MW00344739

Continuous
Univariate
Distributions

Volume 1

WILEY SERIES IN PROBABILITY AND MATHEMATICAL STATISTICS

Established by WALTER A. SHEWHART and SAMUEL S. WILKS

Editors: *Vic Barnett, Ralph A. Bradley, Nicholas I. Fisher, J. Stuart Hunter, J. B. Kadane, David G. Kendall, Adrian F. M. Smith, Stephen M. Stigler, Jozef L. Teugels, Geoffrey S. Watson*

A complete list of the titles in this series appears at the end of this volume

Continuous Univariate Distributions

Volume 1

Second Edition

NORMAN L. JOHNSON
University of North Carolina
Chapel Hill, North Carolina

SAMUEL KOTZ
University of Maryland
College Park, Maryland

N. BALAKRISHNAN
McMaster University
Hamilton, Ontario, Canada

A Wiley-Interscience Publication
JOHN WILEY & SONS, INC.
New York • Chichester • Brisbane • Toronto • Singapore

This text is printed on acid-free paper.

Copyright © 1994 by John Wiley & Sons, Inc.

All rights reserved. Published simultaneously in Canada.

Reproduction or translation of any part of this work beyond
that permitted by Section 107 or 108 of the 1976 United
States Copyright Act without the permission of the copyright
owner is unlawful. Requests for permission or further
information should be addressed to the Permissions Department,
John Wiley & Sons, Inc., 605 Third Avenue, New York, NY
10158-0012.

Library of Congress Cataloging in Publication Data:

Johnson, Norman Lloyd.
    Continuous univariate distributions / Norman L. Johnson, Samuel
Kotz, N. Balakrishnan.—2nd ed.
        p.   cm.—(Wiley series in probability and mathematical
statistics. Applied probability and statistics.)
    "A Wiley-Interscience publication."
    Includes bibliographical references and index.
    ISBN 0-471-58495-9 (v. 1)
    1. Distribution (Probability theory) I. Kotz, Samuel.
II. Balakrishnan, N., 1956– . III. Title. IV. Series.
QA273.6.J6 1994   93-45348
519.2'4—dc20

Printed in the United States of America

10 9 8 7 6 5 4 3 2 1

To
Regina Elandt-Johnson
Rosalie Kotz
Colleen Cutler and Sarah Balakrishnan

Contents

Preface

As a continuation of *Univariate Discrete Distributions*, second edition, this book is the first of two volumes to discuss continuous univariate distributions. The second edition of *Continuous Univariate Distributions* differs from the first, published in 1970, in two important aspects: (1) Professor N. Balakrishnan has joined the two original authors as a coauthor. (2) Because of substantial advances in theory, methodology, and application of continuous distributions, especially gamma, Weibull, and inverse Gaussian during the last two decades, we have decided to move the chapter on extreme value distributions to the next volume. The chapter on gamma distributions has been split into two chapters, one dealing only with chi-squared distributions. Even so, as in the revision of the volume on *Discrete Distributions*, the great amount of additional information accruing since the first edition has led to a substantial increase in length.

In accordance with the principle stated in the original General Preface, we continue to aim at "excluding theoretical minutiae of no apparent practical importance," although we do include material on characterizations that some may regard as being of doubtful practical value. The more general Chapter 12 has been expanded relatively less than the other chapters that deal with specific distributions.

Even with omission of the extreme value distribution chapter, the great amount of new information available has forced us to be very selective for inclusion in the new work. One of our main preoccupations has been to assist the struggle against fragmentation, the necessity for which has been elegantly expressed by Professor A. P. Dawid, new editor of *Biometrika* [*Biometrika*, **80**, 1 (1993)]. We realize that some authors may be affronted at omission of their work, but we hope that it will not be regarded as unfriendly action, or at "best" a consequence of ignorance. These volumes are intended to be useful to readers, rather than an "honor roll" of contributors.

We acknowledge with thanks the invaluable assistance of Mrs. Lisa Brooks (University of North Carolina), Mrs. Cyndi Patterson (Bowling Green State University), and Mrs. Debbie Iscoe (Hamilton, Canada) in their skillful typing of the manuscript. We also thank the Librarians of the University of

North Carolina, Bowling Green State University, McMaster University, University of Waterloo, and the University of Maryland for their help in library research. Samuel Kotz's contribution to this volume was, to a large extent, executed as a Distinguished Visiting Lukacs Professor in the Department of Mathematics and Statistics at Bowling Green State University (Bowling Green, Ohio) during September–December 1992. Likewise N. Balakrishnan's work for this volume was carried out mostly in the Department of Statistics and Actuarial Science at the University of Waterloo where he was on research leave during July 1992—June 1993.

Special thanks are also due to Mrs. Kate Roach and Mr. Ed Cantillon at John Wiley & Sons in New York for their sincere efforts in the fine production of this volume. We also thank Ms. Dana Andrus for all her efforts in copy-editing the long manuscript.

Thanks are offered to the Institute of Mathematical Statistics, the American Statistical Association, the Biometrika Trustees, the Institute of Electrical and Electronics Engineerings, the Association for Computing Machinery, Marcel Dekker, Inc., the Royal Statistical Society, the American Society for Quality Control, the Australian Statistical Society, Gordon and Breach Science Publishers, Blackwell Publishers, and the editors of *Biometrical Journal*, *Sankhyā* and *Tamkang Journal of Mathematics*, for their kind permission to reproduce previously published tables and figures.

Authors of this kind of informational survey—designed primarily for nonspecialists—encounter the problem of having to explain results that may be obvious for a specialist but are not part and parcel of the common knowledge, and still have enough space for information that would be new and valuable for experts. There is a danger of "high vulgarization" (over simplification) on the one hand and an overemphasis on obscure "learned" points—which many readers will not need—on the other hand. We have tried to avoid these pitfalls, to the best of our ability.

It is our sincere hope that these volumes will provide useful facts—coherently stated—for a "lay reader," and also arouse perceptions and memories among well-informed readers, stimulating introspection and further research.

NORMAN L. JOHNSON
SAMUEL KOTZ
N. BALAKRISHNAN

List of Tables

Continuous
Univariate
Distributions

Volume 1

CHAPTER 12

Continuous Distributions (General)

1 INTRODUCTION

Chapters 1 and 2 (of Volume 1) contain some general results and methods that are also useful in our discussion of continuous distributions. In this chapter we will supplement this information with techniques that are relevant to continuous distributions. As in Chapter 2, some general *systems* of (continuous) distributions will be described.

Continuous distributions are generally amenable to more elegant mathematical treatment than are discrete distributions. This makes them especially useful as approximations to discrete distributions. Continuous distributions are used in this way in most applications, both in the construction of models and in applying statistical techniques. Continuous distributions have been used in approximating discrete distributions of discrete statistics in Volume 1. The fact that most uses of continuous distributions in model building are as approximations to discrete distributions may be less widely appreciated but is no less true. Very rarely is it more reasonable, in an absolute sense, to represent an observed value by a continuous, rather than a discrete, random variable. Rather this representation is a convenient *approximation*, facilitating mathematical and statistical analysis.

An essential property of a continuous random variable is that there is zero probability that it takes any specified numerical value, but in general a nonzero probability, calculable as a definite integral of a *probability density function* (see Section 1.4) that it takes a value in specified (finite or infinite) intervals. When an observed value is represented by a continuous random variable, the recorded value is, of necessity "discretized." For example, if measurements are made to the nearest 0.01 units, then all values actually in the interval (8.665, 8.675) will be recorded as 8.67. The data are therefore *grouped*. Adjustments in the estimation of population moments, to correct (on average) for this effect, were proposed by Sheppard (1896). These may be summarized for the case of equal group widths h (with $\mu'_r, {}_g\mu'_r$ denoting the

1

rth original moment and rth grouped moment, respectively):

$$\mu_1' = {}_g\mu_1', \tag{12.1a}$$

$$\mu_2' = {}_g\mu_2' - \frac{1}{12}h^2, \tag{12.1b}$$

$$\mu_3' = {}_g\mu_3' - \frac{1}{4}{}_g\mu_3'h^3, \tag{12.1c}$$

$$\mu_4' = {}_g\mu_4' - \frac{1}{2}{}_g\mu_2'h^2 + \frac{7}{240}h^4. \tag{12.1d}$$

The general formula is [Sheppard (1896); Wold (1934)]

$$\mu_r' = \sum_{j=0}^{r} (2^{1-j} - 1)\binom{r}{j} B_j {}_g\mu_{r-j}'h^j, \tag{12.2}$$

where B_j is the jth Bernoulli number (see Chapter 1, Section A9).

These results depend on the assumption that if the centers of groups are

$$\ldots, a - h, a, a + h, \ldots \qquad (|a| < \tfrac{1}{2}h),$$

then a has a uniform distribution (Chapter 26) between $-\tfrac{1}{2}h$ and $+\tfrac{1}{2}h$ [see also Haitovsky (1983)]. Effects of invalidity of this assumption have been evaluated by Tricker (1984), who showed that the characteristic function (Chapter 1, Section B8) of ${}_gX$ (the grouped variable corresponding to a random variable X) is

$$\varphi_{gX}(t) = \sum_{j=-\infty}^{\infty} \frac{\sin(\tfrac{1}{2}th + j\pi)}{\tfrac{1}{2}th + j\pi} \exp(1 - i \cdot 2\pi jh) \cdot \varphi_X(th + 2\pi j), \tag{12.3}$$

where $i = \sqrt{-1}$ and $\varphi_X(t)$ is the characteristic function of X.

In the same paper, effects of rounding when the distribution of X is normal (Chapter 13), Laplace (Chapter 24), or gamma (Chapter 17) are evaluated numerically for selected cases. For a given group width h, the magnitude of the correction for grouping tends to increase with the magnitude of the skewness, as measured by $|\sqrt{\beta_1}|$; it is quite small for symmetrical distributions.

Some concepts that have great value for discrete distributions are much less valuable in the discussion of continuous distributions. Probability-generating functions, in particular, are little used in this part of the book. Factorial moments also rarely offer the advantages of conciseness and simplicity that they do for discrete distributions, although they can be calculated.

On the other hand, *standardization* (use of the transformed variable

$$\frac{(X - E[X])}{\sqrt{\operatorname{var}(X)}}$$

to produce a distribution with zero mean and unit standard deviation) is much more useful for continuous distributions. In particular, the shape of a distribution can be conveniently summarized by giving standardized values of a number of *quantiles* (i.e., values of the variable for which the cumulative distribution function has specified values). Care should be taken to distinguish between *standardized* and *standard* forms of distributions. The latter are usually convenient ways of writing the mathematical formulas for probability density functions. They *may* happen to be standardized, but this is not essential.

MacGillivray (1992) has introduced the *skewness function*

$$\gamma_X(u) = \frac{F^{-1}(u) + F^{-1}(1 - u) - 2F^{-1}(\tfrac{1}{2})}{F^{-1}(u) - F^{-1}(1 - u)}. \qquad (12.4a)$$

Note that $F^{-1}(\tfrac{1}{2}) = \text{median }(X)$ and $\gamma_X(\tfrac{3}{4})$ is Galton's measure of skewness:

$$\frac{(\text{Upper quartile} - \text{Median}) - (\text{Median} - \text{Lower quartile})}{\text{Interquartile distance}}.$$

MacGillivray proposes

$$\sup_{\tfrac{1}{2} \le u < 1} |\gamma_X(u)| \qquad (12.4b)$$

as "a measure of overall asymmetry for the central $100(1 - 2\alpha)\%$ of the distribution."

At this point we introduce indices that are applied more commonly to continuous than to discrete distributions:

1. *Gini's mean differences*, $\gamma(X)$, for the distribution of X is the expected value, $E[|X_1 - X_2|]$ of the absolute value of the difference between two independent variables X_1, X_2, each distributed as X.
   If $X_1, \ldots, X_n$ are i.i.d., then the statistic

$$g = \binom{n}{2}^{-1} \sum_{i<j}^{n} \sum^{n} |X_i - X_j| \qquad (12.5)$$

is an unbiased estimator of

$$\gamma(X) = E[|X_1 - X_2|]. \qquad (12.6)$$

2. The *Lorenz curve*, for a positive random variable X, is defined as the graph of the ratio

$$L(F_X(x)) = \frac{E[X|X \le x]F_X(x)}{E[X]} \tag{12.7}$$

against $F_X(x)$. If X represents annual income, $L(p)$ is the proportion of total income that accrues to individuals having the $100p\%$ lowest incomes.

It is easy to see that

$$L(p) \le p,$$

$$L(0) = 0,$$

$$L(1) = 1.$$

A typical Lorenz curve is shown in Figure 12.1. If all individuals earn the same income, $L(p) = p$. The area between the line $L(p) = p$ and the Lorenz curve may be regarded as a measure of inequality of income, or more generally, of variability in the distribution of X. There is extensive discussion of Lorenz curves in Gail and Gastwirth (1978), and a concise account of their properties in Dagum (1985).

Gastwirth (1971) has given the following definition of the Lorenz function $L(p)$:

$$L(p) = \{E[X]\}^{-1} \int_0^p F_X^{-1}(t)\, dt, \tag{12.8}$$

where

$$F_X^{-1}(t) = \inf_x \{x : F_X(t) \ge t\}.$$

This equation is equivalent to (12.7) for continuous distributions, but it also applies to discrete distributions.

Lorenz ordering is sometimes used to compare the amounts of inequality in two or more distributions. It is based on comparisons of the values of $L(p)$ for the distributions. If their difference is of the same sign for all p, the distributions are Lorenz ordered appropriately. If the sign changes, Lorenz ordering does not apply. There is a useful description of Lorenz ordering in Arnold (1987); see also Chapter 33.

The area between the line $L(p) = p$ and the actual Lorenz curve is called the *area of concentration*. The *Gini concentration index* $C(X)$ is twice this

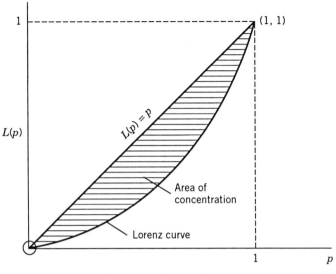

Figure 12.1 Lorenz Curve

area. It is related to *Gini's mean difference*, defined as in (12.5):

$$C(X) = 2\int_0^1 \{p - L(p)\} \, dp = 1 - 2\int_0^1 L(p) \, dp.$$

Now

$$\int_0^1 L(p) \, dp = \frac{1}{E[X]} \int_0^1 E[X|X < x] F_X(x) \, dF_X(x)$$

$$= \frac{1}{E[X]} \int_{-\infty}^{\infty} \left\{ \int_{-\infty}^{x} t p_X(t) \, dt \right\} p_X(x) \, dx$$

$$= \frac{1}{E[X]} E[X_1|X_1 < X_2] \Pr[X_1 < X_2],$$

with X_1, X_2 independently distributed as X. Since X_1 and X_2 are continuous and independent,

$$\Pr[X_1 < X_2] = \Pr[X_1 > X_2] = \tfrac{1}{2};$$

thus

$$C(X) = 1 - \frac{E[X_1|X_1 < X_2]}{E[X]}.$$

Since

$$\tfrac{1}{2}\{E[X_1|X_1 < X_2] + E[X_1|X_1 > X_2]\} = E[X]$$

and

$$\gamma(X) = E[|X_1 - X_2|] = E[X_1|X_1 > X_2] - E[X_1|X_1 < X_2],$$

we have

$$E[X_1|X_1 < X_2] = E[X] - \tfrac{1}{2}\gamma(X).$$

Hence

$$C(X) = \frac{\tfrac{1}{2}\gamma(X)}{E[X]}. \tag{12.9a}$$

The ratio $\gamma(X)/E[X] = E[|X_1 - X_2|]/E[X]$ is analogous to the coefficient of variation. Tziafetas (1989) obtains the alternative form:

$$C(X) = \frac{\text{Cov}(X, F_X(X))}{E[X]}. \tag{12.9b}$$

Order statistics are of much greater use, and simpler in theoretical analysis, for continuous than for discrete distributions. The next section will be devoted to a general discussion of order statistics for continuous variables, with particular reference to their use in statistical analysis.

2 ORDER STATISTICS

If $X_1, X_2, \ldots, X_n$ are random variables, and $X_1' \le X_2' \le \cdots \le X_n'$ are the same variables arranged in ascending order of magnitude [so that $X_1' = \min(X_1, X_2, \ldots, X_n)$, $X_n' = \max(X_1, X_2, \ldots, X_n)$], then $X_1', X_2', \ldots, X_n'$ are called the *order statistics* corresponding to $X_1, X_2, \ldots, X_n$ (see also Chapter 1, Section B10). If it is necessary to indicate the total number of variables explicitly, the symbols $X_{1:n}', \ldots, X_{n:n}'$ will be used.

If the differences $\{X_i - X_j\}$ are continuous random variables, then the events $\{X_i = X_j\}$ all have zero probability. Being finite in number, they can be neglected in probability calculations. We will suppose, from now on, that this is the case, so we can assume that $X_1' < X_2' \cdots < X_n'$, without altering any probabilities that relate to the joint distribution of $X_1', X_2', \ldots, X_n'$.

The cumulative distribution function of X'_n is defined by

$$\Pr[X'_n \leq x] = \Pr\left[\bigcap_{j=1}^{n} (X_j \leq x)\right]. \tag{12.10}$$

If $X_1, X_2, \ldots, X_n$ are mutually independent, then

$$\Pr[X'_n \leq x] = \prod_{j=1}^{n} \Pr[X_j \leq x], \tag{12.11}$$

and the probability density function of X'_n is equal to

$$\left(\begin{array}{c} \text{Cumulative distribution} \\ \text{function of } X'_n \end{array}\right) \times \sum_{j=1}^{n} \left|\begin{array}{c} \text{Probability density} \\ \text{function of } X_j \\ \hline \text{Cumulative distribution} \\ \text{function of } X_j \end{array}\right|. \tag{12.12}$$

If all X_j's have identical distributions with $\Pr[X_j \leq x] = F(x)$, and $dF(x)/dx = p(x)$, then the probability density function of X'_n is

$$n[F(x)]^{n-1}p(x). \tag{12.13a}$$

Similarly (again assuming that all X_j's are independent and identically distributed) the probability density function of X'_1 is

$$n[1 - F(x)]^{n-1}p(x). \tag{12.13b}$$

More generally, in this case, with $1 \leq a_1 < a_2 < \cdots < a_s \leq n$ (and setting $a_0 = 0$, $a_{s+1} = n$, $F(x_{a_0}) = 0$, $F(x_{a_{s+1}}) = 1$), the joint probability density function of $X'_{a_1}, X'_{a_2}, \ldots, X'_{a_s}$ is (in an obvious notation)

$$\frac{n!}{\prod_{j=1}^{s+1}(a_j - a_{j-1})!} \left[\prod_{j=1}^{s+1} \{F(x_{a_j}) - F(x_{a_{j-1}})\}^{a_j - a_{j-1} - 1}\right] \prod_{j=1}^{s} p(x_{a_j})$$

$$(x_{a_1} \leq x_{a_2} \leq \cdots \leq x_{a_s}). \tag{12.14}$$

In particular, the joint probability density function of X'_1 and X'_n is

$$n(n-1)p(x_1)p(x_n)[F(x_n) - F(x_1)]^{n-2} \quad (x_1 \leq x_n). \tag{12.15}$$

From this joint distribution it is possible to evaluate the cumulative distribution function of the range ($W = X'_n - X'_1$). The formulas

$$\Pr[W \leq w] = n \int_{-\infty}^{\infty} p(x)[F(x) - F(x - w)]^{n-1} dx \qquad (12.16)$$

and

$$E[W] = \int_{-\infty}^{\infty} \{1 - [F(x)]^n - [1 - F(x)]^n\} dx \qquad (12.17)$$

are of interest.

If $n = 2m + 1$ is odd (i.e., m is an integer), then X'_{m+1} represents the (*sample*) *median* of $X_1, X_2, \ldots, X_n$. Its probability density function is

$$\frac{(2m + 1)!}{(m!)^2} [F(x)\{1 - F(x)\}]^m p(x). \qquad (12.18)$$

Generally the $100p\%$ sample percentile is represented by $X'_{(n+1)p}$ and is defined only if $(n + 1)p$ is an integer. The median corresponds to $p = \frac{1}{2}$; we have the *lower* and *upper quartile* for $p = \frac{1}{4}, \frac{3}{4}$, respectively.

Often under certain conditions of regularity, it is possible to obtain useful approximations to the moments of order statistics in terms of the common probability density function of the X's. This approach makes use of the fact that the statistics $Y_1 = F(X_1)$, $Y_2 = F(X_2), \ldots, Y_n = F(X_n)$ are independently distributed with common rectangular distribution (see Chapter 26) over the range 0 to 1. The corresponding order statistics $Y'_1, Y'_2, \ldots, Y'_n$ have the joint probability density function

$$p_{Y'_1, \ldots, Y'_n}(y_1, \ldots, y_n) = n! \qquad (0 \leq y_1 \leq y_2 \leq \cdots \leq y_n \leq 1).$$

The joint probability density function of any subset $Y'_{a_1}, \ldots, Y'_{a_s}$ ($1 \leq a_1 < a_2 < \cdots < a_s \leq n$) is [using (12.14)]

$$p_{Y'_{a_1}, \ldots, Y'_{a_s}}(y_{a_1}, \ldots, y_{a_s}) = \frac{n!}{\prod_{j=1}^{s+1}(a_j - a_{j-1})!} \prod_{j=1}^{s+1} (y_{a_j} - y_{a_{j-1}})^{a_j - a_{j-1} - 1},$$

$$(12.19)$$

with $a_0 = 0$, $a_{s+1} = n$; $y_{a_0} = 0$; $y_{a_{s+1}} = 1$, and where $Y'_{a_1}, \ldots, Y'_{a_s}$ is denoted by $\{Y'_a\}$. The moments and product moments of the Y''s are given by the formula

$$E\left[\prod_{j=1}^{s} Y_{a_j}^{\prime r_j}\right] = \frac{n!}{(n + \Sigma_{j=1}^{s} r_j)!} \prod_{j=1}^{s} \left\{\frac{\left(a_j + \sum_{i=1}^{j} r_i - 1\right)!}{(a_j + \Sigma_{i=1}^{j-1} r_i - 1)!}\right\}. \quad (12.20)$$

We now expand X_r', as a function of Y_r', about the value $E[Y_r'] = r/(n + 1)$; thus

$$X_r' = F^{-1}\left(\frac{r}{n+1}\right) + \left(Y_r' - \frac{r}{n+1}\right)\left[\frac{dF^{-1}(y)}{dy}\Bigg|_{y=r/(n+1)}\right]$$

$$+ \frac{1}{2}\left(Y_r' - \frac{r}{n+1}\right)^2\left[\frac{d^2F^{-1}(y)}{dy^2}\Bigg|_{y=r/(n+1)}\right] + \cdots. \quad (12.21)$$

So we can take expected values of each side of (12.13) (using the method of statistical differentials, described in Chapter 1). Note that since

$$y = F(x) = \int_{-\infty}^{x} p(t)\, dt,$$

$$\frac{dF^{-1}}{dy} = \frac{dx}{dy} = \frac{1}{dy/dx} = \frac{1}{p(x)}$$

and

$$\frac{dF^{-1}(y)}{dy}\Bigg|_{y=r/(n+1)} = \frac{1}{p(\xi_r')},$$

where ξ_r' satisfies the equation

$$\frac{r}{n+1} = \int_{-\infty}^{\xi_r'} p(x)\, dx. \quad (12.22)$$

Similarly $d^2F^{-1}/dy^2 = -[p(x)]^{-2}[dp(x)/dy] = -[p(x)]^{-3}[dp(x)/dx]$, and so on.

David and Johnson (1954) found it convenient to arrange the series so obtained in descending powers of $(n + 2)$. Some of their results follow. [In these formulas, $p_s = s/(n + 1)$; $q_s = 1 - p_s$; $(F^{-1})'_r = dF^{-1}/dy|_{y=r/(n+1)}$; $(F^{-1})''_r = d^2F^{-1}/dy^2|_{y=r/(n+1)}$, etc.]

$$E[X'_r] = \xi'_r + \frac{p_r q_r}{2(n+2)}(F^{-1})''_r + \frac{p_r q_r}{(n+2)^2}$$

$$\times \left[\frac{1}{3}(q_r - p_r)(F^{-1})'''_r + \frac{1}{8}p_r q_r(F^{-1})_r^{iv} \right] + \ldots , \qquad (12.23a)$$

$$\mathrm{Var}(X'_r) = \frac{p_r q_r}{n+2}\{(F^{-1})'_r\}^2 + \frac{p_r q_r}{(n+2)^2}\left[2(q_r - p_r)(F^{-1})'_r(F^{-1})''_r \right.$$

$$\left. + p_r q_r\left\{ (F^{-1})'_r(F^{-1})'''_r + \frac{1}{2}\left[(F^{-1})''_r\right]^2 \right\} \right] + \ldots , \qquad (12.23b)$$

$$\mathrm{Cov}(X'_r, X'_s) = \frac{p_r q_s}{n+2}\{(F^{-1})'_r(F^{-1})'_s\}$$

$$+ \frac{p_r q_s}{(n+2)^2}\left[(q_r - p_r)(F^{-1})''_r(F^{-1})'_s + (q_s - p_s)(F^{-1})'_r(F^{-1})''_s \right.$$

$$+ \frac{1}{2}p_r q_r(F^{-1})'''_r(F^{-1})'_s + \frac{1}{2}p_s q_s(F^{-1})'_r(F^{-1})'''_s$$

$$\left. + \frac{1}{2}p_r q_s(F^{-1})''_r(F^{-1})''_s \right] + \ldots , \qquad r < s, \quad (12.23c)$$

$$\mu_3(X'_r) = \frac{p_r q_r}{(n+2)^2}\left[2(q_r - p_r)\{(F^{-1})'_r\}^3 \right.$$

$$\left. + 3p_r q_r\{(F^{-1})'_r\}^2(F^{-1})''_r\} \right] + \ldots , \qquad (12.23d)$$

$$\mu_4(X'_r) = \frac{3p_r^2 q_r^2}{(n+2)^2}\{(F^{-1})'_r\}^4$$

$$+ \frac{p_r q_r}{(n+2)^3}\left[6\{(q_r - p_r)^2 - p_r q_r\}\{(F^{-1})'_r\}^4 \right.$$

$$+ 36 p_r q_r(q_r - p_r)\{(F^{-1})'_r\}^3(F^{-1})'''_r$$

$$\left. + 5p_r^2 q_r^2\{2\{(F^{-1})'_r\}^3(F^{-1})'''_r + 3\{(F^{-1})'_r(F^{-1})''_r\}^2\} \right] + \ldots . \quad (12.23e)$$

By inserting the values of ξ'_r, $(F^{-1})'_r$, $(F^{-1})''_r$, and so on, appropriate to the particular distribution, approximate formulas can be obtained that correspond to any absolutely continuous common distribution of the original independent variables. These formulas generally tend to be more accurate for large n, and for larger $\min(p_r, q_r)$ [with $\text{Cov}(X'_r, X'_s)$ for larger $\min(p_r, p_s, q_r, q_s)$]. David (1981) and Arnold and Balakrishnan (1989) provide detailed discussions on bounds and approximations for moments of order statistics.

If the distribution of X is such that $\Pr[X < x]$ is a function of only $(x - \theta)/\phi$ so that θ and ϕ (> 0) are *location* and *scale* parameters, then it is easy to see that $Z = (X - \theta)/\phi$ has a distribution that does not depend on θ or ϕ. Denoting the order statistics corresponding to independent random variables $Z_1, Z_2, \ldots, Z_n$, each distributed as Z, by $Z'_{1:n}, Z'_{2:n}, \ldots, Z'_{n:n}$ it is easy to see that

$$E[X'_{r:n}] = \theta + \phi E[Z'_{r:n}], \tag{12.24a}$$

and further that

$$\text{Var}(X'_{r:n}) = \phi^2 \text{Var}(Z'_{r:n}),$$

$$\text{Cov}(X'_{r:n}, X'_{s:n}) = \phi^2 \text{Cov}(Z'_{r:n}, Z'_{s:n}). \tag{12.24b}$$

Hence it is possible to obtain *best linear unbiased estimators* of θ and ϕ, based on the order statistics $X'_{1:n}, X'_{2:n}, \ldots, X'_{n:n}$, by minimizing the quadratic form:

$$\sum_r \sum_s c_{rs}(X'_{r:n} - \theta - \phi E[Z'_{r:n}])(X'_{s:n} - \theta - \phi E[Z'_{s:n}]),$$

where the matrix (c_{rs}) is the inverse of the matrix of variances and covariances of the $Z'_{r:n}$'s [Lloyd (1952)].

In later chapters a number of results obtained by this method will be presented. The method is of particular value when not all the $X_{r:n}$'s are used. For example, when data are *censored* (as described in Chapter 1, Section B10), not all the order statistics are available. Even if they were available, we would want to use only a limited number based on robustness considerations. It is useful, in such cases, to know which sets of a fixed number of order statistics will minimize the variance of the best linear unbiased estimator of θ or ϕ (or perhaps some function of these parameters). Exact calculation is usually tedious, but approximate calculation, using only the first terms of formulas (12.23a), (12.23b) is less troublesome.

In using these results, it is desirable to bear in mind that (1) there may be nonlinear estimators that are (in some sense) more accurate, (2) "best" is defined in terms of variance which is not always appropriate, and (3) the constraint of unbiasedness may exclude some good estimators. However, it does appear that the best linear unbiased estimators of location and scale parameters, based on order statistics, usually offer accuracy close to the utmost attainable from the data.

Bennett (1952) developed a general method for determining "asymptotically efficient linear unbiased estimators." These are estimators of the form

$$L_n = \sum_{j=1}^{n} J\left(\frac{j}{n+1}\right) X'_{j:n} \qquad (12.25)$$

where $J(\cdot)$ is a "well-behaved" function; that is, the limiting distribution of $\sqrt{n}\,(L_n - \theta)$ is normal with expected value zero. Bennett's thesis is not easily available. The following results are quoted from Chernoff, Gastwirth, and Johns (1967) who have also demonstrated the asymptotic normality (as $n \to \infty$) of these estimators.

Relatively simple formulas for $J(\cdot)$ are available for the special case when the parameters θ_1, θ_2 are location and scale parameters so that (for each unordered X)

$$\Pr[X \le x] = g\left(\frac{x - \theta_1}{\theta_2}\right), \qquad \theta_2 > 0. \qquad (12.26)$$

The corresponding density function is $\theta_2^{-1} g'[(x - \theta_1)/\theta_2]$, and the Fisher information matrix is

$$\begin{pmatrix} I_{11} & I_{12} \\ I_{21} & I_{22} \end{pmatrix} = \begin{vmatrix} \int_{-\infty}^{\infty} \dfrac{dL_1}{dy} g(y)\,dy & \int_{-\infty}^{\infty} \dfrac{dL_2}{dy} g(y)\,dy \\ \int_{-\infty}^{\infty} y\dfrac{dL_1}{dy} g(y)\,dy & \int_{-\infty}^{\infty} y\dfrac{dL_2}{dy} g(y)\,dy \end{vmatrix} \theta_2^{-2}, \quad (12.27)$$

where

$$L_1(y) = -\frac{g'(y)}{g(y)},$$

$$L_2(y) = -1 - \frac{yg'(y)}{g(y)}.$$

[Note that $I_{21} = I_{12}$ provided that $g''(y)$ exists and that $\lim_{y \to \pm\infty} yg'(y) = 0$.] Then for estimating θ_1 with θ_2 known, we can use

$$J(u) = I_{11}^{-1} L_1'(F^{-1}(u)). \qquad (12.28)$$

To make the estimator unbiased, $I_{11}^{-1}I_{12}\theta_2$ must be subtracted. For estimating θ_2 with θ_1 known, we can use

$$J(u) = I_{22}^{-1}L_2'(F^{-1}(u)).$$

To make the estimator unbiased, $I_{22}^{-1}I_{12}\theta_1$ must be subtracted.
  If neither θ_1 nor θ_2 is known, then for estimating θ_1,

$$J(u) = I^{11}L_1'(F^{-1}(u)) + I^{12}L_2'(F^{-1}(u))$$

and for estimating θ_2,

$$J(u) = I^{12}L_1'(F^{-1}(u)) + I^{22}L_2'(F^{-1}(u)),$$

where

$$\begin{pmatrix} I^{11} & I^{12} \\ I^{12} & I^{22} \end{pmatrix}$$

is the inverse of the matrix **I**. These estimators are unbiased. Chernoff, Gastwirth, and Johns (1967) also obtain formulas to use when the data are censored. Balakrishnan and Cohen (1991) discuss this, and related methods of estimation, in considerable detail.

The *limiting distributions* of order statistics as n tends to infinity have been studied by a number of writers. It is not difficult to establish that if $r - n\omega$ tends to zero as n tends to infinity, the limiting distribution of $n(X_{r:n}' - X_\omega)$ (where $\Pr[X \le X_\omega] = \omega$) is normal with expected value zero and standard deviation $\sqrt{\omega(1 - \omega)}/p(X_\omega)$. However, other limiting distributions are possible. Wu (1966) has shown that a lognormal limiting distribution may be obtained.

Books by Galambos (1987), Resnick (1987), and Leadbetter, Lindgren, and Rootzén (1983) include discussions of asymptotic results for extreme order statistics, while those by Shorack and Wellner (1986) and Serfling (1980) have dealt with central order statistics and linear functions of order statistics. Reiss (1989) has discussed convergence results for all order statistics. Recently Arnold, Balakrishnan, and Nagaraja (1992) have presented a useful summary of many of these developments on the asymptotic theory of order statistics.

Chan (1967) has shown that the distribution function is characterized by either of the sets of values $\{E[X_{1:n}']\}$ or $\{E[X_{n:n}']\}$ (for all n) provided that the expected value of the distribution is finite. Many characterization results involving order statistics are reviewed in Galambos and Kotz (1978) and Arnold, Balakrishnan, and Nagaraja (1992).

3   CALCULUS OF PROBABILITY DENSITY FUNCTIONS

The reader may have noted that many of the results of the preceding section were expressed in terms of probability density functions. Although these are only auxiliary quantities—actual probabilities being the items of real importance—they are convenient in the analysis of continuous distributions. In this section we briefly describe techniques for working with probability density functions that will be employed in later chapters. More detailed discussions, and proofs, can be found in textbooks; see, for example, Mood, Graybill, and Boes (1974), Bickel and Doksum (1977), Hogg and Craig (1978), Dudewicz and Mishra (1988), and Casella and Berger (1990).

If $X_1, X_2, \ldots, X_n$ are independent random variables with probability density functions $p_{X_1}(x_1), p_{X_2}(x_2), \ldots, p_{X_n}(x_n)$, then the joint probability density function may be taken as

$$p_{X_1, X_2, \ldots, X_n}(x_1, x_2, \ldots, x_n) = \prod_{j=1}^{n} p_{X_j}(x_j). \tag{12.29}$$

If the variables are not independent, conditional probability density functions must be used. Then in place of (12.29) we have

$$p(x_1, x_2, \ldots, x_n) = p(x_1)p(x_2|x_1)p(x_3|x_1, x_2) \ldots p(x_n|x_1, x_2, \ldots, x_{n-1}).^* \tag{12.30}$$

Of course (12.30) includes (12.29), since, if $X_1, \ldots, X_n$ are a mutually independent set of variables, then

$$p(x_2|x_1) = p(x_2),$$

$$p(x_3|x_1, x_2) = p(x_3) \ldots .$$

If $p(x_1, \ldots, x_n)$ is known, then the joint probability density function of any subset of n random variables can be obtained by repeated use of the formula

$$\int_{-\infty}^{\infty} p(x_1, \ldots, x_n) \, dx_p = p(x_1, \ldots, x_{p-1}, x_{p+1}, \ldots, x_n). \tag{12.31}$$

To find the joint distribution of n functions of $X_1, \ldots, X_n$ (*statistics*) when $T_1 \equiv T_1(X_1, \ldots, X_n), \ldots, T_n = T_n(X_1, \ldots, X_n)$ and the transformation from $(X_1, \ldots, X_n)$ to $(T_1, \ldots, T_n)$ is one to one, then the formula

$$p_{T_1, \ldots, T_n}(t_1, \ldots, t_n) = p_{X_1, \ldots, X_n}(x_1(t), \ldots, x_n(t)) \left| \frac{\partial(x_1, \ldots, x_n)}{\partial(t_1, \ldots, t_n)} \right| \tag{12.32}$$

---

*In the remainder of this and succeeding sections of this chapter, subscripts following p will often be omitted for convenience. In succeeding chapters the subscripts will usually appear.

may be used [$t_1, \ldots, t_n$ and $x_1, \ldots, x_n$ are related in the same way as $T_1, \ldots, T_n$ and $X_1, \ldots, X_n$; $x_j(t)$ means that x_j is expressed in terms of $t_1, \ldots, t_n$ and $\partial(x_1, \ldots, x_n)/\partial(t_1, \ldots, t_n)$ is the Jacobian of $(x_1, \ldots, x_n)$ with respect to $(t_1, \ldots, t_n)$ in that it is a determinant of n rows and n columns with the element in the ith row and the jth column equal to $\partial x_i/\partial t_j$].

If the transformation is not one to one, the simple formula (12.32) cannot be used. In special cases, however, straightforward modifications of (12.32) can be employed. For example, if k different sets of values of the x's produce the *same* set of values of the t's, it may be possible to split up the transformation into k separate transformations. Then (12.32) is applied to each, and the results added together.

Having obtained the joint distribution of $T_1, T_2, \ldots, T_n$, the joint distribution of any subset thereof can be obtained by using (12.21) repeatedly. The conditional distribution of X_1, given $X_2, \ldots, X_n$ is sometimes called the *array distribution* of X_1 (given $X_2, \ldots, X_n$). The expected value of this conditional distribution (a function of $X_2, \ldots, X_n$) is called the *regression* of X_1 on $X_2, \ldots, X_n$. The variance is called the *array variance* (of X_1, given $X_2, \ldots, X_n$); if it does not depend on $X_2, \ldots, X_n$, the variation is said to be *homoscedastic*.

4   SYSTEMS OF DISTRIBUTIONS

Some families of distributions have been constructed to provide approximations to as wide a variety of observed distributions as possible. Such families are often called *systems* of distributions, or, more often, *systems of frequency curves*. Although theoretical arguments may indicate the relevance of a particular system, their value should be judged primarily on practical, ad hoc considerations. Particular requirements are ease of computation and facility of algebraic manipulation. Such requirements make it desirable to use as few parameters as is possible in defining an individual member of the system. How few we may use, without prejudicing the variety of distributions included, is a major criterion in judging the utility of systems of distributions.

For most practical purposes it is sufficient to use four parameters. There is no doubt that at least three parameters are needed; for some purposes this is enough. Inclusion of a fourth parameter does produce noticeable improvement, but it is doubtful whether the improvement obtained by including a fifth or sixth parameter is commensurate with the extra labor involved. Here we will describe some systems of frequency curves. Among these systems there should be at least one that suffices for practical needs and possibilities in most situations.

4.1   Pearson System

Between 1890 and 1895 Pearson (1895) designed a system whereby for every member the probability density function $p(x)$ satisfies a differential equation of form

$$\frac{1}{p}\frac{dp}{dx} = -\frac{a+x}{c_0 + c_1 x + c_2 x^2}. \tag{12.33}$$

The shape of the distribution depends on the values of the parameters a, c_0, c_1, and c_2. If $-a$ is not a root of the equation

$$c_0 + c_1 x + c_2 x^2 = 0,$$

p is finite when $x = -a$ and dp/dx is zero when $x = -a$. The slope (dp/dx) is also zero when $p = 0$. But, if $x \neq -a$ and $p \neq 0$, then $dp(x)/dx \neq 0$. Since the conditions $p(x) \geq 0$ and

$$\int_{-\infty}^{\infty} p(x)\, dx = 1$$

must be satisfied, it follows from (12.33) that $p(x)$ must tend to zero as x tends to infinity, and likewise must dp/dx. This may not be true of *formal* solutions of (12.33). In such cases the condition $p(x) \geq 0$ is not satisfied, and it is necessary to restrict the range of values of x to those for which $p(x) > 0$ and to assign the value $p(x) = 0$ when x is outside this range.

    The shape of the curve representing the probability density function varies considerably with a, c_0, c_1, and c_2. Pearson classified the different shapes into a number of types. We will give a resumé of his classification. We follow his system of numbering because it is well-established, but it does not have a clear systematic basis.

    The form of solution of (12.33) evidently depends on the nature of the roots of the equation

$$c_0 + c_1 x + c_2 x^2 = 0, \tag{12.34}$$

and the various types correspond to these different forms of solution. We first note that if $c_1 = c_2 = 0$, equation (12.33) becomes

$$\frac{d \log p(x)}{dx} = -\frac{x+a}{c_0},$$

whence

$$p(x) = K \exp\left[-\frac{(x+a)^2}{2c_0}\right],$$

where K is a constant, chosen to make

$$\int_{-\infty}^{\infty} p(x) \, dx = 1.$$

It is clear that c_0 must be positive and that $K = \sqrt{2\pi c_0}$. As a result the corresponding distribution is *normal* with expected value $-a$ and standard deviation $\sqrt{c_0}$. The next chapter is devoted to this distribution.

The normal curve is not assigned to particular type. It is in fact a limiting distribution of all types. From now on we will suppose that the origin of the scale of X has been chosen so that $E[X] = 0$.

Type I corresponds to both roots of (12.33) being real, and of opposite signs. Denoting the roots by a_1, a_2, with

$$a_1 < 0 < a_2,$$

we have

$$c_0 + c_1 x + c_2 x^2 = -c_2 (x - a_1)(a_2 - x).$$

Equation (12.33) can be written

$$\frac{1}{p} \frac{dp}{dx} = \frac{a + x}{(x - a_1)(a_2 - x)}$$

the solution of which is

$$p(x) = K(x - a_1)^{m_1}(a_2 - x)^{m_2}, \qquad (12.35)$$

with

$$m_1 = \frac{a + a_1}{c_2(a_2 - a_1)},$$

$$m_2 = -\frac{a + a_2}{c_2(a_2 - a_1)}.$$

For both $x - a_1$ and $a_2 - x$ to be positive, we must have $a_1 < x < a_2$, so we limit the range of variation to these values of x. Equation (12.35) can represent a proper probability density function provided $m_1 > -1$ and $m_2 > -1$. This is a general form of *beta distribution*, which will be discussed further in Chapter 25. Here we briefly note a few points relating to the function $p(x)$.

The limiting value of $p(x)$, as x tends to a_j is zero or infinite depending on whether m_j is positive or negative (for $j = 1, 2$). If m_1 and m_2 have the same sign, $p(x)$ has a single mode or antimode (according as the m's are positive or negative, respectively). Type I distributions can be subdivided according to the appearance of the graph of $p(x)$ against x. Thus we have

Type I(U): if $m_1 < 0$ and $m_2 < 0$.
Type I(J): if $m_1 < 0$ and $m_2 > 0$, or if $m_1 > 0$ and $m_2 < 0$.

If m_j is zero, then $p(x)$ tends to a nonzero limit as x tends to a_j ($j = 1$ or 2).

The symmetrical form of (12.35), with $m_1 = m_2$, is called a *Type II* distribution. If the common value is negative, the distribution is U-shaped, which is sometimes described as Type II(U). Type III corresponds to the case $c_2 = 0$ (and $c_1 \neq 0$). In this case (12.33) becomes

$$\frac{d \log p(x)}{dx} = -\frac{x + a}{c_0 + c_1 x} = -\frac{1}{c_1} - \frac{a - c_0/c_1}{c_0 + c_1 x},$$

whence

$$p(x) = K(c_0 + c_1 x)^m \exp\left(\frac{-x}{c_1}\right), \tag{12.36}$$

with $m = c_1^{-1}(c_0 c_1^{-1} - a)$. If $c_1 > 0$, we take the range of x as $x > -c_0/c_1$; if $c_1 < 0$, the range is taken to be $x < -c_0/c_1$. Type III distributions are *gamma distributions* and are discussed further in Chapter 17.

Type IV distributions correspond to the case in which the equation

$$c_0 + c_1 x + c_2 x^2 = 0$$

does not have real roots. Then we use the identity

$$c_0 + c_1 x + c_2 x^2 = C_0 + c_2(x + C_1)^2,$$

with $C_0 = c_0 - \frac{1}{4}c_1^2 c_2^{-1}$, $C_1 = \frac{1}{2}c_1 c_2^{-1}$. We write (12.33) as

$$\frac{d \log p(x)}{dx} = \frac{-(x + C_1) - (a - C_1)}{C_0 + c_2(x + C_1)^2}.$$

From this it follows that

$$p(x) = K\left[C_0 + c_2(x + C_1)^2\right]^{-(2c_2)^{-1}} \exp\left[-\frac{a - C_1}{\sqrt{c_2 C_0}} \tan^{-1} \frac{x + C_1}{\sqrt{C_0/c_2}}\right]. \tag{12.37}$$

(Note that since $c_0 + c_1 x + c_2 x^2 = 0$ has no real roots, $c_1^2 < 4c_0 c_2$, and so $c_2 C_0 = c_0 c_2 - \frac{1}{4} c_1^2$ is positive.)

Since no common statistical distributions are of Type IV form, it will not be discussed in a later chapter, but we devote a little space here to this type. Formula (12.37) leads to intractable mathematics if one attempts to calculate values of the cumulative distribution function. As we will see later, it is possible to express the parameters a, c_0, c_1, and c_2 in terms of the first four moments of the distribution, so it is possible to fit by equating actual and fitted moments. Fitting Type IV by maximum likelihood is very difficult (with unknown accuracy in finite-sized samples) and rarely attempted.

The tables in Pearson and Hartley (1972) give standardized quantiles (percentiles) of Pearson system distributions to four decimal places for $\sqrt{\beta_1} = 0.0(0.1)2.0$ and for β_2 increasing by intervals of 0.2. With some interpolation these tables can provide approximate values of the cumulative distribution function, without the need to evaluate K (for Type IV). To evaluate K for Type IV, special tables must be used or a special quadrature of $K^{-1} p(x)$ [according to (12.37)] carried out.

Amos and Daniel (1971) and Bouver and Bargmann (1974) provide more extensive tables of quantiles, to four and five decimal places, respectively. Also Bouver (1973) gives values of the cdf for Type IV distributions, to nine decimal places.

Bowman and Shenton (1979a, b) obtained rational fraction approximations to percentile points of standardized Pearson distributions. Davis and Stephens (1983) utilized these approximations to construct a computer algorithm to determine approximate 0.01, 0.025, 0.05, 0.10, 0.25, 0.50, 0.75, 0.90, 0.95, 0.975, and 0.99 percentiles, for $0 \le |\sqrt{\beta_1}| \le 2$ and values of β_2 (between 1.5 and 15.8) depending on $\sqrt{\beta_1}$ and the percentile being evaluated.

On account of the technical difficulties associated with the use of Type IV distributions, efforts have been made to find other distributions with simpler mathematical forms and with circumstances close enough to Type IV distributions to replace them. More information on this point will be given later in this chapter.

Approximations for the cdfs of Type IV distributions, especially aimed at accuracy in the tails of the distributions, have been developed by Woodward (1976) and Skates (1993). Woodward (1976), taking the pdf in the form

$$p_X(x) = K_1 \left(1 + \frac{x^2}{a^2} \right)^{-m} \exp\left\{ -k \tan^{-1}\left(\frac{x}{a} \right) \right\}, \qquad (12.37)'$$

obtained the approximation

$$F_X(x) \doteq 1 - K_1 a^2 \left(1 + \frac{x^2}{a^2} \right)^{-m+1} \frac{\exp\{ -k \tan^{-1}(x/a) \}}{2mx + ak}. \qquad (12.38)$$

Skates (1993), using techniques for approximating integrals developed by Reid and Skates (1986) and Barndorff-Nielsen (1990), formulated the pdf as

$$
p_X(x) = K_2 \left\{ 1 + \frac{1 + \rho^2}{\nu} \left(x + \rho \sqrt{\frac{\nu}{1 + \rho^2}} \right) \right\}^{-(\nu+1)/2}
$$

$$
\times \exp\left[\nu\rho \tan^{-1}\left\{ \left(x + \rho\sqrt{\frac{\nu}{1 + \rho^2}} \right) \sqrt{\frac{1 + \rho^2}{\nu}} \right\} \right], \quad (12.39)
$$

which can be obtained from (12.37)′ by putting

$$
C_0 = \frac{\nu}{(\nu + 1)(1 + \rho^2)},
$$

$$
C_1 = \rho\sqrt{\frac{\nu}{1 + \rho^2}},
$$

$$
C_2 = (\nu + 1)^{-1},
$$

$$
\rho = \frac{\rho}{\nu + 1}\sqrt{\frac{\nu}{1 + \rho^2}}.
$$

He derived a sequence of "secant approximations," whose initial approximation is

$$
F_X(x) \doteq \Phi(r\sqrt{\nu}) - \phi(r\sqrt{\nu})\left\{ \frac{1}{x}\sqrt{\{1 + (x + \rho)^2\}} - \frac{1}{r} \right\}\frac{1}{\sqrt{\nu}} \quad (12.40)
$$

with

$$
r^2 = \log\left\{ \frac{1 + (x + \rho)^2}{1 + \rho^2} \right\} - 2\rho\{\tan^{-1}(x + \rho) - \tan^{-1}\rho\},
$$

and also an approximation based on Barndorff-Nielsen's (1990) analysis,

$$
F_X(x) \doteq \Phi\left(r\sqrt{\nu} - \frac{1}{r\sqrt{\nu}}\log\left[\frac{r}{x}\sqrt{\{1 + (x + \rho)^2\}} \right] \right). \quad (12.41)
$$

Type V corresponds to the case where $c_0 + c_1 x + c_2 x^2$ is a perfect square $(c_1^2 = 4c_0 c_2)$. Equation (12.33) can be rewritten

$$\frac{d\log p(x)}{dx} = -\frac{x + a}{c_2(x + C_1)^2}$$

$$= -\frac{1}{c_2(x + C_1)} - \frac{a - C_1}{c_2(x + C_1)^2},$$

whence

$$p(x) = K(x + C_1)^{-1/c_2} \exp\left[\frac{a - C_1}{c_2(x + C_1)}\right]. \tag{12.42}$$

If $(a - C_1)/c_2 < 0$, then $x > -C_1$; if $(a - C_1)/c_2 > 0$, then $x < -C_1$. [The inverse Gaussian distribution (Chapter 15) belongs to this family.] If $a = C_1$ and $|c_2| < 1$, then we have the special case

$$p(x) = K(x + C_1)^{-1/c_2},$$

which is sometimes called Type VIII and Type IX provided that $c_2 > 0$ or $c_2 < 0$. From (12.42) it can be seen that $(X + C_1)^{-1}$ has a Type III distribution.

Type VI corresponds to the case when the roots of $c_0 + c_1 x + c_2 x^2 = 0$ are real and of the same sign. If they are both negative (e.g., $a_1 < a_2 < 0$), then an analysis similar to that leading to equation (12.35) can be carried out, with the result written in the form

$$p(x) = K(x - a_1)^{m_1}(x - a_2)^{m_2}. \tag{12.43}$$

Since the expected value is greater than a_2, it is clear that the range of variation of x must be $x > a_2$. [Formula (12.43) can represent a proper probability density function provided that $m_2 < -1$ and $m_1 + m_2 < 0$.]

Finally, Type VII corresponds to the case where $c_1 = a = 0$, $c_0 > 0$, and $c_2 > 0$. In this case equation (12.33) becomes

$$\frac{d \log p(x)}{dx} = -\frac{x}{c_0 + c_2 x^2},$$

whence

$$p(x) = K(c_0 + c_2 x^2)^{-(2c_2)^{-1}}. \tag{12.44}$$

A particularly important distribution belonging to this family is the (*central*) *t distribution*, which will be discussed further in Chapter 28. Distribution

(12.44) can be obtained by a simple multiplicative transformation from a t distribution with "degrees of freedom" (possibly fractional) equal to $c_2^{-1} - 1$.

The parameters a, c_0, c_1, and c_2 in (12.33) can be expressed in terms of the moments of the distribution. Equation (12.33) may be written (after multiplying both sides by x^r)

$$x^r(c_0 + c_1 + c_2 x^2)\frac{dp(x)}{d(x)} + x^r(a + x)p(x) = 0. \qquad (12.45)$$

Integrating both sides of (12.45) between $-\infty$ and $+\infty$, and assuming that $x^r p(x) \to 0$ as $x \to \pm\infty$ for $r \le 5$, we obtain the equation

$$-rc_0\mu'_{r-1} + [-(r+1)c_1 + a]\mu'_r + [-(r+2)c_2 + 1]\mu'_{r+1} = 0. \qquad (12.46)$$

Putting $r = 0, 1, 2, 3$ in (12.46), and noting that $\mu'_0 = 1$ and (in the present context) $\mu'_{-1} = 0$, we obtain four simultaneous linear equations for a, c_0, c_1, and c_2 with coefficients which are functions of μ'_1, μ'_2, μ'_3, and μ'_4. The expected value of the variable can always be arranged (as we have done above) to be zero. If this be done, then $\mu'_1 = 0$ and $\mu'_r = \mu_r$ for $r \ge 2$. The formulas for a, c_0, c_1, and c_2 are then

$$c_0 = (4\beta_2 - 3\beta_1)(10\beta_2 - 12\beta_1 - 18)^{-1}\mu_2, \qquad (12.47a)$$

$$a = c_1 = \sqrt{\beta_1}(\beta_2 + 3)(10\beta_2 - 12\beta_1 - 18)^{-1}\sqrt{\mu_2}, \qquad (12.47b)$$

$$c_2 = (2\mu_4\mu_2 - 3\mu_3^2 - 6\mu_2^3)(10\mu_4\mu_2 - 12\mu_3^2 - 18\mu_2^3)^{-1}$$

$$= (2\beta_2 - 3\beta_1 - 6)(10\beta_2 - 12\beta_1 - 18)^{-1}. \qquad (12.47c)$$

From the definitions of the various types of distributions, it is clear that equations (12.47) yield

Type I: $\kappa = \frac{1}{4}c_1^2(c_0 c_2)^{-1} = \frac{1}{4}\beta_1(\beta_2 + 3)^2(4\beta_2 - 3\beta_1)^{-1}(2\beta_2 - 3\beta_1 - 6)^{-1} < 0$.

Type II: $\beta_1 = 0$, $\beta_2 < 3$.

Type III: $2\beta_2 - 3\beta_1 - 6 = 0$.

Type IV: $0 < \kappa < 1$.

Type V: $\kappa = 1$.

Type VI: $\kappa > 1$.

Type VII: $\beta_1 = 0$, $\beta_2 > 3$.

The division of the (β_1, β_2) plane among the various types is shown in Figure 12.2. (Note that it is impossible to have $\beta_2 - \beta_1 - 1 < 0$.)

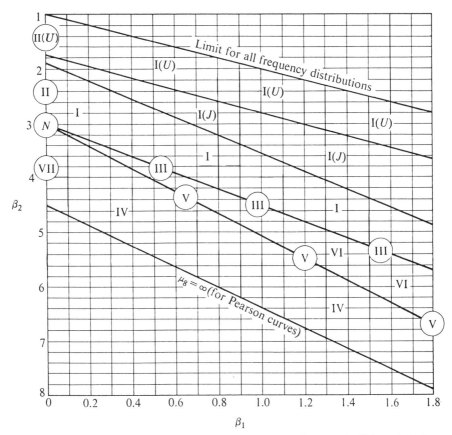

Figure 12.2   A Chart Relating the Type of Pearson Frequency Curve to the Values of β_1, β_2

The upside-down presentation of this figure is in accordance with well-established convention. Note that only Types I, VI, and IV correspond to areas in the (β_1, β_2) diagram. The remaining types correspond to lines and are sometimes called *transition types*. Other forms of diagrams have been proposed by Boetti (1964) and Craig (1936). The latter uses $(2\beta_2 - 3\beta_1 - 6)/(\beta_2 + 3)$ in place of β_2 for one axis.

Examples of fitting Pearson curves to numerical data are given in Elderton and Johnson (1969). Computer programs for producing values of random variables having Pearson type distributions have been described by Cooper et al. (1965). We conclude this subsection by noting a few general properties of Pearson type distributions.

Glänzel (1991) has shown that if $E[X^2]$ is finite, and $E[X|X \geq x]$ and $E[X^2|X \geq x]$ are differentiable functions of x, the distribution of X belongs to the Pearson system if

$$E[X^2|X \geq x] = L_1(x)E[X|X \geq x] + L_2(x)$$

where L_1 and L_2 are linear functions with real coefficients. But this is not a necessary condition. Necessary conditions can be established by addition of the conditions

$$\left(c_0 + c_1 x + c_2 x^2\right) x p(x) \to 0 \qquad [\text{see } (12.33)]$$

as x tends to the upper limit of its support.

Nair and Sankaran (1991) have obtained the following characterization of the Pearson system. If X is a continuous random variable over (a, b) with hazard rate $h_X(x)$, then it has a distribution belonging to the Pearson system if and only if

$$E[X|X > x] = \xi + \left(a_0 + a_1 x + a_2 x^2\right) h_X(x)$$

with real ξ, a_0, a_1, and a_2, for all x in (a, b) where a may be $-\infty$, and b may be $+\infty$. Korwar (1991) gives a rather more complicated characterization of the Pearson system.

By an analysis similar to that leading to equations (12.47), it can be shown that

$$\text{Mean deviation} = 2\left(\frac{1 - 3c_2}{1 - 2c_2}\right) \mu_2 p(\mu_1') \qquad (12.48)$$

for all Pearson type distributions [Pearson (1924); Kamat (1966); Suzuki (1965)]. Note that

$$\frac{1 - 3c_2}{1 - 2c_2} = \frac{4\beta_2 - 3\beta_1}{6(\beta_2 - \beta_1 - 1)} > 0.$$

As mentioned earlier, the derivative $dp(x)/dx$ equals zero at $x = -a$. There is a mode, or an antimode, of the distribution at this value of x, so

$$\text{Mode (or antimode)} - \text{Expected value} = -a.$$

Finally, we have

$$\frac{d^2 p(x)}{dx^2} = -\frac{x + a}{c_0 + c_1 x + c_2 x^2} \frac{dp(x)}{dx} + \frac{c_2 x^2 + 2ac_2 x + ac_1 - c_0}{\left(c_0 + c_1 x + c_2 x^2\right)^2} p(x)$$

$$= \frac{p(x)}{\left(c_0 + c_1 x + c_2 x^2\right)^2}\left[(x + a)^2 + c_2(x + a)^2 + a^2(1 - c_2) - c_0\right],$$

since $c_1 = a$. Provided that $a^2(1 - c_2) < c_0$, there are points of inflexion equidistant from the mode at

$$x = \text{Mode} \pm \sqrt{\{c_0 - a^2(1 - c_2)\}(1 + c_2)^{-1}}.$$

Hansmann (1934) investigated distributions for which the pdf satisfied a modified form of equation (12.33):

$$\frac{1}{p}\frac{dp}{dx} = -\frac{x}{c_0 + c_1 x^2 + c_2 x^4},\qquad (12.49)$$

where the equation $c_0 + c_1 x + c_2 x^2 = 0$ has two positive roots $b^2 \geq a^2$. The pdf's so obtained [including a correction due to Pawula and Rice (1989)] are

$$p(x) = K\left(\frac{b^2 - x^2}{x^2 - a^2}\right)^k,\qquad |x| \leq b;\qquad (12.50a)$$

$$p(x) = \begin{cases} K_1\left(\dfrac{b^2 - x^2}{a^2 - x^2}\right)^k, & |x| \leq a, \\[2mm] K_2\left(\dfrac{b^2 - x^2}{x^2 - a^2}\right)^k & a < |x| \leq b; \end{cases}\qquad (12.50b)$$

$$p(x) = K \exp\left[-\tfrac{1}{2}\{c_2(b^2 - x^2)\}^{-1}\right],\qquad |x| \leq b,\ a = b.\quad (12.50c)$$

Here $k = \frac{1}{2}\{c_2(b^2 - a^2)\}^{-1}$, and K, K_1, K_2 are normalizing constants. Note that K_1 and K_2 have to satisfy the conditions

$$\int p(x)\,dx = 1,$$

$$\int x^2 p(x)\,dx = \sigma_X^2.$$

All the distributions are symmetric about zero.

4.2 Expansions

For a wide class of continuous distributions it is possible to change the values of the cumulants by a simple application of an operator to the probability density function. If $p(x)$ is a probability density function with cumulants $\kappa_1, \kappa_2, \ldots,$ then the function

$$g(x) = \exp\left[\sum_{j=1}^{\infty} \varepsilon_j \left\{ \frac{(-D)^j}{j!} \right\} \right] p(x) \tag{12.51}$$

will have cumulants $\kappa_1 + \varepsilon_1, \kappa_2 + \varepsilon_2, \ldots$. It is necessary to explain the meaning of (12.51) rather carefully. The operator

$$\exp\left[\sum_{j=1}^{\infty} \varepsilon_j \left\{ \frac{(-D)^j}{j!} \right\} \right]$$

is to be understood in the sense described in Chapter 1. That is, the exponential must be formally expanded as

$$\sum_{i=0}^{\infty} \frac{\left[\sum_{j=1}^{\infty} \varepsilon_j \{(-D)^j/j!\}\right]^i}{i!}$$

and then applied to $p(x)$. [As in Chapter 1, Section A4, D is the differentiation operator and $D^j p(x) = d^j p(x)/dx^j$.] It should be clearly understood that $g(x)$ may not satisfy the condition $g(x) \geq 0$ for all x. Note that the cumulants of $g(x)$ are defined as coefficients of $t^r/r!$ in the expansion of

$$\log\left[\int_{-\infty}^{\infty} e^{tx} g(x)\, dx \right]$$

whether or not $g(x) \geq 0$.

Despite this limitation it is often possible to obtain useful approximate representation of a distribution with known moments (and known cumulants) in terms of a known pdf $p(x)$. By far the most commonly used initial family of distributions is the normal distribution. The representations arising from this choice of initial distribution are called *Gram-Charlier series*. From (12.51) we find (formally)

$$g(x) = p(x) - \varepsilon_1 Dp(x) + \frac{1}{2}\left(\varepsilon_1^2 + \varepsilon_2\right) D^2 p(x)$$

$$- \frac{1}{6}\left(\varepsilon_1^3 + 3\varepsilon_1\varepsilon_2 + \varepsilon_3\right) D^3 p(x)$$

$$+ \frac{1}{24}\left(\varepsilon_1^4 + 6\varepsilon_1^2\varepsilon_2 + 4\varepsilon_1\varepsilon_3 + \varepsilon_4\right) D^4 p(x) + \ldots . \tag{12.52}$$

For the approximation to the cumulative distribution function, we have

$$\int_{-\infty}^{x} g(t) = \int_{-\infty}^{x} p(t) \, dt - \varepsilon_1 p(x) + \tfrac{1}{2}(\varepsilon_1^2 + \varepsilon_2)Dp(x) + \cdots . \quad (12.53)$$

In many cases [including the case when $p(x)$ is a normal probability density function]

$$D^j p(x) = P_j(x)p(x),$$

where $P_j(x)$ is a polynomial of degree j in x. Then (12.52) can be written in the form

$$g(x) = \left[1 - \varepsilon_1 P_1(x) + \frac{1}{2}(\varepsilon_1^2 + \varepsilon_2)P_2(x) - \frac{1}{6}(\varepsilon_1^3 + 3\varepsilon_1\varepsilon_2 + \varepsilon_3)P_3(x) \right. $$
$$\left. + \frac{1}{24}(\varepsilon_1^4 + 6\varepsilon_1^2\varepsilon_2 + 4\varepsilon_1\varepsilon_3 + \varepsilon_4)P_4(x) - \cdots \right] p(x) \quad (12.52)'$$

with a corresponding form for (12.53).

If the expected values and standard deviations of $p(x)$ and $g(x)$ have been made to agree, then $\varepsilon_1 = \varepsilon_2 = 0$, and (12.52)′ becomes

$$g(x) = \left[1 - \frac{1}{6}\varepsilon_3 P_3(x) + \frac{1}{24}\varepsilon_4 P_4(x) - \cdots \right] p(x) \quad (12.54)$$

and also

$$\int_{-\infty}^{x} g(t) \, dt = \int_{-\infty}^{x} p(t) \, dt - \left[\frac{1}{6}\varepsilon_3 P_2(x) - \frac{1}{24}\varepsilon_4 P_3(x) + \cdots \right] p(x),$$
$$(12.55)$$

assuming that $P_j(x)p(x) \to 0$ at the extremes of the range of variation of x. A common way of ensuring this agreement in the expected value and the standard deviation is to use standardized variables and to choose $p(x)$ so that the corresponding distribution is standardized. If desired, the actual expected value and standard deviation can be restored by an appropriate linear transformation.

Suppose that we use a standardized variable. In taking $p(x) = (\sqrt{2\pi})^{-1}e^{-x^2/2}$ (normal), we have $(-1)^j P_j(x)$ as the Hermite polynomial $H_j(x)$ described in Chapter 1. Then, since $\kappa_r = 0$ when r is greater than 2 for the normal distribution, $\varepsilon_3, \varepsilon_4, \ldots$, are equal to the corresponding cumulants of the distribution that we want to approximate. Further, since this function is standardized, we have

$$\varepsilon_3 = \alpha_3 = \sqrt{\beta_1},$$

$$\varepsilon_4 = \alpha_4 - 3 = \beta_2 - 3,$$

where the shape factors refer to this distribution. Thus we have

$$g(x) = \left[1 + \frac{1}{6}\sqrt{\beta_1}\, H_3(x) + \frac{1}{24}(\beta_2 - 3) H_4(x) + \dots\right](\sqrt{2\pi})^{-1} e^{-x^2/2},$$

(12.56)

and by integrating both sides of (12.56), we obtain

$$\int_{-\infty}^{x} g(t)\, dt = \Phi(x) - \left[-\frac{1}{6}\sqrt{\beta_1}\, H_2(x) + \frac{1}{24}(\beta_2 - 3) H_3(x) + \dots\right]\phi(x)$$

$$= \Phi(x) - \frac{1}{6}\sqrt{\beta_1}\,(x^2 - 1)\phi(x)$$

$$- \frac{1}{24}(\beta_2 - 3)(x^3 - 3x)\phi(x) + \dots,$$

(12.57)

where

$$\Phi(x) = (\sqrt{2\pi})^{-1} \int_{-\infty}^{x} e^{-t^2/2}\, dt,$$

$$\phi(x) = (\sqrt{2\pi})^{-1} e^{-x^2/2}.$$

(12.58)

Equations (12.56) and (12.57) are known as *Gram-Charlier expansions* (1905); some earlier writers refer to them as Bruns-Charlier expansions (1906). In these expansions the terms occur in sequence determined by the successive derivatives of $\phi(x)$. This is not necessarily in decreasing order of importance, and a different ordering is sometimes used. The ordering is based on the fact that for a sum of n independent, identically distributed standardized random variables, the rth cumulant is proportional to $n^{1-r/2}$ ($r \geq 2$). This means that, in our notation, $\varepsilon_r \propto n^{1-r/2}$. Collecting terms of equal order in $n^{-1/2}$, and rearranging in ascending order, gives an *Edgeworth expansion* [Edgeworth (1896, 1907)] whose leading terms are

$$g(x) = \left[1 + \frac{1}{6}\sqrt{\beta_1}\, H_3(x) + \frac{1}{24}(\beta_2 - 3) H_4(x) + \frac{1}{72}\beta_1 H_6(x) + \dots\right]\phi(x)$$

(12.59)

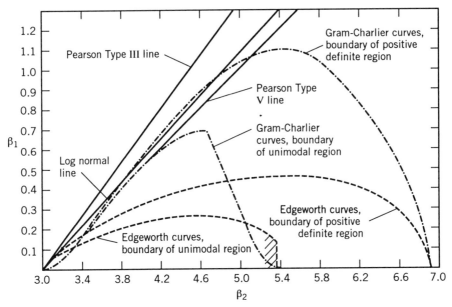

Figure 12.3 β_1, β_2 Plane Showing Regions of Unimodal Curves and Regions of Curves Composed Entirely of Nonnegative Ordinates

from which we obtain

$$\int_{-\infty}^{x} g(t)\, dt = \Phi(x) - \frac{1}{6}\sqrt{\beta_1}\,(x^2 - 1)\phi(x) - \frac{1}{24}(\beta_2 - 3)(x^3 - 3x)\phi(x)$$

$$- \frac{1}{72}\beta_1(x^5 - 10x^3 + 15x)\phi(x) + \dots . \qquad (12.59)'$$

As we noted at the beginning of this subsection, the mathematical expression obtained by applying a cumulant modifying function will not, in general, represent a proper probability density function because there are many intervals throughout which it is negative. This is also true when only a finite number of terms of the expansion is used. Figure 12.3 presents the results of an investigation by Barton and Dennis (1952) and shows the regions in the (β_1, β_2) plane where the expressions (12.56) and (12.59) are never negative. The shaded region should be excluded from the Edgeworth unimodal region, as was shown by Draper and Tierney (1972). [See also Balitskaya and Zolotuhina (1988).]

Figure 12.3 also shows the region where the curves corresponding to (12.56) and (12.59) are unimodal. Multimodality in expansions like (12.56) or (12.59), fitted to empirical data, often indicates an unnecessary fidelity to more or less accidental features of the data, in the form of "humps" in the tails. This kind of phenomenon is more likely to be encountered as more terms are used in the expansion.

In most applications only the first four moments are used, and the following terminating expressions are used:

$$g(x) = \left[1 + \frac{1}{6}\sqrt{\beta_1}(x^3 - 3x) + \frac{1}{24}(\beta_2 - 3)(x^4 - 6x^2 + 3) \right]\phi(x)$$

$$\text{(Gram-Charlier)}, \quad (12.60)$$

or

$$g(x) = \left[1 + \frac{1}{6}\sqrt{\beta_1}(x^3 - 3x) + \frac{1}{24}(\beta_2 - 3)(x^4 - 6x^2 + 3) \right.$$

$$\left. + \frac{1}{72}\beta_1(x^5 - 10x^3 + 15x) \right]\phi(x) \qquad \text{(Edgeworth)}. \qquad (12.61)$$

Note that the Edgeworth has no *general* theoretical superiority over the Gram-Charlier expansion—it depends on a particular assumption about the orders of magnitude of successive cumulants which may, or may not, be a good approximation to actual conditions.

Although the expansions (12.60) and (12.61) terminate, and so the general theory at the beginning of this paragraph does not apply, it can be seen, from the orthogonality (with normal weight function) of Hermite polynomials that they are functions with the correct values for the first four moments, and that they also satisfy the condition $\int_{-\infty}^{\infty} g(x)\,dx = 1$. Since

$$\frac{1}{\sqrt{2\pi}} \int_{-\infty}^{\infty} |x|x^j e^{-x^2/2}\,dx = 0 \qquad \text{for } j \text{ odd,}$$

and for j even

$$\frac{1}{\sqrt{2\pi}} \int_{-\infty}^{\infty} |x|x^j e^{-x^2/2} = \sqrt{\frac{2}{\pi}} \int_0^{\infty} x^{j+1} e^{-x^2/2}\,dx$$

$$= \sqrt{\frac{2}{\pi}}\, 2^{(j+1)/2 - 1/2} \int_0^{\infty} t^{j/2} e^{-t}\,dt,$$

or

$$\frac{1}{\sqrt{2\pi}} \int_{-\infty}^{\infty} |x|x^j e^{-x^2/2}\,dx = \sqrt{\frac{2}{\pi}}\, 2^{j/2} \left(\frac{1}{2}j\right)! \qquad (j \text{ even}),$$

it follows that for the Gram-Charlier (finite term) distribution as given by (12.45) the mean deviation is

$$\sqrt{\frac{2}{\pi}}\left[1 - \frac{1}{24}(\beta_2 - 3)\right] = \sqrt{\frac{2}{\pi}}\,\frac{27 - \beta_2}{24}.$$

This is also the ratio of mean deviation to standard deviation for this Gram-Charlier expansion, with general values for the expected value and variance. Note that for $\beta_2 > 27$ the mean deviation is negative. This is because the probability density function is negative for some values of x.

However, for $1 < \beta_2 < 7$,

$$\frac{5}{6}\sqrt{\frac{2}{\pi}} < \frac{\text{Mean deviation}}{\text{Standard deviation}} < \frac{13}{12}\sqrt{\frac{2}{\pi}}.$$

Similar results can be obtained for the Edgeworth expansion (12.61). For this, Bhattacharjee (1965) and Singh (1967) have obtained the distributions of extreme values and ranges, and they have given numerical values for expected values and variances in random samples of size up to 12. Subrahmaniam (1966) has obtained the distributions of linear functions of independent sample values, and of sample variance for random samples for this distribution.

It is possible to derive expansions of Gram-Charlier form by arguments similar to those used in deriving certain central limit theorems (Chapter 13). Cramér (1928) gives a general discussion; Longuet-Higgins (1964) gives an analysis from a physicist's point of view.

Some theoretical results of Bol'shev (1963) are relevant here. Starting from the normalizing transformation

$$y(x) = \Phi^{-1}[\Pr[S \le x]] \tag{12.62}$$

one can expand the argument $\Pr[S \le x]$ about $\phi(x)$, as a power series in $\Pr[S \le x] - \Phi(x)$. If this difference, in turn, be expanded as a power series in x, then (12.62) gives a power series (in x) expression for $y(x)$. [See also equation (12.21) et seq. of this chapter.)

In the particular case where $S \equiv S_n$ is the standardized sum of n independent identically distributed variables $X_1, X_2, \ldots, X_n$ with finite expected value ξ and standard deviation σ, respectively (and finite cumulants of all orders) [i.e., $S_n = (n\sigma^2)^{-1/2}\sum_{j=1}^{n}(X_j - \xi)$], there is an expansion

$$\Pr[S_n \le x] - \Phi(x) = Q(x), \tag{12.63}$$

where $Q(x)$ is a polynomial with coefficient depending on the common moment ratios of each X_i. Inserting (12.62) in (12.63), we obtain

$$y(x) = x + \sum P_{3j-1}(x) n^{-j/2} + O(n^{-(r-2)/2}), \qquad (12.64)$$

with

$$P_2(x) = \frac{1}{6} \sqrt{\beta_1} (x^2 - 1),$$

$$P_5(x) = \frac{1}{36} \beta_1 (4x^3 - 7x) - \frac{1}{24} (\beta_2 - 3)(x^3 - 3x).$$

Bol'shev shows that of all functions $u(x, n)$ satisfying the conditions that (1) $\partial^{r-2} u / \partial x^{r-2}$ exists and is continuous with respect to n on the line $n = 0$ and (2) $\partial u / \partial x$ exists in a domain

$$|x| < Cn^{-(r-2)}(r - 1), \qquad C > 0, r \geq 3,$$

the only one for which

$$\Pr[u(x, n) \leq u_0] = \phi(u_0) + O(n^{-(r-2)/2})$$

is the function given by (12.64). Bol'shev has applied this result in a number of special cases.

It will be appreciated that it is not essential that $p(x)$ in (12.52)′ must be a normal probability density function. In particular, if $p(x)$ is a standard gamma probability density function, expansions in terms of Laguerre polynomials are obtained. Such expansions have been discussed by Khamis (1958) and applied to approximate the distribution of noncentral F (Chapter 30) by Tiku (1965). If $p(x)$ is a standard beta distribution, expansions in terms of Jacobi polynomials are obtained, but these have not been much used.

Woods and Posten (1968) have made a systematic study of the use of Fourier series expansions in calculating cumulative distribution functions. Their methods are based on the following theorem: If X is a random variable with $F_X(x) = 0, 1$ for $x < 0, x > 1$, respectively, then for $0 \leq x \leq 1$,

$$F_X(x) = 1 - \theta \pi^{-1} - \sum_{j=1}^{\infty} b_j \sin j\theta, \qquad (12.65)$$

where

$$\theta = \cos^{-1}(2x - 1)$$

and

$$b_j = 2(j\pi)^{-1} E\left[\cos(j \cos^{-1}(2X - 1))\right].$$

(The function of which the expected value is to be taken is the jth Chebyshev polynomial $T_j(x)$; see Chapter 1, Section A1.)

Woods and Posten also use a generalized form of this theorem which expresses $F_X(x)$ in terms, of any conveniently chosen "distribution function" $G(x)$, with $G(x) = 0, 1$ for $x < 0$, $x > 1$, respectively. For $0 \leq x \leq 1$,

$$F_X(x) = G(x) - \sum_{j=1}^{\infty} d_j \sin j\theta, \qquad (12.66)$$

with

$$d_j = b_j - 2(j\pi)^{-1} \int_0^1 T_j(x) \, dG(x).$$

Appropriate choice of $G(x)$—usually close to $F_X(x)$—can increase the rate of convergence of the infinite series, though the d's are not so easily computed as the b's. There are similar results for the case when $F_X(x) = 0$, 1 for $x < -1$, $x > 1$.

Computer programs based on these theorems are given in Woods and Posten (1968) for evaluating the cumulative distribution and percentage points of the beta, F and chi-square distributions (Chapters 25, 27, and 18, respectively) and of the noncentral forms of these distributions (Chapters 30 and 29). In Chapter 25 (Section 6) some further details are given in regard to the application of these series to the beta distributions.

A book by Hall (1992) contains an excellent exposition of Edgeworth expansions and their properties. It includes the use of *tilted* Edgeworth expansions, designed for use when greater accuracy is required in certain parts (e.g., in upper or lower tails) of the distribution.

4.3 Transformed Distributions

If the distribution of a random variable X is such that a simple explicit function $f(X)$ has a well-known distribution, it becomes possible to use the results of research on the latter—including published tables—in studying the former distribution. The best known of such distributions is the *lognormal* distribution (Chapter 14) where $\log(X - \xi)$ has a normal distribution. Other well-known families of distributions correspond to cases in which $(X - \xi)^c$ or $e^{-(X-\xi)}$ have exponential distributions [Type II (or Weibull) and Type III extreme value distributions, respectively; see Chapters 21 and 22].

Edgeworth (1916, 1917) considered the possibility of polynomial transformations to normality. To make sure the transformation is monotonic, it is necessary to impose restrictions on the coefficients in the polynomial. This analysis is rather complicated and this kind of transformation is not often used at present.

Plotting on probability paper* will indicate the form of the transformation. One of the earliest papers on this method is Kameda (1928). Further references include Flapper (1967) and Chambers and Fowlkes (1966). Sets of "model plots" of quantiles of various distributions against those of the unit normal distribution contained in the latter reference can be helpful in deciding on suitable transformations.

By analogy with the Pearson system of distributions, it would be convenient if a simple transformation to a normally distributed variable could be found such that, for any possible pair of values $\sqrt{\beta_1}, \beta_2$ there is just one member of the corresponding family of distributions. No such single simple transformation is available, but Johnson (1949, 1954) and Tadikamalla and Johnson (1982, 1989) have described sets of three such transformations which, when combined, do provide one distribution corresponding to each pair of values $\sqrt{\beta_1}$ and β_2. We now describe the sets presented in 1949 in some detail.

One of the three transformations is simply

$$Z = \gamma + \delta \log(X - \xi), \qquad X \geq \xi, \qquad (12.67a)$$

which corresponds to the family of lognormal distributions. The others are

$$Z = \gamma + \delta \log \left\{ \frac{X - \xi}{\xi + \lambda - X} \right\}, \qquad \xi \leq X \leq \xi + \lambda, \qquad (12.67b)$$

$$Z = \gamma + \delta \sinh^{-1} \left\{ \frac{X - \xi}{\lambda} \right\}. \qquad (12.67c)$$

The distribution of Z is, in each case, unit normal. The symbols γ, δ, ξ, and λ represent parameters. The value of λ must be positive, and we conventionally make the sign of δ positive also.

The range of variation of X in (12.67b) is bounded and the corresponding family of distributions is denoted by S_B; in (12.67c) the range is unbounded, and the symbol S_U is used. For lognormal distributions the range is bounded below (if $\delta < 0$, it would be bounded above).

It is clear that the shapes of the distribution of X depends only on the parameters γ and δ (δ only, for lognormal). Writing $Y = (X - \xi)/\lambda$, we have

$$Z = \gamma' + \delta \log Y, \qquad \text{for lognormal } (\gamma' = \gamma - \delta \log \lambda), \qquad (12.68a)$$

$$Z = \gamma + \delta \log \left\{ \frac{Y}{1 - Y} \right\}, \qquad \text{for } S_B, \qquad (12.68b)$$

$$Z = \gamma + \delta \sinh^{-1} Y, \qquad \text{for } S_U, \qquad (12.68c)$$

---

*Probability paper is graph paper designed so that a plot of the cumulative frequency against the variable value would give a linear relation for a specified distribution—often the normal distribution.

and Y must have a distribution of the same shape as X. The moments of Y in (12.68a) are given in Chapter 14.

For S_B, from (12.68b),

$$\mu'_r(Y) = (\sqrt{2\pi})^{-1} \int_{-\infty}^{\infty} [1 + e^{-(z-\gamma)/\delta}]^{-r} e^{-z^2/2} \, dz.$$

Although it is possible to give explicit expressions for $\mu'_r(Y)$ (not involving integral signs) [see Johnson (1949), equations (56) and (57)], they are very complicated.

It is interesting to note that for S_B distributions,

$$\frac{\partial \mu'_r}{\partial \gamma} = \frac{r}{\delta}(\mu'_{r+1} - \mu'_r), \qquad (12.69a)$$

$$\frac{\partial \mu'_r}{\partial \delta} = \frac{r}{\delta^3}(\gamma\delta - r)(\mu'_r - \mu'_{r+1}) + \frac{r(r+1)}{\delta^3}(\mu'_{r+1} - \mu'_{r+2}). \quad (12.69b)$$

For S_U, however, we obtain from (12.68a),

$$\mu'_r(Y) = (\sqrt{2\pi})^{-1} 2^{-r} \int_{-\infty}^{\infty} [e^{(z-\gamma)/\delta} - e^{-(z-\gamma)/\delta}]^r e^{-z^2/2} \, dz \quad (12.70)$$

and this can be evaluated in a straightforward manner, yielding the following values for the expected value, and lower central moments of Y:

$$\mu'_1(Y) = \omega^{1/2} \sinh \Omega, \qquad (12.71a)$$

$$\mu_2(Y) = \tfrac{1}{2}(\omega - 1)(\omega \cosh 2\Omega + 1), \qquad (12.71b)$$

$$\mu_3(Y) = -\tfrac{1}{4}\omega^{1/2}(\omega - 1)^2\{\omega(\omega + 2)\sinh 3\Omega + 3 \sinh \Omega\}, \quad (12.71c)$$

$$\mu_4(Y) = \tfrac{1}{8}(\omega - 1)^2\{\omega^2(\omega^4 + 2\omega^3 + 3\omega^2 - 3)\cosh 4\Omega + 4\omega^2(\omega + 2)$$

$$\times \cosh 2\Omega + 3(2\omega + 1)\}, \quad (12.71d)$$

where $\omega = \exp(\delta^{-2})$, $\Omega = \gamma/\delta$. Note that $\omega > 1$ and that μ_3, so $\alpha_3(= \sqrt{\beta_1})$ has the *opposite* sign to γ. For $\gamma = 0$, the shape factors (for both X and Y) are

$$\alpha_3^2 = \beta_1 = 0,$$

$$\alpha_4 = \beta_2 = \tfrac{1}{2}(\omega^4 + 2\omega^2 + 3).$$

As γ increases both β_1 and β_2 increase, and the (β_1, β_2) point approaches the point with coordinates $[(\omega - 1)(\omega + 2)^2, \omega^4 + 2\omega^3 + 3\omega^2 - 3]$ as $\gamma \to \infty$. The latter point is on the "lognormal line" (see Chapter 14). It corresponds to a lognormal distribution defined by (12.67a).

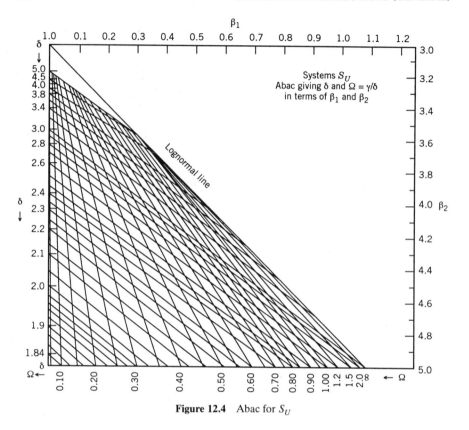

Figure 12.4   Abac for S_U

The variation of (β_1, β_2) with γ and δ is shown diagrammatically in Figure 12.4. This figure can be used to obtain approximate values of γ and δ, for given β_1 and β_2 (note that the sign of γ must be *opposite* to that of $\sqrt{\beta_1}$; see above). More accurate values can be obtained using tables in Johnson (1965), possibly combined with methods of iterative calculation described in that reference. Methods of fitting based on order statistics have been described by Mage (1982), Slifker and Shapiro (1980), and Wheeler (1980). Siekierski (1992) contains a useful comparative study of methods of fitting.

It can be seen from Figure 12.4 (and proved analytically) that for any (β_1, β_2) point "below" the lognormal, line, there is an appropriate S_U distribution. Similarly for any possible point "above" the lognormal line there is an appropriate S_B distribution. In fact the lognormal, S_U and S_B families (or systems) cover the whole possible (β_1, β_2) plane uniquely—namely there is just one appropriate distribution corresponding to each (β_1, β_2) point. The normal distribution, corresponding to the point $(0, 3)$, has to be considered as the limiting form of all three families as $\delta \to \infty$.

Some typical probability density functions belonging to the S_B and S_U families are shown in Figures 12.5a–f (S_B) and 12.6a–c (S_U) [taken from

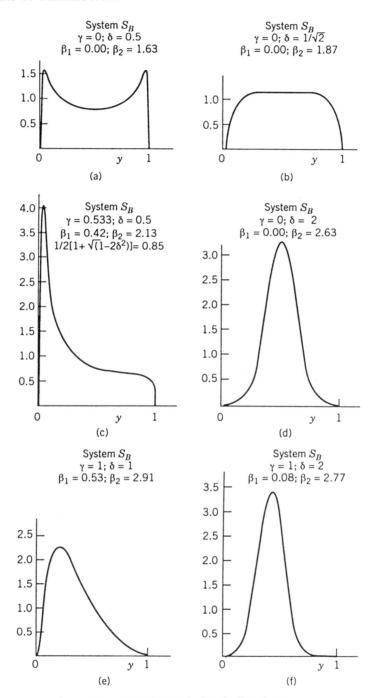

Figure 12.5 System S_B Density Functions

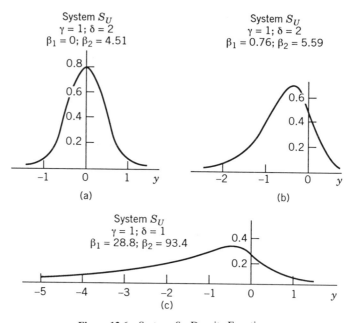

Figure 12.6   System S_U Density Functions

Johnson (1949)]. All S_U curves are unimodal; S_B curves may be unimodal, or they may have two modes, with an antimode between them. The latter case occurs if

$$\delta < 1/\sqrt{2},$$

$$|\gamma| < \delta^{-1} \sqrt{1 - 2\delta^2} - 2\delta \tanh^{-1} \sqrt{1 - 2\delta^2}. \tag{12.72}$$

There are transition cases in which one mode and the antimode coalesce into a point of inflection (see Figures 12.5a, c). If $\gamma = 0$ and $\delta = 1/\sqrt{2}$ (Figure 12.5b), a flat-topped distribution is obtained.

For all S_B and S_U (and also lognormal) distributions there is "high contact" at the extremities of the range of variation. That is to say, not only does the probability density function tend to zero as the extremity is approached, but so do all derivatives. (This applies as $Y \to \pm\infty$, as well as when the extremities are finite.) This property is not shared by all Pearson system distributions.

Numerical comparisons of the cumulative distribution functions of the Pearson Type IV and S_U distributions corresponding to the same pair of values of β_1 and β_2 indicate remarkably close agreement. Distributions of either system may often be used in place of their counterpart in the other system with low risk of errors of practical importance being incurred. Agreement is poorest in the lower tails for large β_1 and/or β_2. Rieck and

Nedelman (1991) introduce a sort of reverse S_U system, wherein X is distributed as a linear function of $\sinh^{-1}((Z - \gamma)/\delta)$, with Z having a unit normal distribution. Thus

$$Z = \gamma + \delta \sinh\left(\frac{X - \xi}{\lambda}\right). \tag{12.73}$$

Analogous sets of three transformations, with Z [in equations (12.68)] having (1) a standard Laplace distribution (S'_L, S'_B, S'_U) or (2) a standard logistic distribution (L_L, L_B, L_U) have been described by Johnson (1954) and Tadikamalla and Johnson (1982), respectively. These two sets possess similar properties to S_L, S_B, S_U. Among these properties are the following:

1. They cover the (β_1, β_2) plane uniquely with $S'_B(L_B)$ "above" and $S'_U(L_U)$ "below" the dividing log-Laplace (log-logistic) line.
2. The S'_B and L_B curves are bimodal for (β_1, β_2) "above" a line in the $S'_B(L_B)$ area, while all the S'_U and L_U curves are unimodal.

Transformations of a standard uniform variable U whose pdf is

$$p_U(u) = 1, \qquad 0 < u < 1$$

(see Chapter 26) have the useful feature that it is very easy to simulate observed values.

Tukey lambda distributions [Hastings et al. (1947); Tukey (1962)] are defined by the transformation

$$X = \frac{aU^\lambda - (1 - U)^\lambda}{\lambda}, \qquad a > 0, \lambda \neq 0; \ -\lambda^{-1} \leq X \leq \lambda^{-1} \text{ for } \lambda > 0,$$
$$\tag{12.74a}$$

with the transitional $\lambda = 0$ transformation

$$X = \log\left\{\frac{U^a}{1 - U}\right\}, \qquad a > 0. \tag{12.74b}$$

Transformations (12.74a) and (12.74b) are each monotonic increasing, so if $\lambda \neq 0$,

$$x = \frac{au^\lambda - (1 - u)^\lambda}{\lambda},$$

then

$$F_X(x) = u,$$

since $F_U(u) = u$, $0 \leq u \leq 1$. Also $F_X(X)$ has a standard uniform distribution, so the transformations (12.74a) and (12.74b) can be regarded as expressions for X corresponding to a specified value F of $F_X(x)$. Expressed differently,

$$x(F) = x = \frac{aF^\lambda - (1-F)^\lambda}{\lambda}, \qquad a > 0, \lambda \neq 0, \qquad (12.74a)'$$

and

$$X = \log\left\{\frac{U^a}{1-U}\right\}, \qquad a > 0. \qquad (12.74b)'$$

Originally attention was mainly directed at the *symmetrical* Tukey lambda distributions $a = 1$, defined by

$$X = \frac{U^\lambda - (1-U)^\lambda}{\lambda}, \qquad \lambda \neq 0, \qquad (12.75a)$$

and

$$X = \log\left\{\frac{U}{1-U}\right\} \qquad \text{(corresponding to } \lambda \to 0\text{)}. \qquad (12.75b)$$

The lower quartile, median and upper quartile for (12.74a) are obtained by putting $U = \frac{1}{4}, \frac{1}{2}$, and $\frac{3}{4}$, which yields

$$(a - 3^\lambda)(4^\lambda\lambda)^{-1}, \quad (a - 1)(2^\lambda\lambda)^{-1} \quad \text{and} \quad (3^\lambda - 1)(4^\lambda\lambda)^{-1},$$

respectively. The value of Galton's skewness index

$$\frac{\text{Lower quartile} + \text{Upper quartile} - 2(\text{Median})}{\text{Upper quartile} - \text{Lower quartile}}$$

is given by

$$\frac{a - 1}{a + 1} \frac{3^\lambda - 2^{\lambda+1} + 1}{3^\lambda - 1}. \qquad (12.76)$$

For (12.74a) the rth moment about zero of X is

$$\mu_r'(X) = \frac{1}{\lambda^r} \int_0^1 \left\{au^\lambda - (1-u)^\lambda\right\}^r du$$

$$= \frac{1}{\lambda^r} \sum_{j=0}^r (-1)^j \binom{r}{j} a^{r-j} B(\lambda(r-j) + 1, \lambda j + 1). \qquad (12.77)$$

In particular, the expected value and variance are

$$E[X] = \lambda^{-1}(\lambda + 1)^{-1}(a - 1) \qquad (12.78a)$$

and

$$Var(X) = (\lambda + 1)^{-1}(2\lambda + 1)^{-1}(a^2 + 1)$$
$$- 2\lambda^{-2}\{B(\lambda + 1, \lambda + 1) - (\lambda + 1)^{-2}\}a. \quad (12.78b)$$

Joiner and Rosenblatt (1971) provide a detailed analysis for the symmetrical distributions (12.75a). They point out that although *explicit* formulas for the pdf of X are not generally available, the values $\lambda = 1$ and $\lambda = 2$ both correspond to uniform distributions for X [over range $(-1, 1), (-\frac{1}{2}, \frac{1}{2})$, respectively]. Also, for appropriate values of λ, very good approximations to normal and t distributions (Chapter 13) can be obtained. They suggest that $\lambda = 0.135$ for normal approximation. (Of course location and scale parameters can be introduced to give any required expected value and standard deviation.) The least possible value of β_2 is 1.75 (approx.). For any value greater than this, there are *two* possible values of λ. Thus for $\beta_2 = 3$ (as for a normal distribution) the values of λ are 0.135 (as suggested above) and *also* 5.2. The two corresponding pdf's (standardized to have expected value zero and standard deviation 1) are shown in Figure 12.7 [taken from Joiner and Rosenblatt (1971)]. The pdf's of X are slightly U-shaped for $1 < \lambda < 2$; they are unimodal for $\lambda < 1$ or $\lambda > 2$.

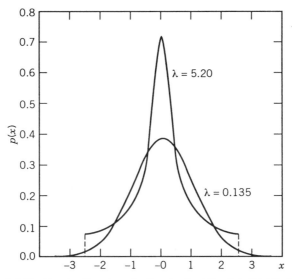

Figure 12.7 Probability Density Functions for Standardized Lambda Distributions, Each Having $\beta_2 = 3.00$ (and $\beta_1 = 0$)

Ramberg (1975) deals with distributions defined by a slight modification of (12.74a), namely

$$X = \lambda_1 + \frac{aU^{\lambda_3} - (1 - U)^{\lambda_3}}{\lambda_2}. \tag{12.74c}$$

He shows that the (β_1, β_2) points cover the whole plane except those "above" a line joining the uniform $(0, 1.8)$ and exponential $(4, 9)$ points [in the Pearson Type I (beta) area]. He gives tables of values of λ_1, λ_2, and λ_3 producing specified values for β_1 and β_2 for a standardized variable.

Skew distributions for X can be obtained in other ways than by putting $a \neq 1$ in (12.74a). Johnson and Kotz (1973) replace U in (12.75a) by T, where T has a standard beta (θ, ϕ) distribution. The resulting pdf of X is

$$p_X(x) = \{B(\theta, \phi)\}^{-1}\{t(x)\}^{\theta-1}\{1 - t(x)\}^{\phi-1}$$
$$\times \left[\{t(x)\}^{\lambda-1} + \{1 - t(x)\}^{\lambda-1}\right]^{-1}, \tag{12.79}$$

where $t(x)$ satisfies the equation

$$x = \frac{\{t(x)\}^{\lambda} - \{1 - t(x)\}^{\lambda}}{\lambda}. \tag{12.80}$$

In general there is no explicit formula for $p_X(x)$, as we have already noted. The rth moment about zero is

$$\mu_r'(X) = \{B(\theta, \phi)\}^{-1}\lambda^{-r} \sum_{j=0}^{r} \binom{r}{j}(-1)^j B(\theta + (r - j)\lambda, \phi + j\lambda). \tag{12.81}$$

Clearly $p_X(x) \to 0$ or ∞ at the extremes of the range of variation of X, according as $p_T(t) \to 0$ or ∞, for $\lambda > 1$.

We now consider the variation in shape of the distribution of X with changes in λ, the parameters θ and ϕ having fixed values. Since $X - (1 - X) = 2X - 1 = X^2 - (1 - X)^2$, we see that the values of $\sqrt{\beta_1}$ and β_2 must be the same for $\lambda = 1$ and $\lambda = 2$. In each one of these two cases we have a beta distribution with the same parameters, though different range, as that of the original variables. By virtue of the continuity of the function involved, β_2 must take a maximum or minimum value for some value of λ between 1 and 2. It is in fact a minimum when $\theta = \phi$, but can be a maximum when $\theta \neq \phi$, and occurs for a value of λ in a remarkably narrow range about $\lambda = 1.45$; see Table 12.1 for a few examples. By considering neighboring loci of $(\sqrt{\beta_1}, \beta_2)$ for varying λ, with slightly different values of θ and ϕ it can be seen that at

Table 12.1 Values of λ Giving Maximum or Minimum β_2

θ	ϕ	λ	$\sqrt{\beta_1}$	β_2	θ	ϕ	λ	$\sqrt{\beta_1}$	β_2
0.5	0.5	1.44	0	1.469[a]	2	2	1.46	0	2.086[a]
0.5	1.0	1.47	0.662	2.134[a]	2	5	1.48	0.661	2.944
0.5	2.0	1.44	1.318	3.987	2	10	1.46	1.035	4.090
0.5	5.0	1.43	2.109	8.171	2	15	1.44	1.195	4.752
0.5	10.0	1.41	2.549	11.69	5	5	1.48	0	2.489[a]
1	1	1.45	0	1.753	5	10	1.51	0.380	2.836
1	2	1.44	0.604	2.401	5	15	1.48	0.551	3.155
1	5	1.45	1.299	4.549	10	10	1.49	0	2.705[a]
1	10	1.43	1.688	6.573	10	15	1.48	0.179	2.804[a]

[a] Minimum.

least for some values of $\sqrt{\beta_1}$ and β_2, there will be more than one set of values (θ, ϕ, λ) giving these values for the shape parameters.

This might be expected, since there are three parameters available to give two specified values $\sqrt{\beta_1}$ and β_2. Another way of looking at the situation is to consider what happens when λ is fixed, but θ and ϕ vary. It is known that for $\lambda = 1$, the region between the boundary $\beta_2 - (\sqrt{\beta_1})^2 - 1 = 0$, and the Type III line $2\beta_2 - 3(\sqrt{\beta_1})^2 - 6 = 0$ is covered. For $\lambda > 0$, generally, the region covered is that between the line $\beta_2 - (\sqrt{\beta_1})^2 - 1 = 0$, and the line [Johnson and Kotz (1972)] corresponding to (Type III variable)$^\lambda$. The latter is approached as $\phi \to \infty$, with θ remaining constant.

From (12.79) we see that $dp_X(x)/dx = 0$ if

$$y^{\lambda - 1} = \frac{1 - \lambda y + (\phi y - \theta)}{y - \lambda - (\phi y - \theta)} = g(y), \text{ say}, \qquad (12.82)$$

where $y = t(x)\{1 - t(x)\}^{-1}$. If $\theta = \phi = 1$, we find that $g(y) = y$. Then the only possible modal values are $y = 0$, 1, or ∞ [i.e., $t(x) = 0$, $\frac{1}{2}$ or 1 corresponding to $x = -\lambda$, 0, or λ].

For $\lambda < 1 < \min(\theta, \phi)$, $g(y)$ increases from 0 to ∞ as y increases from $(\theta - 1)/(\phi - \lambda)$ to $(\theta - \lambda)/(\phi - 1)$, while $y^{\lambda - 1}$ decreases as u increases. Equation (12.82) thus has just one root, since for y outside this range $g(y)$ is negative, and so the distribution of X is unimodal.

A similar situation, in reverse, holds if $1 < \lambda < \min(\theta, \phi)$. If $\phi < \min(1, \lambda)$ and $\theta > \max(1, \lambda)$, then $g(y)$ is always negative for $u > 0$, and (12.82) has no solution. In this case, since $\theta > 1$ and $\phi < 1$, $p_X(x) \to 0$ as $y \to -1$ and $\to \infty$ as $y \to 1$. The density function of X is J-shaped, as is that of T. Related results hold similarly if $\phi > \max(1, \lambda)$ and $\theta < \min(1, \lambda)$. For $\phi > 1$ and $\theta > 1$, $p_Y(y) = 0$ at $t = 0$ or 1; hence, the slope of $p_X(x)$ is also zero at these points.

An interesting special case corresponds to $\lambda = \theta + \phi - 1$. Then (12.82) becomes

$$y = \left(\frac{\theta - 1}{\phi - 1} \right)^{(\theta + \phi - 2)^{-1}}. \tag{12.83}$$

In the symmetrical case $\theta = \phi$, (12.82) is satisfied by $y^{\lambda - 1} = 1$, whence $t = \frac{1}{2}$ and the corresponding modal value for X is 0, as is to be expected. The distribution can, however, be bimodal, with antimode at 0. This is so, for example, if $\lambda > 1$ and $\theta = \phi < \frac{1}{2}(\lambda^2 - \lambda + 2)$. Moments of order statistics are easily obtained. Joiner and Rosenblatt (1971) give formulas for the rth moment of X_i' and for the expected value and variance of the range in random samples of size n.

Other systems that include asymmetric distributions are

$$X = \begin{cases} -\lambda_1 (\lambda_2 - U)^{\lambda_3}, & \lambda_2 \geq U, \\ \lambda_1 (U - \lambda_2)^{\lambda_3}, & \lambda_2 \leq U \end{cases} \tag{12.84}$$

[Schmeiser and Deutsch (1977), quoted by Butterworth (1987)],

$$X = U^{\lambda_1} - (1 - U)^{\lambda_2} \quad (\lambda_1 \neq \lambda_2) \tag{12.85}$$

[Ramberg and Schmeiser (1974); Ramberg et al. (1979)]. As in all cases, location and scale parameters can be introduced to give desired values for expectation and variance. Transformation (12.84) has the attractive feature that it has a simple explicit inverse.

Lam, Bowman, and Shenton (1980) discuss (12.85). They note, as did Joiner and Rosenblatt (1971) for the symmetrical distribution, that the equations for λ_1 and λ_2 obtained by requiring $\sqrt{\beta_1}$ and β_2 to have specified values may not have unique solutions. Figure 12.8, taken from Lam, Bowman, and Shenton (1980), indicates loci of constant $\sqrt{\beta_1}$ and of constant β_2 as functions of λ_2 and λ_1 / λ_2. This paper also includes comparisons of 1%, 5%, 95%, and 99% points for standardized curves of various systems with common values $\sqrt{\beta_1}, \beta_1$, indicating quite close agreement. Similar more extensive comparisons are available in several other papers [e.g., Pearson, Johnson, and Burr (1979)]. The closeness of results is often due to the fact that the different systems can have similar features of smoothness that increase comparability beyond that to be expected from just equality of first four moments. Mallows (1956) provides an interesting study of these possibilities (see also Chapter 33).

Another system of this kind, ascribed to H. A. Thomas by Hosking (1986), quoting Houghton (1978), is defined by the relation

$$X = \xi + \left(\frac{\alpha}{\beta} \right) \{ 1 - (1 - U)^{\beta} \} - \left(\frac{\gamma}{\delta} \right) \{ 1 - (1 - U)^{-\delta} \}. \tag{12.86}$$

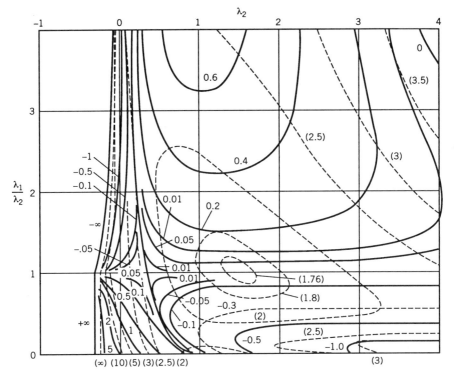

Figure 12.8 Contours of $\sqrt{\beta_1}$ and β_2 for Distribution (12.85). ——————— constant $\sqrt{\beta_1}$
- - - - - - - - - constant β_2

Essentially $X = \phi(1 - U)^{-\delta} - \theta(1 - U)^{\beta} + k$ with $\phi = \gamma/\delta,\ \theta = \alpha/\beta;\ k = \xi + \theta - \phi$.

If $\alpha,\ \beta,\ \gamma$, and δ are each positive, X is already a monotonic increasing function of U. Since (12.86) is unchanged if α and γ, and β and $-\delta$ are interchanged, it can be assumed, without loss of generality, that $\beta + \delta \geq 0$. The following further conditions ensure uniqueness of the relation between X and U [i.e., equation (12.86) has only one solution]:

1. Either $\beta + \delta > 0$, or $\beta = \gamma = \delta = 0$.
2. If $\alpha = 0$, then $\beta = 0$.
3. If $\gamma = 0$, then $\delta = 0$.
4. $\gamma \geq 0$.
5. $\alpha + \gamma \geq 0$.

These distributions have been used in hydrology to model flood and stream flow data. They are called *Wakeby distributions*.

The range of variation of X is from ξ up to infinity if $\delta \geq 0$ and $\gamma > 0$. If *either* $\delta < 0$ or $\gamma = 0$, the range is finite, from ξ to $\xi + \alpha/\beta - \gamma/\delta$

$(= \xi + \theta - \phi)$. The pdf $p(x)$ of X is related to the cdf $F(x)$ by the formula

$$p(x) = \left[\alpha\{1 - F(x)\}^{\beta-1} + \gamma\{1 - F(x)\}^{-\delta-1}\right]^{-1}$$
$$= \{1 - F(x)\}^{\delta+1}(\alpha t + \gamma)^{-1} \tag{12.87}$$

with

$$t = \{1 - F(x)\}^{\beta+\delta}.$$

As $x \to \xi$,

$$p(x) \to \begin{cases} \infty & \text{if } \alpha + \gamma = 0, \\ (\alpha + \gamma)^{-1} & \text{if } \alpha + \gamma > 0. \end{cases}$$

As $x \to \infty$,

$$p(x) \sim \begin{cases} \gamma^{1/\delta}(\delta x)^{-1-1/\delta} & \text{if } \delta > 0, \\ \gamma^{-1} \exp\left(\dfrac{-x}{\gamma}\right) & \text{if } \delta = 0, \gamma > 0. \end{cases} \tag{12.88a}$$

If $\delta < 0$ so that the range is from ξ to $\omega = \xi + \theta - \phi$, then as $x \to \omega$,

$$p(x) \sim (-\delta^{-1})(-\theta)^{\delta-1}(\omega - x)^{-\delta-1-1}. \tag{12.88b}$$

Pdf's of some Wakeby distributions are shown in Figures 12.9, for parameter values shown below ($\xi = 0$ in all cases):

Pdf Number	α	β	γ	δ	Pdf Number	α	β	γ	δ
1	30	20			9	5			
2	1	5	1	0.3	10	0.5	5	1	−0.8
3	1	2			11	0.01			
4	1	0.5			12	−0.5			
5	10	10			13	5			
6	1	3	1		14	0.5	10		
7	1	2		−0.3	15	0.01		1	−1.1
8	1	1	3		16	−0.5	5		
					17	−0.9	1.12	10	
					18	−1.1	1.11	1.11	

The fact that the distributions have *three* disposable shape parameters (plus location and scale parameters) gives the system considerable flexibility. See also generalized Pareto distributions (Chapter 20, Section 8).

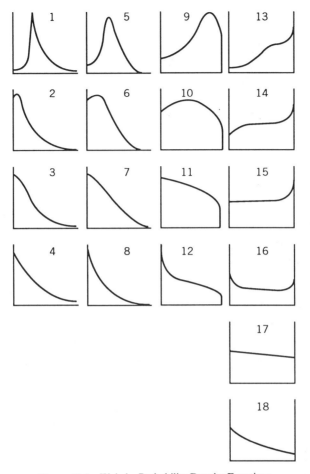

Figure 12.9   Wakeby Probability Density Functions

Hoaglin (1984) describes the *Tukey g- and h-system* obtained by the transformations

$$X = \begin{cases} \xi + \lambda g^{-1} \exp\left(\tfrac{1}{2}hZ^2 + gZ - 1\right), & g \neq 0, \\ \xi + \lambda Z \exp\left(\tfrac{1}{2}hZ^2\right), & g = 0, \end{cases} \tag{12.89}$$

where Z is a unit normal variable. This system covers a large part of the (β_1, β_2) diagram, excluding some symmetrical U-shaped distributions.

Combining the Pearson and Tukey lines of approach, Voit (1992) defines a system of *S-distributions* by the differential equation

$$p(x) = \frac{dF(x)}{dx} = \alpha\left[\{F(x)\}^g - \{F(x)\}^h\right]. \tag{12.90}$$

There are three parameters α, g, and h in (12.90). A fourth (location) parameter ξ is introduced by the condition

$$F(0) = F_X(\xi).$$

α is a scale parameter; g and h determine the shape. This system includes a few standard distributions—notably logistic distributions ($g = 1, h = 2$)—and gives good approximations to many others. It is especially well suited for computational work.

The condition $p(x) \geq 0$ requires that $g < h$ [if $g = h$, we have $p(x) = 0$ for all x] and $\alpha \neq 0$. Voit (1992) notes that the distribution of X is symmetric about zero if

$$\{F(x)\}^g - \{F(x)\}^h = \{1 - F(x)\}^g - \{1 - F(x)\}^h \qquad \text{for all } x,$$

or

$$\{F(x)\}^g - \{1 - F(x)\}^g = \{F(x)\}^h - \{1 - F(x)\}^h \qquad \text{for all } F(x)$$

in $[0, 1]$, which leads to $g = 1$, $h = 2$ (logistic). [Equation clearly holds if $F(x) = \frac{1}{2}$ at the median of X but not, in general, for other values of x.]

At a mode, with $dp(x)/dx = 0$, we have

$$gp(x)\{F(x)\}^{g-1} = hp(x)\{F(x)\}^{h-1}.$$

So, if $p(x) \neq 0$,

$$g\{F(x)\}^{g-1} = h\{F(x)\}^{h-1}. \tag{12.91}$$

Since $\{F(x)\}^{g-h}$ is a monotonic function of $F(x)$, there is a single mode, at $F^{-1}((g/h)^{1/(h-g)})$.

Pdf's of three S-distributions are shown in Figure 12.10, taken from Voit (1992).

Power transformations have generally been the subject of several studies. We will encounter some special cases in later chapters including power transformations of gamma and Pareto variables (generalized gamma and generalized Pareto) in Chapters 17 and 20, respectively. For power transformations to normality, Box and Cox (1964) is a standard reference, though there are several others. They discussed transformations of the form

$$Y = \begin{cases} \dfrac{X^\lambda - 1}{\lambda}, & \lambda \neq 0, \\ \log X, & \lambda = 0. \end{cases} \tag{12.92}$$

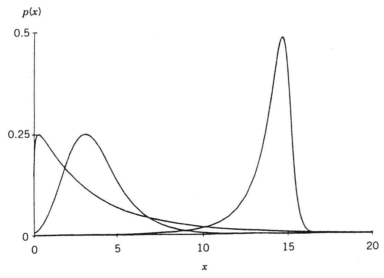

Figure 12.10 Densities of Three *S*-Distributions. From left to right: $g = 0.25$, $h = 0.5$; $g = 0.75$, $h = 1.5$; $g = 1.2$, $h = 5$. In all three cases $\alpha = 1$ and $F(0) = 0.01$.

(The $\lambda = 0$ of course leads to lognormal distributions; see Chapter 14.) This approach is limited to variables X that have support excluding negative values. In particular, X cannot have a normal distribution. Attempts at overcoming this constraint include papers by Moore (1957) and Goto, Inoue, and Tsuchiya (1983, 1987).

Moore (1957) approached the problem by seeking to find numbers r, m, and c such that

$$Z = \frac{X^r - m}{c}$$

is approximately normal, with a very small probability that Z is negative (i.e., m/c is large).

Goto, Matsubara, and Tsuchiya (1983) define a *power-normal* distribution having pdf,

$$\frac{x^\lambda}{\Phi(\text{sgn}(\lambda)|\lambda m + 1|/\lambda\sigma)}\phi\left(\frac{x^\lambda - 1 - \lambda\mu}{\lambda}\right), \qquad \lambda \neq 0, \quad (12.93)$$

where

$$\Phi(u) = (\sqrt{2\pi})^{-1}\int_{-\infty}^{u} \exp(-\tfrac{1}{2}t^2)\,dt;$$

$$\phi(u) = \Phi'(u) = (\sqrt{2\pi})^{-1}\exp(-\tfrac{1}{2}u^2).$$

4.4 Bessel Function Distributions

McKay (1932) described a system of distributions that would provide (at least) one distribution corresponding to any pair of values (β_1, β_2) for which $(\beta_2 - 3)/\beta_1 > 1.5$ (i.e., "below" the Type III line). In fact for the narrow strip

$$1.5 < \frac{\beta_2 - 3}{\beta_1} \leq 1.57735, \qquad (12.94)$$

there are *three* possible distributions of this system corresponding to a single (β_1, β_2) point. Although most of the formulas presented below were derived by McKay, we will first mention a possible genesis of the system constructed by Bhattacharyya (1942). The distributions can in fact be obtained as distributions of $X_1\sigma_1^2 \pm X_2\sigma_2^2$ where X_1, X_2 are mutually independent random variables, each distributed as χ^2 with ν degrees of freedom (see Chapter 18).

The distribution of $Y = X_1\sigma_1^2 + X_2\sigma_2^2$ is the first of McKay's forms. It has probability density function

$$p(y) = \frac{|1 - c^2|^{m+1/2}|y|^m}{\pi^{1/2}2^m b^{m+1}\Gamma(m + \frac{1}{2})} e^{-cy/b} I_m\left(\left|\frac{y}{b}\right|\right), \qquad y > 0, \quad (12.95)$$

with

$$b = 4\sigma_1^2\sigma_2^2(\sigma_1^2 - \sigma_2^2)^{-1},$$

$$c = (\sigma_1^2 + \sigma_2^2)(\sigma_1^2 - \sigma_2^2)^{-1} > 1,$$

$$m = 2\nu + 1.$$

The distribution of $Z = X_1\sigma_1^2 - X_2\sigma_2^2$ is the second of McKay's forms. It has probability density function

$$p_Z(z) = \frac{|1 - c^2|^{m+1/2}|y|^m}{\pi^{1/2}2^m b^{m+1}\Gamma(m + \frac{1}{2})} e^{-cz/b} K_m\left(\left|\frac{z}{b}\right|\right), \qquad (12.96)$$

with

$$b = 4\sigma_1^2\sigma_2^2(\sigma_1^2 + \sigma_2^2)^{-1},$$

$$c = -(\sigma_1^2 - \sigma_2^2)(\sigma_1^2 + \sigma_2^2)^{-1}, \qquad |c| < 1,$$

$$m = 2\nu + 1.$$

In (12.95) and (12.96), $I_m(\cdot)$ and $K_m(\cdot)$ are modified Bessel functions (of the first and second kind, respectively) of order m (see Chapter 1, Section 3).

For *both* kinds of distributions (12.95) and (12.96) the moment-generating function is

$$\left[(1-c^2)\{1-(c-tb)^2\}^{-1}\right]^{m+1/2}$$

(with of course appropriate values of b and c, depending on which kind of distribution is being considered).

It follows that the rth cumulant is

$$\kappa_r = (r-1)!(m+\tfrac{1}{2})b^r(c^2-1)^{-r}\left[(c-1)^r + (c+1)^r\right]. \quad (12.97)$$

(This can be established directly from Bhattacharyya's approach.) In particular

$$\mu_1' = (2m+1)bc(c^2-1)^{-1},$$

$$\mu_2 = (2m+1)b^2(c^2+1)(c^2-1)^{-2},$$

$$\beta_1 = 4c^2(c^2+3)^2(2m+1)^{-1}(c^2+1)^{-3},$$

$$\beta_2 = 3 + 6(c^4+6c^2+1)(2m+1)^{-1}(c^2+1)^{-2}. \quad (12.98)$$

$\sqrt{\beta_1}$ has the opposite sign to c. From the last two equations of (12.98) it can be shown that

$$2c^2(c^2+3)^2\left(\frac{\beta_2-3}{\beta_1}\right) - 3(c^2+1)(c^4+6c^2+1) = 0. \quad (12.99)$$

Regarded as a cubic equation in c^2, (12.99) has a single positive root for

$$\frac{\beta_2-3}{\beta_1} > 1.57735,$$

and three positive roots for $(\beta_2-3)/\beta_1 > 1.5$. [The three positive roots are given by McKay (1932) to 5 significant figures for $(\beta_2-3)/\beta_1 = 1.502(0.002)1.576$.] For the region between the line

$$\beta_2 - 1.57735\beta_1 - 3 = 0 \quad (12.100)$$

(termed the *Bessel line* by McKay) and the axis of β_2 there is a unique "Bessel distribution" corresponding to any given (β_1, β_2) point. Figure 12.11 [taken from Bhattacharyya (1942)] shows how the values of m and c^2 vary over this region, which was called the K-region by McKay [because only distributions of form (12.96) can be used here]. This diagram also shows how

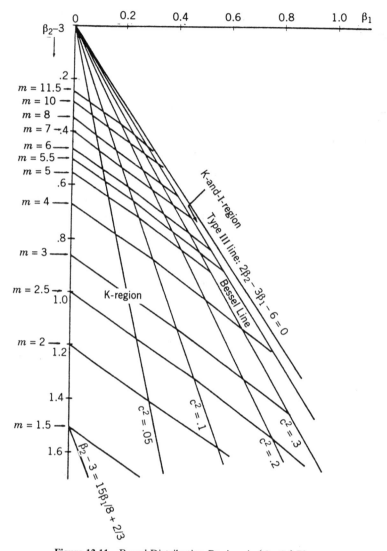

Figure 12.11   Bessel Distribution Regions in (β_1, β_2) Plane

narrow is the K and I *region*, where there are two distributions of form (12.95) and one of form (12.96) for each pair of values (β_1, β_2).

It would seem that the K-form (12.96) is likely to be more generally useful than the I-form (12.95). Indeed, it has been suggested that (12.96) would lead to less troublesome computation than the Pearson Type IV distributions that have the same (β_1, β_2) values. However, the I-form (12.95) has been used [Bose (1938)] for graduating an observed frequency distribution. Also the family of I-distributions include the noncentral χ^2 distributions [see Laha (1953) and Chapter 29] and the distribution of the Mahalanobis D^2-statistic. A generalization of the K-form, with probability density function proportional to $|z|^{m'} K_m(|z/b|)$, $m' \neq m$, has been studied by Sastry (1948).

McNolty (1967) has described the application of Bessel function I-distributions to the distributions of signal or noise in output processed by a radar receiver under various sets of conditions. This paper contains interesting accounts of how the Bessel distribution might be expected to arise in each case. McLeish (1982) suggests the family of symmetrical distributions (of T) generated by $T = G^{1/2}Z$, where G has a gamma $(\alpha, 2)$ distribution (see Chapter 17) and Z is a unit normal variable, independent of G, as "robust alternatives" to the normal distribution, in the sense of possessing robust maximum likelihood estimators. The resulting distribution has pdf,

$$p_T(t) = g(t) = \frac{1}{\sqrt{\pi}\,\Gamma(\alpha)} \left(\frac{|t|}{2} \right)^{\alpha - 1/2} K_{\alpha - 1/2}(t).$$

If α is an integer, we have

$$g_\alpha(t) = \frac{e^{-|t|}}{2^{2\alpha - 1}} \sum_{j=0}^{\alpha - 1} \binom{2\alpha - 1 - j}{\alpha - 1} \frac{|2t|^j}{j!}.$$

Location and scale parameters may be introduced in the usual way. McLeish (1982) notes that this distribution was defined by Pearson, Jeffery, and Elderton (1929) and studied in Pearson, Stouffer, and David (1932), and that a more general family with pdf,

$$p_T(t) = (1 - a^2)^\alpha e^{at} g_\alpha(t)$$

is "closely" related to McKay's Bessel-function distributions.

4.5 Miscellaneous

Burr (1942) has suggested a number of forms of cumulative distribution function that might be useful for purposes of graduation. In choosing one of these forms of distribution, the principal aim is to facilitate the mathematical analysis to which it is to be subjected, while attaining a reasonable approximation.

The forms for the cumulative distribution function ($F_Y(y)$) listed by Burr are shown below (the first corresponds to a uniform distribution and is included because it is in the original list):

  (I) y for $0 < y < 1$.
  (II) $(e^{-y} + 1)^{-k}$.
  (III) $(y^{-c} + 1)^{-k}$ for $0 < y$.
  (IV) $[((c - y)/y)^{1/c} + 1]^{-k}$ for $0 < y < c$.
  (V) $(ce^{-\tan y} + 1)^{-k}$ for $-\pi/2 < y < \pi/2$.
  (VI) $(ce^{-k\sinh y} + 1)^{-k}$.
  (VII) $2^{-k}(1 + \tanh y)^k$.
  (VIII) $[(2/\pi)\tan^{-1} e^y]^k$.
  (IX) $1 - 2/\{c[(1 + e^y)^k - 1] + 2\}$.
  (X) $(1 - e^{-y^2})^k$ for $0 < y$.
  (XI) $[y - (1/2\pi)\sin 2\pi y]^k$ for $0 < y < 1$.
  (XII) $1 - (1 + y^c)^{-k}$ for $0 < y$.

Here k and c are positive parameters. Setting $y = (x - \xi)/\lambda$, we can introduce two extra parameters.

Burr (1942) devoted special attention to the family (XII) of distributions. The probability density function of $Z = Y^c$ is

$$k(1 + z)^{-(k+1)}, \qquad z > 0. \tag{12.101}$$

and the rth moment of X is

$$E[Z^{r/c}] = kB\left(\frac{r}{c} + 1, k - \frac{r}{c}\right), \qquad r < ck. \tag{12.102}$$

Burr gives, for $c = 1(1)10$ and $k = 1(1)11$, tables of mean and standard deviation to 5 decimal places, and of $\sqrt{\beta_1}$ and β_2 to 3 decimal places or 4 significant figures. In Burr (1968) and Burr and Cislak (1968) there is further discussion of properties of this family, with special reference to the distributions of sample median and range. There is a thorough analysis of Type XII distributions in Rodriguez (1977). [See also Vodă (1982).] Wingo (1993) has discussed the maximum likelihood estimation of the parameters of Type XII distributions based on progressively censored samples.

It should be noticed that it is possible to relate some of the above forms of distribution by simple transformations. Thus (III) can be obtained from (II) by replacing y by $c \log y$. Fry (1993) has recently provided a detailed survey of the univariate and multivariate Burr distributions.

Mielke's (1973) *kappa distributions* have cdf's of form

$$\left\{1 - (1 + y^{\alpha\theta})^{-1}\right\}^{1/\alpha}. \tag{12.103}$$

Fattorini and Lemmi (1979) describe applications and provide graphs of some pdf's. Ferreri (1964) has described a system of distributions with probability density functions of form

$$p_X(x) = \sqrt{b} \left[G\left(-\tfrac{1}{2}, a\right) \right]^{-1} \left[\exp\{a + b(x - \xi)^2\} + c \right]^{-1}, \qquad c = \pm 1,$$

$$(12.104)$$

where

$$G(p, a) = \Gamma(p + 1) \sum_{j=1}^{\infty} (-c)^{j-1} e^{-ja} j^{-(p+1)}.$$

This distribution depends on four parameters a, b, c, and ξ. The rth absolute moment about ξ (the mean) is

$$b^{-r/2} \frac{G\left(\tfrac{1}{2}(r - 1), a\right)}{G\left(-\tfrac{1}{2}, a\right)}. \qquad (12.105)$$

For $c = 1$, $\beta_2 < 3$; for $c = -1$, $\beta_2 > 3$. As a increases, the distribution approaches normality.

Among other systems of distributions we note those described by Toranzos (1952) and Laha (1961). Toranzos describes a class of bell-shaped frequency distributions with probability density functions of form

$$(\text{Constant}) \cdot x^c \exp\left[-\tfrac{1}{2}(\alpha + \beta x)^2 \right], \qquad x > 0. \qquad (12.106)$$

Laha considers distributions for which there is a standard form having characteristic function $(1 + |t|^\alpha)^{-1}$ for some value of α in range $0 < \alpha \le 2$. This distribution, known as the *Linnik distribution*, is discussed in Chapter 24.

There are some very broad classes of distributions, including most of the common distributions. We will not devote much attention to these, as we are concerned more with properties of specific distributions than with broad classification. The *exponential-type class* includes all density functions that can be written in the form

$$p(x) = \exp[A(x)B(\theta) + C(x) + D(\theta)], \qquad (12.107)$$

where $A(\cdot)$, $B(\cdot)$, $C(\cdot)$, and $D(\cdot)$ are arbitrary functions. This class was recognized nearly simultaneously by Darmois (1935) and Koopman (1936) (as the form taken by the density function if a single sufficient statistic for θ exists, given values of n independent identically distributed random variables). It is often called the *Darmois-Koopman* (or *Koopman-Darmois*) class. A subclass of these distributions, still very broad, is thoroughly discussed in Morris (1983).

An even broader class is that of the *Pólya-type* distributions, introduced by Karlin (1957). A frequency function $\Lambda(x)$ is said to be a Pólya-type frequency function if for every positive integer n and every pair of sets of increasing numbers $x_1 < x_2 < \cdots < x_n$, $y_1 < y_2 < \cdots < y_n$ the determinant $\|\Lambda(x_i - y_k)\| \geq 0$. A characterization of Pólya-type frequency functions by means of the structure of their characteristic functions was given by Lukacs (1968).

Mathai and Saxena (1966) have pointed out that the formula for the density function

$$p_X(x) = \frac{da^{c/d}\Gamma(\alpha)\Gamma(\beta)\Gamma(r - c/d)}{\Gamma(c/d)\Gamma(r)\Gamma(\alpha - c/d)\Gamma(\beta - c/d)} x^{c-1}{}_2F_1(\alpha, \beta; r; -ax^d)$$

$$(12.108)$$

(with $x > 0$, $c > 0$, $\alpha - c/d > 0$, $\beta - c/d > 0$) and the limiting form obtained by letting α tend to infinity, and a to zero in such a way that a^α tends to a':

$$p_X(x) = \frac{da'^{c/d}\Gamma(\beta)\Gamma(r - c/d)}{\Gamma(c/d)\Gamma(r)\Gamma(\beta - c/d)} x^{c-1}{}_1F_1(\beta; r; -a'x^d) \quad (12.109)$$

can be made to represent a considerable variety of commonly used distributions by appropriate choice of values for the parameters; the confluent hypergeometric functions ${}_2F_1$ and ${}_1F_1$ are defined in Chapter 1, Sections A6 and A7, respectively.

The class of *stable distributions* has very considerable importance in probability theory, though statistical applications appear to be rather limited. Nevertheless, Mandelbrot (1963) and Fama (1965) have applied stable laws to stock market data, though this application was criticized by Nicklin and Paulson (1975). See also Fielitz and Smith (1972), and, more recently, Akgiray, Booth, and Loistl (1989). Holtsmark (1919) applied them to modeling electrical fields; see also DuMouchel (1973). Informally speaking, a location scale family is stable if the convolution of two distributions belonging to the family also belongs to the family. More formally if the pdf of X_i is

$$\frac{1}{\sigma_i} f\left(\frac{x - \theta_i}{\sigma_i}\right), \qquad i = 1, 2,$$

and X_1 and X_2 are independent, then the pdf of $(X_1 + X_2)$ is of form $f[(x - \theta)/\sigma]/\sigma$ for some θ and σ.

An alternative definition [Monrad and Stout (1988)] is that if $X, X_1, \ldots, X_n$ are i.i.d. and there exist constants $a_n > 0$ and b_n such that $\sum_{j=1}^n X_j/a_n - b_n$ has the same distribution as X, the common distribution is stable (excluding the case of degenerate one-point distributions).

All stable distributions have characteristic functions (Chapter 1, Section B8) of form

$$\varphi_x(t) = \exp\left[i\mu t - c|t|^\alpha\{1 + i\beta \, \text{sgn}(t)\omega(|t|,\alpha)\}\right], \qquad (12.110)$$

where

$$\omega(|t|,\alpha) = \begin{cases} \tan(\tfrac{1}{2}\pi\alpha), & \alpha \neq 1, \\ 2n^{-1}\log(t), & \alpha = 1, \end{cases}$$

and $i = \sqrt{-1}$ $(0 < \alpha \le 2; \; -1 \le \beta \le 1)$. μ is a location parameter and c is a scale parameter. If $c = 0$, the distribution is degenerate. The shape of the pdf is determined by the parameters α and β. The latter is a skewness parameter. If $\beta = 0$, the distribution is symmetric. When $\beta = 0$,

- $\alpha = 2$ gives normal distributions (Chapter 13) for any value of β,
- $\alpha = 1$ gives Cauchy distributions (Chapter 16),
- $\alpha = \tfrac{1}{2}$ gives distributions with the same shape as $(\chi^2$ with 1 degree of freedom$)^{-1}$, a special case of Pearson Type III (Section 4.2).

For other values of α (and/or $\beta \neq 0$) we do not have simple expressions for the pdfs. There are a few special cases [Zolotarev (1954)]:

$$\alpha = \frac{1}{3}, \quad \beta = 1 \qquad \text{(the \textit{Macdonald} distribution)}$$

$$\left. \begin{array}{l} \alpha = \dfrac{2}{3}, \quad \beta = 0 \text{ or } 1 \\[2mm] \alpha = \dfrac{3}{2}, \quad \beta = 1 \end{array} \right\} \qquad \text{(in terms of Whittaker functions)}$$

[Kropac (1982) has a discussion of distributions involving Macdonald functions.]

There are explicit formulas for the pdf's of standardized stable distributions as convergent series, although they are rather complicated. For $0 < \alpha < 1$,

$$p_X(x) = \frac{1}{\pi x} \sum_{j=1}^{\infty} \frac{(-1)^{j-1}\Gamma(j\alpha + 1)}{j!} \; \frac{\sin\left[j\{\tfrac{1}{2}\pi\alpha - \tan^{-1}(\beta \tan \tfrac{1}{2}\pi\alpha)\}\right]}{x^{j\alpha}(1 + \beta^2 \tan^2 \tfrac{1}{2}\pi\alpha)^{j/2}}.$$

$$(12.111a)$$

For $1 < \alpha \le 2$,

$$p_X(x) = \frac{1}{\pi x} \sum_{j=1}^{\infty} \frac{(-1)^{j-1} \Gamma(j\alpha_{-1} + 1)}{j!} x^j \left(1 + \beta^2 \tan^2 \frac{1}{2}\pi\alpha\right)^{j/(2\alpha)}.$$

(12.111b)

Chernin and Ibragimov (1959) have shown that all stable pdf's are unimodal.

Figures 12.12a–c show pdf's of stable distributions for $\alpha = 0.5$, 1.0, and 1.5 each for various values of β, taken from Holt and Crow (1973), which further contains figures and tables of the pdf to 4 decimal places, for

$\alpha = 0.25(0.25)1.75$,

$\beta = -1.00(0.25)1.00$,

$x = 0.000(0.001)0.010(0.01)2.00(0.1)M(1)M^*(5 \text{ or } 10)$;

the values of M, M^* were chosen to allow for easy interpolation out to value 0.0001 for the pdf.

Tables of the cdf [to four decimal places, when $x = 0.05(0.05)1.00(0.1)2.0(0.2)4.0(0.4)6.0(1)8$, 10, 15, 20] and percentile points [to 3 decimal places, with $p = 0.52(0.02)0.94(0.01)0.97(0.005)0.995, 0.9975$] for *symmetric* stable distributions ($\beta = 0$) with $\alpha = 1.0(0.1)1.9, 1.95, 2.0$ had already been given by Fama and Roll (1968). Soon after Holt and Crow (1973), additional tables of the cdfs of standardized symmetric distributions were given by Worsdale (1975). Included were values for $\alpha = 0.6(0.1)2.0$ with $x = 0(0.05)3.00$, and also with $\log_{10} x = 0.40(0.05)2.50$. Values were given to five decimal places, though there is a warning of possible round-off error in the fourth place.

Tables calculated by Worsdales's method, for $\alpha = 1.0(0.1)1.9, 1.95, 2.0$ and $x = 0.00(0.05)1.00(0.1)2.0(0.2)4.0(0.4)6.0, 7, 8, 10, 15, 20$ are presented in Panton (1992). There has been some reluctance to include tables for smaller values of α, presumably associated with the somewhat unusual shapes of the corresponding distributions, already becoming apparent in Figure 12.12a (where $\alpha = 0.5$).

Properties of symmetric stable distributions have been studied by Gawronski (1984) and Nagaer and Shkol'nik (1989). Gawronski (1984) showed that all stable densities are "bell-shaped"—that is, $p_X(x)$ is infinitely differentiable (for x real), and $d^k p_X(x)/dx^k$ has exactly k zeros (all simple) on the support of the distribution varying with α and chosen so as to make interpolation feasible. An unusual feature is that for $\alpha \ne 1$, the heavier tail is in the negative direction (negative skewness in the ordinary sense) for $\beta > 0$, and conversely for $\beta < 0$; the situation is reversed for $\alpha = 1$. (See Figure 12.12b).

de Haan and Resnick (1980) presented a simple asymptotic estimate for the index of a stable distribution. Saniga, Pfaffenberger, and Hayya (1975)

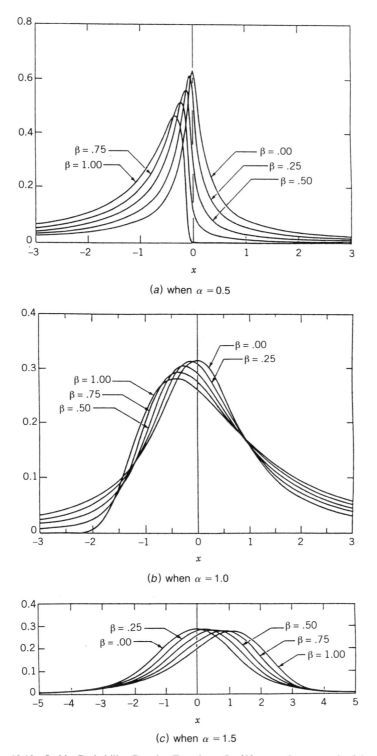

Figure 12.12 Stable Probability Density Functions. In (b) note the reversal of location of heavier tail from that for $\alpha \neq 1$.

discussed small-sample properties of the maximum likelihood estimator of the index α by means of Monte Carlo simulations. These authors also examined the power of various goodness-of-fit tests for stable versus normal and normal versus stable distributions.

Barndorff-Nielsen (1978) observed that from observational data, plots of log (frequency) against value of the variable will often show approximately hyperbolic formula. This implied that the pdf would be of form

$$\left\{2\sqrt{1 + \alpha^2}\, K_1(\zeta)\right\}^{-1} \exp\left[\sqrt{(1 + \alpha^2)(1 + x^2)} - \alpha x\right], \quad (12.112)$$

where $\zeta < 0$ and $K_1(\zeta)$ is the Bessel function (Chapter 1, Section A5). Location and scale parameters, θ and σ, respectively, can be introduced, making four parameters in all. Equation (12.112) defines the *hyperbolic system* of distributions. Barndorff-Nielsen and Blaesild (1983) give a succinct account of this system. The expected value for (12.112) is

$$E[X] = \frac{\alpha K_2(\zeta)}{K_1(\zeta)}, \quad (12.113a)$$

and the variance is

$$\text{Var}(X) = \zeta^{-1}\left\{\frac{K_2(\zeta)}{K_1(\zeta)}\right\} + \frac{\alpha^2\{K_3(\zeta)K_1(\zeta) - K_2^2(\zeta)\}}{K_1^2(\zeta)}. \quad (12.113b)$$

Sargan pdf's [Goldfield and Quandt (1981), ascribed to D. Sargan] are of form

$$\tfrac{1}{2}\left\{\sum_{j=0}^{p}(j!\beta_j)^{-1}\right\}e^{-\alpha|x|}\sum_{j=0}^{p}\beta_j\alpha^j|x|^j, \qquad \beta_0 = 1; \alpha > 0, \beta_j \geq 1. \quad (12.114)$$

This is a *p-order* Sargan density. The distributions are symmetrical about zero. Goldfield and Quandt (1981) recommend (12.114) as a viable alternative to normal distributions in econometric modeling. For $p = 0$ we have Laplace distributions (Chapter 24). Missiakoulis (1983) notes that the pdf of the arithmetic mean of $p + 1$ independent Laplace variables is a p-order Sargan density, and he discusses the choice of order when fitting a Sargan distribution. Tse (1987) also considers this question and indicates a preference for higher orders than those recommended by Missiakoulis (1983).

Azzalini (1985) noted that if X and Y are independent random variables, each having a pdf symmetric about zero, then for any λ,

$$\tfrac{1}{2} = Pr[X - \lambda Y < 0] = \int_{-\infty}^{\infty} p_Y(y) F_X(\lambda y) \, dy.$$

Hence $2 p_Y(y) F_X(\lambda y)$ is a pdf. Taking X and Y each to be unit normal variables, we obtain Azzalini's "*skew-normal*" class of distributions, with pdf's,

$$2\phi(x)\Phi(\lambda x). \tag{12.115}$$

These distributions have a single parameter λ. ($\lambda = 0$ gives a unit normal distribution.) Location and scale parameters can be added, in the usual way. It is interesting to note that if X has pdf (12.115), then X^2 is distributed as χ^2 with one degree of freedom (for all values of λ).

To increase coverage of the (β_1, β_2) plane Azzalini (1985) introduced a further parameter, ξ, and defined densities

$$p_X(x|\lambda, \xi) = \frac{\phi(x)\Phi(\lambda x + \xi)}{\Phi\left\{\xi(1 + \lambda^2)^{-1/2}\right\}}. \tag{12.116}$$

Henze (1986) showed that if X_1 and X_2 are independent standard normal variables ($N(0, 1)$), then

$$X = (1 + \lambda^2)^{-1/2}\{\lambda|X_1| + X_2\}$$

has pdf (12.115). Also, extending Azzalini's original approach, he showed that the distribution of X_2, conditional on $X_1 < \lambda X_2 + \xi$ has pdf (12.116).

Mallows (1983) constructed a system of distributions that would cover the whole (β_1, β_2) plane in a smooth manner, without the sharp transitions of the Pearson and Johnson systems. The system is defined as distributions of the random variable

$$X = \xi + \lambda(\psi Y_\omega + \theta Z_\alpha), \tag{12.117}$$

where ξ, λ, ψ, ω, θ, and α are parameters, Y_ω and Z_α are mutually independent, Y_ω is a continuous variable with pdf

$$\{R(\omega)\}^{-1}(1 + y^2)^{-1}\phi(\omega y),$$

with $R(\omega) = \{1 - \Phi(\omega)\}/\{\phi(\omega)\}$ (Mills's ratio; see Chapter 13). Z_α is a two-point Bernoulli variable with

$$Pr\left[Z_\alpha = \sqrt{\left(\frac{1 - \alpha}{\alpha}\right)}\right] = \alpha,$$

$$Pr\left[Z_\alpha = -\sqrt{\left(\frac{\alpha}{1 - \alpha}\right)}\right] = 1 - \alpha.$$

$[E[Z_\alpha] = 0, \ \mathrm{Var}(Z_\alpha) = 1, \ \mu_3(Z_\alpha) = \{\alpha(1 - \alpha)\}^{-2}(1 - 2\alpha), \ \text{and} \ \mu_4(Z_\alpha) = \{\alpha(1 - \alpha)\}^{-1} - 3.]$

Although there are six parameters in (12.117), the scale factor (λ) is redundant. Mallows suggested fitting by the first four moments, introducing the relationship

$$\{\beta_2(Y_\omega) - \beta_2(X)\}\{\beta_2(Y_\omega) - 3\} = 2\beta_1(X). \tag{12.118}$$

From (12.118) the value of

$$\beta_2(Y_\omega) = \frac{1}{\omega^3 R(\omega)\left[\{\omega R(\omega)\}^{-1}\right]^2} - \frac{1}{\{\omega R(\omega)\}^{-1} - 1}, \tag{12.119}$$

and hence of ω, is determined. Then θ is determined from

$$\beta_2(X) = \theta^{-2}\beta_1(X) + \theta^4 + 6\theta^2(1 - \theta^2) + \beta_2(Y_\omega)(1 - \theta^2)^2. \tag{12.120}$$

Finally, the values

$$\psi = \left\{(1 - \theta^2)\left[\omega R(\omega)\}^{-1} - 1\right]\right\}^{1/2} \tag{12.121}$$

and

$$\alpha = \tfrac{1}{2}\left[1 - \left\{\beta_1(X)(\beta_2(X))^{-1} + 4\theta^6\right\}^{1/2}\right] \tag{12.122}$$

are calculated.

The standard *slash* distribution is the distribution of the ratio of a unit normal variable to an independent standard uniform $(0, 1)$ variable. Its pdf is

$$p(x) = \begin{cases} \dfrac{\{\phi(0) - \phi(x)\}}{x^2}, & x \neq 0, \\ \frac{1}{2}\phi(0), & x = 0, \end{cases} \qquad (12.123)$$

where $\phi(x) = (\sqrt{2\pi})^{-1} \exp(-\frac{1}{2}x^2)$. Equation (12.123) is mainly of use in simulation studies. General slash distributions can be obtained by replacing the unit normal variable by a general normal $N(\xi, \sigma^2)$ random variable. Properties of this family of distributions have been discussed by Rogers and Tukey (1972) and Mosteller and Tukey (1977). Their applications to robustness studies through Monte Carlo simulations have been well illustrated by Andrews et al. (1972) and Gross (1973). The maximum likelihood estimation of location and scale parameters for this family has been described by Kafadar (1973).

Albert, Delampady, and Polasek (1991) have proposed the use of an *extended-power family* of distributions in robustness studies. These are symmetric distributions with pdf,

$$p_X(x) = K\theta^{1/2} \exp\left[-\frac{1}{2}c\lambda^{-1}\left\{\left(1 + \frac{\theta}{c-1}(x-\xi)^2\right)^\lambda - 1\right\}\right],$$

$$\lambda, \theta > 0; c > 1, \quad (12.124a)$$

where K depends on c and λ but not on θ. For $\lambda = 0$,

$$p_X(x) = K\theta^{1/2} \exp\left[-\frac{1}{2}c\log\left\{1 + \frac{\theta}{c-1}(x-\xi)^2\right\}\right] \quad (12.124b)$$

is suggested. This is a Cauchy distribution (Chapter 16), since the right hand side of (12.124b) can be written as

$$K\theta^{1/2}\left\{1 + \frac{\theta}{c-1}(x-\xi)^2\right\}^{-c/2}.$$

5   CORNISH–FISHER EXPANSIONS

If any distribution is fitted by making the first s moments of the fitted and actual distributions agree, it is, in principle, possible to calculate quantiles of the fitted distribution and to regard these as approximations to the corresponding quantiles of the actual distribution. In fact the fitted quantiles are functions of the s fitted moments. So we have estimators of the actual quantiles which are functions of these s moments.

Usually these functions are very complicated and not easily expressible in explicit form. However, in the case of the Gram-Charlier and Edgeworth expansions, described in Section 4.2, it is possible to obtain explicit expansions for standardized quantiles as functions of corresponding quantiles of the unit normal distributions. In these expansions the terms are polynomial functions of the appropriate unit normal quantile, with coefficients that are functions of the moment-ratios of the distribution. We now outline the method of derivation of these expansions, using an argument propounded by Cornish and Fisher (1937). [The argument has been reformulated in Fisher and Cornish (1960) and extended by Finney (1963).]

From (12.51) we have formally

$$\int_{-\infty}^{x} g(t)\, dt = D^{-1} \exp\left[\sum_{j=1}^{\infty} \frac{\varepsilon_j(-D)^j}{j!}\right] p(x), \qquad (12.125)$$

and if $p(x) = \phi(x) = (\sqrt{2\pi})^{-1} e^{-x^2/2}$, then $D^j p(x) = (-1)^j H^j(x)\phi(x)$. Now suppose that X_α and U_α are defined by

$$\int_{-\infty}^{X_\alpha} g(x)\, dx = \alpha = \int_{-\infty}^{U_\alpha} \phi(x)\, dx.$$

Using the expansion (12.125)

$$\int_{-\infty}^{X_\alpha} \phi(x)\, dx + \left[\sum_{i=0}^{\infty} D^{-1} \frac{\left\{\sum_{j=1}^{\infty} \varepsilon_j(-D)^j/j!\right\}^i}{i!}\right] \phi(X_\alpha) = \int_{-\infty}^{U_\alpha} \phi(x)\, dx.$$

$$(12.126)$$

We now expand the right-hand side as

$$\int_{-\infty}^{U_\alpha} \phi(x)\, dx = \int_{0}^{X_\alpha} \phi(x)\, dx + \sum_{j=0}^{\infty} \left\{ \frac{(U_\alpha - X_\alpha)^j}{j!} \right\} D^j \phi(X_\alpha)$$

$$= \int_{0}^{X_\alpha} \phi(x)\, dx + \sum_{j=1}^{\infty} \left\{ \frac{(X_\alpha - U_\alpha)^j}{j!} \right\} H_j(X_\alpha)\phi(X_\alpha). \quad (12.127)$$

Inserting this in (12.126) gives the *identity*

$$
\left[\sum_{i=0}^{\infty} D^{-1} \frac{\left\{ \sum_{j=1}^{\infty} \varepsilon_j (-D)^j / j! \right\}^i}{i!} \right] \phi(X_\alpha)
$$

$$
= \left[\sum_{j=1}^{\infty} \left\{ \frac{(X_\alpha - U_\alpha)^j}{j!} \right\} H_j(X_\alpha) \right] \phi(X_\alpha). \qquad (12.128)
$$

Expanding the left-hand side and dividing both sides by $\phi(X_\alpha)$ gives an identity, of polynomial form, between $(X_\alpha - U_\alpha)$ and X_α. By straightforward, though tedious, algebra it is possible to rearrange (12.118) to obtain either U_α as a function of X_α [i.e., $U_\alpha = U(X_\alpha)$] or X_α as a function of U_α [i.e., $X_\alpha = X(U_\alpha)$].

Cornish and Fisher (1937) gave detailed formulas for $U(X_\alpha)$ and $X(U_\alpha)$, and they extended these in Fisher and Cornish (1960). They collected terms according to Edgeworth's system (see Section 4.3). Based on their formula we have (to order n^{-1} if κ_r is of order $n^{1-r/2}$ with the distribution standardized)

$$
\begin{aligned}
X(U_\alpha) = U_\alpha &+ \frac{1}{6}(U_\alpha^2 - 1)\kappa_3 \\
&+ \frac{1}{24}(U_\alpha^3 - 3U_\alpha)\kappa_4 - \frac{1}{36}(2U_\alpha^3 - 5U_\alpha)\kappa_3^2 \\
&+ \frac{1}{120}(U_\alpha^4 - 6U_\alpha^2 + 3)\kappa_5 - \frac{1}{24}(U_\alpha^4 - 5U_\alpha^2 + 2)\kappa_3\kappa_4 \\
&+ \frac{1}{324}(12U_\alpha^4 - 53U_\alpha^2 + 17)\kappa_3^3 \\
&+ \frac{1}{720}(U_\alpha^5 - 10U_\alpha^3 + 15U_\alpha)\kappa_6 \qquad (12.129) \\
&- \frac{1}{180}(2U_\alpha^5 - 17U_\alpha^3 + 21U_\alpha)\kappa_3\kappa_5 \\
&- \frac{1}{384}(3U_\alpha^5 - 24U_\alpha^3 + 29U_\alpha)\kappa_4^2 \\
&+ \frac{1}{288}(14U_\alpha^5 - 103U_\alpha^3 + 107U_\alpha)\kappa_3^2\kappa_4 \\
&- \frac{1}{7776}(252U_\alpha^5 - 1688U_\alpha^3 + 1511U_\alpha)\kappa_3^4 + \cdots.
\end{aligned}
$$

Also

$$U(X_\alpha) = X_\alpha - \frac{1}{6}(X_\alpha^2 - 1)\kappa_3$$

$$-\frac{1}{24}(X_\alpha^3 - 3X_\alpha)\kappa_4 + \frac{1}{36}(4X_\alpha^3 - 7X_\alpha)\kappa_3^2$$

$$-\frac{1}{120}(X_\alpha^4 - 6X_\alpha^2 + 3)\kappa_5 + \frac{1}{144}(11X_\alpha^4 - 42X_\alpha^2 + 15)\kappa_3\kappa_4$$

$$-\frac{1}{648}(69X_\alpha^4 - 187X_\alpha^2 + 52)\kappa_3^3$$

$$-\frac{1}{720}(X_\alpha^5 - 10X_\alpha^3 + 15X_\alpha)\kappa_6 \qquad\qquad (12.130)$$

$$+\frac{1}{360}(7X_\alpha^5 - 48X_\alpha^3 + 51X_\alpha)\kappa_3\kappa_5$$

$$+\frac{1}{384}(5X_\alpha^5 - 32X_\alpha^3 + 35X_\alpha)\kappa_4^2$$

$$-\frac{1}{864}(111X_\alpha^5 - 547X_\alpha^3 + 456X_\alpha)\kappa_3^2\kappa_4$$

$$+\frac{1}{7776}(948X_\alpha^5 - 3628X_\alpha^3 + 2473X_\alpha)\kappa_3^4 + \cdots.$$

The formulas given in the 1937 and 1960 papers include terms adjusting the mean and variance. In formulas (12.129) and (12.130), however, it has been assumed that the distribution to be fitted has been standardized, so no correction is needed.

Numerical values of the coefficients in formula (12.129) of Fisher and Cornish (1960) are given to 5 decimal places for $\alpha = 0.5$, 0.75, 0.9, 0.95, 0.975, 0.99, 0.995, 0.999, and 0.9995. This paper also gives the values of the first seven Hermite polynomials, to 12 decimal places, for the same values of α. It is especially to be noted that the functional forms $U(\cdot)$ and $X(\cdot)$ *do not* depend on the value of α. The function $U(X)$ may be regarded as a *normalizing transformation* of the random variable X. The function $X(\cdot)$ expresses the quantiles of the (standardized) distribution of X as a function of corresponding quantiles of the unit normal distribution.

In practice only a finite number of terms of the expansions $U(\cdot)$ or $X(\cdot)$ are used. It is important to recognize that the results obtained are not equivalent to those obtained by retaining a similar number (or indeed any specified number) of terms in the Gram-Charlier expansions. They may be (in favorable cases) good approximations to the quantiles of the "distributions" represented by the complete expansions.

Even though the Cornish-Fisher expansions are directly related to the Edgeworth form of distribution, there is a difference in the way these two are used. It is unusual to use moments higher than the fourth in fitting an Edgeworth (or Gram-Charlier) expansion. This is partly because the possibility of negative values (and multimodality) becomes more serious as further terms are added and partly because, with observed data, estimation of higher moments is often of low accuracy. Cornish-Fisher expansions, on the other hand, are more usually applied to theoretically determined distributions (with known moments), and it is quite usual to use moments of order as high as six, or even greater.

As Finney (1963) has pointed out, it would be possible to obtain analogues of Cornish-Fisher expansions by operating on Laguerre series (or other) forms of distribution in the same way as described (for Edgeworth series) at the beginning of this section. Use of Cornish-Fisher expansions to calculate percentiles of Pearson distributions has been reported by Bowman and Shenton (1979a, b) and Davenport and Herring (1979). Modifications of the expansion have been described by McCune (1977). Lee and Lin (1992) present a computer program for extending the Cornish-Fisher expansion to terms of order n^{-9}. (It can be adjusted to give any required order, provided the computer has sufficient capacity.) They note that the expansion is ultimately divergent and suggests, as a practical stopping rule, that the odd order and even order series of terms be considered separately, and that the summation be terminated immediately before either series increases in absolute value.

6 NOTE ON CHARACTERIZATIONS

Characterizations of specific distributions will be described at appropriate places in the text. However, the two following types of characterization, which apply to all distributions in this volume, will *not* usually be mentioned individually, since this would lead to tautological repetition.

If $X_1, \ldots, X_n$ are i.i.d., their common distribution is determined by

1. that of their sum, or, a fortiori their arithmetic mean ($\overline{X}$), or

2. that of any one order statistic $X'_{r:n}$ (the rth smallest of the X's).

In case 1, which applies to all distributions (continuous or discrete), the common characteristic function is equal to the $(1/n)$th power of that of the sum and thus determines the common distribution. In case 2 for absolutely continuous distributions, the cdf of $X'_{r:n}$ is

$$F_{X'_{r:n}}(x) = I_{F(x)}(r, n - r + 1), \tag{12.131}$$

where $F(x)$ is the common cdf of $X_1, \ldots, X_n$, and $I_p(a, b)$ is the incomplete beta function ratio (Chapter 1, Section 5). The right-hand side of (12.131) is a monotonic-increasing function of $F(x)$ and so determines $F(x)$, given $F_{X'_{r:n}}(x)$. [See Johnson and Kotz (1990).]

BIBLIOGRAPHY

Akgiray, V., Booth, G. G., and Loistl, O. (1989). Stable laws are inappropriate for describing German stock returns, *Allgemeine Statistische Archivefe*, **73**, 115–121.

Albert, J., Delampady, M., and Polasek, W. (1991). A class of distributions for robustness studies, *Journal of Statistical Planning and Inference*, **28**, 291–304.

Amos, D. E., and Daniel, S. L. (1971). Tables of percentage points of standardized Pearson distributions, *Record Report* SC-RR-710348, Sandia Laboratories, Albuquerque, New Mexico.

Andrews, D. F., Bickel, P. J., Hampel, F. R., Huber, P. J., Rogers, W. H., and Tukey, J. W. (1972). *Robust Estimates of Location: Survey and Advances*, Princeton: Princeton University Press.

Arnold, B. C. (1987). Majorization and the Lorenz order, *Lecture Notes in Statistics No. 43*, New York and Berlin: Springer-Verlag.

Arnold, B. C., and Balakrishnan, N. (1989). Relations, bounds and approximations for order statistics, *Lecture Notes in Statistics No. 53*, New York: Springer-Verlag.

Arnold, B. C., Balakrishnan, N., and Nagaraja, H. N. (1992). *A First Course in Order Statistics*, New York: Wiley.

Azzalini, A. (1985). A class of distributions which includes the normal ones, *Scandinavian Journal of Statistics*, **12**, 171–178.

Balakrishnan, N., and Cohen, A. C. (1991). *Order Statistics and Inference: Estimation Methods*, San Diego: Academic Press.

Balanda, K. P., and MacGillivray, H. L. (1990). Kurtosis and spread, *Canadian Journal of Statistics*, **18**, 17–30.

Balitskaya, E. O., and Zolotuhina, L. A. (1988). On the representation of a density by an Edgeworth series, *Biometrika*, **75**, 185–187.

Barndorff-Nielsen, O. (1978). Hyperbolic distributions and distributions on hyperbolae, *Scandinavian Journal of Statistics*, **5**, 151–157.

Barndorff-Nielsen, O. E. (1990). A note on the standardized signed log-likelihood ratio, *Scandinavian Journal of Statistics*, **17**, 157–160.

Barndorff-Nielsen, O., and Blaesild, P. (1983). Hyperbolic distributions, *Encyclopedia of Statistical Sciences*, **3**, S. Kotz, N. L. Johnson, and C. B. Read (editors), New York: Wiley.

Barndorff-Nielsen, O. E., and Cox, D. R. (1989). *Asymptotic Techniques for Use in Statistics*, London: Chapman and Hall.

Barton, D. E., and Dennis, K. E. R. (1952). The conditions under which Gram-Charlier and Edgeworth curves are positive definite and unimodal, *Biometrika*, **39**, 425–427.

Bennett, C. A. (1952). *Asymptotic Properties of Ideal Linear Estimators*, Unpublished thesis, Ann Arbor, MI: University of Michigan.

Benson, F. (1949). A note on the estimation of mean and standard deviation from quantiles, *Journal of the Royal Statistical Society, Series B*, **11**, 91–100.

Berry, K. J., Mielke, P. W., and Wong, R. K. W. (1986). Approximate MRPP P-values obtained from four exact moments, *Communications in Statistics—Theory and Methods*, **15**, 581–589.

Bhat, B. R. (1988). On exponential and curved exponential families in stochastic processes, *Mathematical Scientist*, **13**, 121–134.

Bhattacharjee, G. P. (1965). Distribution of range in non-normal samples, *Australian Journal of Statistics*, **7**, 127–141.

Bhattacharyya, B. C. (1942). The use of McKay's Bessel function curves for graduating frequency distributions, *Sankhyā*, **6**, 175–182.

Bhattacharyya, B. C. (1943). On an aspect of Pearsonian system of curves and a few analogies, *Sankhyā*, **6**, 415–418.

Bickel, P. J., and Doksum, K. A. (1977). *Mathematical Statistics: Basic Ideas and Selected Topics*, San Francisco: Holden Day.

Birnbaum, A., and Saunders, S. C. (1969). A new family of life distributions, *Journal of Applied Probability*, **6**, 319–327.

Boetti, G. (1964). Una rappresentazione grafica per la determinazione del tipo di curva, tra le curve del sistema di Pearson, corrispondente a dati valori numerici dei momenti 3° e 4° delle distribuzioni, *Giornale dell' Istituto Italiano degli Attuari*, **27**, 99–121.

Bol'shev, L. N. (1963). Asymptotic Pearsonian transformations, *Teoriya Veroyatnostei i ee Primeneniya*, **8**, 129–155. (In Russian)

Bortolotti, G. (1965). Nuove vedute sulle distribuzioni di frequenze e sui loro rapporti con la legge normale delle probabilita. I & II, *Statistica, Bologna*, **25**, 197–288 and 329–362.

Bose, S. S. (1938). On a Bessel function population, *Sankhyā*, **3**, 253–261.

Bouver, H. (1973). Table of the cumulative standardized Pearson Type IV distribution, Themis Report No. 28, Dept. of Statistics, Athens: University of Georgia.

Bouver, H. (1974). Tables of the standardized percentage points of the Pearson system of curves in terms of β_1 and β_2, Themis Report No. 32, Dept. of Statistics, Athens: University of Georgia.

Bouver, H., and Bargmann, R. E. (1977). The Pearsonian distribution package: An application in curve-fitting, *ASA Proceedings in Statistical Computation*, 127–132.

Bowman, K. O., and Shenton, L. R. (1979a). Approximate percentage points for Pearson distributions, *Biometrika*, **66**, 147–155.

Bowman, K. O., and Shenton, L. R. (1979b). Further approximate Pearson percentage points and Cornish-Fisher, *Communications in Statistics—Simulation and Computation*, **8**, 231–244.

Box, G. E. P., and Cox, D. R. (1964). An analysis of transformations, *Journal of the Royal Statistical Society, Series B*, **26**, 211–243.

Bruns, H. (1906). *Wahrscheinlichkeitsrechnung und Kollektivmasslehre*, Leipzig: Teubner.

Burr, I. W. (1942). Cumulative frequency functions, *Annals of Mathematical Statistics*, **13**, 215–232.

Burr, I. W. (1968). On a general system of distributions III. The sample range, *Journal of the American Statistical Association*, **63**, 636–643.

Burr, I. W., and Cislak, P. J. (1968). On a general system of distributions, I. Its curve-shape characteristics, II. The sample median, *Journal of the American Statistical Association*, **63**, 627–635.

Butterworth, N. J. (1987). A fast and flexible family of distributions for simulation studies, *AIIE Transactions*, **19**, 439–444.

Casella, G., and Berger, R. L. (1990). *Statistical Inference*, Pacific Grove, California: Brooks/Cole.

Castoldi, L. (1963). A continuous analogon of Poisson's distribution, *Rendiconti del Seminario della Facoltà di Scienze della Università di Cagliari*, **33**, 1–5.

Chambers, E., and Fowlkes, E. B. (1966). *A Dictionary of Distributions: Comparisons with the Standard Normal*, Murray Hill, NJ: Bell Telephone Laboratories.

Chan, L. K. (1967). On a characterization of distributions by expected values of extreme order statistics, *American Mathematical Monthly*, **74**, 950–951.

Charlier, C. V. L. (1905). Uber die Darstellung willkürlicher Funktionen, *Arkiv für Matematik, Astronomi och Fysik*, **2**, No. 20, 1–35.

Charlier, C. V. L. (1914). Contributions to the mathematical theory of statistics, *Arkiv für Matematik, Astronomi och Fysik*, **9**, 1–18.

Chernin, K. E., and Ibragimov, I. A. (1959). On the unimodality of stable laws, *Theory of Probability and Its Applications*, **4**, 417–419.

Chernoff, H., Gastwirth, J. L., and Johns, M. V. (1967). Asymptotic distribution of linear combinations of functions of order statistics with applications to estimation, *Annals of Mathematical Statistics*, **38**, 52–72.

Cooper, J. D., Davis, S. A., and Dono, N. R. (1965). Pearson Universal Distribution Generator (PURGE), *Proceedings of the 19th Annual Conference, American Society for Quality Control*, 402–411.

Cornish, E. A., and Fisher, R. A. (1937). Moments and cumulants in the specification of distributions, *Review of the International Statistical Institute*, **5**, 307–320.

Craig, C. C. (1936). A new exposition and chart for the Pearson system of curves, *Annals of Mathematical Statistics*, **7**, 16–28.

Cramér, H. (1928). On the composition of elementary errors, *Skandinavisk Aktuarietidskrift*, **11**, 13–74 and 141–180.

Dagum, C. (1985). Lorenz Curve, *Encyclopedia of Statistical Sciences*, **5**, S. Kotz, N. L. Johnson, and C. B. Read (editors), 156–161, New York: Wiley.

Darmois, G. (1935). Sur les lois de probabilités à estimation exhaustif, *Comptes Rendus de l'Académie des Sciences, Paris*, **200**, 1265–1267.

Davenport, J. M., and Herring, T. A. (1979). On the use of curve fitting to model the error of the Cornish-Fisher expansion of the Pearson Type VI distribution, *Communications in Statistics—Simulation and Computation*, **8**, 311–333.

David, F. N., and Johnson, N. L. (1954). Statistical treatment of censored data, Part I: Fundamental formulae, *Biometrika*, **41**, 228–240.

David, H. A. (1981). *Order Statistics* (Second edition), New York: Wiley.

Davis, C. S., and Stephens, M. A. (1983). Approximate percentage points using Pearson curves, Algorithm AS192, *Applied Statistics*, **32**, 322–327.

de Fériet, J. K. (1966). *The Gram-Charlier approximation of the normal law etc.*, Report 2013, Applied Mathematics Laboratory, David Taylor Model Basin, Department of the Navy, Washington, DC.

de Haan, L., and Resnick, S. I. (1980). A simple asymptotic estimate for the index of a stable distribution, *Journal of the Royal Statistical Society, Series B*, **42**, 83–87.

Dershowitz, A. F. (1966). *Polynomial fit to percentiles of Pearson curves*, T. I. S. Report 66-Ch-SD-511, General Electric Company.

Draper, N. R., and Tierney, D. E. (1972). Regions of positive and unimodal series expansion of the Edgeworth and Gram-Charlier expansion, *Biometrika*, **59**, 463–465.

Draper, N. R., and Tierney, D. E. (1973). Exact formulas for additional terms in some important series expansions, *Communications in Statistics*, **1**, 495–524.

Dudewicz, E. J., and Mishra, S. N. (1988). *Modern Mathematical Statistics*, New York: Wiley.

DuMouchel, W. H. (1971). *Stable Distributions in Statistical Inference*, Ph.D. Thesis, Yale University, New Haven, CT.

DuMouchel, W. H. (1973). 1. Stable distributions in statistical inference: 2. Information from stably distributed sample, *Journal of the American Statistical Association*, **70**, 386–393.

Edgeworth, F. Y. (1896). The asymmetrical probability curve, *Philosophical Magazine*, *5th Series*, **41**, 90–99.

Edgeworth, F. Y. (1907). On the representation of statistical frequency by a series, *Journal of the Royal Statistical Society, Series A*, **70**, 102–106.

Edgeworth, F. Y. (1914). On the use of analytic geometry to represent certain kinds of statistics, *Journal of the Royal Statistical Society, Series A*, **77**, 838–852.

Edgeworth, F. Y. (1916). On the mathematical representation of statistical data, *Journal of the Royal Statistical Society, Series A*, **79**, 455–500.

Edgeworth, F. Y. (1917). On the mathematical representation of statistical data, *Journal of the Royal Statistical Society, Series. A*, **80**, 65–83; **80**, 266–288; **80**, 411–437.

Elderton, W. P., and Johnson, N. L .(1969). *Systems of Frequency Curves*, London: Cambridge University Press.

Fama, E. F. (1965). The behavior of stock market prices, *Journal of Business, University of Chicago*, **38**, 34–105.

Fama, E. F., and Roll, R. (1968). Some properties of symmetric stable distributions, *Journal of the American Statistical Association*, **63**, 817–836.

Fama, E. F., and Roll, R. (1971). Parameter estimates for symmetric stable distributions, *Journal of the American Statistical Association*, **66**, 331–338.

Fattorini, L., and Lemmi, A. (1979). Proposta di un modello alternativo per l'analisi della distribuzione personale del reditto, *Atti delle Giornate di Lavoro AIRO*, 1–28.

Ferreri, C. (1964). A new frequency distribution for single variate analysis, *Statistica (Bologna)*, **24**, 223–251. (In Italian)

Fielitz, B. D., and Smith, E. W. (1972). Asymmetric stable distribution of stock price changes, *Journal of the American Statistical Association*, **67**, 813–814.

Finney, D. J. (1963). Some properties of a distribution specified by its cumulants, *Technometrics*, **5**, 63–69.

Fisher, R. A., and Cornish, E. A. (1960). The percentile points of distributions having known cumulants, *Technometrics*, **2**, 209–226.

Flapper, P. (1967). Transformatie van niet-normale verdelingen, *Statistica Neerlandica*, **21**, 151–155.

Freimer, M., Kollia, G., Mudholkar, G. S., and Lin, C. T. (1988). A study of the generalized Tukey lambda family, *Communications in Statistics—Theory and Methods*, **17**, 3547–3567.

Fry, T. R. L. (1993). Univariate and multivariate Burr distributions: A survey, *Pakistan Journal of Statistics, Series A*, **9**, 1–24.

Gail, M. H., and Gastwirth, J. L. (1978). A scale-free goodness-of-fit test for the exponential distribution based on the Lorenz curve, *Journal of the American Statistical Association*, **73**, 786–793.

Galambos, J. (1987). *The Asymptotic Theory of Extreme Order Statistics* (Second edition), Malabar, FL: Kreiger.

Galambos, J., and Kotz, S. (1978). Characterizations of Probability Distributions, *Lecture Notes in Mathematics No. 675*, New York: Springer-Verlag.

Gastwirth, J. L. (1971). A general definition of the Lorenz curve, *Econometrica*, **39**, 1037–1038.

Gastwirth, J. L. (1972). The estimation of the Lorenz curve and Gini index, *Review of Economics and Statistics*, **54**, 306–316.

Gawronski, W. (1984). On the bell-shape of stable distributions, *Annals of Probability*, **12**, 232–242.

Gawronski, W., and Weissner, M. (1992). Asymptotics and inequalities for the mode of stable laws, *Statistics and Decisions*, **10**, 183–197.

Glänzel, W. (1991). Characterization through some conditional moments of Pearson-type distributions and discrete analogues, *Sankhyā, Series B*, **53**, 17–24.

Goldfield, S. M., and Quandt, R. E. (1981). Econometric modelling with non-normal disturbances, *Journal of Econometrics*, **17**, 141–155.

Goto, M., Inoue, T., and Tsuchiya, Y. (1983). On estimation of parameters for power-normal distribution, *Bulletin of Information and Systematics*, **21**(1-2), 41–53.

Goto, M., Inoue, T., and Tsuchiya, Y. (1987). Double power-normal transformation and its performance—An extensive version of Box-Cox transformation, *Journal of the Japan Statistical Society*, **17**, 149–163.

Goto, M., Matsubara, Y., and Tsuchiya, Y. (1983). Power-normal distribution and its application, *Reports of Statistics Application Research, JUSE*, **30**(3), 8–28.

Groeneveld, R. A. (1986). Measurement of skewness and kurtosis for the generalized Tukey lambda distributions, *Communications in Statistics—Theory and Methods*, **15**, 329–343.

Gross, A. M. (1973). A Monte Carlo swindle for estimators of location, *Applied Statistics*, **22**, 347–353.

Haitovsky, Y. (1983). Grouped Data, *Encyclopedia of Statistical Sciences*, **3**, S. Kotz, N. L. Johnson, and C. B. Read (editors), 527–536, New York: Wiley.

Hall, P. (1992). *The Bootstrap and Edgeworth Expansion*, New York and Berlin: Springer-Verlag.

Hansmann, G. H. (1934). On certain non-normal symmetric frequency distributions, *Biometrika*, **26**, 129–135.

Harris, E. K., and DeMets, D. L. (1972a). Effects of intra- and inter-individual variation on distributions of single measurements, *Clinical Chemistry*, **18**, 244.

Harris, E. K., and DeMets, D. L. (1972b). Estimation of normal ranges and cumulative proportions by transforming observed distributions to Gaussian form, *Clinical Chemistry*, **18**, 605–612.

Harter, H. L. (1968). The use of order statistics in estimation, *Operations Research*, **16**, 783–798.

Hastings, C., Mosteller, F., Tukey, J. W., and Winsor, C. P. (1947). Low moments for small samples: A comparative study of order statistics, *Annals of Mathematical Statistics*, **18**, 413–426.

Henze, N. (1986). A probabilistic representation of the "skew-normal" distribution, *Scandinavian Journal of Statistics*, **13**, 271–275.

Hoadley, A. B. (1968). Use of the Pearson densities for approximating a skew density whose left terminal and first three moments are known, *Biometrika*, **55**, 559–563.

Hoaglin, D. C. (1984). Summarizing shape numerically: The g- and h-distributions, *Data Analysis for Tables, Trends and Shapes: Robust and Exploratory Techniques*, D. C. Hoaglin, F. Mosteller, and J. W. Tukey (editors), New York: Wiley.

Hogg, R. V., and Craig, A. T. (1978). *Introduction to Mathematical Statistics* (Fourth edition), New York: MacMillan.

Holt, D. R., and Crow, E. L. (1973). Tables and graphs of the stable probability density functions, *Journal of Research, National Bureau of Standards*, **77B**, 144–198.

Holtsmark, J. (1919). Über der Verbreitering von Spektrallinien, *Annalen der Physik*, **58**, 577–630.

Houghton, J. E. (1978). Birth of a parent: The Wakeby distribution for modeling flood flows, *Water Resources Research*, **14**, 1105–1110.

Hougaard, P. (1986). Survival models for heterogeneous populations derived from stable distributions, *Biometrika*, **73**, 387–396.

Janssen, A. (1986). Scale invariant exponential families and one-sided test procedures, *Statistics and Decisions*, **4**, 147–174.

Johnson, N. L. (1949). Systems of frequency curves generated by methods of translation, *Biometrika*, **36**, 149–176.

Johnson, N. L. (1954). Systems of frequency curves derived from the first law of Laplace, *Trabajos de Estadistica*, **5**, 283–291.

Johnson, N. L. (1965). Tables to facilitate fitting S_U frequency curves, *Biometrika*, **52**, 547–558.

Johnson, N. L., and Kotz, S. (1973). Extended and multivariate Tukey distributions, *Biometrika*, **60**, 655–661.

Johnson, N. L., and Kotz, S. (1990). Characterization of an absolutely continuous distribution by the distribution of any order statistic, *Indian Journal of Mathematics*, **32**, 193–194.

Johnson, N. L., Nixon, E., Amos, D. E., and Pearson, E. S. (1963). Table of percentage points of Pearson curves, for given β_1 and β_2, expressed in standard measure, *Biometrika*, **50**, 459–498.

Joiner, B. L., and Rosenblatt, J. R. (1971). Some properties of the range in samples from Tukey's symmetric lambda distribution, *Journal of the American Statistical Association*, **66**, 394–399.

Kafadar, K. (1982). A biweight approach to the one-sample problem, *Journal of the American Statistical Association*, **77**, 416–424.

Kafael, M.-A. and Schmidt, P. (1985). On the adequacy of the "Sargan distribution" as an approximation to the normal, *Communications in Statistics—Theory and Methods*, **14**, 509–526.

Kamat, A. R. (1966). A property of the mean deviation for the Pearson type distributions, *Biometrika*, **53**, 287–289.

Kameda, T. (1928). On the reduction of frequency curves, *Skandinavisk Aktuarietidskrift*, **11**, 112–118.

Karlin, S. (1957). Pólya-type distributions II, *Annals of Mathematical Statistics*, **28**, 281–308.

Khamis, S. H. (1958). Incomplete gamma functions, *Bulletin of the International Institute of Statistics*, **37**, 385–396.

Kokonendji, C. C. (1992). Caracterisation des functions variances de Seshadri des familles exponentielles sur $\mathbb{R}$, *Comptes Rendus, Académie des Sciences, Paris*, **314**, 1063–1068.

Koopman, B. O. (1936). On distributions admitting a sufficient statistics, *Transactions of the American Mathematical Society*, **39**, 339–409.

Korwar, R. M. (1991). On characterizations of distributions by mean absolute deviation and variance bounds, *Annals of the Institute of Statistical Mathematics*, **43**, 287–295.

Krieger, A. M. (1979). Bounding moments, the Gini index and Lorenz curve from grouped data for unimodal density functions, *Journal of the American Statistical Association*, **74**, 375–378.

Kropác, O. (1982). Some properties and applications of probability distributions based on the Macdonald function, *Aplikace Matematiky*, **27**, 285–301.

Laha, R. G. (1953). On some properties of the Bessel function distributions, *Bulletin of the Calcutta Mathematical Society*, **46**, 59–72.

Laha, R. G. (1961). On a class of unimodal distributions, *Proceedings of the American Mathematical Society*, **12**, 181–184.

Lam, H.-K., Bowman, K. O., and Shenton, L. R. (1980). Remarks on the generalized Tukey's lambda family of distributions, *Proceedings of the ASA Statistical Computing Section*, 134–139.

Leadbetter, M. R., Lindgren, G., and Rootzén, H. (1983). *Extremes and Related Properties of Random Sequences and Processes*, New York: Springer-Verlag.

Lee, Y. -S., and Lin, T. -K. (1992). Higher-order Cornish-Fisher expansion, Algortihm AS269, *Applied Statistics*, **41**, 233–240. Correction, *Ibid.*, **42**, 268–269.

Leitch, R. A., and Paulson, A. S. (1975). Estimation of stable law parameters: Stock price application, *Journal of the American Statistical Association*, **70**, 690–697.

Lévy, P. (1927). Théorie des erreurs—La loi de Gauss et les lois exceptionelles, *Bulletin de la Société Mathematique de France*, **52**, 49–85.

Linnik, Yu. V. (1953). Linear forms and statistical criteria. II, *Ukrainskii Matematicheskii Zhurnal*, **5**, 247–290. (In Russian)

Lloyd, E. H. (1952). Least squares estimation of location and scale parameters using order statistics, *Biometrika*, **39**, 88–95.

Longuet-Higgins, M. S. (1964). Modified Gaussian distribution for slightly non-linear variables, *Journal of Research of The National Bureau of Standards*, **68D**, 1049–1062.

Lukacs, E. (1968). Contributions to a problem of D. van Dantzig, *Teoriya Veroyatnostei i ee Primeneniya*, **13**, 114–125.

Lukacs, E. (1969). Stable distributions and their characteristic functions, *Jahresbericht der Deutschen Mathematischen Vereinigung*, **71**, 84–114.

MacGillivray, H. L. (1992). Shape properties of the *g*- and *h*- and Johnson families, *Communications in Statistics—Theory and Methods*, **21**, 1233–1250.

Mage, D. T. (1980). An explicit solution for S_B parameters using four percentile points, *Technometrics*, **22**, 247–251.

Mallows, C. L. (1956). Generalizations of Tchebycheff's inequalities, *Journal of the Royal Statistical Society, Series B*, **18**, 139–176.

Mallows, C. L. (1983). A new system of frequency curves, *Contributions to Statistics: Essays in Honour of Norman L. Johnson*, P. K. Sen (editor), 329–338, Amsterdam: North-Holland.

Mandelbrot, B. (1963). The variation of certain speculative prices, *Journal of Business, University of Chicago*, **36**, 394–519.

Mardia, K. V. (1965). Tippett's formulas and other results on sample range and extreme, *Annals of the Institute of Statistical Mathematics*, **17**, 85–91.

Martinez, J., and Iglewicz, B. (1984). Some properties of the Tukey *g*- and *h*-family of distributions, *Communications in Statistics—Theory and Methods*, **13**, 353–359.

Martins, E. G., and Pestansa, D. (1982). The tabulation of stable density functions with small characteristic exponent, *Acta V. Jornade de Luso-España Mathematica*, **9**, 665–668.

Mathai, A. M., and Saxena, R. K. (1966). On a generalized hypergeometric distribution, *Metrika*, **11**, 127–132.

McCune, D. (1977). On Cornish-Fisher expansion as a ratio of determinants, without cumulants, *Communications in Statistics—Theory and Methods*, **6**, 243–250.

McKay, A. T. (1932). A Bessel function distribution, *Biometrika*, **24**, 39–44.

McLeish, D. L. (1982). A robust alternative to the normal distribution, *Canadian Journal of Statistics*, **10**, 89–102.

McNolty, F. (1967). Applications of Bessel function distributions, *Sankhyā, Series B*, **29**, 235–248.

Mehran, F. (1976). Linear measures of income inequality, *Econometrica*, **44**, 805–809.

Mendoza, G. A., and Iglewicz, B. (1983). A comparative study of systems of univariate frequency distributions, *ASA Proceedings of Statistical Computing Section*, 249–254.

Mielke, P. W. (1973). Another family of distributions for describing and analyzing precipitation data, *Journal of Applied Meteorology*, **12**, 275–280.

Missiakoulis, S. (1983). Sargan densities, which one?, *Journal of Econometrics*, **23**, 223–233.

Monrad, D., and Stout, W. (1988). Stable Distributions, *Encyclopedia of Statistical Sciences*, **8**, S. Kotz, N. L. Johnson, and C. B. Read (editors), 617–621, New York: Wiley.

Mood, A. M., Graybill, F. A., and Boes, D. C. (1974). *Introduction to the Theory of Statistics* (Third edition), New York: McGraw-Hill.

Moore, P. G. (1957). Transformations to normality using fractional powers of the variable, *Journal of the American Statistical Association*, **52**, 237–246.

Morris, C. N. (1983). Natural exponential families with quadratic variance function, *Annals of Statistics*, **11**, 515–529.

Mosteller, F., and Tukey, J. W. (1977). *Data Analysis and Regression: A Second Course in Statistics*, Reading, MA: Addison-Wesley.

Müller, P. -H., and Vahl, H. (1976). Pearson's system of frequency curves whose left boundary and first three moments are known, *Biometrika*, **63**, 191–194.

Nagaev, A. V., and Shkol'nik, S. M. (1989). Some properties of symmetric stable distributions close to the normal distribution, *Theory of Probability and Its Applications*, **33**, 139–144.

Nair, N. U., and Sankaran, P. G. (1991). Characterization of the Pearson family of distributions, *IEEE Transactions on Reliability*, **40**, 75–77.

Nicklin, F. H., and Paulson, A. S. (1975). The stable law in security analysis, Research Report 37-75-P2, School of Management, Troy, NY: Rensselaer Polytechnic Institute.

O'Hagan, A., and Leonard, T. (1976). Bayes' estimation subject to uncertainty about parameter constraints, *Biometrika*, **63**, 201–202.

Panton, D. B. (1992). Cumulative distribution function values for symmetric standardized stable distributions, *Communications in Statistics—Simulation and Computation*, **21**, 485–492.

Paulson, A. S., Holcomb, E. A., and Leitch, R. A. (1975). The estimation of the parameters of the stable laws, *Biometrika*, **62**, 163–174.

Pawula, R. F., and Rice, S. O. (1989). A note on Hansmann's 1934 family of distributions, *IEEE Transactions on Information Theory*, **35**, 910–911.

Pearson, E. S., and Hartley, H. O. (eds.) (1972). *Biometrika Tables for Statisticians*, **2**, Cambridge: Cambridge University Press.

Pearson, E. S., Johnson, N. L., and Burr, I. W. (1979). Comparisons of the percentage points of distributions with the same first four moments, chosen from eight different systems of frequency curves, *Communications in Statistics—Simulation and Computation*, **8**, 191–229.

Pearson, K. (1895). Contributions to the mathematical theory of evolution. II. Skew variations in homogeneous material, *Philosophical Transactions of the Royal Society of London*, Series A, **186**, 343–414.

Pearson, K. (1924). On the mean error of frequency distributions, *Biometrika*, **16**, 198–200.

Pearson, K., Jeffery, G. B., and Elderton, E. M. (1929). On the distribution of the first product moment-correlation, in samples drawn from an indefinitely large normal population, *Biometrika*, **21**, 164–193.

Pearson, K., Stouffer, S. A., and David, F. N. (1932). Further applications in statistics of the $T_m(x)$ Bessel function, *Biometrika*, **24**, 292–350.

Proctor, J. W. (1987). Estimation of two generalized curves covering the Pearson system, *Proceedings of ASA Section on Statistical Computing*, 287–292.

Ramberg, J. S. (1975). A probability distribution with applications to Monte Carlo simulation studies, *Statistical Distributions in Scientific Work*, 2 G. P. Patil, S. Kotz, and J. K. Ord (editors), 51–64, Dordrecht: Reidel.

Ramberg, J. S., Dudewicz, E. J., Tadikamalla, P. R., and Mykytka, E. F. (1979). A probability distribution and its use in fitting data, *Technometrics*, **21**, 201–214.

Ramberg, J. S., and Schmeiser, B. W. (1974). An approximate method for generating asymmetric random variables, *Communications of the Association for Computing Machinery*, **17**, 78–82.

Reid, W. H., and Skates, S. J. (1986). On the asymptotic approximation of integrals, *SIAM Journal on Applied Mathematics*, **46**, 351–358.

Reiss, R. D. (1989). *Approximate Distributions of Order Statistics: With Applications to Nonparametric Statistics*, Berlin: Springer-Verlag.

Resnick, S. I. (1987). *Extreme Values, Regular Variation, and Point Processes*, New York: Springer-Verlag.

Rieck, J. R., and Nedelman, J. R. (1991). A log-linear model for the Birnbaum-Saunders distribution, *Technometrics*, **33**, 51–60.

Rigo, P. (1987). Two inequalities for the Gini index, *Statistica* (*Bologna*), **47**, 531–541.

Rocke, D. M. (1993). On the beta transformation family, *Technometrics*, **35**, 72–75.

Rodriguez, R. N. (1977). A guide to the Burr Type XII distributions, *Biometrika*, **64**, 129–134.

Rogers, W. H., and Tukey, J. W. (1972). Understanding some long-tailed symmetric distributions, *Statistica Neerlandica*, **26**, 211–226.

Rust, P. F. (1991). Modeling approximately normal distributions with S-systems, *Proceedings of ASA Section on Statistical Computing*, 162–166.

Saniga, E. M., Pfaffenberger, R. C., and Hayya, J. C. (1975). Estimation and goodness-of-fit tests for symmetric stable distributions, *ASA Proceedings of Business and Economic Statistics Section*, 530–534.

Sastry, K. V. K. (1948). On a Bessel function of the second kind and Wilks' Z-distribution, *Proceedings of the Indian Academy of Sciences*, Series A, **28**, 532–536.

Sathe, Y. S., and Lingres, S. R. (1979). Bounds for the Pearson Type IV tail probabilities, *Communications in Statistics—Theory and Methods*, **8**, 533–541.

Schmeiser, B. W., and Deutsch, S. J. (1977). A versatile four parameter family of probability distributions suitable for simulation, *AIIE Transactions*, **9**, 176–180.

Serfling, R. J. (1980). *Approximation Theorems of Mathematical Statistics*, New York: Wiley.

Shapiro, S. S., and Gross, A. J. (1981). *Statistical Modeling Techniques*, New York: Marcel Dekker.

Shenton, L. R., and Carpenter, J. A. (1965). The Mills ratio and the probability integral for the Pearson Type IV distribution, *Biometrika*, **52**, 119–126.

Sheppard, W. F. (1896). On the calculation of the most probable values of frequency constants for data arranged according to equidistant divisions of a scale, *Proceedings of the London Mathematical Society*, **29**, 353–380.

Shorack, G. R., and Wellner, J. A. (1986). *Empirical Processes with Applications to Statistics*, New York: Wiley.

Siekierski, K. (1992). Comparison and evaluation of three methods of estimation of the Johnson S_B distribution, *Biometrical Journal*, **34**, 879–895.

Singh, C. (1967). On the extreme values and range of samples from non-normal populations, *Biometrika*, **54**, 541–550.

Skates, S. J. (1993). On secant approximations to cumulative distribution functions, *Biometrika*, **80**, 223–235.

Slifker, J. F., and Shapiro, S. S. (1980). The Johnson system: Selection and parameter estimation, *Technometrics*, **22**, 239–246.

Steiner, F. (ed.) (1991). *The Most Frequent Value: Introduction to a Modern Conception of Statistics*, Budapest: Akadámiai Kiadó.

Stok, J. P. van der (1908). On the analysis of frequency curves according to a general method, *Koninklijke Nederlandse Akademie van Wetenschappen, Series A, Mathematical Sciences*, **10**, 799–817.

Subrahmaniam, K. (1966). Some contributions to the theory of non-normality. I. (univariate case), *Sankhyā, Series A*, **28**, 389–406.

Suzuki, G. (1965). A consistent estimator for the mean deviation of the Pearson type distribution, *Annals of the Institute of Statistical Mathematics*, **17**, 271–285.

Tadikamalla, P. R., and Johnson, N. L. (1982). Systems of frequency curves generated by transformations of logistic variables, *Biometrika*, **69**, 461–465.

Thiele, T. N. (1903). *Theory of Observations*, London: Layton.

Thionet, P. (1966). Note sur les mélanges de certaines distributions de probabilités, *Publications de l'Institut de Statistique de l'Université de Paris*, **15**, 61–80.

Tiku, M. L. (1965). Laguerre series forms of non-central chi-squared and F distributions, *Biometrika*, **52**, 415–427.

Tippett, L. H. C. (1925). On the extreme individuals and the range of the samples taken from the normal population, *Biometrika*, **17**, 364–387.

Toranzos, F. I. (1952). An asymmetric bell-shaped frequency curve, *Annals of Mathematical Statistics*, **23**, 467–469.

Tricker, A. R. (1984). Effects of rounding on the moments of a probability distribution, *The Statistician*, **33**, 381–390.

Tse, Y. K. (1987). A note on Sargan densities, *Journal of Econometrics*, **34**, 349–354.

Tukey, J. W. (1962). The future of data analysis, *Annals of Mathematical Statistics*, **33**, 1–67.

Tziafetas, G. N. (1989). A formula for the Gini coefficient and its decomposition, *Biometrical Journal*, **8**, 961–967.

Vodǎ, V. G. (1982). Burr distribution revisited, *Revue Roumaine de Mathématiques Pures et Appliquées*, **27**, 885–893.

Voit, E. O. (1992). The *S*-distribution: A tool for approximation and classification of univariate, unimodal probability distributions, *Biometrical Journal*, **34**, 855–878.

Wheeler, R. E. (1980). Quantile estimators of Johnson curve parameters, *Biometrika*, **67**, 725–728.

Williams, E. J. (1977). Some representations of stable random variables as products, *Biometrika*, **64**, 167–169.

Wingo, D. R. (1993). Maximum likelihood methods for fitting the Burr Type XII distribution to multiply (progressively) censored life test data, *Metrika*, **40**, 203–210.

Winterborn, K. B. (1978). Determining parameters of the Johnson S_U distributions, *Communications in Statistics—Simulation and Computation*, **7**, 223–226.

Wold, H. (1934). Sulla correzione di Sheppard, *Giornale dell' Istituto Italiano degli Attuari*, **5**, 304–314. (In Italian)

Wood, J. T. (1974). An extension of the analysis of transformations of Box and Cox, *Applied Statistics*, **3**, 278–283.

Woods, J. D., and Posten, H. O. (1968). *Fourier series and Chebyshev polynomials in statistical distribution theory*, Research Report No. 37, Department of Statistics, Storrs, CT: University of Connecticut.

Woodward, G. J. (1976). Approximations of Pearson Type IV tail probabilities, *Journal of the American Statistical Association*, **71**, 513–514.

Worsdale, G. J. (1975). Tables of cumulative distribution functions for symmetric stable distributions, *Applied Statistics*, **24**, 123–131.

Wu, C. -Y. (1966). The types of limit distributions for some terms of variational series, *Scientia Sinica*, **15**, 749–762.

Zolotarev, V. M. (1954). Expression of the density of a stable distribution with exponent α greater than one by means of a frequency with exponent $1/\alpha$, *Doklady Akademii Nauk SSSR*, **98**, 735–738. [In Russian. English translation in *Selected Translations AMS-IMS*, **1**, 163–167 (1961).]

Zolotarev, V. M. (1966). On representation of stable laws by integrals, *Selected Translations in Mathematical Statistics and Probability*, **6**, 84–88.

CHAPTER 13

Normal Distributions

1 DEFINITION AND TABLES

A random variable X is *normally* distributed if it has the probability density function:

$$\frac{1}{\sqrt{2\pi}} \exp\left[-\frac{1}{2}\left(\frac{x-\xi}{\sigma}\right)^2\right], \qquad \sigma > 0. \tag{13.1}$$

The probability density function of $U = (X - \xi)/\sigma$ is

$$p_U(u) = \left(\sqrt{2\pi}\right)^{-1} \exp\left(-\tfrac{1}{2}u^2\right), \tag{13.2}$$

which does not depend on the parameters ξ, σ. This is called the *standard form* of normal distribution. (It is also the *standardized form*.) The random variable U is called a *standard*, or *unit*, normal variable.

Since

$$\Pr[X \le x] = \Pr\left[U \le \frac{x-\xi}{\sigma}\right], \tag{13.3}$$

such probabilities can be evaluated from tables of the cumulative distribution function of U, which is

$$\Phi(u) = \Pr[U \le u] = \left(\sqrt{2\pi}\right)^{-1} \int_{-\infty}^{u} e^{-x^2/2}\, dx. \tag{13.4}$$

The notation $\Phi(\cdot)$ is widely used, so it will be used in this book. Further it is convenient to have a systematic notation for the quantiles of the distribution of U. We use the system defined by

$$\Phi(U_\alpha) = \alpha$$

so that $U_{1-\alpha}$ is the upper $100\alpha\%$ point, and U_α $(= -U_{1-\alpha})$ is the lower $100\alpha\%$ point of the distribution.

There are other forms of notation that are much less frequently encountered in statistical work. The parameter in (13.1) is sometimes replaced by the *precision modulus*

$$h = (\sigma\sqrt{2})^{-1}.$$

Other functions are

$$\text{erf}(x) = 2\pi^{-1/2} \int_0^x e^{-t^2}\, dt = 2\Phi(x\sqrt{2}) - 1,$$

$$(13.5)$$

$$\text{erfc}(x) = 1 - \text{erf}(x)$$

(erf is the *error function*, or *Cramp function*, and erfc the *error function complement*). Other names for the distribution are *second law of Laplace*, *Laplace*, *Gaussian*, *Laplace-Gauss*, *de Moivre*; $\Phi(\cdot)$ is also called the *Laplace-Gauss integral*, or simply the *probability integral*, and erf($\cdot$) is also known by this last name and is sometimes called the *error integral*.

Tables relating to the unit normal distribution are a necessary ingredient of any textbook in statistical theory or its applications. This is because for many decades the normal distribution held a central position in statistics. As pointed out, tables of the *unit* normal distribution suffice for calculations relating to *all* normal distributions. Some care is necessary in using these tables; for example, putting

$$(\sqrt{2\pi})^{-1} e^{-x^2/2} = Z(x),$$

$$(13.6)$$

it is necessary to remember the multiplier σ^{-1} in

$$(\sqrt{2\pi}\,\sigma)^{-1} \exp\left[-\frac{1}{2}\left(\frac{x-\xi}{\sigma}\right)^2\right] = \sigma^{-1} Z\left(\frac{x-\xi}{\sigma}\right).$$

$$(13.7)$$

But no real difficulties are presented by the extended use of tables of the unit normal distribution. [The symbols $\varphi(x), \phi(x)$ are often used in place of $Z(x)$.]

In most of the tables only positive values of the variable are given. This is all that is necessary, since

$$Z(x) = Z(-x) \quad \text{and} \quad \Phi(x) = 1 - \Phi(-x).$$

$$(13.8)$$

Here we give a list of only the more easily available tables. Fuller lists are given in the National Bureau of Standards (1952) (up to 1952) and Greenwood and Hartley (1962) (up to 1958). The functions most often tabulated are $\Phi(x)$, $Z(x)$, and U_α, but there are many variants for special uses.

Pearson and Hartley (1948) give tables based on values originally computed by Sheppard (1903, 1907). These contain

1. $\Phi(x)$ and $Z(x)$ to 7 decimal places for $x = 0.00(0.01)4.50$; and to 10 decimal places for $x = 4.50(0.01)6.00$.
2. U_α to 4 decimal places for $\alpha = 0.501(0.001)0.980(0.0001)0.9999$.
3. $Z(U_\alpha)$ to 5 decimal places for $\alpha = 0.500(0.001)0.999$.

Pearson and Hartley (1970) have also provided these tables.

Fisher and Yates (1963) give U_α to 6 decimal places for $\alpha = 0.505(0.005)0.995$, and to 5 decimal places for $1 - \alpha = 0.0'1$ $[r = 2(1)8]$. These tables include values of "probits"—$(5 + U_\alpha)$—to 4 decimal places for $\alpha = 0.001(0.001)0.980(0.0001)0.9999$ and of $Z(u)$ to 4 decimal places for $u = 0.00(0.01)3.00(0.1)3.9$.

Owen (1962) gives $Z(x)$ and $\Phi(x)$ to 6 decimal places, $Z^{(1)}(x)$, $Z^{(2)}(x)$, $Z^{(3)}(x)$, and $\{1 - \Phi(x)\}/Z(x)$ to 5 decimal places, and $\Phi(x)/Z(x)$ to 4 decimal places, for $x = 0.00(0.01)3.99$; also $\{1 - \Phi(x)\}$ to 5 significance figures for $x = 3.0(0.1)6.0(0.2)10.0(1)20(10)100(25)200(50)500$, and U_α and $Z(U_\alpha)$ to 5 decimal places for $\alpha = 0.500(0.001)0.900(0.005)0.990$.

Kelley (1948) gives U_α to 8 decimal places for $\alpha = 0.5000(0.0001)0.9999$. Hald (1952) gives $Z(x)$ and $\Phi(x)$ to 4 significant figures for

$$x = \pm 0.00(0.01)4.99,$$

and probits $(5 + U_\alpha)$ to 3 decimal places for

$$\alpha = 0.0001(0.0001)0.0250(0.001)0.9750(0.0001)0.9999.$$

We next describe some tables containing larger number of decimal places, useful for special calculations.

In Zelen and Severo (1964) there are tables of $Z(x)$, $\Phi(x)$ and $Z^{(1)}(x)$ to 15 decimal places, $Z^{(2)}(x)$ to 10, and $Z^{(r)}(x)$ $(r = 3, 4, 5, 6)$ to 8 decimal places for $x = 0.00(0.02)3.00$. For the values $x = 3.00(0.05)5.00$, $\Phi(x)$ is given to 10 decimal places, $Z(x)$ to 10 significant figures, and $Z^{(r)}(x)$ $(r = 2, \ldots, 6)$ to 8 significant figures. A further table gives $Z^{(r)}(x)$ $(r = 7, \ldots, 12)$ to 8 significant figures for $x = 0.0(0.1)5.0$. There are also tables [based on Kelley (1948)] of U_α and $Z(U_\alpha)$ to 5 decimal places for $\alpha = 0.500(0.001)0.99$, and of U_α to 5 decimal places for $\alpha = 0.9750(0.0001)0.9999$.

In the National Bureau of Standards tables (1953) there are given tables of $Z(x)$ and $2\Phi(x) - 1$ $[= \text{erf}(x/\sqrt{2})]$ to 15 decimal places for $x = 0(0.0001)1.0000(0.001)7.800$; and also of $2[1 - \Phi(x)]$ to 7 significant figures for $x = 6.00(0.01)10.00$.

Pearson (1931) gives tables [calculated by Kondo and Elderton (1931), and Mills and Camp in Mills (1926)] of U_α, $Z(U_\alpha)$, $\alpha/Z(U_\alpha)$, $(1 - \alpha)/Z(U_\alpha)$, $Z(U_\alpha)/\alpha$ and $Z(U_\alpha)/(1 - \alpha)$ to 10 decimal places for $\alpha = 0.500(0.001)0.999$ and of $[1 - \Phi(x)]/Z(x)$ to 5 decimal places for $x = 0.00(0.01)4.00(0.05)5.00(0.1)10.0$. (This last quantity "ratio of area to bounding ordinate" is known as *Mills ratio*.) Wichura (1990) has made some comments on approximating this Mills ratio.

In Emersleben (1951) there are tables of erf x, and $2^{3/2}Z(x\sqrt{2})$ to 10 decimal places for $x = 0(0.01)2.00$, and also of $xe^{x^2}\text{erfc}(x)$ to 7 decimal places for $x = 0.000(0.005)0.250$ and $\text{erfc}(\sqrt{n\pi})$ to 15 decimal places for $n = 1(1)10$.

Bol'shev and Smirnov (1965) give tables of $\Phi(x)$ to six decimal places for $x = 0.000(0.001)3.000$ and to five decimal places for $x = 3.00(0.01)5.00$; $Z(x)$, $Z^{(r)}(x)$ $(r = 1, 2, 3, 4, 5)$ for $x = 0.000(0.004)3.00(0.02)4.00(0.04)5.0(0.1)6.0$ and U_α to 6 decimal places for $x = 0.500(0.001)0.9700(0.0001)0.9990$.

Harvard University Computation Laboratory Tables (1952) give values of $\Phi(x)$ to six decimal places for $x = 0.000(0.004)4.892$; of $Z(x)$ to six decimal places for $x = 0.000(0.004)5.216$ and of $Z^{(r)}(x)$ $(r = 1, 2, \ldots, 20)$ for

$r = 1(1)4$ to six decimal places for $x = 0.000(0.004)6.468$;

$r = 5(1)10$ to six decimal places for $x = 0.000(0.004)8.236$;

$r = 11(1)15$ to six decimal places for $x = 0.000(0.002)6.198$ and seven significant figures for $x = 6.2(0.002)9.61$;

$r = 16(1)20$ to seven significant figures for $x = 0(0.002)8.398$ and to six decimal places for $x = 8.4(0.002)10.902$.

The set of tables of the Advanced Series of Mathematics and Engineering (1961) contains values of $Z(x)$ for

$x = 0.0000(0.001)2.7000$ to 8 decimal places,

$x = 2.7000(0.0001)3.4500$ to 9 decimal places,

$x = 3.4500(0.0001)4.1000$ to 10 decimal places,

$x = 4.1000(0.0001)4.6500$ to 11 decimal places,

$x = 4.6500(0.0001)5.0000$ to 12 decimal places.

Values of $\Phi(x)$ are given according to the same scheme, except that only 7 decimal places are given for $0.0000 \le x < 1.0000$, while 11 places are given for $4.0500 \le x < 4.1000$. There are also values of $Z^{(2)}(x)$, $Z^{(3)}(x)$ and $Z^{(4)}(x)$, to six decimal places, for $x = 0.000(0.001)5.000$. Extremely detailed tables of the error function and its first 20 derivatives have been published in the U.S.S.R.; see Smirnov (1960, 1965) and Tables of Probability Functions (1959).

Table 13.1 Percentile Points of Normal Distribution, as Standardized Deviates (Values of U_α)

α	U_α
0.5	0.000000
0.6	0.253347
0.7	0.524401
0.75	0.674490[a]
0.8	0.841621
0.9	1.281552
0.95	1.644854
0.975	1.959964
0.99	2.326348
0.995	2.575829
0.9975	2.807034
0.999	3.090232

[a]The value of $U_{0.75}$ ($= 0.6745$), the upper quartile of the unit normal distribution, is occasionally called the *probable error* of the distribution, though this nomenclature is seldom used at present. The probable error of distribution (1) is, of course, $U_{0.75}\sigma$.

There are many other publications containing various forms of tables of the normal distribution. Further tables of special functions associated with the normal distribution are used in connection with probit analysis. There is no need for extensive tables of the normal distribution to be given here. We confine ourselves, in Table 13.1, to a few commonly used values of U_α.

Tables of random unit normal deviates (representing values of a random variable having a unit normal distribution) have been constructed from tables of random numbers (representing values of a random variable having a discrete rectangular distribution over the integers 0–9). In 1948 Wold (1948) published a set of 25,000 random unit normal deviates (to 3 decimal places), based on Kendall and Babington Smith's (1942) tables of random numbers. A set of 10,400 random unit normal deviates (also to 3 decimal places), based on Tippett's (1927) table of random numbers, was published by Sengupta and Bhattacharya (1958). These replaced an earlier set of tables, first appearing in 1936 [Mahalanobis et al. (1934)] which were found to contain a number of errors.

A set of 100,000 random unit normal deviates, to 3 decimal places, based on the first half-million random numbers produced in 1947, was published by RAND (1955). In Buslenko et al. (1966) there is a table of 1000 random unit normal deviates, to 4 decimal places. These were calculated from the values of five independent random variables $R_1, \ldots, R_5$ each randomly distributed over the range 0 to 1 (see Chapter 26), using the formulas

$$U = X - 0.01(3X - X^3),$$

where

$$X = \frac{1}{\sqrt{5}} \sum_{j=1}^{5} \sqrt{3}\,(2R_j - 1).$$

[This formula was suggested by Bol'shev (1959). Note that $\sqrt{3}\,(2R_j - 1)$ has a *standardized* rectangular distribution.]

2 HISTORICAL REMARKS

Because of the importance of the normal distribution, considerable attention has been paid to its historical development. The earliest workers regarded the distribution only as a convenient approximation to the binomial distribution. At the beginning of the nineteenth century appreciation of its broader theoretical importance spread with the work of Laplace and Gauss. The normal distribution became widely and uncritically accepted as the basis of much practical statistical work, particularly in astronomy. Around the beginning of the present century, a more critical spirit developed with more attention being paid to systems of "skew (nonnormal) frequency curves" (see Chapter 12). This critical spirit has persisted, but it is offset by developments in both theory and practice. The normal distribution has a unique position in probability theory, and it can be used as an approximation to other distributions. In practice, "normal theory" can frequently be applied, with small risk of serious error, when substantially nonnormal distributions correspond more closely to observed values. This allows us to take advantage of the elegant nature and extensive supporting numerical tables of normal theory.

The earliest published derivation of the normal distribution (as an approximation to a binomial distribution) seems to be that in a pamphlet of de Moivre dated 12 November 1733. This pamphlet was in Latin; in 1738 de Moivre published an English translation, with some additions. [See also Archibald (1926) and Daw (1966).]

In 1774 Laplace obtained the normal distribution as an approximation to hypergeometric distribution, and four years later he advocated tabulation of the probability integral [$\Phi(x)$, in our notation]. The work of Gauss in 1809 and 1816 established techniques based on the normal distribution, which became standard methods used during the nineteenth century.

Most theoretical arguments for the use of the normal distribution are based on forms of *central limit theorems*. These theorems state conditions under which the distribution of standardized sums of random variables tends to a unit normal distribution as the number of variables in the sum increases, that is, with conditions sufficient to ensure an *asymptotic* unit normal distribution. Gauss's (1816) derivation of the normal distribution, as the resultant of a large number of additive independent errors, may be regarded as one of the earliest results of this kind.

Formal rigorous mathematical discussion of central limit theorems (for independent random variables) may be said to start with the work of Lyapunov (1900). A useful theorem associated with his name states that if $X_1, X_2, \ldots, X_n$ are independent, identically distributed random variables with finite mean and standard deviation then the distribution of the standardized sum

$$\frac{\sum_{j=1}^{n} X_j - nE[X]}{\sqrt{n \, \text{Var}(X)}}$$

tends to the unit normal distribution as n tends to infinity. Lyapunov also obtained an upper bound for the magnitude of the difference between the cumulative distribution functions of the standardized sum and the unit normal. This upper bound was of the form $Cn^{-1/2} \log n$, where C is a constant depending on the variances and third moments of the X_i's. It has subsequently been considerably improved by Cramér (1928), Berry (1941), Esseen (1942), Zahl (1966), and Zolotarev (1967). For the case when the variables $\{X_i\}$ are identically distributed the upper bound obtained by Zolotarev (1967) is

$$0.82\left(\frac{\nu_3}{\sigma^3}\right)n^{-1/2}$$

where

$$\sigma^2 = \text{Var}(X_i),$$

$$\nu_3 = E\left[|X_i - E[X_i]|^3\right].$$

This result was an improvement on an earlier result of Wallace (1959) [correcting a result of Berry (1941)]. Zahl (1966) has shown that the upper bound

$$0.65\left(\frac{\nu_3}{\sigma^3}\right)n^{-1/2}$$

can be obtained, provided $\nu_3/\sigma^3 \geq 3/\sqrt{2} = 2.22$.

It can be shown by consideration of particular cases that the upper bound must be at least

$$C\left(\frac{\nu_3}{\sigma^3}\right)n^{-1/2},$$

with

$$C = \frac{\sqrt{13}}{6\sqrt{2\pi}} = 0.40974.$$

Zolotarev (1967) has shown that if the variance and absolute third central moment of X_j are σ_j^2, ν_{3j}, respectively ($j = 1, 2, \ldots, n$), then an upper bound for the magnitude of the difference between cumulative distribution functions is

$$0.9051\left(\sum_{j=1}^{n} \nu_{3j}\right)\left(\sum_{j=1}^{n} \sigma_j^2\right)^{-3/2}.$$

For the general case of independent (but not necessarily identically distributed) variables, Lindeberg (1922) showed that putting $\text{Var}(X_i) = \sigma_i^2$ and

$$\sigma_{(n)}^2 = \sum_{i=1}^{n} \sigma_i^2,$$

if

$$\lim_{n \to \infty} \sigma_{(n)}^{-2} \sum_{i=1}^{n} \left(\Pr\{|X_i - E[X_i]| \geq t\sigma_{(n)}\}\right.$$

$$\left. \times E\left[\{X_i - E[X_i]\}^2 \big| |X_i - E[X_i]| \geq t\sigma_{(n)}\right]\right) = 0$$

for all $t > 0$, the distribution of the standardized sum

$$\sigma_{(n)}^{-1} \sum_{i=1}^{n} (X_i - E[X_i])$$

tends to the unit normal distribution as n tends to infinity. The *necessity* of Lindeberg's condition was established by Feller (1935). Since then attention has moved to consideration of conditions under which a limiting normal distribution applies to sums of nonindependent random variables. An account of some such conditions can be found in a book by Loève (1963).

A comprehensive account of the central limit theorem and related problems (up to the early 1950s) has been given by Gnedenko and Kolmogorov

(1954). Multidimensional extensions of central limit theorems have been investigated by Bergström (1945), Esseen (1958), Sadikova (1966), and Sazanov (1967) among others.

Porter (1985), while discussing the historical details of the concepts of variation and error in Quetelet's statistics, has brought out the vital role that the normal distribution plays in the mathematics of society. Wilf (1988) has commented briefly on the general quest for normality. Read (1985) has provided a fine review of the various important developments on the normal distribution. Stigler (1982) has proposed a new standard for the normal distribution.

As one would expect, there has been a phenomenal development on various aspects of the normal distribution. Consequently several books and monographs have appeared dealing specifically with inference, characterizations, tolerance limits, prediction, goodness-of-fit, and so on. It is therefore neither feasible nor necessary to discuss all these developments in detail. Fortunately there is a handbook prepared by Patel and Read (1981) available on the distribution; the second edition of this book is currently under preparation. We are hopeful and confident that this volume will provide a comprehensive treatment to the distribution, and hence we have concentrated on adding only those results that are primarily of distributional nature (rather than specific inferential aspect). We refer the readers to the above-mentioned handbook and other books/monographs (listed for specific topics) for an elaborate discussion.

3   MOMENTS AND OTHER PROPERTIES

If U has the unit normal distribution, then, since the distribution is symmetrical about $U = 0$,

$$E[U] = 0, \tag{13.9}$$

and so

$$\mu_r = \mu'_r = E(U^r) = \left(\sqrt{2\pi}\right)^{-1} \int_{-\infty}^{\infty} x^r e^{-x^2/2} \, dx. \tag{13.10}$$

If r is odd,

$$\mu_r = 0.$$

If r is even,

$$\mu_r = \left(\sqrt{2/\pi}\right)\int_0^\infty x^r e^{-x^2/2}\, dx$$

$$= \left(\sqrt{2/\pi}\right)2^{(r+1)/2}\int_0^\infty t^{(r-1)/2}e^{-t}\, dt$$

$$= 2^{r/2}\Gamma\left(\tfrac{1}{2}(r+1)\right)/\sqrt{\pi}$$

$$= (r-1)(r-3)\ldots 3.1. \tag{13.11}$$

Hence

$$\mathrm{Var}(U) = \mu_2 = 1,$$

$$\alpha_3(U) = 0,$$

$$\beta_2(U) = \alpha_4(U) = 3.$$

This, as pointed out in Section 1, reveals that the unit normal is also the standardized normal distribution. If X has the general normal distribution (13.1), then

$$X = \xi + \sigma U, \tag{13.12}$$

where U is a unit normal variable.

Some normal probability density functions are shown in Figure 13.1. The nine curves shown correspond to all possible combinations of $\xi = -1, 0, 1$ and $\sigma = \tfrac{1}{2}, 1, 2$. The curve in the center represents the unit normal distribution ($\xi = 0$, $\sigma = 1$). The distribution is symmetrical about $X = \xi$; the probability density function has points of inflexion at $X = \xi \pm \sigma$. The distribution is unimodal with mode at $X = \xi$ (which is also the median of the distribution). The modal value of the probability density function is $(\sqrt{2\pi})^{-1} = 0.3979$.

The moment generating function of X ($= \xi + \sigma U$) is

$$E[e^{tX}] = e^{t\xi + (t^2\sigma^2/2)} \tag{13.13}$$

and the characteristic function is $e^{it\xi - (t^2\sigma^2/2)}$. For all $r > 2$, the cumulants κ_r are zero. This property characterizes normal distributions.

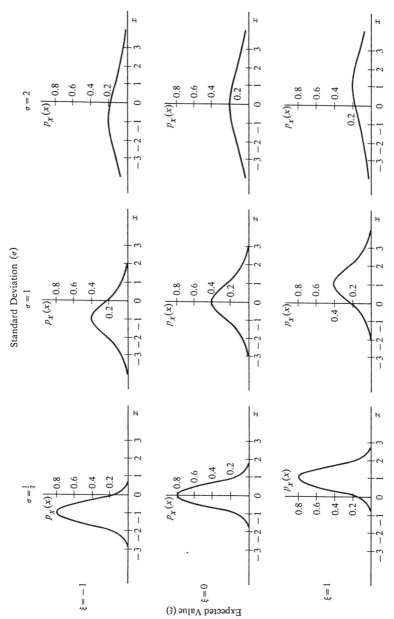

Figure 13.1 Normal Density Functions

The mean deviation of X is $\sigma\sqrt{2/\pi} = 0.798\sigma$. For all normal distributions

$$\frac{\text{Mean deviation}}{\text{Standard deviation}} = \sqrt{\frac{2}{\pi}} = 0.798. \tag{13.14}$$

The information-generating function of X is

$$(\sqrt{2\pi}\,\sigma)^{-u}\left(\frac{\sqrt{2\pi}\,\sigma}{\sqrt{u}}\right) = (\sqrt{2\pi}\,\sigma)^{-(u-1)}u^{-1/2}. \tag{13.15}$$

The entropy is

$$\log(\sqrt{2\pi}\,\sigma) + \tfrac{1}{2}. \tag{13.16}$$

It is of some interest to note that the probability density function (13.1) can be expressed in the numerical form

$$0.3979(0.6065)^{[(x-\xi)/\sigma]^2}. \tag{13.17}$$

The derivatives of the function $Z(\cdot)$ are also of some interest. They are used, for example, in the Gram-Charlier expansion (see Chapter 12). We have already discussed them in Chapter 1, and Section 1 of Chapter 12 contains some references to tables of their numerical values.

If $X_1, X_2, \ldots, X_n$ are independent, normally distributed random variables, then any linear function of these variables is also normally distributed. It is of interest to note that if X_1 and X_2 are independent, and each is normally distributed with zero expected value, then $X_1 X_2(X_1^2 + X_2^2)^{-1/2}$ is also normally distributed. If further var(X_1) = var(X_2), then $(X_1^2 - X_2^2)/(X_1^2 + X_2^2)$ is also normally distributed [Shepp (1964)].

If $X_1, X_2, \ldots, X_n$ are independent random variables each distributed as (13.1), then by applying the transformation

$$\begin{cases} X_1 = \bar{X} + (1\cdot2)^{-1/2}U_2\sigma + (2\cdot3)^{-1/2}U_3\sigma + \cdots + [(n-1)n]^{-1/2}U_n\sigma, \\ X_2 = \bar{X} - (1\cdot2)^{-1/2}U_2\sigma + (2\cdot3)^{-1/2}U_3\sigma + \cdots + [(n-1)n]^{-1/2}U_n\sigma, \\ X_3 = \bar{X} \qquad\qquad\quad -2(2\cdot3)^{-1/2}U_3\sigma + \cdots + [(n-1)n]^{-1/2}U_n\sigma, \\ \vdots \\ X_n = \bar{X} \qquad\qquad\qquad\qquad\qquad\qquad -(n-1)[(n-1)n]^{-1/2}U_n\sigma, \end{cases} \tag{13.18}$$

it can be shown that

1. $\bar{X}\,(= n^{-1}\Sigma_{j=1}^{n}X_j)$ has a normal distribution with expected value ξ and standard deviation $\sigma/\sqrt{n}$.

2. Each U_j $(j = 2, \ldots, n)$ is a unit normal variable.

3. $\bar{X}, U_2, \ldots, U_n$ are a mutually independent set of variables, and hence

4. $\sum_{j=1}^{n}(X_j - \bar{X})^2 = \sigma^2 \sum_{j=2}^{n} U_j^2$ is distributed as σ^2 (χ^2 with $(n - 1)$ degrees of freedom).

This last result was obtained by Helmert in 1875–76. The transformation (13.18) is called *Helmert's transformation*.

Since any function

$$g\left(X_1 - \bar{X}, \ldots, X_n - \bar{X}\right)$$

of the deviations $\{X_j - \bar{X}\}$ alone is a function of $\{U_j\}$ alone, we further note that by (13.18) and by property 3,

5. $\bar{X}$ and any function $g(X_1 - \bar{X}, \ldots, X_n - \bar{X})$ are mutually independent.

This result is helpful in calculating moments and distributions of statistics such as $\bar{X}[\mathrm{Range}(X_1, \ldots, X_n)]^{-1}$; $\bar{X}[n^{-1}\sum_{j=1}^{n}|X_j - \bar{X}|]^{-1}$. It can also be shown that

6. $\sum_{j=1}^{n}(X_j - \bar{X})^2$ and any function of the ratios

$$\left\{(X_i - \bar{X})\left[\sum_{j=1}^{n}\left(X_j - \bar{X}\right)^2\right]^{-1/2}\right\}$$

are mutually independent.

Zehna (1991) has recently given a simple proof for the result that $\bar{X}$ and S^2 are statistically independent. Bondesson (1981) discussed a normal sample with given sample mean and variance. Szekely (1985) established the multiplicative infinite divisibility of a standard normal distribution, while Chernoff (1981) presented an inequality involving the normal distribution function. Berg (1988) showed that the distribution of the cube (or, indeed, any odd power) of a normal random variable is not determined by its moments.

Hawkins (1975) has made a comment on the computation of noncentral t and normal integrals. Hawkins and Wixley (1986) have made some observations on the transformation of chi-squared variables to the normal distribution (see also Chapter 18). Aroian, Taneja, and Cornwell (1978) have derived the mathematical forms of the distribution of the product of two normal variables, while Cornwell, Aroian, and Taneja (1977) have discussed the numerical evaluation of that distribution. Hayya, Armstrong, and Gressis (1975) have considered the distribution of the ratio of two normal variables. Karhunen and Narula (1989) have similarly derived the distribution of the ratio of the absolute values of two normal variables. Anscombe and Glynn

(1983) have discussed the distribution of the kurtosis statistic b_2 based on normal samples. Owen (1980) has presented a valuable table giving formulas for various integrals related to the normal distribution. Some additions to this table have been made by Chou (1981).

Recently, for the U.S. population of men and women aged 17–84 years, Brainard and Burmaster (1992) showed that the normal distribution fits the marginal histograms of height (in cells representing 1-in. intervals) for both genders. Further, they demonstrated that the bivariate histogram for men is fit well by the bivariate normal distribution between the height and the natural logarithm of weight (in cells representing 10-lb intervals).

4 ORDER STATISTICS

Let $X_1' \leq X_2' \leq \cdots \leq X_n'$ be the order statistics obtained from a random sample of size n from the standard normal distribution (13.4). Then the density of X_r' $(1 \leq r \leq n)$ is

$$p_{X_r'}(x) = \frac{n!}{(r-1)!(n-r)!}\{\Phi(x)\}^{r-1}\{1 - \Phi(x)\}^{n-r}Z(x),$$

$$-\infty < x < \infty, \quad (13.19)$$

and the joint density of X_r' and X_s' $(1 \leq r < s \leq n)$ is

$$p_{X_r', X_s'}(x, y)$$

$$= \frac{n!}{(r-1)!(s-r-1)!(n-s)!}\{\Phi(x)\}^{r-1}\{\Phi(y) - \Phi(x)\}^{s-r-1}$$

$$\times \{1 - \Phi(y)\}^{n-s}Z(x)Z(y), \quad -\infty < x < y < \infty. \quad (13.20)$$

From (13.19) and (13.20), the single and the product moments of order statistics may be derived. Derivations of these quantities explicitly in terms of some elementary functions have been attempted by numerous authors, including Jones (1948), Godwin (1949a), Ruben (1954, 1956), Watanabe et al. (1957, 1958), Bose and Gupta (1959), and David (1963). These authors have been successful in their attempts at least for small sample sizes. By adopting a differential equation approach and using the characterizing differential equation (of the standard normal density function) $Z^{(1)}(x) = dZ(x)/dx = -xZ(x)$, Bose and Gupta (1959) have shown, for example, that when $n = 2$,

$$E[X_2'] = -E[X_1'] = \frac{1}{\sqrt{\pi}};$$

when $n = 3$,

$$E[X'_3] = -E[X'_1] = \frac{3}{2\sqrt{\pi}}, \qquad E[X'_2] = 0;$$

when $n = 4$,

$$E[X'_4] = -E[X'_1] = \frac{6}{\pi\sqrt{\pi}} \tan^{-1}\sqrt{2},$$

$$E[X'_3] = -E[X'_2] = \frac{6}{\sqrt{\pi}} - \frac{18}{\pi\sqrt{\pi}} \tan^{-1}\sqrt{2};$$

and when $n = 5$,

$$E[X'_5] = -E[X'_1] = \frac{15}{\pi\sqrt{\pi}} \tan^{-1}\sqrt{2} - \frac{5}{2\sqrt{\pi}},$$

$$E[X'_4] = -E[X'_2] = \frac{10}{\sqrt{\pi}} - \frac{30}{\pi\sqrt{\pi}} \tan^{-1}\sqrt{2}, \qquad E[X'_3] = 0.$$

This method, however, fails to give explicit expressions in terms of elementary functions for expected values of order statistics in samples of size 6 or more. It is of interest to mention that Ruben (1954), who has shown that the single moments of order statistics can be expressed as linear functions of the contents of certain hyperspherical simplices, noted that for dimension greater than three these contents cannot be expressed in terms of elementary functions. This possibly explains why the method of Bose and Gupta fails at sample size 6. Renner (1976) has discussed the evaluation of expected values by means of power series for sample sizes six and seven.

Similar explicit formulas may be derived for the product moments of order statistics from small sample sizes. First, it is clear that when $n = 2$,

$$E[X'_1 X'_2] = E[X_1 X_2] = E[X_1]E[X_2] = 0.$$

Following the differential equation approach, it can be shown that when $n = 3$,

$$E[X'_1 X'_2] = E[X'_2 X'_3] = \frac{\sqrt{3}}{2\pi}, \qquad E[X'_1 X'_3] = -\frac{\sqrt{3}}{\pi},$$

and when $n = 4$,

$$E[X_1'X_2'] = E[X_3'X_4'] = \frac{\sqrt{3}}{\pi}, \qquad E[X_2'X_3'] = \frac{\sqrt{3}}{\pi}(2 - \sqrt{3}),$$

$$E[X_1'X_3'] = E[X_2'X_4'] = -\frac{\sqrt{3}}{\pi}(2 - \sqrt{3}), \qquad E[X_1'X_4'] = -\frac{3}{\pi}.$$

A general approach, given by Godwin (1949a), is to express the product moments in terms of integrals of the form

$$J_n = \int_0^\infty \cdots \int_0^\infty e^{-Q(x_1,\ldots,x_n)} \, dx_1 \ldots dx_n, \qquad (13.21)$$

where $Q(x_1, \ldots, x_n)$ is a quadratic form in the x_i's. For $n = 1, 2, 3, J_n$ can be expressed explicitly in terms of elementary functions as follows:

$$n = 1, \qquad Q(x_1) = a_{11}x_1^2, \qquad J_1 = \frac{\sqrt{\pi}}{2a_{11}},$$

$$n = 2, \qquad Q(x_1, x_2) = a_{11}x_1^2 + a_{22}x_2^2 + 2a_{12}x_1x_2,$$

$$J_2 = \frac{1}{\sqrt{\Delta_2}}\left\{\frac{\pi}{2} - \tan^{-1}\left(\frac{a_{12}}{\sqrt{\Delta_2}}\right)\right\},$$

where

$$\Delta_2 = \begin{vmatrix} a_{11} & a_{12} \\ a_{12} & a_{22} \end{vmatrix} = a_{11}a_{22} - a_{12}^2;$$

$$n = 3, \qquad Q(x_1, x_2, x_3)$$

$$= a_{11}x_1^2 + a_{22}x_2^2 + a_{33}x_3^2 + 2a_{12}x_1x_2 + 2a_{13}x_1x_3 + 2a_{23}x_2x_3,$$

$$J_3 = \frac{\sqrt{\pi}}{4\sqrt{\Delta_3}}\left\{\frac{\pi}{2} + \tan^{-1}\left(\frac{a_{12}a_{13} - a_{11}a_{23}}{\sqrt{a_{11}\Delta_3}}\right) + \tan^{-1}\left(\frac{a_{12}a_{23} - a_{13}a_{22}}{\sqrt{a_{22}\Delta_3}}\right)\right.$$

$$\left. + \tan^{-1}\left(\frac{a_{13}a_{23} - a_{12}a_{33}}{\sqrt{a_{33}\Delta_3}}\right)\right\},$$

where

$$\Delta_3 = \begin{vmatrix} a_{11} & a_{12} & a_{13} \\ a_{12} & a_{22} & a_{23} \\ a_{13} & a_{23} & a_{33} \end{vmatrix}$$

$$= a_{11}a_{22}a_{33} - a_{11}a_{23}^2 - a_{22}a_{13}^2 - a_{33}a_{12}^2 + 2a_{12}a_{13}a_{23}.$$

The values of means of order statistics have been tabulated to five decimal places by Harter (1961a) for $n = 2(1)100(25)250(50)400$, and also by Harter (1970) for some more choices of n. The mean and variance of the ith quasi-range have been tabulated by Harter (1959) for n up to 100. Tippett (1925) has computed the expected value of the sample range for $n \leq 1000$, while Harter (1960) presented tables of the mean, variance, and the coefficients of skewness and kurtosis for $n \leq 100$. Teichroew (1956) has presented tables of means and product moments of order statistics for sample sizes up to 20. By making use of Teichroew's tables, Sarhan and Greenberg (1962) have tabulated the variances and covariances of order statistics (to 10 decimal places) for $n \leq 20$. These tables have been extended by Tietjen et al. (1977) for sample sizes up to 50. The values of the mean and standard deviation of order statistics prepared by Yamauti (1972) for sample sizes up to 50 are contained in the tables of Tietjen et al. (1977). For the largest order statistic X_n', Ruben (1954) has tabulated the first ten moments for $n \leq 50$ and Borenius (1965) has presented the first two moments for $n \leq 120$. Parrish (1992a, b) has presented tables of means, variances and covariances of order statistics (up to 25 decimal places) for some selected sample sizes up to 50. Miyakawa, Tamiya, and Kotani (1985a) have discussed the numerical evaluation of moments of order statistics through an orthogonal inverse expansion. Öztürk and Aly (1991) have proposed some simple approximations for the moments of order statistics.

Royston (1982) has given an algorithm for computing the expected values of normal order statistics. This algorithm will compute and present the exact values for sample sizes up to 1000 and will also present an approximate value for the quantity for larger sample sizes [see Koeniger (1983) for an additional remark on this algorithm]. Balakrishnan (1984) has presented an algorithm, based on an orthogonal inverse expansion, to approximate the sum of squares of normal scores, namely $S = \sum_{i=1}^{n} \{E[X_i']\}^2$. This quantity arises often in nonparametric statistics. Dansie (1986) described the normal order statistics as permutation probability models, while Nelson (1983) discussed the usefulness of normal scores as a transformation.

For the standard normal distribution, an interesting property satisfied by order statistics is that

$$\sum_{j=1}^{n} E[X_i'X_j'] = \sum_{j=1}^{n} \text{Cov}(X_i', X_j') = 1, \qquad 1 \leq i \leq n. \qquad (13.22)$$

That is, every row (or column) of the product-moment matrix or the variance-covariance matrix adds up to 1. This follows easily from the fact that $\bar{X}$ and $X_i' - \bar{X}$ are statistically independent. Some relationships between moments of order statistics have also been established by various authors, and almost all of them have been derived by exploiting the characterizing differential equation $Z^{(1)}(x) = -xZ(x)$. For example, Govindarajulu (1963) has shown that for $1 \le i \le n$,

$$E[X_i'^2] = 1 + n\binom{n-1}{i-1}\sum_{j=0}^{n-i}(-1)^j\binom{n-i}{j}\frac{1}{i+j}E[X_{1:i+j}'X_{2:i+j}']. \quad (13.23)$$

If we set $i = n$ in (13.23), we obtain the relation

$$E[X_1'^2] = E[X_n'^2] = 1 + E[X_1'X_2']. \quad (13.24)$$

The results in (13.22), (13.23), and (13.24) have been used for checking the computation of the product moments. Furthermore Davis and Stephens (1977, 1978) have applied (13.22) and (13.24) to improve the David-Johnson approximation of the variance-covariance matrix (see Chapter 12). Reference may also be made to Shea and Scallon (1988) for further remarks in this regard.

By noting that the condition $Z^{(1)}(x) = -xZ(x)$ is satisfied by both the standard normal and the half-normal (see Section 10) distributions, Joshi and Balakrishnan (1981) established the following results satisfied by order statistics from both these distributions:

$$\sum_{j=i}^{n} E[X_i'X_j'] = 1 + \sum_{j=i}^{n} E[X_{i-1}'X_j'] \qquad \text{for } 1 \le i \le n, \quad (13.25)$$

$$\sum_{j=i+1}^{n} E[X_i'X_j'] = \sum_{j=i+1}^{n} E[X_j'^2] - (n-i) \qquad \text{for } 1 \le i \le n-1, \quad (13.26)$$

$$\sum_{j=1}^{n} E[X_i'X_j'] = 1 + nE[X_{1:1}']E[X_{i-1:n-1}'] \qquad \text{for } 1 \le i \le n, \quad (13.27)$$

and

$$\sum_{j=1}^{n} \mathrm{Cov}(X_i', X_j') = 1 - (n-i+1)E[X_{1:1}']\{E[X_i'] - E[X_{i-1}']\}$$

$$\text{for } 1 \le i \le n. \quad (13.28)$$

Joshi and Balakrishnan (1981) have used these relations to derive a convenient expression for the variance of the selection differential or reach statistic, defined as $\Delta_k = \overline{X}_k - \overline{X}$, where $\overline{X}_k$ is the average of the k largest order statistics, $X'_{n-k+1}, \ldots, X'_n$. For example, Joshi and Balakrishnan (1981) have shown that

$$E\left[\overline{X}_k^2\right] = \frac{1}{k^2} \sum_{i=n-k+1}^{n} (2i - 2n + 2k - 1)E\left[X'^2_i\right] - \frac{k-1}{k}. \quad (13.29)$$

Consequently the mean and variance of $\overline{X}_k$ (and hence of Δ_k) can be determined from the first two raw moments of X'_i alone. These quantities have been tabulated for sample sizes up to 50 by Joshi and Balakrishnan (1981). Some properties of $\overline{X}_k$ have also been discussed by Schaeffer, van Vleck, and Velasco (1970) and Burrows (1972, 1975). In particular, they observed that $k \, \text{Var}(\overline{X}_k)$ remains almost constant for the selected fraction k/n. While Schaeffer, van Vleck, and Velasco (1970) tabulated the values of $k \, \text{Var}(\overline{X}_k)$ for $n \leq 20$ and all choices of k, Burrows (1972, 1975) provided approximations to $E[\overline{X}_k]$ and $k \, \text{Var}(\overline{X}_k)$ for large values of n. But, Joshi and Balakrishnan (1981) have pointed out that Burrows' approximation for $k \, \text{Var}(\overline{X}_k)$ is not satisfactory for small values of k even when n is 50, and that the approximation improves when k increases. The statistic Δ_k is related to Murphy's test statistic for outliers; for details, see Hawkins (1980) and Barnett and Lewis (1994).

The cumulative distribution function of the extreme X'_n was tabulated by Tippett (1925) for $n = 3, 5, 10, 20, 30, 50, 100(100)1000$. Percentage points for $n \leq 30$ were given by Pearson and Hartley (1970), and the cumulative distribution function of X'_n was tabulated by Pearson and Hartley (1972), to 7 decimal places, for $n = 3(1)25(5)60, 100(100)1000$ and for x in steps of 0.1. Gupta (1961) and Govindarajulu and Hubacker (1964) have presented percentage points of all order statistics for $n \leq 10$ and $n \leq 30$, respectively. Eisenhart, Deming, and Martin (1963) have tabulated percentage points of the sample median. Pearson and Hartley (1942, 1970) have provided tables for the distribution function of the sample range $X'_n - X'_1$. Harter and Clemm (1959) have given extensive tables of the cumulative distribution function (to 8 decimal places) and also of the percentage points (to 6 decimal places) of the range for $n = 2(1)20(2)40(10)100$ and the argument in steps of 0.01, and 23 different percentage points for each n, respectively. Besides reproducing these tables, Harter (1970) has presented tables of the probability density function of the sample range (to 8 decimal places) for n up to 16 and the argument in steps of 0.01. The cumulative distribution function of the ith quasi-range, $X'_{n-i+1} - X'_i$, is presented by Harter (1970) to 8 decimal places for $n = 2(1)20(2)40(10)100$, $i = 1, 2, \ldots, 9$, and the argument in steps of 0.05. Harter (1970) has also presented tables of corresponding percentage points to 6 decimal places. Currie (1980) has discussed the distribution of the studentized range. An algorithm for calculating the probability integral of the sample range has been presented by Barnard (1978); see El Lozy (1982) for some additional remarks on this computational algorithm.

David, Kennedy, and Knight (1977) have provided tables of means, variances, and covariances of order statistics arising from a standard normal sample with one outlier. These values are presented for all n up to 20, for the two cases: (1) location-outlier, the outlier being $N(\lambda, 1)$ for $\lambda = 0(0.5)3$, 4; and (2) scale-outlier, the outlier being $N(0, \tau^2)$ for $\tau = 0.5, 2, 3, 4$. These tables have since been used in many robustness studies because they can be utilized to determine exactly the bias and the mean square error of any linear estimator when there is a single outlier present in the sample.

In addition to the numerous tables listed here, there are also several tables pertaining tests for outliers. The recently published third edition of Barnett and Lewis (1994) provides a complete list of all the available tables.

5 RECORD VALUES

Let $X_{U(1)}, X_{U(2)}, \ldots$ be the upper record values arising from a sequence $\{X_i\}$ of i.i.d. standard normal variables. That is, with $T_0 = 0$ and

$$T_n = \min\{j : j > T_{n-1}, X_j > X_{T_{n-1}}\}$$

denoting the upper record times, the record value sequence $\{X_{U(n)}\}_{n=1}^{\infty}$ is defined by $X_{U(n)} = X_{T_{n-1}}$, $n = 1, 2, \ldots$. Then the density of $X_{U(n)}$ is

$$p_{X_{U(n)}}(x) = \frac{1}{(n-1)!}\{-\log(1 - \Phi(x))\}^{n-1}Z(x), \qquad -\infty < x < \infty,$$

$$(13.30)$$

and the joint density of $X_{U(m)}$ and $X_{U(n)}$ is given by

$$p_{X_{U(m)}, X_{U(n)}}(x, y)$$

$$= \frac{1}{(m-1)!(n-m-1)!}\{-\log(1 - \Phi(x))\}^{m-1}\frac{Z(x)}{1 - \Phi(x)}$$

$$\times \{-\log(1 - \Phi(y)) + \log(1 - \Phi(x))\}^{n-m-1}Z(y),$$

$$-\infty < x < y < \infty, 1 \le m < n. \quad (13.31)$$

From (13.30) and (13.31), Houchens (1984) and Balakrishnan and Chan (1994) have determined (by numerical methods) the values of $E[X_{U(n)}]$, $\mathrm{Var}(X_{U(n)})$, and $\mathrm{Cov}(X_{U(m)}, X_{U(n)})$. By making use of these values, these authors have also derived the best linear unbiased estimators of ξ and σ based on the first n upper record values. Balakrishnan and Chan (1994) have also discussed the prediction of a future record and a test for spuriosity of a current record value. They have also established that

$$E\left[X_{U(n-1)}X_{U(n)}\right] = E\left[X_{U(n)}^2\right] - 1 \qquad \text{for } n \geq 2, \qquad (13.32)$$

and consequently that

$$\begin{aligned} \mathrm{Cov}\left(X_{U(n-1)}, X_{U(n)}\right) \\ = \mathrm{Var}\left(X_{U(n)}\right) + E\left[X_{U(n)}\right]\left\{E\left[X_{U(n)}\right] - E\left[X_{U(n-1)}\right]\right\} - 1. \quad (13.33) \end{aligned}$$

Suppose that $X_{L(1)}, X_{L(2)}, \ldots$, denote the lower record values arising from a sequence $\{X_i\}$ of i.i.d. standard normal variables; that is, with $T_0^* = 1$ and

$$T_n^* = \min\left\{j : j > T_{n-1}^*, X_j < X_{T_{n-1}^*}\right\}$$

denoting the lower record times, the lower record value sequence $\{X_{L(n)}\}_{n=1}^{\infty}$ is defined by $X_{L(n)} = X_{T_{n-1}^*}$, $n = 1, 2, \ldots$. Then, due to the symmetry of the standard normal distribution, it may be easily observed that

$$X_{L(n)} \stackrel{d}{=} -X_{U(n)},$$

$$\left(X_{L(m)}, X_{L(n)}\right) \stackrel{d}{=} \left(-X_{U(m)}, -X_{U(n)}\right).$$

With this property, moments of the lower record values (and inference based on the lower record values) can be easily obtained from the corresponding results for the upper record values.

6 CHARACTERIZATIONS

We first summarize normal characterizations presented in the first edition of this book [Johnson and Kotz (1970)]. In all cases $X_1, \ldots, X_n$ are i.i.d. random variables, unless explicitly stated otherwise.

1. $\bar{X} = n^{-1}\sum_{j=1}^{n} X_j$ has a normal distribution. [See Chapter 12, Section 6. Janson (1988) expresses this in the form $d^k \phi_X(t)/dt^k|_{t=0} = 0$ for all $k > 2$, where $\phi_X(t)$ is the characteristic function of the common distribution of the X's.]

    A more sophisticated characterization based on properties of $\bar{X}$ is due to Fieger (1971). If the distribution of the X's belongs to a location family, with $F_X(x) = g(x - \theta)$, if $E[|X_i|]$ is finite, and if $\bar{X}$ is the best translation invariant estimator of θ for any convex loss function

$W(\theta^*, \theta) = W(|\theta^* - \theta|)$ with $W(u) \geq W(0)$ for all u, then the common distribution is normal.

2. $\bar{X}$ and $g(\{X_i - X_j\}, i \neq j)$ mutually independent, given one of three conditions:

   a. $g(\cdot) = \sum_{j=1}^n (X_j - \bar{X})^2 \, (= nS)$. [Lukacs (1942); weakened to $\bar{X}$ and S which have joint pdf of form $h(\bar{X}, S)S^{n-2}$, with $\partial h/\partial \bar{X}$ and $\partial h/\partial S$ existing, by Kaplansky (1943).]

   b. $g(\cdot)$ is a k-statistic [i.e., $g(\cdot)$ is a polynomial in the X's and $E[g(\cdot)] = \kappa_r$, which is the rth cumulant of the common distribution of the X's for some integer $r > 2$]. [Basu and Laha (1954); Lukacs (1955). See also Geary (1936) and Kawata and Sakamoto (1949).]

   c. $g(\cdot) = 0$ if and only if $X_j = \bar{X}$ for all j; and

   $$g(c\underset{\sim}{X}) = |c| g(\underset{\sim}{X})$$

   [Paskevich (1958), Rao (1958), Zinger (1958)].

3. Conditions on conditional expected values:

   a. $E[\bar{X} | \{X_i - X_j\}, i \neq j] = E[\bar{X}]$ (for $n > 2$) [Kagan, Linnik, and Rao (1965, 1973)]. Note that for $n = 3$, this condition is satisfied by *any* symmetric distribution with finite expected value. Rao (1967) extended this condition to

   $$E\left[X_i | X_j - \bar{X}\right] = E[\bar{X}] \qquad \text{for all } i \neq j.$$

   b. The common distribution has zero mean and finite variance, and there exist $(n - 1)$ linearly independent statistics $Y_j = \sum_{i=1}^n a_{ji} X_i$ $(j = 1, \ldots, n - 1)$ such that

   $$E\left[Y_j \left| \sum_{i=1}^n b_i X_i \right.\right] = 0 \qquad \text{for all } j = 1, \ldots, n - 1 \quad (13.34)$$

   and some $\{b_i\}$ ($\{a_{j1}, \ldots, a_{jn}\}$ not all zero). Kagan, Linnik, and Rao (1973, p. 156) improve this result by requiring only $j = 1$ in (13.34) and not finite variance, but restricted to $n \geq 3$. [Cacoullos (1967b)].

   c. $\sum_{j=1}^n a_j b_j = 0$, where $a_j \neq 0$ $(j = 1, \ldots, n)$, $\{b_j\}$ are not all zero, and

   $$E\left[\left(\sum_{j=1}^n b_j X_j\right)^2 \left| \sum_{j=1}^n a_j X_j \right.\right] = E\left[\left(\sum_{j=1}^n b_j X_j\right)^2\right]$$

   [Cacoullos (1967a)].

4. Conditions based on identity of distributions:

   a. $\sum_{j=1}^{n} a_j X_j$ ($a_j \neq 0$, $j = 1, \ldots, n$) and each X_i ($i = 1, \ldots, n$) for some $\{a_j\}$ [Shimizu (1962), referring to Linnik (1952)].

   b. $\sum_{j=1}^{n} a_j X_j$ and $\sum_{j=1}^{n} b_j X_j$, where $a_j b_j \neq 0$ (for all j), and $\{b_j\}$ is not just a rearrangement of $\{a_j\}$, provided the common distribution has finite moments of all order [Marcinkiewicz (1939)]. [Linnik (1952) showed that this result is not valid if some moments are infinite.]

   c. ($n = 2$) of X_1 and $(X_1 + X_2)/\sqrt{2}$ [Pólya (1923), referred to by Bryc (1990)].

5. Characterizations based on order statistics [Govindarajulu (1966)]:

   a. Provided the common distribution has a finite variance, the condition

   $$E\left[X_n'^2 - X_{n-1}' X_n' \right] = \sigma^2$$

   ensure that the common distribution is *either* normal (with variance σ^2) *or* truncated (from above) normal.

   b. If the expected value of the common distribution is zero, the condition

   $$nE\left[X_j' \overline{X} \right] = \sigma^2, \qquad j = 1, \ldots, n; \quad n \geq 2, \qquad (13.35)$$

   ensures that the common distribution is normal. [Note that if it be assumed that the common distribution has zero probability for negative values, condition (13.35) ensures that it is *half-normal* (see Section 10).]

6. Conditions based on distributions of specific statistics:

   a. $\sum_{j=1}^{n} X_j^2$ and $\{X_i^2/(\sum_{j=1}^{n} X_j^2)\}$ ($i = 1, \ldots, n$) are mutually independent [Tamhankar (1967)].

   b. The distribution of $\sum_{j=1}^{n}(X_j + a_j)^2$ depends on the parameters $\{a_j\}$ only through $\sum_{j=1}^{n} a_j^2$ [Kagan and Shalayevskii (1967)].

   c. Provided that $a_j b_j \neq 0$ ($j = 1, \ldots, n$), $\sum_{j=1}^{n} a_j X_j$ and $\sum_{j=1}^{n} b_j X_j$ are mutually independent for some $\{a_j, b_j\}$. [Darmois (1951), Skitovich (1953). The result is known as the *Darmois-Skitovich theorem*. Another proof was given by Tranquilli (1968).]

   This condition and the following one do not require assumption of finite variance—or indeed, identity of distributions of $X_1, \ldots, X_n$ —though mutual independence is still a necessary property.

   d. Extending condition 1 to cases of not necessarily identically distributed variables, if $(X_1 + X_2)$ has a normal distribution, so do each of X_1 and X_2. [Cramér (1936), extended to $n > 2$ random variables by Lukacs (1956).]

e. If each X_j is distributed symmetrically about zero, the condition that the statistics

$$Y_j = \sqrt{j-1}\, X_j \bigg/ \left\{ \sum_{i=1}^{j-1} X_i^2 \right\}^{1/2}, \qquad j = 2, \ldots, n,$$

are mutually independent and distributed as t with $j - 1$ degrees of freedom $(j = 2, \ldots, n)$ is necessary and sufficient to ensure that each X_j has the same normal distribution [Kotlarski (1966)].

Books by Kagan, Linnik, and Rao (1973) and Mathai and Pederzoli (1977), as well as a survey by Kotz (1974), provide a useful, and reasonably adequate, basis for more mathematically inclined readers desiring to understand derivations of the above results. Ghurye and Olkin (1973) include a clear exposition of the work of Pólya (1923) and Linnik (1953), and they illustrate the application of Linnik's results.

In the two decades following publication of the first edition of this book, there has been substantial growth in literature on characterizations, in which characterization of normal distributions has played a prominent part. In fact, to some extent, characterization of normal distributions has become a branch of mathematics, with emphasis on functional equations and characteristic functions but with only limited impact on applied statistics. We are unable to provide a comprehensive, or even fully representative, account due in part to space restriction but also to our feeling that a sizable proportion of the more recent results are of little value in applied work. Exploitation of earlier results has not occurred to the extent that one might have expected, or hoped for. This is true even for results of type 2 above (p. 101).

A possible exception is the following characterization initially due to Csörgö and Seshadri (1971) and Csörgö, Seshadri, and Yalovsky (1975). It was used to develop tests of normality. An early version is as follows: Given $X_1, \ldots, X_n$ ($n = 2k + 3$, $k \geq 2$) which are i.i.d. with expected value ξ and variance σ^2, and

$$Z_j = \frac{\sum_{i=1}^{j} X_i - jX_j}{\sqrt{j(j+1)}} \qquad j = 1, \ldots, n-1,$$

$$(13.36)$$

$$Z_n = \sum_{i=1}^{n} X_i,$$

and given

$$Y_g = Z_{2g-1}^2 + Z_{2g}^2, \qquad g = 1, \ldots, k+1 = \tfrac{1}{2}(n-1),$$

and the statistics

$$Y_r^* = \frac{\sum_{g=1}^r Y_g}{\sum_{g=1}^{k+1} Y_g}, \qquad r = 1, 2, \ldots, k,$$

which have the joint distribution of order statistics for k mutually independent standard uniform variables, then the common distribution of $X_1, \ldots, X_n$ is normal.

A parallel result, applicable for $n = 2k$, if $E[X_j] = \xi$ is known, *and* it can be assumed that the common distribution is symmetrical, is that if the statistics

$$Y_r'^* = \frac{\sum_{g=1}^r Y_g'}{\sum_{g=1}^k Y_g'}, \qquad r = 1, 2, \ldots, k-1,$$

where $Y_g' = (X_{2g-1} - \xi)^2 + (X_{2g} - \xi)^2$ $(g = 1, \ldots, k)$ have the joint distribution of order statistics for $k-1$ mutually independent standard uniform variables, the common distribution of the X's is normal. Further extensions have been obtained in later work.

A typical result of this class due to Pakshirajan and Mohan (1971), which states that if X_1, X_2, and X_3 are mutually independent random variables each symmetric about zero, with cdf continuous at 0 (i.e., $\Pr[X_j = 0] = 0$), then the joint characteristic function of the ratios X_1/X_3 and X_2/X_3 is $\exp(-\sqrt{t_1^2 + t_2^2}\,)$ if and only if the X's have a common normal distribution with zero expected value. However, as noted in the first edition (p. 53), the distribution of the ratio X_1/X_2 of two i.i.d. variables does *not* characterize a normal distribution [Fox (1965)].

Bryc (1990) has extended Pólya's (1923) result (4c above) as follows: If X_1 and X_2 are i.i.d. random variables with finite variance such that, for some α and β (> 0), $(\alpha + \beta X_1)$ and $(X_1 + X_2)$ have identical distributions, then X_1 has a normal distribution.

A further generalization [Wesolowski (1990)] is this: If $X_1, \ldots, X_n$ $(n \geq 2)$ are square integrable random variables and $\{a_1, \ldots, a_n\}$ are real numbers with $a_1 \neq -1$, 0, or 1 and X_1 and $\sum_{j=1}^n a_j X_j$ have identical distributions, then X_1 has a normal distribution [compare 3a; Shimizu (1962) requires a similar condition for *each* X_j]. Arnold and Isaacson (1978) give a simpler proof. See also Lukacs and Laha (1964), who relaxed the condition of square integrability.

Kagan, Linnik, and Rao (1973) contains some extensions and refinements of results already mentioned. These include

1. If $X_1, \ldots, X_n$ are i.i.d. and

$$\sum_{j=1}^{n} a_j X_j \left(\sum_{j=1}^{n} a_j^2 = 1 \right) \quad \text{and} \quad \sum_{j=1}^{n} X_j^2 - \left(\sum_{j=1}^{n} a_j X_j \right)^2$$

are mutually independent, the X's are normally distributed (pp. 105–106) (cf. 2a).

2. If $X_1, \ldots, X_n$ $(n \geq 3)$ are mutually independent, with $E[X_i] = 0$, if

$$L_i - \sum_{j=1}^{n} a_{ij} X_j, \quad i = 1, \ldots, n,$$

are linearly independent, and if

$$E[L_1 | L_2, \ldots, L_n] = 0,$$

then the X's each have a normal distribution (p. 419).

3. If $X_1, \ldots, X_n$ are mutually independent and $\{a_j\}, \{b_j\}$ are nonzero real numbers satisfying the conditions

$$a_j, b_j \neq 0 \quad \text{for all } j,$$

$$a_j b_j^{-1} + a_k b_k^{-1} \neq 0 \quad \text{for all } j \neq k,$$

then each X_j has a normal (possibly degenerate) distribution provided that the conditional distribution of $\sum_{j=1}^{n} a_j X_j$ and $\sum_{j=1}^{n} b_j X_j$ is symmetric.

We also note the following:

4. If X_1 and X_2 are i.i.d., they are unit normally distributed if and only if, for some $a_1, a_2 \neq 0$ each of $(a_1 X_1 + a_2 X_2)^2/(a_1^2 + a_2^2)$ and $(a_1 X_1 - a_2 X_2)^2/(a_2^2 + a_1^2)$ has a χ^2 distribution with 1 degree of freedom [Geisser (1973, pp. 492–494)].

5. Kelker and Matthes (1970) considered location-scale families. These are families with cdf of form

$$F_X(x) = \frac{1}{\sigma} g\left(\frac{x - \theta}{\sigma} \right),$$

where θ, σ are the location and scale parameters, respectively.

According to Kelker and Matthes, "Within the location and scale parameter families $(\overline{X}, S)$ is a sufficient statistic [for (θ, σ)] if and only if the family is the normal distribution family." This is indeed correct, but Bondesson (1977) has pointed out that their proof required $F_X(x)$ to represent a *continuous* distribution. He amended the proof to avoid this assumption. Kelker and Matthes (1970) also showed that

6. If $X_1, \ldots, X_n$ are independent and nondegenerate random variables with location family cdfs $F_X(x) = g_i(x - \theta)$ $(i = 1, \ldots, n)$, then a necessary and sufficient condition for $\sum_{j=1}^{n} b_j X_j$ $(b_1, \ldots, b_n \neq 0)$ to be a sufficient statistic for θ is that X_j have a normal distribution with variance proportional to b_j^{-1}.

7. If $X_1, \ldots, X_n$ $(n = 4)$ are i.i.d. and $(X_1 - X_2)/S$ is independent of $(\overline{X}, S)$, then the common distribution of the X's is normal.

Braverman (1985) has obtained characterizations in terms of *unconditional* expected values. These include the following characterizations:

1. If X_1, X_2, and X_3 are i.i.d. random variables with finite moments of odd order, and there are constants C_a such that

$$E\left[\left|\sum_{j=1}^{3} a_j X_j\right|^p\right] = C_a\left(\sum_{j=1}^{3} a_j^2\right)^{p/2} \tag{13.37}$$

for all $a = (a_1, a_2, a_3)$ and all odd p, then each of the three variables has a normal distribution.

This result is not true for even integers p.

2. If X_1 and X_2 are i.i.d., with a symmetric common distribution, with

$$E\left[\exp\left(\lambda |X_j|^2\right)\right] < \infty \qquad \text{for some } \lambda > 0$$

and

$$E\left[|X_j|^s\right] \neq 0 \qquad \text{for all } s$$

real or complex, with Re $s > 0$, and

$$E[|a_1 X_1 + a_2 X_2|] = C\left(a_1^2 + a_2^2\right)^{1/2} \tag{13.38}$$

for suitable C, then the common distribution is normal.

Braverman (1989) has also shown that

3. If X_1 and X_2 are i.i.d. (not necessarily symmetric) and there are two odd numbers, p_1 and p_2, such that

$$E[|a_1 X_1 + a_2 X_2|^{p_j}] = C_j\left(a_1^2 + a_2^2\right)^{p_j}, \qquad j = 1, 2, \tag{13.39}$$

for all real a_1, a_2 and suitable C_j, then the common distribution of the X's is normal.

Next we present a few notes on "stability" (or "robustness") of characterizations—namely, how far small departures from exact characterization conditions can affect size of departure from the characterized form.

Defining the distance between two cdfs as

$$\delta(G(x), F(x)) = \sup_x |G(x) - F(x)|$$

Meshalkin (1968) described the two distributions as *ε-coincident* if $\delta(G(x), F(x)) \leq \varepsilon$, and termed a random variable X, with cdf $F_X(x)$, *ε-normal* if

$$\delta\left[F_X(x), \Phi\left(\frac{x - \xi}{\sigma}\right)\right] \leq \varepsilon$$

for some pair (ξ, σ).

Sapogov (1956) showed that if X_1 and X_2 are independent, with

$$F_{X_1}(0) = \tfrac{1}{2},$$

$$\int_{-g(\varepsilon)}^{g(\varepsilon)} x \, dF_{X_1}(x) = \xi_1,$$

$$\int_{-g(\varepsilon)}^{g(\varepsilon)} x^2 \, dF_{X_1}(x) = \sigma_1^2 + \xi_1^2,$$

where

$$g(\varepsilon) = (-2 \log \varepsilon)^{1/2} + 1, \qquad 0 < \varepsilon < 1,$$

then, if $X_1 + X_2$ is ε-normal, X_1 is $\beta(\varepsilon)$-normal with

$$\beta(\varepsilon) = C\sigma_1^{-3}(-\log \varepsilon)^{-1/2},$$

where C is some constant (not depending on ε), because

$$\sup_x \left| F_{X_1}(x) - \Phi\left(\frac{x - \xi_1}{\sigma_1}\right)\right| < C\sigma_1^{-3}(-\log \varepsilon)^{-1/2}. \qquad (13.40)$$

Using the concept of ε-*independence* of X_1 and X_2, defined by

$$\left| \int_{a_j x_1 + b_j x_2 < c_j \ (j=1,2)} \delta\big(F_{X_1, X_2}(x_1, x_2) - F_{X_1}(x_1) F_{X_2}(x_2)\big) dx_1 dx_2 \right| \le \varepsilon$$

for all (a_j, b_j, c_j), Nye (1966) showed that

1. if X_1 and X_2 are mutually independent, and $(X_1 + X_2)$ and $(X_1 - X_2)$ are ε-independent, X_1 and X_2 are both $\beta_1(\varepsilon)$-normal,
2. if $X_1, \ldots, X_n$ are i.i.d. random variables and $\overline{X}$ and S^2 are ε-independent, then the common distribution of the X_i's is $\beta_2(\varepsilon)$-normal.

Both $\beta_1(\varepsilon)$ and $\beta_2(\varepsilon)$ were of order $(-\log \varepsilon)^{-1/2}$, and so was $\beta(\varepsilon)$.

Meshalkin (1968) obtained an improved order of magnitude in the following results: If X_1 and X_2 are i.i.d., with expected value zero and variance 1, and $E[|X_i|^2]$ is bounded, then (1') ε-independence of $(X_1 + X_2)$ and $(X_1 - X_2)$ implies $\beta_3 \varepsilon^{1/2}$-normality of the common distribution, and (2') ε-coincidence of the distribution functions of $(X_1 + X_2)/\sqrt{2}$ and X_i implies $\beta_4 \varepsilon^{1/2}$-normality of X_1 and X_2. The multipliers β_3 and β_4 depend on the bound of $E[|X_i|^3]$, but not on ε.

Yanushkyavichyus (1989) in his doctoral dissertation at Vilnius University provided the following stability theorem, improving on the above results: Let $X_1, \ldots, X_n$ be independent identically distributed random variables. If $\overline{X}$ and S^2, defined by the formulas,

$$\overline{X} = \frac{1}{n} \sum_{j=1}^{n} X_j,$$

$$S^2 = \frac{1}{n} \sum_{j=1}^{n} \left(X_j - \overline{X}\right)^2, \qquad n \ge 2$$

are (δ, ε)-independent [i.e., $\delta(F_{(X,Y)}, F_X F_Y) \le \varepsilon$, where $F_{(X,Y)}(x, y) = \Pr(X < x, Y < y)$, $F_X(x) = \Pr(X < x)$], then

$$\delta(X_j, N) \le C \left(\ln \frac{1}{\varepsilon} \right)^{-1/2} \tag{13.41}$$

[i.e., there exists a normal r.v. N, such that (13.41) is valid, where C is an absolute constant].

Yanushkyavichyus (1989) also improved on Nye (1966) and Meshalkin's result by showing that if X_1 and X_2 are independent identically distributed random variables and if $S = X_1 + X_2$ and $T = X_1 - X_2$ are (δ, ε)-independent, then there exists a normal r.v. N, such that

$$\delta(X_j, N) \leq C\varepsilon^{1/3}\left(\ln \frac{1}{\varepsilon}\right)^{(2/3)\log_2 3}.$$

Note that the loss of order from ε to $\varepsilon^{1/3}$ with a logarithm factor is due to passage from the functional equations, in terms of distribution functions $F_{(S,D)}(x, y) = F_S(x)F_D(y) + r(x, y)$, $(|r(x, y)| \leq \varepsilon$, for all $(x, y) \in R^2)$ to a functional equation in terms of characteristic functions. Yanushkyavichyus's results do not involve any restrictions on moments or on possible degeneracy of the r.v. under consideration.

Stability in regression models was studied by Lukacs (1942) and Yanushkyavichyus (1989), among others. Yanushkyavichyus's result is as follows: Let μ be a distance defined on the space of random variables. A r.v. X (with a finite expectation $E[X]$) has (μ, ε)-*regression on r.v.* Y if

$$\mu|E[X|Y], E[X]| \leq \varepsilon.$$

Yanushkyavichyus (1989) selects $\mu|Z_1, Z_2| = E|Z_1 - Z_2|$ and proves the following theorem: If X_1, X_2 are i.i.d. random variables with $E[X_j] = 0$, $E[X_j^2] = \sigma^2$, $E[|X_j|^{2+\delta}] \leq M < \infty$ for some $\delta > 0$ and $L_1 = a_1X_1 + a_2X_2$ has (μ, ε)-regression on $L_2 = b_1X_1 + b_2X_2$ (and $a_1b_1 + a_2b_2 \neq 0$; $|b_2|$, $|b_1| < 1$), then there is a normal distribution $[G(x)]$ with parameter $(0, \sigma^2)$ such that

$$\sup_x |F(x) - G(x)| \leq C\sqrt{-\log \varepsilon},$$

where $F(x) = \Pr[X_j \leq x]$ and C depends on M and (a_1, a_2, b_1, b_2) only.

Numerous other characterizations of normal are available in the literature. Gabovich (1974) has discussed the stability of certain characterization results. Khatri (1975a, b) has characterized the normal distribution by the constancy of regression; see Gyires (1975). Sampson (1975) has characterized the general exponential family of distributions by moment generating functions. Ruben (1974, 1975) has presented some characterizations through the distribution of the sample variance (see also Chapter 18). A characterization of normal distribution, through the general linear model, has been given by Ruben (1976). Parthasarathy (1976) characterized the normal law through the local independence of some statistics. Several weak sense analogues of characteristic properties were given by Kagan (1976). Zinger (1977) presented a characterization through identically distributed linear statistics. Many of these characterizations have been reviewed in the book on this topic by Mathai and Pederzoli (1977).

Some characterization results were also developed based on some properties of estimators. For example, Klebanov and Melamed (1976) provided a characterization through properties of Bayesian estimators. Klebanov and Melamed (1978) also characterized the normal distribution through properties of Fisher information amount. Fieger (1976) characterized the distribution based on a homogeneous Pitman estimator. Bondesson (1976), by starting with the question when the sample mean is BLUE, established a characterization result. Bondesson (1974, 1975, 1978) also proved some characterizations of the normal law through properties of distributions of some statistics. In particular, in the 1978 paper Bondesson has shown that the sample variance, when properly normalized, is distributed as chi-square only for the normal distribution (also see Chapter 18). In an interesting note, Goel and DeGroot (1980) established that only normal distributions have linear posterior expectations in linear regression. Ahmad (1978) and Wesolowski (1987) have provided further characterizations based on regression. Bischoff, Cremers, and Fieger (1987) used the sufficiency of the least-squares estimation for characterization. Eberl (1986) has characterized the normal distribution in translation classes through properties of Bayes estimators. Stadje (1988) provided a characterization through maximum likelihood estimation. Klebanov and Neupokoeva (1990) recently proved an interesting characterization by a property of the expected values of order statistics. A characterization of the normal law in the Gauss-Markov model has been given by Stepniak (1991).

Fieger (1977) discussed transformations that characterize the normal distribution. Arnold and Isaacson (1978) presented some characterizations through the distribution of linear forms (assuming finite variance). Prakasa Rao (1979) used some identities for characterizations. Lukacs (1976, 1977, 1980) presented stability theorems for many characterizations. Some more assorted characterization results for the normal distribution are due to Talwalker (1980), Lajko (1980), Joshi (1982), Borovkov and Utev (1983), Ramasubramanian (1985), Viskov (1985), Ahsanullah and Hamedani (1988), and Ahsanullah (1990). Letac (1981) proved some interesting characterizations via the concepts of isotropy and sphericity. Findeisen (1982) has discussed Gauss's characterization of the normal distribution. Hombas (1985) has characterized the normal density function as the solution of a differential equation. Ahsanullah (1989) has used properties of linear statistics as well as chi-squared in order to characterize the normal distribution. It is important to mention that the finiteness of the variance is a critical assumption in many of the above mentioned characterization results; see Lancaster (1987) for some comments in this regard. Quite recently, Cacoullos, Papathanasiou, and Utev (1993) discussed a characterization of the normal distribution and also presented a proof of the central limit theorem connected with it.

7 APPROXIMATIONS AND ALGORITHMS

The most common use of the normal distribution is as an approximation where either normality is ascribed to a distribution in the construction of a model or a known distribution is replaced by a normal distribution with the same expected value and standard deviation. Examples of such replacement are the Fisher and Wilson-Hilferty approximations to the χ^2-distribution (Chapter 18), the normal approximation to the (central) t-distribution (Chapter 28), and the use of normal distribution to approximate the distribution of the arithmetic mean of a number (often not very large, around 8 or more) of independent and identically distributed random variables. But now we are concerned with approximations *to* the normal distribution. It is possible to regard the distributions that are approximated *by* the normal distribution as being, themselves, approximations *to* normal distributions. However, they are usually more complex than the normal distribution, and we would like to study approximations that are simpler than the normal distribution.

From the point of view of replacement of a normal distribution by another distribution we note that:

1. A lognormal distribution can give a good representation of a normal distribution that has a small absolute value (say, less than 0.25) of the coefficient of variation.

2. A particular form of logistic distribution is very close to a normal distribution (see Chapter 23).

3. A form of the Weibull distribution with the shape parameter ≈ 3.25 is almost identical with the unit normal distribution (see Chapter 21).

4. Raab and Green (1961) have suggested that the distribution with probability density function

$$(2\pi)^{-1}(1 + \cos x), \qquad -\pi < x < \pi, \qquad (13.42)$$

can be used to replace a normal distribution. The correspondence is not very precise (see Table 13.2 comparing standardized percentile deviates of the two distributions) but will sometimes give useful analytical results. The replacement would only be used if substantial simplification in analysis were effected thereby.

The expected value and standard deviation of a random variable with distribution (13.42) are zero and $(\frac{1}{3}\pi^2 - 2)^{1/2} = 1.14$. The standardized range of the distribution $(-\pi, \pi)$ is thus from -2.77 to $+2.77$ standard deviations, and obviously the replacement gives a poor fit in the tails.

5. Bell (1962) has described even simpler approximations, using triangular distributions (Chapter 26). He pointed out that such approximations can be regarded as the second stage in a series of approximations

Table 13.2   Standardized Percentile Points of Distribution (13.42) and the Normal Distribution

| Cumulative | Standardized Value | |
Probability	Normal	Distribution (13.42)
0.5	0.000	0.000
0.6	0.253	0.279
0.75	0.674	0.732
0.9	1.282	1.334
0.95	1.645	1.649
0.975	1.960	1.888
0.99	2.326	2.124
β_2	3.000	2.406

by distributions of means of increasing numbers of independent rectangularly distributed variables (see the method of construction of "random normal deviates" used by Buslenko et al. (1966) described in Section 1).

Chew (1968) includes 2, 4, and 5 in a list of five possible replacements for normal distributions. The two other distributions he suggests are uniform (Chapter 26) and Laplace (Chapter 24). These also will be very crude approximations.

6. Hoyt (1968) has suggested using the distribution of the sum of three mutually independent random variables each uniformly distributed over the interval -1 to $+1$ as an approximation to the unit normal distribution. The density function is

$$\tfrac{1}{8}(3 - x^2) \qquad \text{for } |x| \le 1,$$

$$\tfrac{1}{16}(3 - |x|)^2 \qquad \text{for } 1 \le |x| \le 3.$$

This gives an error not exceeding 0.01 in the cumulative distribution function.

7. Steffensen (1937) has suggested the use of the distribution of a multiple of a chi random variable (i.e., $c\chi_\nu$), with ν sufficiently large. He called this a "semi-normal" distribution.

8. A different kind of approximation has been developed in connection with calculation of the functions $\Phi(\cdot)$, $Z(\cdot)$ in computers. These approximations usually employ polynomial expressions. They give quite high accuracy, sometimes only within definite limits on the values of the variable. Outside these limits they may give quite poor approximation.

Zelen and Severo (1964) quote, among other things, the following formulas, which are based on formulas given by Hastings (1955):

$$\Phi(x) \doteq 1 - \left(a_1 t + a_2 t^2 + a_3 t^3\right) Z(x), \qquad (13.43)$$

with $t = (1 + 0.33267x)^{-1}$, $a_1 = 0.4361836$, $a_2 = -0.1201676$, and $a_3 = 0.9372980$. The error in $\Phi(x)$, for $x \geq 0$, is less than 1×10^{-5}.

$$\Phi(x) \doteq 1 - \tfrac{1}{2}\left(1 + a_1 x + a_2 x^2 + a_3 x^3 + a_4 x^4\right)^{-4}, \quad (13.44)$$

with $a_1 = 0.196854$, $a_2 = 0.115194$, $a_3 = 0.000344$, and $a_4 = 0.019527$. The error in $\Phi(x)$, for $x \geq 0$, is less than 2.5×10^{-4}.

$$Z(x) \doteq \left(a_0 + a_2 x^2 + a_4 x^4 + a_6 x^6\right)^{-1}, \qquad (13.45)$$

with $a_0 = 2.490895$, $a_2 = 1.466003$, $a_4 = -0.024393$, and $a_6 = 0.178257$. The error in $Z(x)$ is less than 2.7×10^{-3}.

Very accurate results can be obtained with the formula [Hart (1966)]

$$1 - \Phi(x) \doteq \left(x\sqrt{2\pi}\right)^{-1}\left[\exp\left(-x^2/2\right)\right]$$

$$\times \left[1 - \frac{(1 + bx^2)^{1/2}}{(1 + ax^2)}\left\{x\sqrt{\frac{\pi}{2}}\right.\right.$$

$$\left.\left. + \left[\frac{1}{2}\pi x^2 + \frac{(1 + bx^2)^{1/2}}{(1 + ax^2)}e^{-x^2/2}\right]^{1/2}\right\}\right]^{-1} \qquad (13.46)$$

with

$$a = \frac{1}{2\pi}\left[1 + (1 + 6\pi - 2\pi^2)^{1/2}\right] = 0.212024,$$

$$b = \frac{1}{2\pi}\left[1 + (1 + 6\pi - 2\pi^2)^{1/2}\right]^2 = 0.282455.$$

For $x > 2$, Schucany and Gray (1968) have constructed the simpler formula

$$1 - \Phi(x) \doteq \left[(x^2 + 2)\sqrt{2\pi}\right]^{-1}$$

$$\times x\left[\exp\left(\frac{-x^2}{2}\right)\right]\frac{x^6 + 6x^4 + 14x^2 - 28}{x^6 + 5x^4 - 20x^2 - 4}, \qquad (13.47)$$

which is even better than (13.46) for $x > 3$. [The proportionate error of (13.46) for $5 \leq x \leq 10$ is about 0.5×10^{-5}; that of (13.47) decreases from 0.39×10^{-5} for $x = 5$ to 0.42×10^{-7} for $x = 10$.]

By use of rather elaborate formulas, quite remarkable accuracy can be attained. Strecock (1968) gives formulas for values of $\mathrm{erf}(x)$ [see (13.5)] and of the inverse function [$\mathrm{inverf}(y)$ where $\mathrm{erf}(\mathrm{inverf}(y)) = y$], correct to 22 decimal places for $|x|$ (or $|\mathrm{inverf}(y)|$) less than 7.85.

9. Burr (1967) has considered approximations to $\Phi(x)$ of form

$$G(x) = 1 - \left[1 + (\alpha + \beta x)^c\right]^{-k}.$$

He suggests taking $\alpha = 0.644693$, $\beta = 0.161984$, $c = 4.874$, and $k = -6.158$. An even better approximation is obtained by using

$$H(x) = \tfrac{1}{2}\left[G(x) + 1 - G(-x)\right],$$

which is a symmetrical function of x. The discrepancy $|H(x) - \Phi(x)|$ reaches its maximum value of about 0.00046 when $x \doteq \pm 0.6$.

10. McGillivray and Kaller (1966) have considered the discrepancy between $\Phi(x)$ and $\Phi(x) + a_{2r} Z(x) H_{2r-1}(x)$, where $H_{2r-1}(x)$ is the Hermite polynomial of order $2r - 1$ and a_{2r} is a constant chosen so that $1 + a_{2r} H_{2r}(x)$ cannot be negative. This means that a_{2r} must be between zero and

$$A_{2r} = \left| \inf_x H_{2r}(x) \right|^{-1}.$$

The second function $\Phi(x) + a_{2r} Z(x) H_{2r-1}(x)$ is the cumulative distribution function of a symmetrical distribution having the first r even central moments (and of course all odd moments), the same as those for a unit normal distribution. The discrepancy cannot exceed

$$A_{2r} \sup_x \left\{ Z(x) | H_{2r-1}(x) | \right\}.$$

The values of this quantity, for $r = 2$, 3, and 4 are 0.10, 0.03, and 0.005, respectively. [Of course other distributions with the same (zero) odd central moments and first r even central moments *might* have greater discrepancies, but these results do give a useful idea of the accuracy obtained by equating moments.]

11. Riffenburgh (1967) has suggested that a symmetrical truncated unit normal distribution be approximated by the density function

$$\frac{\frac{1}{2}(Z(x) - Z(c))}{\Phi(c) - \frac{1}{2} - cZ(c)}, \qquad -c \le x \le c,$$

where $-c, c$ are the points of truncation. Use of this approximation is recommended only when c exceeds 1 (preferably $c \ge 1.5$). Tables of the variance (to 3 decimal places) of the approximate distribution are given by Riffenburgh (1967) for $c = 0.8(0.1)1.2(0.05)4.00$, and also of $\Pr[X \le x] - \frac{1}{2}$ (to 4 decimal places) for $c = 1.2(0.1)3.0$ and x at intervals of 0.05. (Riffenburgh has also developed test procedures based on this distribution.)

We now discuss some *bounds* on the value of $\Phi(x)$. Various inequalities for Mills's ratio can also be interpreted as bounds for $\Phi(x)$ or $Z(x)$. Using a simple geometrical argument (based on the joint distribution of two independent unit normal variables) it can be shown that

$$\frac{1}{2}\left[1 + (1 - e^{-x^2/2})^{1/2}\right] \le \Phi(x) \le \frac{1}{2}\left[1 + (1 - e^{-x^2})^{1/2}\right] \quad (13.48)$$

[e.g., see D'Ortenzio (1965)]. By a refinement of the argument, on the left-hand side of (13.48), $(1 - e^{-x^2/2})$ can be replaced by

$$1 - e^{-x^2/2} + \left(2\pi^{-1} - \frac{1}{2}\right)^2 e^{-x^2},$$

and on the right-hand side, $(1 - e^{-x^2})$ can be replaced by

$$1 - e^{-x^2} - (1 - 2\pi^{-1})^2 e^{-x^2}.$$

The approximation

$$\Phi(x) \doteq \frac{1}{2}\left[1 + \left\{1 - \exp\left(\frac{-2x^2}{\pi}\right)\right\}^{1/2}\right] \quad (13.49)$$

was obtained by Pólya (1945). This has a maximum error of 0.003, when $x = 1.6$. Cadwell (1951) modified (13.49) to

$$\Phi(x) \doteq \frac{1}{2}\left[1 + \left\{1 - \exp\left(-2\pi^{-1}x^2 - \frac{2}{3}\pi^{-2}(\pi - 3)x^4\right)\right\}^{1/2}\right]. \quad (13.50)$$

Over the range $0 < x < 3.5$, the maximum error of (13.50) is 0.0007, when $x = 2.5$. Formula (13.50) should not be used for large values of x. Cadwell suggested, on empirical grounds, the addition of the terms

$$-0.0005x^6 + 0.00002x^8$$

to the exponent in (13.50). This reduces the maximum error to 0.00005.

Carta (1975) developed approximations, similar to that of Hastings in (13.44), of the form

$$\Phi(x) \doteq 1 - \tfrac{1}{2}\left(a_1 + a_2 x + \cdots + a_n x^{n-1}\right)^{-2^q}, \quad \text{for } x \geq 0. \quad (13.51)$$

Note the added flexibility of a variable first coefficient, as opposed to the fixed value of 1 in (13.44). For different choices of n and q, Carta (1975) has presented the coefficients a_i's that yield the minimum absolute error. For example, from Carta's table, we have the coefficients corresponding to $n = 6$ and $q = 4$ as

0.9999998582,   0.0487385796,   0.02109811045,

0.003372948927,   −0.00005172897742,   0.00008569579420.

The absolute error for this approximation (for all $x \geq 0$) is less than 1.2×10^{-6}. Note that in this case (and also for some other choices of n, q), the leading coefficient is very nearly 1 [as in (13.44)]. Carta also has presented similar approximations for x restricted in the intervals $[0, 3.09]$, $[0, 4.00]$, and $[0, 5.20]$.

Badhe (1976) presented the following approximation which is easily implementable on a hand calculator:

$$\Phi(x) \doteq 1 - \frac{Z(x)}{x}\left[1 - \frac{1}{Y}\left(1 + \frac{1}{Y}\left\{7 + \frac{1}{Y}\left[55 + \frac{1}{Y}\left(445 + \frac{3745 Q_1(x)}{Y}\right)\right]\right\}\right)\right],$$

$$(13.52)$$

where $Y = x^2 + 10$ and

$$Q_1(x) = 8.5(x^2 - 0.4284639753x^{-2} + 1.240964109)^{-1} + 1.$$

Badhe (1976) has pointed out that the approximation in (13.52) is good when $x > 4$, but certainly not suitable when $x \leq 2$. For the case where $x \leq 2$, Badhe has presented a seventh degree polynomial approximation (obtained by Chebyshev economization) given by

$$\Phi(x) \doteq \tfrac{1}{2} + x\{a + Y[b + Y(c + Y\{d + Y[e + Y(f + Y(g + hY))]\})]\},$$
(13.53)

where

$Y = x^2/32,$
$a = 0.3989422784,$
$b = -2.127690079,$
$c = 10.2125662121,$
$d = -38.8830314909,$
$e = 120.2836370787,$
$f = -303.2973153419,$
$g = 575.073131917,$
$h = -603.9068092058.$

The maximum absolute error for this approximation, for $x \in [0, 2]$, is 0.2×10^{-8}.

By making use of the Hermite expansion

$$\sqrt{2\pi} \int_0^x Z(t)\, dt = 2e^{-x^2/8} \sum_{n=0}^{\infty} \frac{1}{(2n+1)!} (x/2)^{2n+1} H_{2n}(x/2), \quad (13.54)$$

where $H_n(x)$ is the nth Hermite polynomial, and the known recurrence relation

$$H_{n+1}(x) = xH_n(x) - nH_{n-1}(x) \qquad \text{for } n = 1, 2, \ldots,$$

with $H_0(x) \equiv 1$ and $H_1(x) = x$, Kerridge and Cook (1976) suggested using the series

$$\sqrt{2\pi} \int_0^x Z(t)\, dt = xe^{-x^2/8} \sum_{n=0}^{\infty} \frac{1}{2n+1} \theta_{2n}(x/2) \qquad (13.55)$$

for computing $\Phi(x)$ on a computer. In (13.55), $\theta_n(x) = x^n H_n(x)/n!$, which is easily computed using the recurrence relation

$$\theta_{n+1} = \frac{x^2(\theta_n - \theta_{n-1})}{n+1}, \qquad n = 1, 2, \ldots.$$

Beasley and Springer (1977) have provided an algorithm for computing the percentage point x_p, for a specified value of p, such that

$$p = \int_{-\infty}^{x_p} Z(t) \, dt.$$

Their Fortran subroutine replaces p by $q = p - \frac{1}{2}$ and then compares $|q|$ with 0.42; if $|q| \le 0.42$, x_p is determined by a rational approximation

$$x_p = \frac{qA(q^2)}{B(q^2)}, \tag{13.56}$$

where A and B are polynomials of degrees 3 and 4, respectively, while if $|q| > 0.42$ an auxiliary variable, $r = \{\ln(\frac{1}{2} - |q|)\}^{1/2}$ is first formed and then x_p as

$$x_p = \pm \frac{C(r)}{D(r)}, \tag{13.57}$$

where C and D are polynomials of degrees 3 and 2, respectively, the sign being taken that of q. See also an earlier algorithm given by Odeh and Evans (1974).

Page (1977) considered simple approximations of the form

$$\Phi(x) \doteq \frac{e^{2y}}{1 + e^{2y}}, \tag{13.58}$$

where $y = a_1 x(1 + a_2 x^2)$, and determined $a_1 = 0.7988$ and $a_2 = 0.04417$ to provide an approximation with maximum error 0.14×10^{-3}. Page has also presented a similar simple approximation for the percentage point that gives two decimal accuracy.

Derenzo (1977) provided an approximation to the unit normal cdf as

$$\Phi(x) \doteq 1 - \frac{1}{2} \exp\left\{ -\frac{(83x + 351)x + 562}{(703/x) + 165} \right\}, \qquad x > 0, \tag{13.59}$$

with a maximum absolute error of 0.042% for $x \in (0, 5.5]$. Another approximation that Derenzo presented is

$$\Phi(x) \doteq 1 - \frac{1}{\sqrt{2\pi} x} \exp\left(-\frac{x^2}{2} - \frac{0.94}{x^2} \right) \tag{13.60}$$

with a maximum absolute error of 0.040% for $x \ge 5.5$. Derenzo (1977) also provided an approximation for the percentage point x_p (for a given p) as

$$x_p \doteq \left\{ \frac{[(4y + 100)y + 205] y^2}{[(2y + 56)y + 192] y + 131} \right\}^{1/2}, \tag{13.61}$$

where $y = -\log(1 - p/2)$, with a maximum absolute error of 1.3×10^{-4} for $x \in (0, 5.2)$. For $x \in [5.2, 22.6)$, Derenzo has given the approximation

$$x_p \doteq \left\{ \frac{[(2y + 280)y + 572]y}{(y + 144)y + 603} \right\}^{1/2} \qquad (13.62)$$

with a maximum absolute error of 4×10^{-4}.

Some more simple (but not necessarily more accurate for all x) approximations for the cumulative distribution function $\Phi(x)$ have been given by Parsonson (1978), Easingwood (1979), Heard (1979), Martynov (1981), Monahan (1981), Allasia and Giordano (1982), Hawkes (1982), Fearn (1983), Edgeman (1988), and Abernathy (1988). Pugh (1989) has made a survey of many of the algorithms available for computing the cdf $\Phi(x)$.

Moran (1980), by slightly modifying a formula of Strecock (1968) given for the error function, came up with the approximation

$$\Phi(x) \doteq \frac{1}{2} + \frac{1}{\pi} \sum_{n=0}^{12} \left(n + \frac{1}{2} \right)^{-1} e^{-(n+(1/2))^2/9} \sin\left\{ \frac{\sqrt{2}}{3} \left(n + \frac{1}{2} \right) x \right\}. \quad (13.63)$$

Compared to the 38-term approximation given by Strecock, this approximation is simpler to use and is also accurate to nine decimal places for $|x| \leq 7$.

Shore (1982), by using a "betalike" cumulative distribution $G(x)$ [with $G(-\infty) = 0$, $G(\infty) = 1$, $G(0) = \frac{1}{2}$, and

$$\frac{dG}{dx} = \begin{cases} cG^{k_1}(1 - G)^{k_2}, & 0 \leq G \leq \frac{1}{2}, \\ c(1 - G)^{k_1}G^{k_2}, & \frac{1}{2} \leq G \leq 1, \end{cases}$$

where $k_1 > 0$] to approximate $\Phi(x)$, derived the following three approximations for the percentage point x_p:

$$x_p \doteq -5.5310\left\{ \left(\frac{1 - p}{p} \right)^{0.1193} - 1 \right\}, \qquad p > \frac{1}{2}, \quad (13.64)$$

$$x_p \doteq -0.4115\left\{ \frac{1 - p}{p} + \ln\left(\frac{1 - p}{p} \right) - 1 \right\}, \qquad p \geq \frac{1}{2}, \quad (13.65)$$

$$x_p \doteq -a\ln\left(\frac{1 - p}{p} \right) + b, \qquad p \geq \frac{1}{2}, \quad (13.66)$$

where a and b should satisfy

$$b = [1 - 1.3682a^2]^{1/2} - 1.3862a$$

or its approximate simpler form $b = 1.3086 - 2.3735a$ (which is easier to use when a value of a is sought that gives the best fit for any desired range of x).

The approximation in (13.64) is the most accurate of the three with a maximum absolute difference of 0.0073 (0.5%) for $0 \le x \le 2.3$. It is interesting to note that for $b = 0$, we have $a = 0.5513 = \sqrt{3}/\pi$, in which case the simple approximation in (13.66) becomes the logistic approximation (with location parameter 0 and shape parameter $\sqrt{3}/\pi$); see Chapter 23. For $-0.5 \le x \le 2.2$, the best approximation of (13.66) is with $a = 0.495$ and $b = 0.1337$; for $x > 2.2$, the best approximation is obtained with $a = 0.4506$ and $b = 0.2252$.

Shore (1982) has also obtained a good approximation for the standard normal pdf as

$$Z(x) \doteq 1.4184(1-p)^{0.8632}p,$$
$$p = \Phi(x) \ge \tfrac{1}{2}. \tag{13.67}$$

This approximation immediately yields an approximation for the Mills's ratio (hazard rate–see Chapter 33, Section 2) as

$$\frac{Z(x)}{1-\Phi(x)} \doteq \begin{cases} \left[1.4184p^{0.8632}\right]^{-1}, & p \le \tfrac{1}{2}, \\ \left[1.4184(1-p)^{-0.1368}p\right]^{-1}, & p \ge \tfrac{1}{2}. \end{cases} \tag{13.68}$$

Shore has also discussed the accuracy of this approximation.

Shah (1985) suggested the following approximation for $\Phi(x) - \tfrac{1}{2}$, $x \ge 0$:

$$\begin{cases} \dfrac{x(4.4-x)}{10}, & 0 \le x \le 2.2 \\ 0.49, & 2.2 < x < 2.6 \\ 0.50, & x \ge 2.6. \end{cases} \tag{13.69}$$

Even though this approximation is simple to use, it is clear from (13.69) that it is not designed to approximate right tail areas for $x \ge 2.6$. For this reason Norton (1989) proposed the approximations for $1 - \Phi(x)$ as

$$\tfrac{1}{2}e^{-(x^2+x)/2}, \qquad 0 \le x \le 2.6, \tag{13.70}$$

$$\tfrac{1}{2}e^{-(x^2+1.2x^{0.8})/2}, \qquad 0 \le x \le 2.7, \tag{13.71}$$

$$Z(x)/x, \qquad x > 2.7. \tag{13.72}$$

A number of comments and criticisms have been made on these approximations given by Norton; see Hoaglin (1989), Pratt (1989), Cox (1989), Shore

(1990), Sharp (1990), Revfeim (1990), and McConnell (1990), and also the replies to these comments by Norton (1990a, b).

Schonfelder (1978) discussed further on Chebyshev expansions for the error and related functions, including $\Phi(x)$. Hamaker (1978) presented simple approximations for the cdf $\Phi(x)$ as well as for the percentage point x_p. Lin (1988) discussed alternatives to Hamaker's approximations. Schmeiser (1979) gave easy approximations for the percentage point x_p that could be used on hand calculators. Bailey (1981) proposed alternatives to Hastings's approximation to x_p. Lew (1981) presented an approximation to $\Phi(x)$ with simple coefficients. Wichura (1988) and Guirguis (1991) have discussed approximations to the percentage point x_p. While Heffernan (1988) gave a simple formula for the cdf $\Phi(x)$, Lin (1989) presented approximations for the normal tail probability as well as its inverse which are easy to use even on a hand calculator. Lin (1990) has also proposed a simpler logistic approximation to $\Phi(x)$ and to x_p [also see the comment after (13.66)]. Even though all these approximations are quick and easy to use, their relative accuracies may vary dramatically, and therefore their use should be adjudged for the particular situation one intends to use. The modern personal computer with its great power and memory does make some of the work (if not all) unnecessary, as the following remarks indicate.

Fleming (1989) has advocated the use of numerical integration for approximating $\Phi(x)$, for it can be easily demonstrated using a spreadsheet program and a personal computer. For example, $\Phi(x)$ can be computed using Lotus 1-2-3 by approximating the area under the standard normal pdf $Z(x)$. Fleming (1989) has noted that a rectangular integration gives results accurate to four decimal places for $0 \leq x \leq 3$, with the interval width being taken as 0.01. As he has pointed out, other methods of integration (like the trapezoidal rule, Simpson's rule, Newton's three-eighths rule) are also easily adaptable and will be even more accurate in evaluating $\Phi(x)$. When a powerful personal computer with a spreadsheet program is available, it should be put to use rather than relying on simple-to-use approximations.

A mechanical method of drawing a normal probability density curve has been described by Edwards (1963). *Normal probability paper* is graph paper with a natural scale in the horizontal (abscissa) direction, while the distances on the vertical (ordinate) scale are proportional to the corresponding normal deviates. The vertical scale is usually marked in percentages. Thus 50% correspond to the horizontal axis, 25% and 75% are at distances 0.6745 below and above this line, 5% and 95% are at distances 1.9600 below and above this line, and so on (see Figure 13.2). Barnett (1976) has discussed convenient probability plotting positions, while Nelson (1976) has elaborated the construction of normal probability paper. Recently Nelson (1989) put forward a stabilized normal probability plotting technique; Rouncefield (1990), among many others, explained how one could use the normal probability paper to assess the validity of the assumption of normal distribution for a sample.

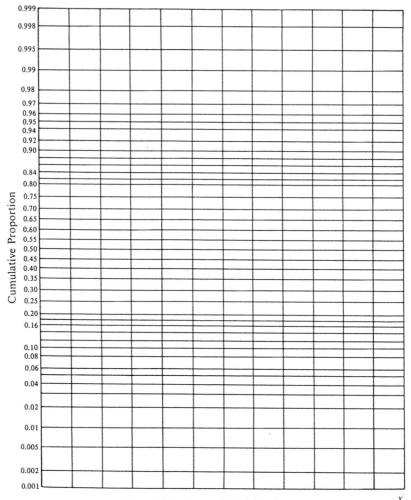

Figure 13.2 Normal Probability Paper

If X has the distribution (13.1) and $\Pr[X \leq x]$ is plotted (as the ordinate) against x (as the abscissa), then a straight line is obtained. The slope of this line is σ^{-1} and its intercept on the horizontal axis is at $x = \xi$. If observed frequencies of the events $(X \leq x)$ are used in place of the actual probabilities, an approximately straight-line plot may be expected. A straight line fitted to these observed points gives estimates of σ and ξ. Such graphical methods of estimation can give good practical accuracy.

If the horizontal scale is logarithmic, we have *lognormal probability paper* (see Chapter 14). *Half-normal probability paper* is simply normal probability paper with negative abscissas omitted. It is used in connection with analysis of variance techniques developed by Daniel (1959).

8   ESTIMATION

The theory of estimation of ξ and σ has been fully worked out. To facilitate comprehension, this section is divided into four subsections. Subsections 8.1 through 8.3 describe techniques primarily appropriate to a complete sample (though some apply also to censored data) corresponding to values of n independent random variables each having distribution (13.1). Subsection 8.4 describes techniques suitable when the data have been censored by omission of certain order statistics. (*Truncated* normal distributions will be discussed in Section 10.) Subsection 8.5 lists various books/monographs available on specific topics relating to inference for the normal distribution.

The variety of applicable methods can be bewildering. To judge rapidly between them, it is necessary to bear in mind accuracy, sensitivity to variations from normality, and ease of calculation. The relative importance of these factors varies with circumstance, but they should always be taken into account. As mentioned earlier in Section 2, there are numerous volumes available on specific topics relating to inference for the normal distribution (as listed in Section 8.5). This is understandable of course, due to the volume of literature. Because of the availability of all these volumes and the upcoming revised edition of "Handbook of the Normal Distribution" by Patel and Read, we only present briefly the recent developments, referring the readers to appropriate volumes for a comprehensive treatment of the topic of interest.

8.1   Estimation of ξ

The arithmetic mean $\overline{X} = n^{-1}\sum_{j=1}^{n}X_j$ and the mean square deviation $S^2 = n^{-1}\sum_{j=1}^{n}(X_j - \overline{X})^2$ are jointly sufficient for ξ and σ, and $\overline{X}$ is sufficient for ξ alone. For most practical purposes $\overline{X}$ is the best estimator for ξ, whether or not σ is known. It is the maximum likelihood estimator and is unbiased.

The only circumstances under which this estimator would not be used are (1) when not all observations are available (as will be discussed more fully in Section 8.4), or (2) when the accuracy of some values (e.g., outlying values) is doubtful. In either case estimation may be based on a central block of order statistics. As an extreme case (when n is odd), a single order statistic, the median, may be used to estimate ξ. This is an unbiased estimator of ξ, and has standard deviation approximately equal to

$$\tfrac{1}{2}\pi\sigma^2 n^{-1} = 1.5708\sigma^2 n^{-1}$$

compared with

$$\mathrm{Var}(\overline{X}) = \sigma^2 n^{-1}.$$

The efficiency of the median, relative to $\overline{X}$, is thus approximately

$$100 \times (1.5708)^{-1}\% = 63.7\%.$$

Alternatively, the jth *Winsorized mean*

$$\tilde{\xi}_{(j)} = n^{-1}\left[jX'_{j+1} + \sum_{i=j+1}^{n-j} X'_i + jX'_{n-j}\right] \tag{13.73}$$

may be used $(j < [(n-1)/2])$. It can be seen that $\tilde{\xi}_{(j)}$ is obtained by replacing each of $X'_1, X'_2, \ldots, X'_j$ by X'_{j+1} and $X'_{n-j+1}, \ldots, X'_n$ by X'_{n-j}. This is also an unbiased estimator of ξ. It is interesting to note that Chernoff, Gastwirth, and Johns (1967) obtain a formula of the type (13.73), if only $X'_{j+1}, \ldots, X'_{n-j}$ are available, with the multipliers (j) of X'_{j+1} and X'_{n-j} replaced by

$$a = \frac{[Z(U_\varepsilon)]^2 \varepsilon^{-1} + U_\varepsilon Z(U_\varepsilon)}{1 - 2\varepsilon + 2U_\varepsilon Z(U_\varepsilon) + 2[Z(U_\varepsilon)]^2 \varepsilon^{-1}},$$

where $\varepsilon = j/(n+1)$, and n^{-1} replaced by $(n - 2j + 2a)^{-1}$. In fact a is slightly less than j.

Rather than replace the values of extreme observations by more central values, we can simply omit them. The resulting unbiased estimator of ξ is the jth *trimmed mean*

$$\tilde{\xi}'_{(j)} = (n - 2j)^{-1} \sum_{i=j+1}^{n-j} X'_i. \tag{13.74}$$

Some relative efficiencies of $\tilde{\xi}'_{(j)}$, compared with $\overline{X}$, are shown in Table 13.3. (For efficiencies of $\tilde{\xi}_{(j)}$, see Table 13.9.)

Table 13.3 Efficiency of Trimmed Means, Relative to $\overline{X}$

n	j	Relative Efficiency of $\tilde{\xi}'_{(j)}$ (%)
5	2	69
10	2	89
10	3	81
15	2	92
15	4	83
15	6	73

It is apparent that the Winsorized mean $\tilde{\xi}_{(j)}$ is more efficient than $\tilde{\xi}'_{(j)}$. In fact, compared with the best linear unbiased estimator *using the same order statistics*, its efficiency never falls below 99.9% for $n \le 20$ [Sarhan and Greenberg (1962)]. Meeden (1976) has pointed out a special property of linear estimates of ξ, while Mehran (1975) has derived relationships between the UMVUEs of the mean and median of a function of a normal distribution.

Knowledge of σ is of no help in calculating point estimators of ξ. It is, however, used in calculating the standard deviations of such estimators and in constructing confidence intervals for ξ. If σ is known, $100(1 - \alpha)\%$ confidence limits for ξ are

$$\bar{X} \pm \frac{U_{1-\alpha/2}\sigma}{\sqrt{n}}. \tag{13.75}$$

Although the similar formulas

$$\tilde{\xi}_{(j)} \pm U_{1-\alpha/2} \sqrt{\operatorname{Var}\left(\tilde{\xi}_{(j)}\right)},$$

$$\tilde{\xi}'_{(j)} \pm U_{1-\alpha/2} \sqrt{\operatorname{Var}\left(\tilde{\xi}'_{(j)}\right)} \tag{13.76}$$

do not give *exact* limits (since $\tilde{\xi}_{(j)}, \tilde{\xi}'_{(j)}$ do not have normal distributions), they give limits that are useful provided that n is not too small (e.g., $n \ge 15$).

If σ is not known, the above formulas cannot be used. It is natural to replace σ by an estimator of σ. If the sample size is large and a good (efficient) estimator of σ is used, this can be done with little serious effect on the confidence coefficient. The "estimator" of σ most often employed is

$$(1 - n^{-1})^{-1/2}S = \left[(n - 1)^{-1} \sum_{j=1}^{n} \left(X_j - \bar{X}\right)^2\right]^{1/2},$$

although this is not an unbiased estimator of σ. If this estimator is used, then $U_{1-\alpha/2}$ in (13.75) should be replaced by $t_{n-1, 1-\alpha/2}$, the upper $50\alpha\%$ point of the t distribution with $(n - 1)$ degrees of freedom (see Chapter 28). The effect of replacement of σ by $(1 - n^{-1})^{-1/2}S$ in formulas (13.76) is not so clear, but there should be comparable increase in the multiplying factor $U_{1-\alpha/2}$.

It can be shown (see Section 3) that $\bar{X}$ and *any* function of the deviations $\{X_j - \bar{X}\}$, only, are mutually independent. This facilitates computation of percentage points of distributions of statistics of form

$$\frac{\sqrt{n}\,(\bar{X} - \xi)/\sigma}{f\left(X_1 - \bar{X}, \ldots, X_n - \bar{X}\right)} = T_{\{f\}},$$

say, with various functions $f(X_1 - \overline{X}, \ldots, X_n - \overline{X})$ in the denominator, subject to the restrictions that $f(\cdot)$ is positive with probability one and that

$$f(ay_1, ay_2, \ldots, ay_n) = af(y_1, y_2, \ldots, y_n)$$

for any $a \geq 0$. For example, we might have

$$f(X_1 - \overline{X}, \ldots, X_n - \overline{X}) = \text{range}$$
$$= \max(X_1 - \overline{X}, \ldots, X_n - \overline{X})$$
$$- \min(X_1 - \overline{X}, \ldots, X_n - \overline{X})$$
$$= W$$

or

$$f(X_1 - \overline{X}, \ldots, X_n - \overline{X}) = \text{mean deviation}$$
$$= n^{-1} \sum_{j=1}^{n} |X_j - \overline{X}|$$
$$= M.$$

Indeed, any of the estimators of σ to be described in Section 8.2 might be used as $f(\cdot)$.

Under the conditions stated, the distribution of $\sigma^{-1}f(X_1 - \overline{X}, \ldots, X_n - \overline{X})$ does not depend on σ. The distribution of $T_{\{f\}}$ therefore does not depend on ξ or σ. As a result it is possible, in principle, to construct tables of percentage points $T_{\{f\}, \alpha}$ of this distribution, defined by $f(\cdot)$ and α alone. The relation

$$\Pr\left[T_{\{f\}, \alpha_1} < \frac{\sqrt{n}\,(\overline{X} - \xi)/\sigma}{f(X_1 - \overline{X}, \ldots, X_n - \overline{X})} < T_{\{f\}, 1-\alpha_2} \,\Big|\, \xi, \sigma \right] = 1 - (\alpha_1 + \alpha_2)$$

can be arranged to show that the limits

$$\overline{X} - \frac{\sigma T_{\{f\}, \alpha_1} f(X_1 - \overline{X}, \ldots, X_n - \overline{X})}{\sqrt{n}},$$

$$\overline{X} - \frac{\sigma T_{\{f\}, 1-\alpha_2} f(X_1 - \overline{X}, \ldots, X_n - \overline{X})}{\sqrt{n}},$$

$$(13.77)$$

form a $100(1 - \alpha_1 - \alpha_2)\%$ confidence interval for ξ.

To calculate such limits it is necessary to have tables of percentage points of the distribution of $T_{(f)}$, among which the following are available:

For $f(\cdot) = (1 - n^{-1})^{-1/2}S$, as described in Chapter 28 (tables of distribution).

For $f(\cdot) = W$ [in Lord (1947)].

For $f(\cdot) = M$ [in Herrey (1965)].

If such tables are not available, approximations may be used, for example, the approximations to the distributions of M and W to be described in Section 8.2; alternatively, they may be determined through Monte Carlo simulations.

8.2 Estimation of σ

The maximum likelihood estimator of σ (ξ not being known) is

$$S = \left[n^{-1} \sum_{j=1}^{n} (X_j - \bar{X})^2 \right]^{1/2}. \tag{13.78}$$

If ξ is known, the maximum likelihood estimator is

$$\left[n^{-1} \sum_{j=1}^{n} (X_j - \xi)^2 \right]^{1/2} = \left[S^2 + (\bar{X} - \xi)^2 \right]^{1/2}. \tag{13.79}$$

It is, however, very unusual to know ξ exactly, and we will not discuss this estimator further, except to note that neither (13.78) nor (13.79) is an unbiased estimator of σ. In fact

$$E[S] = \frac{\sigma(2/n)^{1/2}\Gamma(n/2)}{\Gamma[(n-1)/2]}$$

$$= \frac{\sigma}{a_n} \tag{13.80}$$

and

$$E\left[\left\{ S^2 + (\bar{X} - \xi)^2 \right\}^{1/2} \right] = \frac{\sigma(2/n)^{1/2}\Gamma[(n+1)/2]}{\Gamma(n/2)}.$$

To obtain an unbiased estimator of σ, we must multiply S by a_n. A few values of a_n are shown in Table 13.4. Values of $a'_n = a_n\sqrt{1 - n^{-1}}$, such that $a'_n E[\{(n-1)^{-1}\Sigma(X_j - \bar{X})^2\}^{1/2}] = \sigma$, are also shown.

Table 13.4 Multipliers a_n, a'_n Such That $E[a_n S] = \sigma = E[a'_n \sqrt{V}]$

n	a_n	a'_n
2	1.77245	1.25331
3	1.38198	1.12838
4	1.25331	1.08540
5	1.18942	1.06385
6	1.15124	1.05094
7	1.12587	1.04235
8	1.10778	1.03624
9	1.09424	1.03166
10	1.08372	1.02811

For n greater than 10, the formulas

$$a_n \doteq 1 + \tfrac{3}{4}(n-1)^{-1},$$

$$a'_n \doteq 1 + \tfrac{1}{4}(n-1)^{-1}$$

give quite good approximations.

$$V = n(n-1)^{-1} S^2 = (n-1)^{-1} \sum_{j=1}^{n} \left(X_j - \bar{X} \right)^2 \qquad (13.81)$$

is an unbiased estimator of σ^2. Jarrett (1968) has given an interesting historical account of tables of these multiplying factors. [See also Cureton (1968) and Bolch (1968).] Note that the value of a minimizing the *mean square error* of aS^2 as an estimator of σ^2 is $(n^{-1} + 1)^{-1}$. The value of b minimizing the mean square error of bS as an estimator of σ is a'_{n+1} [Markowitz (1968)]. Iliescu and Voda (1974) have discussed the estimation of σ in detail.

The variance of the unbiased estimator, $a_n S$, of σ is

$$\mathrm{Var}(a_n S) = \left[a_n^2 (1 - n^{-1}) - 1 \right] \sigma^2. \qquad (13.82)$$

The variance of V is

$$\mathrm{Var}(V) = 2(n-1)^{-1} \sigma^4. \qquad (13.83)$$

Unbiased estimator of σ can also be obtained by multiplying the mean deviation (M) and the range (W) by appropriate factors (depending of course on n). The resulting unbiased estimators $b_n M, c_n W$ are identical with $a_n S$ for $n = 2$, and have greater variances than $a_n S$ for $n > 2$. Values of b_n

can be calculated from the simple formula $b_n = \sqrt{(\pi/2)n(n - 1)}^{-1}$. Values of c_n for $n = 2(1)20$ are given in Pearson and Hartley (1948).

Relative efficiencies (inverse ratio of variances) of $b_n M, c_n W$ and other unbiased estimators of σ compared with $a_n S$, are shown in Table 13.7 on page 136. From this table it can be seen that the estimator based on range is (slightly) more efficient than that based on mean deviation for $n = 4, 5$, but less efficient for $n \geq 6$. (For $n = 2, 3$ the two estimators give identical estimators of σ.)

The formula for b_n, quoted above, follows from

$$E[M] = \sigma \sqrt{\left(\frac{2}{\pi}\right)(1 - n^{-1})} . \qquad (13.84)$$

The variance of M is

$$\text{Var}(M) = \frac{2\sigma^2}{n\pi}\left(1 - \frac{1}{n}\right)\left\{\frac{1}{2}\pi + \sqrt{n(n - 2)} - n + \sin^{-1}\frac{1}{n - 1}\right\}. \qquad (13.85)$$

For $n \geq 5$, a very good approximation (error no more than about 0.00001) is

$$\sigma^2 \text{Var}(M) \doteq n^{-1}(1 - 2\pi^{-1})(1 - 0.12n^{-1}). \qquad (13.86)$$

Approximate formulas for the moment ratios of M are

$$\beta_1(M) \doteq 1.05n^{-1},$$

$$\beta_2(M) = 3 + 0.91n^{-1}.$$

Godwin and Hartley (1945) calculated a table of the probability integral of the distribution of M, giving $\Pr[M \leq m\sigma]$ to 5 decimal places for $m = 0.00(0.02)3.00$ and $n = 2(1)10$, and also multipliers M_α for percentile points such that

$$\Pr[M \leq M_\alpha \sigma] = \alpha$$

to 3 decimal places, for $n = 2(1)10$ and $\alpha = 0.001, 0.005, 0.01, 0.025, 0.05, 0.1, 0.9, 0.95, 0.975, 0.99, 0.995,$ and 0.999. For $n = 10$ there are also given approximate values calculated from the formula

$$M_\alpha \doteq E\left[\frac{M}{\sigma}\right] + U_\alpha\left(\sqrt{\text{Var}\left(\frac{M}{\sigma}\right)}\right). \qquad (13.87)$$

Although the upper and lower 2.5%, 5% and 10% values are not too inaccurate, approximation is poor for the more extreme values. A better approximation was obtained by Cadwell (1953), by regarding $(M/\sigma)^{1.8}$ as having (approximately) a $c\chi_\nu^2$ distribution with

$$\log c = -\log 2 - 1.8\left[\log\left(\frac{5}{9} + \frac{\nu}{2}\right) - \log \Gamma\left(\frac{\nu}{2}\right)\right.$$

$$\left. - \frac{1}{2}\log\left\{\left(\frac{2}{\pi}\right)(1 - n^{-1})\right\}\right],$$

(13.88)

$$\nu = \nu_0 + 0.196 - 0.159\nu_0^{-1},$$

$$\nu_0 = \frac{0.617[E[M]]^2}{\mathrm{Var}(M)}.$$

The approximation was obtained by first finding values of λ, c, and ν to make $(M/\sigma)^\lambda$ and $c\chi_\nu^2$ have the same first three moments. The values of course depend on n; they are shown in Table 13.5 [based on Cadwell (1953)]. This table also gives the results of similar calculations for the range W. It can be seen that for M, the values of λ do not vary much, once n exceeds 5, say. An "average" value of 1.8 was chosen; the values of ν and c were then adjusted to make the first *two* moments of $(M/\sigma)^{1.8}$ and $c\chi_\nu^2$ agree. It might

Table 13.5 Values of ν, λ, and log c Such That First Three Moments of $(T / \sigma)^\lambda$ and $c\chi_\nu^2$ Agree

	T = Range			T = Mean Deviation		
n	ν	λ	$\log_{10} c$	ν	λ	$\log_{10} c$
2	1.00	2.00	0.3010	1.00	2.00	$\bar{1}$.6990
3	2.05	1.96	0.2368	2.05	1.96	$\bar{1}$.4234
4	3.20	1.90	0.1547	3.35	1.84	$\bar{1}$.2370
5	4.5	1.83	0.0607	4.6	1.80	$\bar{1}$.1149
6	6.0	1.75	$\bar{1}$.9588	5.9	1.77	$\bar{2}$.0177
7	7.7	1.67	$\bar{1}$.8574	7.2	1.74	$\bar{2}$.9388
8	9.5	1.60	$\bar{1}$.7685	8.4	1.74	$\bar{2}$.8765
9	12.0	1.51	$\bar{1}$.6457	9.6	1.74	$\bar{2}$.8222
10	14.0	1.46	$\bar{1}$.5785	11.0	1.72	$\bar{2}$.7674
12	19.0	1.36	$\bar{1}$.4252	13.0	1.74	$\bar{2}$.6964
14	26.0	1.24	$\bar{1}$.2479	16.0	1.70	$\bar{2}$.6126
16	34.0	1.14	$\bar{1}$.0964	18.0	1.72	$\bar{2}$.5617
18	46.0	1.03	$\bar{2}$.9142	21.0	1.70	$\bar{2}$.4988
20	60.0	0.94	$\bar{2}$.7569	23.0	1.71	$\bar{2}$.5491

be thought that $\lambda = 1.7$ would have been a better choice, but the chosen value of λ ($= 1.8$) does give closer approximation for smaller values of n, without affecting the accuracy too severely for larger values of n.

In the same paper Cadwell discussed approximations to the distributions of the arithmetic mean of a number (k) of (independent) (M/σ)'s or (W/σ)'s. He also considered the distributions of the ratios

$$\frac{\max_j(M_j)}{\min_j(M_j)} \quad \text{and} \quad \frac{\max_j(W_j)}{\min_j(W_j)},$$

which are analogous to the ratios of maximum to minimum of a number of independent mean square deviations ($S_1^2, S_2^2, \ldots, S_k^2$), each based on the same number of observed values. Cadwell gives tables of approximate upper 5% and 1% points of these statistics for $n = 3(1)10$ and $k = 2(1)12$ (for the 5% points, $n = 12, 15, 20, 30, 60$ are also included for the mean deviation ratio, and $n = 12, 15, 20$ for the range ratio). Similar tables for the statistics

$$\frac{\max_j(S_j^2)}{\min_j(S_j^2)}$$

are given by Pearson and Hartley (1948). They also provide some useful tables of values connected with the distribution of (M/σ), including the expected value, variance, β_1, and β_2 for $n = 2(1)20, 30, 60$, and the upper and lower 0.1, 0.5, 2.5, 5, and 10 percentage points for $n = 2(1)10$.

Among the few simple exact results concerning the distribution of range, we note the following:

For $n = 2$, $E[W] = 2\sigma/\sqrt{\pi}$; $\text{Var}(W) = 2\sigma^2(1 - 2\pi^{-1})$.
For $n = 3$, $E[W] = 3\sigma/\sqrt{\pi}$; $\text{Var}(W) = \sigma^2[2 - (9 - 3\sqrt{3})\pi^{-1}]$.
For $n = 4$, $E[W] = (3\sigma/\sqrt{\pi})(1 + 2\pi^{-1}\sin^{-1}(\tfrac{1}{3}))$.

Godwin (1949a) gives a number of other exact values of first and second moments.

Subsequently, quite extensive tables of the distribution and moments of (W/σ) have become available. A historical survey (up to 1960) of these tables has been given by Harter (1960), who also provided tables of percentage point multipliers, W_α, to 6 decimal places for $n = 2(1)20(2)40(10)100$ and $\alpha = 0.0001, 0.0005, 0.001, 0.005, 0.01, 0.025, 0.05, 0.1(0.1)0.9, 0.95, 0.975,$ $0.99, 0.995, 0.999, 0.9995,$ and 0.9999. There are also tables of the expected value, variance, and β_1 and β_2 of (W/σ) to 8 (or more) significant figures for $n = 2(1)100$ in Harter (1960). Pearson and Hartley (1948) give tables of $\Pr[W \le w\sigma]$ to 4 decimal places for $n = 2(1)20$ and $w = 0.00(0.05)7.25$. They also give the upper and lower 0.1, 0.5, 2.5, 5, and 10 percentage points of the distribution of (W/σ) to 2 decimal places, and expected value and variance (5 decimal places), β_1 (4 decimal places), and β_2 (3 decimal places) for $n = 2(1)20$, 30, and 60. More elaborate tables are also available in this context, and some numerical algorithms have also been developed; see Section 4 for details.

From Table 13.5 it can be seen that a single value of λ might not be found such that $(W/\sigma)^\lambda$ is well approximated by a distribution of $c\chi_\nu^2$ (for suitably chosen c and ν) for a range of values of $n \leq 20$. Pearson (1952) and Cox (1949) have investigated this kind of approximation in some detail. From their investigations it appears that for smaller values of n (e.g., $n \leq 8$), an approximation of the form $c\chi_\nu$ is preferable (indeed, it is exact for $n = 2$); an approximation of the form $c\chi_\nu^2$ is better for larger values of n.

Using tables of percentage points of (M/σ), (W/σ), and χ^2 (see Chapter 18), it is possible to construct confidence intervals for σ by rewriting the equation

$$\Pr\left[T_{\alpha_1}\sigma < T < T_{1-\alpha_2}\sigma \mid \sigma\right] = 1 - \alpha_1 - \alpha_2$$

in the form

$$\Pr\left[\frac{T}{T_{1-\alpha_2}} < \sigma < \frac{T}{T_{\alpha_1}} \mid \sigma\right] = 1 - \alpha_1 - \alpha_2, \tag{13.89}$$

which shows that $(T/T_{1-\alpha_2}, T/T_{\alpha_1})$ is a $100(1 - \alpha_1 - \alpha_2)\%$ confidence interval for σ. Here T can be replaced by M, W, or $\sqrt{n}\,S$, and T_α by $(M/\sigma)_\alpha$, $(W/\sigma)_\alpha$, or $\sqrt{n}\,\chi_{n-1,\,\alpha}$, respectively.

The maximum likelihood estimator S of σ (like its unbiased counterpart $a_n S$) is not a linear function of the observed values of $X_1, X_2, \ldots, X_n$. It is, however, possible to construct a *best linear unbiased estimator* of σ, using the order statistics $X_1', X_2', \ldots, X_n'$. Such estimators (using all the sample values) are of form

$$D = \sum_{j=1}^{[n/2]} \alpha_j \left(X_{n-j+1}' - X_j'\right). \tag{13.90}$$

Values of α_j have been calculated for $n = 2(1)20$; they have been published in Sarhan and Greenberg (1962). Balakrishnan (1990) has presented similar tables for $n = 21(1)30(5)40$. The efficiency of D relative to $a_n S$ is always greater than 98%. Although this is very satisfactory, these estimators are not used often because it is just as easy to calculate $a_n S$ if *all* values are to be used, and this does not require such extensive auxiliary tables. If a linear estimator is desired (e.g., to reduce effects of inaccurate outlying observations), there are other linear estimators, nearly as efficient as D, but with simpler formulas for the coefficients.

We take especial note of *Gini's mean difference*

$$G = \binom{n}{2}^{-1} \sum_{i<j}^{n} \sum^{n} |X_i - X_j|$$

$$= \frac{4}{n(n-1)} \sum_{j=1}^{[n/2]} \left\{ \frac{1}{2}(n+1) - j \right\} (X'_{n-j+1} - X'_j). \qquad (13.91)$$

We have

$$E[G] = \left(\frac{2}{\sqrt{\pi}} \right) \sigma,$$

$$\operatorname{Var}[G] = 4[n(n-1)]^{-1} \left[\frac{1}{3}(n+1) + 2\pi^{-1}(n-2)\sqrt{3} \right.$$

$$\left. - 2\pi^{-1}(2n-3) \right] \sigma^2.$$

The statistic $\frac{1}{2}\sqrt{\pi}\,G$ is an unbiased estimator of σ. The first three lines of Table 13.7 [taken from Nair (1949)] show the efficiencies of $M\sqrt{\pi/2}$, $\frac{1}{2}\sqrt{\pi}\,G$ and D, the best linear unbiased estimator of σ, for $n = 2(1)10$. It can be seen that $\frac{1}{2}\sqrt{\pi}\,G$ is very nearly as efficient as D. As n tends to infinity, the efficiency of $\frac{1}{2}\sqrt{\pi}\,G$, relative to $a_n S$, tends to

$$\left(\tfrac{2}{3} + 4\sqrt{3} - 8 \right)^{-1} = 97.8\%.$$

The asymptotically efficient estimator obtained by the method of Chernoff, Gastwirth, and Johns (1967) (see Section 2 of Chapter 12) is obtained by putting α_j in (13.90) equal to $U_{j/(n+1)}$ for all j.

We also mention an estimator of σ suggested by Gupta (1952) as an approximation to D. This is obtained by replacing the coefficients α_j in (13.90) by

$$\alpha'_j = \frac{E[U'_{n-j+1}]}{\sum_{j=1}^{n} \{E[U'_j]\}^2}, \qquad (13.92)$$

where $U'_1 \le U'_2 \le \cdots \le U'_n$ are order statistics corresponding to n independent unit normal variables so that the estimator is

$$D' = \frac{\sum_{j=1}^n X_j' E[U_j']}{\sum_{j=1}^n \{E[U_j']\}^2}. \tag{13.93}$$

For large n this estimator is very nearly the same as the asymptotically efficient estimator just described. Shapiro and Wilk (1965) described estimators similar to D', except that the ratio of α_1' to the remaining α''s is modified.

In view of the accuracy attainable with $\frac{1}{2}\sqrt{\pi}\,G$, it does not seem necessary to consider the use of D'. However, we note an estimator proposed by Mead (1966) which is based on the same general idea as Gupta's estimator and may be convenient to apply in special circumstances (e.g., using certain kinds of rapid measuring apparatus). Suppose that the number of observations (n) is a multiple of m, say, km. The data are then sorted into k groups so that the m least values are in the first group, the next m least values in the second group, and so on; the last group consists of the m greatest values. If the unit normal distribution is truncated between u_{i-1} and u_i ($> u_{i-1}$), then (see Section 10.1) the expected value of the truncated distribution is

$$\lambda_{(i)} = \frac{Z(u_i) - Z(u_{i-1})}{\Phi(u_i) - \Phi(u_{i-1})}.$$

Mead's estimator is then

$$\frac{\sum_{i=1}^n \lambda_{(i)}[\text{Mean of the } i\text{th group}]}{\sum_{i=1}^k \lambda_{(i)}^2}, \tag{13.94}$$

with

$$u_0 = -\infty,$$

$$u_i = U_{i/k}, \qquad i = 1, \ldots, (k-1),$$

$$u_k = \infty.$$

(Note that $\Phi(u_i) - \Phi(u_{i-1}) = k^{-1}$ for all i.) Denoting the mean of the ith group by $\bar{Y}_i$, the estimators are as follows: for $k = 2$, $0.62666\,(\bar{Y}_2 - \bar{Y}_1)$; for $k = 3$, $0.45838\,(\bar{Y}_3 - \bar{Y}_1)$; for $k = 4$, $0.36927\,(\bar{Y}_4 - \bar{Y}_1) + 0.09431\,(\bar{Y}_3 - \bar{Y}_2)$; for $k = 5$, $0.31213\,(\bar{Y}_5 - \bar{Y}_1) + 0.11860\,(\bar{Y}_4 - \bar{Y}_2)$; and for $k = 6$, $0.27177\,(\bar{Y}_6 - \bar{Y}_1) + 0.12373\,(\bar{Y}_5 - \bar{Y}_2) + 0.03844\,(\bar{Y}_4 - \bar{Y}_3)$. Mead obtained these values for the asymptotic efficiency of this estimator (n large) relative to $a_n S$:

k	2	3	4	5	6
Asymptotic Efficiency (%)	87.6	93.3	95.6	96.8	97.8

Yet another class of linear estimators has been studied by Oderfeld and Pleszczyńska (1961) and Pleszczyńska (1965). These are linear functions of the order statistics $Y'_1, \ldots, Y'_n$ corresponding to the random variables $Y_j = |X_j - \overline{X}|$. These authors estimated the values of the coefficients in

$$\sum_{j=1}^{n} \alpha_j Y'_j$$

on an empirical basis, using result of sampling experiments. In the first paper the only value used for n was 5, and the suggested estimator was

$$-0.065 Y'_1 + 0.150 Y'_2 + 0.175 Y'_3 + 0.312 Y'_4 + 0.405 Y'_5.$$

In the second paper the smallest absolute deviate was omitted (i.e., α_1 taken equal to zero). Coefficients (α_j) were estimated for $n = 3(1)10$. In all cases the largest coefficient was α_n, indicating the relative importance of extreme observations in estimating σ. These estimators appear (from empirical evidence) to have quite high (over 95%) efficiencies relative to $a_n S$, though they are no higher than, for example, estimators based on Gini's mean difference.

For $n \leq 10$ (at least) there are even simple unbiased estimators of σ that are not much less efficient than $\frac{1}{2}\sqrt{\pi}\, G$ or D. These are based on the *thickened ranges*

$$J_{(r)} = \sum_{j=1}^{r} \left(X'_{n-j+1} - X'_j \right) \tag{13.95}$$

[Jones (1946)]. Values of the multiplying factor to be applied to $J_{(2)}$ to make it an unbiased estimator of σ are shown in Table 13.6 [taken from Jones (1946)]. Relative efficiencies of these estimators (compared with $a_n S$) are also shown in Table 13.7. It can be seen that if $J_{(1)}$ ($\equiv W$) be used for $n \leq 5$ and $J_{(2)}$ for $6 \leq n \leq 10$, the relative efficiency never falls below 98%.

For large n Prescott (1968) has given the approximate formulas

$$E[J_{(r)}] = \frac{2r\sigma Z(U_p)}{p},$$

$$\text{Var}(J_{(r)}) = \frac{2r^2\sigma^2}{np^2}\left[p + \{pU_p - Z(U_p)\}\{(1-2p)U_p + 2Z(U_p)\} \right],$$

Table 13.6 Multiplying Factor for $J_{(2)}$

n	Factor
4	0.37696
5	0.30157
6	0.26192
7	0.23702
8	0.21970
9	0.20684
10	0.19684

where $p = r/n$ is not too small. For n large the maximum efficiency (96.65%) is attained with $p = 0.225$. Prescott suggests using $p = \frac{1}{6}$, since the efficiency is still over 90% and the easily remembered quantity

$$\frac{1}{3r}J_{(r)}$$

is very nearly an unbiased estimator of σ.

Dixon (1960) has considered estimators of the form

$$k' \sum_j W_{(j)} = k' \sum_j (X'_{n+j+1} - X'_j), \tag{13.96}$$

where the summation Σ_j is over some set of values of j. The statistic $W_{(j)}$ is

Table 13.7 Relative Efficiencies of Unbiased Estimators of σ

Estimator Based on		n = 2	3	4	5	6	7	8	9	10
Mean deviation		100.0	99.19	96.39	94.60	93.39	92.54	91.90	91.4	91.0
Gini mean difference		100.0	99.19	98.75	98.50	98.34	98.24	98.16	98.1	98.1
Best linear		100.0	99.19	98.92	98.84	98.83	98.86	98.90	98.9	99.0
Range ($J_{(1)} \equiv W$)		100.0	99.19	97.52	95.48	93.30	91.12	89.00	86.9	85.0
Thickened range	$J_{(2)}$	—	—	91.25	93.84	95.71	96.67	96.97	96.8	96.4
	$J_{(3)}$	—	—	—	—	90.25	91.78	93.56	95.0	95.9
	$J_{(4)}$	—	—	—	—	—	—	89.76	90.7	92.2
	$J_{(5)}$	—	—	—	—	—	—	—	—	89.4

Table 13.8   Unbiased Estimators of σ

n	Estimator	Efficiency (%) Relative to $a_n S$
11	$0.1608(W + W_{(2)} + W_{(4)})$	96.7
12	$0.1524(W + W_{(2)} + W_{(4)})$	97.2
13	$0.1456(W + W_{(2)} + W_{(4)})$	97.5
14	$0.1399(W + W_{(2)} + W_{(4)})$	97.7
15	$0.1352(W + W_{(2)} + W_{(4)})$	97.7
16	$0.1311(W + W_{(2)} + W_{(4)})$	97.5
17	$0.1050(W + W_{(2)} + W_{(3)} + W_{(5)})$	97.8
18	$0.1020(W + W_{(2)} + W_{(3)} + W_{(5)})$	97.8
19	$0.09939(W + W_{(2)} + W_{(3)} + W_{(5)})$	97.9
20	$0.10446(W + W_{(2)} + W_{(4)} + W_{(6)})$	98.0

sometimes called the jth *quasi-range*, occasionally the jth *subrange*. Evidently

$$J_{(r)} = \sum_{j=1}^{r} W_{(j)} \qquad \left(\text{and } W_{(1)} \equiv W\right). \qquad (13.97)$$

Dixon found that for $n \leq 10$ the most efficient estimators of form (13.96) are those based on the range ($W \equiv W_{(1)}$), or thickened range $J_{(2)}$ just described. For $n = 11 - 20$ he obtained the most efficient unbiased estimators [in the class (13.96)] given in Table 13.8. The efficiencies compare quite well with those of D, though they are not generally as high as those of Gini's mean difference.

Note that those unbiased linear estimators which are optimal (in various restricted classes) give large relative weight to the extreme observations. [Even S, which appears to be symmetrical, can be regarded (very roughly) as using weights proportional to the absolute magnitude of deviation from the sample mean.] Although we have obtained high efficiencies with these estimators, the calculations are all based on the complete validity of the normal distribution as applied to the data. Distributions of extreme order statistics are likely to be especially sensitive to departures from normality, and it is sometimes more important to guard against this possibility than to squeeze the last drop of formal "efficiency" from the data.

The simplest method of this kind uses only a single pair of symmetrically placed order statistics, in the form of a quasi-range, giving an estimator

$$\beta_r W_{(r)} = \beta_r \left(X'_{n-r+1} - X'_r \right).$$

For n large it is best to take $r \doteq 0.069n$. The efficiency (relative to $a_n S$) of the corresponding unbiased estimator of σ is about 65% [Pearson (1920)]. [For estimating ξ by a statistic of form

$$\beta_{r/2}(X'_{n-r+1} + X'_r),$$

the best choice is $r \doteq 0.270n$, and the efficiency is about 67%.]

Kulldorff (1963, 1964) has studied the construction of estimators of this kind in some detail. He found that if estimators of form

$$\sum_{i=1}^{k} \beta_{r_i} W_{(r_i)} \tag{13.98}$$

were considered, quite good results could be obtained by taking β_{r_i} proportional to i, giving estimators of form

$$\gamma \sum_{i=1}^{k} i W_{(r_i)}. \tag{13.99}$$

For $k = 2$ Kulldorff found that the best values of r_1 and r_2 [subject to an estimator of form (13.99) being used] to take are $0.0235n$ and $0.1279n$, respectively, with $\gamma = 0.1174$; for $k = 3$ optimal values are $r_1 = 0.0115n$, $r_2 = 0.0567n$, and $r_3 = 0.1704n$, with $\gamma = 0.0603$. The corresponding relative efficiencies are approximately 82% for $k = 2$ and 89% for $k = 3$. Note that these results apply to "large" samples (large values of n) and cannot be expected to apply when $n \leq 20$, as in the discussion of other estimators.

Using the large-sample approximations, Eisenberger and Posner (1965) have constructed "best linear unbiased estimators" of mean and standard deviation using only a fixed number (k) of quantiles, and excluding quantiles more extreme than 1 and 99%, or 2.5 and 97.5%, for $k = 2(2)20$. They also give (for the same values of k) pairs of linear estimators minimizing

(Variance of estimator of mean)

$+ \lambda$(Variance of estimator of standard deviation)

for $\lambda = 1, 2,$ and 3. Formulas appropriate for censored samples (described in Section 8.4) can also be used for complete samples if certain observed values must be ignored.

Ogawa (1976) has commented on the optimal spacing of the systematic statistics for the estimation of normal parameters. Ogawa (1977) later presented optimal spacings for the simultaneous estimation of ξ and σ based on selected two sample quantiles. Cheng (1980) discussed the asymptotic best linear unbiased estimator of ξ from a censored sample. Fattorini (1978) gave percentile estimators for ξ and σ, while Cheng and Ling (1983) discussed the best linear unbiased estimators based on incomplete samples. Miyakawa, Tamiya, and Kotani (1985b) addressed the question of whether optimal spacings (in the case of small samples) for the estimation of ξ and σ should be symmetric. Balakrishnan and Cohen (1991) have presented a detailed account of the optimal linear estimation problem.

For the linear estimators based on order statistics (apart from those based on W alone) there are no easily available tables of percentage points. Such tables would be needed to construct confidence intervals for σ by rearranging the relation

$$\Pr\left[T_{\alpha_1}\sigma < T < T_{1-\alpha_2}\sigma \mid \sigma\right] = 1 - \alpha_1 - \alpha_2$$

in the form

$$\Pr\left[\frac{T}{T_{1-\alpha_2}} < \sigma < \frac{T}{T_{\alpha_1}} \middle| \sigma\right] = 1 - \alpha_1 - \alpha_2$$

[see (13.89)]. Even for those cases where such tables are available (M and W) only symmetrical intervals (with $\alpha_1 = \alpha_2$) are practically useful. Except in connection with intervals based on S, no attempt has been made to construct shortest confidence intervals for σ. Nair (1947) suggested calculation of upper and lower 1% and 5% points of the distribution of the second thickened range $J_{(2)}$, but such tables have not been published. In many cases an approximation using a distribution of the form of that of a multiple of a chi random variable ($c\chi_\nu$) (Chapter 18) may give results that are not seriously inaccurate for practical purposes.

8.3   Estimation of Functions of ξ and σ

Certain functions of both ξ and σ are sometimes the primary target of statistical estimation. Among these we particularly note the $100\alpha\%$ percentile point ($\xi + U_\alpha\sigma$) and the proportion (of population) less than a fixed number x,

$$P_x = \left(\sqrt{2\pi}\,\sigma\right)^{-1} \int_{-\infty}^{(x-\xi)/\sigma} e^{-u^2/2}\, du. \tag{13.100}$$

These quantities may be estimated by general methods, not using the special form of distribution. However, when the validity of the assumption of normality has been clearly established, it is to be expected that more accurate estimates can be obtained by using this knowledge of the form of distribution.

Evidently, if $\hat{\xi}'$, $\hat{\sigma}'$ are any unbiased estimators of ξ, σ, respectively, then ($\hat{\xi}' + U_\alpha\hat{\sigma}'$) is an unbiased estimator of $\xi + U_\alpha\sigma$. If $\overline{X}$ is used as the estimator of ξ and if $\hat{\sigma}'$ is any of the unbiased estimators of σ described in

Section 8.2, then, since $\bar{X}$ and $\hat{\sigma}'$ are independent,

$$\mathrm{Var}\left(\bar{X} + U_\alpha \hat{\sigma}'\right) = \sigma^2 n^{-1} + U_\alpha^2 \,\mathrm{Var}(\hat{\sigma}'). \tag{13.101}$$

Combination of $\bar{X}$ with the best linear unbiased estimator of σ (or one of the other, nearly as efficient, linear estimators of σ) will give a good linear estimator of $\xi + U_\alpha \sigma$.

If $a_n S$ is used as an estimator of σ, the distribution of the estimator $(\bar{X} + U_\alpha a_n S)$ may be evaluated in the following way:

$$\Pr\left[\bar{X} + U_\alpha a_n S \le K\right]$$

$$= \Pr\left[\left(\xi + \frac{U\sigma}{\sqrt{n}}\right) + \left(\frac{U_\alpha a_n \sigma}{\sqrt{n}}\right)\chi_{n-1} \le K\right]$$

$$= \Pr\left[\left\{U + \frac{\sqrt{n}\,(\xi - K)}{\sigma}\right\}\left(\frac{\chi_{n-1}}{\sqrt{n-1}}\right)^{-1} \le -U_\alpha a_n \sqrt{n-1}\right]$$

$$= \Pr\left[t'_{n-1}\left(\frac{\sqrt{n}\,(\xi - K)}{\sigma}\right) \le -U_\alpha a_n \sqrt{n-1}\right], \tag{13.102}$$

where $t'_{n-1}(\lambda)$ denotes a noncentral t variable (see Chapter 31) with $n - 1$ degrees of freedom and a noncentrality parameter λ.

If other estimators of σ are used, approximate results of similar form can be obtained by approximating the distribution of $\hat{\sigma}'$ by that of $c\chi_\nu$, with suitable values of c and ν. It will usually be troublesome to assess the accuracy of these approximations. If effects of unreliable outlying observations are to be specially avoided, then estimators of ξ and σ not using such observations may be used. However, if the reason for this precaution is that lack of normality is suspected, it is doubtful that $\xi + U_\alpha \sigma$ should be estimated at all.

Coming now to the estimation of quantities like

$$\Pr[X \le x] = \left(\sqrt{2\pi}\right)^{-1}\int_{-\infty}^{(x-\xi)/\sigma} e^{-u^2/2}\,du,$$

it is clear that the maximum likelihood estimator is obtained by replacing ξ by $\bar{X}$, and σ by S. The resulting estimator is, in general, biased. (It is unbiased if it so happens that $x = \xi$.)

To obtain the minimum variance unbiased estimator, the Blackwell-Rao theorem may be used. The estimator

$$T = \begin{cases} 1 & \text{if } X_1 \le x, \\ 0 & \text{if } X_1 > x \end{cases} \tag{13.103}$$

is an unbiased estimator of $\Pr[X \le x]$ and $\bar{X}$ and S are jointly complete sufficient statistics for ξ and σ. Hence the minimum variance unbiased estimator of $\Pr[X \le x]$ is

$$E[T|\bar{X}, S] = \Pr\left[X_1 \le x | \bar{X}, S\right]$$

$$= \Pr\left[\frac{X_1 - \bar{X}}{S} \le \frac{x - \bar{X}}{S} \Big| \bar{X}, S\right].$$

Since the conditional distribution of $(X_1 - \bar{X})/S$ is independent of both $\bar{X}$ and S, it is the same as the unconditional distribution of $(X_1 - \bar{X})/S$. Making an orthogonal transformation with one new variable $\sqrt{n/(n-1)}\,(X_1 - \bar{X})$, and one equal to $\sqrt{n}\,\bar{X}$, it can be seen that $(X_i - \bar{X})/S$ is distributed symmetrically about zero as the signed square root of $(n-1)$ times a beta variable with parameters $\frac{1}{2}, \frac{n}{2} - 1$ (see Chapter 25). Hence the minimum variance unbiased estimator of $\Pr[X \le x]$ is

$$\begin{cases} \left[B\left(\frac{1}{2}, \frac{n}{2} - 1\right)\right]^{-1} \int_{-1}^{(x-\bar{X})/(S\sqrt{n-1})} (1 - v^2)^{(n-4)/2}\, dv & \text{for } |x - \bar{X}| \le S\sqrt{n-1}, \\ 0 & \text{for } x < \bar{X} - S\sqrt{n-1}, \\ 1 & \text{for } x > \bar{X} + S\sqrt{n-1}. \end{cases} \tag{13.104}$$

(Numerical evaluation can be effected using tables of the incomplete beta function, as described in Chapter 25.)

At this point we note that if X_{n+1} is independent of, and has the same distribution as each X_j, $(X_{n+1} - \bar{X})/S$ is distributed as $[(n+1)/(n-1)]^{1/2}$ times t with $n - 1$ degrees of freedom. Hence the interval

$$\left(\bar{X} + t_{n-1,\alpha_1}\left[\frac{n+1}{n-1}\right]^{1/2} S,\ \bar{X} + t_{n-1,1-\alpha_2}\left[\frac{n+1}{n-1}\right]^{1/2} S\right)$$

contains on average a proportion $1 - \alpha_1 - \alpha_2$ of the population values. It is thus a form of *tolerance interval* for the normal distribution. Unlike the tolerance intervals described in Chapter 12, the construction of this interval makes use of knowledge of the form of population distribution. It cannot be used for other populations without the possibility of introducing bias.

Wald and Wolfowitz (1946) have shown that a good approximation to *tolerance limits*, such that there is a probability equal to $1 - \alpha$ that the limits include *at least* a specified proportion γ of the population, is

$$\bar{X} \pm \frac{\lambda_\gamma \sqrt{n}\, S}{\chi_{n-1,\alpha}}, \qquad (13.105)$$

where λ_γ satisfies the equation

$$\Phi\left(n^{-1/2} + \lambda_\gamma\right) - \Phi\left(n^{-1/2} - \lambda_\gamma\right) = \gamma.$$

The construction of exact *one-sided* tolerance limits can be simply effected, using the noncentral t distribution (see Chapter 31). We note that the population proportion less than $(\bar{X} + kS)$ is

$$\Phi\left(\frac{\bar{X} + kS - \xi}{\sigma}\right),$$

and this is at least γ if

$$\frac{\bar{X} + kS - \xi}{\sigma} \geq U_\gamma.$$

This inequality can be rearranged in the form

$$\frac{\sqrt{n}\left\{(\bar{X} - \xi)/\sigma\right\} - \sqrt{n}\, U_\gamma}{\sqrt{n}\,(S/\sigma)/\sqrt{n-1}} \geq -k\sqrt{n-1}. \qquad (13.106)$$

The statistic on the left-hand side of (13.106) has a noncentral t distribution with $(n - 1)$ degrees of freedom and noncentrality parameter $(-\sqrt{n}\, U_\gamma)$. In order that the probability that at least a proportion γ of the population is less than $(\bar{X} + kS)$, should be equal to $(1 - \alpha)$, we make $(-k\sqrt{n-1})$ equal to the lower $100\alpha\%$ point of the noncentral t distribution, i.e.

$$k = \frac{-t'_{n-1,\alpha}\left(-\sqrt{n}\, U_\gamma\right)}{\sqrt{n-1}}. \qquad (13.107)$$

Nelson (1977) has discussed tolerance factors for the normal distribution. Odeh and Owen (1980) have presented elaborate tables of normal tolerance limits. Gerisch, Struck, and Wilke (1987) determined one-sided tolerance limit factors in the case of censored samples through Monte Carlo simulations. Mee (1988) discussed the estimation of the percentage of a normal distribution lying outside a specified interval. Eberhardt, Mee, and Reeve (1989) determined factors for exact two-sided tolerance limits for the normal distribution.

Sometimes it is desired to estimate the *mean square error* $(\xi - \xi_0)^2 + \sigma^2$, where ξ_0 is a specified number. The mean square

$$n^{-1} \sum_{j=1}^{n} (X_j - \xi_0)^2$$

is an unbiased estimator of this quantity. It is distributed as

$$n^{-1}\sigma^2 \times \left(\begin{array}{l} \text{Noncentral } \chi^2 \text{ with} \\ n \text{ degrees of freedom and} \\ \text{noncentrality parameter} \\ n(\xi - \xi_0)^2/\sigma^2 \end{array} \right)$$

(see Chapter 29) and has variance

$$2n^{-2}\sigma^4 \left[n + \frac{2n(\xi - \xi_0)^2}{\sigma^2} \right] = 2n^{-1}\sigma^4 \left[1 + \frac{2(\xi - \xi_0)^2}{\sigma^2} \right].$$

A natural estimate of the coefficient of variation (σ/ξ) is the ratio $a_n S/\overline{X}$, or more generally $a'_n S/\overline{X}$, with a'_n being suitably chosen. Since the expected value of $S/\overline{X}$ is infinite, it is not possible to obtain an unbiased estimator of this form. We can, however, construct an *approximate* confidence interval for σ/ξ. We will suppose that $\Pr[\overline{X} < 0]$ can be neglected (i.e., σ/ξ sufficiently small—less than $\frac{1}{4}$, say). Then, since $\overline{X}/S$ is distributed as $(n - 1)^{-1/2}$ times noncentral t with $n - 1$ degrees of freedom and noncentrality parameter $\sqrt{n}\,\xi/\sigma$, it follows that (in the notation of Chapter 31)

$$\Pr\left[t'_{n-1,\alpha_1}\left(\frac{\sqrt{n}\,\xi}{\sigma} \right) \le \frac{\sqrt{n-1}\,\overline{X}}{S} \le t'_{n-1,1-\alpha_2}\left(\frac{\sqrt{n}\,\xi}{\sigma} \right) \right] = 1 - \alpha_1 - \alpha_2$$

or

$$\Pr\left[n^{-1/2}g_{1-\alpha_2}\left(\frac{\sqrt{n-1}\,\overline{X}}{S} \right) \le \frac{\xi}{\sigma} \le n^{-1/2}g_{\alpha_1}\left(\frac{\sqrt{n-1}\,\overline{X}}{S} \right) \right] \doteq 1 - \alpha_1 - \alpha_2,$$

$$(13.108)$$

where $g_\alpha(z)$ is the solution (for g) of the equation

$$t'_{n-1,\alpha}(g) = z \qquad (13.109)$$

(assuming that $\overline{X}$ is not too small). Assuming now that $\xi > 0$, (13.108) can be rewritten

$$\Pr\left[\sqrt{n}\,/g_{\alpha_1}\!\left(\frac{\sqrt{n-1}\,\overline{X}}{S}\right) < \frac{\sigma}{\xi} < \sqrt{n}\,/g_{1-\alpha_2}\!\left(\frac{\sqrt{n-1}\,\overline{X}}{S}\right)\right] \doteq 1 - \alpha_1 - \alpha_2.$$

It is necessary to use tables of the noncentral t distribution (see Chapter 31, Section 7) to calculate even these approximate limits.

More easily calculable, but rather rough, approximate limits are obtained from the formula

$$\text{Lower limit} = V\left[1 - n^{-1/2}U_{\alpha_2}\sqrt{\tfrac{1}{2} + V^2}\right]^{-1},$$
$$\qquad\qquad (13.110)$$
$$\text{Upper limit} = V\left[1 - n^{-1/2}U_{1-\alpha_1}\sqrt{\tfrac{1}{2} + V^2}\right]^{-1},$$

where $V = S/\overline{X}$. These are based on the assumption that $(S - k\overline{X})$ is approximately normally distributed with expected value $(\sigma - k\xi)$ and variance $n^{-1}\sigma^2(1 + \tfrac{1}{2}k^2)$ so that (since $\xi \gg \sigma$)

$$\Pr\left[\frac{S}{\overline{X}} < k\right] \doteq \Phi\left(\sqrt{\frac{n}{1 + \tfrac{1}{2}k^2}}\left\{1 - \frac{k\xi}{\sigma}\right\}\right),$$

that is, $\sqrt{n/\left(1 + \tfrac{1}{2}V^2\right)}\,(1 - V\xi/\sigma)$ has approximately a unit normal distribution. A similar argument indicates that if X_1, X_2 are independent normal random variables and $E[X_j] = \xi_j$, $\text{Var}(X_j) = \sigma_j^2$ $(j = 1, 2)$ with $\xi_2 \gg \sigma_2$, then putting $X_1/X_2 = R$, the distribution of

$$\frac{R\xi_2 - \xi_1}{\left(R^2\sigma_2^2 + \sigma_1^2\right)^{1/2}}$$

is approximately unit normal.

Koopmans, Owen, and Rosenblatt (1964) have pointed out that if the distribution of each of the independent variables is *lognormal* (see Chapter 14), then construction of *exact* confidence intervals for the coefficient of variation is straightforward. Since it is possible to approximate a normal distribution quite closely by a lognormal distribution (see Chapter 14, Section 3), it is likely that the same formulas will give good results for normal variables (though they will not of course give *exactly* specified values for

confidence coefficients). The (approximate) confidence limits, in terms of the
original variables $X_1, \ldots, X_n$ obtained by this method, are

$$
\left[\exp\left\{ \frac{\sum_{j=1}^{n}\left(\log X_j - \overline{\log X}\right)^2}{\chi^2_{n-1,1-\alpha_1}} \right\} - 1 \right]^{1/2}
\tag{13.111}
$$

and

$$
\left[\exp\left\{ \frac{\sum_{j=1}^{n}\left(\log X_j - \overline{\log X}\right)^2}{\chi^2_{n-1,\alpha_2}} \right\} - 1 \right]^{1/2},
$$

where $\overline{\log X} = n^{-1}\sum_{j=1}^{n} \log X_j$.

The cumulative distribution function of the rth quasi-range for random
samples from a unit normal distribution is [Jones et al. (1969)]

$$
F_{W_{(r)}}(w) = \sum_{i=0}^{r} \frac{n^{(2r-i+1)}}{r!(r-i)!} \sum_{j=0}^{r-i} \sum_{k=0}^{n-2r+i-1} (-1)^{n-2r+i-1-j+k} P, \qquad w > 0,
$$

where

$$
P = \binom{r-i}{j}\binom{n-2r+i-1}{k} \Pr\left[\bigcap_{l=1}^{n-r+i+j-1} \left(Y_l \le \frac{\delta_{j+h-l}w}{\sqrt{2}} \right) \right]
$$

and $\delta_h = 0, 1$ for $h < 0, \ge 0$, respectively; the Y's are standardized multi-normal variables (Chapter 35) with all correlations equal to $\frac{1}{2}$.

Estimates for the reliability function of the normal distribution has been
considered by Hurt (1980). Sinha (1985) has discussed Bayes's estimations of
the reliability function.

Owen and Hua (1977) have presented tables of confidence limits on the
tail area of the normal distribution. Fertig and Mann (1977) have discussed
one-sided prediction intervals for at least p out of m future observations.
Nelson and Schmee (1981) have given prediction limits for the last failure
time of a normal sample from early failures. Lingappaiah (1983) has discussed prediction problems in normal samples. Along the lines of Fertig and
Mann (1977), Chou and Owen (1986) have studied one-sided simultaneous
lower prediction intervals for l future samples from a normal distribution.
Whitmore (1986) has made some interesting comments on the prediction
limits for a single normal observation. Odeh (1989a) has developed simultaneous two-sided prediction intervals to contain at least l out of k future
means. Similarly Odeh (1989b) has also discussed simultaneous one-sided
prediction intervals to contain all of k future means.

8.4   Estimation from Censored Data

We will consider situations in which the r_1 least, and r_2 greatest, observations are censored (i.e., not recorded) leaving only $X'_{r_1+1}, \ldots, X'_{n-r_2}$. Best linear unbiased estimators, based on these order statistics, are particularly useful in these circumstances, since the maximum likelihood estimators of ξ and σ are much more difficult to calculate than they are for complete samples. We will, first, discuss maximum likelihood estimators and then possible approximations.

The joint probability density function of $X'_{r_1+1}, \ldots, X'_{n-r_2}$ is

$$p(x_{r_1+1}, \ldots, x_{n-r_2}) = \frac{n!}{r_1! r_2!} \left[\Phi\left(\frac{x_{r_1+1} - \xi}{\sigma} \right) \right]^{r_1} \left[1 - \Phi\left(\frac{x_{n-r_2} - \xi}{\sigma} \right) \right]^{r_2}$$

(13.112)

$$\times \, \sigma^{-(n-r_1-r_2)} \prod_{j=r_1+1}^{n-r_2} Z\left(\frac{x_j - \xi}{\sigma} \right).$$

The maximum likelihood estimators $\hat{\xi}, \hat{\sigma}$ of ξ, σ, respectively satisfy the following equations [using the notation $\hat{U}'_j = (X'_j - \hat{\xi})/\hat{\sigma}$]:

$$r_1 \left[-\frac{Z(\hat{U}'_{r_1+1})}{\Phi(\hat{U}'_{r_1+1})} \right] + \sum_{j=r_1+1}^{n-r_2} \hat{U}'_j$$

$$+ \, r_2 \left[\frac{Z(\hat{U}'_{n-r_2})}{1 - \Phi(\hat{U}'_{n-r_2})} \right] = 0,$$

(13.113)

$$r_1 \left[1 - \frac{\hat{U}'_{r_1+1} Z(\hat{U}'_{r_1+1})}{\Phi(\hat{U}'_{r_1+1})} \right] + \sum_{j=r_1+1}^{n-r_2} \hat{U}'^2_j$$

$$+ \, r_2 \left[1 + \frac{\hat{U}'_{n-r_2} Z(\hat{U}'_{n-r_2})}{1 - \Phi(\hat{U}'_{n-r_2})} \right] = n.$$

(13.114)

From equations (13.115) below it can be seen that in (13.113) the censored "observations" are replaced by the expected value of the appropriate tail of the normal distribution truncated at X'_{r_1+1} or X'_{n-r_2}, as the case may be, and that in (13.114) the squared standardized deviates of these "observations" are replaced by the corresponding expected values for the tails. With this in mind, we obtain approximate solutions of equations (13.113) and (13.114) by replacing $(X'_{r_1+1} - \hat{\xi})/\hat{\sigma}$ and $(X'_{n-r_2} - \hat{\xi})/\hat{\sigma}$ by U_{α_1} and $U_{1-\alpha_2}$, respectively (except in the summations) with $\alpha_j = (r_j + 1)/(n + 1)$ for $j = 1, 2$. By introducing the following notation for the moments of the tails (of singly truncated normal distributions, Section 10),

$$-\frac{Z(U_\alpha)}{\Phi(U_\alpha)} = \mu'_{1,(\alpha-)}, \quad \frac{Z(U_\alpha)}{1 - \Phi(U_\alpha)} = \mu'_{1,(\alpha+)},$$

$$1 - \frac{U_\alpha Z(U_\alpha)}{\Phi(U_\alpha)} = \mu'_{2,(\alpha-)}, \quad 1 + \frac{U_\alpha Z(U_\alpha)}{1 - \Phi(U_\alpha)} = \mu'_{2,(\alpha+)},$$

(13.115)

we obtain the approximate equations

$$\overline{X}' + (n - r_1 - r_2)^{-1}(r_1\mu'_{1,(\alpha_1-)} + r_2\mu'_{1,(1-\alpha_2+)})\hat{\sigma} \doteq \hat{\xi}, \quad (13.116)$$

$$\left[\sum_{j=r_1+1}^{n-r_2} \left(X'_j - \overline{X}'\right)^2\right]\left[n - r_1\mu'_{2,(\alpha_1-)} - r_2\mu'_{2,(1-\alpha_2+)}\right.$$

$$\left. - (n - r_1 - r_2)^{-1}(r_1\mu'_{1,(\alpha_1-)} + r_2\mu'_{1,(1-\alpha_2+)})^2\right]^{-1} = \hat{\sigma}^2, \quad (13.117)$$

with $\overline{X}' = (n - r_1 - r_2)^{-1}\sum_{j=r_1+1}^{n-r_2}X'_j$. Values of $\mu'_{1,\alpha\pm}$ and $\mu'_{2,\alpha\pm}$ can be obtained from tables mentioned in Section 1 [also in Harter and Moore (1966)].

Having obtained first approximations to $\hat{\xi}$ and $\hat{\sigma}$ from (13.116) and (13.117), these can be used to calculate new values of $\mu'_{j,\pm\alpha}$, now using $(X'_{r_1+1} - \hat{\xi})/\hat{\sigma}$ and $(X'_{n-r_2} - \hat{\xi})/\hat{\sigma}$ in place of U_{α_1} and $U_{1-\alpha_2}$, respectively. Then (13.113) and (13.114) give new approximations to $\hat{\xi}, \hat{\sigma}$, and so on. Cohen (1957) has given a chart that may be used in the solution of (13.113) and (13.114).

Approximations (for large values of n) to the variances and covariance of $\hat{\xi}$ and $\hat{\sigma}$ are the same as for the corresponding truncated distributions, with truncation points corresponding to $\alpha_j = r_j/n$ $(j = 1, 2)$. Some numerical values are given in Section 10. Detailed discussions of these developments and corresponding tables and graphs may be found in the volumes of Harter (1970), Nelson (1982), Schneider (1986), Cohen and Whitten (1988), Cohen (1991), and Balakrishnan and Cohen (1991). Although solution of the maximum likelihood equations is practicable if the techniques described above or variants thereof are used, it is often convenient to use simpler estimating formulas.

Making the further approximation $r_j/n \doteq \Phi(U_{\alpha_j})$ $(j = 1, 2)$, equations (13.116) and (13.117) can be written

$$\bar{X}' + \hat{\sigma}\big[Z(U_{1-\alpha_2}) - Z(U_{\alpha_1})\big][1 - \alpha_1 - \alpha_2]^{-1} = \hat{\xi}, \qquad (13.118)$$

$$\left[n^{-1} \sum_{j=r_1+1}^{n-r_2} \left(X_j' - \bar{X}'\right)^2 \right]$$

$$\times \left[1 - \alpha_1 - \alpha_2 + U_{\alpha_1} Z(U_{\alpha_1}) - U_{1-\alpha_2} Z(U_{1-\alpha_2}) \right.$$

$$\left. - \left\{ Z(U_{\alpha_1}) - Z(U_{1-\alpha_2}) \right\}^2 (1 - \alpha_1 - \alpha_2)^{-1} \right]^{-1} = \hat{\sigma}^2. \quad (13.119)$$

Tiku (1967) has suggested that the approximate formulas

$$\frac{Z(x)}{\Phi(x)} = a_1 + b_1 x,$$

$$\frac{Z(x)}{1 - \Phi(x)} = a_2 + b_2 x$$

be used to simplify equations (13.113) and (13.114). (The values of a_1, b_1, a_2, and b_2 are chosen to give good fits over the range $U_{\alpha_1} \le x \le U_{1-\alpha_2}$.) This leads to the following equations for the estimators of ξ and σ (denoted here by $\hat{\xi}', \hat{\sigma}'$):

$$\hat{\xi}' = K + L\hat{\sigma}',$$

$$(1 - \alpha_1 - \alpha_2)\hat{\sigma}'^2 - \left\{ \alpha_2 a_2 X_{n-r_2}' - \alpha_1 a_1 X_{r_1+1}' - (\alpha_2 a_2 - \alpha_1 a_1) K \right\} \hat{\sigma}'^2$$

$$- \left\{ n^{-1} \sum_{j=r_1+1}^{n-r_2} X_j'^2 + \alpha_2 b_2 X_{n-r_2}'^2 - \alpha_1 b_1 X_{r_1+1}'^2 \right\}$$

$$- (1 - \alpha_1 - \alpha_2 + \alpha_2 b_2 + \alpha_1 b_1) K^2 = 0,$$

with

$$K = \frac{n^{-1}\sum_{j=r_1+1}^{n-r_2} X_j' + \alpha_2 b_2 X_{n-r_2}' - \alpha_1 b_1 X_{r_1+1}'}{[1 - \alpha_1 - \alpha_2 + \alpha_2 b_2 - \alpha_1 b_1]}$$

and

$$L = \frac{\alpha_2 a_2 - \alpha_1 a_1}{1 - \alpha_1 - \alpha_2 + \alpha_2 b_2 - \alpha_1 b_1}$$

(the quadratic in $\hat{\sigma}'$ has only one positive root).

For symmetrical censoring ($\alpha_1 = \alpha_2 = \alpha$), we have $a_2 = a_1 = a$, say, and $b_2 = -b_1 = b$, leading to

$$\hat{\xi}' = \frac{n^{-1}\sum_{j=r+1}^{n-r}X_j' + \alpha b(X_{r+1}' + X_{n-r}')}{1 - 2\alpha + 2\alpha b}$$

and

$$(1 - 2\alpha)\hat{\sigma}'^2 - \alpha a(X_{n-r}' - X_{r+1}')\hat{\sigma}'$$

$$- \left[n^{-1}\sum_{j=r+1}^{n-r} X_j'^2 + \alpha b(X_{n-r}'^2 + X_{r+1}'^2) - (1 - 2\alpha + 2\alpha b)K^2\right] = 0.$$

Note that $\hat{\xi}'$ is of similar form to the Winsorized mean (13.73). Tiku gives tables to assist in obtaining good values for a and b. These simplified estimators, referred to as the *modified maximum likelihood estimators*, have been utilized to develop several robust inference procedures (robust to departures from normality); see Tiku, Tan, and Balakrishnan (1986).

For symmetrical censoring, the Winsorized mean (13.73) described in Section 8.1 is a natural choice as estimator of ξ. For a moderate degree of asymmetry in censoring (i.e., $|r_1 - r_2|$ small), it may be worthwhile to set aside the excess observations and use the Winsorized mean of the largest available symmetrically censored set of sample values. Table 13.9 [from

Table 13.9 Efficiency (%) of Winsorized Mean Based on $X_{j+1}', \ldots, X_{n-j}'$ Relative to the Best Linear Unbiased Estimator of ξ Based on $X_j', X_{j+1}', \ldots, X_{n-j}'$

n \ j	1	2	3	4	5	6
3	100.0					
4	96.2					
5	96.4	100.0				
6	96.9	96.4				
7	97.3	96.3	100.0			
8	97.7	96.7	96.7			
9	98.0	97.1	96.5	100.0		
10	98.2	97.4	96.8	97.1		
12	98.6	97.9	97.4	97.0	97.4	
14	98.8	98.3	97.9	97.5	97.1	97.6
16	99.0	98.6	98.2	97.9	97.5	97.3
18	99.1	98.8	98.5	98.2	97.9	97.6
20	99.2	98.9	98.7	98.4	98.2	98.0

Dixon (1960)] gives the efficiency of the Winsorized mean

$$\xi_{(j)} = n^{-1}\big[(j+1)X'_{j+1} + X'_{j+2} + \cdots + X'_{n-j-1} + (j+1)X'_{n-j}\big]$$

relative to the best linear unbiased estimator of ξ based on $X'_j, X'_{j+1}, \ldots, X'_{n-j}$. It can be seen that the loss in efficiency from ignoring the value X'_j is trifling.

Even if one of r_1 and r_2 is zero, the Winsorized mean can still be used with little loss of efficiency. Even this can be reduced by using a modified Winsorized mean of form (for $r_1 = 0$, $r_2 = r$)

$$(n + a - 1)^{-1}\big[aX'_1 + X'_2 + \cdots + (r+1)X'_{n-r}\big] \qquad (13.120)$$

with a chosen to make this an unbiased estimator of ξ. Minimum values of the efficiency of this estimator (relative to the best linear unbiased estimator) are shown below:

r	Minimum Relative Efficiency
1	99.8%
2	99.2%
3	98.5%
4	97.7%
5	96.9%
6	96.0%

Values of a are given in Dixon (1960, table II), but they can easily be calculated from tables of expected values of ordered normal variables (see Section 4).

Estimation of σ is more seriously, and adversely, affected by omission of extreme values, than is estimation of ξ. However, there is the compensating feature that the estimates are generally less sensitive to departures from exact normality of distribution (as already noted in Section 8.2). Rhiel (1986), for example, has examined the effect of nonnormality on the use of the sample range in estimating the population standard deviation σ. The simplest estimators of σ that may be used when analyzing censored samples are those based on the *quasi-ranges*

$$W_{(j)} = X'_{n-j+1} - X'_j. \qquad (13.121)$$

Dixon found that for symmetrical censoring with $r_1 = r_2 = r \leq 6$, and $n \leq 20$, estimators based on

1. $W_{(r+1)} + W_{(r+2)}$ for $r = 1$, $11 \leq n \leq 15$, and $r = 2$, $16 \leq n \leq 19$,
2. $W_{(r+1)} + W_{(r+3)}$ for $r = 1$, $16 \leq n \leq 20$, and $r = 2$, $n = 20$,
3. $W_{(r+1)}$ for all other $r \leq 6$, $n \leq 20$,

have efficiencies relative to the best linear unbiased estimators of σ of at least 96.5%. A similar situation exists when $|r_1 - r_2| = 1$. Dixon also gives simple formulas for linear unbiased estimators for *single* censoring with $r_1 = 0$, $r_2 \leq 6$, $n \leq 20$, that have a minimum efficiency of 93.7% relative to the best linear unbiased estimators.

Selvin (1976) has discussed a graphical estimate of ξ based on censored samples. Custer and Pam (1976) have provided correction for bias in the maximum likelihood estimator of σ in a right-censored sample. Persson and Rootzen (1977) have proposed some simple and highly efficient estimators for ξ and σ when the available sample is Type-I censored. Healy (1978) has given a mean difference estimator of σ in the case of symmetrically censored samples; see Prescott (1979) for a similar estimator of σ in the case of asymmetrically censored samples. A linear estimator of σ has also been provided by Healy (1982) for symmetrically censored samples. Schneider (1984) presented some simple and very efficient estimators for both ξ and σ for censored samples; also see Schneider (1986) for a detailed discussion on various estimators of ξ and σ under censored sampling. Wolynetz (1979) has developed an algorithm for determining the maximum likelihood estimates of ξ and σ from truncated and censored data. Hill (1987) and Swan (1977) have discussed algorithms for the maximum likelihood estimation from grouped and censored data. Schader and Schmid (1984) compared some algorithms available for the computation of the maximum likelihood estimates of ξ and σ from a grouped sample. Schader and Schmid (1988) have also discussed small-sample properties of these maximum likelihood estimates from a grouped data.

Sobel and Tong (1976) have described the estimation of a normal percentile by grouping. Kabe (1976) has constructed confidence bands for the quantiles. Dyer, Keating, and Hensley (1977) have compared different point estimators of normal percentiles. Durrant (1978) has presented a nomogram for confidence limits on quantiles of the normal distribution. In continuation of the work of Dyer, Keating, and Hensley (1977), a further look at the comparison of different point estimators of normal percentiles has been given by Dyer and Keating (1979). Hassanein, Saleh, and Brown (1986) have tabulated the best linear unbiased estimators of normal quantiles for sample sizes up to 20. Oppenlander, Schmee, and Hahn (1988) have proposed some simple robust estimators of normal tail percentiles and then examined their performance.

Rukhin (1986) has provided an interesting discussion on the estimation of normal tail probabilities. Gertsbakh and Winterbottom (1991) have examined both point and interval estimation of normal tail probabilities.

Shapiro (1986) has elaborated on how to test for normality, while D'Agostino and Stephens (1986) have presented different types of goodness-of-fit tests for the normal distribution, examined their comparative performance, and presented necessary tables of percentage points for these tests. Barnett and Lewis (1994) have discussed in some detail various methods of testing for the presence of outliers in normal samples and have also presented several tables to facilitate the use of such tests.

9 SIMULATIONAL ALGORITHMS

Many algorithms for generating pseudorandom numbers from a normal population have been developed over the years. Of course any uniform random number generating algorithm can be successfully used with the inverse cumulative distribution function (or an efficient approximation of it, as discussed in Section 7) to generate pseudorandom normal observations. But simpler, more efficient, and faster methods of simulation have been developed, and we describe a few of them here.

9.1 Box–Muller Method

By starting with two independent standard normal variables X_1 and X_2, Box and Muller (1958) considered the transformation

$$Y_1 = \exp\left\{-\frac{1}{2}(X_1^2 + X_2^2)\right\} \quad \text{and} \quad Y_2 = \frac{1}{2\pi}\tan^{-1}\left(\frac{X_1}{X_2}\right) \quad (13.122)$$

and showed that the variables Y_1 and Y_2 are independently uniformly distributed in $(0,1)$. Since Y_2 is uniformly distributed in $(0,1)$, it readily follows that X_1/X_2 has a standard Cauchy distribution (see Chapter 16). Hence, after generating two pseudorandom uniform $(0,1)$ observations Y_1 and Y_2, (13.122) may be used to obtain the required pseudorandom normal observations X_1 and X_2.

Alternatively, if we consider the polar transformation

$$X_1 = r\sin\theta \quad \text{and} \quad X_2 = r\cos\theta, \quad (13.123)$$

it can be easily verified that the joint density of r and θ is

$$p(r,\theta) = \frac{1}{2\pi}e^{-r^2/2}r, \qquad 0 \le r < \infty, 0 \le \theta \le 2\pi. \quad (13.124)$$

By factorization theorem, we immediately have the variables r and θ to be statistically independent. Further the variables

$$U_1 = e^{-r^2/2} \quad \text{and} \quad U_2 = \frac{1}{2\pi}\theta \qquad (13.125)$$

are independently uniformly distributed in $(0, 1)$. [These are exactly same as Y_1 and Y_2 in (13.122).] Inverting the transformation, we get

$$\begin{aligned} X_1 &= \sqrt{-2\ln U_1}\,\cos(2\pi U_2), \\ X_2 &= \sqrt{-2\ln U_1}\,\sin(2\pi U_2) \end{aligned} \qquad (13.126)$$

to be a pair of pseudorandom standard normal observations; see also Golder and Settle (1976).

9.2 Marsaglia–Bray's Improvement

Marsaglia and Bray (1964) improved Box–Muller method as follows: Let U_1 and U_2 be uniformly distributed on $(-1, 1)$, subject to the condition that $U_1^2 + U_2^2 \leq 1$; that is, $p_{U_1, U_2}(u_1, u_2)$ is a disc of uniform height and radius 1, centered at the origin, and is given by

$$p_{U_1, U_2}(u_1, u_2) = \frac{1}{\pi}, \qquad -1 \leq u_1, u_2 \leq 1, 0 \leq u_1^2 + u_2^2 \leq 1.$$

Then by direct transformation of variables it can be verified that the variables

$$\begin{aligned} X_1 &= U_1 \left\{ \frac{-2\ln(U_1^2 + U_2^2)}{U_1^2 + U_2^2} \right\}^{1/2}, \\ X_2 &= U_2 \left\{ \frac{-2\ln(U_1^2 + U_2^2)}{U_1^2 + U_2^2} \right\}^{1/2} \end{aligned} \qquad (13.127)$$

are independently distributed as standard normal. After simulating a pseudorandom observation (u_1, u_2) from a uniform circular disc of radius 1, centered at the origin, (13.127) can be used to generate a pair of pseudorandom standard normal observations. Atkinson and Pearce (1976) have demonstrated that this method is faster than the Box–Muller method.

9.3 Acceptance–Rejection Method

The acceptance–rejection method is a fast method (certainly faster than the inverse cumulative distribution function method mentioned earlier) of generating pseudorandom normal observations; see Kinderman and Ramage (1976). In this method the normal density is represented as a mixture of densities over which a variety of acceptance–rejection methods due to Marsaglia (1964), Marsaglia and Bray (1964), and Marsaglia, MacLaren, and Bray (1964) are applied.

Suppose that the pdf $p(x)$ is bounded above by a function $cq(x)$, where $c > 0$ and $q(x)$ is a density function for all x. Generally $q(x)$ is taken to be an easily computable function. The "rejection method" generates pseudorandom numbers from $p(x)$ as follows:

1. Provides a pair of independent uniform $(0, 1)$ random numbers, u_1 and u_2.
2. Provides a random number from $q(x)$ using u_1, say, x_q.
3. Accepts x_q as a pseudorandom observation from $p(x)$ if

$$u_2 \leq \frac{p(x_q)}{cq(x_q)}.$$

4. If x_q is rejected, returns to step (1).

The constant c is to be selected as the smallest number such that $p(x) \leq cq(x)$ for all x. For example, with $p(x)$ being the standard normal density function and $q(x)$ being the Laplace density function (Chapter 24)

$$q(x) = \frac{\alpha}{2} e^{-\alpha |x|},$$

x_q may be generated (by inverse cdf method) as

$$x_q = \begin{cases} \dfrac{1}{\alpha} \log(2u_1) & \text{if } 0 < u_1 \leq \tfrac{1}{2}, \\[2ex] -\dfrac{1}{\alpha} \log\{2(1 - u_1)\} & \text{if } \tfrac{1}{2} < u_1 < 1. \end{cases} \qquad (13.128)$$

The inequality in step 3 becomes in this case

$$\log u_2 \leq c_1 + \alpha |x_q| - \frac{x_q^2}{2}, \qquad (13.129)$$

where $c_1 = \tfrac{1}{2} \log\{2/(\pi c^2 \alpha^2)\}$. If we consider the Laplace distribution with mean 0 and variance 1, then $\alpha^2 = 2$, and in this case we have $c = e/\sqrt{\pi}$ and $c_1 = -1$. Consequently (13.129) reduces to

$$\log u_2 \leq -1 + \sqrt{2}\,|x_q| - \frac{x_q^2}{2} = -\frac{1}{2}\left(|x_q| - \sqrt{2}\right)^2. \qquad (13.130)$$

9.4 Ahrens–Dieter Method

From the Box–Muller transformation in (13.122) or (13.126), observe that the variables

$$C = \frac{X_2}{X_1} \quad \text{and} \quad S = X_1^2 + X_2^2 \qquad (13.131)$$

are independently distributed as Cauchy and exponential with mean 2, respectively. Thence the inverse of this transformation yields

$$X_1 = \sqrt{\frac{S}{1 + C^2}} \quad \text{and} \quad X_2 = C\sqrt{\frac{S}{1 + C^2}} \,. \qquad (13.132)$$

The snag in using this transformation is that the signs of the square roots are undetermined. However, as Ahrens and Dieter (1988) have pointed out, these signs must be positive and negative independently and with equal probability $\frac{1}{2}$. So X_1 may receive a random sign, and sign(X_2) becomes sign(C). Thus the pseudorandom standard normal observations X_1 and X_2 may be generated from (13.132) by using directly the Cauchy generator (see Chapter 16) and the exponential generator (see Chapter 19) with $S = E + E$. For this algorithm Ahrens and Dieter (1988) have pointed out that (for generating a pair of pseudorandom standard normal observations) 1.023195 is the average consumption of uniform (0, 1) random variables (with 1.039921 uniform variables for the exponential algorithm and 1.006469 for the Cauchy algorithm).

Several other simulational algorithms are also available for generating normal observations. Kinderman, Monahan, and Ramage (1975), Marsaglia, Ananthanarayanan, and Paul (1976), Atkinson and Pearce (1976), Sakasegawa (1978), Best (1979), and Deak (1980, 1981) have all presented simple and economical methods of generating normal observations. Gates (1978) presented a modified Butler algorithm for this purpose. Burford and Willis (1978) carried out a comparative study of unit normal variables generated by the Box–Muller algorithm using alternative standard uniform generators. Kronmal, Peterson, and Hutchinson (1979) discussed the generation of normal random variables using the uniform alias-rejection mixture method. Schuster (1983) made some comments on the method of generating normal variates by summing three uniforms. Dagpunar (1988) has presented a method of generating random variates from the tail of a normal distribution. Dagpunar (1987) has also given a nomogram to enable the manual sampling of random variates from a normal distribution.

Generating a normal sample with given sample mean and variance has been considered by Pullin (1979), Marsaglia (1980), and Jones (1985).

10   RELATED DISTRIBUTIONS

In Section 8 we provided references for the distributions of arithmetic mean, median, variance, range, mean deviation, and so on, in "random samples from normal populations." Many chapters of this volume discuss other distributions related to the normal distribution. In the present section we will confine ourselves to discussion of *truncated* normal distributions and *mixtures* of normal distributions, together with some brief references to some other distributions related to the normal.

10.1   Truncated Normal Distributions

A random variable X has a *doubly truncated* normal distribution if its probability density function is

$$\frac{1}{\sqrt{2\pi}\,\sigma}e^{-(x-\xi)^2/2\sigma^2}\left[\frac{1}{\sqrt{2\pi}\,\sigma}\int_A^B e^{-(t-\xi)^2/2\sigma^2}\,dt\right]^{-1}$$

$$= \sigma^{-1}Z\left(\frac{x-\xi}{\sigma}\right)\left[\Phi\left(\frac{B-\xi}{\sigma}\right) - \Phi\left(\frac{A-\xi}{\sigma}\right)\right]^{-1}, \qquad A \leq x \leq B.$$

$$(13.133)$$

The *lower* and *upper truncation points* are A, B, respectively; *the degrees of truncation* are $\Phi((A-\xi)/\sigma)$ (from below) and $1 - \Phi((B-\xi)/\sigma)$ (from above). If A is replaced by $-\infty$, or B by ∞, the distribution is *singly truncated* from *above*, or *below*, respectively. More elaborate forms of truncation (e.g., omission of a number of intervals of X) will not be considered here.

Some typical doubly and singly truncated normal probability density functions are shown in Figure 13.3. These are classified according to the degrees of truncation. It can be seen that when the truncations are large, the distribution bears little resemblance to a normal distribution. It is indeed more like a rectangular or trapezoidal distribution (Chapter 26). The case $A = \xi$, $B = \infty$ produces a *half-normal* distribution. This is actually the distribution of $\xi + \sigma|U|$ where U is a unit normal variable. We will discuss, in detail, only doubly truncated normal distributions. Treatment of singly truncated normal distributions follows on similar lines.

The expected value of X [from (13.133)] is given by

$$E[X] = \xi + \frac{Z((A-\xi)/\sigma) - Z((B-\xi)/\sigma)}{\Phi((B-\xi)/\sigma) - \Phi((A-\xi)/\sigma)}\sigma, \qquad (13.134)$$

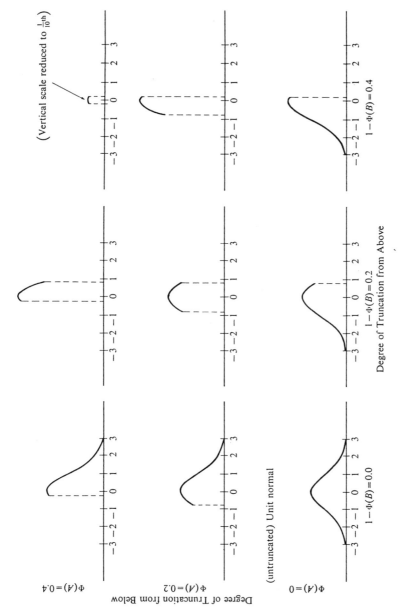

Figure 13.3 Density Functions of Some Truncated Unit Normal Distributions, Points of Truncation, $A < B$

157

and the variance of X by

$\text{Var}(X)$

$$
= \left[1 + \frac{((A-\xi)/\sigma)Z((A-\xi)/\sigma) - ((B-\xi)/\sigma)Z((B-\xi)/\sigma)}{\Phi((B-\xi)/\sigma) - \Phi((A-\xi)/\sigma)} \right.
$$

$$
\left. - \left\{ \frac{Z((A-\xi)/\sigma) - Z((B-\xi)/\sigma)}{\Phi((B-\xi)/\sigma) - \Phi((A-\xi)/\sigma)} \right\}^2 \right] \sigma^2. \quad (13.135)
$$

Note that if $A - \xi = -(B - \xi) = -\delta\sigma$, then

$$
\begin{cases}
E[X] = \xi, \\
\text{Var}(X) = \left[1 - \dfrac{2\delta Z(\delta)}{2\Phi(\delta) - 1} \right] \sigma^2.
\end{cases} \quad (13.136)
$$

The mean deviation of X is

$$
2 \left[Z(J) - Z\left(\frac{B-\xi}{\sigma} \right) - \left\{ \Phi\left(\frac{B-\xi}{\sigma} \right) - \Phi(J) \right\} J \right]
$$

$$
\times \left[\Phi\left(\frac{B-\xi}{\sigma} \right) - \Phi\left(\frac{A-\xi}{\sigma} \right) \right]^{-1}, \quad (13.137)
$$

where

$$
J = \frac{Z((A-\xi)/\sigma) - Z((B-\xi)/\sigma)}{\Phi((B-\xi)/\sigma) - \Phi((A-\xi)/\sigma)}.
$$

Some values of $E[X]$, $\sqrt{\text{Var}(X)}$ and the ratio of the mean deviation to the standard deviation are given in Table 13.10.

The moments of the truncated distribution can also be expressed in terms of the Hh functions [Fisher (1930)] defined by

$$
Hh_n(y) = (n!)^{-1} \int_y^\infty (t-y)^n e^{-t^2/2} \, dt. \quad (13.138)
$$

(Note that $Hh_n(y) = \int_y^\infty Hh_{n-1}(t)\, dt$; $Hh_0(y) = \sqrt{2\pi}(1 - \Phi(y))$; $Hh_{-1}(y) = e^{-y^2/2}$.)

Sugiura and Gomi (1985) have given Pearson diagrams (of the coefficients of skewness and kurtosis) for the truncated normal distributions. Nakamura (1980) has discussed the moments of positively truncated normal distribution. Hall (1979) has also derived inverse moments for a class of truncated normal distributions.

Table 13.10 Expected Value, Standard Deviation, and (Mean Deviation) / (Standard Deviation) for Truncated Normal Distributions

| Degrees of Truncation | | | | |
$q_1 = \Phi\left(\dfrac{A - \xi}{\sigma}\right)$ (Below)	$q_2 = 1 - \Phi\left(\dfrac{B - \xi}{\sigma}\right)$ (Above)	$[-\xi + $ (Expected Value)] $\times \sigma^{-1}$	(Standard Deviation) $\times \sigma^{-1}$	(Mean Deviation)/ (Standard Deviation)
0	0.1	-0.195	0.844	0.814
0	0.2	-0.350	0.764	0.812
0	0.3	-0.497	0.702	0.809
0	0.4	-0.644	0.650	0.805
0	0.5	-0.798	0.603	0.801
0	0.6	-0.966	0.558	0.795
0	0.7	-1.159	0.514	0.791
0	0.8	-1.400	0.468	0.785
0	0.9	-1.755	0.411	0.777
0.1	0.1	0.000	0.662	0.844
0.1	0.2	-0.149	0.566	0.850
0.1	0.3	-0.287	0.489	0.853
0.1	0.4	-0.422	0.420	0.854
0.1	0.5	-0.559	0.355	0.856
0.2	0.2	0.000	0.463	0.856
0.2	0.3	-0.135	0.382	0.859
0.2	0.4	-0.266	0.309	0.861
0.2	0.5	-0.397	0.239	0.863
0.3	0.3	0.000	0.297	0.862
0.3	0.4	-0.129	0.222	0.864
0.3	0.5	-0.256	0.151	0.865
0.4	0.4	0.000	0.146	0.865
0.4	0.5	-0.126	0.073	0.866

The values of A and B are usually known, at least fairly closely. If A and B are known, the maximum likelihood estimation of ξ and σ is equivalent to estimation by equating first and second sample and population moments. The equations satisfied by the estimators $\hat{\xi}$ and $\hat{\sigma}$ are similar in form to those for censored samples, with $X'_{r_1 + 1}$, $X'_{n - r_2}$ replaced by A, B, respectively, and with the multipliers $r_1 n^{-1}, r_2 n^{-1}$ replaced by the degrees of truncation $\Phi((A - \xi)/\sigma)$, $1 - \Phi((B - \xi)/\sigma)$, respectively.

Harter and Moore (1966) have given a table of asymptotic variances of, and correlation between, $\hat{\xi}$ and $\hat{\sigma}$ (for large n). Table 13.11 is based on their figures and includes the asymptotic variance of the maximum likelihood estimator of ξ when σ is known and of σ when ξ is known. If $\Phi((A - \xi)/\sigma)$ is greater than $[1 - \Phi((B - \xi)/\sigma)]$, the variable X can be replaced by $-X$,

Table 13.11 Asymptotic Values of $n \times$ (Variance) $/ \sigma^2$ for Maximum Likelihood Estimators

Proportions Censored		ξ and σ Unknown			σ Known	ξ Known
		Variance of		Correlation between	Variance of	Variance of
$\Phi\left(\dfrac{A-\xi}{\sigma}\right)$	$1-\Phi\left(\dfrac{B-\xi}{\sigma}\right)$	$\hat{\xi}$	$\hat{\sigma}$	$\hat{\xi}, \hat{\sigma}$	$\hat{\xi}$	$\hat{\sigma}$
0.0	0.0	1.00	0.50	0.000	1.00	0.50
0.0	0.1	1.02	0.59	0.053	1.02	0.58
0.0	0.2	1.06	0.69	0.125	1.05	0.68
0.0	0.3	1.14	0.82	0.214	1.09	0.78
0.0	0.4	1.27	0.99	0.320	1.14	0.89
0.0	0.5	1.52	1.24	0.441	1.22	1.00
0.0	0.6	1.99	1.62	0.572	1.34	1.09
0.0	0.7	3.02	2.25	0.700	1.53	1.14
0.0	0.8	5.78	3.54	0.822	1.87	1.15
0.0	0.9	17.79	7.51	0.918	2.78	1.18
0.1	0.1	1.04	0.70	0.000	1.04	0.70
0.1	0.2	1.07	0.85	0.075	1.06	0.84
0.1	0.3	1.14	1.04	0.172	1.11	1.01
0.1	0.4	1.27	1.32	0.293	1.17	1.20
0.1	0.5	1.54	1.74	0.417	1.25	1.41
0.1	0.6	2.13	2.46	0.597	1.37	1.58
0.1	0.7	3.67	3.95	0.757	1.57	1.69
0.1	0.8	9.77	8.66	0.896	1.94	1.71
0.2	0.2	1.10	1.05	0.000	1.10	1.05
0.2	0.3	1.15	1.34	0.103	1.14	1.32
0.2	0.4	1.28	1.78	0.239	1.20	1.68
0.2	0.5	1.56	2.54	0.413	1.29	2.10
0.2	0.6	2.30	4.09	0.618	1.42	2.53
0.2	0.7	5.18	8.93	0.827	1.64	2.82
0.3	0.3	1.19	1.79	0.000	1.19	1.80
0.3	0.4	1.29	2.57	0.150	1.26	2.51
0.3	0.5	1.57	4.16	0.368	1.35	3.59
0.3	0.6	2.69	9.04	0.665	1.50	5.04
0.4	0.4	1.33	4.17	0.000	1.33	4.17
0.4	0.5	1.57	9.09	0.286	1.44	8.35

ξ by $-\xi$, A by $-B$ and B by $-A$, then $\Phi((-B+\xi)/\sigma)$ will be less than $[1 - \Phi((-A+\xi)/\sigma)]$, and the tables may be applied.

The tables by Harter and Moore (1966) give 6 decimal places. Earlier tables by Gupta (1952) give the first three columns to 5 decimal places for singly truncated distributions ($A = -\infty$ and so $\Phi((A-\xi)/\sigma) = 0$), for $\Phi((B-\xi)/\sigma) = 0.05(0.05)0.95(0.01)0.99$. The values given in these tables also give the asymptotic variances for estimators of ξ and σ from singly and doubly censored sets of independent identically distributed normal variables.

They are, however, only *asymptotic* values, and some care is needed in using them when the sample size n is not large (e.g., less than 50).

Sampling experiments [Harter and Moore (1966)] with $n = 10$ and $n = 20$ have indicated that there is a negative bias in both $\hat{\xi}$ and $\hat{\sigma}$ that increases with the degree of truncation. For a moderate degree of truncation (or censoring) knowledge of either parameter does not result in much reduction of variance of the estimator of other parameter. Occasionally, the points of truncation (A and B) are not known and must be estimated from the data, as well as ξ and σ. With sufficiently large samples it will probably be good enough to take A equal to a value slightly less than the least observed value, and B slightly larger than the greatest observed value, but sometimes more elaborate methods may be needed.

The maximum likelihood estimation of the four parameters A, B, ξ, and σ has been described by Cohen (1950a). An estimation using the first four moments has been described by Shah and Jaiswal (1966). There are marked similarities between the equations obtained by the two methods. Cohen gives tables of auxiliary quantities useful in solving the maximum likelihood equations. These tables have been extended by Lifsey (1965).

Cohen (1950b) also has described how the parameters ξ and σ can be estimated when there is a single truncation (i.e., $A = -\infty$ or $B = \infty$), using the first three sample moments. As the point of truncation is supposed known, it is not really necessary to use the third moment. However, by introducing the third moment, simple explicit formulas are obtained. Supposing that $B = \infty$ (i.e., left truncation only), the estimators of ξ and σ^2 are

$$\begin{cases} A - \left(2m_1'^2 - m_2'\right)^{-1}\left(2m_1'm_2' - m_3'\right), \\ \left(2m_1'^2 - m_2'\right)^{-1}\left(m_1'm_3' - m_2'^2\right), \end{cases} \tag{13.139}$$

respectively, where m_r' is the rth sample moment about the point of truncation A. Cohen found that the asymptotic efficiency of these estimators (relative to maximum likelihood estimators) is never less than 77% for ξ, 72% for σ^2. For very small or very large degrees of truncation the efficiencies are greater.

The use of the third moment may be expected to introduce unnecessary inaccuracies. Pearson and Lee (1908) gave formulas from which estimators can be obtained, equating only first and second sample and population moments; the estimators are also maximum likelihood estimators. The equations for the estimators $\hat{\xi}, \hat{\sigma}$ are (again assuming $B = \infty$)

$$\hat{\sigma} = \left(\frac{Z(\hat{\delta})}{1 - \Phi(\hat{\delta})} - \hat{\delta}\right)^{-1} \bar{X}, \tag{13.140}$$

$$\left(\frac{\hat{\sigma}}{\bar{X}}\right)\left(\frac{\hat{\sigma}}{\bar{X}} - \hat{\delta}\right) = \frac{1}{n\bar{X}^2}\sum_{j=1}^{n} X_j^2, \tag{13.141}$$

where $\hat{\delta} = (A - \hat{\xi})/\hat{\sigma}$ (= estimate of lower truncation point as a standardized deviate) and $\bar{X} = n^{-1}\sum_{j=1}^{n}X_j$. Using (13.140), (13.141) can be solved for $\hat{\delta}$, and then $\hat{\sigma}$ is calculated from (13.140).

Cohen and Woodward (1953) give tables of the functions

$$Y(\delta) = \left(\frac{Z(\delta)}{1 - \Phi(\delta)} - \delta\right)^{-1},$$

$$\tfrac{1}{2}Y(\delta)[Y(\delta) - \delta]$$

[to 8 significant figures for $\delta = -4.0(0.1)3.0$ to aid in the solution of (13.140) and (13.141)]. Since (13.141) can be written in the form

$$\frac{1}{2n\bar{X}^2}\sum_{j=1}^{n}X_j^2 = \frac{1}{2}Y(\hat{\delta})[Y(\hat{\delta}) - \hat{\delta}],$$

inverse interpolation suffices to determine $\hat{\delta}$, and then the table of $Y(\delta)$ is used to calculate $\hat{\sigma}$.

Crain (1979) has discussed further the estimation of parameters of a truncated normal distribution. Chiu and Leung (1981) have presented a graphical method of estimating the parameters, and Mittal and Dahiya (1987) and Hegde and Dahiya (1989) have provided further discussions on the estimation of parameters from a doubly truncated normal distribution. Detailed discussions on the estimation of parameters from truncated normal distributions may be found in the volumes by Harter (1970), Schneider (1986), Cohen and Whitten (1988), Cohen (1991), and Balakrishnan and Cohen (1991).

DePriest (1983) has used the singly truncated normal distribution to analyze satellite data. Levedahl (1989) has found an application for the truncated normal distribution in evaluating the effect of commodity donation programs. Ord and Bagchi (1983) have suggested the truncated normal–gamma mixture as a distribution for lead time demand.

Distributions of functions of truncated normal variables cannot usually be expressed in mathematically elegant forms. In Francis (1946) there are some tables (calculated by Campbell) of the distribution of the sums of n independent identically distributed truncated normal variables. The tables are not extensive, but they do give a useful idea of the effects of truncation on the distribution of arithmetic means, even for nonnormal distributions. Values of the cumulative distribution, to 4 decimal places, are given for values of the argument at intervals of 0.1, and for $n = 2$ and 4.

As an example of special problems that can sometimes arise, we mention the derivation of the distribution of the sum of two independent random variables, one normal and the other truncated normal. This has been discussed in Weinstein (1964).

10.2 Mixtures

Compound normal distributions are formed by ascribing a distribution to one, or both, of the parameters ξ, σ in (13.1). There are two distinct kinds of distributions: those obtained by treating ξ and/or σ as continuous random variables are of methodological and theoretical interest, while when (ξ, σ) takes only a finite (usually small) number of possible values, the fitting of corresponding distributions is usually regarded as the "dissection" of a heterogeneous population into more homogeneous "parts."

Using the notation introduced in Chapter 8, we remark that the distribution*

$$\text{Normal}(\xi, \sigma) \bigwedge_{\xi} \text{Normal}(\mu, \sigma')$$

is also a normal distribution, with expected value μ and standard deviation $\sqrt{\sigma^2 + \sigma'^2}$. This can be demonstrated by direct integration, or simply by regarding the compound distribution as that of $(\mu + U'\sigma') + U\sigma$, where U, U' are independent unit normal variables.

It can also be shown that

$$\text{Normal}(\xi, \sigma) \bigwedge_{\sigma^{-2}} \text{Gamma}(c\chi_\nu^2)$$

is equivalent to a Pearson Type VII distribution (Chapter 12, Section 4.1). In fact

$$(2c)^{-\nu/2}\left[\Gamma\left(\frac{\nu}{2}\right)\right]^{-1}\int_0^\infty \left[\sqrt{2\pi}\,\sigma\right]^{-1}(\sigma^{-2})^{(\nu/2)-1}$$

$$\times \exp\left[-(2c\sigma^2)^{-1} - (2\sigma^2)^{-1}(x-\xi)^2\right]d\sigma^{-2} \quad (13.142)$$

$$= \frac{c^{1/2}}{B(1/2, \nu/2)}\left[1 + \frac{(x-\xi)^2}{c}\right]^{-(\nu+1)/2}.$$

(It may be noted that the distribution ascribed to σ^{-2} is of a kind sometimes called *fiducial*, which is obtained by formal inversion of the statement "V is distributed as $\chi_\nu^2\sigma^2$" to become "σ^{-2} is distributed as $V^{-1}\chi_\nu^2$.")

Teichroew (1957) has studied the distribution

$$\text{Normal}(0, \sigma^2) \bigwedge_{\sigma^2} \text{Gamma}(c\chi_\nu^2).$$

*In the notation introduced in Chapter 8, ξ is a random variable.

The distribution has a complicated form, although its characteristic function is simply $(1 + ct^2)^{-\nu/2}$.

The distribution

$$\text{Normal}(\xi, \sigma) \bigwedge_{\xi} \text{Rectangular}$$

has been studied by Bhattacharjee et al. (1963). Clow, Hansen, and McNolty (1974) have considered this distribution as the distribution for the pulse-plus-stationary Gaussian noise (under a constant amplitude).

Coming now to mixtures of a finite number (k) of normal *components*, a general form for the probability density function is

$$\sum_{t=1}^{k} \omega_t \left(\sqrt{2\pi}\, \sigma_t\right)^{-1} \exp\left[-\frac{1}{2}\left\{\frac{(x - \xi_t)^2}{\sigma_t^2}\right\}\right]. \tag{13.143}$$

The quantities $\omega_1, \omega_2, \ldots, \omega_k$ $(0 < \omega_t; \Sigma_{t=1}^{k}\omega_t = 1)$ are called the *weights* of the component normal distributions. We will consider in detail only the case $k = 2$. With increasing number of components the general case rapidly becomes extremely complicated, though simplifications (e.g., supposing all σ_t's to be equal) can sometimes be used to render the analysis more manageable.

With $k = 2$, and ξ_1 and ξ_2 sufficiently different, it is possible for (13.143) to represent bimodal distributions [Helguero (1904); Prasad (1955); Teichroew (1957)]. A systematic study of conditions under which this is so has been made by Eisenberger (1964) [see also Wessels (1964)]. He summarizes his results as follows:

1. If

$$(\xi_1 - \xi_2)^2 < \frac{27\sigma_1^2\sigma_2^2}{4(\sigma_1^2 + \sigma_2^2)},$$

the distribution cannot be bimodal (in particular, if $\xi_1 = \xi_2$).

2. If

$$(\xi_1 - \xi_2)^2 > \frac{8\sigma_1^2\sigma_2^2}{\sigma_1^2 + \sigma_2^2},$$

there are values of ω_1 and ω_2 $(= 1 - \omega_1)$ for which the distribution is bimodal.

3. For *any* set of values ξ_1, ξ_2, σ_1, and σ_2 there are values of ω_1 and ω_2 for which the distribution is unimodal. (This is fairly evident, on considering that if $\omega_1 = 0$ or $\omega_1 = 1$, a normal distribution, which is unimodal, is obtained.)

Tables of moments of (13.143) with $k = 2$ [(standard deviation) σ^{-1}, α_3 and $(\alpha_4 - 3)$ to 3 decimal places for $\omega_2/\omega_1 = 0.1$, 0.9, and 1.0; $\sigma_2/\sigma_1 = 1.0(0.5)3.0$ and $(\xi_2 - \xi_1)/\sigma_1 = 0.0(0.5)3.0$] have been given by Linders (1930). If ω_1 is nearly 1, and so ω_t (for $t > 1$) is small, the mixture distribution (13.143) is sometimes called a *contaminated* normal distribution [Tukey (1949)]. It has been used as a model to assess tests for rejection of outlying observations obtained in samples from a supposedly normal distribution, and also in robustness studies for various inference procedures.

In the general case (13.143), the rth moment of X about zero is

$$\mu'_r(X) = \sum_{t=1}^{k} E[X^r|\xi_t, \sigma_t] = \sum_{t=1}^{k} E[(\xi_t + U\sigma_t)^r]$$

$$= \sum_{j=0}^{[r/2]} \binom{r}{2j} E[U^{2j}] \sum_{t=1}^{k} \xi_t^{r-2j}\sigma_t^{2j}, \tag{13.144}$$

where U is a unit normal variable (remember that $E[U^j] = 0$ if j is odd).

Suppose that k is equal to 2, and that we want to estimate the 5 parameters $\omega_1 (= 1 - \omega_2)$, ξ_1, ξ_2, σ_1, and σ_2 by the method of moments. Five moments will be needed. From the equations

$$\mu'_1 = \omega_1\xi_1 + (1 - \omega_1)\xi_2,$$

$$\mu'_2 = \omega_1(\xi_1^2 + \sigma_1^2) + (1 - \omega_1)(\xi_2^2 + \sigma_2^2),$$

$$\mu'_3 = \sum_{t=1}^{2} \omega_t(\xi_t^3 + 3\xi_t\sigma_t^2),$$

$$\mu'_4 = \sum_{t=1}^{2} \omega_t(\xi_t^4 + 6\xi_t^2\sigma_t^2 + 3\sigma_t^4),$$

$$\mu'_5 = \sum_{t=1}^{2} \omega_t(\xi_t^5 + 10\xi_t^3\sigma_t^2 + 15\xi_t\sigma_t^4),$$

$$\tag{13.145}$$

we try to find values for ω_1, ξ_1, ξ_2, σ_1, and σ_2.

This problem was considered by Pearson in 1894. Subsequently a number of improvements have been effected [e.g., Charlier and Wicksell (1924); Helguero (1905)]. Papers by Molenaar (1968) and Cohen (1967) give a useful account of this work. The following summary is based on information in these papers.

Putting $\theta_j = \xi_j - \mu'_1$ $(j = 1, 2)$ the *central* moments μ_2, μ_3, μ_4, and μ_5 are obtained from (13.145) by replacing ξ_t by θ_t $(t = 1, 2)$. From the resulting equations can be derived an equation of ninth degree for $\phi = \theta_1\theta_2$:

$$\sum_{j=0}^{9} a_j\phi^j = 0, \tag{13.146}$$

where

$$a_0 = -24\mu_3^6,$$

$$a_1 = -96\mu_3^4\kappa_4,$$

$$a_2 = -63\mu_3^2\kappa_4^2 - 72\mu_3^3\kappa_5,$$

$$a_3 = 288\mu_3^4 - 108\mu_3\kappa_4\kappa_5 + 27\kappa_4^3,$$

$$a_4 = 444\mu_3^2\kappa_4 - 18\kappa_5^2,$$

$$a_5 = 90\kappa_4^2 + 72\mu_3\kappa_5,$$

$$a_6 = 36\mu_3^2,$$

$$a_7 = 84\kappa_4,$$

$$a_8 = 0,$$

$$a_9 = 24,$$

with

$$\kappa_4 = \mu_4 - 3\mu_2^2,$$

$$\kappa_5 = \mu_5 - 10\mu_2\mu_3.$$

In application, values of μ_r are replaced by sample values of these moments.

Since equation (13.146) might have as many as nine roots, there may be difficulty in choosing the "right" root. Since μ_1' lies between ξ_1 and ξ_2, it follows (unless $\xi_1 = \xi_2$) that θ_1 and θ_2 are of opposite signs, and so $\phi = \theta_1\theta_2 < 0$. Hence only negative roots of (13.146) need be considered. For ease of computation the following method appears to be convenient.

If the value of $\phi' = \theta_1 + \theta_2$ is known then ϕ is a negative root of the cubic equation

$$6\phi^3 - 2\phi'^2\phi^2 + (3\kappa_4 - 4\phi'\mu_3)\phi + \mu_3^2 = 0, \tag{13.147}$$

and this equation has only one such root. Using ϕ', and the value of ϕ obtained from (13.147), values of θ_1 and θ_2 can be determined, and from these ξ_1, ξ_2 are estimated as

$$\xi_j = \bar{X} + \theta_j, \qquad j = 1, 2 \tag{13.148}$$

($\bar{X}$ being the sample mean), and ω_1 as

$$\omega_1 = \frac{\theta_2}{\theta_2 - \theta_1}. \tag{13.149}$$

Finally, we have

$$\sigma_j^2 = \frac{1}{3}\theta_j\left(2\phi' - \frac{\mu_3}{\phi}\right) + \mu_3 - \theta_j^2, \qquad j = 1, 2. \qquad (13.150)$$

Using the parameter values so obtained, a value for μ_5 can be calculated such as $\mu_5(\phi')$. By inverse interpolation from a series of such values of $\mu_5(\phi')$, a value for ϕ' (and hence a set of values for all five parameters) can be estimated. [This does not exclude the possibility that more than one value of ϕ' may make $\mu_5(\phi')$ equal to the sample value, so it may still be necessary to distinguish among such values. Pearson (1894) suggested that the value giving closest agreement between sample and population sixth moments be chosen. It is easy, however, to think of other criteria, e.g., choosing the value giving the least value of χ^2 or some other goodness-of-fit criterion.]

In view of the likely inaccuracy in estimating the sixth central moment, it seems preferable to use the first and third absolute central moments, ν_1 (mean deviation) and ν_3, together with the variance μ_2. From the equations

$$\nu_1 = \left[\omega_1\sigma_1 + (1 - \omega_1)\sigma_2\right]\sqrt{\frac{2}{\pi}},$$

$$\mu_2 = \omega_1\sigma_1^2 + (1 - \omega_1)\sigma_2^2, \qquad (13.151)$$

$$\nu_3 = \left[\omega_1\sigma_1^3 + (1 - \omega_1)\sigma_2^3\right]\left(2\sqrt{\frac{2}{\pi}}\right),$$

we obtain σ_1, σ_2 as roots of the equation

$$\left(\mu_2 - \frac{1}{2}\pi\nu_1^2\right)z^2 - \left(\frac{1}{2}\sqrt{\frac{\pi}{2}}\,\nu_3 - \sqrt{\frac{\pi}{2}}\,\nu_1\mu_2\right)z + \frac{1}{4}\pi\nu_1\nu_3 - \mu_2^2 = 0.$$

$$(13.152)$$

Sometimes simpler procedures can give adequate results. If the difference between the means $|\xi_1 - \xi_2|$ is large enough, then the left-hand and right-hand tails of the distribution come almost entirely from single (and different) components of the mixture. Figure 13.4 typifies such a situation, with $\xi_1 < \xi_2$ (and $\sigma_1 > \sigma_2$). In such cases a truncated normal distribution may be fitted to each tail separately (as described, for single truncation, in Section 1). This gives estimates of ξ_1, ξ_2, σ_1, and σ_2. Finally, ω_1 is determined from the equation

$$\omega_1\xi_1 + (1 - \omega_1)\xi_2 = \bar{X}.$$

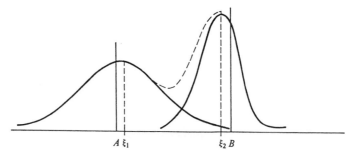

Figure 13.4 Data in the Region Below A Used for Estimation of the Component with Expected Value ξ_1, and Data above B Used for Estimation of the Component with Expected Value ξ_2

A major difficulty in this method is the choice of points of truncation. Some check is possible by moving the points inward, toward the bulk of the distribution, so long as the estimators of ξ_j and σ_j remain "reasonably" consistent. A number of graphical methods of estimation have been developed [e.g., Harding (1949), Molenaar (1965), Taylor (1965), and Wiechselberger (1961)]. Rivest (1981) has discussed the sum of two contaminated normals.

If $\xi_1 = \xi_2$, then the distribution is symmetrical for any value of ω_1. A symmetrical distribution is also obtained (even with $\xi_1 \neq \xi_2$) if $\sigma_1 = \sigma_2$ and $\omega_1 = \frac{1}{2}$. This can be distinguished from the symmetrical distribution obtained with $\xi_1 = \xi_2$ (and $\sigma_1 \neq \sigma_2$), since in the former case $\kappa_4 < 0$ while in the latter case $\kappa_4 > 0$. Moment estimators for the case $\xi_1 = \xi_2$ have been discussed in detail by Agard (1961). A maximum likelihood estimation procedure, for the case $\sigma_1 = \sigma_2$, was described by Molenaar (1965), who also constructed a computer program for this procedure.

Since 1970 extensive work has gone on with regard to the inference, modeling, and application aspects of the mixture-normal distributions. Books by Titterington, Smith, and Makov (1985) and McLachlan and Basford (1987) provide elaborate discussions on all these developments.

10.3 Other Related Distributions

If X_1 and X_2 are independent unit normal variables, and $F = (X_1/X_2)^2$, then F is distributed as $F_{1,1}$ (see Chapter 27). Equivalently, $t = X_1/X_2$ is distributed as t_1 (see Chapter 28), and so has a Cauchy distribution (see Chapter 16).

Among statistics used in testing normality are the following:

1. The sample skewness,

$$\sqrt{b_1} = \left[n^{-1} \sum_{j=1}^{n} \left(X_j - \bar{X} \right)^2 \right]^{-3/2}.$$

2. The sample kurtosis,

$$b_2 = \left[n^{-1} \sum_{j=1}^{n} \left(X_j - \overline{X} \right)^4 \right] \left[n^{-1} \sum_{j=1}^{n} \left(X_j - \overline{X} \right)^2 \right]^{-2}.$$

[Considerable attention has been devoted to obtaining formulas for the higher moments (up to eight) of these two statistics. See Fisher (1930), Geary (1933, 1947), Geary and Worlledge (1947), Hsu and Lawley (1940), and Wilkinson (1961). Tables of approximate percentiles are available in Pearson (1965).]

3. The ratio (sample mean deviation)/(sample standard deviation) [Geary (1935, 1936)].

4. The ratio (best linear unbiased estimator of σ) $\times$ (sample variance)$^{-1}$ [Shapiro and Wilk (1965)].

5. Ratios of symmetrical differences between order statistics,

$$\frac{X'_{n-r_1+1} - X'_{r_1}}{X'_{n-r_2+1} - X'_{r_2}}$$

[David and Johnson (1954)].

6. The ratio (sample range)/(sample standard deviation) [Pearson, Hartley, and David (1954); Pearson and Stephens (1964)].

As mentioned earlier in Section 8, a complete account of the goodness-of-fit tests for the normal distribution (along with detailed tables of percentage points, approximations for large samples, comparative power studies, etc.) can be had from the books by Shapiro (1986) and D'Agostino and Stephens (1986). In the construction of tests for outlying observations, distributions of the following statistics have been studied:

1. $(X'_n - \overline{X})/\sigma$, $(\overline{X} - X'_1)/\sigma$ and the same statistics with σ replaced by the sample standard deviation [David (1956); Nair (1947); Sarhan (1954); Pearson and Chandra Sekar (1936)].

2. Ratios of order statistics of form

$$\frac{X'_r - X'_1}{X'_{r+1} - X'_1}; \quad \frac{X'_n - X'_{n-r-1}}{X'_n - X'_{n-r}} \qquad \text{for } r = 2, 3, 4, \ldots$$

[Dixon (1950)].

Since 1970 the literature on outliers and related issues has expanded significantly. This is clearly evident from the third edition of Barnett and Lewis (1994). Many new tests have been proposed, their performance have been

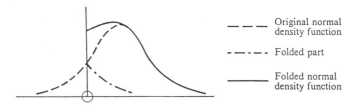

Figure 13.5   Folded Normal Distribution

evaluated, comparative studies have been carried out, and several tables have been constructed to facilitate the users. All these developments have been discussed in great detail in Barnett and Lewis (1994).

If X has a normal distribution as in (13.2), then $|X|$ is said to have a *folded normal distribution*. As the name suggests, the distribution can be regarded as being formed by folding the part corresponding to negative X about the vertical axis (see Figure 13.5) and then adding it to the positive part.

The folded normal distribution is identical with the distribution of

$$\sigma \times \left(\begin{array}{l} \text{Noncentral } \chi \text{ with} \\ \text{one degree of freedom and} \\ \text{noncentrality parameter } (\xi/\sigma)^2 \end{array} \right),$$

and, as such, will be described in Chapter 29. If the folding is about the mean, then $\xi = 0$ and a *central* χ is obtained. This is also the *half-normal* distribution referred to earlier in this chapter (Section 10.1). Leone, Nelson and Nottingham (1961) and Nelson (1980) have discussed various properties and applications of the folded normal distribution. Sinha (1983) has derived Bayesian estimates for the parameters of this distribution.

In Chapter 12 we discussed *Johnson's system of distributions* (as three different transformations of normal random variables). Shenton and Bowman (1975) have used the S_U-distributions to examine the sample skewness and kurtosis statistics. Hill (1976) has given an algorithm for normal-Johnson and Johnson-normal transformations. Olsson (1979) has explained the fitting of Johnson's S_B and S_U systems of curves using the maximum likelihood method. Hill and Wheeler (1981) have commented on the algorithm of Hill (1976) and also on the fitting of Johnson curves by the method of moments; also see Dodgson and Hill (1983). The *logistic-normal distribution*, in particular, has received considerable attention from many authors. Aitchison and Shen (1980) have discussed some properties and uses of this distribution, while Lenk (1988) used this distribution for Bayesian and nonparametric predictive densities. Titterington (1989) has provided a review of the logistic-normal distribution. Crouch and Spiegelman (1990) recently considered integrals of the form $\int_{-\infty}^{\infty} f(t) e^{-t^2} dt$ and applied them to the study of logistic-normal models.

Azzalini's (1985) *skewed-normal distributions* was described in Chapter 12. This family of distributions, including the normal (when $\lambda = 0$) and the half-normal (when $\lambda = \infty$), has been discussed further by Azzalini (1986). Henze (1986) has provided a probabilistic representation of this family of distributions. Kotz and Johnson (1988) have provided a brief review of this distribution, while Liseo (1990) has recently discussed inferential aspects of this distribution from a Bayesian viewpoint.

A robust alternative to the normal distribution proposed by McLeish (1982) has already been pointed out in Chapter 12. Tardiff (1980) and Canal (1981) have discussed a generalized normal distribution, while Vianelli (1983) has defined families of normal and lognormal distributions of order r. Goto and Hatanaka (1985) have studied power-normal transformations. The *Sargan distribution* was introduced in Chapter 12. Kafael and Schmidt (1985) have examined the adequacy of the Sargan distribution as an approximation to the normal distribution.

When X has a $N(\xi, \sigma^2)$ distribution, then it is easily seen that $Y = 1/X$ has the density function

$$p_Y(y|\xi, \sigma) = \frac{1}{\sqrt{2\pi}\,\sigma y^2} e^{-\{(1/y)-\xi\}^2/(2\sigma^2)}. \tag{13.153}$$

Robert (1991), in generalizing the density in (13.153), has proposed the family of *generalized inverse normal distributions* with density

$$p_Y(y|\alpha, \xi, \sigma) = \frac{K(\alpha, \xi, \sigma)}{|y|^\alpha} e^{-\{(1/y)-\xi\}^2/(2\sigma^2)}, \qquad \alpha > 1, \sigma > 0, \tag{13.154}$$

where $K(\alpha, \xi, \sigma)$ is the normalizing constant. The distribution is always bimodal, with modes at

$$y_1 = -\frac{\xi + \sqrt{\xi^2 + 4\alpha\sigma^2}}{2\alpha\sigma^2} \quad \text{and} \quad y_2 = \frac{\sqrt{\xi^2 + 4\alpha\sigma^2} - \xi}{2\alpha\sigma^2}.$$

The constant $K(\alpha, \xi, \sigma)$ can be expressed explicitly in terms of confluent hypergeometric functions. The kth moment exists only if $\alpha > k + 1$. For example, when $\alpha > 2$, the mean of the distribution is given by

$$\frac{\xi}{\sigma^2} \frac{{}_1F_1((\alpha - 1)/2; 3/2; \xi^2/2\sigma^2)}{{}_1F_1((\alpha - 1)/2; 1/2; \xi^2/2\sigma^2)}.$$

Kanji (1985) and Jones and McLachlan (1990) have considered *Laplace-normal mixture distribution* with density function

$$p(x; \xi, \sigma, \alpha, \beta) = \frac{\alpha}{2\beta} e^{-|(x-\xi)/\beta|} + \frac{1 - \alpha}{\sqrt{2\pi}\,\sigma} e^{-(x-\xi)^2/(2\sigma^2)} \tag{13.155}$$

and applied the distribution to fit wind shear data.

A random variable X is said to have a *circular normal distribution* if its probability density function is [Kotz and Johnson (1982)]

$$p_X(x) = \frac{1}{2\pi I_0(k)} e^{k\cos(x-\theta)}, \qquad 0 \le x < 2\pi, 0 < \theta < 2\pi, \quad (13.156)$$

where $I_0(k) = \sum_{j=0}^{\infty}(k/2)^{2j}/(j!)^2$ is the modified Bessel function of first kind and order 0. The parameters of the distribution are θ and k; as k increases, the distribution becomes more and more concentrated around the pole $x = \theta$. Sometimes a bivariate normal distribution with equal marginal standard deviations is also referred as a circular normal distribution; see Dyer (1974). In fact Dyer considered a truncated form of that distribution, discussed an estimation method for the parameters, and illustrated the method with ballistic applications.

A random variable X is said to have a *lineo-normal distribution* if its cdf is given by

$$F_X(x) = \sqrt{\frac{2}{\pi\sigma^2}} \int_{-\infty}^{x} \int_0^1 t^{1/2} e^{-y^2/(2t\sigma^2)} \, dt \, dy \qquad (13.157)$$

[Kotz and Johnson (1985a)]. The distribution is symmetric about 0 and has variance $2\sigma^2/3$. This distribution is a member of the *modified normal distributions* constructed by Romanowski. The cdf of this family is [Kotz and Johnson (1985c)]

$$F_X(x) = \frac{a+1}{\sqrt{2\pi}\sigma} \int_{-\infty}^{x} \int_0^1 t^{a-1/2} e^{-y^2/(2t\sigma^2)} \, dt \, dy. \qquad (13.158)$$

Formula (13.158) is in fact a compound (or mixture) distribution obtained by ascribing the power function distribution with density $(a+1)t^a$, $0 \le t \le 1$ and $a \ge -1$, to the distribution $N(0, \sigma^2)/\sigma^2$. This distribution is also symmetric about 0, has variance $\sigma^2(a+1)/(a+2)$ and kurtosis $3(a+2)^2/\{(a+1)(a+3)\}$. The lineo-normal distribution in (13.157) is a special case of (13.158) with $a = 1$. When $a = 0$, the distribution in (13.158) is referred to as *equi-normal distribution*. When $a = \frac{1}{2}$, (13.158) is called as a *radico-normal distribution*. When $a = 2$, (13.158) is called as a *quadri-normal distribution*.

The *lagged normal density function* is the convolution of a normal density function and a descending exponential function, and is given by

$$q(t) = \int_{-\infty}^{t} p_1(\tau) p_2(t-\tau) \, d\tau, \qquad (13.159)$$

where $p_1(x)$ is the density function of $N(\xi, \sigma^2)$ and $p_2(x)$ is the exponential density $\lambda \exp(-\lambda x)$, $x \ge 0$. The mean and variance of the density (13.159) are simply $\xi + 1/\lambda$ and $\sigma^2 + 1/\lambda^2$, respectively. A *generalized lagged normal density function* may similarly be defined by convoluting the normal density function with two or more exponentials. Realizing that the convolution of two or more exponentials is the generalized Erlangian distribution, the generalized lagged normal density function simply becomes the convolution of a normal density function and a generalized Erlangian density function. Davis and Kutner (1976) have applied this generalized lagged normal density function to fit indicator-dilution curves.

The *alpha distribution* has its cdf as

$$F_X(x) = \Phi\left(\alpha - \frac{\beta}{x}\right)\bigg/\Phi(\alpha), \qquad \alpha, \beta > 0, \qquad (13.160)$$

and pdf as

$$p_X(x) = \frac{\beta}{\sqrt{2\pi}\,\Phi(\alpha)x^2} e^{-(\alpha-\beta/x)^2/2}. \qquad (13.161)$$

The distribution has its mode at $\beta(\sqrt{\alpha^2 + 8} - \alpha)/4$, and it moves to the left (right) as $\alpha(\beta)$ increases. This distribution has been applied to tool wear problems and has also been suggested for use in modeling lifetimes under accelerated test condition. Salvia (1985) has provided a characterization of the distribution, some other inferential results, and further pointed out the reliability application of the distribution. Note the similarity between (13.161) and the inverse normal density function (13.153). The similarity is due to the fact that (13.161) is nothing but the density function of $X = 1/Y$ when Y has a $N(\xi, \sigma^2)$, truncated to the left of 0, distribution (with $\alpha = \xi/\sigma$ and $\beta = 1/\sigma$).

A random variable X is said to have a *two-piece normal distribution* with parameters ξ, σ_1 and σ_2 if it has its pdf,

$$p_X(x) = \begin{cases} Ae^{-(x-\xi)^2/(2\sigma_1^2)}, & x \le \xi, \\ Ae^{-(x-\xi)^2/(2\sigma_2^2)}, & x > \xi, \end{cases} \qquad (13.162)$$

where $A = \sqrt{2}\,(\sigma_1 + \sigma_2)^{-1}/\sqrt{\pi}$. John (1982) investigated the fitting of this distribution by the method of moments as well as the maximum likelihood method, and he discussed a number of properties of the distribution. In addition John has suggested a test of normality against a two-piece normal alternative. The distribution was originally introduced by Gibbons and Mylroie (1973) by joining two half-normal distributions; they then applied it to fit impurity profiles data in ion-implantation research. Kimber (1985) has examined this distribution further, and discussed methods of estimation and testing. Kimber has also considered the truncated two-piece normal distribution and derived maximum likelihood estimates of its three parameters. Kimber and Jeynes (1985) have applied the truncated two-piece normal distribution to an analysis of the depth measurements of arsenic implants in silicon.

BIBLIOGRAPHY

Abernathy, R. W. (1988). Finding normal probabilities with a hand-held calculator, *Mathematics Teacher*, **81**, 651–652.

Agard, J. (1961). Mélange de deux populations normales et étude de quelques fonctions $f(x, y)$ de variables normales x, y, *Revue de Statistique Appliquée*, **9**, No. 4, 53–70.

Ahmad, R. (1978). A characterization of the normal distribution based on the sample mean and the residual vector, *Trabajos de Estadistica*, **29/2**, 77–80.

Ahrens, J. H., and Dieter, U. (1988). Efficient table-free sampling methods for the exponential, Cauchy, and normal distributions, *Communications of the ACM*, **31**, 1330–1337.

Ahsanullah, M. (1989). Characterizations of the normal distribution by properties of linear statistics and chi-squared, *Pakistan Journal of Statistics, Series B*, **5**, 267–276.

Ahsanullah, M. (1990). Some characteristic properties of normal distribution, *Computational Statistics and Data Analysis*, **10**, 117–120.

Ahsanullah, M., and Hamedani, G. G. (1988). Some characterizations of normal distribution, *Bulletin of the Calcutta Statistical Association*, **37**, 95–99.

Aitchison, J., and Shen, S. M. (1980). Logistic-normal distributions: Some properties and uses, *Biometrika*, **67**, 261–272.

Aitkin, M. (1966). The correlation between variate-values and ranks in a doubly truncated normal distribution, *Biometrika*, **53**, 281–282.

Alam, K. (1971). A characterization of normality, *Annals of the Institute of Statistical Mathematics*, **23**, 523–525.

Allasia, G., and Giordano, C. (1982). Bilateral approximation of the normal distribution function, *Statistica*, **42**, 389–396.

Anscombe, F. J., and Glynn, W. J. (1983). Distribution of the kurtosis statistic b_2 for normal samples, *Biometrika*, **70**, 227–234.

Archibald, T. C. (1926). A rare pamphlet of Moivre and some of his discoveries, *Isis*, **8**, 671.

Arnold, B. C., and Isaacson, D. L. (1978). On normal characterizations by the distribution of linear forms, assuming finite variance, *Stochastic Processes and Their Applications*, **7**, 227–230.

Aroian, L. A., Taneja, V. S., and Cornwell, L. W. (1978). Mathematical forms of the distribution of the product of two normal variables, *Communications in Statistics —Theory and Methods*, **7**, 165–172.

Athreya, K. B. (1986). Another conjugate family for the normal distribution, *Statistics & Probability Letters*, **4**, 61–64.

Atkinson, A. C., and Pearce, M. C. (1976). The computer generation of beta, gamma and normal random variables (with discussion), *Journal of the Royal Statistical Society, Series A*, **139**, 431–460.

Azzalini, A. (1985). A class of distributions which includes the normal ones, *Scandinavian Journal of Statistics*, **12**, 171–178.

Azzalini, A. (1986). Further results on a class of distributions which includes the normal ones, *Statistica*, **46**, 199–208.

Badhe, S. K. (1976). New approximation of the normal distribution function, *Communications in Statistics—Simulation and Computation*, **5**, 173–176.

Bailey, B. J. R. (1981). Alternatives to Hastings' approximation to the inverse of the normal cumulative distribution function, *Applied Statistics*, **30**, 275–276.

Baker, G. A. (1930). Random sampling from non-homogeneous populations, *Metron*, **8**, 67–87.

Balakrishnan, N. (1984). Approximating the sum of squares of normal scores, *Applied Statistics*, **33**, 242–245.

Balakrishnan, N. (1990). Best linear unbiased estimates of the mean and standard deviation of normal distribution for complete and censored samples of sizes 21(1)30(5)40, *McMaster University Report*, Hamilton, Canada.

Balakrishnan, N., and Chan, P. S. (1994). On the normal record values and associated inference, *Submitted for publication*.

Balakrishnan, N., and Cohen, A. C. (1991). *Order Statistics and Inference: Estimation Methods*, San Diego: Academic Press.

Barnard, J. (1978). Probability integral of the normal range, *Applied Statistics*, **27**, 197–198.

Barnett, V. (1976). Convenient probability plotting positions for the normal distribution, *Applied Statistics*, **25**, 47–50.

Barnett, V., and Lewis, T. (1994). *Outliers in Statistical Data* (Third edition), New York: Wiley.

Barton, D. E. (1961). Unbiased estimation of a set of probabilities, *Biometrika*, **48**, 227–229.

Basu, D., and Laha, R. G. (1954). On some characterizations of the normal distribution, *Sankhyā*, **13**, 359–362. (Addendum, **14**, 180).

Beasley, J. D., and Springer, S. G. (1977). The percentage points of the normal distribution, *Applied Statistics*, **26**, 118–121.

Bell, S. (1962). *Approximating the normal distribution with the triangular*, Sandia Corporation Report No. 494.

Benson, F. (1949). A note on the estimation of mean and standard deviation from quantiles, *Journal of the Royal Statistical Society, Series B*, **11**, 91–100.

Berg, C. (1988). The cube of a normal distribution is indeterminate, *Annals of Probability*, **16**, 910–913.

Bergström, H. (1945). On the central limit theorem in the space R^k, $k > 1$, *Skandinavisk Aktuarietidskrift*, **28**, 106–127.

Bernstein, S. N. (1941). On a characteristic property of the normal law, *Trudy Leningrad Polytekhn. Instituta*, **3**, 21–22.

Berry, A. C. (1941). The accuracy of the Gaussian approximation to the sum of independent variates, *Transactions of the American Mathematical Society*, **49**, 122–136.

Best, D. J. (1979). Some easily programmed pseudo-random normal generators, *Australian Computer Journal*, **11**, 60–62.

Bhattacharjee, G. P., Pandit, S. N. N., and Mohan, R. (1963). Dimensional chains involving rectangular and normal error-distributions, *Technometrics*, **5**, 404–406.

Bhattacharya, N. (1959). An extension of Hald's table for one-sided censored normal distribution, *Sankhyā*, **21**, 377–380.

Billingsley, P. (1963). Limit theorems for randomly selected partial sums, *Annals of Mathematical Statistics*, **33**, 85–92.

Bischoff, W., Cremers, H., and Fieger, W. (1987). A characterization of the normal distribution by sufficiency of the least squares estimation, *Metrika*, **34**, 259–273.

Bland, R. P., Gilbert, R. D., Kapadia, C. H., and Owen, D. B. (1966). On the distributions of the range and mean range for samples from a normal distribution, *Biometrika*, **53**, 245–248.

Bliss, C. I., and Stevens, W. L. (1937). The calculation of the time-mortality curve, *Annals of Applied Biology*, **24**, 815–852.

Blum, J. R. (1956). On a characterization of the normal distribution, *Skandinavisk Aktuarietidskrift*, **39**, 59–62.

Bolch, B. W. (1968). More on unbiased estimation of the standard deviation, *The American Statistician*, **22**, No. 3, 27.

Bol'shev, L. N. (1959). On the transformations of random variables, *Teoriya Veroyatnostei i ee Primeneniya*, **4**, 136–149. (In Russian. English translation, 129–141.)

Bol'shev, L. N., and Smirnov, N. V. (1965). *Tables of Mathematical Statistics*, Moscow: Nauka.

Bondesson, L. (1974). A characterization of the normal law, *Sankhyā*, *Series A*, **36**, 321–324.

Bondesson, L. (1975). Normal distribution, gamma distribution, and quadratic polynomial statistics, *Scandinavian Journal of Statistics*, **2**, 138–144.

Bondesson, L. (1976). When is the sample mean BLUE? *Scandinavian Journal of Statistics*, **3**, 116–120.

Bondesson, L. (1977a). The sample variance, properly normalized, is χ^2-distributed for the normal law only, *Sankhyā*, *Series A*, **39**, 303–304.

Bondesson, L. (1977b). A note on sufficiency and independence, *Technical Report* NFMS-3046, Department of Statistics, Lund University, Sweden.

Bondesson, L. (1981). A normal sample with given sample mean and variance, *Journal of Statistical Computation and Simulation*, **14**, 51–52.

Borenius, G. (1958). On the distribution of extreme values in a sample from a normal distribution, *Skandinavisk Aktuarietidskrift*, **41**, 131–166.

Borenius, G. (1965). On the limit distribution of an extreme value in a sample from a normal distribution, *Skandinavisk Aktuarietidskrift*, **48**, 1–16.

Borovkov, A. A., and Utev, S. A. (1983). On an inequality and a related characterization of the normal distribution, *Theory of Probability and Its Applications*, **28**, 219–228.

Bose, R. C., and Gupta, S. S. (1959). Moments of order statistics from a normal population, *Biometrika*, **46**, 433–440.

Box, G. E. P., and Muller, M. E. (1958). A note on the generation of random normal deviates, *Annals of Mathematical Statistics*, **29**, 610–611.

Brainard, J., and Burmaster, D. E. (1992). Bivariate distributions for height and weight of men and women in the United States, *Risk Analysis*, **12**, 267–275.

Braverman, M. S. (1985). Characteristic properties of normal and stable distributions, *Theory of Probability and Its Applications*, **30**, 465–474.

Braverman, M. S. (1989). On a characterization of the normal distribution, Preprint, Khabarovsk University, Khabarovsk, Russia.

Breakwell, J. V. (1953). On estimating both mean and standard deviation of a normal population from the lowest *r* out of *n* observations (Abstract), *Annals of Mathematical Statistics*, **24**, 683.

Bryc, W. (1990). Normal distributions and characterizations, Unpublished manuscript, Cincinnati, OH: University of Cincinnati.

Bryc, W., and Szablowski, P. J. (1990). Some characteristics of normal distribution by conditional moments, *Bulletin of the Polish Academy of Sciences*: *Mathematics*, **38**, 209–218.

Burford, R. L., and Willis, J. E. (1978). The comparative quality of unit normal variates generated by the Box-Muller algorithm using alternative unit uniform generators, *ASA Proceedings of Statistical Computing Section*, 84–89.

Burr, I. W. (1967). A useful approximation to the normal distribution function, with application to simulation, *Technometrics*, **9**, 647–651.

Burrows, P. M. (1972). Expected selection differentials for directional selection, *Biometrics*, **28**, 1091–1100.

Burrows, P. M. (1975). Variances of selection differentials in normal samples, *Biometrics*, **31**, 125–133.

Buslenko, N. P., Golenko, D. I., Shreider, Yu. A., Sobol, I. M., and Sragovich, V. G. (1966). *The Monte Carlo Method*, Oxford: Pergamon. (Original Russian edition published 1962.)

Cacoullos, T. (1967a). Some characterizations of normality, *Sankhyā*, *Series A*, **29**, 399–404.

Cacoullos, T. (1967b). Characterizations of normality by constant regression of linear statistics on another linear statistic, *Annals of Mathematical Statistics*, **38**, 1894–1898.

Cacoullos, T., Papathanasiou, V., and Utev, S. (1993). Another characterization of the normal law and a proof of the central limit theorem connected with it, *Theory of Probability and Its Applications*, **37**, 581–588.

Cadwell, J. H. (1951). The bivariate normal integral, *Biometrika*, **38**, 475–479.

Cadwell, J. H. (1953a). Approximating to the distributions of measures of dispersion by a power of χ^2, *Biometrika*, **40**, 336–346.

Cadwell, J. H. (1953b). The distribution of quasi-ranges in samples from a normal population, *Annals of Mathematical Statistics*, **24**, 603–613.

Cadwell, J. H. (1954). The statistical treatment of mean deviation, *Biometrika*, **41**, 12–18.

Canal, L. (1981). On the generalized normal distribution, *Quaderni di Statistica e Matematica Applicata alle Scienze Economico-Sociali*, **4**, 97–107. (In Italian)

Carta, D. G. (1975). Low-order approximations for the normal probability integral and the error function, *Mathematics of Computation*, **29**, 856–862.

Charlier, C. V. L., and Wicksell, S. D. (1924). On the dissection of frequency functions, *Arkiv för Matematik, Astronomi och Fysik*, **18**, No. 6, 1–64.

Cheng, S. W. (1980). On the ABLUE of the normal mean from a censored sample, *Journal of Statistical Planning and Inference*, **4**, 259–265.

Cheng, S. W., and Ling, K. D. (1983). On the BLUE's of location and scale parameters based on incomplete samples, *Soochow Journal of Mathematics*, **9**, 35–45.

Chernoff, H. (1981). A note on an inequality involving the normal distribution, *Annals of Probability*, **9**, 533–535.

Chernoff, H., Gastwirth, J. L., and Johns, M. V. (1967). Asymptotic distribution of linear combinations of order statistics with application to estimation, *Annals of Mathematical Statistics*, **38**, 52–71.

Chew, V. (1968). Some useful alternatives to the normal distribution, *The American Statistician*, **22**, No. 3, 22–24.

Chiu, W. K., and Leung, M. P. Y. (1981). A graphical method for estimating the parameters of a truncated normal distribution, *Journal of Quality Technology*, **13**, 42–45.

Chou, Y.-M. (1981). Additions to the table of normal integrals, *Communications in Statistics—Simulation and Computation*, **10**, 537–538.

Chou, Y.-M., and Owen, D. B. (1986). One-sided simultaneous lower prediction intervals for *l* future samples from a normal distribution, *Technometrics*, **28**, 247–251.

Chu, J. T. (1955). On bounds for the normal integral, *Biometrika*, **42**, 263–265.

Chu, J. T., and Ya'Coub, K. (1966). Quadratic order estimates and moments of normal order statistics, *Annals of the Institute of Statistical Mathematics*, **18**, 337–341.

Clark, C. E. (1966). *Random Numbers in Uniform and Normal Distribution*, San Francisco: Chandler.

Clow, R., Hansen, E., and McNolty, F. (1974). Bayesian density functions for Gaussian pulse shapes in Gaussian noise, *Proceedings of the IEEE*, **62**, 134–136.

Cohen, A. C. (1949). On estimating the mean and standard deviation of truncated normal distributions, *Journal of the American Statistical Association*, **44**, 518–525.

Cohen, A. C. (1950a). Estimating the mean and variance of normal populations from singly truncated and doubly truncated samples, *Annals of Mathematical Statistics*, **21**, 557–569.

Cohen, A. C. (1950b). On estimating the mean and variance of singly truncated normal frequency distributions from the first three sample moments, *Annals of the Institute of Statistical Mathematics*, **3**, 37–44.

Cohen, A. C. (1955). Censored samples from truncated normal distributions, *Biometrika*, **42**, 516–519.

Cohen, A. C. (1957). On the solution of estimating equations for truncated and censored samples from normal populations, *Biometrika*, **44**, 225–236.

Cohen, A. C. (1961). Tables for maximum likelihood estimates: Singly truncated and singly censored samples, *Technometrics*, **3**, 535–540.

Cohen, A. C. (1963). Progressively censored samples in life testing, *Technometrics*, **5**, 327–339.

Cohen, A. C. (1967). Estimation in mixtures of two normal distributions, *Technometrics*, **9**, 15–28.

Cohen, A. C. (1991). *Truncated and Censored Samples: Theory and Applications*, New York: Marcel Dekker.

Cohen, A. C., and Whitten, B. J. (1988). *Parameter Estimation in Reliability and Life Span Models*, New York: Marcel Dekker.

Cohen, A. C., and Woodward, J. (1953). Tables of Pearson-Lee-Fisher functions of singly truncated normal distributions, *Biometrics*, **9**, 489–497.

Cornwell, L. W., Aroian, L. A., and Taneja, V. S. (1977). Numerical evaluation of the distribution of the product of two normal variables, *ASA Proceedings of Statistical Computing Section*, 163–168.

Cox, D. R. (1949). Use of range in sequential analysis, *Journal of the Royal Statistical Society, Series B*, **11**, 101–114.

Cox, D. R. (1960). Notes on the analysis of mixed frequency distributions, *British Journal of Mathematical and Statistical Psychology*, **19**, 39–47.

Cox, S. H. (1989). Comment on "Pocket-calculator approximation for areas under the standard normal curve," *The American Statistician*, **43**, 290.

Craig, C. C. (1928). An application of Thiele's semi-invariants to the sampling problem, *Metron*, **7**, No. 4, 3–74.

Crain, B. R. (1979). Estimating the parameters of a truncated normal distribution, *Applied Mathematics and Computation*, **5**, 149–156.

Cramér, H. (1928). On the composition of elementary errors, *Skandinavisk Aktuarietidskrift*, **11**, 13–74, 141–180.

Cramér, H. (1936). Über eine Eigenschaft der normalen Verteilungsfunktion, *Mathematische Zeitschrift*, **41**, 405–414.

Crouch, E. A. C., and Spiegelman, D. (1990). The evaluation of integrals of the form $\int_{-\infty}^{\infty} f(t) \exp(-t^2)\, dt$: Application to logistic-normal models, *Journal of the American Statistical Association*, **85**, 464–469.

Csörgö, M., and Seshadri, V. (1971). Characterizing the Gaussian and exponential law via mappings onto the unit interval, *Zeitschrift für Wehrscheinlichkeitstheorie und ihre Verwendte Gebiete*, **18**, 333–339.

Csörgö, M., Seshadri, V., and Yalovsky, M. (1975). Application of characterizations in the area of goodness-of-fit, In *Statistical Distributions in Scientific Work* (G. P. Patil, ed.), Dordrecht, Netherlands: Reidel, **2**, 79–90.

Cureton, E. E. (1968). Unbiased estimation of the standard deviation, *The American Statistician*, **22**, No. 1, 22. (Priority correction, *Ibid.*, **22**, No. 3, 27.)

Currie, I. D. (1980). On the distribution of the studentized range in a single normal sample, *Scandinavian Journal of Statistics*, **7**, 150–154.

Custer, S. W., and Pam, S. H. (1976). Correction for bias in maximum likelihood estimators of σ in a right-censored normal distribution, *Communications in Statistics—Simulation and Computation*, **5**, 15–22.

D'Agostino, R. B., and Stephens, M. A. (editors) (1986). *Goodness-of-fit Techniques*, New York: Marcel Dekker.

Dagpunar, J. S. (1987). Nomograms for the manual sampling of random variates from gamma, normal and associated distributions, *The Statistician*, **36**, 31–36.

Dagpunar, J. S. (1988). Computer generation of random variates from the tail of t and normal distributions, *Communications in Statistics—Simulation and Computation*, **17**, 653–661.

Daly, J. F. (1946). On the use of the sample range in an analogue of Student's t-test, *Annals of Mathematical Statistics*, **17**, 71–74.

Daniel, C. (1959). Use of half-normal plots in interpreting factorial two-level experiments, *Technometrics*, **1**, 311–341.

Dansie, B. R. (1986). Normal order statistics as permutation probability models, *Applied Statistics*, **35**, 269–275.

Darmois, G. (1951). Sur une propriété caractéristique de la loi de probabilité de Laplace, *Comptes Rendus de l' Académie des Sciences, Paris*, **232**, 1999–2000.

Darmois, G. (1953). Analyse générale des liasons stochastiques. Étude particuliaire de l'analyse factorielle linéaire, *Revue de l'Institut Internationale de Statistique*, **21**, 2–8.

David, F. N., and Johnson, N. L. (1954). Tests for skewness and kurtosis with ordered variables, *Proceedings of the International Congress of Mathematics*, **2**, 286.

David, H. A. (1956). Revised upper percentage points of the extreme studentized deviate from the sample mean, *Biometrika*, **43**, 449–451.

David, H. A. (1957). Estimation of means of normal populations from observed minima, *Biometrika*, **44**, 282–286.

David, H. A. (1968). Gini's mean difference rediscovered, *Biometrika*, **55**, 573–575.

David, H. A., Kennedy, W. J., and Knight, R. D. (1977). Means, variances, and covariances of the normal order statistics in the presence of an outlier, *Selected Tables in Mathematical Statistics*, **5**, 75–204.

David, H. T. (1963). The sample mean among the extreme normal order statistics, *Annals of Mathematical Statistics*, **34**, 33–55.

Davis, C. S., and Stephens, M. A. (1977). The covariance matrix of normal order statistics, *Communications in Statistics—Simulation and Computation*, **6**, 75–82.

Davis, C. S., and Stephens, M. A. (1978). Approximating the covariance matrix of normal order statistics, *Applied Statistics*, **27**, 206–211.

Davis, G. C., Jr. and Kutner, M. H. (1976). The lagged normal family of probability density functions applied to indicator-dilution curves, *Biometrics*, **32**, 669–675.

Daw, R. H. (1966). Why the normal distribution, *Journal of the Institute of Actuaries Students Society*, **18**, 2–15.

Deak, I. (1980). A fast normal random number generator, *Alkalmazott Matematikai Lapok*, **6**, 83–92. (In Hungarian)

Deak, I. (1981). An economical method for random number generation and a normal generator, *Computing*, **27**, 113–121.

DePriest, D. J. (1983). Using the singly truncated normal distribution to analyze satellite data, *Communications in Statistics—Theory and Methods*, **12**, 263–272.

Derenzo, S. E. (1977). Approximations for hand calculators using small integer coefficients, *Mathematics of Computation*, **31**, 214–225.

Dixon, W. J. (1950). Analysis of extreme values, *Annals of Mathematical Statistics*, **21**, 488–506.

Dixon, W. J. (1960). Simplified estimation from censored normal samples, *Annals of Mathematical Statistics*, **31**, 385–391.

Dixon, W. J. (1962). Rejection of observations, *Contributions to Order Statistics*, A. E. Sarhan and B. G. Greenberg (editors), 299–342, New York: Wiley.

Dodgson, J. H., and Hill, I. D. (1983). A remark on Algorithm AS100: Normal-Johnson and Johnson-normal transformations, *Applied Statistics*, **32**, 345.

Doeblin, W. (1937). *Sur les propriétés asymptotiques de mouvements regis par certains types de chaînes simples*, Thesis, University of Paris.

Doornbos, R. (1956). Significance of the smallest of a set of estimated normal variances, *Statistica Neerlandica*, **10**, 117–126.

D'Ortenzio, R. J. (1965). Approximating the normal distribution function, *Systems Design*, **9**, 4–7.

Durrant, N. F. (1978). A nomogram for confidence limits on quantiles of the normal distribution with application to extreme value distributions, *Journal of Quality Technology*, **10**, 155–158.

Dyer, D. D. (1974). Estimation in a truncated circular normal distribution with ballistic applications, *Operations Research*, **22**, 197–201.

Dyer, D. D., and Keating, J. P. (1979). A further look at the comparison of normal percentile estimators, *Communications in Statistics—Theory and Methods*, **8**, 1–16.

Dyer, D. D., Keating, J. P., and Hensley, O. L. (1977). Comparison of point estimators of normal percentiles, *Communications in Statistics—Simulation and Computation*, **6**, 269–284.

Easingwood, T. (1979). Approximation to the normal distribution function, *Mathematical Gazette*, **63**, 40–42.

Eaton, M. L. (1966). Characterization of distributions by the identical distribution of linear forms, *Journal of Applied Probability*, **3**, 481–494.

Eberhardt, K. R., Mee, R. W., and Reeve, C. P. (1989). Computing factors for exact two-sided tolerance limits for a normal distribution, *Communications in Statistics—Simulation and Computation*, **18**, 397–413.

Eberl, W., Jr. (1986). Characterizations of normality in translation classes by properties of Bayes estimators, *Metrika*, **33**, 111–119.

Edgeman, R. L. (1988). Normal distribution probabilities and quantiles without tables, *Mathematics of Computer Education*, **22**, 95–99.

Edwards, A. W. F. (1963). A linkage for drawing the normal distribution, *Applied Statistics*, **12**, 44–45.

Eisenberger, I. (1964). Genesis of bimodal distributions, *Technometrics*, **6**, 357–363.

Eisenberger, I., and Posner, E. C. (1965). Systematic statistics used for data compression in space telemetry, *Journal of the American Statistical Association*, **60**, 97–133.

Eisenhart, C., Deming, L. S., and Martin, C. S. (1963). Tables describing small-sample properties of the mean, median, standard deviation, and other statistics in sampling from various distributions, National Bureau of Standards, Technical Note No. 191.

Elfving, G. (1947). The asymptotical distribution of range in samples from a normal population, *Biometrika*, **34**, 111–119.

El Lozy, M. (1982). A remark on Algorithm AS 126: Probability integral of the normal range, *Applied Statistics*, **31**, 99.

Emersleben, O. (1951). Numerische Werte des Fehlerintegrals für $\sqrt{n\pi}$, *Zeitshrift für Angewandte Mathematik und Mechanik*, **31**, 393–394.

Esseen, C. -G. (1942). On the Liapounoff limit of error in the theory of probability, *Arkiv för Matematik, Astronomi och Fysik*, **28A**, 1–19.

Esseen, C. -G. (1945). Fourier analysis of distribution functions. A mathematical study of the Laplace-Gauss law, *Acta Mathematica*, **77**, 1–125.

Esseen, C. -G. (1958). On mean central limit theorems, *Kunglige Tekniska Högskolans Handlingar*, No. 121, 1–30.

Evans, I. G. (1964). Bayesian estimation of the variance of a normal distribution, *Journal of the Royal Statistical Society, Series B*, **26**, 63–68.

Fattorini, L. (1978). Percentile estimators for the parameters of the normal distribution, *Metron*, **36/3**, 151–158.

Fearn, D. (1983). Elementary methods for approximating the cumulative distribution of beta, *F*, *t*, and normal variables, *Computer Science and Statistics: Proceedings of the Fifteenth Symposium on the Interface*, J. E. Gentle (editor), 344–346, Amsterdam: North-Holland.

Feller, W. (1935). Über den zentralen Grenzwertsatz der Wahrscheinlichkeitsrechnung, *Mathematische Zeitschrift*, **40**, 521–559.

Feller, W. (1945). The fundamental limit theorems in probability, *Bulletin of the American Mathematical Society, Series 2*, **51**, 800–832.

Feller, W. (1966). *An Introduction to Probability Theory*, **2**, New York: Wiley.

Fertig, K. W., and Mann, N. R. (1977). One-sided prediction intervals for at least *p* out of *m* future observations from a normal population, *Technometrics*, **19**, 167–178.

Fieger, W. (1971). A statistical characterization of the normal distribution, *Zeitschrift für Wahrscheinlichkeitstheorie und Ihre Verwendte Gebiete*, **19**, 330–344.

Fieger, W. (1976). On a characterization of the normal law based on a homogeneous Pitman estimator, *Sankhyā, Series A*, **38**, 394–396.

Fieger, W. (1977). Transformations that characterize the normal distribution, *Metrika*, **24**, 7–22. (In German)

Fields, R. I., Kramer, C. Y., and Clunies-Ross, C. W. (1962). Joint estimation of the parameters of two normal populations, *Journal of the American Statistical Association*, **57**, 446–454.

Findeisen, P. (1982). Gauss' characterization of normality, *Metrika*, **29**, 55–63.

Fisher, R. A. (1925). Theory of statistical estimation, *Proceedings of the Cambridge Philosophical Society*, **22**, 700–706.

Fisher, R. A. (1930). The moments of the distributions of normal samples of measures of departure from normality, *Proceedings of the Royal Society of London*, **130**, 16–28.

Fisher, R. A. (1931). The truncated normal distribution, *British Association for the Advancement of Science, Mathematical Tables*, **5**, xxxiii–xxxiv.

Fisher, R. A., and Yates, F. (1963). *Statistical Tables for Biological, Agricultural and Medical Research*, London and Edinburgh: Oliver & Boyd.

Fleming, N. S. (1989). Approximating the normal cumulative distribution function using a spreadsheet program, *The American Statistician*, **43**, 68. Reply, *Ibid.*, **43**, 290–291.

Fox, C. (1965). A family of distributions with the same ratio property as normal distribution, *Canadian Mathematical Bulletin*, **8**, 631–636.

Francis, V. J. (1946). On the distribution of the sum of n sample values drawn from a truncated normal population, *Journal of the Royal Statistical Society, Series B*, **8**, 223–232.

Gabovich, Yu. R. (1974). On the stability of certain characteristic properties of the normal distribution, *Theory of Probability and Its Applications*, **19**, 365–370. (In Russian)

Galambos, J., and Kotz, S. (1978). Characterizations of Probability Distributions, *Lecture Notes in Mathematics*, **675**, Heidelberg: Springer-Verlag.

Gates, C. E. (1978). Generating random normal deviates using a modified Butler algorithm, *ASA Proceedings of Statistical Computing Section*, 111–114.

Gauss, C. F. (1809). *Theoria Motus Corporum Coelestium*, Hamburg: Perthes & Besser. (English translation by C. H. Davis, published 1857, Boston: Little, Brown.)

Gauss, C. F. (1816). Bestimmung der Genauigkeit der Beobachtungen, *Zeitschrift Astronomi*, **1**, 185–197.

Gautschi, W. (1964). Error function and Fresnel integrals, *Handbook of Mathematical Functions*, M. Abramowitz and I. A. Stegun (editors) 295–309, U.S. Department of Commerce, *Applied Mathematics Series*, **55**.

Geary, R. C. (1933). A general expression for the moments of certain symmetrical functions of normal samples, *Biometrika*, **25**, 184–186.

Geary, R. C. (1935). The ratio of the mean deviation to the standard deviation as a test of normality, *Biometrika*, **27**, 310–332.

Geary, R. C. (1936). Moments of the ratio of the mean deviation to the standard deviation for normal samples, *Biometrika*, **28**, 295–305.

Geary, R. C. (1947). The frequency distribution of $\sqrt{b_1}$ for samples of all sizes drawn at random from a normal population, *Biometrika*, **34**, 68–97.

Geary, R. C. (1947). Testing for normality, *Biometrika*, **34**, 209–242.

Geary, R. C., and Worlledge, J. P. G. (1947). On the computation of universal moments of tests of statistical normality derived from samples drawn at random from a normal universe. Application to the calculation of the seventh moment of b_2, *Biometrika*, **34**, 98–110. (Correction, *Ibid.*, **37**, 189.)

Geisser, S. (1956). A note on the normal distribution, *Annals of Mathematical Statistics*, **27**, 858–859.

Geisser, S. (1973). Normal characterization via the squares of random variables, *Sankhyā, Series A*, **35**, 492–494.

Gerisch, W., Struck, W., and Wilke, B. (1987). One-sided Monte Carlo tolerance limit factors for normal distributions in the case of censored samples, *Computational Statistics Quarterly*, **3**, 265–288.

Gertsbakh, I., and Winterbottom, A. (1991). Point and interval estimation of normal tail probabilities, *Communications in Statistics—Theory and Methods*, **20**, 1497–1514.

Ghosh, J. K. (1969). Only linear transformations preserve normality, *Sankhyā*, *Series A*, **31**, 309–312.

Ghurye, S. G., and Olkin, I. (1973). Identically distributed linear forms and the normal distribution, *Advances in Applied Probability*, **5**, 138–152.

Gibbons, J. F., and Mylroie, S. (1973). Estimation of impurity profiles in ion-implanted amorphous targets using joined half-Gaussian distributions, *Applied Physics Letters*, **22**, 568–569.

Gjeddebaek, N. F. (1949). Contribution to the study of grouped observations. Applications of the method of maximum likelihood in case of normally distributed observations, *Skandinavisk Aktuarietidskrift*, **42**, 135–150.

Glänzel, W. (1988). A characterization of the normal distribution, *Studia Scientiarum Mathematicum Hungarica*, **23**, 89–91.

Gnedenko, B. V. (1948). On a theorem of S. N. Bernstein, *Izvestiya Akademii Nauk SSSR*, *Seria Matematicheskaya*, **12**, 97–100. (In Russian)

Gnedenko, B. V., and Kolmogorov, A. N. (1954). *Limit distributions for sums of independent random variables*, Reading, MA: Addison-Wesley.

Godwin, H. J. (1945). On the distribution of the estimate of mean deviation obtained from samples from a normal population, *Biometrika*, **33**, 254–256.

Godwin, H. J. (1948). A further note on the mean deviation, *Biometrika*, **35**, 304–309.

Godwin, H. J. (1949a). On the estimation of dispersion by linear systematic statistics, *Biometrika*, **36**, 92–100.

Godwin, H. J. (1949b). Some low moments of order statistics, *Annals of Mathematical Statistics*, **20**, 279–285.

Godwin, H. J., and Hartley, H. O. (1945). Tables of the probability integral and the percentage points of the mean deviation in samples from a normal population, *Biometrika*, **33**, 254–265.

Goel, P. K., and DeGroot, M. H. (1980). Only normal distributions have linear posterior expectations in linear regression, *Journal of the American Statistical Association*, **75**, 895–900.

Golder, E. R., and Settle, J. G. (1976). The Box-Muller method for generating pseudo-random normal deviates, *Applied Statistics*, **25**, 12–20.

Gosslee, D. G., and Bowman, K. O. (1967). *Evaluation of maximum likelihood estimates of parameters in mixtures of normal distributions*, Oak Ridge National Laboratory, Report ORNL-TM-2110.

Goto, M., and Hatanaka, S. (1985). On power-normal transformation diagnostics, *Japanese Journal of Applied Statistics*, **14**, 69–80. (In Japanese)

Govindarajulu, Z. (1963). On moments of order statistics and quasi-ranges from normal populations, *Annals of Mathematical Statistics*, **34**, 633–651.

Govindarajulu, Z. (1966). Characterizations of normal and generalized truncated normal distributions using order statistics, *Annals of Mathematical Statistics*, **37**, 1011–1015.

Govindarajulu, Z., and Hubacker, N. W. (1964). Percentiles of order statistics in samples from uniform, normal, chi (1 d.f.) and Weibull populations, *Reports in Statistical Applied Research, JUSE*, **11**, 64–90.

Graybill, F. A., and Kingman, A. (1970). A survey of characterizations of normal distribution by identical distribution of linear forms from Pólya (1923), *Annals of Mathematical Statistics*, **41**, 1889–1895.

Greenwood, J. A., and Hartley, H. O. (1962). *Guide to Tables in Mathematical Statistics*, Princeton, NJ: Princeton University Press.

Grubbs, F. E. (1950). Sample criteria for testing outlying observations, *Annals of Mathematical Statistics*, **21**, 27–58.

Grubbs, F. E., and Weaver, C. L. (1947). The best unbiased estimate of population standard deviation based on group ranges, *Journal of the American Statistical Association*, **42**, 224–241.

Grundy, P. M. (1952). The fitting of grouped truncated and grouped censored normal distribution, *Biometrika*, **39**, 252–259.

Guirguis, G. H. (1991). A rational approximation of the inverse normal probability function, *Computational Statistics and Data Analysis*, **11**, 199–201.

Gupta, A. K. (1952). Estimation of the mean and standard deviation of a normal population from a censored sample, *Biometrika*, **39**, 260–273.

Gupta, S. S. (1961). Percentage points and modes of order statistics from the normal distribution, *Annals of Mathematical Statistics*, **32**, 888–893.

Gupta, S. S. (1962). Life-test sampling plans for normal and lognormal distributions, *Technometrics*, **4**, 151–175.

Guttman, I. (1960). Tests for the scale parameter of the truncated normal, *Canadian Mathematical Bulletin*, **3**, 225–236.

Gyires, B. (1975). A characterization of the normal distribution, *Statistical Distributions in Scientific Work*, **3**, 211–220, Dordrecht: D. Reidel.

Hahn, G. J. (1969). Finding an interval for the next observation from a normal distribution, *Journal of Quality Technology*, **1**, 168–171.

Hahn, G. J. (1970a). Statistical intervals for a normal population, Part I. Tables, examples and applications, *Journal of Quality Technology*, **2**, 115–125.

Hahn, G. J. (1970b). Statistical intervals for a normal population, Part II. Formulas, assumptions, some derivations, *Journal of Quality Technology*, **2**, 195–206.

Hald, A. (1949). Maximum likelihood estimation of the parameters of a normal distribution which is truncated at a known point, *Skandinavisk Aktuarietidskrift*, **32**, 119–134.

Hald, A. (1952). *Statistical Tables and Formulas*, New York: Wiley.

Hall, R. L. (1979). Inverse moments for a class of truncated normal distributions, *Sankhyā, Series B*, **41**, 66–76.

Halperin, M. (1952). Maximum likelihood estimation in truncated samples, *Annals of Mathematical Statistics*, **23**, 226–238.

Halperin, M. (1952). Estimation in the truncated normal distribution, *Journal of the American Statistical Association*, **47**, 457–465.

Halperin, M. (1961). Confidence intervals from censored samples, *Annals of Mathematical Statistics*, **32**, 828–837.

Halperin, M. (1966). Confidence intervals from censored samples, **II**, *Technometrics*, **8**, 291–301.

Hamaker, H. C. (1978). Approximating the cumulative normal distribution and its inverse, *Applied Statistics*, **27**, 76–77.

Harding, J. P. (1949). The use of probability paper for the graphical analysis of polynomial frequency distributions, *Journal of Marine Biology Association, United Kingdom*, **28**, 141–153.

Harris, H., and Smith, C. A. B. (1949). The sib-sib age of onset correlation among individuals suffering from a hereditary syndrome produced by more than one gene, *Annals of Eugenics, London*, **14**, 309–318.

Hart, R. G. (1966). A close approximation related to the error function, *Mathematics of Computation*, **20**, 600–602.

Harter, H. L. (1959). The use of sample quasi-ranges in estimating population standard deviation, *Annals of Mathematical Statistics*, **30**, 980–999. Correction, *Ibid.*, **31**, 228.

Harter, H. L. (1960). Tables of range and studentized range, *Annals of Mathematical Statistics*, **31**, 1122–1147.

Harter, H. L. (1961). Expected values of normal order statistics, *Biometrika*, **48**, 151–157.

Harter, H. L. (1963). Percentage points of the ratio of two ranges and power of the associated test, *Biometrika*, **50**, 187–194.

Harter, H. L. (1964). Criteria for best substitute interval estimators, with an application to the normal distribution, *Journal of the American Statistical Association*, **59**, 1133–1140.

Harter, H. L. (1970). *Order Statistics and Their Use in Testing and Estimation*, **1** and **2**, Washington, DC: Government Printing Office.

Harter, H. L., and Clemm, D. S. (1959). The probability integrals of the range and of the Studentized range—probability integrals, percentage points, and moments of the range, **1**, Wright Air Development Center Technical Report No. 58-484.

Harter, H. L., and Moore, A. H. (1966). Iterative maximum likelihood estimation of the parameters of normal populations from singly and doubly censored samples, *Biometrika*, **53**, 205–213.

Hartley, H. O. (1942). Numerical evaluation of the probability integral (*of range*), *Biometrika*, **32**, 309–310.

Hartley, H. O. (1945). Note on the calculation of the distribution of the estimate of mean deviation in normal samples, *Biometrika*, **33**, 257–258.

Hartley, H. O., and Pearson, E. S. (1951). Moment constants for the distribution of range in small samples, *Biometrika*, **38**, 463–464.

Hassanein, K. M., Saleh, A. K. Md. E., and Brown, E. F. (1986). Best linear unbiased estimators for normal distribution quantiles for sample sizes up to 20, *IEEE Transactions on Reliability*, **35**, 327–329.

Hasselblad, V. (1966). Estimation of parameters for a mixture of normal distributions, *Technometrics*, **8**, 431–444. (Discussion by A. C. Cohen, 445–446.)

Hastings, C. (1955). *Approximations for Digital Computers*, Princeton, NJ: Princeton University Press.

Hastings, C., Mosteller, F., Tukey, J. W., and Winsor, C. P. (1947). Low moments for small samples: A comparative study of order statistics, *Annals of Mathematical Statistics*, **18**, 413–426.

Hawkes, A. G. (1982). Approximating the normal tail, *The Statistician*, **31**, 231–236.

Hawkins, D. M. (1975). From the noncentral *t* to the normal integral, *The American Statistician*, **29**, 42.

Hawkins, D. M. (1980). *Identification of Outliers*, London: Chapman & Hall.

Hawkins, D. M., and Wixley, R. A. J. (1986). A note on the transformation of chi-squared variables to normality, *The American Statistician*, **40**, 296–298.

Hayya, J., Armstrong, D., and Gressis, N. (1975). A note on the ratio of two normally distributed variables, *Management Sciences*, **21**, 1338–1341.

Healy, M. J. R. (1978). A mean difference estimator of standard deviation in symmetrically censored normal samples, *Biometrika*, **65**, 643–646.

Healy, M. J. R. (1982). A linear estimator of standard deviation in symmetrically trimmed normal samples, *Applied Statistics*, **31**, 174–175.

Heard, T. J. (1979). Approximation to the normal distribution function, *Mathematical Gazette*, **63**, 39–40.

Heath, D. F. (1967). Normal or lognormal: Appropriate distributions (Letter), *Nature, London*, **213**, 1159–1160.

Heffernan, P. M. (1988). New measures of spread and a simple formula for the normal distribution, *The American Statistician*, **42**, 100–102.

Hegde, L. M., and Dahiya, R. C. (1989). Estimation of the parameters in a truncated normal distribution, *Communications in Statistics—Theory and Methods*, **18**, 4177–4195.

Helguero, F. de (1904). Sui massima delle curve dimorfiche, *Biometrika*, **3**, 84–98.

Helguero, F. de (1905). Per la risoluzione delle curve dimorfiche, *Biometrika*, **4**, 230–231.

Helmert, F. R. (1875). Über die Berechnung des wahrscheinlichen Fehlers aus einer endlichen Anzahl wahrer Beobachtungsfehler, *Zeitschrift für angewandte Mathematik und Physik*, **20**, 300–303.

Helmert, F. R. (1876). Über die Wahrscheinlichkeit der Potenzsummen der Beobachtungsfehler und über einige damit im Zusammenhänge stehende Frangen, *Zeitschrift für angewandte Mathematik und Physik*, **21**, 192–218.

Henze, N. (1986). A probabilistic representation of "skew-normal" distribution, *Scandinavian Journal of Statistics*, **13**, 271–275.

Herrey, E. M. J. (1965). Confidence intervals based on the mean absolute deviation of a normal sample, *Journal of the American Statistical Association*, **60**, 257–269.

Hill, B. M. (1963). Information for estimating the proportions in mixtures of exponential and normal distributions, *Journal of the American Statistical Association*, **58**, 918–932.

Hill, I. D. (1976). Normal-Johnson and Johnson-normal transformations, *Applied Statistics*, **25**, 190–192.

Hill, I. D. (1987). A remark on Algorithm AS 16. Maximum likelihood estimation from grouped and censored normal data, *Applied Statistics*, **36**, 119–120.

Hill, I. D., and Wheeler, R. E. (1981). A remark on Algorithms AS99: Fitting Johnson curves by moments and AS100: Normal-Johnson and Johnson-normal transformations, *Applied Statistics*, **30**, 106.

Hoaglin, D. C. (1989). Comment on "Pocket-calculator approximation for areas under the standard normal curve," *The American Statistician*, **43**, 289.

Hojo, T. (1931). Distribution of the median, quartiles and interquartile distance in samples from a normal population, *Biometrika*, **23**, 315–360.

Hombas, V. C. (1985). Characterizing the normal density as a solution of a differential equation, *Statistica Neerlandica*, **39**, 387–388.

Houchens, R. L. (1984). Record value theory and inference, Doctoral dissertation, Riverside, CA: University of California.

Hoyt, J. P. (1968). A simple approximation to the standard normal probability density function, *The American Statistician*, **22**, No. 2, 25–26.

Hsu, C. T., and Lawley, D. N. (1940). The derivation of the fifth and sixth moments of the distribution of b_2 in samples from a normal population, *Biometrika*, **31**, 238–248.

Hurt, J. (1980). Estimates of reliability for the normal distribution, *Aplikace Matematiky*, **25**, 432–444.

Hyrenius, H. (1949). Sampling distributions from a compound normal parent population, *Skandinavisk Aktuarietidskrift*, **32**, 180–187.

Ibragimov, I. A. (1965). On the rate of convergence to normality, *Doklady Akademii Nauk SSSR*, **161**, 1267–1269. (In Russian)

Iliescu, D. V., and Voda, V. Gh. (1974). On the estimation of standard deviation for a normal population, *Trabajos de Estadistica*, **25**, 71–98.

Irwin, J. O. (1925). The further theory of Francis Galton's individual-difference problem, *Biometrika*, **17**, 100–128.

Janson, S. (1988). Normal convergence by higher semi-invariants with applications to sums of dependent variables and random graphs, *Annals of Probability*, **16**, 305–312.

Jarrett, R. F. (1968). A minor exercise in history, *The American Statistician*, **22**, No. 3, 25–26.

Jilek, M., and Likař, O. (1960). Tolerance limits of the normal distribution with known variance and unknown mean, *Australian Journal of Statistics*, **2**, 78–83.

John, S. (1982). The three-parameter two-piece normal family of distributions and its fitting, *Communications in Statistics—Theory and Methods*, **11**, 879–885.

Jones, A. E. (1946). A useful method for the routine estimation of dispersion in large samples, *Biometrika*, **33**, 274–282.

Jones, G. M., Kapadia, C. H., Owen, D. B., and Bland, R. P. (1969). On the distribution of the quasi-range and mid-range for samples from a normal population, Technical Report No. 20, THEMIS Contract, Department of Statistics, Dallas, TX: Southern Methodist University.

Jones, H. L. (1948). Exact lower moments of order statistics in small samples from a normal distribution, *Annals of Mathematical Statistics*, **19**, 270–273.

Jones, M. C. (1985). Generating normal variates with given sample mean and variance, *Applied Statistics*, **34**, 295.

Jones, P. N., and McLachlan, G. J. (1990). Laplace-normal mixtures fitted to wind shear data, *Journal of Applied Statistics*, **17**, 271–276.

Joshi, P. C., and Balakrishnan, N. (1981). An identity for the moments of normal order statistics with applications, *Scandinavian Actuarial Journal*, 203–213.

Joshi, V. M. (1982). A characterization of the normal population, *Journal of the Indian Statistical Association*, **20**, 85–88.

Kabe, D. G. (1976). On confidence bands for quantiles of a normal population, *Journal of the American Statistical Association*, **71**, 417–419.

Kac, M. (1939). On a characterization of the normal distribution, *American Journal of Mathematics*, **61**, 726–728.

Kafael, M. A., and Schmidt, P. (1985). On the adequacy of the "Sargan distribution" as an approximation to the normal, *Communications in Statistics—Theory and Methods*, **14**, 509–526.

Kagan, A. M. (1976). Several weak sense analogues of characteristic properties of the normal distribution, *Journal of Soviet Mathematics*, **61**, 59–67. (In Russian)

Kagan, A. M., Linnik, Yu. V., and Rao, C. R. (1965). On a characterization of the normal law based on a property of the sample average, *Sankhyā*, Series A, **27**, 405–406.

Kagan, A. M., Linnik, Yu. V., and Rao, C. R. (1973). *Characterization Problems of Mathematical Statistics*, New York: Wiley.

Kagan, A. M., and Shalayevskii, O.V. (1967). Characterization of the normal law by a property of the non-central χ^2-distribution, *Lietuvos Matematikos Rinkinys*, **7**, 57–58. (In Russian)

Kamat, A. R. (1953). The third moment of Gini's mean difference, *Biometrika*, **40**, 451–452.

Kamat, A. R. (1954a). Distribution theory of two estimates for standard deviation based on second variate differences, *Biometrika*, **41**, 1–11.

Kamat, A. R. (1954b). Moments of the mean deviation, *Biometrika*, **41**, 541–542.

Kamat, A. R., and Sathe, Y. S. (1957). Approximations to the distributions of some measures of dispersion based on successive differences, *Biometrika*, **44**, 349–359.

Kanji, G. K. (1985). A mixture model for wind shear data, *Journal of Applied Statistics*, **12**, 49–58.

Kaplansky, I. (1943). A characterization of the normal distribution, *Annals of Mathematical Statistics*, **14**, 197–198.

Karhunen, P. J., and Narula, S. C. (1989). The probability distribution of the ratio of the absolute values of two normal variables, *Journal of Statistical Computation and Simulation*, **33**, 173–182.

Katti, S. K., and Khan, B. A. (1988). A proof of a characterization of normality, *Revista Brasilea Probabilitade Estadistica*, (*REBRAPE*), **2**, 1–5.

Kawata, R., and Sakamoto, H. (1949). On the characterization of the normal population by the independence of the sample mean and the sample variance, *Journal of the Mathematical Society of Japan*, **1**, 111–115.

Kelker, D., and Matthes, R. K. (1970). A sufficient statistics characterization of the normal distribution, *Annals of Mathematical Statistics*, **41**, 1086–1090.

Kelley, T. L. (1948). *The Kelley Statistical Tables*, Cambridge: Harvard University Press.

Kendall, M. G., and Babington Smith, B. (1942). Random Sampling Numbers, *Tracts for Computers*, xxiv. Cambridge: Cambridge University Press.

Kerridge, D. F., and Cook, G. W. (1976). Yet another series for the normal integral, *Biometrika*, **63**, 401–403.

Keyfitz, N. (1938). Graduation by a truncated normal, *Annals of Mathematical Statistics*, **9**, 66–67.

Khatri, C. G. (1975a). Characterization of normal law by constancy of regression, *Statistical Distributions in Scientific Work*, **3**, 199–210, Dordrecht: D. Reidel.

Khatri, C. G. (1975b). A characteristic property of normal distribution, *Gujarat Statistical Review*, **2/2**, 24–27.

Kimber, A. C. (1985). Methods for the two-piece normal distribution, *Communications in Statistics—Theory and Methods*, **14**, 235–245.

Kimber, A. C., and Jeynes, C. (1985). An application of the truncated two-piece normal distribution to the measurement of depths of arsenic implants in silicon, Technical Report No. 51, Department of Mathematics, University of Surrey, England.

Kinderman, A. J., Monahan, J. F., and Ramage, J. G. (1975). Computer generation of random variables with normal and Student's *t* distributions, *ASA Proceedings of Statistical Computing Section*, 128–131.

Kinderman, A. J., and Ramage, J. G. (1990). Computer generation of normal random variables, *Journal of the American Statistical Association*, **71**, 893–896. (Correction, *Ibid.*, **85**, 272.)

King, E. P. (1953). Estimating the standard deviation of a normal population, *Industrial Quality Control*, **10**, No. 2, 30–33.

Kingman, A., and Graybill, F. A. (1970). A non-linear characterization of the normal distribution, *Annals of Mathematical Statistics*, **41**, 1889–1895.

Klebanov, L. B., and Melamed, I. A. (1976). The characterization of the normal and gamma distributions by properties of their Bayesian estimates, *Litovskii Matematicheskii Sbornik*, **16/1**, 123–137. (In Russian)

Klebanov, L. B., and Melamed, I. A. (1978). Characterization of normal and gamma distributions by properties of Fisher information amount, *Journal of Soviet Mathematics*, **9**, 881–886.

Klebanov, L. B., and Melamed, I. A. (1982). Stability of some problems of characterization of the normal and related distributions, *Studia Scientiarum Mathematicarum Hungarica*, **17**, 367–378.

Klebanov, L. B., and Neupokoeva, M. V. (1990). A characterization of a distribution by a property of the mean values of order statistics, *Theory of Probability and Its Applications*, **34**, 715–720.

Koeniger, W. (1983). A remark on AS177. Expected normal order statistics (exact and approximate), *Applied Statistics*, **32**, 223–224.

Kondo, T., and Elderton, E. M. (1931). Table of functions of the normal curve to ten decimal places, *Biometrika*, **22**, 368–376.

Koopmans, L. H., Owen, D. B., and Rosenblatt, J. I. (1964). Confidence intervals for the coefficient of variation for the normal and log-normal distributions, *Biometrika*, **51**, 25–32.

Kotlarski, I. (1966). On characterizing the normal distribution by Student's law, *Biometrika*, **53**, 603–606.

Kotz, S. (1974). Characterizations of statistical distributions: A supplement to recent surveys, *International Statistical Review*, **42**, 39–45.

Kotz, S., and Johnson, N. L. (1982). Circular normal distribution, *Encyclopedia of Statistical Sciences*, **1**, S. Kotz, N. L. Johnson, and C. B. Read (editors), 479, New York: Wiley.

Kotz, S., and Johnson, N. L. (1985a). Lineo-normal distribution, *Encyclopedia of Statistical Sciences*, **5**, S. Kotz, N. L. Johnson, and C. B. Read (editors), 73, New York: Wiley.

Kotz, S., and Johnson, N. L. (1985b). Normal-gamma (prior) density, *Encyclopedia of Statistical Sciences*, **6**, S. Kotz, N. L. Johnson, and C. B. Read (editors), 362, New York: Wiley.

Kotz, S., and Johnson, N. L. (1985c). Modified normal distributions, *Encyclopedia of Statistical Sciences*, **5**, S. Kotz, N. L. Johnson, and C. B. Read (editors), 590–591, New York: Wiley.

Kotz, S., and Johnson, N. L. (1986). Radico-normal distribution, *Encyclopedia of Statistical Sciences*, **7**, S. Kotz, N. L. Johnson, and C. B. Read (editors), 500–501, New York: Wiley.

Kotz, S., and Johnson, N. L. (1988). Skew-normal distributions, *Encyclopedia of Statistical Sciences*, **8**, S. Kotz, N. L. Johnson, and C. B. Read (editors), 507, New York: Wiley.

Kronmal, R. A., Peterson, A. V., Jr., and Hutchinson, F. (1979). Generating normal random variables using the uniform alias-rejection mixture method, *ASA Proceedings of Statistical Computing Section*, 250–255.

Krutchkoff, R. G. (1966). The correct use of the sample mean absolute deviation in confidence intervals for a normal variate, *Technometrics*, **8**, 663–673.

Kulldorff, G. (1958a). Maximum likelihood estimation of the mean of a normal random variable when the sample is grouped, *Skandinavisk Aktuarietidskrift*, **41**, 1–17.

Kulldorff, G. (1958b). Maximum likelihood estimation of the standard deviation of a normal random variable when the sample is grouped, *Skandinavisk Aktuarietidskrift*, **41**, 18–36.

Kulldorff, G. (1963, 1964). On the optimum spacing of sample quantiles from a normal distribution, Part I, *Skandinavisk Aktuarietidskrift*, **46**, 143–161. Part II, *Skandinavisk Aktuarietidskrift*, **47**, 71–87.

Laha, R. G. (1956). On the stochastic independence of a homogeneous quadratic statistic and the sample mean, *Vestnik Leningrad University*, **1**, 23–32.

Laha, R. G. (1957). On a characterization of the normal distribution from properties of suitable linear statistics, *Annals of Mathematical Statistics*, **28**, 126–139.

Lajko, K. (1980). A characterization of generalized normal and gamma distributions, *Analytic Function Methods in Probability Theory*, No. 21, 199–225, Amsterdam: North-Holland.

Lancaster, H. O. (1987). Finiteness of the variances in characterizations of the normal distribution, *Australian Journal of Statistics*, **29**, 101–106.

Laplace, P. S. (1774). Determiner le milieu que l'on doit prendre entre trois observations données d'un même phénomené, *Mémoires de Mathématique et Physique presentées à l'Académie Royale des Sciences par divers Savans*, **6**, 621–625.

Laplace, P. S. (1778). Mémoire sur les probabilités, *Histoire de l'Academie Royale de Sciences, Année 1778* (published 1781), 227–332.

Lee, A. (1914). Tables of the Gaussian "tail" functions when the "tail" is longer than the body, *Biometrika*, **10**, 208–214.

Lenk, P. J. (1988). The logistic normal distribution for Bayesian, nonparametric, predictive densities, *Journal of the American Statistical Association*, **83**, 509–516.

Leone, F., Nelson, L. S., and Nottingham, R. B. (1961). The folded normal distribution, *Technometrics*, **3**, 543–550. Correction, *Ibid.*, **22**, 452.

Leslie, R. T., and Brown, B. M. (1966). Use of range in testing heterogeneity of variance, *Biometrika*, **53**, 221–227.

Letac, G. (1981). Isotropy and sphericity: Some characterizations of the normal distribution, *Annals of Statistics*, **9**, 408–417.

Levedahl, J. W. (1989). Evaluating the effect of commodity donation programs using the truncated normal distribution, *Proceedings of the Kansas State University Conference on Applied Statistics in Agriculture*, G. A. Milliken and J. R. Schwenke (editors), 176–185, Lawrence: Kansas State University.

Levy, H., and Roth, L. (1936). *Elements of Probability*, Oxford: Oxford University Press.

Lévy, P. (1935). Propriétés asymptotiques des sommes de variables aléatoires independantes ou enchaînées, *Journal de Mathématiques Pures et Appliquées*, **14**, 347–402.

Lew, R. A. (1981). An approximation to the cumulative normal distribution with simple coefficients, *Applied Statistics*, **30**, 299–301.

Lifsey, J. D. (1965). Auxiliary estimating functions for doubly truncated normal samples, NASA Technical Memo. TM X-53221.

Lin, J. -T. (1988). Alternatives to Hamaker's approximations to the cumulative normal distribution and its inverse, *The Statistician*, **37**, 413–414.

Lin, J. -T. (1989). Approximating the normal tail probability and its inverse for use on a pocket calculator, *Applied Statistics*, **38**, 69–70.

Lin, J. -T. (1990). A simpler logistic approximation to the normal tail probability and its inverse, *Applied Statistics*, **39**, 255–257.

Lindeberg, J. W. (1922). Einie neue Herleitung des Exponentialgesetzes in der Wahrscheinlichkeitsrechnung, *Mathematische Zeitschrift*, **15**, 211–225.

Linders, F. J. (1930). On the addition of two normal frequency curves, *Nordic Statistical Journal*, **2**, 63–73.

Lingappaiah, G. S. (1983). Prediction in samples from a normal population, *Journal of Statistical Research*, **17**, 43–50.

Linnik, Yu. V. (1952). Linear statistics and the normal distribution law, *Doklady Akademii Nauk SSSR*, **83**, 353–355. (In Russian. English translation published by American Mathematical Society, 1961.)

Liseo, B. (1990). The skew-normal class of densities: Inferential aspects from a Bayesian viewpoint, *Statistica*, **50**, 59–70. (In Italian)

Lloyd, E. H. (1952). Least squares estimation of location and scale parameters using order statistics, *Biometrika*, **39**, 88–95.

Loève, M. (1950). Fundamental limit theorems of probability theory, *Annals of Mathematical Statistics*, **21**, 321–338.

Loève, M. (1963). *Probability Theory* (Third edition), New York: Van Nostrand.

Looney, S. W., and Gulledge, T. R., Jr. (1985). Probability plotting positions and goodness of fit for the normal distribution, *The Statistician*, **34**, 297–303.

Lord, E. (1947). The use of range in place of standard deviation in the *t*-test, *Biometrika*, **34**, 41–67. Correction, *Ibid.*, **39**, 442.

Lukacs, E. (1942). A characterization of the normal distribution, *Annals of Mathematical Statistics*, **13**, 91–93.

Lukacs, E. (1955). Applications of Faà di Bruno's formula in mathematical statistics, *American Mathematical Monthly*, **62**, 340–348.

Lukacs, E. (1956). Characterization of populations by properties of suitable statistics, *Proceedings of the Third Berkeley Symposium on Mathematical Statistics and Probability*, **2**, 195–214.

Lukacs, E. (1960). *Characteristic Functions*, London: Griffin.

Lukacs, E. (1962). On tube statistics and characterization problems, *Zeitschrift für Wahrscheinlichkeitstheorie und verwandte Gebiete*, **1**, 116–125.

Lukacs, E. (1976). On the stability of a characterization of the normal distribution, *Osterreichische Akademie de Wissenschaften*, **185**, 167–179.

Lukacs, E. (1977). Stability theorems for characterizations of the normal and of the degenerate distributions, In *Recent Developments in Statistics*, Proceedings of the 1976 European Meeting of Statisticians, September 6–11, 1976, Grenoble, France, pp. 535–536, Amsterdam: North-Holland.

Lukacs, E. (1980). Stability theorems for characterizations of the normal and of the degenerate distributions, *Asymptotic Theory of Statistical Tests and Estimation*, 205–229, New York: Academic Press.

Lukacs, E., and King, E. P. (1954). A property of the normal distribution, *Annals of Mathematics Statistics*, **25**, 389–394.

Lukacs, E., and Laha, R. G. (1964). *Applications of Characteristic Functions*, New York: Hafner.

Lyapunov, A. (1900). Sur une proposition de la théorie des probabilités, *Izvestiya Akademii Nauk SSSR*, Series *V*, **13**, 359–386. [Also Lyapunov, A. (1954). *Collected Works*, **1**, 125–151, Moscow: Akademia Nauk SSSR. (In Russian)]

Lyapunov, A. (1901). Nouvelle forme du théorème sur la limite de probabilité, *Mémoire Académia St. Petersbourg*, Series *VII*, **12**, 1–24. [Also Lyapunov, A. (1954). *Collected Works*, **1**, 157–176. Moscow: Akademia Nauk SSSR. (In Russian)]

Mahalanobis, P. C., Bose, S. S., Roy, P. R., and Banerjee, S. K. (1934). Tables of random samples from a normal distribution, *Sankhyā*, **1**, 289–328.

Marcinkiewicz, J. (1939). Sur une propriété de la loi de Gauss, *Mathematische Zeitschrift*, **44**, 612–618.

Markowitz, E. (1968). Minimum mean-square-error estimation of the standard deviation of the normal distribution, *The American Statistician*, **22**, No. 3, 26.

Marsaglia, G. (1964). Generating a variable from the tail of a normal distribution, *Technometrics*, **6**, 101–102.

Marsaglia, G. (1980). Generating a normal sample with given sample mean and variance, *Journal of Statistical Computation and Simulation*, **11**, 71–73.

Marsaglia, G., Ananthanarayanan, K., and Paul, N. J. (1976). Improvements on fast methods for generating normal random variables, *Information Processing Letters*, **5**, 27–30.

Marsaglia, G., and Bray, T. A. (1964). A convenient method for generating normal variables, *SIAM Review*, **6**, 260–264.

Marsaglia, G., MacLaren, D., and Bray, T. A. (1964). A fast procedure for generating normal random variables, *Communications of the ACM*, **7**, 4–10.

Martynov, G. V. (1981). Evaluation of the normal distribution function, *Journal of Soviet Mathematics*, **17**, 1857–1876.

Mathai, A. M., and Pederzoli, G. (1977). *Characterizations of the Normal Probability Law*, New York: Wiley.

McConnell, C. R. (1990). Comment on "Pocket-calculator approximation for areas under the standard normal curve" (**43**, 24–26), *The American Statistician*, **44**, 63–64.

McGillivray, W. R., and Kaller, C. L. (1966). A characterization of deviation from normality under certain moment assumptions, *Canadian Mathematical Bulletin*, **9**, 509–514.

McKay, A. T. (1933a). The distribution of $\sqrt{\beta_1}$ in samples of four from a normal universe, *Biometrika*, **25**, 204–210.

McKay, A. T. (1933b). The distribution of β_2 in samples of four from a normal universe, *Biometrika*, **25**, 411–415.

McKay, A. T. (1935). The distribution of the difference between the extreme observation and the sample mean in samples of n from a normal universe, *Biometrika*, **27**, 466–471.

McLachlan, G. J., and Basford, K. E. (1987). *Mixture Models: Inference and Applications to Clustering*, New York: Marcel Dekker.

McLeish, D. L. (1982). A robust alternative to the normal distribution, *Canadian Journal of Statistics*, **10**, 89–102.

Mead, R. (1966). A quick method of estimating the standard deviation, *Biometrika*, **53**, 559–564.

Mee, R. W. (1988). Estimation of the percentage of a normal distribution lying outside a specified interval, *Communications in Statistics—Theory and Methods*, **17**, 1465–1479.

Meeden, G. (1976). A special property of linear estimates of the normal mean, *Annals of Statistics*, **4**, 649–650.

Meeker, W. Q., and Hahn, G. J. (1982). Sample sizes for prediction intervals, *Journal of Quality Technology*, **14**, 201–206.

Mehran, F. (1975). Relationships between the UMVU estimators of the mean and median of a function of a normal distribution, *Annals of Statistics*, **3**, 457–460.

Meshalkin, L. D. (1968). On the robustness of some characterizations of the normal distribution, *Annals of Mathematical Statistics*, **39**, 1747–1750.

Mills, J. F. (1926). Table of the ratio: Area to bounding ordinate for any portion of normal curve, *Biometrika*, **18**, 395–400.

Mittal, M. M., and Dahiya, R. C. (1987). Estimating the parameters of a doubly truncated normal distribution, *Communications in Statistics—Simulation and Computation*, **16**, 141–159.

Miyakawa, K., Tamiya, T., and Kotani, K. (1985a). Some comments on the numerical evaluation of moments of order statistics from the normal distribution: Applications of the orthogonal inverse expansion, *TRU Mathematics, Science University of Tokyo*, **21/1**, 147–159.

Miyakawa, K., Tamiya, T., and Kotani, K. (1985b). Are the optimum spacings symmetric? In the case of normal distribution for small samples, *TRU Mathematics, Science University of Tokyo*, **21/1**, 161–171.

Moivre, A. de (1733). Approximatio ad Summam Ferminorum Binomii $(a + b)^n$ in Seriem expansi, *Supplementum II to Miscellanae Analytica*, 1–7.

Moivre, A. de (1738). *The Doctrine of Chances* (Second edition). (First edition 1718, third edition 1756; new impression of second edition, with some additional material, 1967. London: Frank Cass and Co.)

Molenaar, W. (1965). Survey of separation methods for two normal distributions , *Statistica Neerlandica*, **19**, 249–263. (In Dutch)

Monahan, J. F. (1981). Approximating the log of the normal cumulative, *Computer Science and Statistics: Proceedings of the Thirteenth Symposium on the Interface* W. F. Eddy (editor), 304–307, New York: Springer-Verlag.

Monk, D. T., and Owen, D. B. (1957). *Tables of the Normal Probability Integral*, Sandia Corporation. (Publications of the U.S. Department of Commerce, Office of Technical Services.)

Moore, P. G. (1956). The estimation of the mean of a censored normal distribution by ordered variables, *Biometrika*, **43**, 482–485.

Moran, P. A. P. (1980). Calculation of the normal distribution function, *Biometrika*, **67**, 675–676.

Muhammad, F., Ahmad, S., and Abiodullah, M. (1993). Use of probability weighted moments in the analysis of means, *Biometrical Journal*, **35**, 371–378.

Muller, M. E. (1959). A comparison of methods for generating normal deviates on digital computers, *Journal of the Association of Computing Machinery*, **6**, 376–383.

Nair, K. R. (1947). The distribution of the extreme deviate from the sample mean and its studentized form, *Biometrika*, **34**, 118–144.

Nair, K. R. (1949). The efficiency of Gini's mean difference, *Bulletin of the Calcutta Statistical Association*, **2**, 129–130.

Nair, K. R. (1950). Efficiencies of certain linear systematic statistics for estimating dispersion from normal samples, *Biometrika*, **37**, 182–183.

Nair, K. R. (1952). Tables of percentage points of the studentized extreme deviate from the sample mean, *Biometrika*, **39**, 189–191.

Nair, U. S. (1936). The standard error of Gini's mean difference, *Biometrika*, **28**, 428–438.

Nakamura, T. (1980). On the moment of positively truncated normal distribution, *Journal of the Japan Statistical Society*, **10**, 139–144.

National Bureau of Standards (1952). *A Guide to Tables of the Normal Probability Integral*, U.S. Department of Commerce, *Applied Mathematics Series*, **21**.

National Bureau of Standards (1953). *Tables of Normal Probability Functions*, U.S. Department of Commerce, *Applied Mathematics Series*, **23**. (Original edition, 1942.)

Nelson, L. S. (1976). Constructing normal probability paper, *Journal of Quality Technology*, **8**, 56–57.

Nelson, L. S. (1977). Tolerance factors for normal distributions, *Journal of Quality Technology*, **9**, 198–199.

Nelson, L. S. (1980). The folded normal distribution, *Journal of Quality Technology*, **12**, 236–238.

Nelson, L. S. (1983). Expected normal scores, a useful transformation, *Journal of Quality Technology*, **15**, 144–146.

Nelson, L. S. (1989). A stabilized normal probability plotting technique, *Journal of Quality Technology*, **21**, 213–215.

Nelson, W. (1982). *Applied Life Data Analysis*, New York: Wiley.

Nelson, W., and Schmee, J. (1981). Prediction limits for the last failure time of a (log) normal sample from early failures, *IEEE Transactions on Reliability*, **30**, 461–463.

Newman, D. (1939). The distribution of range in samples from a normal population, expressed in terms of an independent estimate of standard deviation, *Biometrika*, **31**, 20–30.

Norton, R. M. (1989). Pocket-calculator approximation for areas under the standard normal curve, *The American Statistician*, **43**, 24–26. Reply, *Ibid.*, **43**, 290.

Norton, R. M. (1990a). Reply to "Comments on 'Pocket-calculator approximation for areas under the standard normal curve' " (**43**, 24–26), *The American Statistician*, **44**, 64.

Norton, R. M. (1990b). Reply to "Comments on 'Pocket-calculator approximation for areas under the standard normal curve' " (**43**, 24–26), *The American Statistician*, **44**, 262.

Nye, H. H. (1966). On the stability of certain theorems characterizing normal distribution, *International Congress of Mathematicians: Information Bulletin*, **6**, 5.

Odeh, R. E. (1989a). Simultaneous two-sided prediction intervals to contain at least ℓ out of k future means from a normal distribution, *Communications in Statistics—Simulation and Computation*, **18**, 429–457.

Odeh, R. E. (1989b). Simultaneous one-sided prediction intervals to contain all of k future means from a normal distribution, *Communications in Statistics—Simulation and Computation*, **18**, 1557–1585.

Odeh, R. E., and Evans, J. O. (1974). Algorithm AS 70. The percentage points of the normal distribution, *Applied Statistics*, **23**, 96–97.

Odeh, R. E., and Owen, D. B. (1980). *Tables for Normal Tolerance Limits, Sampling Plans, and Screening*, New York: Marcel Dekker.

Oderfeld, J., and Pleszczyńska, E. (1961). Liniowy estymator odchylenia średniego w populacji normalnej, *Zastosowania Matematyki*, **6**, 111–117.

Ogawa, J. (1951). Contributions to the theory of systematic statistics, I, *Osaka Mathematics Journal*, **3**, 175–213.

Ogawa, J. (1962). Determination of optimum spacings in the case of normal distributions, *Contributions to Order Statistics*, A. E. Sarhan and B. G. Greenberg (editors), 272–283, New York: Wiley.

Ogawa, J. (1976). A note on the optimal spacing of the systematic statistics—Normal distribution, *Essays in Probability and Statistics*, pp. 467–474, Tokyo, Japan: Shinko Tsuko Company.

Ogawa, J. (1977). Optimal spacings for the simultaneous estimation of the location and scale parameters of a normal distribution based on selected two sample quantiles, *Journal of Statistical Planning and Inference*, **1**, 61–72.

Olsson, D. M. (1979). Fitting Johnson's S_B and S_U systems of curves using the method of maximum likelihood, *Journal of Quality Technology*, **11**, 211–217.

Oppenlander, J. E., Schmee, J., and Hahn, G. J. (1988). Some simple robust estimators of normal distribution tail percentiles and their properties, *Communications in Statistics—Theory and Methods*, **17**, 2279–2301.

Ord, J. K., and Bagchi, U. (1983). The truncated normal-gamma mixture as a distribution for lead time demand, *Naval Research Logistics Quarterly*, **30**, 359–365.

Owen, D. B. (1962). *Handbook of Statistical Tables*, Reading, MA: Addison-Wesley.

Owen, D. B. (1980). The table of normal integrals, *Communications in Statistics—Simulation and Computation*, **9**, 389–419. Correction, *Ibid.*, **10**, 541.

Owen, D. B., and Hua, T. A. (1977). Tables of confidence limits on the tail area of the normal distribution, *Communications in Statistics—Simulation and Computation*, **7**, 183–201.

Öztürk, A., and Aly, E. E. A. A. (1991). Simple approximation for the moments of normal order statistics, *The Frontiers of Statistical Computation, Simulation, and Modeling*, **1**, P. R. Nelson, E. J. Dudewicz, A. Öztürk, and E. C. van der Meulen (editors), pp. 151–170, Syracuse, NY: American Sciences Press.

Pachares, J. (1959). Tables of the upper 10% points of the studentized range, *Biometrika*, **46**, 461–466.

Page, E. (1977). Approximations to the cumulative normal function and its inverse for use on a pocket calculator, *Applied Statistics*, **26**, 75–76.

Pakshirajan, P., and Mohan, N. C. (1971). A characterization of the normal law, *Annals of the Institute of Statistical Mathematics*, **21**, 529–532.

Parrish, R. S. (1992a). Computing expected values of normal order statistics, *Communications in Statistics—Simulation and Computation*, **21**, 57–70.

Parrish, R. S. (1992b). Computing variances and covariances of normal order statistics, *Communications in Statistics—Simulation and Computation*, **21**, 71–101.

Parsonson, S. L. (1978). An approximation to the normal distribution function, *Mathematical Gazette*, **62**, 118–121.

Parthasarathy, K. R. (1976). Characterization of the normal law through the local independence of certain statistics, *Sankhyā, Series A*, **38**, 174–178.

Paskevich, V. S. (1953). On a property of control charts used in quality control, *Prikladnaya Matematika i Mekhanika*, **17**, 49–57. (In Russian)

Patel, J. K., and Read, C. B. (1981). *Handbook of the Normal Distribution*, New York: Marcel Dekker.

Patil, G. P. (1964). On certain compound Poisson and compound binomial distribution, *Sankhyā, Series A*, **26**, 293–294.

Patil, S. A. (1961). Monotonicity property of indices of information functions of samples censored below and samples censored above from normal distribution, *Bulletin of the Calcutta Statistical Association*, **10**, 153–162.

Pearson, E. S. (1930). A further development of tests for normality, *Biometrika*, **22**, 239–249.

Pearson, E. S. (1931). Note on tests for normality, *Biometrika*, **22**, 423–424.

Pearson, E. S. (1932). The percentage limits for the distribution of range in samples from a normal population, *Biometrika*, **24**, 404–417.

Pearson, E. S. (1936). Note on probability levels for $\sqrt{b_1}$, *Biometrika*, **28**, 306–307.

Pearson, E. S. (1942). The probability integral of the range in samples of n observations from a normal population. Part I: Foreword and tables, *Biometrika*, **32**, 301–308.

Pearson, E. S. (1945). The probability integral of the mean deviation (Editorial Note), *Biometrika*, **33**, 252–253.

Pearson, E. S. (1952). Comparison of two approximations to the distribution of the range in small samples from normal populations, *Biometrika*, **39**, 130–136.

Pearson, E. S. (1965). Tables of percentage points of $\sqrt{b_1}$ and b_2 in normal samples; a rounding off, *Biometrika*, **52**, 282–285.

Pearson, E. S., and Chandra Sekar, C. (1936). The efficiency of statistical tools and a criterion for the rejection of outlying observations, *Biometrika*, **28**, 308–320.

Pearson, E. S., and Hartley, H. O. (1942). The probability integral of the range in samples of n observations from a normal population, *Biometrika*, **32**, 301–310.

Pearson, E. S., and Hartley, H. O. (1943). Tables of the probability integral of the 'Studentized' range, *Biometrika*, **33**, 89–99.

Pearson, E. S., and Hartley, H. O. (1948). *Biometrika Tables for Statisticians* (Second edition), **1**, Cambridge: Cambridge University Press.

Pearson, E. S. and Hartley, H. O. (1970). *Biometrika Tables for Statisticians* (Third edition), **1**, Cambridge: Cambridge University Press.

Pearson, E. S., and Hartley, H. O. (1972). *Biometrika Tables for Statisticians*, **2**, Cambridge: Cambridge University Press.

Pearson, E. S., Hartley, H. O., and David, H. A. (1954). The distribution of the ratio, in a single normal sample, of range to standard deviation, *Biometrika*, **41**, 482–493.

Pearson, E. S., and Stephens, M. A. (1964). The ratio of range to standard deviation in the same normal sample, *Biometrika*, **51**, 484–487.

Pearson, K. (1894). Contributions to the mathematical study of evolution, *Philosophical Transactions of the Royal Society of London, Series A*, **185**, 71–110.

Pearson, K. (1920). On the probable errors of frequency constants, Part III, *Biometrika*, **13**, 112–132.

Pearson, K. (1924). Historical note on the origin of the normal curve of errors, *Biometrika*, **16**, 402–404.

Pearson, K. (ed.) (1931). *Tables for Statisticians and Biometricians*, Part II, Cambridge: Cambridge University Press.

Pearson, K. (1931). Historical note on the distribution of the standard deviations of samples of any size drawn from an indefinitely large normal parent population, *Biometrika*, **23**, 416–418.

Pearson, K., and Lee, A. (1908). On the generalized probable error in multiple normal correlation, *Biometrika*, **6**, 59–68.

Pepper, J. (1932). The sampling distribution of the third moment coefficient—an experiment, *Biometrika*, **24**, 55–64.

Persson, T., and Rootzén, H. (1977). Simple and highly efficient estimators for a type I censored normal sample, *Biometrika*, **64**, 123–128.

Pillai, K. C. S. (1952). On the distribution of "studentized" range, *Biometrika*, **39**, 194–195.

Pleszczyńska, E. (1963). Tabela wag liniowego estymatora odchylenia średniego w populacji normalnej, *Zastosowania Matematyki*, **7**, 117–124.

Plucinska, A. (1965). O pewnych zagadieniach zwiazanych z podzialem populacji normalnej na czesci, *Zastosowania Matematyki*, **8**, 117–125.

Pólya, G. (1932). Verleitung des Gauss'chen Fehlergesetzes aus einer Funktional-gleichung, *Mathematische Zeitschrift*, **18**, 96–108.

Pólya, G. (1945). Remarks on computing the probability integral in one and two dimensions, *Proceedings of the 1st Berkeley Symposium on Mathematical Statistics and Probability*, 63–78.

Porter, T. M. (1985). The mathematics of society: variation and error in Quetelet's statistics, *The British Journal for the History of Science*, **18**, 51–69.

Prakasa Rao, B. L. S. (1979). Characterizations of distributions through some identities, *Journal of Applied Probability*, **16**, 902–908.

Prasad, A. (1955). Bi-modal distributions derived from the normal distribution, *Sankhyā*, **14**, 369–374.

Pratt, J. W. (1989). Comment on "Pocket-calculator approximation for areas under the standard normal curve," *The American Statistician*, **43**, 289–290.

Prescott, P. (1968). A simple method of estimating dispersion from normal samples, *Applied Statistics*, **17**, 70–74.

Prescott, P. (1979). A mean difference estimator of standard deviation in asymmetrically censored normal samples, *Biometrika*, **66**, 684–686.

Proschan, F. (1953). Confidence and tolerance intervals for the normal distribution, *Journal of the American Statistical Association*, **48**, 550–564.

Pugh, G. A. (1989). Computing with Gaussian distribution: A survey of algorithms, *ASQC Technical Conference Transactions*, 496–501.

Pullin, D. I. (1979). Generation of normal variates with given sample mean and variance, *Journal of Statistical Computation and Simulation*, **9**, 303–309.

Raab, D. H., and Green, E. H. (1961). A cosine approximation to the normal distribution, *Psychometrika*, **26**, 447–450.

RAND Corporation (1955). *A Million Random Digits with 100,000 Normal Deviates*, Glencoe, IL: Free Press.

Ramasubramanian, S. (1985). A characterization of the normal distribution, *Sankhyā*, *Series A*, **47**, 410–414.

Rao, C. R. (1967). On some characterizations of the normal law, *Sankhyā*, *Series A*, **29**, 1–14.

Rao, C. R., Mitra, S. K., and Mathai, A. (1966). *Formulae and Tables for Statistical Work*, Calcutta: Statistical Publishing Society.

Rao, J. N. K. (1958). A characterization of the normal distribution, *Annals of Mathematical Statistics*, **29**, 914–919.

Rao, M. M. (1963). Characterizing normal law and a nonlinear integral equation, *Journal of Mathematics and Mechanics*, **12**, 869–880.

Read, C. B. (1973). An application of a result of Watson to estimation of the normal standard deviation, *Communications in Statistics*, **1**, 183–185.

Read, C. B. (1985). Normal distribution, *Encyclopedia of Statistical Sciences*, **6**, S. Kotz, N. L. Johnson, and C. B. Read (editors), 347–359, New York: Wiley.

Renner, R. M. (1976). Evaluation by power series of the means of normal order statistics of samples of sizes six and seven, *Mathematical Chronicle*, **4**, 141–147.

Revfeim, K. J. A. (1990). Comment on "Pocket-calculator approximation for areas under the standard normal curve" (**43**, 24–26), *The American Statistician*, **44**, 63.

Rhiel, G. S. (1986). The effect of non-normality on the use of the range in estimating the population standard deviation, *Journal of Statistical Computation and Simulation*, **24**, 71–82.

Riffenburgh, R. H. (1967). Transformation for statistical distribution approximately normal but of finite sample range, Report NUWC-TP-19, Naval Undersea Warfare Center, San Diego, CA. [Abbreviated version *Technometrics*, **11**, (1969), 47–59.]

Rivest, L. -P. (1981). On the sum of contaminated normals, *The American Statistician*, **35**, 155–156.

Robert, C. (1991). Generalized inverse normal distributions, *Statistics & Probability Letters*, **11**, 37–41.

Rouncefield, M. (1990). Is it normal? How to use probability (or normal) graph paper, *Teaching Statistics*, **12**, 6–8.

Royston, J. P. (1982). Expected normal order statistics (exact and approximate), *Applied Statistics*, **31**, 161–165.

Ruben, H. (1954). On the moments of order statistics in samples from normal populations, *Biometrika*, **41**, 200–227.

Ruben, H. (1956). On the moments of the range and product moments of extreme order statistics in normal samples, *Biometrika*, **43**, 458–460.

Ruben, H. (1960). On the geometrical significance of the moments of order statistics, and of deviations of order statistics from the mean in samples from Gaussian populations, *Journal of Mathematics and Mechanics*, **9**, 631–638.

Ruben, H. (1962). The moments of the order statistics and of the range in samples from normal populations, *Contributions to Order Statistics*, A. E. Sarhan and B. G. Greenberg (editors), 165–190, New York: Wiley.

Ruben, H. (1974). A new characterization of the normal distribution through the sample variance, *Sankhyā, Series A*, **36**, 379–388.

Ruben, H. (1975). A further characterization of normality through the sample variance, *Sankhyā, Series A*, **37**, 72–81.

Ruben, H. (1976). A characterization of normality through the general linear model, *Sankhyā, Series A*, **38**, 186–189.

Rukhin, A. L. (1986). Estimating normal tail probabilities, *Naval Research Logistics Quarterly*, **33**, 91–99.

Sadikova, S. M. (1966). On two-dimensional analogs of an inequality of C. -G. Esseen and their application to the central limit theorem, *Teoriya Veroyatnostei i ee Primeneniya*, **11**, 370–380. (In Russian)

Sakasegawa, H. (1978). On a generation of normal pseudo-random numbers, *Annals of the Institute of Statistical Mathematics*, **30**, 271–280.

Salvia, A. A. (1985). Reliability application of the alpha distribution, *IEEE Transactions on Reliability*, **34**, 251–252.

Sampson, A. R. (1975). Characterizing exponential family distributions by moment generating functions, *Annals of Statistics*, **3**, 747–753.

Sandelius, M. (1957). On the estimation of the standard deviation of a normal distribution from a pair of percentiles, *Skandinavisk Aktuarietidskrift*, **40**, 85–88.

Sapogov, N. A. (1956). Stability problem for Cramér's theorem, *Leningrad Universitet Vestnik (II)*, 60–64.

Sarhan, A. E. (1954). Estimation of the mean and standard deviation by order statistics, Part I, *Annals of Mathematical Statistics*, **25**, 317–328.

Sarhan, A. E., and Greenberg, B. G. (1962). The best linear estimates for the parameters of the normal distribution, *Contributions to Order Statistics*, A. E. Sarhan and B. G. Greenberg, (editors), 206–269, New York: Wiley.

Sarkadi, K. (1960). On testing for normality, *Matematikai Kutato Intezetenet Közlemenyei, Hungarian Academy of Science*, **5**, 269–275.

Saw, J. G., and Chow, B. (1966). The curve through the expected values of ordered variates and the sum of squares of normal scores, *Biometrika*, **53**, 252–255.

Sazanov, V. V. (1967). On the rate of convergence in the multidimensional central limit theorem, *Teoriya Veroyatnostei i ee Primeneniya*, **12**, 82–95. (In Russian)

Schader, M., and Schmid, F. (1984). Computation of maximum likelihood estimates for μ and σ from a grouped sample of a normal population: A comparison of algorithms, *Statistische Hefte*, **25**, 245–258.

Schader, M., and Schmid, F. (1988). Small sample properties of the maximum likelihood estimators of the parameters μ and σ from a group sample of a normal population, *Communications in Statistics—Simulation and Computation*, **17**, 229–239.

Schaeffer, L. R., van Vleck, L. D., and Velasco, J. A. (1970). The use of order statistics with selected records, *Biometrics*, **26**, 854–859.

Schay, G. (1965). Approximation of sums of independent random variables by normal random variables, *Zeitschrift für Wahrscheinlichkeitstheorie und verwandte Gebiete*, **4**, 209–216.

Schmeiser, B. W. (1979). Approximations to the inverse cumulative normal function for use on hand calculators, *Applied Statistics*, **28**, 175–176.

Schneider, H. (1984). Simple and highly efficient estimators for censored normal samples, *Biometrika*, **71**, 412–414.

Schneider, H. (1986). *Truncated and Censored Samples from Normal Populations*, New York: Marcel Dekker.

Schonfelder, J. L. (1978). Chebyshev expansions for the error and related functions, *Mathematics of Computation*, **32**, 1232–1240.

Schucany, W. R., and Gray, H. L. (1968). A new approximation related to the error function, *Mathematics of Computation*, **22**, 201–202.

Schuster, E. F. (1983). Generating normal variates by summing three uniforms, *Computer Science and Statistics: Proceedings of the Fifteenth Symposium on the Interface*, J. E. Gentle (editor), 324–327, Amsterdam: North-Holland.

Selvin, S. (1976). A graphical estimate of the population mean from censored normal data, *Applied Statistics*, **25**, 8–11.

Sengupta, J. M., and Bhattacharya, N. (1958). Tables of random normal deviates, *Sankhyā*, **20**, 250–286.

Shah, A. K. (1985). A simpler approximation for areas under the standard normal curve, *The American Statistician*, **39**, 80. Correction, *Ibid.*, **39**, 327.

Shah, S. M., and Jaiswal, M. C. (1966). Estimation of parameters of doubly truncated normal distribution from first four sample moments, *Annals of the Institute of Statistical Mathematics*, **18**, 107–111.

Shapiro, S. S. (1986). *How to Test Normality and Other Distributional Assumptions*, ASQC Press, Milwaukee, Wisconsin.

Shapiro, S. S., and Wilk, M. B. (1965). An analysis of variance test for normality (complete samples), *Biometrika*, **52**, 591–611.

Sharp, W. E. (1990). Comment on "Pocket-calculator approximation for areas under the standard normal curve" (**43**, 24–26), *The American Statistician*, **44**, 261.

Shea, B. L., and Scallon, A. J. (1988). Approximating the covariance matrix of normal order statistics, *Applied Statistics*, **37**, 151–155.

Shenton, L. R., and Bowman, K. O. (1975). Johnson's S_U and the skewness and kurtosis statistics, *Journal of the American Statistical Association*, **70**, 220–228.

Shepp, L. (1964). Normal functions of normal random variables, *SIAM Review*, **6**, 459–460.

Sheppard, W. F. (1903). New tables of the probability integral, *Biometrika*, **2**, 174–190.

Sheppard, W. F. (1907). Table of deviates of the normal curve, *Biometrika*, **5**, 404–406.

Sheppard, W. F. (1939). The Probability Integral, *British Association for the Advancement of Science, Mathematical Tables*, **7**.

Shimizu, R. (1961). A characterization of the normal distribution, *Annals of the Institute of Statistical Mathematics*, **13**, 53–56.

Shimizu, R. (1962). Characterization of the normal distribution, Part II, *Annals of the Institute of Statistical Mathematics*, **14**, 173–178.

Shore, H. (1982). Simple approximations for the inverse cumulative function, the density function and the loss integral of the normal distribution, *Applied Statistics*, **31**, 108–114.

Shore, H. (1990). Comment on "Pocket-calculator approximation for areas under the standard normal curve" (**43**, 24–26), *The American Statistician*, **44**, 261.

Sinha, S. K. (1983). Folded normal distribution: A Bayesian approach, *Journal of the Indian Statistical Association*, **21**, 31–34.

Sinha, S. K. (1985). Bayes estimation of the reliability function of normal distribution, *IEEE Transactions on Reliability*, **34**, 360–362.

Skitovich, V. P. (1953). On a property of the normal distribution, *Doklady Akademii Nauk SSSR*, **89**, 217–219. (In Russian)

Skitovich, V. P. (1954). Linear Forms of Independent Random Variables and the Normal Distribution, *Izvestiya Akademii Nauk SSSR, Series Mathematics 18*, 185–200.

Smirnov, N. V. (ed.) (1960). *Tables of the Normal Integral, Normal Density and Its Normalized Derivatives*, Moscow: Akademia Nauk SSSR. (In Russian)

Smirnov, N. V. (ed.) ([1960] 1965). *Tables of the Normal Probability Integral, Normal Density and Its Normalized Derivatives*, New York: Macmillan. (Translation)

Sobel, M., and Tong, Y. L. (1976). Estimation of a normal percentile by grouping, *Journal of the American Statistical Association*, **71**, 189–192.

Stadje, W. (1988). A generalized maximum likelihood characterization of the normal distribution, *Metrika*, **35**, 93–97.

Steffensen, J. F. (1937). On the semi-normal distribution, *Skandinavisk Aktuarietidskrift*, **20**, 60–74.

Stepniak, C. (1991). On characterization of the normal law in the Gauss-Markov model, *Sankhyā, Series A*, **53**, 115–117.

Stigler, S. M. (1982). A modest proposal: A new standard for the normal, *The American Statistician*, **36**, 137–138.

Strecock, A. J. (1968). On the calculation of the inverse of the error function, *Mathematics of Computation*, **22**, 144–158.

Subramanya, M. T. (1965). Tables of two-sided tolerance intervals for normal distribution $N(\mu, \sigma)$ with μ unknown and σ known, *Journal of the Indian Statistical Association*, **3**, 195–201.

Sugiura, N., and Gomi, A. (1985). Pearson diagrams for truncated normal and truncated Weibull distributions, *Biometrika*, **72**, 219–222.

Swamy, P. S. (1963). On the amount of information supplied by truncated samples of grouped observations in the estimation of the parameters of normal populations, *Biometrika*, **50**, 207–213.

Swan, A. V. (1977). A remark on Algorithm AS16: Maximum likelihood estimation from grouped and censored normal data, *Applied Statistics*, **26**, 122.

Szekely, G. (1985). Multiplicative infinite divisibility of standard normal and gamma distribution, *Proceedings of the Seventh Conference on Probability Theory*, 1982, Brasov, Romania, Editura Academiei, Bucharest IVNU Science Press, Utrecht, pp. 579–581.

Tables of the Error Function and Its First Twenty Derivatives (1952). Cambridge: Harvard University Computation Laboratory.

Tables of Normal Distribution (1961). *Advanced Series of Mathematics and Engineering Tables*, **3**, Tokyo: Corona.

Tables of Probability Functions (1959). **2**, Computing Center, Moscow: Akademia Nauk SSSR.

Tallis, G. M., and Young, S. S. M. (1962). Maximum likelihood estimation of parameters of the normal, log-normal, truncated normal and bivariate normal distributions from grouped data, *Australian Journal of Statistics*, **4**, 49–54.

Talwalker, S. (1980). On characterizing the normal and Poisson distributions, *Communications in Statistics—Theory and Methods*, **9**, 347–354.

Tamhankar, M. V. (1967). A characterization of normality, *Annals of Mathematical Statistics*, **38**, 1924–1927.

Tardiff, R. M. (1980). Generalized normal distribution, *Stochastica*, **4**, 221–225.

Taylor, B. J. R. (1965). The analysis of polymodal frequency distributions, *Journal of Animal Ecology*, **34**, 445–452.

Teicher, H. (1961). Maximum likelihood characterization of distributions, *Annals of Mathematical Statistics*, **32**, 1214–1222.

Teichroew, D. (1956). Tables of expected values of order statistics and products of order statistics for samples of size twenty and less from the normal distribution, *Annals of Mathematical Statistics*, **27**, 410–426.

Teichroew, D. (1957). The mixture of normal distributions with different variances, *Annals of Mathematical Statistics*, **28**, 510–512.

Teichroew, D. (1962). Tables of lower moments of order statistics for samples from the normal distribution, *Contributions to Order Statistics*, A. E. Sarhan and B. G. Greenberg, (editors), New York: Wiley.

Thompson, W. R. (1935). On a criterion for the rejection of observations and the distribution of the ratio of deviation to sample standard deviation, *Annals of Mathematical Statistics*, **6**, 214–219.

Tietjen, G. L., Kahaner, D. K., and Beckman, R. J. (1977). Variances and covariances of the normal order statistics for sample sizes 2 to 50, *Selected Tables in Mathematical Statistics*, **5**, 1–74.

Tiku, M. L. (1967). Estimating the mean and standard deviation from a censored normal sample, *Biometrika*, **54**, 155–165.

Tippett, L. H. C. (1925). On the extreme individuals and the range of samples taken from a normal population, *Biometrika*, **17**, 364–387.

Tippett, L. H. C. (1927). *Random Sampling Numbers* (Tracts for Computers XV), Cambridge: Cambridge University Press.

Titterington, D. M. (1989). Logistic-normal distribution, *Encyclopedia of Statistical Sciences*, Supplementary volume, S. Kotz, N. L. Johnson, and C. B. Read (editors), 90–91, New York: Wiley.

Titterington, D. M., Smith, A. F. M., and Makov, U. E. (1985). *Statistical Analysis of Finite Mixture Distributions*, New York: Wiley.

Tranquilli, G. B. (1966). Sul teorema di Basu-Darmois, *Giornale dell'Istituto Italiano degli Attuari*, **29**, 135–152.

Trotter, H. F. (1959). An elementary proof of the central limit theorem, *Archiv der Mathematik*, **10**, 226–234.

Truax, D. R. (1953). An optimum slippage test for the variances of k normal distributions, *Annals of Mathematical Statistics*, **24**, 669–674.

Tukey, J. W. (1949). Comparing individual means in the analysis of variance, *Biometrics*, **5**, 99–114.

Tukey, J. W. (1962). The future of data analysis, *Annals of Mathematical Statistics*, **33**, 1–67.

Vianelli, S. (1983). The family of normal and lognormal distributions of order r, *Metron*, **41/1**, 3–10.

Viskov, O. V. (1985). An algebraic characterization of the normal distribution, *Theory of Probability and Its Applications*, **30**, 141–143.

Wald, A., and Wolfowitz, J. (1946). Tolerance limits for a normal distribution, *Annals of Mathematical Statistics*, **17**, 208–215.

Walker, H. M. (1924). *Studies in the History of Statistical Methods*, Baltimore: William and Wilkins.

Wallace, D. (1959). A corrected computation of Berry's bound for the central limit theorem error, *Statistics Research Center, University of Chicago*.

Watanabe, Y., Isida, M., Taga, S., Ichijô, Y., Kawase, T., Niside, G., Takeda, Y., Horisuzi, A., and Kuriyama, I. (1957). Some contributions to order statistics, *Journal of Gakugei, Tokushima University*, **8**, 41–90.

Watanabe, Y., Yamamoto, T., Satô, T., Fujimoto, T., Inoue, M., Suzuki, T., and Uno, T. (1958). Some contributions to order statistics (continued), *Journal of Gakugei, Tokushima University*, **9**, 31–86.

Weichselberger, K. (1961). Über ein graphisches Verfahren zur Ternnung von Mischverteilungen und zur Identifikation kapierter Normalverteilungen bei grossem Stichprobenumfang, *Metrika*, **4**, 178–229.

Weinstein, M. A. (1964). The sum of values from a normal and a truncated normal distribution (Answer to Query), *Technometrics*, **6**, 104–105. (See also answers by M. Lipow, N. Mantel, and J. W. Wilkinson, *Technometrics*, **6**, 469–471.)

Weissberg, A., and Beatty, G. H. (1960). Tables of tolerance-limit factors for normal distributions, *Technometrics*, **2**, 483–500.

Wesolowski, J. (1987). A regressional charcterization of the normal law, *Statistics & Probability Letters*, **6**, 11–12.

Wesolowski, J. (1990). Personal communication (to W. Bryc).

Wessels, J. (1964). Multimodality in a family of probability densities, with application to a linear mixture of two normal densities, *Statistica Neerlandica*, **18**, 267–282.

Wetherill, G. B. (1965). An approximation to the inverse normal function suitable for the generation of random normal deviates on electronic computers, *Applied Statistics*, **14**, 201–205.

White, J. S. (1970). Tables of normal percentile points, *Journal of the American Statistical Association*, **65**, 635–638.

Whitmore, G. A. (1986). Prediction limits for a univariate normal observation, *The American Statistician*, **40**, 141–143.

Wichura, M. J. (1988). The percentage points of the normal distribution, *Applied Statistics*, **37**, 477–484.

Wichura, M. J. (1990). A note on Das's approximation to Mill's ratio, *Statistics & Probability Letters*, **10**, 297–299.

Wilf, H. S. (1988). The quest for normality, *Educational and Psychological Measurement*, **48**, 711–712.

Wilkinson, G. N. (1961). The eighth moment and cumulant of the distribution for normal samples of the skewness coefficient, $\gamma = k_3 k_2^{-3/2}$, *Australian Journal of Statistics*, **3**, 108–109.

Wishart, J. (1930). The derivation of certain high order sampling product moments from a normal population, *Biometrika*, **22**, 224–238.

Wold, H. (1948). *Random Normal Deviates* (Tracts for Computers, XXV), Cambridge: Cambridge University Press. (Also *Statistica Uppsala*, **3**.)

Wolfowitz, J. (1946). Confidence limits for the fraction of a normal population which lies between two given limits, *Annals of Mathematical Statistics*, **17**, 483–488.

Wolynetz, M. S. (1979). Maximum likelihood estimation from confined and censored normal data, *Applied Statistics*, **28**, 185–195.

Yamauti, Z. (ed.) (1972). *Statistical Tables and Formulas with Computer Applications*, JSA-1972, Tokyo, Japan: Japanese Standard Association.

Yanushkyavichyus, R. V. (1988). Convolution equations in problems of the stability of characterization of probability laws, *Teoriya Veroyatnostei i ee Primeneniya*, **33**, 720–734. (English translation: *Theory of Probability and Its Applications*, **33**, 668–680.

Yanushkyavichyus, R. V. (1989). Doctoral thesis, *Abstracts*, *Vilnius, Lithuaniya*: Vilnius University.

Zackrisson, U. (1959). The distribution of "Student's" t in samples from individual nonnormal populations, *Publications of the Statistical Institute, Gothenburg, University of Gothenburg*, No. 6.

Zacks, S. (1966). Unbiased estimation of the common mean of two normal distributions based on small samples of equal size, *Journal of the American Statistical Association*, **61**, 467–476.

Zahl, S. (1966). Bounds for the central limit theorem error, *SIAM Journal of Applied Mathematics*, **14**, 1225–1245.

Zehna, P. W. (1991). On proving that $\bar{X}$ and S^2 are independent, *The American Statistician*, **45**, 121–122.

Zeigler, R. K. (1965). A uniqueness theorem concerning moment distributions, *Journal of the American Statistical Association*, **60**, 1203–1206.

Zelen, M., and Severo, N. C. (1964). Probability functions, *Handbook of Mathematical Functions*, M. Abramowitz and I. A. Stegun (editors), 925–995, U. S. Department of Commerce, *Applied Mathematics Series*, **55**.

Zinger, A. A. (1958). New results on independent statistics, *Trudy Soveshchaniya Teorii Veroyatnostei i Matematicheskoi Statistike, Erevan, AN ArmSSR*, 103–105. (In Russian)

Zinger, A. A. (1977). On the characterization of the normal law by identically distributed linear statistics, *Sankhyā, Series A*, **39**, 232–242.

Zinger, A. A., and Linnik, Yu. V. (1964). On a characterization of the normal distribution, *Teoriya Veroyatnostei i ee Primeneniya*, **9**, 692–695. (In Russian)

Zolotarev, V. M. (1967). A sharpening of the inequality of Berry-Esseen, *Zeitschrift für Wahrscheinlichkeitstheorie und verwandte Gebiete*, **8**, 332–342.

CHAPTER 14

Lognormal Distributions

1 INTRODUCTION

The idea of a transformation such that the transformed variable is normally distributed was encountered in Chapter 12. In Section 4.3 of that chapter some specific transformations were introduced. Of these the most commonly used, and the only one of sufficient importance to merit a separate chapter, is the simple logarithmic transformation.

If there is a number θ such that $Z = \log(X - \theta)$ is normally distributed, the distribution of X is said to be *lognormal*. For this to be the case, it is clearly necessary that X take any value exceeding θ but have zero probability of taking any value less than θ. The term "lognormal" can also be applied to the distribution of X if $\log(\theta - X)$ is normally distributed, X having zero probability of exceeding θ. However, since replacement of X by $-X$ (and θ by $-\theta$) reduces this situation to the first, we will consider only the first case.

The distribution of X can be defined by the equation

$$U = \gamma + \delta \log(X - \theta) \tag{14.1}$$

where U is a unit normal variable and γ, δ and θ are parameters. From (14.1) it follows that the probability density function of X is

$$p_X(x) = \delta\left[(x - \theta)\sqrt{2\pi}\right]^{-1} \exp\left[-\tfrac{1}{2}\{\gamma + \delta \log(x - \theta)\}^2\right], \qquad x > \theta. \tag{14.2}$$

(We take $\delta > 0$, without loss of generality, since $-U$ has the same distribution as U.) An alternative notation replaces γ and δ by the expected value ζ and standard deviation σ of $Z = \log(X - \theta)$. The two sets of parameters are related by the equations

$$\zeta = -\frac{\gamma}{\delta}, \qquad \sigma = \delta^{-1},$$

so that (14.1) becomes

$$U = \frac{\log(X - \theta) - \zeta}{\sigma}, \qquad (14.1)'$$

and (14.2) becomes

$$p_X(x) = \left[(x - \theta)\sqrt{2\pi}\,\sigma\right]^{-1} \exp\left[-\frac{1}{2}\frac{\{\log(x - \theta) - \zeta\}^2}{\sigma^2}\right](x > \theta). \quad (14.2)'$$

The lognormal distribution is sometimes called the *antilognormal* distribution. This name has some logical basis in that it is not the distribution of the logarithm of a normal variable (this is not even always real) but of an exponential—that is, antilogarithmic—function of such a variable. However, "lognormal" is most commonly used, and we will follow this practice. The minor variants *logarithmic-* or *logarithmico*-normal have been used, as have the names of pioneers in its development, notably *Galton* (1879) and *McAlister* (1879), *Kapteyn* (1903), *van Uven* (1917a), and *Gibrat* (1930) (see Section 2). When applied to economic data, particularly production functions, it is sometimes called the *Cobb-Douglas* distribution [e.g., Dhrymes (1962)].

It can be seen that a change in the value of the parameter θ affects only the location of the distribution. It does not affect the variance or the shape (or any property depending only on differences between values of the variable and its expected value). It is convenient to assign θ a particular value for ease of algebra, with the understanding that many of the results so obtained can be transferred to the more general distribution. In many applications θ is "known" to be zero (so that $\Pr[X \leq 0] = 0$ or X is a "positive random variable"). This important case has been given the name *two-parameter lognormal distribution* (parameters γ, δ or ζ, σ). For this distribution (14.1) becomes

$$U = \gamma + \delta \log X \qquad (14.3)$$

and (14.1)' becomes

$$U = \frac{\log X - \zeta}{\sigma}. \qquad (14.4)$$

The general case (with θ not necessarily zero) can be called the *three-parameter lognormal distribution* (parameters γ, δ, θ or ζ, σ, θ).

A so-called four parameter lognormal distribution has been defined by

$$U = \gamma + \delta \log\left\{\frac{X - \theta}{\lambda}\right\}. \tag{14.5}$$

Since (14.5) can be rewritten

$$U = \gamma' + \delta \log(X - \theta) \tag{14.5}'$$

with $\gamma' = \gamma - \delta \log \lambda$, it is really only a three-parameter lognormal that is defined by (14.1).

2 HISTORICAL REMARKS

Galton (1879) pointed out that, if $X_1, X_2, \ldots, X_n$ are independent positive random variables and

$$T_n = \prod_{j=1}^{n} X_j,$$

then

$$\log T_n = \sum_{j=1}^{n} \log X_j,$$

and if the independent random variables $\log X_j$'s are such that a central limit type of result applies, then the standardized distribution of $\log T_n$ would tend to a unit normal distribution as n tends to infinity. The limiting distribution of T_n would then be (two-parameter) lognormal. In an accompanying paper McAlister (1879) obtained expressions for the mean, median, mode, variance, and certain percentiles of the distribution.

   Subsequent to this, little material was published relating to the lognormal distribution until 1903 when Kapteyn again considered its genesis on the lines described above. [Fechner (1897) mentioned the use of the distribution in the description of psychophysical phenomena but gave little emphasis to this topic.] Kapteyn and van Uven (1916) gave a graphical method (based on the use of sample quantiles) for estimating parameters, and in the following years there was a considerable increase in published information on the lognormal and related distributions. Wicksell (1917) obtained formulas for the higher moments (in a study of the distribution of ages at first marriage), while van Uven (1917a, b) considered transformations to normality from a more general point of view; Nydell (1919) obtained approximate formulas for the standard deviations of estimators obtained by the method of moments. Estimation from percentile points was described by Davies (1925, 1929) and tables to facilitate estimation from sample moments were published by Yuan (1933). Unbiased estimators based on sample moments were constructed by Finney (1941).

From 1930 onward fields of application of the lognormal distribution have increased steadily. Gibrat (1930, 1931) found the distribution usefully representing the distribution of size for varied kinds of 'natural' economic units. Gibrat's "law of proportional effect" can be expressed as

$$X_j = X_{j-1}(1 + Z_j), \tag{14.6}$$

where $X_0, X_1, X_2, \ldots, X_n$ is a sequence of random variables and $\{Z_j\}$ is a set of mutually independent random variables, statistically independent of $\{X_j\}$.

Formula (14.6) leads to

$$X_n = X_0 \prod_{j=1}^{n} (1 + Z_j).$$

For Z_j small compared to 1, we have

$$\log X_n \doteq \log X_0 + \sum_{j=1}^{n} Z_j.$$

Using a multiplicative analogue to the additive central limit theorem, we arrive at the conclusion that X_n is asymptotically lognormally distributed with two parameters. Variations and extensions of Gibrat's argument have appeared in the literature. Kalecki (1945) assumed that the regression of $\log(1 + Z_j)$ on $\log X_{j-1}$ is linear, with

$$\log(1 + Z_j) = -\alpha_j \log X_{j-1} + Y_j \quad \text{with} \quad E[Y_j] = 0,$$

leading to

$$\log X_n = \log\left\{ X_0 \prod_{j=1}^{n} (1 - \alpha_j) + Y_1 \prod_{j=2}^{n} (1 - \alpha_j) + \cdots + Y_{n-1}(1 - \alpha_n) + Y_n \right\}.$$

Soon after Gibrat's work, Gaddum (1933) and Bliss (1934) found that distributions of critical dose (dose just causing reaction) for a number of drugs could be represented with adequate accuracy by a (two-parameter) lognormal distribution. On the basis of these observations, a highly developed method of statistical analysis of "quantal" (all-or-none) response data has been elaborated. (The term "probit analysis" has been given to such analyses, although "probit" is often understood to apply to a special transformed value, $\Phi^{-1}(\hat{p}) + 5$, where $\hat{p}$ is an observed proportion.)

Lognormal distributions have also been found to be applicable to distributions of particle size in naturally occurring aggregates [Hatch (1933), Hatch and Choute (1929), and Krumbein (1936), with Bol'shev, Prohorov, and Rudinov (1963), Herdan (1960), Kalinske (1946), Kolmogorov (1941), Kottler (1950), and Wise (1952) among many others]. It is possible to give a general theoretical basis for this application on lines similar to those of Gibrat's arguments. Consider a quantity A, subjected to a large number of successive independent subdivisions, the jth resulting in a proportion X_j of the quantity surviving, so that after n divisions the surviving quantity is $AX_1X_2 \cdots X_n$. If $X_1, \ldots, X_n$ are each uniformly distributed (see Chapter 26) over the interval 0 to 1, then the distribution of this quantity will be approximately lognormal. [See Halmos (1944) and Herdan (1960) for more complete accounts of this argument.]

Further applications, in agricultural, entomological, and even literary research were described by Cochran (1938), Williams (1937, 1940), Grundy (1951), Herdan (1958, 1966), and Pearce (1945). Koch (1966, 1969) discussed mechanisms that might generate lognormal distributions in a variety of biological and pharmacological situations.

This broadening of areas of practical usefulness of the distribution was followed by (and presumably associated with) renewed interest in estimation of its parameters. (Section 4 of this chapter contains technical accounts of estimation procedures.) Even in 1957 there were so many doubtful points in this regard that Aitchison and Brown (1957) devoted a substantial part of their book to discussion of problems of estimation, on account of "unre-solved difficulties." Many of these difficulties, especially those concerning maximum likelihood estimation, have now been solved, or at least clarified. A recent paper [Nakamura (1991)] contains a description of construction of an adequate theoretical framework. Wu (1966) has shown that lognormal distri-butions can arise as limiting distributions of order statistics when order and sample size increase in certain relationships.

The book by Aitchison and Brown (1957) was very useful to us when organizing material for this chapter in the first edition. It can still be recommended for supplementary reading together with a more recent com-pendium on lognormal distributions, edited by Crow and Shimizu (1988), containing contributions from several experts. In revising the present chapter, we have tended to include topics which are less thoroughly covered in this compendium.

3 MOMENTS AND OTHER PROPERTIES

Most of the following discussion will be in terms of the two-parameter distribution, using (14.4) rather than (14.3). The rth moment of X about zero is

$$\mu'_r = E[X^r] = E[\exp r(\zeta + U\sigma)] = \exp(r\zeta + \tfrac{1}{2}r^2\sigma^2). \quad (14.7)$$

The expected value of X is

$$\mu_1' = \exp\left(\zeta + \tfrac{1}{2}\sigma^2\right) = e^\zeta \omega^{1/2}, \tag{14.8a}$$

where $\omega = \exp(\sigma^2)$, and the variance is

$$\mu_2 = \mu_2' - \mu_1'^2 = e^{2\zeta}\left(\omega^2 - \left(\omega^{1/2}\right)^2\right) = e^{2\zeta}\omega(\omega - 1). \tag{14.8b}$$

The rth central moment of X [Wartmann (1956)] is

$$\mu_r = E\left[(X - \mu_1')^r\right] = \sum_{j=0}^{r} (-1)^j \binom{r}{j} \mu_{r-j}' \mu_1'^j$$

$$= \sum_{j=0}^{r} (-1)^j \binom{r}{j} \exp\left\{(r-j)\zeta + \frac{1}{2}(r-j)^2\sigma^2 + j\zeta + \frac{1}{2}j\sigma^2\right\}$$

$$= \sum_{j=0}^{r} (-1)^j \binom{r}{j} \exp\left[r\zeta + \frac{1}{2}\{(r-j)^2 + j\}\sigma^2\right]$$

$$= e^{r\zeta} \sum_{j=0}^{r} (-1)^j \binom{r}{j} \omega^{\{(r-j)^2 + j\}/2}$$

$$= \omega^{r/2}\left\{\sum_{j=0}^{r} (-1)^j \binom{r}{j} \omega^{(r-j)(r-j-1)/2}\right\} e^{r\zeta}. \tag{14.8c}$$

In particular

$$\mu_3 = \omega^{3/2}(\omega - 1)^2(\omega + 2)e^{3\zeta}, \tag{14.8d}$$

$$\mu_4 = \omega^2(\omega - 1)^2(\omega^4 + 2\omega^3 + 3\omega^2 - 3)e^{4\zeta}. \tag{14.8e}$$

The shape factors are

$$\alpha_3 = \sqrt{\beta_1} = (\omega - 1)^{1/2}(\omega + 2) \tag{14.9a}$$

and

$$\alpha_4 = \beta_2 = \omega^4 + 2\omega^3 + 3\omega^2 - 3. \tag{14.9b}$$

Neither of these depends on ζ. Note that $\alpha_3 > 0$ and $\alpha_4 > 3$—that is, the distributions are positively skewed and are leptokurtic.

Equations (14.9a) and (14.9b) may be regarded as parametric equations of a curve in the (β_1, β_2) plane. This curve is called the *lognormal* line and is shown in Figure 12.1 of Chapter 12, where it separates the regions for Johnson S_U and S_B distributions. The coefficient of variation is $(\omega - 1)^{1/2}$. It also does not depend on ζ. The distribution of X is unimodal; the mode is at

$$\text{Mode } (X) = \omega^{-1}e^{\zeta}. \tag{14.10}$$

From (14.4) it is clear that the value x_α, such that $\Pr[X \le x_\alpha] = \alpha$, is related to the corresponding percentile u_α of the unit normal distribution by the formula:

$$x_\alpha = \exp(\zeta + u_\alpha\sigma). \tag{14.11a}$$

In particular

$$\text{Median } (X) = x_{0.5} = e^{\zeta} \quad (\text{since } u_{0.5} = 0). \tag{14.11b}$$

Comparison of (14.8a), (14.10), and (14.11b) shows that

$$E[X] > \text{Median } (X) > \text{Mode } (X) \tag{14.12a}$$

and

$$\frac{\text{Mode } (X)}{E[X]} = \omega^{-3/2} = \left[\frac{\text{Median } (X)}{E[X]}\right]^3. \tag{14.12b}$$

The Theil coefficient is

$$E\left[\frac{X}{E[X]}\log\left\{\frac{X}{E[X]}\right\}\right] = \frac{1}{2}\sigma^2. \tag{14.13}$$

Since the coefficient of variation is $(\omega - 1)^{1/2}$, the asymmetry index is (from (14.9a))

$$\frac{\sqrt{\beta_1}}{\text{Coefficient of variation}} = \omega + 2 = (\text{Coefficient of variation})^2 + 3. \tag{14.14}$$

In hydrological literature special attention has been devoted to relations between μ_i, Median (X) and $\sqrt{\beta_1}$ [Burges, Lettenmeier, and Bates (1975); Charbeneau (1978); Burges and Hoshi (1978)].

The *standardized $100\alpha\%$ deviate* is, for example,

$$x'_\alpha = \frac{x_\alpha - \mu'_1(X)}{\sqrt{\mu_2(X)}}$$

$$= \frac{\omega^{-1/2}\exp(u_\alpha\sigma) - 1}{(\omega - 1)^{1/2}} = g(u_\alpha) \tag{14.15}$$

[compare (14.11a), (14.8a) and (14.8b)]. Some values of x'_α are shown in Table 14.1.

Slifker and Shapiro (1980) have noted that for lognormal distributions

$$\frac{\{g(3u_\alpha) - g(u_\alpha)\} - \{g(-u_\alpha) - g(-3u_\alpha)\}}{\{g(u_\alpha) - g(-u_\alpha)\}^2} = 1 \quad \text{for all } \alpha, \tag{14.16}$$

Table 14.1 Standardized 100α% Points (x'_α) of Lognormal Distributions ($\sigma = \delta^{-1}$)

α (Upper Tail) σ (Lower Tail)	99.95 0.05	99.9 0.1	99.75 0.25	99.5 0.5	99 1	97.5 2.5	95 5	90 10	75 25	50 (Median)
0.02	3.39	3.18	2.88	2.63	2.37	1.99	1.66	1.29	0.669	
	−3.19	−3.01	−2.75	−2.52	−2.28	−1.93	−1.63	−1.27	−0.680	−0.010
0.04	3.49	3.27	2.95	2.69	2.42	2.02	1.68	1.29	0.663	
	−3.10	−2.92	−2.67	−2.46	−2.24	−1.90	−1.61	−1.27	−0.680	−0.020
0.06	3.60	3.36	3.02	2.75	2.46	2.04	1.69	1.30	0.657	
	−3.01	−2.84	−2.61	−2.41	−2.20	−1.87	−1.59	−1.26	−0.689	−0.030
0.08	3.71	3.45	3.09	2.81	2.50	2.07	1.71	1.30	0.650	
	−2.29	−2.76	−2.54	−2.36	−2.15	−1.85	−1.57	−1.25	−0.693	−0.040
0.10	3.82	3.54	3.17	2.87	2.55	2.10	1.72	1.31	0.643	
	−2.83	−2.69	−2.48	−2.30	−2.11	−1.82	−1.56	−1.24	−0.697	−0.050
0.12	3.93	3.64	3.24	2.93	2.60	2.13	1.74	1.31	0.635	
	−2.75	−2.61	−2.42	−2.25	−2.07	−1.79	−1.54	−1.23	−0.701	−0.060
0.14	4.05	3.74	3.32	2.99	2.64	2.15	1.75	1.31	0.628	
	−2.67	−2.54	−2.36	−2.20	−2.03	−1.76	−1.52	−1.23	−0.704	−0.069
0.16	4.17	3.84	3.40	3.05	2.69	2.18	1.77	1.32	0.620	
	−2.59	−2.47	−2.30	−2.15	−1.98	−1.73	−1.50	−1.22	−0.706	−0.079
0.18	4.29	3.95	3.48	3.11	2.73	2.21	1.78	1.32	0.611	
	−2.51	−2.40	−2.24	−2.10	−1.94	−1.70	−1.48	−1.21	−0.708	−0.089
0.20	4.42	4.05	3.56	3.17	2.78	2.23	1.79	1.32	0.603	
	−2.44	−2.33	−2.18	−2.05	−1.90	−1.67	−1.46	−1.14	−0.714	−0.098
0.30	5.10	4.61	3.97	3.49	3.00	2.35	1.84	1.32	0.555	
	−2.10	−2.03	−1.92	−1.82	−1.71	−1.53	−1.36	−1.14	−0.714	−0.143
0.40	5.86	5.23	4.41	3.81	3.22	2.45	1.88	1.30	0.402	
	−1.81	−1.76	−1.68	−1.61	−1.53	−1.39	−1.25	−1.07	−0.709	−0.185
0.50	6.71	5.89	4.86	4.13	3.42	2.54	1.89	1.27	0.444	
	−1.56	−1.52	−1.47	−1.42	−1.36	−1.25	−1.15	−1.00	−0.695	−0.220
1.00	11.66	9.41	6.90	5.32	3.98	2.52	1.63	0.904	0.145	
	−0.746	−0.742	−0.735	−0.728	−0.718	−0.698	−0.674	−0.634	−0.527	−0.300

and they have suggested using this as a basis for deciding whether a lognormal distribution is appropriate.

From Table 14.1 we see that as σ tends to zero (or δ to infinity), the standardized lognormal distribution tends to a unit normal distribution, and $x'_\alpha \to u_\alpha$. For σ small,

$$x'_\alpha \doteq u_\alpha + \tfrac{1}{2}\left(u_\alpha^2 - 1\right)\sigma + \tfrac{1}{12}\left(2u_\alpha^3 - 9u_\alpha\right)\sigma^2. \qquad (14.17)$$

It is to be expected that

1. x'_α will change relatively slowly with σ (for σ small) if $u_\alpha^2 \doteq 1$ (i.e., $\alpha \doteq 0.84$ or 0.16),
2. the standardized inter-$100\alpha\%$ distance,

$$x'_\alpha - x'_{1-\alpha} \doteq 2u_\alpha + \tfrac{1}{6}\left(2u_\alpha^3 - 9u_\alpha\right)\sigma^2,$$

will change relatively slowly if $u_\alpha^2 \doteq 9/2$ (i.e., $\alpha \doteq 0.983$ or 0.017). These features also are indicated in Table 14.1. It is also clear from this table that as σ increases the lognormal distribution rapidly becomes markedly nonnormal. Table 14.2, which gives values of α_3 and α_4, also indicates how rapidly the skewness and kurtosis increase with σ. It is for this reason that only relatively small values of σ are used in Table 14.1.

It can be seen in Table 14.1 that for larger σ, there is a high probability density below the expected value, leading to small numerical standardized deviations. Conversely there is a long positive tail with large standardized deviations for upper percentiles. Figure 14.1 shows some typical probability density functions (standardized in each case so that the expected value is zero and the standard deviation is 1).

Table 14.2 Values of α_3 and α_4 for Distribution (14.2)'

σ	α_3	α_4
0.1	0.30	3.16
0.2	0.61	3.68
0.3	0.95	4.64
0.4	1.32	6.26
0.5	1.75	8.90
0.6	2.26	13.27
0.7	2.89	20.79
0.8	3.69	34.37
0.9	4.75	60.41
1.0	6.18	113.94

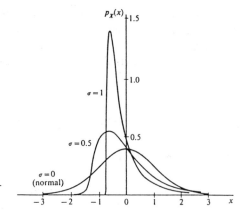

Figure 14.1   Standardized Lognormal Distribution

Wise (1952) has shown that the probability density function of the two-parameter distribution has two points of inflection at

$$x = \exp\left[\zeta - \frac{3\sigma^2}{2} \pm \sigma\sqrt{1 + \frac{1}{4}\sigma^2}\right]. \qquad (14.18)$$

Just as the sum of two independent normal variables is also normally distributed, so the product of two independent (two-parameter) lognormally distributed variables is also lognormally distributed. If

$$U_j = \gamma_j + \delta_j \log X_j, \qquad j = 1, 2,$$

are independent unit normal variables, then

$$\log(X_1 X_2) = \delta_1^{-1} U_1 + \delta_2^{-1} U_2 - \left(\gamma_1 \delta_1^{-1} + \gamma_2 \delta_2^{-1}\right),$$

and so

$$\left(\delta_1^{-2} + \delta_2^{-2}\right)^{-1/2}\left[\log(X_1 X_2) + \gamma_1 \delta_1^{-1} + \gamma_2 \delta_2^{-1}\right]$$

is a unit normal variable. In particular, if $X_1, X_2, \ldots, X_n$ is a random sample from the distribution (14.2)' (with $\theta = 0$) so that $\log X_j$ is distributed normally with expected value ζ and variance σ^2, then $n^{-1}\sum_{j=1}^{n} \log X_j = \log(G(x))$, where $G(x)$ is the sample geometric mean of $X_1, \ldots, X_n$, is distributed normally with expected value ζ and variance σ^2/n. The sample geometric mean $G(x)$ therefore has a lognormal distribution with parameters $\zeta, n^{-1}\sigma^2$ [in (14.2)'], or $\gamma = -\sqrt{n}\,\zeta/\sigma$, $\delta = \sqrt{n}\,/\sigma$.

We introduce the notation $\Lambda(\zeta, \sigma^2)$ to denote a distribution defined by (14.4) so that $X_1 X_2$ has the distribution $\Lambda(\zeta_1 + \zeta_2, \sigma_1^2 + \sigma_2^2)$ if X_j has the distribution $\Lambda(\zeta_j, \sigma_j^2)$ (for any j) and X_1, X_2 are mutually independent. Then the distribution of $a_j X_j$, where a_j is any positive constant, is

$\Lambda(\zeta_j + \log a_j, \sigma_j^2)$. Clearly, if $X_1, X_2, \ldots, X_k$ are a mutually independent set, then $\prod_{j=1}^k a_j X_j$ is distributed as $\Lambda(\sum_{j=1}^k \{\zeta_j + \log a_j\}, \sum_{j=1}^k \sigma_j^2)$.

Distributions of sums of independent random variables, each having lognormal distribution, have received special attention in the literature. These distributions are of importance in many areas of telecommunication, statistical detection problems, multihop scatter systems, and so on.

Fenton (1960) approximates the distribution of sum by a lognormal distribution with the same first two moments. Barakat (1976) considers the sum $S = \sum_{j=1}^N X_j$ of N i.i.d. X_j's distributed as $\Lambda(\zeta, \sigma^2)$. He expresses the characteristic function of each X_j in the form

$$\phi_X(t) = \frac{e^{i\sigma t}}{\sigma\sqrt{2\pi}} \int_{-\infty}^{\infty} e^{-y^2/2\sigma^2} e^{i\sigma t(e^y - y - 1)} e^{ity} \, dy.$$

He then uses the expansion

$$\exp\{i\sigma t(e^y - y - 1)\} = \sum_{n=0}^{\infty} a_n(i\sigma t) \frac{y^n}{n!}$$

to obtain the characteristic function in the form

$$\phi_X(t) = \exp\left\{i\sigma t - \frac{1}{2}(\sigma t)^2\right\} \sum_{n=0}^{\infty} \frac{(i\sigma)^n}{n!} a_n(i\sigma t) h_n(i\sigma t). \quad (14.19)$$

Values of the coefficients $a_n(s)$ are given in Barakat's Table 1; the $h_n(x)$ are "quasi-Hermite" polynomials. The pdf is obtained by inverting (14.19) using quadrature.

The somewhat less ambitious target of obtaining the distribution of the sum of two independent lognormal variables was considered by Naus (1969). He obtained the moment-generating function of the distribution of the sum of two independent $\Lambda(0, \sigma^2)$ variables. Schwartz and Yeh (1982) consider the same problem when the two lognormal variables are not necessarily identically distributed.

The following (condensed) analysis of the problem indicates the difficulties involved: Suppose that X_j is distributed $\Lambda(\zeta_j, \sigma_j^2)$ $(j = 1, 2)$ and that X_1 and X_2 are mutually independent. The joint pdf of X_1 and X_2 is

$$p_{X_1, X_2}(x_1, x_2) = \frac{1}{2\pi\sigma_1\sigma_2 x_1 x_2} \exp\left\{-\frac{1}{2}\sum_{j=1}^{2}\left(\frac{\log x_j - \zeta_j}{\sigma_j}\right)^2\right\}, \qquad 0 \le x_1, x_2.$$

Changing to variables $Y_1 = X_1 + X_2$, $Y_2 = X_2$, we have the Jacobian

$$\left| \frac{\partial(X_1, X_2)}{\partial(Y_1, Y_2)} \right| = \begin{vmatrix} 1 & -1 \\ 0 & 1 \end{vmatrix} = 1,$$

and the joint pdf of Y_1 and Y_2 is

$$\frac{1}{2\pi\sigma_1\sigma_2(y_1 - y_2)y_2} \exp\left[-\frac{1}{2}\left(\frac{\log(y_1 - y_2) - \zeta_1}{\sigma_1} \right)^2 \right.$$

$$\left. -\frac{1}{2}\left(\frac{\log y_2 - \zeta_2}{\sigma_2} \right)^2 \right], \qquad 0 \le y_2 \le y_1.$$

Hence

$$p_{Y_1}(y_1) = \int_{-\infty}^{\infty} p_{Y_1, Y_2}(y_1, y_2)\, dy_2 = \int_{0}^{y_1} p_{Y_1, Y_2}(y_1, y_2)\, dy_2$$

$$= \frac{1}{2\pi\sigma_1\sigma_2} \int_0^{y_1} \frac{1}{(y_1 - y_2)y_2} \exp\left[-\frac{1}{2}\left(\frac{\log(y_1 - y_2) - \zeta_1}{\sigma_1} \right)^2 \right.$$

$$\left. -\frac{1}{2}\left(\frac{\log y_2 - \zeta_2}{\sigma_2} \right)^2 \right] dy_2.$$

Setting $y_2 = y_1 t$, we obtain

$$p_{Y_1}(y_1) = \frac{1}{2\pi\sigma_1\sigma_2 y_1} \int_0^1 \frac{1}{t(1 - t)} \exp\left[-\frac{1}{2}\left(\frac{\log(1 - t) + \log y_1 - \zeta_1}{\sigma_1} \right)^2 \right.$$

$$\left. -\frac{1}{2}\left(\frac{\log t + \log y_1 - \zeta_2}{\sigma_2} \right)^2 \right] dt.$$

$$(14.20)$$

The integral in (14.20) may be evaluated by quadrature.

The preceding discussion has been limited to variables with *two-parameter* lognormal distributions. The extra parameter, θ in (14.2) or (14.2)' moves ("translates") the whole distribution by an amount θ in the direction of increasing X. Many properties—particularly variance and shape—remain unchanged, but the reproductive property does not hold if $\theta \neq 0$ for either X_1 or X_2.

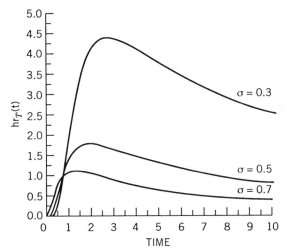

Figure 14.2 Hazard Rate of the Lognormal Distribution, with $E[\log T] = 0$; $\sigma = $ S.D.$(\log T) = 0.3, 0.5, 0.7$

The information-generating function corresponding to (14.2) is

$$\left(\frac{\delta}{\sqrt{2\pi}}\right)^{u-1} \exp\left[\frac{1}{2}(u-1)^2(u\delta^2)^{-1} + (u-1)\frac{\gamma}{\delta}\right]. \qquad (14.21)$$

The entropy is

$$\frac{1}{2} - \left(\frac{\gamma}{\delta}\right) - \log\left(\frac{\delta}{\sqrt{2\pi}}\right). \qquad (14.22)$$

A recent reference for the characteristic function of $\Lambda(\zeta, \sigma^2)$ is Leipnik (1991). The distribution is not determined by its moments. A result of Bondesson (1979) is that any distribution with pdf of form

$$cx^{\beta-1}g(x), \qquad \beta > 0, g(0) = 1,$$

with $g(\cdot)$ completely monotone on $(0, \infty)$, satisfying some mild regularity conditions is infinitely divisible (see Chapter 1). This result enables us to deduce that lognormal distributions are infinitely divisible.

Sweet (1990) has studied the hazard rate of lognormal distributions. Figure 14.2, taken from that paper, exhibits some of his results for $\Lambda(\zeta, \sigma^2)$. From this figure it can be seen that $h(0) = 0$ and that $h(t)$ runs to a maximum and thereafter decreases slowly to zero. The value t_M of t that maximizes $h(t)$ is

$$t_M = \exp(\zeta + z_M\sigma), \qquad (14.23)$$

where $(z_M + \sigma)^{-1} = \{1 - \Phi(z_M)\}/\phi(z_M) = R(z_M)$, which is Mills's ratio (see Chapter 13).

From (14.23), $-\sigma < z_M < -\sigma + \sigma^{-1}$, and hence

$$e^{\zeta - \sigma^2} < t_M < e^{\zeta - \sigma^2 + 1}. \tag{14.24}$$

As $\sigma \to \infty$, $t_M \to \exp(\zeta - \sigma^2)$, and so for large σ,

$$\max_t h(t) \doteq \frac{\exp(\zeta - \sigma^2/2)}{\sigma\sqrt{2\pi}}. \tag{14.25a}$$

As $\sigma \to 0$, $t_M \to \exp(\zeta - \sigma^2 + 1)$, and so for small σ,

$$\max_t h(t) \doteq \{\sigma^2 \exp(\zeta - \sigma^2 + 1)\}^{-1}. \tag{14.25b}$$

4 ESTIMATION

4.1 θ Known

If the value of the parameter θ is known (it is often possible to take θ equal to zero), then estimation of the parameters ζ and σ (or γ and δ) presents few difficulties beyond those already discussed in connection with the normal distribution. In fact, by using values of $Z_i = \log(X_i - \theta)$, the problem is reduced to that of estimation of parameters of a normal distribution. Many specialized problems, such as those relating to truncated distributions or censored samples, or the use of order statistics, may be reduced to corresponding problems (already discussed in Chapter 13) for the normal distribution. Maximum likelihood estimation is exactly equivalent to maximum likelihood estimation for normal distributions, so that the maximum likelihood estimators $\hat{\zeta}, \hat{\sigma}$ for ζ and σ, respectively, are

$$\hat{\zeta} = \bar{Z}, \tag{14.26a}$$

$$\hat{\sigma} = \left[n^{-1} \sum_{j=1}^{n} \left(Z_j - \bar{Z} \right)^2 \right]^{1/2}, \tag{14.26b}$$

where $\bar{Z} = n^{-1}\sum_{j=1}^{n} Z_j$ (assuming of course that $X_1, X_2, \ldots, X_n$ are independent random variables each having the same lognormal distribution with θ known).

There are a few situations where techniques appropriate to normal distributions cannot be applied directly. One is the construction of best *linear* unbiased estimators, using order statistics of the original variables X_j. Another is that where the original values $\{X_j\}$ are not available, but the data are in groups of equal width. While the results can immediately be written down in the form of grouped observations from a normal distribution, the groups appropriate to the transformed variables will not be of equal width. In such circumstances methods of the kind described in Chapter 12, Section 2, might be applied.

Sometimes it may be desired to estimate not ζ or σ but rather the expected value $\exp(\zeta + \sigma^2/2)$ and the variance $e^{\sigma^2}(e^{\sigma^2} - 1)e^{2\zeta}$ of the variable X. Since $\bar{Z}$ and $S^2 = (n - 1)^{-1}\sum_{j=1}^{n}(Z_j - \bar{Z})^2$ are jointly complete and sufficient statistics for ζ and σ, if σ^2 is known the UMVU estimator of $\theta_{a,b} = \exp(a\zeta + b\sigma^2)$ is

$$\hat{\theta}_u = \exp\left\{a\bar{Y} + b\sigma^2 - \frac{a^2\sigma^2}{2n}\right\}. \tag{14.27}$$

If a function $f(\bar{Z}, S^2)$ has expected value $h(\zeta, \sigma)$, it is the minimum variance unbiased estimator of $h(\zeta, \sigma)$. Finney (1941) obtained such estimators of the expected value and variance in the form of infinite series

$$M = \exp(\bar{Z})g\left(\frac{1}{2}S^2\right) \tag{14.28a}$$

$$V = \exp(2\bar{Z})\left\{g(2S^2) - g\left(\frac{(n-2)S^2}{n-1}\right)\right\}, \tag{14.28b}$$

respectively, where

$$g(t) = 1 + \frac{n-1}{n}t + \sum_{j=2}^{\infty} \frac{(n-1)^{2j-1}}{n^j(n+1)(n+3)\ldots(n+2j-3)} \frac{t^j}{j!}. \tag{14.29}$$

Unfortunately, the series in (14.29) converges slowly (except for very small values of t). Finney recommends using the approximation

$$g(t) \doteq e^t\left\{1 - \frac{t(t+1)}{n} + \frac{t^2(3t^2 + 22t + 21)}{6n^2}\right\} \tag{14.30}$$

which (he states) should be safe for $n > 50$ in (14.28a) and $n > 100$ in (14.28b) provided that the coefficient of variation $[(e^{\sigma^2} - 1)^{1/2}]$ is less than 1 (corresponding to $\sigma < 0.83$).

Finney showed that [with $\omega = \exp(\sigma^2)$]

$$n\,\mathrm{Var}(M) \doteq \sigma^2 e^{2\zeta}\omega\left(1 + \tfrac{1}{2}\sigma^2\right), \tag{14.31a}$$

$$n\,\mathrm{Var}(V) \doteq 2\sigma^2 e^{4\zeta}\omega^2\left\{2(\omega - 1)^2 + \sigma^2(2\omega - 1)^2\right\}. \tag{14.31b}$$

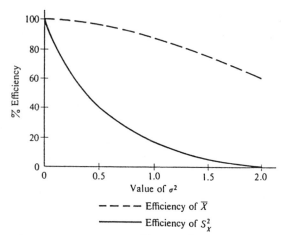

Figure 14.3   The Efficiency of $\overline{X}$ and of S_X^2 in Large Samples

By comparison, for the unbiased estimators of the expected value and variance, $\overline{X}$ and $S_X^2 = (n - 1)^{-1}\sum_{j=1}^n (X_j - \overline{X})^2$, respectively,

$$n \, \text{Var}(\overline{X}) = e^{2\zeta}\omega(\omega - 1), \qquad (14.32a)$$

$$n \, \text{Var}(S_X^2) \doteq e^{4\zeta}\omega^2(\omega - 1)^2(\omega^4 + 2\omega^3 + 3\omega^2 - 4). \qquad (14.32b)$$

Figure 14.3 [taken from Finney (1941)] shows approximate values of the "efficiency" ratios $100 \times \text{Var}(M)/\text{Var}(\overline{X})$ and $100 \times \text{Var}(V)/\text{Var}(S_X^2)$ as a function of σ^2. It will be noted that while $\overline{X}$ is reasonably efficient compared with M, considerable reduction in variance is achieved by using V in place of S_X^2. [See also Oldham (1965).] Peters (1963) has constructed best *quadratic* estimators of $\log \mu_r' = r\zeta + r^2\sigma^2/2$, using order statistics based on $\log X_j$'s.

Confidence limits for ζ and/or σ can of course be constructed from the transformed values $Z_1, Z_2, \ldots, Z_n$ using techniques in Chapter 13. In particular (when $\theta = 0$), since the coefficient of variation of X is $(e^{\sigma^2} - 1)^{1/2}$, confidence limits (with confidence coefficient $100\alpha\%$) for this quantity are $(\exp[(n - 1)S^2/\chi_{n-1,1-\alpha/2}^2] - 1)^{1/2}$ and $(\exp[(n - 1)S^2/\chi_{n-1,\alpha/2}^2] - 1)^{1/2}$, where $\chi_{\nu,\varepsilon}^2$ denotes the $100\varepsilon\%$ point of the χ^2 distribution with ν degrees of freedom, to be described in Chapter 18. [See Koopmans, Owen, and Rosenblatt (1964).] Similar arguments could be used to construct confidence limits for any other monotonic function of σ, such as $\sqrt{\beta_1}$ and β_2.

4.2 θ Unknown

Estimation problems present considerable difficulty when θ is not known. As might be expected, estimation of θ is particularly inaccurate. This parameter is a "threshold value," below which the cumulative distribution function is

Table 14.3 Percentile Points of Standardized Lognormal Distributions

		Lower			Upper	
σ	θ	10%	25%	Median	25%	10%
0.02	-49.995	-1.275	-0.680	-0.010	0.669	1.288
0.04	-24.990	-1.268	-0.685	-0.020	0.663	1.293
0.06	-16.652	-1.260	-0.689	-0.030	0.657	1.299
0.08	-12.480	-1.252	-0.693	-0.040	0.650	1.303
0.10	-9.975	-1.244	-0.697	-0.050	0.643	1.307
0.12	-8.303	-1.235	-0.700	-0.060	0.635	1.311
0.14	-7.108	-1.225	-0.703	-0.069	0.628	1.314
0.16	-6.210	-1.216	-0.706	-0.079	0.620	1.316
0.18	-5.511	-1.206	-0.708	-0.089	0.611	1.318
0.20	-4.950	-1.195	-0.710	-0.098	0.603	1.320

zero and above which it is positive. Such values are often—one might say "usually" difficult to estimate.

However, estimation of parameters is often not as important as estimation of probabilities—in particular, of cumulative distribution functions. Table 14.3 shows the median and upper and lower 10% and 25% points of lognormal distributions with widely differing values of θ, but with ζ and σ so chosen that each distribution is standardized (i.e., has zero expected value and unit standard deviation; see also Table 14.1).

There can be considerable variation in θ with little effect on the percentiles and little effect on values of the cumulative distribution functions for fixed values of X. Insensitivity to variation in θ is most marked for large negative values of θ. These observations and Table 14.3 should correct any feeling of depression caused by the inaccuracy of estimation of θ. Of course there can be situations where the accuracy of estimation of θ itself is a primary consideration. In these cases special techniques, both of experimentation and analysis, may be needed. These will be briefly described at the end of this section.

Maximum likelihood estimation of θ, ζ, and σ might be expected to be attained by the following tedious, but straightforward, procedure. Since for given θ, the likelihood function is maximized by taking

$$\hat{\zeta} = \hat{\zeta}(\theta) = n^{-1} \sum_{j=1}^{n} \log(X_j - \theta), \qquad (14.33a)$$

$$\hat{\sigma} = \hat{\sigma}(\theta) = \left[n^{-1} \sum_{j=1}^{n} \left\{ \log(X_j - \theta) - \hat{\zeta}(\theta) \right\}^2 \right]^{1/2}, \qquad (14.33b)$$

$$[\text{cf. } (14.26a), (14.26b)],$$

one might take a sequence of values of θ, calculate the maximized likelihood corresponding to each, and then try to estimate numerically the value $\hat{\theta}$ of θ by maximizing the maximized likelihood. However, Hill (1963) has shown that as θ tends to $\min(X_1, X_2, \ldots, X_n)$,* the maximized likelihood tends to infinity. Formally this would seem to indicate that we should accept the "estimates": $\hat{\theta} = \min(X_1, X_2, \ldots, X_n)$, $\hat{\zeta} = -\infty$, $\hat{\sigma} = \infty$. Hill resolved this difficulty by introducing a prior joint distribution for θ, ζ, and σ, and then using Bayes's theorem. This leads to the conclusion that solutions of the *formal* maximum likelihood equations should be used, with $\hat{\theta}$ satisfying

$$\sum_{j=1}^{n} \left(X_j - \hat{\theta} \right)^{-1} + \left[\hat{\sigma}(\hat{\theta}) \right] \left\{ \sum_{j=1}^{n} \left(X_j - \hat{\theta} \right)^{-1} Z_j' \right\} = 0, \qquad (14.34)$$

where $Z_j' = [\log(X_j - \hat{\theta}) - \hat{\zeta}(\hat{\theta})]/\hat{\sigma}(\hat{\theta})$ and $\hat{\zeta}(\hat{\theta})$, $\hat{\sigma}(\hat{\theta})$ satisfy (14.33), with θ replaced by $\hat{\theta}$.

Formal calculation of limiting variances gives

$$n \operatorname{Var}(\hat{\theta}) \doteq \sigma^2 e^{2\zeta} \omega^{-1} \left[\omega(1 + \sigma^2) - 2\sigma^2 - 1 \right]^{-1}, \qquad (14.35a)$$

$$n \operatorname{Var}(\hat{\zeta}) \doteq \sigma^2 \left[\omega(1 + \sigma^2) - 2\sigma^2 \right] \left[\omega(1 + \sigma^2) - 2\sigma^2 - 1 \right]^{-1}, \quad (14.35b)$$

$$n \operatorname{Var}(\hat{\sigma}) \doteq \sigma^2 \left[\omega(1 + \sigma^2) - 1 \right] \left[\omega(1 + \sigma^2) - 2\sigma^2 - 1 \right]^{-1}. \qquad (14.35c)$$

Formulas equivalent to these have been given by Cohen (1951) and Hill (1963), but (14.35c) is unlikely to be valid [see Hill (1963) and below].

We may note that Harter and Moore (1966) carried out a sampling experiment, with $\zeta = 4$, $\sigma = 2$, and $\theta = 10$. Their results can be represented approximately, for sample size n in the range 50 to 200 by

$$n \operatorname{Var}(\hat{\zeta}) \doteq 4.2,$$

$$n \operatorname{Var}(\hat{\sigma}) \doteq 3.2,$$

$$n^2 \operatorname{Var}(\hat{\theta}) \doteq 1800$$

(note the factor n^2 in the last formula). If θ is known, their results give $n \operatorname{Var}(\hat{\zeta}) \doteq 4.1$—not much less than if θ is not known—but $n \operatorname{Var}(\hat{\sigma})$ is now only about 2.1.

Tiku (1968) has suggested an approximate linearization of the maximum likelihood equations based on the approximate formula

$$Z(x) [1 - \Phi(x)]^{-1} \doteq \alpha + \beta x \qquad (14.36)$$

---

*It is clear that we must take $\theta \le \min(X_1, X_2, \ldots, X_n)$.

for Mills's ratio. Appropriate values of α and β depend on the range of values of x. (See discussion of a similar technique in Chapter 13, Section 8.4.)

Over the last 20 years there has been further consideration of maximum likelihood estimation for the three-parameter lognormal and its difficulties. We are inclined to agree with the opinion expressed by Griffiths (1980), that "the method of maximum likelihood has been wrongly discredited because of supposed computational difficulties and theoretical uncertainties."

Calitz (1973) suggests using Cohen's (1951) method to solve the maximum likelihood equations. He solves (14.34) numerically using the Newton-Raphson method, which will converge provided the initial value of θ is chosen so that $\partial \log L(\theta)/\partial\theta < 0$. Using the value $\theta = \max(0.6X_{(1)}, X_{(1)} - 0.1)$, where $X_{(1)}$ is the least of the observed values of X, provided satisfactory results. Lambert's (1964) failure to solve this problem—which triggered many denunciations of the maximum likelihood method—was due to his choice of method of iteration. He tried to iterate on *all three* maximum likelihood equations simultaneously.

Wingo (1976) proposed using penalty functions, maximizing the function

$$-n\{\hat{\zeta}(\theta) + \log \hat{\sigma}(\theta)\},$$

with $\hat{\zeta}(\theta), \hat{\sigma}(\theta)$ given by (14.33), subject to constraints of form

$$\begin{cases} \theta + c > 0, \\ \theta < \min(X_1, \ldots, X_n). \end{cases} \tag{14.37}$$

The constant c is chosen to be a large positive number (although its value can be problem dependent). Hawkins (1991) points out that a necessary condition for the maximum likelihood equations to "degenerate" so that they cannot be solved is that

$$2S_1 + \log(S_2 - S_1^2),$$

where $S_i = \sum_{j=1}^{n}[\{\log(X_j + \alpha)\}^i/n]$ $(i = 1, 2)$ be a decreasing function of α.

Noting that all data are in fact subject to grouping (and so are, in reality, discrete), Giesbrecht and Kempthorne (1976) have shown that the problem of an unbounded likelihood function can be removed by taking a suitable discretized model. Griffiths (1980) utilizes the discretization procedure to obtain a confidence interval for the threshold parameter θ. Provided that the group width is sufficiently small, the likelihood will be practically of the same form for both the (approximate) continuous and (correct) discrete models, except in the neighborhood of the singularity in the continuous model. Griffiths (1980) also discusses construction of approximate confidence intervals for θ.

Dahiya and Guttman (1982) construct a "shortest prediction interval" of level $1 - \alpha$ for a two-parameter lognormal variable. [This is not an estimated interval, it is simply the interval (a, b) such that $b - a$ is minimized subject to $\Pr[a < X < b] = 1 - \alpha$.] Clearly we must have $a = \exp(\zeta + A\sigma)$ and $b = \exp(\zeta + B\sigma)$, with

$$\Phi(B) - \Phi(A) = 1 - \alpha \qquad (14.38)$$

(to provide level $1 - \alpha$). Subject to (14.38) we have to minimize $b - a$, or equivalently

$$e^{B\sigma} - e^{A\sigma}. \qquad (14.39)$$

From (14.38), $\phi(B)(\partial B / \partial A) - \phi(A) = 0$, whence

$$\frac{\partial B}{\partial A} = \frac{\phi(A)}{\phi(B)}. \qquad (14.40a)$$

To minimize (14.39), we equate

$$\frac{\partial}{\partial A}(e^{B\sigma} - e^{A\sigma}) = \sigma\left(e^{B\sigma}\frac{\partial B}{\partial A} - e^{A\sigma}\right) \qquad (14.40b)$$

to zero.

Inserting the value of $\partial B / \partial A$ from (14.40a), we obtain

$$e^{B\sigma} \cdot \phi(A) = e^{A\sigma}\phi(B),$$

whence $B\sigma - A^2/2 = A\sigma - B^2/2$; that is,

$$A + B = -2\sigma. \qquad (14.41)$$

The values of A and B are easily determined from (14.38) and (14.41), but Dahiya and Guttman (1982) provide tables to facilitate the process. These authors also consider construction of shortest *confidence* intervals for the median (e^ζ). [Of course, if it is known that $\theta = 0$, confidence intervals for ζ ($= E[\log X]$) are easily constructed using normal theory results (see Chapter 13).]

Since

$$\overline{\log X} = n^{-1} \sum_{i=1}^{n} \log X_i$$

and

$$S^2 = (n - 1)^{-1} \sum_{i=1}^{n} \left(\log X_i - \overline{\log X}\right)^2$$

are sufficient for the parameters of $\Lambda(\zeta, \sigma^2)$, a minimum variance unbiased estimator of

$$\mu'_1 = \log E[X] = \zeta + \tfrac{1}{2}\sigma^2$$

is

$$\tilde{\mu}'_1 = \overline{\log X} + \tfrac{1}{2}S^2; \tag{14.42a}$$

a minimum variance unbiased estimator of the variance of $\tilde{\mu}'_1$,

$$\mathrm{Var}(\tilde{\mu}'_1) = \mathrm{Var}(\overline{\log X}) + \tfrac{1}{4}\mathrm{Var}(S^2),$$

(since $\overline{\log X}$ and S^2 are independent) is [Land (1972)]

$$\tilde{\gamma}^2 = \widetilde{\mathrm{Var}}(\tilde{\mu}'_1) = n^{-1}S^2 + \tfrac{1}{2}(n+1)^{-1}S^4. \tag{14.42b}$$

Among other proposed modifications of maximum likelihood estimation we note the suggestions of Cohen and Whitten (1980) based on introduction of order statistics: They propose replacing the maximum likelihood equation $\partial \log L/\partial\theta = 0$ by

$$\log(X'_r - \theta) = \zeta + \sigma E[U'_r] \tag{14.43}$$

where X'_r is the rth order statistic among $X_1, \ldots, X_n$ and U'_r is the rth order statistic among n mutually independent unit normal variables, or by

$$\log(X'_r - \theta) = \zeta + \sigma \Phi^{-1}\left(\frac{r}{n+1}\right). \tag{14.44}$$

Other modifications include: the equation is replaced by

$$E[X] = \overline{X}$$

or by

$$\mathrm{Var}(X) = S^2 \quad \text{with} \quad S^2 = \frac{\sum_{i=1}^{n}(X_i - \overline{X})^2}{n-1}.$$

The actual estimating equations for θ are

$$\theta = \overline{X} - e^\zeta\sqrt{\omega}$$

and

$$S^2 = e^{2\zeta}\omega(\omega - 1).$$

Nevertheless, uncertainties and difficulties in the use of (14.33) and (14.34) lead us to consider methods other than maximum likelihood.

If the first three sample moments are equated to the corresponding population values, we obtain the formulas

$$\bar{X} = \tilde{\theta} + \exp\left(\tilde{\zeta} + \tfrac{1}{2}\tilde{\sigma}^2\right), \tag{14.45a}$$

$$m_2 = n^{-1} \sum_{j=1}^{n} \left(X_j - \bar{X}\right)^2 = e^{2\tilde{\zeta}+\tilde{\sigma}^2}(e^{\tilde{\sigma}^2} - 1), \tag{14.45b}$$

$$m_3 = n^{-1} \sum_{j=1}^{n} \left(X_j - \bar{X}\right)^3 = e^{3\tilde{\zeta}+3\tilde{\sigma}^2/2}(e^{\tilde{\sigma}^2} - 1)^2(e^{\tilde{\sigma}^2} + 2), \tag{14.45c}$$

whence

$$b_1 = m_3^2 m_2^{-3} = (e^{\tilde{\sigma}^2} - 1)(e^{\tilde{\sigma}^2} + 2)^2$$

$$= (\tilde{\omega} - 1)(\tilde{\omega} + 2)^2. \tag{14.46}$$

From (14.46), $\tilde{\omega}$ and then $\tilde{\sigma}$ can be determined. From (14.45b),

$$\tilde{\zeta} = \tfrac{1}{2} \log\left[m_2 \tilde{\omega}^{-1}(\tilde{\omega} - 1)^{-1}\right],$$

and finally $\tilde{\theta}$ can be determined from (14.45a).

Yuan (1933) provided a table of values of the right-hand side of (14.46) to aid in the solution of that equation. Aitchison and Brown (1957) give a table for $\tilde{\sigma}^2$ directly to four decimal places for $\sqrt{b_1} = 0.0(0.2)10.0(1)24$. Without using such tables, an iterative method using equation (14.46) in the form

$$\tilde{\omega} = 1 + (\tilde{\omega} + 2)^{-2}b_1 \tag{14.46}'$$

is quite easy to apply, or the explicit solution

$$\tilde{\omega} = \left[1 + \tfrac{1}{2}b_1 + \sqrt{\left(1 + \tfrac{1}{2}b_1\right)^2 - 1}\right]^{1/3}$$

$$+ \left[1 + \tfrac{1}{2}b_1 - \sqrt{\left(1 + \tfrac{1}{2}b_1\right)^2 - 1}\right]^{1/3} - 1$$

can be used.

This method is easy to apply but liable to be inaccurate because of sampling variation in b_1. [However, Sinha (1981) found it to be quite reliable.] Even when θ is known, estimation of population variance by sample variance is relatively inaccurate (see Figure 14.3). Approximate formulas for the variances of the estimators were obtained by Nydell (1919). They are

$$n \operatorname{Var}(\tilde{\zeta}) \doteq \tfrac{1}{6}[4u^{-1} + 54 + 306u + 895u^2 + 1598u^3 + 1753.5u^4 + 1242.7u^5],$$

$$n \operatorname{Var}(\tilde{\sigma}) \doteq \tfrac{1}{2}u\sigma^{-2}\{8 + 106u + 332u^2 + 479u^3 + 404u^4 + 216u^5\},$$

where $u = \omega - 1 = e^{\sigma^2} - 1 = $ (square of coefficient of variation).

A modification of the method of moments is to estimate θ by this method and then use Finney's estimators M, V [see (14.28a) and (14.28b)] of expected value and variance, applied to the variables $\{X_i - \tilde{\theta}\}$.

Alternatively, if one is prepared to accept some loss in efficiency, it might be possible to use a relatively simple method of calculating estimates. Certain special forms of the method of percentile points give very simple formulas for the estimators. Using the relationship [see (14.1) and (14.11a)]

$$u_{\alpha_j} = \gamma + \delta \log(x_{\alpha_j} - \theta), \qquad j = 1, 2, 3, \tag{14.47}$$

the following formula can be obtained:

$$\frac{u_{\alpha_1} - u_{\alpha_2}}{u_{\alpha_2} - u_{\alpha_3}} = \frac{\log[(x_{\alpha_1} - \theta)/(x_{\alpha_2} - \theta)]}{\log[(x_{\alpha_2} - \theta)/(x_{\alpha_3} - \theta)]}. \tag{14.48}$$

If estimated values $\hat{X}_{\alpha_j}$ of the x_{α_j}'s are inserted in (14.48), the solution of the resulting equation for θ must be effected by trial and error. However, if we choose $\alpha_2 = \tfrac{1}{2}$ (corresponding to the median) and $\alpha_1 = 1 - \alpha_3$, then (noting that $u_{1/2} = 0$, and $u_{1-\alpha_1} = -u_{\alpha_1}$) the equations are (asterisk denoting "estimated value"):

$$\hat{X}_{\alpha_1} = \theta^* + e^{-\gamma^*/\delta^*} \exp\left[\frac{u_{\alpha_1}}{\delta^*}\right], \tag{14.49a}$$

$$\hat{X}_{1/2} = \theta^* + e^{-\gamma^*/\delta^*}, \tag{14.49b}$$

$$\hat{X}_{1-\alpha_1} = \theta^* + e^{-\gamma^*/\delta^*} \exp\left[-\frac{u_{\alpha_1}}{\delta^*}\right], \tag{14.49c}$$

whence

$$\frac{\hat{X}_{\alpha_1} - \hat{X}_{1/2}}{\hat{X}_{1/2} - \hat{X}_{\alpha_3}} = \exp\left[\frac{u_{\alpha_1}}{\delta^*}\right]$$

and

$$\delta^* = u_{\alpha_1}\left[\log\left\{\frac{\hat{X}_{\alpha_1} - \hat{X}_{1/2}}{\hat{X}_{1/2} - \hat{X}_{1-\alpha_1}}\right\}\right]^{-1} \tag{14.50}$$

so that no solution by trial and error is needed. Aitchison and Brown (1957) suggest taking $\alpha_1 = 0.95$, but it is likely that any value between 0.90 and 0.95 will give about the same accuracy. Lambert (1970) suggested using values of the parameters obtained by this method as a starting point for an iterative formal solution of the maximum likelihood equations. He used the value $\alpha_1 = 61/64 \doteq 0.953$.

It is possible for (14.50) to give a negative value of δ^*, but this happens only if

$$\hat{X}_{\alpha_1} - \hat{X}_{1/2} < \hat{X}_{1/2} - \hat{X}_{1-\alpha_1},$$

which is very unlikely to be the case if the distribution has substantial positive skewness.

If the sample median $\hat{X}_{1/2}$ is replaced by the sample arithmetic mean $\overline{X}$ and equation (14.49b) by

$$\overline{X} = \theta^* + e^{-\gamma^*/\delta^*} \exp\left(\tfrac{1}{2}\delta^{*-2}\right), \tag{14.49d}$$

then from (14.49a), (14.49c), and (14.49d),

$$\frac{\overline{X} - \hat{X}_{1-\alpha_1}}{\hat{X}_{\alpha_1} - \overline{X}} = \frac{\exp\left(\tfrac{1}{2}\delta^{*-2}\right) - \exp\left(-u_{\alpha_1}\delta^{*-1}\right)}{\exp\left(u_{\alpha_1}\delta^{*-1}\right) - \exp\left(\tfrac{1}{2}\delta^{*-2}\right)}. \tag{14.51}$$

From this equation δ^* can be found numerically. Aitchison and Brown (1957) give graphs of the function on the right-hand side of (14.51), for $\alpha_1 = 0.95, 0.90$, and 0.80, which helps to give an initial value for the solution. This method of estimation is called *Kemsley's method* (1952). It can also give a negative value for δ^*, though this is unlikely to occur.

A method suggested by Boswell, Ord, and Patil (1979) is easy to apply: Consider the 100αth lower, 50th and 100αth upper percentiles of the *normal variable* $Z = \log(X - \theta)$. These are $\zeta - z\sigma$, ζ, and $\zeta + z\sigma$, respectively, where $\alpha = \Phi(-z) = 1 - \Phi(z)$. The corresponding (unknown) percentiles of X will be $x_{(1)} = \exp(\zeta - z\sigma) + \theta$, $x_{(2)} = \exp(\zeta) + \theta$, and $x_{(3)} = \exp(\zeta + z\sigma) + \theta$. Solving for θ, from these three equations we obtain

$$\theta = \frac{x_{(1)}x_{(3)} - x_{(2)}^2}{x_{(1)} - 2x_{(2)} + x_{(3)}}. \tag{14.52}$$

Estimating $x_{(1)}$, $x_{(2)}$, and $x_{(3)}$ from the corresponding percentiles of lognormal data we obtain an estimator of θ.

Boswell, Ord, and Patil (1979) recommended choosing z in the range of 1.5 to 2. Aitchison and Brown (1957) made an experimental comparison of results of using the method of moments (14.46), the method of quantiles

(14.50), and Kemsley's method (14.51) to estimate θ. They came to the conclusion that (14.50) with $\alpha_1 = 0.95$ is slightly better than (14.51) and that both are considerably better than the method of moments.

Other methods of estimating θ have been suggested. *Cohen's method* (1951) is based on the idea that in a large sample at least one observed value of X should be not much greater than the threshold value θ. The least sample value $\hat{X}_{min}$ is equated to the formula for the $100(n + 1)^{-1}\%$ point of the distribution, giving the relationship

$$\hat{X}_{min} = \theta* + e^{-\gamma}* /^{\delta}* \exp\left[\frac{u_{1/(n+1)}}{\delta*}\right]$$

or

$$\hat{X}_{min} = \theta* + e^{\zeta*} \exp[u_{1/(n+1)}\sigma*]. \tag{14.53}$$

This is then combined with the first two maximum likelihood equations (14.33).

There are natural modifications of this method, to allow for cases when there are several equal (or indistinguishable, as in grouped data) minimum values. Using a variant of this method, an initial value of θ (for use in some iterative process) may be chosen as $\hat{X}_{min}$ *minus* some arbitrary (usually rather small) value.

Cohen, Whitten, and Ding's (1985) *modified moment* estimation method uses the equations (in an obvious notation)

$$E[X] = \bar{X}, \tag{14.54a}$$

$$\text{Var}(X) = S^2, \tag{14.54b}$$

$$\theta + \exp\{\zeta + \sigma E[U_{1,n}]\} = \hat{X}_{min}. \tag{14.54c}$$

Equation (14.54c) is obtained from (14.53) by replacing $u_{1/(n+1)} = \Phi^{-1}((n + 1)^{-1})$ by $E[U_{1,n}]$. Table 14.4 gives values of $-u_{1/(n+1)}$ and $-E[U_{1,n}]$. [More detailed tables of $E[U_{1,n}]$ are available in Harter (1961); see also Tippett (1925) and Chapter 13, Section 4.]

Table 14.4 Values of $-u_{1/(n+1)}$ and $-E(U_{1,n})$

n	5	10	20	30	40	50	60	70
$-u_{1/(n+1)}$	0.967	1.340	1.671	1.848	1.970	2.062	2.135	2.195
$-E(U_{1,n})$	1.163	1.539	1.867	2.043	2.161	2.249	2.319	2.377

n	80	90	100	150	200	300	400	1000
$-u_{1/(n+1)}$	2.246	2.291	2.330	2.477	2.578	2.714	2.808	3.093
$-E(U_{1,n})$	2.427	2.470	2.508	2.649	2.746	2.878	2.968	3.241

The explicit form of the estimating equations is

$$\theta^* + e^{\zeta^*}\omega^{*1/2} = \overline{X}, \tag{14.54a}'$$

$$e^{2\zeta^*}\omega^*(\omega^* - 1) = S^2, \tag{14.54b}'$$

$$\theta^* + e^{\zeta^*}\exp\{\sqrt{\log \omega^*}\, E[U_{1,n}]\} = \hat{X}_{\min}. \tag{14.54c}'$$

The above three equations give

$$f(\omega^*) = \frac{\omega^*(\omega^* - 1)}{\left[\sqrt{\omega^*} - \exp\{\sqrt{\log \omega^*}\, E[U_{1,n}]\}\right]^2} = \frac{S^2}{\left(\overline{X} - \hat{X}_{\min}\right)^2}. \tag{14.55}$$

From (14.55), ω^* (and so $\sigma^* = \sqrt{\log \omega^*}$) can be determined numerically. Cohen, Whitten, and Ding (1985) and Cohen and Whitten (1988) provide graphs of $f(\omega^*)$ as a function of σ^* for $n = 10, 15, 20, 30, 50, 100$, and 400.

A pioneering paper by Cheng and Amin (1982) introduced a new method of estimation—maximum product-of-spacings (MPS). This overcomes difficulties involved in maximum likelihood estimation for the three-parameter lognormal distribution. The authors claim that Giesbrecht and Kempthorne's (1976) grouping and discretization method does not appear to identify the essential difficulty. "Whether it is possible in practice or not, it is still legitimate to hypothesize a situation where the observations are truly continuously distributed, when discretization is merely an approximation to the actual situation." The authors contend that in maximum likelihood estimation for a continuous density $p_X(x|\phi)$, one maximizes, approximately

$$\prod_{i=1}^{n} \Pr\{X_i \leq x_i \leq X_i + (\delta X_i)|\phi\} = \prod_{i=1}^{n} \int_{X_i}^{X_i + (\delta X_i)} p_X(x|\phi)\, dx, \tag{14.56}$$

replacing the probabilities on the right-hand side by the first-order approximations

$$\int_{X_i}^{X_i + (\delta X)} p_X(x|\phi)\, dx = p_X(X_i|\phi)(\delta X_i).$$

The quantities

$$D_i = Y_i - Y_{i-1} = \int_{X'_{i-1}}^{X'_i} p_X(x|\phi)\, dx, \qquad i = 1, 2, \ldots, n + 1, \tag{14.57}$$

where $X'_0 = -\infty < X'_1 < X'_2 < \cdots < X'_n < X'_{n+1} = \infty$ constitute an ordered random sample of size n, are called *spacings* of $\{Y\}$. The MPS method chooses ϕ that maximizes

$$G_n(\phi, \mathbf{X}) = \left\{ \prod_{i=1}^{n+1} D_i \right\}^{1/(n+1)}. \tag{14.58}$$

Under certain assumptions on $p_X(x|\phi)$ the MPS estimators are consistent. For the three-parameter lognormal distribution, if the true parameter $\phi = (\zeta, \sigma, \theta)$ lies in any set in $\{\phi: -\infty < \theta, \zeta < \infty, 0 < \sigma < \infty\}$, then the estimators are consistent as $n \to \infty$, which is a stronger result than is available for maximum likelihood estimators.

For the lognormal model we have to maximize

$$h(\zeta, \sigma, \theta | X_1, \ldots, X_n) = \sum \log\{\Phi(Z_i) - \Phi(Z_{i-1})\}, \tag{14.59}$$

where $Z_i = \sigma^{-1}\{\log(X_i - \theta) - \zeta\}$, $Z_0 = -\infty$, $Z_{n+1} = +\infty$. Instead of direct maximization Cheng and Amin (1982) obtained the restricted maximum of $h(\cdot)$ with respect to ζ and σ for a sequence of values of θ, and then maximized with respect to θ. They used $\zeta_{a+1} = \overline{Z}_a = n^{-1}\sum_{i=1}^n Z_{ai}$ and $\sigma_{a+1} = \{n^{-1}\sum(Z_{ai} - \overline{Z}_a)^2\}^{1/2}$, with $Z_{ai} = \sigma_a^{-1}\{\log(X_i - \theta) - \zeta_a\}$, in an iterative procedure, and reported fast convergence, usually in three or four iterations, using the stopping rule

$$\left| h(\zeta_{a+1}, \sigma_{a+1}, \theta) - h(\zeta_a, \sigma_a, \theta) \right| < 10^{-6}.$$

Evans and Shaban (1974) discuss estimation of parameters of the form

$$\gamma = \exp(a\zeta + b\sigma^2),$$

where a and b are arbitrary constants. Other papers on this subject by Neyman and Scott (1960), Mehran (1973), Bradu and Mundlak (1970), Likeš (1980), Shimizu and Iwase (1981), and Shimizu (1983) are summarized in Crow and Shimizu (1988).

Zellner (1971), Sinha (1979), and Rukhin (1986) studied Bayesian estimation of parameters of the two-parameter lognormal distribution. The first two authors utilized improper priors $p(\zeta, \sigma) \propto 1/\sigma^c$, $c > 0$. Rukhin utilizes generalized prior density which is also "uniform" in ζ. Wang, Ma, and Shi (1992) discuss, among other things, Bayesian estimation of the mean $\theta = \exp(\zeta + \sigma^2/2)$ of lognormal distribution $\Lambda(\zeta, \sigma^2)$. Using $g(\zeta, \sigma) \propto \sigma^{-1}$ as a joint noninformative prior on (ζ, σ), they derive the marginal posterior for θ

in the form

$$g(\theta|x) \propto \theta^{(n-2)/2} \int_0^{+\infty} \sigma^{n-1}$$

$$\times \exp\left\{-\left[\frac{(n-1)s_Z^2 + n(\log\theta - \overline{Z})^2}{2\sigma^2} + \frac{n\sigma^2}{8}\right]\right\} d\sigma,$$

$$0 < \theta + \infty,$$

where $\overline{Z} = (1/n)\Sigma \log X_i$ and $s_Z^2 = \Sigma(\log X_i - \overline{Z})^2/(n-1)$ [which is Zellner's (1971) result].

For *censored samples* $X_1' \leq \cdots \leq X_r'$ $(r < n)$ with σ known and using prior normal distribution $N(\lambda_0, \psi_0^2)$ on ζ, they obtain

$$h(\zeta|X_r') = \left[\Phi\left(\frac{\zeta - \log X_r'}{\sigma}\right)\right]^{n-r}$$

$$\times \exp\left[-\frac{r^2 + \sigma^2/\psi_0^2}{2\sigma^2}\left(X_r' - \frac{X_r' + \lambda_0\sigma^2/\psi_0^2}{r + \sigma^2/\psi_0^2}\right)^2\right] - I^{-1},$$

where

$$I = \int_{-\infty}^{+\infty} \left[\Phi\left(\frac{\zeta - \log X_r'}{\sigma}\right)\right]^{n-r}$$

$$\times \exp\left[-\frac{r^2 + \sigma^2/\psi_0^2}{2\sigma^2}\left(\zeta - \frac{X_r' + \lambda_0\sigma^2/\psi_0^2}{r + \sigma^2/\psi_0^2}\right)^2\right] d\zeta.$$

The integral I has no closed form solution and must be numerically evaluated. These authors use two-parameter lognormal distributions $\Lambda(\zeta_i, \sigma_i^2)$ for independent variables X_i $(i = 1, 2)$ with independent "noninformative prior" distributions for ζ_i's and σ_i's in analysis of environmental data.

The *reliability function* for two parameter lognormal distributions is given by

$$R(t; \zeta, \sigma) = 1 - \Phi\left[\frac{\log t - \zeta}{\sigma}\right], \qquad t > 0.$$

The Bayesian estimator based on a sample of observations $X_1, X_2, \ldots, X_n$ is given by [cf. Padgett and Wei (1977)]:

$$\widetilde{R_B}(t) = 1 - \Pr\left[t_{2\alpha'} < \sqrt{\frac{\alpha'}{\beta'}} \frac{(\log t - m')}{\sqrt{1 + 1/(\tau^{-1} + n)}}\right],$$

where $t_{2\alpha'}$ is a random variable with t-distribution with $2\alpha'$ degrees of freedom, provided that (ζ, σ^2) has the "normal-gamma prior" [i.e., prior $(\zeta|\sigma)$ is $N(m, \tau\sigma^2)$ and prior (σ^{-2}) is $G(\alpha, \beta)$, with density $g(\nu) =$

$\{\beta^{\alpha}/\Gamma(\alpha)\}\nu^{\alpha-1}e^{-\beta\nu}, \nu \geq 0]$. The joint prior is

$$p(\zeta, \sigma) \propto \sigma^{1-\alpha} \exp\left[-\beta\sigma^2 - \frac{(\zeta - m)^2}{2\tau\sigma^2}\right].$$

Here

$$m' = \frac{\tau^{-1}m + n\bar{Y}}{\tau^{-1} + n},$$

$$\beta' = \beta + \frac{1}{2}\sum_{i=1}^{n}\left(Y_i - \bar{Y}\right)^2 + \frac{\tau^{-1}n(\bar{Y} - m)^2}{2(\tau^{-1} + n)},$$

where $Y_i = \log X_i \sim N(\zeta, \sigma^2)$ and $\alpha' = \alpha + n/2$. As $\alpha \to 0$, $\beta \to 0$, and $\tau^{-1} \to 0$, corresponding to Jeffreys's "vague" prior, we have

$$\Pr\left[t_{2\alpha'} < \sqrt{\frac{\alpha'}{\beta'}} \cdot \frac{\log t - m'}{\sqrt{1 + 1/(\tau^{-1} + n)}}\right]$$

$$\to \Pr\left[t_n < \frac{\log t - \bar{y}}{\sqrt{1 + 1/n}}\sqrt{\frac{n}{\sum_{i=1}^{n}(x_i - \bar{x})^2}}\right]. \qquad (14.60)$$

Sinha (1989) obtains a very similar result directly assuming the joint prior: $p(\zeta, \sigma) \propto 1/\sigma$.

Comparison with the maximum likelihood estimator

$$\widehat{R_{ML}}(t) = 1 - \Phi\left[\frac{\log t - \hat{\zeta}}{\hat{\sigma}}\right]$$

and with the minimum variance unbiased estimator

$$\widehat{R_{MVU}}(t) = \begin{cases} 1, & w \leq 0, \\ 0, & w \geq 1, \\ 1 - I_w\left(\dfrac{n-2}{2}, \dfrac{n-2}{2}\right), & 0 < w < 1, \end{cases}$$

where $I_z(a, b)$ is the incomplete beta function ratio and $w = \frac{1}{2} + \sqrt{n}\left[(\log t - \bar{X})/\{2S\sqrt{n-1}\}\right], t > 0$, with $\bar{X} = \sum_{i=1}^{n}X_i/n$ and $S^2 = \sum_{i=1}^{n}(X_i - \bar{X})^2/(n-1)$, shows, as might be expected, that the Bayes's estimator has the smallest variance when the assumed priors are actually the true priors. The estimator is, however, slightly biased.

4.3   Graphical Estimation

Wise (1952) has put forward the following interesting suggestion for a graphical method of estimating the parameters. It is particularly useful when the "rate of increase" of probability (i.e., probability density) is observed rather than actual frequencies, though it can also be used in the latter case, with sufficiently extensive data. He starts by observing that the tangents at the two points of inflexion of the probability density function remain close to the curve for a considerable length, and so should be estimable graphically with fair accuracy. The modulus of the ratio of the slopes (lower/upper) is

$$\exp\left[2\sinh^{-1}\tfrac{1}{2}\sigma + \sigma\sqrt{1 + \tfrac{1}{4}\sigma^2}\right]. \tag{14.61}$$

Wise (1966a) provides tables of the logarithm of this quantity to four decimal places for $\sigma = 0(0.01)0.80$, to aid in estimating σ from the observed slopes of the inflection tangents. The initial point θ is estimated from the values x_1, x_2 of x at the points where these tangents cut the horizontal axis (see Figure 14.4). If $x_1 < x_2$, then θ is estimated from the formula

$$\tfrac{1}{2}(x_1 + x_2) - \tfrac{1}{2}(x_2 - x_1)L^{-1}, \tag{14.62}$$

where

$$L = \left[1 + \phi\tanh\left(\sigma\sqrt{1 + \tfrac{1}{4}\sigma^2}\right)\right]\left[\tanh\left(\sigma\sqrt{1 + \tfrac{1}{4}\sigma^2}\right) + \phi\right]^{-1}$$

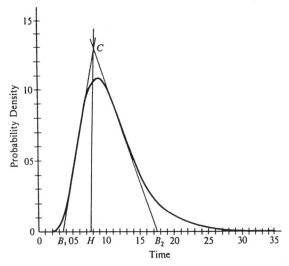

Figure 14.4   Geometrical Method of Estimating Lognormal Parameters

with

$$\phi = \left(1 + \tfrac{1}{2}\sigma^2\right)\left[\sigma\sqrt{1 + \tfrac{1}{4}\sigma^2}\right]^{-1}.$$

Wise (1966a) also provides values of L to four decimal places for $\sigma = 0.(0.01)0.80$.

Finally ζ is estimated from the formula

$$\zeta = \tfrac{3}{2}\sigma^2 + \tfrac{1}{2}\log[(x_1 - \theta)(x_2 - \theta)]. \tag{14.63}$$

Sequential estimation and testing have been discussed by Zacks (1966) and Tomlinson (1957).

Issues of testing normality versus lognormality has received attention in the literature for the last 20 years. See Kotz (1973), Klimko, Rademaker, and Antle (1975), and more recently Al-Khalidi and Hwang (1991).

5 TABLES AND GRAPHS

Aitchison and Brown (1957) give values of the coefficient of variation, α_3, $(\alpha_4 - 3)$, the ratios [from (14.12b)] of mean to median ($e^{\sigma^2/2}$) and mean to mode ($e^{3\sigma^2/2}$), and the probability that X does not exceed $E[X]$ for

$$\sigma = 0.00(0.005)1.00(0.1)3.0.$$

(When $\sigma = 0$, values appropriate to the unit normal distribution are shown.) All values [except for coefficient of variation and $(\alpha_4 - 3)$ when $\sigma = 0.05$] are given to four significant figures.

Tables of percentile points of (two-parameter) lognormal distributions have been published by Moshman (1953): Upper and lower 10%, 5%, 2.5%, 1%, and 0.5% values to 4 decimal places, for $\alpha_3 = 0(0.05)3.00$; and by Broadbent (1956): ratios of upper and lower 5% and 1% values to the expected value, to 4 decimal places, for coefficient of variation $[(e^{\sigma^2} - 1)^{1/2}]$ equal to $0(0.001)0.150$. (Note that this corresponds to $0 \leq \sigma \leq 0.15$ approximately.)

Tables of *random lognormal deviates* have been published by Hyrenius and Gustafsson (1962). These were derived from tables of random normal deviates (see Chapter 13) and are given to 2 decimal places for distributions with skewness $\alpha_3^2 = 0.2, 0.5, 1.0,$ and 2.0 (corresponding approximately to $\sigma = 0.006, 0.16, 0.31,$ and 0.55).

Graph paper with the horizontal (abscissa) scale logarithmic and the vertical (ordinate) scale normal [marked with proportion P at a distance from the origin equal to x where $P = \Phi(x)$] is called *lognormal probability paper*. If X has a two-parameter lognormal distribution, plotting $P = \Pr[X \leq x]$ as ordinate against x as abscissa will give a straight-line plot. The slope of the line is δ ($= \sigma^{-1}$); it meets the horizontal axis at $-\gamma/\delta$ ($= \zeta$). From sample plots these parameters may be estimated by observing the slope of a fitted line and its intersection with the horizontal axis. If a third parameter (θ) is needed, the plotted points tend to curve up, away from a straight line, as x decreases. It is possible, with some patience, to obtain a graphical estimate of θ by trial and error [as suggested by Gibrat (1930)], changing the value of θ until a plot using ($x - \theta$) in place of x can be most closely fitted by a straight line. For a subjective method of this kind, it is not possible to obtain even an approximate formula for the standard deviation of the estimator of θ. However, the method seems to give quite useful results and can certainly be used to obtain initial values for use in iterative processes.

Graph paper of the kind described above is useful in probit analysis of quantal response data. Moments, product moments, and percentage points of various order statistics for standard lognormal distribution $\Lambda(0, 1)$ have been computed and tabulated for all samples of size 20 or less by Gupta, McDonald, and Galarneau (1974).

6 APPLICATIONS

Section 2 of this chapter indicates several fields in which lognormal distributions have been found applicable, and the references in the present section contain further examples. Application of the distribution is not only based on empirical observation but can, in some cases, be supported by theoretical argument—for example, in the distribution of particle sizes in natural aggregates (see Section 2), and in the closely related distribution of dust concentration in industrial atmospheres [Kolmogorov (1941); Tomlinson (1957); Oldham (1965)]. Geological applications have been described by Ahrens (1954–57), Chayes (1954), Miller and Goldberg (1955), and Prohorov (1963). Thébault (1961) gives a number of examples of such applications and includes some further references.

The three-parameter lognormal distribution was introduced to geology by Krige (1960) for modeling gold and uranium grades, and it is now widely regarded as the "natural" parametric model for low-concentration deposits. The lognormal model for minerals present in low concentrations has been experimentally verified for many minerals [Krige (1971); Harbaugh and Ducastaing (1981)]. Its use for gold deposits was pioneered by Sichel (1947). Further applications, mentioned by Oldham (1965), include duration of sickness absence and physicians' consultation time. Wise (1966) has described application to dye-dilution curves representing concentration of indicator as a function of time. Hermanson and Johnson (1967) found that it gives a good representation of flood flows, although extreme value distributions (see Chapter 22) are more generally associated with this field. Medical applications are summarized by Royston (1992), who mentions, in particular, modeling the weights of children [Rona and Altman (1977)] and construction of age-specific reference ranges for clinical variables [Royston (1991)]. Royston (1992) fits lognormal distributions to observations of antibody concentration in μg/ml of type III group B streptococcus (BGS).

Leipnik (1991) mentions applications of sums of independent lognormal variables in telecommunication and studies of effects of the atmosphere on radar signals. Molecular particles are modeled in radar noise theory as having normal velocity distribution, while dust particles of a given type are assigned lognormal distributions.

The lognormal distribution has also been found to be a serious competitor to the Weibull distribution (Chapter 21) in representing lifetime distributions (for manufactured products). Among our references, Adams (1962), Epstein (1947, 1948), Feinlieb (1960), Goldthwaite (1961), Gupta (1962), and Nowick and Berry (1961) refer to this topic. Other applications in quality control are described by Ferrell (1958), Morrison (1958), and Rohn (1959). Aitchison and Brown (1957) cite various applications such as the number of persons in a census occupation class and the distribution of incomes in econometrics, the distribution of stars in the universe, and the distribution of the radical component of Chinese characters.

The 1974 distribution of wealth in the United Kingdom was studied by Chester (1979) who fitted a two-parameter lognormal distribution to data reported in the *Inland Revenue Statistics* (1977). Theoretical grounds for this application of lognormal distributions are developed by Sargan (1957) and Pestieau and Possen (1979) [see also Gibrat (1930, 1931)].

The two-parameter distribution is, in at least one important respect, a more realistic representation of distributions of characters like weight, height, and density than is the normal distribution. These quantities cannot take negative values, but a normal distribution ascribes positive probability to such events, while the two-parameter lognormal distribution does not. Furthermore, by taking σ small enough, it is possible to construct a lognormal distribution closely resembling any normal distribution. Hence, even if a normal distribution is felt to be really appropriate, it might be replaced by a suitable lognormal distribution. Such a replacement is convenient when obtaining confidence limits for the coefficient of variation. Koopmans, Owen, and Rosenblatt (1964) pointed out that if the normal distribution is replaced by a lognormal distribution, then confidence limits for the coefficient of variation are easily constructed (as described in Section 4 of this chapter). Wise (1966) has pointed out marked similarities in shape between appropriately chosen lognormal distributions and inverse Gaussian (Chapter 15) and gamma (Chapter 17) distributions.

The lognormal distribution is also applied, in effect, when certain approximations to the distribution of Fisher's $z = \frac{1}{2} \log F$ are used (see Chapter 27). It is well known that the distribution of z is much closer to normality than is that of F [e.g., Aroian (1939)]. Logarithmic transformations are also often used in attempts to "equalize variances" [Curtiss (1943); Pearce (1945)]. If the standard deviation of a character is expected to vary (e.g., from locality to locality) in such a way that the coefficient of variation remains roughly constant (i.e., with standard deviation roughly proportional to expected value), then application of the method of statistical differentials (as described in Chapter 1) indicates that by use of logarithms of observed values, the dependence of standard deviation on expected value should be substantially reduced. Very often the transformed variables also have a distribution more nearly normal than that of the original variables.

Ratnaparkhi and Park (1986)—starting from Yang's (1978) deterministic model for the rate of change in the residual strength with respect to the number of fatigue cycles

$$\frac{dX(n)}{dn} = - \frac{f(n)}{c(X(n))^{c-1}}, \qquad c \neq 0, \qquad (14.64)$$

where $X(n)$ is the residual strength at cycle n, $f(n)$ is a nonnegative function of n, and c is a physical parameter—proposed using a lognormal distribution for $X(0)$, the initial (or "ultimate") strength. They deduced from this that the random variable N (the fatigue failure cycle) has a three-parameter lognormal distribution.

The relationship between leaving a company and employees tenure has been described by two-parameter lognormal distributions with great success [Young (1971); McClean (1976)]. It was found that the lognormal model accurately defines the leaving behavior for any "entry cohort" (defined as a group of people of about the same quality, performing roughly the same work and joining the company at about the same time). Agrafiotis (1985) estimated the early leavers for each entry cohort of a company based on the lognormal hypothesis by fitting an appropriate unweighted least-squares regression.

O'Neill and Wells (1972) point out that recent work in analyzing automobile insurance losses has shown that the lognormal distribution can be effectively used to fit the distribution for individual insurance claim payments. Many applications in biochemistry, including mechanisms generating the lognormal distribution can be found in Masuyama (1984) and its references.

Recently, for the U.S. population of men and women aged 17–84 years, Brainard and Burmaster (1992) showed that the lognormal distribution fits the marginal histograms of weight (in cells representing 10-lb intervals) for both genders.

7   CENSORING, TRUNCATED LOGNORMAL
AND RELATED DISTRIBUTIONS

As pointed out in Section 4, estimation for the two-parameter lognormal distribution, and the three-parameter distribution with known value for θ, presents no special difficulties beyond those already encountered with normal distributions in Chapter 13. The same is true for censored samples from truncated forms of these distributions [see, e.g., Gajjar and Khatri (1969)].

If $\log X$ has a normal distribution with expected value ζ and standard deviation σ, truncated from below at $\log X_0$, then the rth moment of X about zero is

$$(r\text{th moment if not truncated})\left(\frac{1 - \Phi(U_0 - r\sigma)}{1 - \Phi(U_0)}\right), \qquad (14.65)$$

with $U_0 = (\log X_0 - \zeta)/\sigma$ [Quensel (1945)]. For the three-parameter distribution, with θ not known, when censoring or truncation has been applied, estimation presents considerable difficulty.

Tiku (1968) has used his approximate linearization formula (described in Section 4) to simplify the maximum likelihood equations for the truncated lognormal; Tallis and Young (1962) have considered estimation from grouped data.

Thompson (1951) gives details of fitting a "truncated lognormal distribution" that is a mixed distribution defined by

$$\Pr[\log X = a] = \Phi\left(\frac{a - \zeta}{\sigma}\right)$$

$$\Pr[a < \log X \leq z] = \Phi\left(\frac{z - \zeta}{\sigma}\right) - \Phi\left(\frac{a - \zeta}{\sigma}\right) \qquad \text{for } z > a. \quad (14.66)$$

Tables are provided to assist estimation by moments.

Harter and Moore (1966) gave the results of some sampling experiments in which values of parameters were estimated for lognormal distributions censored by omission of proportions q_1 at the lower and q_2 at the upper limits of the range of variation, with $q_1 = 0$ or 0.01 and $q_2 = 0$ or 0.5. (The case $q_1 = q_2 = 0$ corresponds to a complete, untruncated lognormal distribution. These results have already been noted in Section 4.)

Taking $\zeta = 4$, $\sigma = 2$, and $\theta = 10$, Table 14.5 shows the values that were obtained for the variances of estimators determined by solving the maximum likelihood equations. Although the results are applicable to a censored sample from lognormal distribution they should give a useful indication of the accuracy to be expected with truncated lognormal distributions. The substantial increase in $\text{Var}(\hat{\theta})$ with even a small amount of censoring ($q_1 = 0.01$) at the lower limits is notable. It would seem reasonable to suppose that variances and covariances are approximately inversely proportional to sample size, if estimates are needed for sample sizes larger than 100. The arithmetic means of the estimators were also calculated. They indicated a positive bias

Table 14.5 Variances and Covariances of Maximum Likelihood Estimators for Censored Three-Parameter Lognormal Distribution

q_1	q_2	$\text{Var}(\hat{\zeta})$	$\text{Var}(\hat{\sigma})$	$\text{Var}(\hat{\theta})$	$\text{Cov}(\hat{\sigma}, \hat{\theta})$	$\text{Cov}(\hat{\zeta}, \hat{\theta})$	$\text{Cov}(\hat{\zeta}, \hat{\sigma})$
0.00	0.5	0.0600	0.0961	0.2109	0.0628	0.0146	0.0296
0.01	0.0	0.0428	0.0351	0.2861	0.0064	0.0126	−0.0047
0.01	0.5	0.0628	0.1244	0.3011	0.0199	0.0239	0.0375
0.00[a]	0.0[a]	0.0416	0.0312	0.1733	0.0232	−0.0015	−0.0032

Note: In the sampling experiment with sample size of 100, $\zeta = 4$, $\sigma = 2$, and $\theta = 10$ (the value of θ was not used in the analysis).

[a] No censoring.

in $\hat{\theta}$ of about 0.8–0.9 when $q_1 = 0.01$ (whether or not $q_2 = 0$) and a positive bias of about 0.3 when $q_2 = 0.5$ with $q_1 = 0$. There was also a positive bias of about 0.5 in $\hat{\sigma}$ when $q_1 = 0.01$ and $q_2 = 0.5$.

It is of interest to compare the figures in Table 14.5 with corresponding values in Table 14.6, where the value 10 of θ was supposed known and used in the analysis. The variance of $\hat{\sigma}$ is considerably reduced (by comparison with Table 14.5), but that of $\hat{\zeta}$ is not greatly changed. The effect of varying q_1 is much smaller.

Progressively censored sampling of three-parameter lognormal distributions has been studied by Gajjar and Khatri (1969) and Cohen (1976). Let N designate the total sample size of items subject to a life or fatigue testing and n the number that fail (resulting in completely determined life spans). Suppose that censoring (i.e., removal) occurs in k stages at times $T_j > T_{j-1}$, $j = 1, 2, \ldots, k$, and that r_j surviving items are removed (censored) from further observation at the jth stage. Thus $N = n + \sum_{i=1}^{k} r_i$.

Cohen (1976) discusses Type I censoring, with T_j's fixed, and the number of survivors at these times are represented by random variables. For k-stage Type I progressively censored sampling the likelihood function, in an obvious notation, is

$$C \prod_{i=1}^{n} p_X(X_i) \prod_{j=1}^{k} \left(1 - F_X(T_j)\right)^{r_j}.$$

Table 14.6 Maximum Likelihood Estimators for Censored Three-Parameter Lognormal Distribution Where Value of θ Is Supposed Known

q_1	q_2	$\text{Var}(\hat{\zeta}|\theta)$	$\text{Var}(\hat{\sigma}|\theta)$	$\text{Cov}(\hat{\zeta}, \hat{\sigma}|\theta)$
0.00	0.5	0.0576	0.0460	0.0198
0.01	0.0	0.0406	0.0224	0.0000
0.01	0.5	0.0574	0.0473	0.0197
0.00[a]	0.0[a]	0.0416	0.0218	0.0002

[a] No censoring.

For the lognormal distribution (14.2)' the local maximum likelihood estimating equations are

$$
\begin{cases}
\sum_{1}^{n} [\log(X_i - \theta) - \zeta] + \sigma \sum_{1}^{k} r_i Z_i = 0, \\[2ex]
\sum_{1}^{n} [\log(X_i - \theta) - \zeta]^2 + \sigma^2 \left[\sum_{1}^{k} r_i \left(\frac{T_i - \zeta}{\sigma} \right) Z_i - n \right] = 0, \\[2ex]
\sum_{1}^{n} \left[\frac{\log(X_i - \theta) - \zeta}{X_i - \theta} \right] + \sigma^2 \sum_{1}^{n} \left(\frac{1}{X_i - \theta} \right) + \sigma \sum_{1}^{k} \left(\frac{r_i Z_i}{T_i - \theta} \right) = 0,
\end{cases}
\qquad (14.67)
$$

where $Z_i = Z(T_i^*) = \phi(T_i^*)/[1 - \Phi(T_i^*)]$, with

$$
T_i^* = \{\log(T_i - \theta) - \zeta \sigma/\theta\}/\sigma.
$$

Cohen suggests selecting trial values θ_i for θ, solving the first two equations with $\theta = \theta_i$ for ζ_i and σ_i, using standard Newton-Raphson procedure, and then substituting these values into the third equation. If the third equation is not satisfied for any value of θ in the permissible interval $\theta \leq \hat{x}_{\min}$, he recommends application of a modified maximum likelihood method (which has proved satisfactory in many applications) in which the third equation is replaced by

$$
\theta = X_r' - \exp\left[\zeta + \sigma \Phi^{-1} \left\{ \frac{r}{(N + 1)} \right\} \right],
$$

where $\Phi(\xi_r) = r/(N + 1)$, and X_r' is the rth order statistic among the X's for some $r > 1$.

The discretized form of the (truncated) lognormal distribution has been found to offer a competitive alternative to logarithmic series distributions in some practical situations (see Chapter 7). Quensel (1945) has described the *logarithmic Gram-Charlier* distribution in which log X has a Gram-Charlier distribution.

Kumazawa and Numakunai (1981) introduced a *hybrid lognormal distribution* $H(\alpha, \beta, \gamma)$ defined by the cdf,

$$
H_X(x|\alpha, \beta, \gamma) = \int_0^x \frac{1}{\sqrt{2\pi}} \left(\frac{\alpha}{x} + \beta \right) \exp\left[-\frac{(\alpha \log x + \beta x + \gamma)^2}{2} \right] dx,
$$

$$
0 < x < \infty, \ \alpha \geq 0, \ \beta \geq 0, \ \alpha + \beta > 0. \quad (14.68)
$$

For $\alpha = 0$, the equation (14.68) represents a normal distribution with parametrs $-\gamma\beta^{-1}$, β^{-2}, and for $\beta = 0$ it gives the lognormal distribution

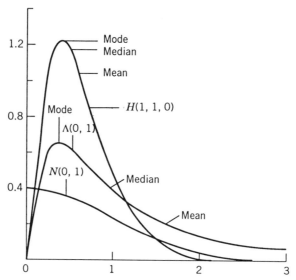

Figure 14.5 Frequency Curves of the Normal, Lognormal, and Hybrid Lognormal Distribution, $N(0, 1)$, $\Lambda(0, 1)$, and $H(1, 1, 0)$, Respectively

$\Lambda(-\gamma\alpha^{-1}, \alpha^{-2})$ [or $N(\log x| -\gamma\alpha^{-1}, \alpha^{-2})$]. Thus (14.68) can be viewed as the cdf of a hybrid normal-lognormal distribution. Figure 14.5 provides comparison of the three curves for standard values of these parameters.

For small $x \ll 1$ the curve of $H(1, 1, 0)$ approaches that of $\Lambda(0, 1)$, and for $x > 1.5$ it decreases faster than both $\Lambda(0, 1)$ and $N(0, 1)$. Data of occupational radiation exposures, as reported in the late 1970s by the U.S. Nuclear Regulatory Commission, are shown to be far better fitted by the hybrid lognormal distribution than by lognormal or normal.

Based on the relation

$$H(x|\alpha, \beta, \gamma) = N\left(\log \rho x + \rho x|\mu, \sigma^2\right), \qquad (14.69)$$

where $\mu = \log(\beta/\alpha) - (\gamma/\alpha)$, $\sigma^2 = 1/\alpha^2$, and $\rho = \beta/\alpha$, Kumazawa and Numakunai (1981) designed hybrid lognormal probability paper that provides a quick method of judging whether given data can be fitted by a hybrid lognormal distribution.

The S_U and S_B systems of distributions [Johnson (1949)], discussed in Chapter 12, are related to the lognormal distribution in that the lognormal distribution is a limiting form of either system and in that the "lognormal line" is the border between the regions in the (β_1, β_2) plane corresponding to the two systems.

The distribution corresponding to the S_B transformation, defined by

$$U = \gamma + \delta \log\left(\frac{X - \xi}{\xi + \lambda - X}\right) \qquad (14.70)$$

which has four parameters γ, δ, ξ, and λ, was considered by van Uven as early as 1917. It has been termed the *four-parameter lognormal* distribution by Aitchison and Brown (1957). This name is not so well recognized as the two-parameter or three-parameter lognormal nomenclature. In view of the existence of other four-parameter transformations (e.g., (14.5)], the usefulness of this name is doubtful.

Lambert (1970) uses the form of pdf,

$$p_X(x) = \frac{(\theta - \tau)}{(x - \tau)(\theta - x)\sqrt{2\pi\beta}} \exp\left[-\frac{1}{2\beta}\left\{\log\frac{x - \tau}{\theta - x} - \mu\right\}^2\right],$$

$$\tau < x < \theta,$$

which is a reparametrization of Johnson's S_B distribution (Chapter 12, Section 4.3). The random variable $\log[(X - \tau)/(\theta - X)]$ has the $N(\mu, \beta)$ distribution. Johnson (1949) carried out estimation of θ and τ using quantiles and a method based on quantiles and points of the range. (See also Chapter 12.)

Estimation becomes a difficult problem when all four parameters must be estimated from observations. Lambert (1970) tackles maximum likelihood estimation of the parameters. The difficulties associated with the behavior of the likelihood function at $\tau = \min X_i$ or $\theta = \max X_i$ are overcome by assuming that observations are multiples of some unit δ (i.e., a recorded observation may differ from its true value by up to $\delta/2$). The likelihood equation is

$$L_1(X_1, X_2, \ldots, X_n | \mu, \beta, \tau, \theta)$$

$$= \begin{cases} \dfrac{(\theta - \tau)^n}{\Pi(X_i - \tau)\Pi(\theta - X_i)(2\pi\beta)^{n/2}} \\ \times \exp\left[-\dfrac{1}{2\beta}\Sigma\left\{\log\dfrac{X_i - \tau}{\theta - X_i} - \mu\right\}^2\right] & \text{if } \tau < X_1 \text{ and } X_n < \theta, \\ 0 & \text{if either } X_1 \le \tau \text{ or } X_n \ge \theta. \end{cases}$$

$$(14.71)$$

(The parameter space is defined by $-\infty < \mu < \infty$; $0 < \beta < \infty$; $-\infty < \tau < X_1$;

$X_n < \theta < \infty$.) At any maximum of L_1 the following equations are satisfied:

$$
\begin{cases}
0 = \dfrac{\partial \log L_1}{\partial \mu} = \dfrac{1}{\beta} \sum \{ \log(X_i - \tau) - \log(\theta - X_i) - \mu \}, \\[2mm]
0 = \dfrac{\partial \log L_1}{\partial \beta} = -\dfrac{n}{2\beta} + \dfrac{1}{2\beta^2} \sum \{ \log(X_i - \tau) - \log(\theta - X_i) - \mu \}^2, \\[2mm]
0 = \dfrac{\partial \log L_1}{\partial \tau} = -\dfrac{n}{\theta - \tau} + \sum \dfrac{1}{X_i - \tau} + \dfrac{1}{\beta} \sum \dfrac{\log(X_i - \tau) - \log(\theta - X_i) - \mu}{X_i - \tau}, \\[2mm]
0 = \dfrac{\partial \log L_1}{\partial \theta} = \dfrac{n}{\theta - \tau} - \sum \dfrac{1}{\theta - X_i} + \dfrac{1}{\beta} \sum \dfrac{\log(X_i - \tau) - \log(\theta - X_i) - \mu}{\theta - X_i}.
\end{cases}
$$

$$(14.72)$$

Solving the first two equations for μ and β yields

$$
\hat{\mu}(\tau, \theta) = \frac{\sum \{ \log(X_i - \tau) - \log(\theta - X_i) \}}{n},
$$

$$
\hat{\beta}(\tau, \theta) = \frac{\sum \{ \log(X_i - \tau) - \log(\theta - X_i) - \mu(\tau, \theta) \}^2}{n}.
$$

Then

$$
L_1^{**}(\tau, \theta) \equiv \sup_{\mu, \beta} L(\mu, \beta, \tau, \theta)
$$

$$
= \frac{(\theta - \tau)^n}{\Pi(X_i - \tau)\Pi(\theta - X_i)\{\hat{\beta}(\tau, \theta)\}^{n/2}} \times \frac{1}{(2\pi e)^{n/2}}.
$$

If the values $\hat{\tau}$ and $\hat{\theta}$ that maximize $L_1^{**}(\tau, \theta)$ are found, then the set $(\hat{\mu}, \hat{\beta}, \hat{\tau}, \hat{\theta})$ will be the solution of equations (14.72), and as long as $\hat{\tau} < X_1 - \delta/2$ and $\hat{\theta} > X_n + \delta/2$, these may be taken as the maximum likelihood estimates.

To locate the greatest value of the likelihood, Lambert (1970) recommends that "the function $\log L_1^{**}(\tau, \theta)$ be plotted over a range of values of τ and θ. Examination of such a plot enables one either to locate a region in which the greatest value of $\log L_1^{**}(\tau, \theta)$ lies or to observe in which direction to shift to find this value; in this case the function was recalculated over a suitable new region. Once a region containing the greatest value is found, changing the intervals in τ and θ at which $\log L_1^{**}$ is calculated enables one to locate the greatest value with any desired accuracy and so to obtain $\hat{\tau}$ and $\hat{\theta}$." Lambert concludes that "although techniques have been developed for estimating, from the likelihood, the parameters of the four-parameter lognormal distribution, and these appear to provide satisfactory estimates with *artificial samples*, the method has not so far yielded useful estimates."

We are unaware of any further development along these lines in estimating the Johnson four-parameter lognormal distribution using maximum likelihood method. Wasilewski (1988) studied mixtures of several two parameter lognormal distributions $\Lambda(\zeta_i, \sigma^2)$ $(i = 1, \ldots, k)$, with pdf

$$
\begin{aligned}
p_X&(x | \sigma, \zeta_1, \ldots, \zeta_k, p_1, \ldots, p_k) \\
&= p_X(x | \sigma, \zeta, p) \\
&= \frac{1}{\sqrt{2\pi}\,\sigma x} \sum_{i=1}^{k} p_i \exp\left(\frac{-(\log x - \zeta_i)^2}{2\sigma^2} \right), \qquad x > 0, \quad (14.73)
\end{aligned}
$$

where $p_i > 0$, $\sum_{i=1}^{k} p_i = 1$, and $0 < \zeta_1 < \cdots < \zeta_k$.

The rth moment about zero of (14.73) is

$$
E[X^r] = \sum_{i=1}^{k} p_i \exp\left(r\zeta_i + \tfrac{1}{2}r^2\sigma^2\right) = e^{r^2\sigma^2/2} \sum_{i=1}^{k} p_i e^{r\zeta_i}. \qquad (14.74)
$$

Wasilewski (1988) discusses estimation of the parameters (σ, ζ, p) by the method of moments, equating for $r = -k, -k + 1, \ldots, k - 1, k$ the values of $E[X^r]$ to the observed values

$$
m_r = n^{-1} \sum_{i=1}^{n} X_i^r.
$$

8   CONVOLUTION OF NORMAL AND LOGNORMAL DISTRIBUTIONS

Hawkins (1991) investigated, in detail, convolutions of normal and lognormal distributions. His work was motivated by application of lognormal distributions—namely, in measuring lognormally distributed quantities with instruments that give quick and inexpensive readings but have substantial random measurement errors. If $\log(Z + \alpha)$ is distributed as $N(\zeta, \sigma^2)$ and X is a measurement of Z with an unbiased normally distributed error Y [$Y \sim N(0, \tau^2)$] independent of Z so that $X = Z + Y$ and $X | Z \sim N(Z, \tau^2)$, then

$$
\begin{aligned}
p_X&(x | \alpha, \tau, \zeta, \sigma) \\
&= \frac{1}{2\pi\sigma\tau} \int_0^\infty z^{-1} \exp\left[-\frac{1}{2}\left\{ \left(\frac{\log z - \zeta}{\sigma} \right)^2 + \left(\frac{x - z + \alpha}{\tau} \right)^2 \right\} \right] dz \\
&= \frac{1}{2\pi\sigma\tau} \int_{-\infty}^\infty \exp\left[-\frac{1}{2}\left\{ \left(\frac{t - \zeta}{\sigma} \right)^2 + \left(\frac{x - e^t + \alpha}{\tau} \right)^2 \right\} \right] dt \qquad (14.74)'
\end{aligned}
$$

(using the transformation $t = \log z$). The parameters α and τ are the location and scale parameters of the convolution.

The canonical form ($\alpha = 0$, $\tau = 1$) is

$$p_X(x|0, 1, \zeta, \sigma) = \int_{-\infty}^{\infty} \frac{1}{2\pi\sigma} \exp\left\{-\frac{1}{2}\left[\frac{(t-\zeta)^2}{\sigma^2} + (x - e^t)^2\right]\right\} dt.$$

$$(14.75)$$

The moments of (14.74)′ are easily calculated from the relation

$$E(X^k) = \sum_{j=0}^{k} \binom{k}{j} E(Z^j) E(U^{k-j}),$$

where U denotes a unit normal variable. In particular,

$$E[X] = -\alpha + \exp\left(\zeta + \tfrac{1}{2}\sigma^2\right), \qquad (14.76)$$

$$\text{Var}(X) = \tau^2 + \exp(2\zeta + \sigma^2)\{e^{\sigma^2} - 1\}. \qquad (14.77)$$

Compare these equations with (14.8a) and (14.8b). The sequence of moments of X increases even more rapidly than that of Z, and (similarly to the lognormal distribution) the distribution of X is also not determined by its moments. Any normal distribution $N(\zeta, \delta^2)$, say, can be obtained as a degenerate limiting case of the convolution, letting $\alpha \to \infty$, $\sigma \to 0$, and keeping the mean and the variance equal to ζ and δ^2, respectively.

Unlike the three-parameter lognormal distribution, the likelihood of the lognormal-normal convolution has no singularities, and the distribution is infinitely differentiable with respect to all its parameters. However, it is still possible that data may conform better to a normal than to a lognormal-normal distribution. Hawkins (1991) finds that his nondegeneracy test (see Section 4) is equally applicable in the lognormal-normal case with Z_i replaced by X_i.

Moment estimation is straightforward. Let $\bar{X} = \Sigma X_i/n$ and $S^2 = \Sigma(X_i - \bar{X})^2/n$, and let $M_3 = \Sigma(X_i - \bar{X})^3/n$ and $B_1 = M_3^2/S^6$, then utilizing equation (14.46)′ or its explicit solution when estimating moments of the three-parameter lognormal, we have

$$\hat{\sigma} = \log \tilde{\omega},$$

$$\hat{\zeta} = \frac{1}{2} \log\left[\frac{S^2}{\tilde{\omega}(\tilde{\omega} - 1)}\right], \qquad (14.78)$$

$$\hat{\alpha} = \exp\left(\hat{\zeta} + \tfrac{1}{2}\hat{\sigma}^2\right) - \bar{X}.$$

However, Hawkins (1991) finds that this modified method of moments in the case of lognormal-normal distribution is by far less effective than the conventional one. Moment estimators turn out to be highly biased, but still they may provide a useful starting point for maximum likelihood estimation. This approach is a very time-consuming routine operation for the lognormal-normal distribution, since it involves numerous evaluations of the density and its derivatives for each value in the sample. Especially undesirable are situations where the initial α is large and the initial σ close to zero (cases close to degeneracy). Hawkins recommends a hybrid algorithm behaving like steepest descent method in the initial steps and like Newton's method close to the optimum.

BIBLIOGRAPHY

Adams, J.-D. (1962). Failure time distribution estimation, *Semiconductor Reliability*, **2**, 41–52.

Agrafiotis, G. K. (1985). Fitting the lognormal distribution to censored labour wastage data, *Mathematics of Social Science*, **10**, 269–279.

Ahrens, L. H. (1954–57). The lognormal distribution of the elements, *Geochimica et Cosmochimica Acta*, **5**, 49–73; **6**, 121–131; **11**, 205–212.

Aitchison, J., and Brown, J. A. C. (1957). *The Lognormal Distribution*, Cambridge: Cambridge University Press.

Al-Khalidi, H. R., and Hwang, L. J. (1991). Some pitfalls of tests of separate families of hypotheses: Normality vs. lognormality, *Communications in Statistics—Theory and Methods*, **20**, 2505–2528.

Aroian, L. A. (1939). A study of Fisher's z distribution, *Annals of Mathematical Statistics*, **12**, 429–438.

Bain, A. D. (1964). *The Growth of Television Ownership in the United Kingdom since the War*, Cambridge: Cambridge University Press.

Balakrishnan, N., and Cohen, A. C. (1991). *Order Statistics and Inference: Estimation Methods*, San Diego: Academic Press.

Barakat, R. (1976). Sums of independent lognormally distributed variables, *Journal of the Optimization Society of America*, **66**, 211–216.

Bliss, C. I. (1934). The method of probits, *Science*, **79**, 38–39.

Bol'shev, L. N., Prohorov, Yu. V., and Rudinov, D. A. (1963). On the logarithmic-normal law in geology (Abstract), *Teoriya Veroyatnostei i ee Primeneniya*, **8**, 114. (In Russian. English translation, 107.)

Bondesson, L. (1979). A general result on infinite divisibility, *Annals of Probability*, **7**, 965–979.

Börges, R. (1966). A characterization of the normal distribution, *Zeitschrift für Wahrscheinlichkeitstheorie und Verwandte Gebiete*, **5**, 244–246.

Boswell, M. T., Ord, J. K., and Patil, G. P. (1979). Normal and lognormal distributions as models of size, *Statistical Distributions in Ecological Work*, J. K. Ord, G. P.

Patil, and C. Taillie (editors), 72–87, Fairland, MD: International Cooperative Publishing House.

Bradu, D., and Mundlak, Y. (1970). Estimation in lognormal linear models, *Journal of the American Statistical Association*, **65**, 198–211.

Brainard, J., and Burmaster, D. E. (1992). Bivariate distributions for height and weight of men and women in the United States, *Risk Analysis*, **12**, 267–275.

Broadbent, S. R. (1956). Lognormal approximation to products and quotients, *Biometrika*, **43**, 404–417.

Bulmer, M. G. (1974). On fitting the Poisson lognormal distribution to species-abundance data, *Biometrics*, **30**, 101–110.

Burges, S. J., and Hoshi, K. (1978). Approximation of a normal distribution by a three-parameter log normal distribution, *Water Resources Research*, **14**, 620–622.

Burges, S. J., Lettenmeier, D. P., and Bates, C. L. (1975). Properties of the three-parameter lognormal probability distribution, *Water Resources Research*, **11**, 229–235.

Calitz, F. (1973). Maximum likelihood estimation of the parameters of the three parameter lognormal distribution—A reconsideration, *Australian Journal of Statistics*, **15**, 229–235.

Camp, B. H. (1938). Notes on the distribution of the geometric mean, *Annals of Mathematical Statistics*, **9**, 221–226.

Carelman, T. (1926). *Les Fonctions Quasi-analytiques*, Paris: Gauthier-Villars.

Charbeneau, R. J. (1978). Comparison of the two- and three-parameters lognormal distributions used in stream-flow analysis, *Water Resources Research*, **14**, 149–160.

Chayes, F. (1954). The lognormal distribution of the elements: A discussion, *Geochimica et Cosmochimica Acta*, **6**, 119–120.

Cheng, R. C. H., and Amin, N. A. K. (1982). Maximum product-of-spacings estimation with application to the lognormal distribution, *Journal of the Royal Statistical Society, Series B*, **44**, 394–403.

Cheng, R. C. H., and Iles, T. C. (1990). Embedded models in three parameter models and their estimation, *Journal of the Royal Statistical Society, Series B*, **52**, 135–149.

Chester, A. (1979). Efficient estimation of income and wealth distribution, Discussion paper, Faculty of Commerce and Social Science, University of Birmingham, England.

Cochran, W. G. (1938). Some difficulties in the statistical analysis of replicated experiments, *Empire Journal of Experimental Agriculture*, **6**, 157–163.

Cohen, A. C. (1951). Estimating parameters of logarithmic-normal distributions by maximum likelihood, *Journal of the American Statistical Association*, **46**, 206–212.

Cohen, A. C. (1963). Progressively censored samples in life-testing, *Technometrics*, **18**, 99–103.

Cohen, A. C., and Whitten, B. J. (1980). Estimation in the three-parameter lognormal distribution, *Journal of the American Statistical Association*, **75**, 399–404.

Cohen, A. C. and Whitten, B. J. (1988). *Parameter Estimation in Reliability and Life Span Models*, New York: Marcel Dekker.

Cohen, A. C., Whitten, B. J., and Ding, Y. (1985). Modified moment estimation for the three-parameter lognormal distribution, *Journal of Quality Technology*, **17**, 92–99.

Crow, E. L. (1988). Applications in Atmospheric Sciences, Chapter 13 in *Lognormal Distributions: Theory and Applications*, E. L. Crow and K. Shimizu, (editors), pp. 331–356, New York: Marcel Dekker.

Crow, E. L., and Shimizu, K. (eds.) (1988). *Lognormal Distributions: Theory and Applications*, New York: Marcel Dekker.

Curtiss, J. H. (1943). On transformations used in the analysis of variance, *Annals of Mathematical Statistics*, **14**, 107–122.

Dahiya, R. C., and Guttman, I. (1982). Shortest confidence and prediction intervals for the log-normal, *Canadian Journal of Statistics*, **10**, 277–291.

Davies, G. R. (1925). The logarithmic curve of distribution, *Journal of the American Statistical Association*, **20**, 467–480.

Davies, G. R. (1929). The analysis of frequency distributions, *Journal of the American Statistical Association*, **24**, 349–366.

Dennis, B., and Patil, G. P. (1988). Applications in Ecology, Chapter 12 in *Lognormal Distributions: Theory and Applications*, E. L. Crow and K. Shimizu (editors), New York: Marcel Dekker.

Dhrymes, P. J. (1962). On devising unbiased estimators for the parameters of the Cobb-Douglas production function, *Econometrica*, **30**, 297–304.

Éltetö, Ö. (1965). Large sample lognormality tests based on new inequality measures, *Bulletin of the International Statistical Institute*, **41**(1), 382–385.

Epstein, B. (1947). The mathematical description of certain breakage mechanisms leading to the logarithmico-normal distribution, *Journal of the Franklin Institute*, **244**, 471–477.

Epstein, B. (1948). Statistical aspects of fracture problems, *Journal of Applied Physics*, **19**, 140–147.

Evans, I. G., and Shaban, S. A. (1974). A note on estimation in lognormal models, *Journal of the American Statistical Association*, **69**, 779–781.

Evans, I. G., and Shaban, S. A. (1976). New estimators (of smaller M.S.E.) for parameters of a lognormal distribution, *Biometrische Zeitschrift*, **18**, 453–466.

Fechner, G. T. (1897). *Kollektivmasslehre*, Leipzig: Engelmann.

Feinlieb, M. (1960). A method of analyzing log-normally distributed survival data with incomplete follow-up, *Journal of the American Statistical Association*, **55**, 534–545.

Fenton, L. F. (1960). The sum of log-normal probability distributions in scatter transmission systems, *IRE Transactions in Communication Systems*, **CS8**, 57–67.

Ferrell, E. B. (1958). Control charts for log-normal universes, *Industrial Quality Control*, **15**(2), 4–6.

Finney, D. J. (1941). On the distribution of a variate whose logarithm is normally distributed, *Journal of the Royal Statistical Society, Series B*, **7**, 155–161.

Gaddum, J. H. (1933). Reports on biological standards III. Methods of biological assay depending on a quantal response, *Special Report Series, Medical Research Council*, London, **183**.

Gaddum, J. H. (1945). Lognormal distributions, *Nature*, **156**, 463–466.

Gajjar, A. V., and Khatri, C. G. (1969). Progressively censored samples from lognormal and logistic distributions, *Technometrics*, **11**, 793–803.

Galton, F. (1879). The geometric mean in vital and social statistics, *Proceedings of the Royal Society of London*, **29**, 365–367.

Gibrat, R. (1930). Une loi des répartitions économiques: L'effet proportionelle, *Bulletin de Statistique Géneral, France*, **19**, 469ff.

Gibrat, R. (1931). *Les Inegalités Economiques*, Paris: Libraire du Recueil Sirey.

Giesbrecht, F., and Kempthorne, O. (1976). Maximum likelihood estimation in the three-parameter lognormal distribution, *Journal of the Royal Statistical Society, Series B*, **38**, 257–264.

Goldthwaite, L. R. (1961). Failure rate study for the lognormal lifetime model, *Proceedings of the Seventh National Symposium on Reliability and Quality Control in Electronics*, 208–213.

Griffiths, D. A. (1980). Interval estimation for the three-parameter lognormal distribution via the likelihood function, *Applied Statistics*, **29**, 58–68.

Grundy, P. M. (1951). The expected frequencies in a sample of an animal population in which the abundances of species are log-normally distributed, Part I, *Biometrika*, **38**, 427–434.

Gupta, S. S. (1962). Life test sampling plans for normal and lognormal distributions, *Technometrics*, **4**, 151–175.

Gupta, S. S., McDonald, G. C., and Galarneau, D. I. (1974). Moments, product moments and percentage points of the order statistics from the lognormal distribution for samples of size twenty and less, *Sankhyā, Series B*, **36**, 230–260.

Halmos, P. R. (1944). Random alms, *Annals of Mathematical Statistics*, **15**, 182–189.

Harbaugh, J., and Ducastaing, M. (1981). Historical changes in oil-field populations as method of forecasting field sizes of undiscovered populations. A comparison of Kansas, Wyoming and California, *Subsurface Geology Series*, Kansas Geological Survey, Lawrence, KS.

Harter, H. L. (1961). Expected values of normal order statistics, *Biometrika*, **48**, 151–166.

Harter, H. L., and Moore, A. H. (1966). Local-maximum-likelihood estimation of the parameters of three-parameter log-normal populations from complete and censored samples, *Journal of the American Statistical Association*, **61**, 842–851.

Hatch, T. (1933). Determination of average particle size from the screen-analysis of non-uniform particulate substances, *Journal of the Franklin Institute*, **215**, 27–37.

Hatch, T., and Choute, S. P. (1929). Statistical description of the size properties of non-uniform particulate substances, *Journal of the Franklin Institute*, **207**, 369–388.

Hawkins, D. M. (1991). The convolution of the normal and lognormal distributions, *South African Statistical Journal*, **25**, 99–128.

Herdan, G. (1958). The relation between the dictionary distribution and the occurrence distribution of word length and its importance for the study of quantitative linguistics, *Biometrika*, **45**, 222–228.

Herdan, G. (1960). *Small Particle Statistics* (Second edition), London: Butterworth.

Herdan, G. (1966). *The Advanced Theory of Language as Choice and Chance*, New York: Springer-Verlag.

Hermanson, R. E., and Johnson, H. P. (1967). Generalized flood-frequency relationships, *Iowa State Journal of Science*, **41**, 247–268.

Heyde, C. C. (1963). On a property of the lognormal distribution, *Journal of the Royal Statistical Society, Series B*, **25**, 392–393.

Hill, B. M. (1963). The three-parameter lognormal distribution and Bayesian analysis of a point-source epidemic, *Journal of the American Statistical Association*, **58**, 72–84.

Hyrenius, H., and Gustafsson, R. (1962). *Tables of Normal and Log-normal Random Deviates: I, II*, Stockholm: Almqvist and Wiksell.

Iwase, K., and Kanefuji, K. (1992). Estimation for 3-parameter lognormal distribution with unknown shifted origin, *Technical Report No. 92-D5*, Department of Applied Mathematics, Hiroshima University, Japan.

Johnson, N. L. (1949). Systems of frequency curves generated by methods of translation, *Biometrika*, **36**, 149–176.

Jones, C. F. (1971). A confidence interval for the lognormal hazard, *Technometrics*, **13**, 885–888.

Kalecki, M. (1945). On the Gibrat distribution, *Econometrica*, **13**, 161–170.

Kalinske, A. A. (1946). On the logarithmic probability law, *Transaction American Geophysical Union*, **27**, 709–711.

Kapteyn, J. C. (1903). *Skew Frequency Curves in Biology and Statistics*, Groningen: Astronomical Laboratory, Noordhoft.

Kapteyn, J. C., and van Uven, M. J. (1916). *Skew Frequency Curves in Biology and Statistics*, Groningen: Hotsema Brothers, Inc.

Kemsley, W. F. F. (1952). Body weight at different ages and heights, *Annals of Eugenics, London*, **16**, 316–334.

Klimko, L. A., Rademaker, A., and Antle, C. E. (1975). A test for normality versus lognormality, *Communications in Statistics*, **4**, 1009–1019.

Koch, A. L. (1966). The logarithm in biology, I. Mechanisms generating the lognormal distributions exactly, *Journal of Theoretical Biology*, **12**, 276–290.

Koch, A. L. (1969). The logarithm in biology, II. Distributions simulating the lognormal, *Journal of Theoretical Biology*, **23**, 251–268.

Kolmogorov, A. N. (1941). Über das logarithmisch normale Verteilungsgesetz der Dimensionen der Teilchen bei Zerstückelung, *Doklady Akademii Nauk SSSR*, **31**, 99–101.

Koopmans, L. H., Owen, D. B., and Rosenblatt, J. I. (1964). Confidence intervals for the coefficient of variation for the normal and lognormal distributions, *Biometrika*, **51**, 25–32.

Kottler, F. (1950). The distribution of particle sizes, I, *Journal of the Franklin Institute*, **250**, 339–356.

Kotz, S. (1973). Normality vs. lognormality with applications, *Communications in Statistics*, **1**, 113–132.

Krige, D. G. (1960). On the departure of one value distributions from the lognormal model in South African gold mines, *Journal of the South African Institute for Mining and Metallurgy*, **61**, 231–244.

Krige, D. G. (1971). Geostatistical case studies of the advantages of lognormal-deWijsian Kriging with mean for a base metal mine and a gold mine, *Journal of the International Association for Mathematical Geology*, **14**, 547–555.

Krumbein, W. C. (1936). Application of logarithmic moments to size frequency distributions of sediments, *Journal of Sedimentary Petrology*, **6**, 35–47.

Kumazawa, S., and Numakunai, T. (1981). A new theoretical analysis of occupational dose distributions indicating the effect of dose limits, *Health Physics*, **41**, 465–475.

Lambert, J. A. (1964). Estimation of parameters in the three-parameter lognormal distribution, *Australian Journal of Statistics*, **6**, 29–32.

Lambert, J. A. (1970). Estimation of parameters in the four parameter lognormal distribution, *Australian Journal of Statistics*, **12**, 33–43.

Land, C. E. (1972). An evaluation of approximate confidence interval estimation methods for lognormal means, *Technometrics*, **14**, 145–158.

Laurent, A. G. (1963). The lognormal distribution and the translation method: description and estimation problems, *Journal of the American Statistical Association*, **58**, 231–235. (Correction, *Ibid.*, **58**, 1163.)

Lawrence, R. J. (1988a). The lognormal as event-time distribution, Chapter 8 in *Lognormal Distributions: Theory and Applications*, E. L. Crow and K. Shimizu (editors), pp. 211–228, New York: Marcel Dekker.

Lawrence, R. J. (1988b). Applications in economics and business, Chapter 9 in *Lognormal Distributions: Theory and Applications*, E. L. Crow and K. Shimizu (editors), pp. 229–266, New York: Marcel Dekker.

Leipnik, R. B. (1991). Lognormal random variables, *Journal of the Australian Mathematical Society, Series B*, **32**, 327–347.

Likes, J. (1980). Variance of the MVUE for lognormal variance, *Technometrics*, **22**, 253–258.

McAlister, D. (1879). The law of the geometric mean, *Proceedings of the Royal Society of London*, **29**, 367–375.

McClean, S. (1976). A comparison of the lognormal and transition models for wastage, *The Statistician*, **25**, 281–294.

Masuyama, M. (1984). A measure of biochemical individual variability, *Biometrical Journal*, **26**, 327–346.

Mehran, F. (1973). Variance of the MVUE for the lognormal mean, *Journal of the American Statistical Association*, **68**, 726–727.

Michelini, C. (1972). Convergence pattern of the scoring method in estimating parameters of a lognormal function, *Journal of the American Statistical Association*, **67**, 319–323.

Miller, R. L., and Goldberg, E. D. (1955). The normal distribution in geochemistry, *Geochimica et Cosmochimica Acta*, **8**, 53–62.

Morrison, J. (1958). The lognormal distribution in quality control, *Applied Statistics*, **7**, 160–172.

Moshman, J. E. (1953). Critical values of the log-normal distribution, *Journal of the American Statistical Association*, **48**, 600–609.

Mosimann, J. E. (1970). Size allometry: Size and shape variables with characterizations of the lognormal and generalized gamma distributions, *Journal of the American Statistical Association*, **65**, 930–945.

Mosimann, J. E., and Campbell, G. (1988). Applications in biology: Simple growth models, Chapter 11 in *Lognormal Distributions: Theory and Applications*, E. L. Crow and K. Shimizu (editors), pp. 287–302, New York: Marcel Dekker.

Nakamura, T. (1991). Existence of maximum likelihood estimates for interval-censored data from some three-parameter models with a shifted origin, *Journal of the Royal Statistical Society, Series B*, **53**, 211–220.

Naus, J. I. (1969). The distribution of the logarithm of the sum of two log-normal variates, *Journal of the American Statistical Association*, **64**, 655–659.

Nelson, W., and Schmee, J. (1979). Inference for (log) normal life distributions from small singly censored samples and BLUEs, *Technometrics*, **21**, 43–54.

Neyman, J., and Scott, E. (1960). Correction for bias introduced by a transformation of variables, *Annals of Mathematical Statistics*, **31**, 643–655.

Nowick, A. S., and Berry, B. S. (1961). Lognormal distribution function for describing anelastic and other relaxation processes, *IBM Journal of Research and Development*, **5**, 297–311, 312–320.

Nydell, S. (1919). The mean errors of the characteristics in logarithmic-normal distribution, *Skandinavisk Aktuarietidskrift*, **1**, 134–144.

Oldham, D. D. (1965). On estimating the arithmetic means of lognormally distributed populations, *Biometrics*, **21**, 235–239.

O'Neill, B., and Wells, W. T. (1972). Some recent results in lognormal parameter estimation using grouped and ungrouped data, *Journal of the American Statistical Association*, **69**, 76–79.

Padgett, W. J., and Wei, L. J. (1977). Bayes estimation of reliability for the two-parameter lognormal distribution, *Communications in Statistics—Theory and Methods*, **6**, 443–457.

Patterson, R. L. (1969). Difficulties involved in the estimation of a population mean using transformed sample data, *Technometrics*, **8**, 535–537.

Pearce, S. C. (1945). Lognormal distribution, *Nature*, **156**, 747.

Pestieau, P., and Possen, U. M. (1979). A model of wealth distribution, *Econometrica*, **47**, 701–722.

Peters, S. (1963). Multi-quantile estimates of the moments of a lognormal distribution, Working Memo No. 142, Arthur D. Little, Inc.

Preston, F. W. (1948). The commonness, and rarity of species, *Ecology*, **29**, 254–283.

Prohorov, Yu. V. (1963). On the lognormal distribution in geo-chemistry, *Teoriya Veroyatnostei i ee Primeneniya*, **10**, 184–187. (In Russian)

Quensel, C.-E. (1945). Studies of the logarithmic normal curve, *Skandinavisk Aktuarietidskrift*, **28**, 141–153.

Ratnaparkhi, M. V., and Park, W. J. (1986). Lognormal distribution—Model for fatigue life and residual strength of composite material, *IEEE Transactions on Reliability*, **R-35**, 312–315.

Reid, D. D. (1981). The Poisson lognormal distribution and its use as a model of plankton aggregation, *Statistical Distributions in Scientific Work*, C. O. Taillie, G. P. Patil, and B. A. Baldessari (editors), 303–316, Dordrecht: Reidel.

Rendu, J.-M. M. (1988). Applications in geology, Chapter 14 in *Lognormal Distributions: Theory and Applications*, E. L. Crow and K. Shimizu (editors), pp. 357–366, New York: Marcel Dekker.

Rohn, W. B. (1959). Reliability prediction for complex systems, *Proceedings of the Fifth National Symposium on Reliability and Quality Control in Electronics*, 381–388.

Rona, R. J., and Altman, D. G. (1977). National study of health and growth: Standards of attained height, weight and triceps skinfold in English children 5 to 11 years old, *Annals of Human Biology*, **4**, 501–523.

Royston, P. (1991). Constructing time-specific reference ranges, *Statistics in Medicine*, **10**, 675–690.

Royston, P. (1992). Estimation, reference ranges and goodness of fit for the three-parameter log-normal distribution, *Statistics in Medicine*, **11**, 897–912.

Rukhin, A. L. (1986). Improved estimation in lognormal models, *Journal of the American Statistical Association*, **81**, 1041–1049.

Sargan, J. D. (1957). The distribution of wealth, *Econometrica*, **25**, 568–590.

Sartwell, P. E. (1950). The distribution of incubation periods of infectious diseases, *American Journal of Hygiene*, **51**, 310–318.

Schwartz, S. C., and Yeh, Y. (1982). On the distribution function and moments of power-sums with log-normal components, *Bell System Technical Journal*, **61**, 1441–1462.

Severo, N. C., and Olds, E. G. (1956). A comparison of tests on the mean of a logarithmico-normal distribution, *Annals of Mathematical Statistics*, **27**, 670–686.

Shaban, S. A. (1988a). Poisson–lognormal distributions, Chapter 7 in *Lognormal Distributions: Theory and Applications*, E. L. Crow and K. Shimizu (editors), pp. 195–210, New York: Marcel Dekker.

Shaban, S. A. (1988b). Applications in industry, Chapter 10 in *Lognormal Distributions: Theory and Applications*, E. L. Crow and K. Shimizu (editors), pp. 267–286, New York: Marcel Dekker.

Shimizu, K. (1986). Estimation in lognormal distribution, *Proceedings of the Second Japan-China Symposium on Statistics*, 237–240, Kyushi University, Fukanoka, Japan.

Shimizu, K., and Iwase, K. (1981). Uniformly minimum variance unbiased estimation in lognormal and related distributions, *Communications in Statistics—Theory and Methods*, **10**, 1127–1147.

Sichel, H. S. (1947). Experimental and theoretical investigations of the bias error in mine sampling with special reference to narrow gold reefs, *Transactions of the Institute of Mining and Metallurgy*, **56**, 403–473.

Sichel, H. S. (1967). Application of statistical techniques to the evaluation of mineral deposits, *Bulletin of the International Statistical Institute*, **42**, 245–268.

Sinha, S. K. (1979). A note on Bayes estimators and robustness of lognormal parameters, *Journal of the Indian Society of Agricultural Statistics*, **31**(3), 49–53.

Sinha, S. K. (1981). On the moment estimation of lognormal parameters, *Journal of the Indian Association for Productivity, Quality and Reliability* (*IAPQR Transactions*), **6**, No. 2, 83–88.

Sinha, S. K. (1989). Bayesian inference about the prediction/credible intervals and reliability function for lognormal distribution, *Journal of the Indian Statistical Association*, **27**, 73–78.

Sinnott, E. W. (1937). The relation of gene to character in quantitative inheritance, *Proc. National Academy of Science, Washington*, **23**, 224–227.

Slifker, J. F., and Shapiro, S. S. (1980). The Johnson system: Selection and parameter estimation, *Technometrics*, **22**, 239–246.

Sweet, A. L. (1990). On the hazard rate of the lognormal distribution, *IEEE Transactions on Reliability*, **39**, 325–328.

Szajnowski, W. J. (1977). Estimators of log-normal distribution parameters, *IEEE Transactions on Aerospace and Electronic Systems*, **AES-13**, 533–536.

Tallis, G. M., and Young, S. S. Y. (1962). Maximum likelihood estimation of parameters of the normal, the log-normal, truncated normal and bivariate normal distributions from grouped data, *Australian Journal of Statistics*, **4**, 49–54.

Thébault, J. Y. (1961). Distribution lognormale de certains caractères de quelques phénomènes géologiques et ses applications, *Revue de Statistique Appliquée*, **9**(2), 37–87.

Thompson, H. R. (1951). Truncated lognormal distributions, I. Solution by moments, *Biometrika*, **38**, 414–422.

Thöni, H. (1969). A table for estimating the mean of a lognormal distribution, *Journal of the American Statistical Association*, **64**, 632–636.

Thorin, O. (1977). On the infinite divisibility of the lognormal distribution, *Scandinavian Actuarial Journal*, 121–148.

Tiku, M. L. (1968). Estimating the parameters of log-normal distribution from censored samples, *Journal of the American Statistical Association*, **63**, 134–140.

Tippett, L. H. C. (1925). On the extreme individuals and the range of samples taken from a normal population, *Biometrika*, **17**, 364–387.

Tokoko, K. (1966). On the mode and median of Gibrat distribution, *Bulletin of the Faculty of Arts and Sciences, Ibaraki University*, (*Natural Sciences*), **17**, 11–15.

Tomlinson, R. C. (1957). A simple sequential procedure to test whether average conditions achieve a certain standard, *Applied Statistics*, **6**, 198–207.

Uven, M. J. van (1917a). Logarithmic frequency distributions, *Proceedings of the Royal Academy of Sciences, Amsterdam*, **19**, 533–546.

Uven, M. J. van (1917b). Logarithmic frequency distributions, *Proceedings of the Royal Academy of Sciences, Amsterdam*, **19**, 670–694.

Wang, H. Z., Ma, B.-H., and Shi, J.-S. (1992). Estimation of environmental factors for the log normal distribution, *Microelectronics & Reliability*, **32**, No. 5, 679–685.

Wartmann, R. (1956). Anwendung der logarithmischen Normalverteilung, *Mitteilungsblatt für Mathematische Statistik*, **8**, 83–91.

Wasilewski, M. (1988). Estimation of the parameters of the mixture ($k \geq 1$) of logarithmic-normal variables, *Trabajos de Estadística*, **3**, 167–175.

Weiss, L. L. (1957). A nomogram for log-normal frequency analysis, *Transactions of the American Geophysical Union*, **38**, 33–37.

Wicksell, S. D. (1917). On the genetic theory of frequency, *Arkiv för Mathematik, Astronomi och Fysik*, **12**, No. 20.

Williams, C. B. (1937). The use of logarithms in the interpretation of certain entomological problems, *Annals of Applied Biology*, **24**, 404–414.

Williams, C. B. (1940). A note on the statistical analysis of sentence length, *Biometrika*, **31**, 356–361.

Wilson, E. G., and Worcester, J. (1945). The normal logarithmic transformation, *Review of Economics and Statistics*, **27**, 17–22.

Wingo, D. R. (1976). Moving truncations barrier-function methods for estimation in three-parameter lognormal models, *Communications in Statistics—Simulation and Computation*, **1**, 65–80.

Wise, M. E. (1952). Dense random packing of unequal spheres, *Philips Research Reports*, **7**, 321–343.

Wise, M. E. (1966a). The geometry of log-normal and related distributions and an application to tracer-dilution curves, *Statistica Neerlandica*, **20**, 119–142.

Wise, M. E. (1966b). Tracer-dilution curves in cardiology and random walk and lognormal distributions, *Acta Physiologica Pharmacologica Neerlandica*, **14**, 175–204.

Wu, C.-Y. (1966). The types of limit distribution for some terms of variational series, *Scientia Sinica*, **15**, 745–762.

Yang, J. N. (1978). Fatigue and residual strength degradation for graphite/epoxy composite under tension-compression cyclic loadings, *Journal of Composite Materials*, **12**, 19–39.

Young, A. (1971). Demographic and ecological models for manpower planning, *Aspects of Manpower Planning*, D. J. Bartholomew and R. B. Morris (editors), London: English Universities Press.

Yuan, P. T. (1933). On the logarithmic frequency distribution and the semi-logarithmic correlation surface, *Annals of Mathematical Statistics*, **4**, 30–74.

Zacks, S. (1966). Sequential estimation of the mean of a log-normal distribution having a prescribed proportional closeness, *Annals of Mathematical Statistics*, **37**, 1688–1695.

Zellner, A. (1971). Bayesian and non-Bayesian analysis of the lognormal distribution and lognormal regression, *Journal of the American Statistical Association*, **66**, 327–330.

CHAPTER 15

Inverse Gaussian (Wald) Distributions

1 INTRODUCTION

The name "inverse Gaussian" was first applied to a certain class of distributions by Tweedie (1947), who noted the inverse relationship between the cumulant generating functions of these distributions and those of Gaussian distributions. The same class of distributions was derived by Wald (1947) as an asymptotic form of distribution of average sample number in sequential analysis. The name "Wald distribution" is also used for members of this class. Wasan (1968b) calls these distributions *first passage time distributions of Brownian motion with positive drift* and uses the abbreviation *T.B.M.P. distribution*.

Inverse Gaussian distributions have attracted considerable attention during the last 20 years. A major influence was the Ph.D. dissertation of R. S. Chhikara, under J. L. Folks at Oklahoma State University, in 1972. This work was followed by several joint publications by Chhikara and Folks culminating in a review paper [Folks and Chhikara (1978)]. This paper attracted many discussants—including probabilists and both theoretical and applied statisticians—and continuing interest in the topic, resulting eventually in a book devoted to inverse Gaussian distributions [Chhikara and Folks (1989)]. In the early 1980s B. Jørgensen's M.Sc. thesis, under the direction of O. Barndorff-Nielsen, appeared as Jørgensen (1982). Tables of percentile points of inverse Gaussian distributions became available in a *CRC Handbook* (1988). Books by A. C. Cohen and B. J. Whitten (1989) and N. Balakrishnan and A. C. Cohen (1991) contain a chapter devoted to estimation of parameters in a three-parameter inverse Gaussian distributions.

In view of this partial consolidation of recent research on inverse Gaussian distributions, and to keep the length of this chapter with reasonable limits, we include here mostly results that are not emphasized in the above references and concentrate on post-1988 results as much as possible. Readers are advised to consult Chhikara and Folks (1989) and Jørgensen (1982) for

259

additional details. The bibliography of this chapter is, however, more comprehensive and contains some references that are not explicitly mentioned in the text but might be viewed as important contributions in regard to particular aspects of the distributions.

2 GENESIS

Suppose that a particle moving along a line tends to move with a uniform velocity v. Suppose also that the particle is subject to linear Brownian motion, which causes it to take a variable amount of time to cover a fixed distance d. It can be shown that the time X required to cover the distance is a random variable with probability density function

$$p_X(x) = \frac{1}{\sqrt{2\pi\beta x^3}} d e^{-(d-vx)^2/(2\beta x)}, \qquad 0 < x, \qquad (15.1)$$

where β is a diffusion constant [Schrödinger (1915)]. Several heuristic and some more or less rigorous derivations of this result have been published [Huff (1974); Whitmore and Seshadri (1987, and references therein)].

Alternatively, when the time x is fixed, the distance D over which the particle travels is a random variable with the normal distribution

$$p_D(d) = \frac{1}{\sqrt{2\pi\beta x}} e^{-(d-vx)^2/(2\beta x)}. \qquad (15.2)$$

While examining the cumulant generating functions of (15.1) and (15.2), Tweedie (1956) noticed the inverse relationship between these functions and suggested the name "inverse Gaussian" for the distribution (15.1).

Distributions of this kind were derived by Wald (1947) as a limiting form of distribution of sample size in certain sequential probability ratio tests. In a more general form, the distribution can be obtained as a solution to the following problem: Given that $Z_1, Z_2, \ldots$ are independent random variables, each having the same distribution with finite expected value $E[Z] > 0$ and nonzero variance $V(Z)$, what is the limiting distribution of the random variable N, defined by

$$\left[\bigcap_{j=1}^{N-1} \left(\sum_{i=1}^{j} Z_i < K \right) \right] \cap \left(\sum_{i=1}^{N} Z_i \geq K \right), \qquad \text{with } K > 0?$$

Note that $\sum_{i=1}^{N} Z_i$ is the first of the sums $Z_1, Z_1 + Z_2, Z_1 + Z_2 + Z_3, \ldots$, to be not less than K, with $K > 0$.

It can be shown that $E[N] = K/E[Z]$ and that

$$\lim_{E(N) \to \infty} \Pr[N \leq xE(N)]$$

is given by the integral up to x of

$$p_X(x) = \sqrt{\frac{\phi}{2\pi}}\, e^\phi x^{-3/2} \exp\left[-\frac{1}{2}\phi(x + x^{-1})\right], \qquad x > 0, \quad (15.3)$$

with

$$\phi = \frac{KE[Z]}{V(Z)} > 0.$$

This is the *standard form* of probability density function of the *Wald distribution*. It is a special inverse Gaussian distribution.

3   DEFINITION

On substituting $\nu = d/\lambda$ and $\beta = d^2/\lambda$ into (15.1), we obtain the standard ("canonical") form of the *two-parameter* inverse Gaussian distribution:

$$
\begin{aligned}
p_X(x|\mu, \lambda) &= \left[\frac{\lambda}{2\pi x^3}\right]^{1/2} \exp\left\{-\frac{\lambda}{2\mu^2 x}(x - \mu)^2\right\} \\
&= \left[\frac{\lambda}{2\pi x^3}\right]^{1/2} \exp\left\{-\frac{\lambda}{2\mu}\left(\frac{x}{\mu} - 2 + \frac{\mu}{x}\right)\right\}, \qquad x > 0. \quad (15.4a)
\end{aligned}
$$

We will denote this distribution by $IG(\mu, \lambda)$. As we will see below, μ is the expected value and μ^3/λ is the variance of $IG(\mu, \lambda)$; μ/λ is the square of the coefficient of variation.

Alternatively, the distribution can be written in any of the three following equivalent forms [Tweedie (1957a)]:

$$p_X(x|\mu, \phi) = \left[\frac{\mu\phi}{2\pi x^3}\right]^{1/2} e^\phi \exp\left\{-\frac{1}{2}\phi\left(\frac{x}{\mu} + \frac{\mu}{x}\right)\right\}, \qquad IG(\mu, \phi\mu),$$

$$(15.4b)$$

$$p_X(x|\phi, \lambda) = \left[\frac{\lambda}{2\pi x^3}\right]^{1/2} e^\phi \exp\left\{-\frac{1}{2}\left(\frac{\phi^2 x}{\lambda} + \frac{\lambda}{x}\right)\right\}, \qquad IG\left(\frac{\lambda}{\phi}, \lambda\right),$$

$$(15.4c)$$

$$p_X(x|\alpha, \lambda) = \left[\frac{\lambda}{2\pi x^3}\right]^{1/2} \exp\left[-\frac{\lambda}{2}\left\{\alpha x - (2\alpha)^{1/2} + \frac{1}{2x}\right\}\right], \qquad IG\left(\frac{\alpha^2}{2}, \lambda\right).$$

$$(15.4d)$$

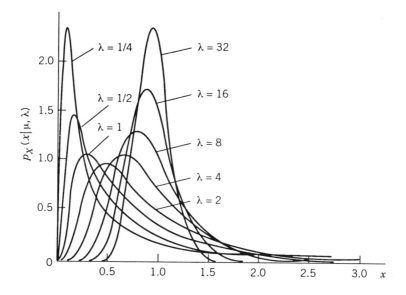

Figure 15.1 Inverse Gaussian Density Functions, $E[X] = \mu = 1$

In these equations $\phi = \lambda/\mu$, $\alpha = \sqrt{(2\mu)}$; λ, μ, ϕ, and α are all positive. The Wald distribution (15.3) is obtained by taking $\mu = 1$ (and $\lambda = \phi$); in our notation it is IG(1, ϕ). Our distinction between standard Wald (one-parameter) and standard inverse Gaussian (two parameters) is arbitrary, but we find it to be convenient.

Note that if X is distributed as IG(μ, λ), then aX ($a > 0$) is distributed as IG($a\mu$, $a\lambda$). Some graphs of $p_X(x|\mu, \lambda)$ are shown in Figure 15.1. The cdf corresponding to (15.4a) is

$$F_X(x|\mu, \lambda) = \Phi\left\{\sqrt{\frac{\lambda}{x}}\left(\frac{x}{\mu} - 1\right)\right\} + e^{2\lambda/\mu}\Phi\left\{-\sqrt{\frac{\lambda}{x}}\left(\frac{x}{\mu} + 1\right)\right\} \quad (15.4e)$$

where $\Phi(\cdot)$ is the standard normal distribution function [Shuster (1968); Chhikara and Folks (1977)].

4 MOMENTS

From (15.4a) we obtain the cumulant generating function

$$\Psi_X(t; \mu, \lambda) = \frac{\lambda}{\mu}\left\{1 - \left(1 - \frac{2\mu^2 t}{\lambda}\right)^{1/2}\right\}. \quad (15.5a)$$

The corresponding characteristic function is

$$\exp\left[\frac{\lambda}{\mu}\left\{1 - \left(1 - \frac{2i\mu^2 t}{\lambda}\right)^{1/2}\right\}\right].$$

The cumulant generating functions corresponding to (15.4b), (15.4c), and (15.4e) are

$$\Psi_X(t;\mu,\phi) = \phi\left\{1 - \left(1 - \frac{2\mu t}{\phi}\right)^{1/2}\right\} \tag{15.5b}$$

$$\Psi_X(t;\phi,\lambda) = \phi\left\{1 - \left(1 - \frac{2\lambda t}{\phi^2}\right)^{1/2}\right\} \tag{15.5c}$$

$$\Psi_X(t;\alpha,\lambda) = 2^{1/2}\lambda\left\{\alpha^{1/2} - \left(\alpha - \frac{t}{\lambda}\right)^{1/2}\right\} \tag{15.5d}$$

respectively.

The first four cumulants corresponding to (15.5a) are

$$\begin{cases} \kappa_1 = \mu & (= E[X]), \\[2mm] \kappa_2 = \dfrac{\mu^3}{\lambda} & (= \mathrm{Var}(X)), \\[2mm] \kappa_3 = \dfrac{3\mu^5}{\lambda^2}, \\[2mm] \kappa_4 = \dfrac{15\mu^7}{\lambda^3}. \end{cases} \tag{15.6}$$

Generally, for integer $r \geq 2$,

$$\kappa_r = 1 \times 3 \times 5 \ldots \times (2r - 3)\frac{\mu^{2r-1}}{\lambda^{r-1}} \tag{15.7}$$

[Tweedie (1957a)]. The first two moment-ratios are

$$\sqrt{\beta_1} = \alpha_3 = 3\sqrt{\frac{\mu}{\lambda}} \qquad (= 3\phi^{-1/2}), \tag{15.8a}$$

$$\beta_2 = \alpha_4 = 3 + \frac{15\mu}{\lambda} \qquad (= 3 + 15\phi^{-1}). \tag{15.8b}$$

In the Pearson (β_1, β_2) plane the inverse Gaussian points fall on the straight line $\beta_2 = 3 + 5\beta_1/3$ which lies between the gamma (Type III) and lognormal lines.

The first three negative moments about zero of $IG(\mu, \lambda)$ are

$$\mu'_{-1} = \mu^{-1} + \lambda^{-1}, \tag{15.9a}$$

$$\mu'_{-2} = \mu^{-2} + 3\mu^{-1}\lambda^{-1} + 3\lambda^{-2}, \tag{15.9b}$$

$$\mu'_{-3} = \mu^{-3} + 6\mu^{-2}\lambda^{-1} + 15\mu^{-1}\lambda^{-2} + 15\lambda^{-3}. \tag{15.9c}$$

Note that for the Wald distribution, $\kappa_1 = 1$ whatever the value of ϕ, and that

$$\kappa_r = 1 \times 3 \times 5 \ldots \times (2r - 3)\phi^{-(r-1)}.$$

From (15.4b) the density function of $Y = X^{-1}$ is

$$p_Y(y|\mu, \phi) = \left[\frac{\mu\phi}{2\pi y^3} \right]^{-1/2} y \exp\left\{ -\frac{\mu\phi y}{2} + \phi - \frac{\phi}{2\mu y} \right\}.$$

It follows that

$$E[X^{-r}|\mu, \phi] = E[Y^r|\mu, \phi] = \mu E[X^{r+1}|\mu^{-1}, \phi].$$

In particular for the standard Wald distribution ($\mu = 1$), we have the remarkable relation between negative and positive moments

$$\mu'_{-r} = \mu'_{r+1}. \tag{15.10}$$

For the standard Wald distribution, the mean deviation is

$$4e^{2\phi}\Phi\left(-2\sqrt{\phi}\right). \tag{15.11}$$

For ϕ large (whatever be the value of μ)

$$\frac{\text{Mean deviation}}{\text{Standard deviation}} = 4\sqrt{\phi}\, e^{2\phi}\Phi\left(-2\sqrt{\phi}\right)$$

$$\doteq 4\sqrt{\phi}\, e^{2\phi}\left[\left(\sqrt{2\pi}\right)^{-1} e^{-2\phi}\left(2\sqrt{\phi}\right)^{-1}\left\{1 - \left(2\sqrt{\phi}\right)^{-2}\right\}\right]$$

$$\doteq \sqrt{\frac{2}{\pi}}\left(1 - \frac{1}{4}\phi^{-1}\right). \tag{15.12}$$

Since taking $\mu = 0$ (to make $E[X] = 0$) also makes $\text{Var}(X) = 0$, a standardized form of the distribution has to be obtained circuitously. By taking the variable

$$Y = X + \xi, \qquad \xi < 0,$$

so that $E[Y] = \mu + \xi$; $\text{Var}(Y) = \mu^3/\lambda$; $\sqrt{\beta_1(Y)} = 3\sqrt{(\lambda/\mu)}$ and then taking $\mu = -\xi$, $\lambda = -\xi^3$, we have

$$E[Y] = 0,$$

$$\text{Var}(Y) = 1,$$

$$\sqrt{\beta_1(Y)} = 3|\xi|.$$

The cdf of Y is

$$F_Y\left(y|\sqrt{\beta_1}\right) = \Phi\left(y + \frac{1}{3}\sqrt{\beta_1}\right)^{-1/2}$$

$$+ \exp\left(\frac{18}{\beta_1}\right)\Phi\left(-\left(y + 6/\sqrt{\beta_1}\right)\left(1 + \frac{1}{3}\sqrt{\beta_1}\right)^{-1}\right). \quad (15.13)$$

The distribution of Y is a *three-parameter* inverse Gaussian $\text{IG}(\mu, \lambda, \xi)$. Its pdf is given in (15.45).

A one-parameter inverse Gaussian [different from Wald (15.3)], obtained by taking $\lambda = \mu^2$ in (15.4a), has been discussed by Vodǎ (1973). [See also Iliescu and Vodǎ (1977, 1981).] Its pdf is

$$p_X(x|\mu) = \mu(2\pi x^3)^{-1/2}\exp\left\{-\tfrac{1}{2}x^{-1}(x - \mu)^2\right\}, \qquad x > 0, \mu > 0. \quad (15.14)$$

In our notation it is an $\text{IG}(\mu, \mu^2)$ distribution.

Yet another one-parameter inverse Gaussian distribution is obtained as a limiting form of $\text{IG}(\mu, \lambda)$ as $\mu \to \infty$. The pdf is

$$p_X(x|\lambda) = \left(\frac{\lambda}{2\pi x^3}\right)^{1/2}\exp\left(-\frac{\lambda}{2x}\right), \qquad x > 0, \lambda > 0. \quad (15.15)$$

The expected value and variance (and indeed $E[X^\alpha]$ for $\alpha \geq \tfrac{1}{2}$) are infinite.

This is the distribution of first passage time of drift-free Brownian motion, which is the one-sided stable law (see Chapter 12) with exponent $\tfrac{1}{2}$. Letac and Seshadri (1989) point out that this distribution is also a member of a natural exponential family.

In the discussion following Folks and Chhikara (1978), Whitmore (1978) [see also Whitmore (1979)] introduced a class of modified inverse Gaussian distributions, which he termed *defective inverse Gaussian* distributions. They have densities of form

$$p_X(x) = a^2(2\pi x^3 \nu)^{-1/2} \exp\left\{ -\frac{(a - \delta x)^2}{2\nu x} \right\} \qquad \text{for } 0 < x < \infty, \delta < 0,$$

(15.16)

together with a nonzero probability

$$\Pr[X = \infty] = 1 - \exp\left(\frac{-2a\delta}{\nu} \right).$$

Note the reparametrization $\mu = a/|\delta|$, $\lambda = a^2/\nu$.

The genesis of distribution (15.16) is as follows: Consider a Wiener process $\{W(t); t \geq 0\}$ with $W(0) = 0$, having drift and volatility parameters δ (< 0) and ν, respectively—namely for any interval of time $(t_1, t_2), W(t_2) - W(t_1)$ is normally distributed with expected value $\delta(t_2 - t_1)$ and variance $\nu(t_2 - t_1)$. There is an absorbing barrier at a (> 0). The probability of eventual absorption is $\exp(2a\delta/\nu)$, and the distribution of first passage time, *conditional* on absorption occurring in finite time is $IG(a/|\delta|, a^2/\nu)$.

Over part of the parameter space (for $a^2|\delta|/\nu \geq 10$, $\delta < 0$) the inverse Gaussian distribution is approximately lognormal (see Chapter 14). However, when $a^2|\delta|/\nu$ is small, the two distributions differ.

5 PROPERTIES

As ϕ tends to infinity (with μ fixed) the *standardized* distribution tends to a unit normal distribution (see Figure 15.1). More precisely, there is the following relation between corresponding quantiles of the standard Wald distribution $X_p(\phi)$ and the normal distribution U_p, obtained by Sigangirov (1962):

$$1 + \frac{U_{p-\varepsilon}}{\sqrt{\phi}} + \frac{U_{p-\varepsilon}^2}{2\phi} + \frac{U_{p-\varepsilon}^3}{8\phi\sqrt{\phi}} - \frac{U_{p-\varepsilon}^5}{128\phi^2\sqrt{\phi}}$$

$$< X_p(\phi) < 1 + \frac{U_p}{\sqrt{\phi}} + \frac{U_p^2}{2\phi} + \frac{U_p^3}{8\phi\sqrt{\phi}},$$

(15.17)

where

$$\varepsilon = \Phi\left(-2\sqrt{\phi} \right)e^{2\phi} < \frac{1}{2\sqrt{2\pi\phi}}.$$

For small ϕ ($\phi < U_p^2/16$),

$$\frac{\phi}{U_{p/2}^2 e^{2\phi}} < X_p(\phi) < \frac{\phi}{U_{p/2}^2}. \tag{15.17}'$$

As μ tends to infinity (λ remaining fixed), the distribution of $Y = X^{-1}$ tends to the gamma distribution:

$$p_Y(y) = \lambda(2\pi)^{-1/2} y^{-1/2} \exp\left(-\tfrac{1}{2}\lambda y\right), \qquad y > 0 \tag{15.18}$$

[Wasan and Roy (1967)].

Krapivin (1965) notes the following equations, satisfied by the Wald probability density function $p_X(x)$ [as given in (15.3)]:

$$W_\phi(x) = F_X(x) = \int_0^x p_X(t)\, dt = \int_{1/x}^\infty t p_X(t)\, dt$$

$$= 1 - \int_0^{1/x} t p_X(t)\, dt. \tag{15.19}$$

This enables one to calculate the expected value of a Wald distribution from tables of $W_\phi(x)$. From (15.4e) with $\mu = 1$ and $\lambda = \phi$, we have the cdf

$$F_X(x) = \Phi\left((x-1)\sqrt{\frac{\phi}{x}}\right) + e^{2\phi}\Phi\left(-(x+1)\sqrt{\frac{\phi}{x}}\right). \tag{15.20a}$$

For x large

$$F_X(x) \doteq 1 - e^{-\phi x/2} e^\phi (\log x). \tag{15.20b}$$

If $X_1, \ldots, X_n$ are independent $IG(\mu_i, \lambda_i)$ variables ($i = 1, \ldots, n$), then the distribution of

$$\sum_{i=1}^n \left(\mu_i^{-2}\lambda_i X_i\right)$$

is $IG(\mu, \lambda)$ with

$$\mu = \sum_{i=1}^n \left(\frac{\lambda_i}{\mu_i}\right), \qquad \lambda = \mu^2.$$

It is thus a one-parameter inverse Gaussian distribution of the type described by Vodă (1973) with pdf given in (15.14).

If $\mu_i = \mu$ and $\lambda_i = \lambda$ for all $i = 1, \ldots, n$, we have

$$\frac{\lambda}{\mu^2} \sum_{i=1}^{n} X_i = \frac{\lambda}{\mu^2} n\bar{X}$$

distributed as $\mathrm{IG}(n\lambda/\mu, n^2\lambda^2/\mu^2)$, whence

$$\bar{X} \text{ is distributed as } \mathrm{IG}(\mu, n\lambda) \tag{15.21}$$

[since aX has an $\mathrm{IG}(a\mu, a\lambda)$ distribution if X has an $\mathrm{IG}(\mu, \lambda)$ distribution; see Section 3, following (15.4d)]. Property (15.21) is termed a *reproducibility* property.

The inverse Gaussian distribution may be characterized by the fact that if $X_1, X_2, \ldots, X_n$ are mutually independent $\mathrm{IG}(\mu, \lambda)$ variables, then $\bar{X} = (\sum_{i=1}^{n} X_i)/n$ and $V = n^{-1}\sum_{i=1}^{n}(X_i^{-1} - \bar{X}^{-1})$ are statistically independent [Tweedie (1957a)]. The converse is also true; that is, if the expected values of X, X^2, $1/X$, and $(1/\sum_{i=1}^{n} X_i)$ exist and are different from zero, and if $\bar{X}$ and $n^{-1}\sum_{i=1}^{n}(X_i^{-1} - \bar{X}^{-1})$ are independently distributed, then the distribution of each variate X_i is inverse Gaussian [Khatri (1962)].

Letac and Seshadri (1985) have provided a short proof of this characterization, replacing the hypothesis on moments by the assumption that X is positive. Seshadri (1983) extended Khatri's result by imposing constancy of expected values of suitable statistics, conditional on $\bar{X}$. One of his results is as follows: If $X_1, X_2, \ldots, X_n$ are independent and identically distributed, and $E[X_i]$, $E[X_i^2]$, $E[X_i^{-1}]$ for all $i = 1, \ldots, n$, $E[(\sum_{i=1}^{n} X_i)^{-1}]$ all exist and are nonzero, and if $E[\sum_{i=1}^{n} a_i X_i^{-1} - n(\sum_{i=1}^{n} a_i)(\sum_{i=1}^{n} X_i)^{-1}|\sum_{i=1}^{n} X_i]$ $(\sum_{i=1}^{n} a_i > 0)$ does not depend on $\sum_{i=1}^{n} X_i$, then the common distribution of the X_i's is inverse Gaussian. Taking $a_i = 1$ $(i = 1, \ldots, n)$, we obtain Khatri's result.

Letac, Seshadri, and Whitmore (1985) have shown that if $X_1, X_2, \ldots, X_n$ are mutually independent $\mathrm{IG}(\mu, \lambda)$ variables, then

$$Q_1 = \frac{n\lambda(\bar{X} - \mu)^2}{\mu^2 \bar{X}}$$

and

$$Q - Q_1 = \lambda\left\{\sum_{i=1}^{n} X_i^{-1} - n\bar{X}^{-1}\right\}$$

are mutually independent and $Q - Q_1$ is distributed as χ^2 with $n - 1$ degrees of freedom.

The distribution is unimodal and its shape depends only on the value of $\phi = \lambda/\mu$ [see Figure 15.1]. The location of the mode is

$$\mathrm{Mode}(X) = \mu\left\{\left(1 + \frac{9}{4\phi^2}\right)^{1/2} - \frac{3}{2\phi}\right\}.$$

Shuster (1968) has pointed out that if X has distribution (15.4a), then $\lambda(X - \mu)^2(\mu^2 X)^{-1}$ has a χ^2 distribution with one degree of freedom (Chapter 18). Some further details are given by Wasan (1968a, b).

Folks and Chhikara (1978) suggest the algebraic formulas

$$\frac{1}{2}\theta\mu^{-2}\sum_{i=1}^{n} X_i^{-1}(X_i - \mu)^2 = \frac{1}{2}\theta\sum_{i=1}^{n}\left(X_i^{-1} - \bar{X}^{-1}\right) + \frac{1}{2}n\theta\mu^{-2}\bar{X}^{-1}\left(\bar{X} - \mu\right)^2,$$

(15.22a)

$$\sum_{i=1}^{k}\sum_{j=1}^{n_i}\left(X_{ij}^{-1} - \bar{X}_{..}^{-1}\right) = \sum_{i=1}^{k} n_i\left(\bar{X}_{i.}^{-1} - \bar{X}_{..}^{-1}\right) + \sum_{i=1}^{k}\sum_{j=1}^{n_i}\left(X_{ij}^{-1} - \bar{X}_{i.}^{-1}\right),$$

(15.22b)

where $\bar{X}_{i.} = n_i^{-1}\sum_{j=1}^{n_i} X_{ij}$ and $\bar{X}_{..} = (\sum_{i=1}^{k} n_i)^{-1}\sum_{i=1}^{k} n_i\bar{X}_i$. These are analogues, for IG parent populations, to the standard decompositions of sums of squares, used in analysis of variance for data in one-way classification by groups for normal parent populations.

They show that for the analogue of a standardized normal variable

$$U = (n\theta)^{1/2}(\bar{X} - \mu)(\mu\bar{X}^{1/2})^{-1},$$

(15.23)

U^2 is distributed as χ^2 with one degree of freedom but U is not normally distributed.

More recently there has been some attention in the literature to the "length biased" pdf of IG(μ, λ), which is generally defined as

$$p_Y(y) = \frac{y p_X(y)}{E[X]}.$$

(15.24)

For the length-biased inverse Gaussian—denoted LBIG(μ, λ)—we have

$$p_Y(y) = \left(\frac{\lambda}{2\pi y\mu^2}\right)^{1/2}\exp\left\{-\frac{\lambda(y - \mu)^2}{2\mu^2 y}\right\}.$$

(15.25)

We note that if X is distributed as IG(μ, λ), then μ^2/X is distributed as LBIG(μ, λ); conversely, if Y is distributed as LBIG(μ, λ), then μ^2/Y is distributed as IG(μ, λ).

If $Y_1, Y_2, \ldots, Y_n$ are mutually independent LBIG(μ, λ) variables, then their harmonic mean

$$\tilde{Y} = n\left(\sum_{i=1}^{n} Y_i^{-1}\right)^{-1}$$

is distributed as

$$n\mu^2 \left(\sum_{i=1}^{n} X_i \right)^{-1} = \frac{\mu^2}{\overline{X}},$$

with $X_1, \ldots, X_n$ each distributed as IG(μ, λ). Since $\overline{X}$ is distributed as IG$(\mu, n\lambda)$, $\tilde{Y}$ is distributed as LBIG$(\mu, n\lambda)$.

Whitmore and Yalovsky (1978) approximate the IG(μ, λ) distribution by ascribing a standard normal distribution to

$$\frac{1}{2}\sqrt{\frac{\mu}{\lambda}} + \sqrt{\frac{\lambda}{\mu}} \log\left(\frac{X}{\mu}\right). \tag{15.26}$$

This is an improvement over the simpler approximation

$$\sqrt{\frac{\lambda}{\mu}} \left(\mu^{-1}X - 1\right) = N(0,1).$$

Sibuya (1992) has examined six rough approximations for the quantile function of the inverse Gaussian distribution. Among these approximations, four are based on fitting approximate normal, log-normal, gamma, and Birnbaum–Saunders distributions. The other two approximations make use of the specific features of the inverse Gaussian distribution. Through numerical comparisons, Sibuya has noted that no single approximation is uniformly better than the others and that approximating the upper tail of the distribution with a large coefficient of variation becomes particularly difficult. Consequently, these approximations may be used as starting value for the Newton method to numerically solving the equation equating the distribution function to a specified p-value.

6 ESTIMATION OF PARAMETERS

The statistic $(\sum_{i=1}^{n} X_i, \sum_{i=1}^{n} X_i^{-1})$ is minimal sufficient for the IG(μ, λ) distribution. Maximum likelihood estimators (MLEs) have been obtained [Tweedie (1957a)] for the following general case: let $X_1, \ldots, X_n$ be a series of observations on n distinct inverse Gaussian distributions IG(μ, λ_i) ($i = 1, \ldots, n$). Also let $\lambda_i = w_i \lambda_0$, where λ_0 is unknown but w_i is a known positive value. The MLE's of μ and λ_0 satisfy the equations

$$\hat{\mu} = \overline{X} = \left[\sum_{i=1}^{n} w_i X_i \right]\left[\sum_{i=1}^{n} w_i \right]^{-1}, \tag{15.27a}$$

$$\frac{1}{\hat{\lambda}_0} = \frac{1}{n}\sum_{i=1}^{n} w_i\left(X_i^{-1} - \overline{X}^{-1}\right). \tag{15.27b}$$

When the w_i's are all equal to unity, (15.27a) and (15.27b) become

$$\hat{\mu} = \bar{X} = \frac{\sum_{i=1}^{n} X_i}{n} \tag{15.28a}$$

$$\frac{1}{\hat{\lambda}_0} = V = n^{-1} \sum_{i=1}^{n} \left(X_i^{-1} - \bar{X}^{-1} \right). \tag{15.28b}$$

It is apparent that in this case $\bar{X}$ is a sufficient statistic for μ. The distribution of $\bar{X}$ is $IG(\mu, n\lambda_0)$. Tweedie (1957a) has shown that $\bar{X}$ and $\hat{\lambda}_0$ are independent. An elementary proof of this basic property is given by Schwarz and Samanta (1991). It can be shown that

$$V\left(= \hat{\lambda}_0^{-1}\right) \text{ is distributed as } (\lambda_0 n)^{-1} \times \left(\chi^2 \text{ with } n - 1 \text{ degrees of freedom}\right).$$
$$\tag{15.29}$$

Combining (15.27b) and (15.29), the following relation is obtained:

$$\sum_{i=1}^{n} w_i \left(X_i^{-1} - \bar{X}^{-1} \right) \text{ is distributed as}$$

$$\lambda_0^{-1} \times \left(\chi^2 \text{ with } n - 1 \text{ degrees of freedom}\right). \tag{15.30}$$

Confidence intervals for λ_0 may be constructed using (15.30) [Tweedie (1957a)]. We also note that

$$(n - 1)^{-1} \sum_{i=1}^{n} w_i \left(X_i^{-1} - \bar{X}^{-1} \right) \tag{15.31}$$

is an unbiased estimator of $1/\lambda_0$. It is in fact the uniform minimum variance unbiased estimator [Roy and Wasan (1968)]. Note that the distribution of this estimator is the same form as that of the usual unbiased estimator of variance of a normal distribution [Tweedie (1957a)].

An *approximately* unbiased estimator of $1/\lambda_0$ is $S^2/\bar{X}^3$, where

$$S^2 = \frac{\sum_{i=1}^{n} \left(X_i - \bar{X} \right)^2}{n - 1}. \tag{15.32}$$

It is a consistent estimator and has asymptotic efficiency of $\phi/(\phi + 3)$. Some further notes on estimators of the cumulants will be found in Section 9.

If the value of μ is known, we may, by a suitable choice of scale, take it to be equal to 1. We are then in the situation of having to estimate the single parameter ϕ, of a standard Wald distribution.

If $X_1, X_2, \ldots, X_n$ are independent random variables, each having distribution (15.3), then the maximum likelihood estimator $\hat{\phi}$ of ϕ satisfies the equation

$$\frac{n}{2}\hat{\phi}^{-1} + n - \frac{1}{2}\sum_{i=1}^{n}(X_i + X_i^{-1}) = 0,$$

whence

$$\hat{\phi} = \left[\frac{1}{n}\sum_{i=1}^{n}(X_i + X_i^{-1}) - 2\right]^{-1}.$$

For n large

$$n\,\mathrm{Var}(\hat{\phi}) \doteq 2\phi^2.$$

We note that if ϕ be estimated by equating observed and expected values of $m_2 = n^{-1}\sum_{i=1}^{n}(X_i - \bar{X})^2$, the resultant estimator is

$$\tilde{\phi} = m_2^{-1}. \tag{15.33}$$

For n large

$$n\,\mathrm{Var}(\tilde{\phi}) \doteq 2\phi^2 + 15\phi. \tag{15.34}$$

The asymptotic efficiency of $\tilde{\phi}$, relative to $\hat{\phi}$, as an estimator of ϕ is

$$\left(1 + 7.5\phi^{-1}\right)^{-1}.$$

From (15.10) it follows that

$$E(\hat{\phi}^{-1}) = E[X + X^{-1} - 2] = 1 + (1 + \phi^{-1}) - 2 = \phi^{-1},$$

that is, $\hat{\phi}^{-1}$ is an unbiased estimator of ϕ^{-1} and m_2 is also an unbiased estimator ϕ^{-1}. Further

$$\mathrm{Var}(\hat{\phi}^{-1}) = 2\phi^{-2}n^{-1}, \tag{15.35a}$$

$$\mathrm{Var}(m_2) = (2\phi^{-2} + 15\phi^{-3})n^{-1}. \tag{15.35b}$$

It is of interest to compare these results with those when μ is unknown. The maximum likelihood estimators are then

$$\hat{\mu} = n^{-1}\sum_{i=1}^{n}X_i = \bar{X}, \tag{15.36a}$$

$$\hat{\phi} = \left[n^{-2}\left(\sum_{i=1}^{n}X_i\right)\left(\sum_{i=1}^{n}X_i^{-1}\right) - 1\right]^{-1}. \tag{15.36b}$$

Evidently

$$\text{Var}(\hat{\mu}) = \mu^2 \phi^{-1} n^{-1}. \tag{15.37a}$$

Also

$$n\,\text{Var}(\hat{\phi}) \doteq 2\phi^2 + \phi, \tag{15.37b}$$

$$\text{Corr}(\hat{\mu}, \hat{\phi}) \doteq -(1 + 2\phi)^{-1/2}. \tag{15.37c}$$

The increase in variance of $\hat{\phi}$, arising from ignorance of the value of μ, should be noted.

Moment estimators are

$$\tilde{\mu} = \overline{X}, \tag{15.38a}$$

$$\tilde{\phi} = \frac{n\overline{X}^2}{\sum_{i=1}^{n}(X_i - \overline{X})^2} = (\text{sample coefficient of variation})^{-2}. \tag{15.38b}$$

Note that $\phi^{-1/2}$ is the coefficient of variation. The variance of $\tilde{\mu}$ is of course the same as that of $\hat{\mu}$ and

$$n\,\text{Var}(\tilde{\phi}) \doteq 10\phi^2 + 19\phi \qquad [\text{cf. (15.34)}]. \tag{15.39}$$

More recently moment and maximum likelihood estimation have been discussed by Padgett and Wei (1979) and maximum likelihood estimation by Cheng and Amin (1981), with special attention to consistency and asymptotic efficiency. The two methods of estimation have been compared by Jones and Cheng (1984) who find maximum likelihood to be clearly superior. Modified maximum likelihood and moment estimation, in which the equation obtained by equating expected and observed values of the first (least) order statistic replaces the maximum likelihood equation equating $\partial \log (\text{likelihood})/\partial \xi$ to zero, or the moment equation between sample and population values of the third moment about zero, are discussed by Chan, Cohen, and Whitten (1984). Cohen and Whitten [(1985), (1988)] and Balakrishnan and Cohen (1991) provide tables to assist in computation of the solution of the newly introduced equation for ξ in terms of μ and λ (for the three-parameter case discussed in Section 4).

A uniform minimum variance unbiased estimator of the variance ($\sigma^2 = \mu^3/\lambda$) of $IG(\mu, \lambda)$ was derived by Korwar (1989) as

$$\hat{\sigma}^2 = (n-1)^{-1}\overline{X}^3 V \,_2F_1\left(1, \frac{3}{2}; \frac{1}{2}(n+1); \overline{X}V\right), \tag{15.40}$$

where

$$_2F_1(a, b; c; z) = \sum_{j=0}^{\infty} \frac{a^{[j]}b^{[j]}}{c^{[j]}} \frac{z^j}{j!}$$

is the Gaussian hypergeometric function discussed in Chapter 1, and V is defined in (15.28b).

Hsieh and Korwar (1990) have shown that

$$\tilde{\sigma}^2 = \frac{(n-1)\hat{\sigma}^2}{n+1} \tag{15.41}$$

has mean square error never greater than that of $\hat{\sigma}^2$, as an estimator of σ^2. The maximum likelihood estimator of σ^2 is $V\bar{X}^3/n$; the mean square error of $V\bar{X}^3/(n+1)$ is never greater than that of $V\bar{X}^3/n$.

In further work Hsieh, Korwar, and Rukhin (1990) have shown that the mean square error of $\bar{X}$, the maximum likelihood estimator of μ, is never *less* than that of an estimator of form

$$\bar{X}\left[1 - 2\varepsilon \exp\left\{-\frac{an}{2\,\bar{X}V}\right\}\right] \tag{15.42}$$

for suitably chosen (positive) values of ε and a. In fact, for any given a (> 0), this will be true for

$$0 < \varepsilon \le \min\left(\inf_x g(x), 0.5\right),$$

where

$$g(x) = \frac{E\left[\left(\sqrt{1 + aY^{-1}} + x - 1\right)\exp\left\{-\left(\sqrt{1 + aY^{-1}} - 1\right)/\gamma\right\}\right]}{E\left[\left(\sqrt{1 + 2aY^{-1}} + x\right)\exp\left\{-\left(\sqrt{1 + 2aY^{-1}} - 1\right)/\gamma\right\}\right]},$$

with Y distributed as $(n-1)^{-1} \times (\chi^2$ with $n-1$ degrees of freedom) and $\gamma = \mu/(n\lambda)$. To obtain an "optimal" choice for the values of ε and a, it is suggested to use $\varepsilon = \frac{1}{2}$, and the following table, with $\hat{\gamma} = \hat{\mu}/(n\hat{\lambda})$, for a:

$\hat{\gamma}$	0.1	0.5	1	5	10	20	50
a	0.46	1.75	3.19	13.84	26.64	51.80	126.25

A uniform minimum variance unbiased estimator of the squared coefficient of variation (μ/λ) is

$$T = n(n-1)^{-1}\bar{X}V. \tag{15.42}$$

Seshadri (1989) derives the distribution of T in terms of modified Bessel function of the third kind. Despite the complicated form of the distribution the variance of T is quite a simple expression. Since $\bar{X}$ and V are mutually independent and $\bar{X}$ has a $N(\mu, n^{-1}\mu^3/\lambda)$ distribution, while V has a

$n\lambda^{-1}\chi^2_{n-1}$ distribution,

$$E[T^2] = n^2(n-1)^{-2}E[\bar{X}^{-2}]E[V^2]$$

$$= n^2(n-1)^{-2}(\mu^2 + n^{-1}\mu^3\lambda^{-1})n^2\lambda^{-2}\{2(n-1) + (n-1)^2\}$$

and

$$\text{Var}(T) = E[T^2] - \{E[T]\}^2 = E[T^2] - \mu^2\lambda^{-2}$$

$$= (n-1)^{-1}\left(\frac{\mu}{\lambda}\right)^2\left\{2 + \frac{n+1}{n}\left(\frac{\mu}{\lambda}\right)\right\}. \tag{15.43}$$

For large n, $\sqrt{n-1}(T - \mu\lambda^{-1})$ has an approximately normal distribution with zero expected value and variance $[2 + (\mu/\lambda)](\mu/\lambda)^2$.

Seto and Iwase (1985) obtained a complicated formula for an UMVU estimator of the mode of an IG(μ, λ) distribution. For $n = 3$ this estimator is

$$Y = \bar{X}\{{}_1F_2(-\tfrac{1}{2}; \tfrac{1}{2}, 1; -\tfrac{1}{4}W^2) - W\}, \tag{15.44}$$

where

$$W = \frac{3}{2}\left\{\left(1 + \frac{1}{3}\bar{X}V\right)^{1/2} - 1\right\},$$

$${}_1F_2(a; b, c; z) = \sum_{j=0}^{\infty} \frac{a^{[j]}}{b^{[j]}c^{[j]}} \frac{z^j}{j!}.$$

Estimation of the parameters of the three-parameter IG(μ, λ, ξ) distribution with pdf

$$p_Y(y|\mu, \lambda, \xi) = \left\{\frac{\lambda}{2\pi(y-\xi)^3}\right\}^{1/2} \exp\left\{-\frac{\lambda(y-\xi-\mu)^2}{2(y-\xi)\mu^2}\right\},$$

$$y > \xi; \mu, \lambda > 0, \tag{15.45}$$

has been discussed by Padgett and Wei (1979), Cheng and Amin (1981), Jones and Cheng (1984), Chan, Cohen, and Whitten (1984), Cohen and Whitten (1985, 1988), and Balakrishnan and Cohen (1991). Putting $\lambda = \mu^3/\sigma^2$, (15.45) can be written as

$$p_Y(y|\mu, \sigma, \xi) = \frac{1}{\sigma\sqrt{2\pi}}\left(\frac{\mu}{y-\xi}\right)^{3/2} \exp\left\{-\frac{1}{2}\left(\frac{\mu}{y-\xi}\right)\left(\frac{y-\xi-\mu}{\sigma}\right)^2\right\},$$

$$y > \xi; \mu, \sigma > 0. \tag{15.46}$$

Recent books by Cohen and Whitten (1988) and Balakrishnan and Cohen (1991) discuss estimation problems for distribution (15.45) in great detail. We do not present these in full here. It should, however, be noted that MLEs are not available if the sample third central moment is negative.

For the special case when the value of $a = E[X]/\text{Var}(X)$ is known—the corresponding distribution being IG$(\mu, a\mu^2)$—procedures for statistical inference on μ have been constructed by Iwase (1987) and Joshi and Shah (1991). Iwase (1987) showed that the uniform minimum variance unbiased estimator of μ, based on n independent IG$(\mu, c\mu^2)$ variables $X_1, \ldots, X_n$ is

$$\tilde{\mu} = T \frac{\psi\left(\frac{1}{2}(n-1), \frac{3}{2}; \frac{1}{2}naT\right)}{\psi\left(\frac{1}{2}(n-1), \frac{1}{2}; \frac{1}{2}naT\right)}, \qquad n \geq 2, \tag{15.47}$$

where $T = \{n^{-1}\sum_{i=1}^{n} X_i^{-1}\}^{-1}$ is the harmonic mean of the X's, and

$$\psi(a, c; x) = x^{a-2} \sum_{j=0}^{\infty} (-1)^j \frac{a^{[j]}(a-c+1)^{[j]}}{j!} x^{-j} \tag{15.48}$$

is the confluent hypergeometric function. (See Chapter 1, Section A7.)

The maximum likelihood estimator of μ is

$$\hat{\mu} = \frac{1}{2}T\left[1 + \left\{1 + 4(cT)^{-1}\right\}^{1/2}\right], \qquad n \geq 1. \tag{15.49}$$

Iwase (1987) showed that

$$\tilde{\mu} = \hat{\mu}\{1 + o(n^{-1})\} \tag{15.50}$$

as $n \to \infty$. Joshi and Shah (1991) consider (from a Bayesian viewpoint) another special case, namely IG$(\mu, c^{-2}\mu)$, where $c = E[X]/\sqrt{\text{Var}(X)}$ is the coefficient of variation of X, and the value of c is known.

Bayesian analysis for the general IG(μ, λ) distribution has been considered by Betro and Rotondi (1991) and by Banerjee and Bhattacharyya (1979). Joshi and Shah (1991) consider Bayesian estimation of the expected value μ of IG$(\mu, c^{-2}\mu)$, where the value of c (the coefficient of variation) is known.

7 TRUNCATED DISTRIBUTIONS—ESTIMATION OF PARAMETERS

It is frequently desirable to estimate the parameters of a truncated inverse Gaussian distribution. Estimators of the parameters and their asymptotic variances have been obtained [Patel (1965)] for both the singly and doubly truncated cases.

7.1 Doubly Truncated Distribution

The density function of the doubly truncated distribution may be written as

$$p_X(x|\mu, \lambda; x_1, x_2) = K x^{-3/2} \exp\left(-\frac{\lambda x}{2\mu^2} - \frac{\lambda}{2x}\right), \qquad (15.51)$$

where

$$K = \left[\int_{x_1}^{x_2} x^{-3/2} \exp\left\{-\frac{\lambda x}{2\mu^2} - \frac{\lambda}{2x}\right\} dx\right]^{-1}, \qquad x_1 \le x \le x_2.$$

The values of x_1 and x_2 denote the lower and upper points of truncation, respectively, of the distribution of X $(0 < x_1 < x_2)$; the parameters μ and λ are positive.

A recurrence relation for the moments may be obtained by differentiating (15.51) partially with respect to x and multiplying both sides by x^r to give, after integration by parts:

$$a\mu'_r - \lambda\mu'_{r-2} - 2(x_1^r p_1 - x_2^r p_2) = (2r - 3)\mu'_{r-1}, \qquad (15.52)$$

where

$a = \lambda/\mu^2$,
$p_i = p_X(x_i|\mu, \lambda; x_1, x_2)$ for $i = 1, 2$,
$\mu'_r = \int_{x_1}^{x_2} x^r p_X(x|\mu, \lambda; x_1, x_2) dx$ for $r = 0, \pm 1, \pm 2, \ldots$.

p_1 and p_2 are of course values of the density at the truncation points x_1 and x_2.

If the population moments (μ'_r) are replaced by the sample moments (m'_r) in the recurrence relation (15.52), the following set of equations results:

$$\mathbf{M}\hat{\mathbf{h}} = \mathbf{c}, \qquad (15.53)$$

where

$$\hat{\mathbf{h}} = \begin{pmatrix} \hat{a} \\ \hat{\lambda} \\ \hat{p}_1 \\ \hat{p}_2 \end{pmatrix}, \quad \mathbf{M} = \begin{pmatrix} 1 & -m'_{-2} & -2 & 2 \\ m'_1 & -m'_{-1} & -2x_1 & 2x_2 \\ m'_2 & -1 & -2x_1^2 & 2x_2^2 \\ m'_3 & -m'_1 & -2x_1^3 & 2x_2^3 \end{pmatrix}, \quad \mathbf{c} = \begin{pmatrix} -3m'_{-1} \\ -1 \\ m'_1 \\ 3m'_2 \end{pmatrix}.$$

$(\hat{\mathbf{h}}$ is a vector of estimators of functions of the parameters.) Provided that $\mathbf{M}$ is nonsingular, (15.53) may be solved as

$$\hat{\mathbf{h}} = \mathbf{M}^{-1}\mathbf{c} \qquad (15.54)$$

from which $\hat{\mu}$ may be derived as $\sqrt{\hat{\lambda}/\hat{a}}$.

The asymptotic variances and covariances of the estimators ($\hat{\mathbf{h}}$) of the parameters may be obtained in matrix form by using a differential method [Patel (1965)]. The elements of this symmetric matrix are as follows:

$$n\,\mathrm{Var}(\hat{a}) \doteq (14 + \lambda a)\mu'_{-2} + a\mu'_{-1} - P_{-1} - \lambda P_{-2}, \qquad (15.55a)$$

$$n\,\mathrm{Cov}(\hat{a}, \hat{\lambda}) \doteq 2\lambda\mu'_{-2} - \lambda P_{-1} + 3a, \qquad (15.55b)$$

$$n\,\mathrm{Cov}(\hat{a}, \hat{p}_1) \doteq \lambda\mu'_{-1} + 3a\mu'_1 - a\lambda - 3, \qquad (15.55c)$$

$$n\,\mathrm{Cov}(\hat{a}, \hat{p}_2) \doteq \lambda^2\mu'_{-1} - (9 + a\lambda)\mu'_1 + 3a\mu'_2, \qquad (15.55d)$$

$$n\,\mathrm{Var}(\hat{\lambda}) \doteq \lambda^2\mu'_{-2} + a^2\mu'_2 - 2a\lambda, \qquad (15.55e)$$

$$n\,\mathrm{Cov}(\hat{\lambda}, \hat{p}_1) \doteq a^2\mu'_3 - a\mu'_2 - a\lambda\mu'_1 + \lambda, \qquad (15.55f)$$

$$n\,\mathrm{Cov}(\hat{\lambda}, \hat{p}_2) \doteq 2a\mu'_3 - a\lambda\mu'_2 + 3\lambda\mu'_1 + aP_4 + \lambda^2, \qquad (15.55g)$$

$$n\,\mathrm{Var}(\hat{p}_1) \doteq 3a\mu'_3 + (1 + a\lambda)\mu'_2 + aP_4, \qquad (15.55h)$$

$$n\,\mathrm{Cov}(\hat{p}_1, \hat{p}_2) \doteq 18\mu'_3 + \mu'_2 + 3P_4 + aP_5, \qquad (15.55i)$$

$$n\,\mathrm{Var}(\hat{p}_2) \doteq \frac{30}{a}(5\mu'_3 + \lambda\mu'_2 + P_4) + 4\lambda\mu'_3 - \lambda P_4 + 3P_5 + aP_6, \qquad (15.55j)$$

where $P_r = (2x_1^r p_1 - x_2^r p_2)$.

Two other useful general relations are

$$\mathrm{Var}(\hat{\mu}) = \frac{1}{4\mu^2 a^4}\left[a^2\,\mathrm{Var}(\hat{\lambda}) + \lambda^2\,\mathrm{Var}(\hat{a}) - 2a\lambda\,\mathrm{Cov}(\hat{a}, \hat{\lambda})\right], \qquad (15.56)$$

$$\mathrm{Cov}(\hat{\lambda}, \hat{\mu}) = \frac{a\,\mathrm{Var}(\hat{\lambda}) - \lambda\,\mathrm{Cov}(\hat{a}, \hat{\lambda})}{2\mu a^2}. \qquad (15.57)$$

7.2 Truncation of the Lower Tail Only

If the truncation point is at x_0 where $x_0 > 0$, then the estimators $\hat{\mathbf{h}}_1$ of the parameters are

$$\hat{\mathbf{h}}_1 = \mathbf{M}_1^{-1}\mathbf{c}_1, \qquad (15.58)$$

where $\mathbf{M}_1$ is a matrix of order 3 (obtained by deleting the last row and column of $\mathbf{M}$ and putting $x_1 = x_0$). The vectors $\hat{\mathbf{h}}_1$ and $\mathbf{c}_1$ are obtained from $\hat{\mathbf{h}}$ and $\mathbf{c}$, respectively, by deleting the last element and putting $p_1 = p_0$.

The asymptotic variance of $\hat{\lambda}$ can be obtained from (15.55e) and that of $\hat{\mu}$ and the covariance between $\hat{\lambda}$ and $\hat{\mu}$ from (15.56) and (15.57), respectively.

7.3 Truncation of the Upper Tail Only

This case may be dealt with in a similar manner. The general relations (15.53) of the first case are again applicable with $x_1 = 0$, $x_0 = x_2$, $p_0 = p_2$, and omitting the last row and column of $\mathbf{M}$. The asymptotic variances and covariances of the estimators are obtained in the same way as in the second case.

8 CONDITIONAL EXPECTATIONS OF THE ESTIMATORS OF THE CUMULANTS

We now return to study the untruncated distribution. The section is of less general interest than Section 7, but contains some useful results. It may be shown that, given the arithmetic mean $\bar{X}$, the conditional expectation of *any* unbiased estimator $\bar{k}_r$ of the rth cumulant must be

$$E\left[\bar{k}_r | \bar{X}\right] = 2\bar{X}\left(\frac{1}{2}\lambda n^2\right)^{r-1} \exp\left(\frac{1}{2}G\right) \int_1^{\bar{X}} (u-1)^{2r-3}$$

$$\times \exp\left(-\frac{1}{2}Gu^2\right) \frac{du}{(r-2)!}, \qquad (15.59)$$

where $G = \lambda n/\bar{X}$. Since the distribution of $\bar{X}$ has the same form as (15.4c) with ϕ replaced by ϕn, then the probability density function of G is

$$p_G(g) = \frac{\exp\left[-(\theta^2/2g) + \theta - (g/2)\right]}{\sqrt{2\pi g}}, \qquad g > 0, \qquad (15.60)$$

where $\theta = \phi n$. The moments of G may be obtained using the formulas developed for the inverse Gaussian variate X [see (15.7) and (15.10)].
   The first few moments of G are

$$E[G] = \theta + 1,$$

$$E[G^2] = \theta^2 + 3\theta + 3,$$

$$E[G^3] = \theta^3 + 6\theta^2 + 15\theta + 15,$$

$$E[G^4] = \theta^4 + 10\theta^3 + 45\theta^2 + 105\theta + 105,$$

$$E[G^5] = \theta^5 + 15\theta^4 + 105\theta^3 + 420\theta^2 + 945\theta + 945,$$

$$E[G^6] = \theta^6 + 21\theta^5 + 210\theta^4 + 1260\theta^3 + 4725\theta^2 + 10395\theta + 10395.$$

$$(15.61)$$

It can be shown [Tweedie (1957b)] that conditional expectations of unbiased estimators of the cumulants, calculated from (15.59) are

$$E\left[\bar{k}_2|\bar{X}; \lambda, n\right] = \lambda^2 n^3 G^{-2} J, \tag{15.62a}$$

$$E\left[\bar{k}_3|\bar{X}; \lambda, n\right] = \tfrac{1}{2}n^2\bar{X}^3\{(G + 3)J - 1\}, \tag{15.62b}$$

$$E\left[\bar{k}_4|\bar{X}; \lambda, n\right] = (\tfrac{1}{8})n^3\bar{X}^4\{(G^2 + 10G + 15)J - G - 7\}, \tag{15.62c}$$

where

$$J = 1 - Ge^{G/2}\int_1^\infty e^{-Gu^2/2}\, du. \tag{15.63}$$

J has a maximum value of 1 and decreases monotonically to 0 as G increases to ∞.

An asymptotic series expansion of (15.63) yields

$$J = G^{-1} - 3G^{-2} + \cdots + (-1)^{r+1}3 \times 5 \ldots \times (2r - 1)G^{-r}$$

$$+ (-1)^r 3 \times 5 \ldots \times (2r - 1) \times (2r + 1)G^{-r}e^{-G/2}\int_1^\infty u^{-2r-2}e^{-Gu^2/2}\, du \tag{15.64a}$$

[Tweedie (1957b)]. A relatively large number of significant figures is required in J in order to yield accurate results when evaluating (15.62) numerically.

Tables of J and $e^{G/2}\int_1^\infty e^{-Gu^2/2}\, du\ (=I)$ were compiled [National Bureau of Standards (1942)]. Tweedie (1957b) gives tables of I and J for $G = 1(1)10$. Tweedie also gives "a useful expansion for moderately large values of G" in the form of the continued fraction

$$J = \cfrac{1}{G + 3 + \cfrac{1 + 2}{G + 5 + \cfrac{1 + 4}{G + \ldots,}}} \tag{15.64b}$$

where the sequence of positive integers in the partial numerators is 1; 3, 2; 5, 4; 7, 6; 9, 8; and so on. For large values of G, asymptotic expansions of

Hermite polynomials may be used, giving

$$E\left[\bar{k}_r|\bar{X}; \lambda, n\right] \sim \frac{4\lambda}{\bar{X}} \left(\frac{\bar{X}^2}{2\lambda}\right)^r \sum_{i=0}^{\infty} \frac{(2i + 2r - 3)!}{i!(r - 2)!(-2G)^i}. \qquad (15.65)$$

On using (15.65), or by applying the asymptotic series expansion (15.64a) of J to (15.62), the expected values of the estimators of the cumulants become

$$E\left[\bar{k}_2|\bar{X}; \lambda, n\right] \sim \lambda^{-1}\bar{X}^3(1 - 3G^{-1} + 15G^{-2} - 105G^{-3} + 945G^{-4} - \dots), \qquad (15.66a)$$

$$E\left[\bar{k}_3|\bar{X}; \lambda, n\right] \sim 3\lambda^{-2}\bar{X}^5(1 - 10G^{-1} + 105G^{-2} - 1260G^{-3}$$
$$+ 17325G^{-4} - \dots), \qquad (15.66b)$$

$$E\left[\bar{k}_4|\bar{X}; \lambda, n\right] \sim 15\lambda^{-3}\bar{X}^7(1 - 21G^{-1} + 378G^{-2} - 6930G^{-3}$$
$$+ 135135G^{-4} - \dots) \qquad (15.66c)$$

[Tweedie (1957b)]. Experimental data involving the inverse Gaussian distribution exhibit a high correlation between changes in sample means and changes in sample variances in different sets of data. This fact can be accounted for theoretically by the relationship between the first and second cumulants $[k_2 = k_1^3/\lambda$, as can be seen from (15.6)]. The regression function of the sample variance S^2 on the sample mean $\bar{X}$ is

$$E\left[S^2|\bar{X}; \lambda, n\right] = \frac{GJ\bar{X}^3}{\lambda} = Jn\bar{X}^2 \qquad (15.67)$$

[Tweedie (1957b)].

9 RELATED DISTRIBUTIONS

9.1 Reciprocal of an Inverse Gaussian Variate

Sometimes it may be convenient to use the reciprocal of a variate X having an inverse Gaussian distribution. [In terms of the moving particle (see Section 2) this corresponds to average velocity.] The probability density function of $Y (= 1/X)$ is (see also Section 3)

$$p_Y(y) = \exp\left\{-\frac{\lambda y}{2} + \frac{\lambda}{\mu} - \frac{\lambda}{2\mu^2 y}\right\}\left[\frac{\lambda}{2\pi y}\right]^{1/2}, \qquad y > 0. \quad (15.68)$$

This distribution is called the *random walk distribution* [Wise (1966); Wasan (1968b)].

The positive moments about zero for Y are the same as the negative moments about zero for the inverse Gaussian variate X and are given by (15.9) and (15.10). The cumulant generating function of Y is

$$\psi_Y(t;\mu,\lambda) = \lambda\mu^{-1}\{1 - (1 + 2t\lambda^{-1})^{1/2}\} - \tfrac{1}{2}\log(1 + 2t\lambda^{-1}). \quad (15.69)$$

The first two cumulants are

$$\kappa_1(Y) = \frac{1}{\mu} + \frac{1}{\lambda}, \quad (15.70a)$$

$$\kappa_2(Y) = \frac{1}{\lambda\mu} + \frac{2}{\lambda^2}. \quad (15.70b)$$

The bias of Y as an estimator of μ^{-1} is λ^{-1}. The mean square error in using Y as an estimator of $1/\mu$ is

$$E\left[\left(Y - \frac{1}{\mu}\right)^2\right] = (\lambda\mu)^{-1} + 3\lambda^{-2}. \quad (15.71)$$

The mode of the density function of Y is located at

$$\text{Mode}(Y) = \frac{1}{\mu}\left[\left(1 + \frac{\mu^2}{4\lambda^2}\right)^{1/2} - \frac{\mu}{2\lambda}\right]. \quad (15.72)$$

Wise (1966) has shown that the density function has two points of inflection, at values of y, satisfying the equation

$$u^4 + 2u^3 + u = \left(\frac{\lambda}{\mu}\right)^2 + \frac{1}{4},$$

where

$$u = \frac{1}{2} + \frac{1}{2}\left(\frac{\lambda}{\mu}\right)\left(\frac{y}{\mu} - \frac{\mu}{y}\right).$$

Figure 15.2 presents plots of the density function of Y for different values of λ [reproduced from Tweedie (1957a)].

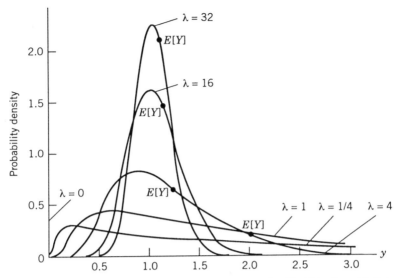

Figure 15.2 Density Functions of Reciprocals of Inverse Gaussian Variables ($\mu = 1$). Note that $\lim_{\lambda \to \infty} E[Y] = 1$.

9.2 Birnbaum–Saunders Distribution

Birnbaum and Saunders (1969a) have proposed the distribution with pdf,

$$
p_X(x; \alpha, \beta) = \frac{1}{2\alpha\beta} \left(\frac{x}{\beta}\right)^{-1/2} \left\{1 + \left(\frac{x}{\beta}\right)^{-1}\right\}
$$

$$
\times \frac{1}{\sqrt{(2\pi)}} \exp\left[-\frac{1}{2\alpha^2}\left(\frac{x}{\beta} - 2 + \frac{\beta}{x}\right)\right], \qquad x > 0;\ \alpha, \beta > 0,
$$

$$(15.73)$$

to represent the distribution of lifetimes of components under specified condition of wear (a "fatigue life" distribution). Desmond (1986) has pointed out that formula (15.73) can be rewritten as

$$
p_X(x; \alpha, \beta) = \frac{1}{3}\left[\left(\frac{\alpha^{-2}\beta}{2\pi x^3}\right)^{1/2} \exp\left\{-\frac{1}{2\alpha^2\beta}(x - 2\beta + \beta^2 x^{-1})\right\}\right.
$$

$$
\left. + \left(\frac{\alpha^{-2}\beta^{-1}}{2\pi x}\right)^{1/2} \exp\left\{-\frac{\beta}{2\alpha^2}(x^{-1} - 2\beta^{-1} + \beta^{-2}x)\right\}\right]. \quad (15.73)'
$$

Comparison with equations (15.4a) and (15.68) shows that the Birnbaum–Saunders distribution is a mixture, in equal proportions, of an $IG(\beta, \alpha^{-2}\beta)$ distribution and a reciprocal $IG(\beta, \alpha^{-2}\beta^{-1})$ distribution. Desmond (1985) provided a more general derivation of this distribution based on a biological model. In his derivation, Desmond has strengthened the physical justification for the use of this distribution by relaxing some of the assumptions made by Birnbaum and Saunders (1969a). Birnbaum and Saunders (1969b) derived the maximum likelihood estimators for the parameters α and β. They also presented a mean mean estimator for β. Ahmad (1988) discussed a jackknife estimator for β based on the mean mean estimator. Tests of hypotheses and interval estimation of parameters, based on the maximum likelihood estimates, have been developed by Engelhardt, Bain, and Wright (1981). Rieck and Nedelman (1991) discussed a log-linear model for this distribution, which is applicable in accelerated life-testing. Estimation of parameters based on symmetrically Type-II censored samples has been considered by Rieck (1994). Bhattacharyya and Fries (1982) have also studied reciprocal inverse Gaussian distributions. They emphasize the exponential family structure of the distributions. More details on the Birnbaum–Saunders distribution may be found in Chapter 33, Section 2.

9.3 Generalized Inverse Gaussian Distributions

An important generalization of inverse Gaussian distributions, introduced by Good (1953) and studied by Wise (1971, 1975), has attracted considerable attention, exemplified especially by an exhaustive and interesting monograph by Jørgensen (1982). He uses parameters ψ, χ, and θ, and defines the pdf of a generalized inverse Gaussian distribution [GIG(ψ, χ, θ)] as

$$p_X(x|\psi, \chi, \theta) = \frac{(\psi/\chi)^{\theta/2}}{2K_\theta(\sqrt{(\psi\chi)})} x^{\theta-1} \exp\left\{-\frac{1}{2}(\chi x^{-1} + \psi x)\right\},$$

$$x > 0; \psi, \chi > 0, \quad (15.74)$$

where $K_\theta(\cdot)$ is a modified Bessel function of the third kind. If $\theta = -\frac{1}{2}$ we have an $IG(\sqrt{(\psi/\chi)}, \chi)$ pdf. Other special cases are

$\theta = \frac{1}{2}$ − reciprocal IG( ) distribution,
$\chi = 0, \theta > 0$ − gamma distribution (Chapter 17),
$\psi = 0, \theta < 0$ − reciprocal gamma distribution (Chapter 17)
$\theta = 0$ − hyperbolic distribution of Barndorff-Nielsen, Blaesild, and Halgreen (1978) and Rukhin (1974) (Chapter 12).

Jørgensen (1982) also uses an alternative form for (15.74), more symmetrical in appearance, obtained by putting $\omega = \sqrt{(\psi\chi)}$ and $\eta = \sqrt{(\chi/\psi)}$. This gives

$$p_X(x|\omega, \eta, \theta) = \left\{2\eta^\theta K_\theta(\omega)\right\}^{-1} x^{\theta-1} \exp\left\{-\tfrac{1}{2}\omega(\eta^{-1}x + \eta x^{-1})\right\},$$

$$x > 0, \omega, \eta > 0. \quad (15.74)'$$

If $\chi = 0$ or $\psi = 0$ so that $\omega = 0$, equation (15.4a) for $IG(\mu, \lambda)$ corresponds to (15.74)', with $\theta = -\tfrac{1}{2}$, $\omega = \lambda/\mu$, and $\eta = \mu$ (and $K_{-1/2}(\omega)$ $= \sqrt{\tfrac{1}{2}\pi\omega^{-1/2}e^{-\omega}}$). Jørgensen (1982) also reports an approximation based on supposing that

$$-\theta^{3/2}\chi^{-1}(2X + \theta^{-1}\chi) \qquad (15.75)$$

has a standard normal $[N(0, 1)]$ distribution and suggests that even better approximation might be secured by taking $\log X$ to have an approximate normal distribution (i.e., by approximating the GIG distribution by a lognormal distribution). He also shows that GIG distributions can be regarded as limiting cases of generalized hyperbolic distributions.

Barndorff-Nielsen, Blaesild, and Halgreen (1978) have shown that GIG distribution with $\theta < 0$ can arise as first passage times of a time-homogeneous diffusion process. The well-known analogy between analysis of data based on normal and IG parent distributions (see Section 6)—wherein the maximum likelihood estimators of μ and λ for $IG(\mu, \lambda)$ are

$$\hat{\mu} = \overline{X},$$

$$\hat{\lambda} = (\overline{X} - \overline{X}^{-1})^{-1} \qquad (\text{see } (15.27a), (15.27b)),$$

with $\hat{\mu}$ having an $IG(\mu, n\lambda)$ distribution and $n\lambda/\hat{\lambda}$ having a χ^2_{n-1} distribution—*does not* extend to GIG distributions. However, the set

$$\left(\sum_{i=1}^{n} X_i, \ \sum_{i=1}^{n} X_i^{-1}, \ \sum_{i=1}^{n} \log X_i\right) \qquad (15.76)$$

is minimally sufficient for GIG distributions. Embrechts (1983) has shown that the GIG distribution belongs to a class of subexponential distributions.

9.4 Mixtures of $IG(\mu, \lambda)$ with Its Complementary Reciprocal

If X_1 is distributed as $IG(\mu, \sigma^{-2})$ and X_2^{-1} is distributed as $IG(\mu^{-1}, \sigma^{-2}\mu^2)$, we call X_2 the *complementary reciprocal* of X_1. Jørgensen, Seshadri, and Whitmore (1991) introduced a mixture of the distributions of X_1 and X_2 in proportions $1 - p$ and p, respectively $(0 < p < 1)$. They denote this distribution by M-IG(μ, σ^{-2}, p). The pdf of this distribution is

$$p_X(x|\mu, \sigma^{-2}, p) = (2\pi\sigma^2 x^3)^{-1/2}\left(1 - p + \frac{px}{\mu}\right)\exp\left\{-\frac{(x-\mu)^2}{2\mu^2\sigma^2 x}\right\},$$

$$x > 0;\ \mu, \sigma > 0;\ 0 < p < 1. \quad (15.77)$$

We have

$$E[X] = \mu + p\sigma^2\mu^2, \qquad (15.78a)$$

$$\mathrm{Var}(X) = \sigma^2\mu^3 + p(3 - p)\sigma^4\mu^4. \qquad (15.78b)$$

Introducing $\gamma = \mu(1 - p)/p$, (15.77) can be written as

$$p_X(x|\mu, \sigma^{-2}, p) = (2\pi\sigma^2 x^3)^{-1/2}\left(\frac{\gamma + x}{\gamma + \mu}\right)\exp\left\{-\frac{(x-\mu)^2}{2\mu^2\sigma^2 x}\right\}. \quad (15.77)'$$

If γ is known, this is a two parameter exponential family. The cdf is

$$F_X(x|\mu, \sigma^{-2}, p) = \Phi\left(\frac{x - \mu}{\sigma\mu\sqrt{x}}\right) + \exp\left(\frac{2}{\sigma^2\mu}\right)\Phi\left(\frac{x + \mu}{\sigma\mu\sqrt{x}}\right). \quad (15.79)$$

For both IG(μ, σ^{-2}) and M-IG(μ, σ^{-2}, p), the statistic $(X - \mu)^2/(\mu^2 X)$ has a χ_1^2 distribution; (see page 269).

A M-IG(μ, σ^{-2}, p) random variable X can be represented as

$$X = X_1 + Y, \qquad (15.80)$$

where X_1 and Y are mutually independent, X_1 has an IG(μ, σ^{-2}) distribution, and $\Pr[Y = 0] = 1 - p$, while Y is distributed as $\mu^2\sigma^2\chi_1^2$ if $Y \neq 0$. Thus M-IG(μ, σ^{-2}, p) is a convolution of an inverse Gaussian and a compound Bernoulli distribution.

Also if X is distributed as M-IG(μ, σ^{-2}, p), then

$$cX \quad \text{is distributed as} \quad M\text{-IG}(\mu, c\sigma^{-2}, p)$$

and

$$X^{-1} \quad \text{is distributed as} \quad M\text{-IG}(\mu^{-1}, \mu^2\sigma^{-2}, 1 - p).$$

Thus the M-IG family is "closed under reciprocation."

Figure 15.3 from Jørgensen, Seshadri, and Whitmore (1991) provides graphs of some M-IG$(1, 1, p)$ pdf's, and Figure 15.4 plots graphs of M-IG distribution with common expected value 1, $\gamma = 1$ [i.e., $\mu = p/(1 - p)$],

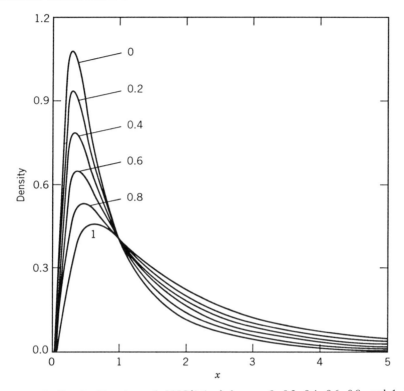

Figure 15.3 Density Functions of M-IG$(1, 1, p)$ for $p = 0$, 0.2, 0.4, 0.6, 0.8, and 1. From Jørgensen, Seshadri, and Whitmore (1991).

$\mu = 0.1(0.2)0.9, 0.95$, and σ chosen so that the variance is 1 [i.e., $(1 - p)p^3\sigma^2 + (3 - p)p^5\sigma^4 = 1$]. The Birnbaum-Saunders distribution is a particular case of the M-IG(μ, σ^2, p).

Abd-El-Hakim and Ahmad (1992) have recently considered a mixture of two inverse Gaussian distributions and discussed its modal properties.

9.5   Other Related Distributions

Al-Hussaini and Abd-El-Hakim (1990) studied estimation of parameters of a inverse Gaussian-Weibull mixture of the form

$$f(x, \theta) = p\left(\frac{\lambda}{2\pi x^3}\right)^{1/2} \exp\left[-\frac{\lambda(x - \mu)^2}{2\mu^2 x}\right] + (1 - p)\left(\frac{c}{\alpha}\right)\left(\frac{x}{\alpha}\right)^{c-1} \exp\left[-\left(\frac{x}{\alpha}\right)^c\right].$$

$$(15.81)$$

They derived equations for both maximum likelihood and weighted maximum likelihood estimators. In an earlier paper (1989) the same authors proposed this mixture as a failure model and investigated its failure rate function.

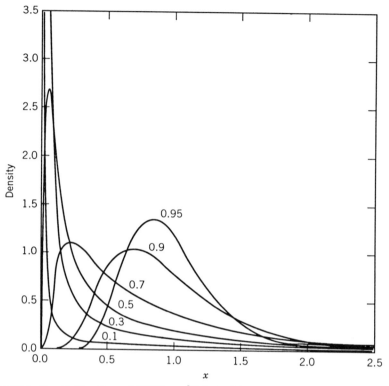

Figure 15.4   Density Functions of M-IG(μ, σ^2) Having Mean Value $m = 1$, $\gamma = 1$, and $\mu = 0.1$, 0.3, 0.5, 0.7, 0.9, and 0.95. From Jørgensen, Seshadri, and Whitmore (1991).

Holla (1966) has described a (discrete) compound Poisson distribution:

$$\text{Poisson}(\theta) \underset{\theta}{\bigwedge} \text{Inverse Gaussian (see Chapter 9)}$$

Robert (1991) proposes another form of *generalized inverse normal* as a generalization of the distribution of $1/X$, when X is distributed as $N(0, \sigma^2)$. The pdf of $Z = 1/X$ is

$$p_Z(z|\theta, \sigma) = \frac{1}{\sqrt{2\pi}\,\sigma}\frac{1}{z^2}\exp\left\{-\frac{1}{2\sigma^2}\left(\frac{1}{z} - \theta\right)^2\right\}.$$

Robert's family $TN(\alpha, \tau, \mu)$ is defined by

$$p_Z(z|\alpha, \mu, \tau) = \frac{K(\alpha, \mu, \tau)}{|z|^\alpha}\exp\left\{-\frac{(z^{-1} - \mu)^2}{2\tau^2}\right\}, \qquad \alpha > 1, \tau > 0,$$

where K is a normalizing constant expressed in terms of a confluent hypergeometric function. The distribution is *bimodal* with modes at

$$z_1 = -\frac{\mu + \sqrt{\mu^2 + 4\alpha\tau^2}}{2\alpha\tau^2}$$

and

$$z_2 = \frac{\sqrt{\mu^2 + 4\alpha\tau^2} - \mu}{2\alpha\tau^2}.$$

For $\alpha = \mu = 2$ $(\tau = 1)$ the distribution is indistinguishable from a distribution with positive support. The case $\alpha = 2$ corresponds to the original inverse normal transformation. The kth moment is defined only if $\alpha > k + 1$. This distribution generates a conjugate prior family for the parameters of $N(\theta, \sigma^2\theta^2)$ distribution.

10 TABLES

Tables of the cumulative distribution function corresponding to (15.4a) have been constructed by Wasan and Roy (1967). They give values of $\Pr[X \le x]$ to four decimal places for $\mu = 5$ and

$$\lambda = 0.25(0.25)1.00(1)10, 16(8)32.$$

For $\lambda < 0.25$ they suggest using a reciprocal gamma approximation, and for $\lambda > 32$, a normal approximation (see section 5). If it is desired to evaluate $\Pr[X \le x]$ for a value of μ different from 5, then one can evaluate

$$\Pr\left[Y = \tfrac{1}{5}\mu X \le \tfrac{1}{5}\mu x\right],$$

since Y has distribution (15.4a) with μ equal to 5, and λ equal to $5 \times$ (original λ)/μ. The intervals of x in the tables vary between 0.005 and 0.5. They were chosen so that the differences between successive tabulated values of $\Pr[X \le x]$ rarely exceed 0.005.

Tables of the standard Wald probability density function (15.3) and cumulative distribution function (15.16) have been published by Krapivin (1965). They contain values of each function to six significant figures for

$$\phi = 0.01(0.01)0.1(0.1)4.0(0.2)5.0(0.5)10$$

and for various values of x. The coverage of x decreases with the increase in the values of ϕ. For example, when $\phi = 0.01$, one has

$$x = 0(0.0001)0.01(0.01)1(0.05)6(0.1)31(0.5)81(1)231(25)1181,$$

and when $\phi = 2.0$, $x = 0(0.01)2(0.05)7(0.2)12.0$. These tables also contain the modal values to six significant figures for

$$\phi = 0.0001(0.0001)0.0100(0.01)1.00(0.1)6.0(1)55,$$

$$60(10)550, 800(100)5700, 60000(10000)550000.$$

Chan, Cohen, and Whitten (1983) give tables of $F_X(x)$ to 6 decimal places for $x = -3.0(0.1)5.9$ with $\alpha_3 = 0.0(0.1)1.2$ and $x = -1.5(0.1)7.4$ with $\alpha_3 = 1.3(0.1)2.5$. Extensive tables of percentage points of the $IG(\lambda, \lambda^2)$ distribution indexed by parameter λ for

$$\lambda = 0.02(0.02)4(0.04)6(0.2)35(1)100(2)150(4)250(10)300(20)$$

$$600(40)2000(80)4000$$

and for cumulative probability $0.001, 0.005, 0.01(0.01)0.99, 0.995, 0.999$, are presented to eight significant figures in a *CRC Handbook* (1989) (J. A. Koziol, editor).

11 APPLICATIONS

Inverse Gaussian distributions have been used to explain the motion of particles influenced by Brownian motion [Wasan (1968b)]. In particular, they have been used to study the motion of particles in a colloidal suspension under an electric field [Tweedie (1947)].

It has been suggested [Tweedie (1957a)] that because of the statistical independence between $\overline{X}$ and $1/\hat{\lambda}$, an analogue of the analysis of variance for nested classifications can be performed. This analogue uses the tables of χ^2 and F developed for the analysis of variance where the measure of dispersion is given by (15.22b).

The main, and the most familiar, applications of Wald distributions are in sequential analysis [e.g., Wald (1947); Bartlett (1956)]. Some authors have used this distribution in various physical applications connected with diffusion processes with boundary conditions. It appears in the calculation of the distribution of time of first hitting the boundary in a random walk.

In Russian literature on electronics and radiotechnique, the Wald distribution is often used, and several studies on this distribution have been published in these journals [e.g., Basharinov, Fleishman, and Samochina (1959); Sigangirov (1962)]. [The Krapivin (1965) extensive tables described above have also been published by the Soviet Institute of Radiotechnique and Electronics.]

Marcus (1975) has suggested that an inverse Gaussian distribution might be used in place of a lognormal distribution when a longtailed distribution is expected to be appropriate—for example, in distribution of particle size in aggregates [see Bardsley (1980) for a related application in meteorology and also Lawrence (1984) in business applications]. If distributions of sums of independent random variables (convolutions) are to be studied, it may be useful to remember that if the mutually independent random variables X_i $(i = 1, \ldots, n)$ are distributed as $IG(\mu_i, c\mu_i^2)$, respectively, then $(X_1 + X_2 + \cdots + X_n)$ has an $IG(\sum_{i=1}^{n} \mu_i, c(\sum_{i=1}^{n} \mu_i)^2)$ distribution (see page 267).

In recent years the inverse Gaussian distribution has played versatile roles in models of stochastic processes including the theory of generalized linear models [McCullagh and Nelder (1983)], reliability and lifetime data analysis [Padgett and Tsai (1986)], accelerated life testing [Bhattacharyya and Fries (1982)], and repair time distributions, especially in cases of preponderance of early failures [Chhikara and Folks (1977)]. In industrial quality control Edgeman (1989a, b) has developed control charts and sampling plans based on inverse Gaussian process distributions.

More recently Dhulesia, Bernicet, and Deheuvels (1991) have provided theoretical justification, based on the properties of Brownian motion with drift, for preferring inverse Gaussian to lognormal or gamma distributions to describe distributions of slug lengths in pipelines, and verified this conclusion by fitting data. (The fit often breaks down for very short slug lengths, which are likely to be produced by special physical phenomena.)

Hougaard (1984) and Feaganes and Suchindran (1991) advocates the use of the inverse Gaussian distribution for distribution of "frailty" [a concept recently developed in survival analysis; e.g., see Hougaard (1984)]. Frailty occurs in models of *heterogeneity*. Ordinary life table analyses implicitly assume that population is homogeneous. If z is frailty, then the hazard rate (or death rate) at age t for a person with frailty z is assumed to be of the multiplicative form

$$\theta(t; z) = z\theta(t),$$

where $\theta(t)$ is independent of z and describes the age effect.

An advantage of using inverse Gaussian to represent the distribution of frailty is its closure property. If frailty in the population is distributed as an inverse Gaussian at the formation of the cohort (e.g., birth), then frailty will still have an inverse Gaussian distribution among survivors at any time after the formation of the cohort. In fact, if the frailty is distributed as $IG(\mu, \lambda)$, then the distribution of z given $T \geq t$ is $IG(\mu_t, \lambda)$, where $\mu_t = \mu\{1 - 2\lambda\mu^{-2}\int_0^t \theta(u)\,du\}^{-1/2}$.

Another advantage is that the coefficient of variation for this conditional distribution of heterogeneity decreases with age (unlike the case of the gamma distribution where it remains constant); that is to say, the survivors become more identical with the passage of time.

BIBLIOGRAPHY

Abd-El-Hakim, N. S., and Ahmad, K. E. (1992). On the modes of a mixture of two inverse Gaussian distributions, *Egyptian Statistical Journal*, **36**, 226–233.

Ahmad, I. A. (1988). Jackknife estimation for a family of life distributions, *Journal of Statistical Computation and Simulation*, **29**, 211–223.

Akman, O., and Gupta, R. C. (1992). A comparison of various estimators of the mean of an inverse Gaussian distribution, *Journal of Statistical Computation and Simulation*, **40**, 71–81.

Al-Hussaini, E. K., and Abd-El-Hakim, N. S. (1989). Failure rate of the inverse Gaussian-Weibull mixture model, *Annals of the Institute of Statistical Mathematics*, **41**, 617–622.

Al-Hussaini, E. K., and Abd-El-Hakim, N. S. (1990). Estimation of parameters of the inverse Gaussian-Weibull mixture model, *Communications in Statistics—Theory and Methods*, **19**, 1607–1622.

Al-Hussaini, E. K., and Abd-El-Hakim, N. S. (1992). Efficiency of schemes from the inverse Gaussian–Weibull mixture model, *Communications in Statistics—Theory and Methods*, **21**, 3143–3169.

Balakrishnan, N., and Cohen, A. C. (1991). *Order Statistics and Inference: Estimation Methods*, San Diego: Academic Press.

Banerjee, A. K., and Bhattacharyya, G. K. (1979). Bayesian results for the inverse Gaussian distribution, with an application, *Technometrics*, **21**, 247–251.

Bardsley, W. E. (1980). Note on the use of the inverse Gaussian distribution for wind energy applications, *Journal of Applied Meteorology*, **19**, 1126–1130.

Barndorff-Nielsen, O. E., Blaesild, P., and Halgreen, C. (1978). First hitting time models for the generalized inverse Gaussian distribution, *Stochastic Processes and their Applications*, **7**, 49–54.

Barndorff-Nielsen, O. E., and Jørgensen, B. (1989). Some parametric models on the simplex, *Journal of Multivariate Analysis*, **39**, 106–116.

Bartlett, M. S. (1956). *Introduction to Stochastic Processes*, Cambridge: Cambridge University Press.

Basharinov, A. E., Fleishman, B. S., and Samochina, M. A. (1959). Binary accumulating systems with two threshold analyzers, *Radiotekhnika i Elektronika*, **4**, 1419–1426. (In Russian).

Betro, B., and Rotondi, R. (1991). On Bayesian inference for the inverse Gaussian distribution, *Statistics & Probability Letters*, **11**, 219–224.

Bhattacharyya, G. K., and Fries, A. (1982). Fatigue failure models—Birnbaum-Saunders vs. Inverse Gaussian, *IEEE Transactions in Reliability*, **31**, 439–440.

Birnbaum, Z. W., and Saunders, S. C. (1969a). A new family of life distributions, *Journal of Applied Probability*, **6**, 319–327.

Birnbaum, Z. W., and Saunders, S. C. (1969b). Estimation for a family of life distributions with applications to fatigue, *Journal of Applied Probability*, **6**, 328–347.

Chan, M. Y., Cohen, A. C., and Whitten, B. J. (1983). The standardized inverse Gaussian distribution-tables of the cumulative distribution function, *Communications in Statistics—Simulation and Computation*, **12**, 423–442.

Chan, M. Y., Cohen, A. C., and Whitten, B. J. (1984). Modified maximum likelihood and modified moment estimators for the three-parameter inverse Gaussian distribution, *Communications in Statistics—Simulation and Computation*, **13**, 47–68.

Cheng, R. C. H., and Amin, N. A. K. (1981). Maximum likelihood estimation of parameters in the inverse Gaussian distribution with unknown origin, *Technometrics*, **23**, 257–263.

Chhikara, R. S. (1972). Statistical Inference Related to the Inverse Gaussian Distribution, Ph.D. dissertation, Stillwater, OK: Oklahoma State University.

Chhikara, R. S., and Folks, J. L. (1977). The inverse Gaussian distribution as a lifetime model, *Technometrics*, **19**, 461–468.

Chhikara, R. S., and Folks, J. L. (1989). *The Inverse Gaussian Distribution*, New York: Marcel Dekker.

Cohen, A. C., and Whitten, B. J. (1985). Modified moment estimation for the three-parameter inverse Gaussian distribution, *Journal of Quality Technology*, **17**, 147–154.

Cohen, A. C., and Whitten, B. J. (1988). *Parameter Estimation in Reliability and Life Span Models*, New York: Marcel Dekker.

CRC Handbook. (1988). *Percentile Points of the Inverse Gaussian Distribution*, J. A. Koziol (editor), Boca Raton, FL: CRC Press.

Desmond, A. F. (1985). Stochastic models of failure in random environments, *Canadian Journal of Statistics*, **13**, 171–183.

Desmond, A. F. (1986). On relationship between two fatigue-life models, *IEEE Transactions on Reliability*, **35**, 167–169.

Dhulesia, H., Bernicet, M., and Deheuvels, P. (1991). Statistical analysis and modelling of slug lengths, *Multi-phase Production*, 80–112, A. P. Burns (editor), New York: Elsevier.

Edgeman, R. L. (1989a). Inverse Gaussian control charts, *Australian Journal of Statistics*, **31**, 294–296.

Edgeman, R. L. (1989b). Control of inverse Gaussian processes, *Quality Engineering*, **1**, 265–276.

Edgeman, R. L. (1989c). Sample Plans for the Inverse Gaussian Distribution, College of Business Working Papers Series, No. 89-10, Fort Collins, CO: Colorado State University.

Embrechts, P. (1983). A property of the generalized inverse Gaussian distribution with some applications, *Journal of Applied Probability*, **20**, 537–544.

Engelhardt, M., Bain, L. J., and Wright, F. T. (1981). Inferences on the parameters of the Birnbaum-Saunders fatigue life distribution based on maximum likelihood estimation, *Technometrics*, **23**, 251–256.

Feaganes, J. R., and Suchindran, C. M. (1991). Weibull regression with unobservable heterogeneity, an application, *ASA Proceedings of Social Statistics Section*, pp. 160–165.

Folks, J. L., and Chhikara, R. S. (1978). The inverse Gaussian distribution and its statistical applications—A review, *Journal of the Royal Statistical Society, Series B*, **40**, 263–289.

Good, I. J. (1953). The population frequencies of the species and the estimation of population parameters, *Biometrika*, **40**, 237–260.

Holla, M. S. (1966). On a Poisson-inverse Gaussian distribution, *Metrika*, **11**, 115–121.

Hougaard, P. (1984). Life table methods for heterogeneous populations: Distributions describing the heterogeneity, *Biometrika*, **71**, 75–83.

Hsieh, H. K. (1990). Inferences on the coefficient of variation of an inverse Gaussian distribution, *Communications in Statistics—Theory and Methods*, **19**, 1589–1605.

Hsieh, H. K., and Korwar, R. M. (1990). Inadmissibility of the uniformly minimum variance unbiased estimator of the inverse Gaussian variance, *Communications in Statistics—Theory and Methods*, **19**, 2509–2516.

Hsieh, H. K., Korwar, R. M., and Rukhin, A. L. (1990). Inadmissibility of the maximum likelihood estimator of the inverse Gaussian mean, *Statistics & Probability Letters*, **9**, 83–90.

Huff, B. W. (1974). A comparison of sample path properties for the inverse Gaussian and Bessell processes, *Scandinavian Actuarial Journal*, 157–166.

Iliescu, D. V., and Vodă, V. G. (1977). Some inferences on alpha distribution, *Revue Roumaine de Mathematiques Pures et Appliquees*, **22**, 1433–1439.

Iliescu, D. V., and Vodă, V. G. (1981). On the inverse Gaussian distribution, *Bulletin Mathematique de la Societe des Sciences Mathematiques de la Republique Socialiste de Roumaine* (*N.S.*), **25**, 281–293.

Iwase, K. (1987). UMVU estimation for the inverse Gaussian distribution $I(\mu, c\mu^2)$ with known c, *Communications in Statistics—Theory and Methods*, **16**, 1315–1320.

Iwase, K., and Kanefuji, K. (1991). Estimation for 3-parameter inverse Gaussian distribution with unknown origin, Technical Report No. 92-04, Department of Applied Mathematics, Hiroshima University, Japan.

Iwase, K., and Seto, N. (1983). Uniformly minimum variance unbiased estimation for the inverse Gaussian distribution, *Journal of the American Statistical Association*, **78**, 660–663.

Jones, G., and Cheng, R. C. H. (1984). On the asymptotic efficiency of moment and maximum likelihood estimators in the three-parameter inverse Gaussian distribution, *Communications in Statistics—Theory and Methods*, **13**, 2307–2314.

Jørgensen, B. (1982). Statistical Properties of the Generalized Inverse Gaussian Distribution, *Lecture Notes in Statistics*, No. 9, New York: Springer-Verlag.

Jørgensen, B., Seshadri, V., and Whitmore, G. A. (1991). On the mixture of the inverse Gaussian distribution with its complementary reciprocal, *Scandinavian Journal of Statistics*, **18**, 77–89.

Joshi, S., and Shah, M. (1991). Estimating the mean of an inverse Gaussian distribution with known coefficient of variation, *Communications in Statistics—Theory and Methods*, **20**, 2907–2912.

Khatri, C. G. (1962). A characterization of the inverse Gaussian distribution, *Annals of Mathematical Statistics*, **33**, 800–803.

Kingman, J. F. C. (1978). Discussion of "The inverse Gaussian distribution and its statistical applications—a review" by J. L. Folks and R. S. Chhikara, *Journal of the Royal Statistical Society, Series B*, **40**, 263–289.

Korwar, R. M. (1980). On the uniformly minimum variance unbiased estimators of the variance and its reciprocal of an inverse Gaussian distribution, *Journal of the American Statistical Association*, **75**, 734–735.

Krapivin, V. F. (1965). *Tables of Wald's Distribution*, Moscow: Nauka. (In Russian).

Lancaster, A. (1978). A stochastic model for the duration of a strike, *Journal of the Royal Statistical Society, Series A*, **135**, 257–271.

Lawrence, R. J. (1984). The log normal distribution of the duration of strikes, *Journal of the Royal Statistical Society, Series A*, **147**, 464–483.

Letac, G., and Seshadri, V. (1985). On Khatri's characterization of the inverse Gaussian distribution, *Canadian Journal of Statistics*, **13**, 249–252.

Letac, G., and Seshadri, V. (1989). The expectation of X^{-1} as a function of $E(X)$ for an exponential family on the positive line, *Annals of Statistics*, **17**, 1735–1741.

Letac, G., Seshadri, V., and Whitmore, G. A. (1985). An exact chi-squared decomposition theorem for inverse Gaussian variates, *Journal of the Royal Statistical Society, Series B*, **47**, 476–481.

Marcus, A. H. (1975). Power-sum distributions: An easier approach using the Wald distribution, *Journal of the American Statistical Association*, **71**, 237–238.

McCullagh, P., and Nelder, J. A. (1983). *Generalized Linear Models*, London: Chapman and Hall.

National Bureau of Standards (1942). *Tables of Probability Functions*, **2**, Washington, DC: Government Printing Office.

Newby, M. (1985). A further note on the statistics of strike duration, *Journal of the Royal Statistical Society, Series A*, **148**, 350–356.

Nonnenmacher, T. F. (1990). Fractional and differential equations for a class of Levy-type probability densities, *Journal of Physics A23*, L697S–L700S.

O'Reilly, F. J., and Rueda, R. (1992). Fit for the inverse Gaussian, *Canadian Journal of Statistics*, **20**, 387–398.

Padgett, W. J., and Tsai, S. K. (1986). Prediction intervals for future observations from the inverse Gaussian distribution, *IEEE Transactions on Reliability*, **35**, 406–408.

Padgett, W. J., and Wei, L. J. (1979). Estimation for the three-parameter inverse Gaussian distribution, *Communications in Statistics*, **8**, 129–137.

Patel, R. C. (1965). Estimates of parameters of truncated inverse Gaussian distribution, *Annals of the Institute of Statistical Mathematics*, **17**, 29–33.

Rieck, J. R. (1994). Estimation of parameters of the Birnbaum–Saunders distribution based on complete and symmetrically censored samples (Preprint).

Rieck, J. R., and Nedelman, J. (1991). A log-linear model for the Birnbaum–Saunders distribution, *Technometrics*, **33**, 51–60.

Robert, C. (1991). Generalized inverse normal distributions, *Statistics & Probability Letters*, **11**, 37–41.

Roy, L. K. (1977). Estimation of parameters of a truncated log-inverse Gaussian, *Trabajos de Estadistica*, **28**, 161–165.

Roy, L. K., and Wasan, M. T. (1968). The first passage time distribution of Brownian motion with positive drift, *Mathematical Biosciences*, **3**, 191–204.

Schrödinger, E. (1915). Zür Theorie der Fall—und Steigversuche an Teilchenn mit Brownsche Bewegung, *Physikalische Zeitschrift*, **16**, 289–295.

Schwarz, C. J., and Samanta, M. (1991). An inductive proof of the sampling distribution for the MLEs of parameters in an inverse Gaussian distribution, *The American Statistician*, **45**, 223–225.

Seal, H. L. (1978). From aggregate claims distribution to probability of ruin, *ASTIN Bulletin*, **10**, 47–53.

Seshadri, V. (1983). The inverse Gaussian distribution: Some properties and characterizations, *Canadian Journal of Statistics*, **11**, 131–136.

Seshadri, V. (1989). A U-Statistic and estimation for the inverse Gaussian distribution, *Statistics & Probability Letters*, **7**, 47–49.

Seshadri, V. (1992). General exponential models on the unit simplex and related multivariate inverse Gaussian distributions, *Statistics & Probability Letters*, **14**, 385–391.

Seto, N., and Iwase, K. (1985). UMVU estimators of the mode and limits of an interval for the inverse Gaussian distribution, *Communications in Statistics—Theory and Methods*, **14**, 1151–1161.

Shuster, J. (1968). On the inverse Gaussian distribution function, *Journal of the American Statistical Association*, **63**, 1514–1516.

Sibuya, M. (1992). Numerical calculation of quantiles of the inverse Gaussian distribution, *Japanese Journal of Applied Statistics*, **22**, 113–127. (In Japanese.)

Sichel, H. S. (1974). On a distribution representing sentence-length in written prose, *Journal of the Royal Statistical Society, Series A*, **137**, 25–34.

Sichel, H. S. (1975). On a distribution law for word frequencies, *Journal of the American Statistical Association*, **70**, 542–547.

Sigangirov, K. Sh. (1962). Representation of Wald's distribution by means of a normal distribution, *Radiotekhnike Elektronika*, **7**, 164–166. (In Russian)

Tweedie, M. C. K. (1947). Functions of a statistical variate with given means, with special reference to Laplacian distributions, *Proceedings of the Cambridge Philosophical Society*, **43**, 41–49.

Tweedie, M. C. K. (1956). Some statistical properties of inverse Gaussian distributions, *Virginia Journal of Science (New Series)*, **7**, 160–165.

Tweedie, M. C. K. (1957a). Statistical properties of inverse Gaussian distributions, I, *Annals of Mathematical Statistics*, **28**, 362–377.

Tweedie, M. C. K. (1957b). Statistical properties of inverse Gaussian distributions, II, *Annals of Mathematical Statistics*, **28**, 696–705.

Vodă, V. G. (1973). A note on the one-parameter inverse Gaussian distribution, *Revista de Ciências Matemáticas, Universidade de Louringo Marques, Série A*, **4**, 47–55.

Wald, A. (1947). *Sequential Analysis*, New York: Wiley.

Wasan, M. T. (1968a). On an inverse Gaussian process, *Skandinavisk Aktuarietidskrift*, 69–96.

Wasan, M. T. (1968b). *First Passage Time Distribution of Brownian Motion*, Monograph, Department of Mathematics, Kingston, Ontario: Queen's University.

Wasan, M. T., and Roy, L. K. (1967). Tables of inverse Gaussian probabilities (Abstract), *Annals of Mathematical Statistics*, **38**, 299.

Whitmore, G. A. (1976). Management applications of inverse Gaussian distributions, *International Journal of Management Science*, **4**, 215–223.

Whitmore, G. A. (1978). Discussion of "The inverse Gaussian distribution and its statistical applications—a review" by J. L. Folks and R. S. Chhikara, *Journal of the Royal Statistical Society, Series B*, **40**, 263–289.

Whitmore, G. A. (1979). An inverse Gaussian model for labour turnover, *Journal of the Royal Statistical Society, Series A*, **142**, 468–478.

Whitmore, G. A. (1983). A regression method for censored inverse Gaussian data, *Canadian Journal of Statistics*, **11**, 305–315.

Whitmore, G. A., and Seshadri, V. (1987). A heuristic derivation of the inverse Gaussian distribution, *The American Statistician*, **41**, 280–281.

Whitmore, G. A., and Yalovsky, M. (1986). Inverse Gaussian ratio estimation, *Applied Statistics*, **35**, 8–15.

Wise, M. E. (1971). Skew probability curves with negative powers of the time and related random walks in series, *Statistica Neerlandica*, **25**, 159–180.

Wise, M. E. (1975). Skew distributions in biomedicine, including some with negative powers of time, *Statistical Distributions in Scientific Work*, **2**, G. P. Patil, S. Kotz, and J. K. Ord (editors), 241–262, Dordrecht: Reidel.

CHAPTER 16

Cauchy Distribution

1 HISTORICAL REMARKS

The density function $\pi^{-1}(1 + x^2)^{-1}$, that is commonly referred to as the Cauchy density, or rather curves proportional to $(x^2 + a^2)^{-1}$—have been studied in the mathematical world for over three centuries. An excellent historical account of the distribution has been prepared by Stigler (1974). As he points out, the curve seems to have appeared first in the works of Pierre de Fermat in the midseventeenth century and was subsequently studied by many including Sir Isaac Newton, Gottfried Leibniz, Christian Huygens, Guido Grandi, and Maria Agnesi; interestingly the curve had acquired the name the "Witch of Agnesi" derived from the name of the last-mentioned Italian mathematician who lived from 1718 to 1799. She had discussed the properties of the Cauchy curve and referred to it as *la Versiera* (Italian for "witch"), following the term introduced by Grandi (1718). Grandi's construction and interpretation of the Cauchy density is very simply the locus of points C in Fig. 16.1 as K goes from $-\infty$ to ∞ with the angle at C remaining a right angle.

Siméon Denis Poisson (1824) was the first to notice that the distribution with density $\pi^{-1}(1 + x^2)^{-1}$ has some peculiar properties and could provide counterexamples to some generally accepted results and concepts in statistics. For example, by presenting the characteristic function of the Cauchy distribution, Poisson showed that a proof given by Laplace for the large-sample justification for Legendre's principle of least squares through the central limit theorem breaks down as the second term in Laplace's expansion of the characteristic function would not be negligible for large n (the sample size). As Eisenhart (1971) indicated, it is this simple form of the characteristic function that may have led Poisson to consider this heavy-tailed density function. But this distribution, known as the "Cauchy distribution," became associated with Augustin Louis Cauchy (1853) when he, in response to an article by I. J. Bienaymé (1853) criticizing a method of interpolation suggested earlier by Cauchy, considered the density function and showed that Legendre's least-squares does not provide the "most probable results" in this case as it does in the case of normally distributed errors. More elaborate historical details in this regard are in Stigler (1974).

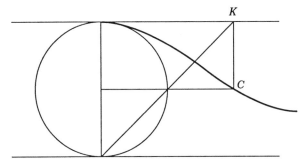

Figure 16.1 Grandi's 1718 Construction of the "Cauchy" Density.

2 DEFINITION AND PROPERTIES

The special form of the Pearson Type VII distribution, with probability density function

$$(\pi\lambda)^{-1}\left[1 + \left\{\frac{x - \theta}{\lambda}\right\}^2\right]^{-1}, \qquad \lambda > 0, \tag{16.1}$$

is called the *Cauchy distribution* [Cauchy (1853) eq. (25), p. 206]. The cumulative distribution function is

$$\frac{1}{2} + \pi^{-1}\tan^{-1}\left[\frac{x - \theta}{\lambda}\right]. \tag{16.2}$$

The parameters θ and λ are location and scale parameters, respectively. The distribution is symmetrical about $x = \theta$. The median is θ; the upper and lower quartiles are $\theta \pm \lambda$. The probability density function has points of inflexion at $\theta \pm \lambda/\sqrt{3}$. It may be noted that the values of the cumulative distribution at the points of inflexion are 0.273 and 0.727, as compared with the corresponding values 0.159 and 0.841 for the normal distribution.

If X has pdf (16.1), then the probability density function of $Y = A + BX$ is of the same form as that of X, with θ replaced by $A + B\theta$ and λ replaced by $|B|\lambda$. The distribution does not possess finite moments of order greater than or equal to 1, and so does not possess a finite expected value or standard deviation. (However, θ and λ are location and scale parameters, respectively, and may be regarded as being analogous to mean and standard deviation.) Goria (1978) derived expressions for the fractional absolute moments of a Cauchy variable. The most notable difference between the normal and Cauchy distributions is in the longer and flatter tails of the latter. These differences are illustrated in Tables 16.1 and 16.2.

It should be mentioned here that Balanda (1987) carried out a kurtosis comparison of the Cauchy and Laplace distributions. Although the coefficient

Table 16.1   Comparison of Cauchy and Normal Distributions

	$\Pr[X \leq x]$			$\Pr[X \leq x]$	
x	Cauchy	Normal	x	Cauchy	Normal
0.0	0.5000	0.5000	2.2	0.8642	0.9311
0.2	0.5628	0.5537	2.4	0.8743	0.9472
0.4	0.6211	0.6063	2.6	0.8831	0.9602
0.6	0.6720	0.6571	2.8	0.8908	0.9705
0.8	0.7148	0.7053	3.0	0.8976	0.9785
1.0	0.7500	0.7500	3.2	0.9036	0.9845
1.2	0.7789	0.7919	3.4	0.9089	0.9891
1.4	0.8026	0.8275	3.6	0.9138	0.9924
1.6	0.8222	0.8597	3.8	0.9181	0.9948
1.8	0.8386	0.8876	4.0	0.9220	0.9965
2.0	0.8524	0.9113			

Note: The Cauchy distribution is in standard form (16.3); the normal distribution has expected value zero and standard deviation $(0.67445)^{-1} = 1.4827$. The two distributions have the same median ($x = 0$) and upper and lower quartiles ($x = \pm 1$). Since both distributions are symmetrical about $x = 0$, there is no need to tabulate negative values of x.

of kurtosis β_2 is infinite for Cauchy and 6 for the double exponential, Balanda noted that this moment-based comparison is inadequate, since it fails to recognize the dominant features of the two distributions: the Cauchy's long tail and the double exponential's dramatic peak (see Chapter 24). For example, Horn (1983) classified the double exponential as being more peaked than the Cauchy distribution, while Rosenberger and Gasko (1983) classified the Cauchy as having heavier tails than the double exponential distribution. Since the moment-based orderings are not helpful because none of the Cauchy's moments are finite as mentioned earlier, Balanda made a kurtosis comparison using kurtosis orderings following the lines of van Zwet (1964) (see Chapter 33).

**Table 16.2   Percentage Points
of Standard Cauchy Distribution**

$\Pr[X \leq x]$	x
0.5	0.0000
0.6	0.3249
0.7	0.7265
0.75	1.0000
0.8	1.3764
0.85	1.9626
0.9	3.0777
0.95	6.3138
0.975	12.7062

There is no standardized form of the Cauchy distribution, as it is not possible to standardize without using (finite) values of mean and standard deviation, which do not exist in this case. However, a *standard form* is obtained by putting $\theta = 0$, $\lambda = 1$. The standard pdf is

$$\pi^{-1}(1 + x^2)^{-1} \tag{16.3}$$

and the standard cdf is

$$\frac{1}{2} + \pi^{-1} \tan^{-1}x. \tag{16.4}$$

(Note that this is the *t-distribution* with 1 degree of freedom; see Chapter 28.) Values of (16.4) are shown in Table 16.1. Percentage points [values X_α such that $\Pr[X \leq X_\alpha] = \alpha$] are shown in Table 16.2.

The characteristic function corresponding to (16.1) is

$$E[e^{itX}] = \exp[it\theta - |t|\lambda]. \tag{16.5}$$

If $X_1, X_2, \ldots, X_n$ are independent random variables with pdf's

$$p_{X_j}(x) = (\pi\lambda_j)^{-1}\left[1 + \left\{\frac{x - \theta_j}{\lambda_j}\right\}^2\right]^{-1}, \qquad \lambda_j > 0; j = 1, 2, \ldots, n, \tag{16.6}$$

then the characteristic function of $S_n = \sum_{j=1}^{n} X_j$ is

$$\exp\left[it \sum_{j=1}^{n} \theta_j - |t| \sum_{j=1}^{n} \lambda_j\right].$$

Hence S_n has a Cauchy probability density function, as in (16.1), with $\lambda = \sum_{j=1}^{n}\lambda_j$; $\theta = \sum_{j=1}^{n}\theta_j$. More generally, $\sum_{j=1}^{n}a_jX_j$ has a Cauchy probability density function, as in (16.1) with $\lambda = \sum_{j=1}^{n}|a_j|\lambda_j$; $\theta = \sum_{j=1}^{n}a_j\theta_j$. In particular, putting

$$\lambda_1 = \lambda_2 = \cdots = \lambda_n = \lambda,$$

$$\theta_1 = \theta_2 = \cdots = \theta_n = \theta,$$

$$a_1 = a_2 = \cdots = a_n = n^{-1},$$

we see that the arithmetic mean $n^{-1}\sum_{j=1}^{n}X_j$ of n independent random variables each having probability density function (16.1) has the *same* Cauchy distribution as each of the X_j's. It should be pointed out here that Dwass (1985), Nelsen (1985), and Blyth (1986) derived the distribution of a sum of independent Cauchy random variables using the convolution formula (instead of the more advanced characteristic function approach described above).

Thus the Cauchy distribution is a "stable" distribution and is also "infinitely divisible." Puglisi (1966) has shown that a variable with a Cauchy distribution can be represented as the sum of two independent random variables, each having infinitely divisible, but *not* stable, distributions. Dugué (1941) presented a decomposition of Cauchy using Raikov theorem. Bondesson (1987) has shown that the half-Cauchy (or folded-Cauchy) distribution is infinitely divisible. Similarly Rohatgi, Steutel, and Székely (1990) have proved that the product and quotient of two i.i.d. standard Cauchy random variables, with density as in (16.3), are infinitely divisible too. Springer and Thompson (1966) have obtained explicit formulas for the probability density function of the product of n (≤ 10) independent and identically distributed Cauchy variables (each with $\theta = 0$, $\lambda = 1$).

If X_1 and X_2 are independent, and each has a standard Cauchy distribution, then $(b_1 X_1 + b_2 X_2)$ has a Cauchy distribution and is in fact distributed as $(|b_1| + |b_2|)X_1$. Pitman and Williams (1967) have extended this result, obtaining [in a form suggested by Williams (1969)]:

"Suppose that the sets of numbers $\{w_j\}$ and $\{a_j\}$ are such that $w_j \geq 0$, $\sum_{j=1}^{\infty} w_j = 1$, and $\{a_j\}$ possesses no limit point. Then, if X has a standard Cauchy distribution, so does

$$\sum_{j=1}^{\infty} w_j (1 + a_j X)(a_j - X)^{-1}."$$

Similarly, when X has a standard Cauchy distribution, Kotlarski (1977) has shown that for $n = 2, 3, \ldots$, the random variables

$$Z_2 = \frac{2X}{1 - X^2}, \quad Z_3 = \frac{3X - X^3}{1 - 3X^2}, \quad Z_4 = \frac{4X - X^3}{1 - 6X^2 + X^4},$$

$$\ldots, \quad Z_n = \frac{\binom{n}{1}X - \binom{n}{3}X^3 + \binom{n}{5}X^5 - \cdots}{1 - \binom{n}{2}X^2 + \binom{n}{4}X^4 - \binom{n}{6}X^6 + \cdots}$$

all have the same standard Cauchy distribution.

Gupta and Govindarajulu (1975) derived the distribution of the quotient of two independent Hotelling's T^2-variates in terms of a hypergeometric function of the second kind, which includes the distribution of the quotient of two independent Cauchy variates as a special case. Eicker (1985) proved that sums of independent squared Cauchy variables grow quadratically, that is,

$$\Pr\left(n^2 o(1) < \sum_{i=1}^{n} X_i^2 < \frac{n^2}{o(1)}\right) \to 1 \quad \text{as} \quad n \to \infty$$

for any positive $o(1)$ sequence; see Bose (1987) for some further remarks in this direction. Zhang (1990) has discussed the distribution of the squared sum of independent Cauchy variables. Salama (1983) has discussed additional problems associated with the Cauchy distribution.

3   ORDER STATISTICS

If $X_1, X_2, \ldots, X_n$ are independent random variables each with standard Cauchy density function (16.3), and $X_1' \le X_2' \le \cdots X_n'$ are the corresponding order statistics, then the density function of X_r' $(1 \le r \le n)$ is

$$p_{X_r'}(x) = \frac{n!}{(r-1)!(n-r)!} \left(\frac{1}{2} + \frac{1}{\pi} \tan^{-1} x\right)^{r-1} \left(\frac{1}{2} - \frac{1}{\pi} \tan^{-1} x\right)^{n-r}$$

$$\times \frac{1}{\pi(1+x^2)}, \qquad -\infty < x < \infty, \tag{16.7}$$

and the joint density function of X_r' and X_s' $(1 \le r < s \le n)$ is

$$p_{X_r', X_s'}(x, y) = \frac{n!}{(r-1)!(s-r-1)!(n-s)!} \left(\frac{1}{2} + \frac{1}{\pi} \tan^{-1} x\right)^{r-1}$$

$$\times \left(\frac{1}{\pi} \tan^{-1} y - \frac{1}{\pi} \tan^{-1} x\right)^{s-r-1} \left(\frac{1}{2} - \frac{1}{\pi} \tan^{-1} y\right)^{n-s}$$

$$\times \frac{1}{\pi^2(1+x^2)(1+y^2)}, \qquad -\infty < x < y < \infty. \tag{16.8}$$

Since the standard Cauchy random variable does not possess finite moments of order ≥ 1, an application of Sen's (1959) theorem immediately reveals that expected values of X_1' and X_n', and the variances of X_1', X_2', X_{n-1}' and X_n' are infinite; interested readers may also refer to Stoops and Barr (1971) for some insight into this existence-of-moments problem.

By employing numerical integration methods, Barnett (1966b) computed from (16.7) and (16.8) values of means, variances, and covariances of order statistics for sample size $n = 5(1)16(2)20$. Explicit expressions (involving infinite series) can be derived for the single and the product moments of order statistics as shown recently by Vaughan (1993). For example, one may write from (16.7) that

$$E[X'_r] = \int_{-\infty}^{\infty} y p_{X'_r}(y) \, dy$$

$$= \frac{n!}{(r-1)!(n-r)!\pi^n} \int_{-\pi/2}^{\pi/2} \left(\frac{\pi}{2}+y\right)^{r-1} \left(\frac{\pi}{2}-y\right)^{n-r} \tan y \, dy$$

$$= \frac{n!}{(r-1)!(n-r)!\pi^n}(I_1 + I_2), \tag{16.9}$$

where I_1 and I_2 are integrals over the ranges $[0, \pi/2]$ and $[-\pi/2, 0]$ with the same integrand. Using Gradshteyn and Ryzhik's (1965) integral formula that for $|y| < \pi$ and for any $a \geq 1$,

$$\int y^a \cot y \, dy = \sum_{i=0}^{\infty} (-1)^i \frac{2^{2i} B_{2i} y^{a+2i}}{(a+2i)(2i)!},$$

where B_{2i} is the Bernoulli number given by

$$B_0 = 1,$$

$$B_{2i} = (-1)^{i-1}(2i)! \zeta(2i)/(2^{2i-1}\pi^{2i}), \qquad i \geq 1,$$

with $\zeta(2i) = \sum_{j=1}^{\infty} j^{-2i}$ [see Gradshteyn and Ryzhik (1965)], we get

$$I_1 = \sum_{j=0}^{r-1} (-1)^j \binom{r-1}{j} \pi^{r-1-j} \sum_{i=0}^{\infty} \frac{(-1)^i 2^{2i} B_{2i} (\pi/2)^{n-r+j+2i}}{(n-r+j+2i)(2i)!}. \tag{16.10}$$

Similarly we find

$$I_2 = -\sum_{j=0}^{n-r} (-1)^j \binom{n-r}{j} \pi^{n-r-j} \sum_{i=0}^{\infty} \frac{(-1)^i 2^{2i} B_{2i} (\pi/2)^{r-1+j+2i}}{(r-1+j+2i)(2i)!}. \tag{16.11}$$

Substitution of the expressions of I_1 and I_2 in (16.10) and (16.11), respectively, into (16.9) yields

$$E[X'_r] = \frac{n!}{(r-1)!(n-r)!\pi}$$

$$\times \left\{ \sum_{j=0}^{r-1} (-1)^j \binom{r-1}{j} 2^{-(n-r+j)} \sum_{i=0}^{\infty} \frac{(-1)^i B_{2i} \pi^{2i}}{(n-r+j+2i)(2i)!} \right.$$

$$\left. - \sum_{j=0}^{n-r} (-1)^j \binom{n-r}{j} 2^{-(r-1+j)} \sum_{i=0}^{\infty} \frac{(-1)^i B_{2i} \pi^{2i}}{(r-1+j+2i)(2i)!} \right\}. \tag{16.12}$$

Similar explicit expressions for the second moment and the product moment have been derived by Vaughan (1993).

For large n, of course, one may use David and Johnson's (1954) approximation to derive approximate expressions for the means, variances and covariances of order statistics as

$$E[X'_r] \doteq \tan\left\{\pi\left(\frac{r}{n} - \frac{1}{2}\right)\right\} = -\cot\left(\frac{\pi r}{n}\right), \qquad (16.13)$$

$$\mathrm{Var}(X'_r) \doteq \frac{1}{n}\pi^2\left(\frac{r}{n}\right)\left(1 - \frac{r}{n}\right)\mathrm{cosec}^4\left(\frac{r\pi}{n}\right), \qquad (16.14)$$

and

$$\mathrm{Cov}(X'_r, X'_s) \doteq \frac{1}{n}\pi^2\left(\frac{r}{n}\right)\left(1 - \frac{s}{n}\right)\mathrm{cosec}^2\left(\frac{r\pi}{n}\right)\mathrm{cosec}^2\left(\frac{s\pi}{n}\right). \quad (16.15)$$

Writing, for example,

$$x(u) = F_X^{-1}(u) = \cot\{\pi(1 - u)\} = \sum_{i=0}^{\infty}(-1)^i 2^{2i}\pi^{2i-1}\frac{B_{2i}(1 - u)^{2i-1}}{(2i)!}$$

$$= \sum_{i=0}^{\infty} b_{2i}(1 - u)^{2i-1}, \qquad (16.16)$$

Hall (1978) has derived an asymptotic formula for mean of X'_{n-r} (for $1 \le r \le n - 2$) as

$$E[X'_{n-r}] = \frac{n!}{r!}\sum_{i=0}^{N} b_{2i}\frac{(r + 2i - 1)!}{(n + 2i - 1)!} + \varepsilon_N; \qquad (16.17)$$

the error term ε_N is always negative and is bounded by

$$|\varepsilon_N| \le b\frac{n!(r + 2N + 1)!}{r!(n + 2N)!(n - r - 1)}\left\{1 - \frac{r + 2N + 2}{2(n + 2N + 1)}\right\}$$

where $b = \sup_{i \ge N+1}|b_{2i}|$. By using inequalities for B_{2i} [see Abramowitz and Stegun (1965, p. 208)], Hall has also shown that

$$\frac{2}{\pi} \le -b_{2i} \le \frac{2}{\pi(1 - 2^{1-2i})} \qquad \text{for } i \ge 1,$$

so that $b_{2i} \to -2/\pi$ as $i \to \infty$, and the error $|b_{2i} + 2/\pi|$ is less than 4×10^{-7} if $i > 10$. Approximate expressions for the second single moment and the product moment similar to that in (16.17) have also been derived by Hall (1978). Mention should also be made of Joshi (1969), who has modified the orthogonal inverse expansion method of Sugiura (1962) to derive bounds and approximations for moments of order statistics from the Cauchy distribution.

The asymptotic distribution of the largest order statistic X'_n can be obtained from the general formulas of Fréchet (1927), Fisher and Tippett (1928), and von Mises (1936) [see Galambos (1987)]. The Cauchy distribution belongs to the class for which

$$\lim_{x \to \infty} \{1 - F_X(x)\}x^2 = A > 0. \tag{16.18}$$

The asymptotic distribution of the extremal quotient ($= X'_n / |X'_1|$) has been discussed by Gumbel and Keeney (1950).

4 METHODS OF INFERENCE

It is not possible to use the method of moments to estimate θ and λ by equating sample and population first and second moments. (It may be remembered, incidentally, that it was shown in Section 2 of this chapter that the arithmetic mean, $n^{-1}\sum_{j=1}^{n} X_j = \overline{X}$, has the same distribution as any one of the X's. Thus $\overline{X}$ is no more informative than any single one of the X'_js, although one might feel it could be a useful estimator of θ.) It is possible to derive methods of estimation using moments of fractional order, but these are not used here.

4.1 Methods Based on Order Statistics

The simple form of cumulative distribution function (16.2) makes it possible to obtain simple estimators by equating population percentage points (quantiles) and sample estimators thereof. The $100p\%$ quantile X_p of the distribution satisfies the equation

$$\Pr[X \leq X_p] = p;$$

a convenient estimator from the sample is the rth-order statistic, X'_r, with $r = (n + 1)p$. We will denote this estimator by $\hat{X}_p$. The value of X_p depends on θ and λ.

If p_1 and p_2 are distinct numbers between 0 and 1, then the equations

$$\hat{X}_{p_j} = X_{p_j}, \qquad j = 1, 2,$$

lead to the estimators

$$\tilde{\lambda} = \left(\hat{X}_{p_1} - \hat{X}_{p_2} \right)\left(\cot \pi p_2 - \cot \pi p_1 \right)^{-1}, \tag{16.19}$$

$$\tilde{\theta} = \left(\hat{X}_{p_1} \cot \pi p_2 - \hat{X}_{p_2} \cot \pi p_1 \right)\left(\cot \pi p_2 - \cot \pi p_1 \right)^{-1}. \tag{16.20}$$

[Note that $\tan[\pi(p - \tfrac{1}{2})] = -\cot \pi p$ as in (16.13).]
  In the symmetrical case with $p_1 = p > 1/2 > 1 - p = p_2$,

$$\tilde{\lambda} = \tfrac{1}{2}\left(\hat{X}_p - \hat{X}_{1-p} \right)\tan[\pi(1 - p)], \tag{16.21}$$

$$\tilde{\theta} = \tfrac{1}{2}\left(\hat{X}_p + \hat{X}_{1-p} \right). \tag{16.22}$$

If the available observations can be represented by values of n independent random variables, $X_1, X_2, \ldots, X_n$, each with probability density function of form (16.1), and $\hat{X}_p, \hat{X}_{1-p}$ are appropriate order statistics [so that $E(\hat{X}_p) \doteq X_p$, $E(\hat{X}_{1-p}) \doteq X_{1-p}$], then

$$n \operatorname{Var}(\tilde{\lambda}) \doteq \lambda^2 \left[2\pi^2(1 - p)(2p - 1) \right]\operatorname{cosec}^2 2\pi p, \tag{16.23}$$

$$n \operatorname{Var}(\tilde{\theta}) \doteq \lambda^2 \left[\tfrac{1}{2}\pi^2(1 - p) \right]\operatorname{cosec}^4 \pi p. \tag{16.24}$$

The estimators $\tilde{\lambda}$ and $\tilde{\theta}$ are uncorrelated; $\tilde{\lambda}$ is an unbiased estimator of θ. The approximate variance [by (16.24)] of $\tilde{\theta}$ is minimized by taking $p = 55.65\%$; the corresponding value of the right-hand side of (16.24) is $2.33\lambda^2$. The Cramér-Rao lower bound for the variance of an unbiased estimator of θ is $2\lambda^2/n$, so the asymptotic efficiency of $\tilde{\theta}$ is 86%. (The quoted lower bound applies whether λ is known or not.) Putting $r = n/2$ in (16.14), we see that the variance of the median is approximately $\tfrac{1}{4}\pi^2\lambda^2/n = 2.47\lambda^2/n$. The median is an unbiased estimator of θ with asymptotic efficiency 81%.

  Rothenberg, Fisher, and Tilanus (1964) obtained an approximation to the variance of the symmetrical censored arithmetic mean, $\tilde{\theta}'$, say, where the censoring causes omission of the $\tfrac{1}{2}pn$ (approx.) lowest and $\tfrac{1}{2}pn$ (approx.) highest observed values in a random sample of size n. The formula they obtained is

$$n \operatorname{Var}(\tilde{\theta}') \doteq \lambda^2 \Big[p(1 - p)^{-2}\cot^2(\tfrac{1}{2}\pi p)$$
$$+ 2\pi^{-1}(1 - p)^{-2}\cot(\tfrac{1}{2}\pi p) - (1 - p)^{-1} \Big]. \tag{16.25}$$

The statistic $\tilde{\theta}'$ is an unbiased estimator of θ. By taking $p \doteq 0.76$ the right-hand side of (16.25) attains its minimum value $2.28\lambda^2$. The asymptotic efficiency of the estimator $\tilde{\theta}'$ is then 88%. [See also Bloch (1966).]

This "optimum" $\tilde{\theta}'$ uses the values of only the central 24% of the observations, while the "best" pair of order statistics to use in $\tilde{\theta}$ are separated by only the central 11% of observations. For the normal distribution (see Chapter 13, Section 8) considerably larger central groups of observations give "optimum" results. This reflects the greater variability in the "tails" of the Cauchy distribution.

Bloch (1966) discussed the asymptotic best linear unbiased estimation of θ based on k optimally selected quantiles, by following the method of Ogawa (1951). Let $X'_{n_1} \le X'_{n_2} \le \cdots \le X'_{n_k}$ be the k-order statistics of interest, and let $X_{\xi_1}, X_{\xi_2}, \dots, X_{\xi_k}$ be the corresponding population quantiles, that is, $F(X_{\xi_i}) = \xi_i$ for $i = 1, 2, \dots, k$. It is then known from Ogawa (1951) that the asymptotic relative efficiency of the BLUE of θ is

$$\varepsilon = \frac{K_1}{E\left[\{p'(X)/p(X)\}^2\right]} = 2K_1, \qquad (16.26)$$

where

$$K_1 = \sum_{i=1}^{k+1} \frac{(p_i - p_{i-1})^2}{\xi_i - \xi_{i-1}}, \qquad (16.27)$$

with $p_i = p_X(X_{\xi_i})$, $p_0 = p_{k+1} = 0$, $\xi_0 = 0$, $\xi_{k+1} = 1$. The optimal value of $\boldsymbol{\xi} = (\xi_1, \dots, \xi_k)$ may be obtained by minimizing K_1 in (16.27) with respect to $\xi_1, \dots, \xi_k$. The corresponding ABLUE is then given by

$$\theta^{**} = \sum_{i=1}^{k} a_i X'_{n_i}, \qquad (16.28)$$

$$a_i = \frac{p_i}{K_1}\left\{\left(\frac{p_i - p_{i-1}}{\xi_i - \xi_{i-1}}\right) - \left(\frac{p_{i+1} - p_i}{\xi_{i+1} - \xi_i}\right)\right\}, \qquad i = 1, 2, \dots, k. \quad (16.29)$$

We note here that Kariya, Sinha, and Subramanyam (1981) derived nearly efficient estimators for θ based on 3 or 5 optimally determined order statistics. Chan (1970) discussed the asymptotic best linear unbiased estimation of the parameters θ and λ based on k optimally selected order statistics, when one or both parameters are unknown. For example, by assuming the spacing to be symmetric, he determined the optimal spacings for the estimation of θ when λ is known. Chan also showed that the asymptotic best linear unbiased estimates of θ and λ, when both are unknown, determined by the spacings $\xi_i = i/(k + 1)$, $i = 1, 2, \dots, k$, have joint asymptotic relative efficiency of at least 65% for all $k > 2$.

Balmer, Boulton, and Sack (1974) showed however that in the case of the estimation of θ, when λ is known, optimal k-spacings (corresponding to the maximum asymptotic relative efficiency) are asymmetric when $k = 4m - 1$, which is in contrast to Chan's assumption of symmetric spacing. By combining a dynamic programming procedure and Newton-Raphson iteration methods, they determined accurately the optimal spacings for $k = 3$, 7, and 11. They also proved that the spacings $\xi_i = i/(k + 1)$, $i = 1, 2, \ldots, k$, is in fact the optimal spacing for the joint estimation of θ and λ; see also Chan, Chan and Mead (1973) and Cane (1974). Furthermore the joint asymptotic relative efficiency of the ABLUEs θ^{**} and λ^{**}, based on the aforementioned k optimal spacings, is given by

$$\text{JARE} = \left\{ \frac{k + 1}{\pi} \sin\left(\frac{\pi}{k + 1} \right) \right\}^4, \tag{16.30}$$

and θ^{**} and λ^{**} are given by

$$\theta^{**} = \sum_{i=1}^{k} a_i X'_{n_i} \quad \text{and} \quad \lambda^{**} = \sum_{i=1}^{k} b_i X'_{n_i}, \tag{16.31}$$

where

$$a_i = -\frac{4}{k + 1} \sin^2\left(\frac{\pi i}{k + 1} \right) \cos\left(\frac{2\pi i}{k + 1} \right),$$

$$b_i = -\frac{4}{k + 1} \sin^2\left(\frac{\pi i}{k + 1} \right) \sin\left(\frac{2\pi i}{k + 1} \right).$$

[For details, see Sarhan and Greenberg (1962).] Tests of hypotheses concerning $\theta = \theta_0$ and $(\theta, \lambda) = (\theta_0, \lambda_0)$ have been considered by Chan and Cheng (1971) and Saleh, Hassanein, and Brown (1985), respectively, based on these k-optimal ABLUEs.

Among estimators of form $\sum_{j=1}^{n} \alpha_j X'_j$ (i.e., linear functions of order statistics) the following values of α_j give asymptotically optimum estimators:

For θ,

$$\alpha_j = \frac{\sin 4\pi \left(j(n + 1)^{-1} - \frac{1}{2} \right)}{\tan \pi \left(j(n + 1)^{-1} - \frac{1}{2} \right)}.$$

For λ,

$$\alpha_j = \frac{8 \tan \pi \left(j(n + 1)^{-1} - \frac{1}{2} \right)}{\sec^4 \pi \left(j(n + 1)^{-1} - \frac{1}{2} \right)}.$$

[See Chernoff, Gastwirth and Johns (1967).] Barnett (1966b) determined the

BLUE of θ given by

$$\theta^* = \frac{\mathbf{1}^T \Sigma^{-1} \mathbf{X}'}{\mathbf{1}^T \Sigma^{-1} \mathbf{1}} = \sum_{i=3}^{n-2} c_i X_i', \tag{16.32}$$

and its variance given by

$$\frac{\text{Var}(\theta^*)}{\lambda^2} = \frac{1}{\mathbf{1}^T \Sigma^{-1} \mathbf{1}}, \tag{16.33}$$

where $\mathbf{1}^T = (1, 1, \ldots, 1)$ and Σ is the variance-covariance matrix of $n - 4$ central order statistics. He has tabulated the coefficients c_i in (16.32) and $\text{Var}(\theta^*)$ in (16.33) for sample sizes up to 20. Weiss and Howlader (1988) extended Barnett's work by tabulating the BLUEs of θ and λ for sample sizes up to 20 (since θ^* and λ^* are uncorrelated due to the symmetry of the Cauchy distribution, they had to present tables only for λ^*). Weiss and Howlader (1988) also presented a simple linear estimator for λ to be

$$\tilde{\lambda} = \sum_{i \in I_\lambda} c_i X_i' = -\frac{1}{r} \sum_{i \in I_\lambda} \tan(\pi \tilde{q}_i) X_i', \tag{16.34}$$

where $r = n - 4$ for even n ($I_\lambda = \{3, 4, \ldots, n - 2\}$) and $r = n - 5$ for odd n ($I_\lambda = \{3, 4, \ldots, (n - 1)/2, (n + 3)/2, \ldots, n - 2\}$) and $\tilde{q}_i = (i - 2 + \frac{1}{2})/(n - 1)$ for $i < n/2$ with $\tilde{q}_{n-i+1} = 1 - \tilde{q}_i$. They show that the estimator $\tilde{\lambda}$ in (16.34) has almost minimum mean squared error even for small samples. Some simplified linear estimators for θ and λ based on quasi-midranges and quasi-ranges, respectively, were proposed earlier by Raghunandanan and Srinivasan (1972) for the complete sample case as well as in the case of symmetrically Type-II censored samples [also see Haq (1971)].

4.2 Maximum Likelihood Inference

Based on n observations $x_1, x_2, \ldots, x_n$ from the Cauchy density in (16.1), we have the likelihood equations for θ and λ to be

$$\frac{\partial \log L}{\partial \theta} = \sum_{i=1}^{n} \frac{2(x_i - \theta)}{\lambda^2 + (x_i - \theta)^2} = 0 \tag{16.35}$$

and

$$\frac{\partial \log L}{\partial \lambda} = \frac{n}{\lambda} - \sum_{i=1}^{n} \frac{2\lambda}{\lambda^2 + (x_i - \theta)^2} = 0. \tag{16.36}$$

For the problem of estimating θ when λ is known, the likelihood function is occasionally multimodal. In fact, for a sufficiently small λ, it will have n

local maxima—one close to each of the n observations [see Edwards (1972)]. In general, the MLE of θ when λ is fixed is a root of a polynomial of degree $2n - 1$ [refer also to Reeds (1985)]. Barnett (1966a) discussed a numerical method of finding a global maximum of θ. For this global optimization problem, given an easily derived upper bound on the second derivative of the negative of the log-likelihood function, Wingo (1983) prescribed the usage of Brent's (1973) univariate numerical global optimization method to locate the absolute maximum among several relative maxima, instead of performing an exhaustive search over the entire real line. In this case, even though the likelihood equation for θ has multiple roots, Bai and Fu (1987) proved that the maximum likelihood estimator $\hat{\theta}$ (the global maximum) converges to θ exponentially and is an asymptotically efficient estimator in the Bahadur sense. Haas, Bain, and Antle (1970) developed numerical methods for solving the likelihood equations in (16.35) and (16.36). Hinkley (1978) conducted large-scale simulations (based on 40,000 samples of size 20 and 20,000 samples of size 40) and very rarely observed the Newton-Raphson procedure to fail to converge. Copas (1975) and Gabrielsen (1982) showed that the joint likelihood function for θ and λ is unimodal. Hence the two-parameter situation is easier to handle than just the location-parameter case in that even simple schemes, such as the steepest ascent, will enable to determine the MLEs of θ and λ.

For a sample of size 2, Haas, Bain, and Antle (1970) showed that the MLE of (θ, λ) is not unique and that all MLEs $(\hat{\theta}, \hat{\lambda})$ for observations x_1 and x_2 with $x_1 < x_2$ are given by

$$x_1 \le \hat{\theta} \le x_2 \quad \text{and} \quad \hat{\lambda}^2 = (\hat{\theta} - x_1)(x_2 - \hat{\theta}). \tag{16.37}$$

For an arbitrary sample size, Copas (1975) showed that the only time the MLE of (θ, λ) is not unique is when half the observations are at a point x_1 and the other half at a point x_2, when the MLE $(\hat{\theta}, \hat{\lambda})$ is again given by (16.37). When otherwise half or more of the observations are at a point x_1, the maximum of the likelihood function occurs at $\hat{\theta} = x_1$ and $\hat{\lambda} = 0$, a singular point for the likelihood equations in (16.35) and (16.36). However, in all other cases, the MLE of (θ, λ) is the unique root of the likelihood equations in (16.35) and (16.36). Curiously enough, Ferguson (1978) showed that closed-form expressions exist for the MLE $(\hat{\theta}, \hat{\lambda})$ for samples of size 3 and 4.

Based on a sample of size 3 denoted by x_1, x_2 and x_3 with $x_1 < x_2 < x_3$, Ferguson (1978) presented the MLEs of θ and λ as

$$\hat{\theta} = \frac{x_1(x_3 - x_2)^2 + x_2(x_3 - x_1)^2 + x_3(x_2 - x_1)^2}{(x_3 - x_2)^2 + (x_3 - x_1)^2 + (x_2 - x_1)^2}, \tag{16.38}$$

$$\hat{\lambda} = \frac{\sqrt{3}(x_3 - x_2)(x_3 - x_1)(x_2 - x_1)}{(x_3 - x_2)^2 + (x_3 - x_1)^2 + (x_2 - x_1)^2}. \tag{16.39}$$

Observe that $\hat{\theta}$ in (16.38) is a weighted average of the three observations with weights proportional to the squares of the differences of the other two. A direct substitution of the expressions of $\hat{\theta}$ and $\hat{\lambda}$ in (16.38) and (16.39) into formulas (16.35) and (16.36) will readily reveal that these estimates are in fact the MLEs (due to the uniqueness of the MLE established by Copas).

Based on a sample of size 4 denoted by x_1, x_2, x_3, and x_4 with $x_1 < x_2 < x_3 < x_4$, Ferguson (1978) similarly presented the MLEs of θ and λ as

$$\hat{\theta} = \frac{x_2 x_4 - x_1 x_3}{x_4 - x_3 + x_2 - x_1},$$ (16.40)

$$\hat{\lambda}^2 = \frac{(x_4 - x_3)(x_3 - x_2)(x_2 - x_1)(x_4 - x_1)}{(x_4 - x_3 + x_2 - x_1)^2}.$$ (16.41)

Observe that $\hat{\theta}$ in (16.40) is a weighted average of the two middle-order statistics with weights proportional to the distance of the opposite outlier from its neighbor, if we write $\hat{\theta} = \alpha x_2 + (1 - \alpha)x_3$, where $\alpha = (x_4 - x_3)/\{(x_4 - x_3) + (x_2 - x_1)\}$. Once again, a direct substitution of the expressions of $\hat{\theta}$ and $\hat{\lambda}$ in (16.40) and (16.41) into formulas (16.35) and (16.36) will readily reveal that $(\hat{\theta}, \hat{\lambda})$ is in fact the MLE.

The fact that $\hat{\theta}$ is a weighted average of the sample values can be established easily for any sample size n. To do this, let us first rewrite the likelihood equations in (16.35) and (16.36) as

$$\sum_{i=1}^{n} \frac{1}{1 + \left\{(x_i - \hat{\theta})/\hat{\lambda}\right\}^2} = \frac{n}{2},$$ (16.42)

$$\sum_{i=1}^{n} \frac{x_i}{1 + \left\{(x_i - \hat{\theta})/\hat{\lambda}\right\}^2} = \frac{n}{2}\hat{\theta}.$$ (16.43)

Upon dividing (16.43) by (16.42), we see immediately that $\hat{\theta}$ is a weighted average of the sample observations x_i (a result we have seen already explicitly for $n = 3$ and 4).

An interesting property of θ in (16.38) and (16.40) is that outlying observations have a negative effect. To be precise, changes in the values of observations in the lower and upper extreme values have the effect of pushing the estimate in the opposite direction. As Ferguson pointed out, this occurs even for equally spaced observations in a sample of size 3. To see this effect, let $x_1 = -1$, $x_2 = 0$ and $x_3 > 0$ in (16.38) so that

$$\hat{\theta}(x_3) = \frac{x_3(1 - x_3)}{2(x_3^2 + x_3 + 1)}.$$

At $x_3 = 1$, we have $\hat{\theta}(x_3) = 0$ as it should be, due to the symmetry, but for $x_3 > 1$ we see that $\hat{\theta}(x_3) < 0$. In fact $(d/dx_3)\hat{\theta}(x_3) < 0$ for all $x_3 > (\sqrt{3} - 1)/2 = 0.366\ldots$. This interesting property had been noticed earlier by

Barnett (1966b), Bloch (1966), Chan (1970) and Balmer, Boulton, and Sack (1974) in the best linear unbiased estimator of θ.

Quite recently McCullagh (1992a, b, c; 1993) established some remarkable results on properties of the MLEs. First of all, he considers n i.i.d. variables from a Cauchy population with density function

$$p_X(x;\eta) = \frac{|\lambda|}{\pi|x - \eta|^2} \tag{16.44}$$

in which x is a real number and $\eta = \theta + i\lambda$ is the Cauchy parameter represented as a complex number. The parameter space here should be taken as the complex plane in which complex conjugate pairs of points are identified; this set is isomorphic to the upper half-plane. He then proved that for any function $h(\cdot)$ analytic on the upper half-plane, $h(\hat\eta)$ is unbiased for $h(\eta)$ provided that the expectation exists. For example, it immediately implies that $E(\hat\eta) = \eta$ for $n \geq 3$, $E(\hat\eta^k) = \eta^k$ for integer $k \leq n - 2$, $E(\log \hat\eta) = \log \eta$ for $n \geq 3$, and so on. Furthermore the Cauchy family has the property (as mentioned earlier in Section 2) that if $X \to \text{Cauchy}(\eta)$, then

$$\frac{aX + b}{cX + d} \to \text{Cauchy}\left(\frac{a\eta + b}{c\eta + d}\right) \tag{16.45}$$

for all real numbers a, b, c, and d with $ad - bc \neq 0$. That is, the Cauchy family of distributions is closed under the action of the real Möbius group G, and the parameter is equivariant. Let $T = T_1 + iT_2$ be the MLE of $\eta = \theta + i\lambda$ based on sample observations x_i, $i = 1, \ldots, n$. It then immediately follows from (16.45) that

$$T\left(\frac{ax + b}{cx + d}\right) = \frac{aT(x) + b}{cT(x) + d}. \tag{16.46}$$

That is, if the components of x are transformed by $h \in G$, the MLE derived from hx will be the same as the transformation h applied to the MLE based on x. In other words, T is equivariant under G. Upon exploiting this equivariance feature, McCullagh proved that the joint density of T has the form

$$p(t_1, t_2; \eta) = \frac{1}{4\pi t_2^2} p_n(\chi), \tag{16.47}$$

where $\chi = |t - \eta|^2/(4t_2\lambda)$. He derived explicit expressions for $p_3(\chi)$ and $p_4(\chi)$ and also the asymptotic large-sample limit. Note that in this derivation of the distribution of T, explicit closed-form expression of T is not required although the closed form expressions of the MLEs given earlier for $n = 3$ and 4 offer no help in obtaining the joint density function. It is the form of the density in (16.47) that is used by McCullagh to establish the remarkable property that $E[h(T)] = h(\eta)$ if $h(\cdot)$ is harmonic and the expectation is finite. As an immediate consequence, of course, we have the result that both components of the MLE are unbiased for $n \geq 3$.

Further, by using the explicit expressions of

$$p_3(x) = \frac{3\sqrt{3}}{\pi\{1 + 3x(1 + x)\}},$$

$$p_4(x) = \frac{12\log(1 + 2x)}{\pi^2 x(x + 1)(2x + 1)},$$

McCullagh derived fairly simple forms for the marginal densities of $T_1 = \hat{\theta}$ and $T_2 = \hat{\lambda}$ for sample sizes 3 and 4. For example, when $n = 3$,

$$p_{T_1}(t_1) = \frac{3}{\pi\left(1 + t_1^2\right)\sqrt{4 + 3t_1^2}},$$

$$p_{T_2}(t_2) = \frac{2\sqrt{3}}{\pi\left(1 + t_2^2\right)^{3/2}} \times \frac{\sqrt{2}}{\sqrt{1 + \varepsilon}\left\{1 + \sqrt{1 + \varepsilon}\right\}^{1/2}}, \qquad t_2 > 0,$$

where $\varepsilon = 4t_2^2/\{3(1 + t_2^2)^2\}$. Interestingly the density of T_2 given above is approximately the positive half of a Student's t-distribution on two degrees of freedom (see Chapter 28), since the second factor is bounded between 0.83 and 1.0 as ε is between 0 and $\frac{1}{3}$.

4.3 Conditional Inference

Suppose that $X_1, X_2, \ldots, X_n$, are i.i.d. random variables from Cauchy population with density as in (16.1). Fisher (1934) observed that the configuration statistic **A** with components $a_i = (X_i - \hat{\theta})/\hat{\lambda}$ is ancillary in the sense that the distribution of **A** does not depend on the parameter (θ, λ) as is evident from the invariance of **A** under location and scale transformations. Here $\hat{\theta}$ and $\hat{\lambda}$ need not be the maximum likelihood estimators. In fact, any pair of equivariant estimators for (θ, λ) will suffice. The basic philosophy behind this method of inference [Fisher (1934)] is that the precision with which the parameters θ and λ are estimated depends on the configuration observed, some configurations being more informative than others; as a result the appropriate distribution that should be used for inferential purposes is the conditional distribution of $(\hat{\theta}, \hat{\lambda})$ given the observed value of the configuration ancillary.

    Lawless (1972) discussed conditional confidence interval estimation procedures for θ and λ. He discussed three cases separately: (1) θ unknown, λ known; (2) θ known, λ unknown; and (3) both θ and λ unknown. After discussing the computational aspects of using this conditional inference approach, he made some numerical comparisons of the conditional confidence interval procedures and the unconditional confidence interval procedures. We will describe this method of inference by considering case (3). First, we have the conditional joint density function of $\hat{\theta}$ and $\hat{\lambda}$, given $\mathbf{A} = \mathbf{a}$, as

$$p\left(\hat{\theta}, \hat{\lambda} \mid a_1, \ldots, a_{n-2}\right) = C \frac{\hat{\lambda}^{n-2}}{\lambda^n} \prod_{i=1}^{n} \left[\frac{1}{\pi \left\{ 1 + \left[(\hat{\theta} - \theta + a_i \hat{\lambda})/\lambda \right]^2 \right\}} \right], \quad (16.48)$$

where $C = C(a_1, \ldots, a_{n-2})$ is the nomalizing constant. It is clear that $z_1 = (\hat{\theta} - \theta)/\lambda$ and $z_2 = \hat{\lambda}/\lambda$ are pivotal quantities, with joint conditional density

$$p\left(z_1, z_2 \mid a_1, \ldots, a_{n-2}\right) = C z_2^{n-2} \prod_{i=1}^{n} \left[\frac{1}{\pi \left\{ 1 + (z_1 + a_i z_2)^2 \right\}} \right]. \quad (16.49)$$

From (16.49) it is of interest to consider the joint conditional density of $z_3 = z_1/z_2$ and z_2, which is given by

$$p\left(z_3, z_2 \mid a_1, \ldots, a_{n-2}\right) = C z_2^{n-1} \prod_{i=1}^{n} \left[\frac{1}{\pi \left\{ 1 + z_2^2 (a_i + z_3)^2 \right\}} \right]. \quad (16.50)$$

To make inferences about θ when λ is unknown, z_2 should be integrated out of (16.50) to obtain the marginal distribution of $z_3 = (\hat{\theta} - \theta)/\hat{\lambda}$, conditional on the observed $a_1, \ldots, a_{n-2}$. Similarly, to make inferences about λ when θ is unknown, z_3 should be integrated out of (16.50) to obtain the marginal distribution of $z_2 = \hat{\lambda}/\lambda$, conditional on the observed $a_1, \ldots, a_{n-2}$. Computing probabilities for the marginal distributions of z_3 and z_2 is too complicated and quite time-consuming. So Lawless (1972) suggested either looking at a plot of contours of (16.50) or at plots of the conditional densities $p(z_3 \mid z_2)$ and $p(z_2 \mid z_3)$ for a few selected values of the conditioning variable. Recently McCullagh (1992a, b) gave a nice discussion on the choice of ancillary in reference to the Cauchy case. He specifically showed that the configuration ancillary is not unique and that the choice of ancillary has some effect on conditional probability calculations. For moderate deviations he showed that the choice of ancillary has very little effect on probability calculations with the asymptotic effect being at most of order $O_p(n^{-1})$. In case of large deviations, the relative or logarithmic difference between the two conditional densities is $O_p(n^{1/2})$ and unbounded in probability as $n \to \infty$. McCullagh (1992b) showed that there exist moderate deviation approximations with absolute error $O(n^{-1})$ that are independent of the choice of ancillary, given that $(\hat{\eta}, \mathbf{A})$ is sufficient.

4.4 Bayesian Inference

Starting with the "vague prior"

$$g(\theta, \lambda) \propto \frac{1}{\lambda} \quad (16.51)$$

and combining it with the likelihood function based on the sample $x_1, \ldots, x_n$, we obtain the joint posterior density of θ and λ as

$$\Pi(\theta, \lambda | \mathbf{x}) \propto \lambda^{n-1} \prod_{i=1}^{n} \frac{1}{\lambda^2 + (x_i - \theta)^2}. \tag{16.52}$$

Then, the Bayesian estimators of θ and λ, under the squared-error loss function, are the posterior means given by

$$\hat{\theta}_B = E(\theta | \mathbf{x}) = \int \int \theta \Pi(\theta, \lambda | \mathbf{x}) d(\theta, \lambda)$$

$$= \frac{\int \int \theta \lambda^{n-1} \prod_{i=1}^{n} \left\{ \lambda^2 + (x_i - \theta)^2 \right\}^{-1} d(\theta, \lambda)}{\int \int \lambda^{n-1} \prod_{i=1}^{n} \left\{ \lambda^2 + (x_i - \theta)^2 \right\}^{-1} d(\theta, \lambda)}, \tag{16.53}$$

$$\hat{\lambda}_B = \frac{\int \int \lambda^{n} \prod_{i=1}^{n} \left\{ \lambda^2 + (x_i - \theta)^2 \right\}^{-1} d(\theta, \lambda)}{\int \int \lambda^{n-1} \prod_{i=1}^{n} \left\{ \lambda^2 + (x_i - \theta)^2 \right\}^{-1} d(\theta, \lambda)}. \tag{16.54}$$

Note that these estimators, unlike in the case of a location parameter θ alone, do not minimize the squared-error loss among equivariant estimates. One needs to start with the prior density $d\theta \, d\lambda / \lambda^3$ in order to derive such estimates.

In deriving a best invariant test for the normal distribution against a Cauchy alternative, Franck (1981) evaluated integrals similar to those in (16.53) and (16.54). Spiegelhalter (1985) obtained closed-form expressions for the Bayesian estimators of θ and λ which, for odd values of n larger than 3, are

$$\hat{\theta}_B = \frac{1}{2} \frac{\Sigma \Sigma_{i<j} w_{ij} (x_i + x_j) \log d_{ij}}{\Sigma \Sigma_{i<j} w_{ij} \log d_{ij}}, \tag{16.55}$$

$$\hat{\lambda}_B = \frac{\pi}{4} \frac{\Sigma \Sigma_{i<j} w_{ij} d_{ij}}{\Sigma \Sigma_{i<j} w_{ij} \log d_{ij}}, \tag{16.56}$$

where

$$d_{ij} = |x_i - x_j|,$$

$$w_{ij} = \frac{(-1)^{i+j} d_{ij}^{m-1}}{\prod_{k=1}^{n} d_{ik} d_{jk}}.$$

Refer to Spiegelhalter (1985) for similar expressions for the Bayesian estimators of θ and λ for even values of n.

Howlader and Weiss (1988a) pointed out that the exact formulas given above are difficult to compute and require great computational precision as these estimates are very unstable and often blow up in values. Through an empirical Monte Carlo study that they carried out, they also observed that the exact method often grossly overestimated λ, especially for small values of n. It is for this reason that Howlader and Weiss (1988a) derived some approximate Bayesian estimators by using a method of approximating ratios of integrals [e.g., (16.53) and (16.54)] due to Lindley (1980). They then showed that these approximate Bayesian estimators perform very well in comparison to the MLEs. It should be mentioned here that Lindley's (1980) expansion, a method similar to the one employed by Hinkley (1978) in correcting for the bias of the MLE and the method of Efron (1975) on second-order efficiency, has been compared by Howlader and Weiss (1987) to another method of approximating ratios of integrals due to Tierney and Kadane (1986). In comparing the performance of the two methods under the Cauchy model, Howlader and Weiss (1987) observed that the former (even though is computationally more tedious) remains quite stable even for small sample sizes (unlike the latter). It is of interest to add here that the Bayesian estimation of the reliability function $R_X(t)$ of the Cauchy density in (16.1) has been discussed by Howlader and Weiss (1988b).

4.5 Other Developments in Inference

Efron and Hinkley (1978), when making an excellent assessment of the normal approximations to the distribution of the maximum likelihood estimator in one-parameter families (with the variance approximation by either the expected total Fisher information or the observed information), used Cauchy distribution as an example. Barnard (1974), in response to an article by Bailey (1973), pointed out that in the problem of data reduction one's interest is typically to reduce a number of independent observations whose distribution involves an unknown parameter θ to an equivalent single observation (ESO), whose distribution also involves θ, in such a way that the information about θ in the set of observations is exactly (or at least nearly) equivalent to that in the ESO. When θ is the location parameter, it so happens that the form of the distribution of the ESO is the same (apart from a scaling factor) as that of the original observations in the case of normal and uniform distributions. However, Barnard (1974) mentioned that this property is rather exceptional and cited the Cauchy distribution in which case for $n = 2$ the distribution differs from Cauchy in being bimodal.

Higgins and Tichenor (1977) proposed "window estimates" for θ and λ, and also established the consistency and asymptotic normality of these estimates. They further showed that the asymptotic distribution of these window estimates is the same as that of the MLEs. Through an empirical comparative study, Higgins and Tichenor (1978) showed that the window estimate of θ is nearly optimal for $n \geq 40$ and is as efficient as many other estimates for $n \geq 20$; further they displayed that the window estimate of λ is nearly optimal even for samples of size 10.

Kadiyala and Murthy (1977) considered the problem of estimation of a linear regression with Cauchy disturbances. They compared the performance of the ML method and the minimum sum of absolute errors method for this specific estimation problem.

Koutrouvelis (1982) discussed a method of estimation of θ and λ using the empirical characteristic function. Interestingly enough, the determination of optimum values at which the empirical characteristic function needs to be evaluated for this method of estimation reduces to the determination of the asymptotically optimum quantiles for the parameter estimation of an exponential distribution by linear functions of order statistics; the latter has a known solution by Ogawa (1960) (also see Chapter 19). These estimators are shown to be asymptotically independent and normally distributed, achieving high asymptotic efficiencies. Feuerverger and McDunnough (1981a, b) also considered this empirical characteristic function approach but restricted it to the determination of the optimum uniform spacing between the points.

Beckman and Johnson (1987) discussed the fitting of the Cauchy distribution (Student t-distribution, more generally) to grouped experimental data by means of a grouped MLE approach. A final mention has to be made to the asymptotically optimal tests proposed by Philippou (1988) based on the concept of contiguity.

5 GENESIS AND APPLICATIONS

The standard Cauchy distribution with $\theta = 0$, $\lambda = 1$, as in (16.3), is the distribution of central t with one degree of freedom (Chapter 28). It is thus the distribution of the ratio U/V, where U and V are independent unit normal variables. An interesting application of this distributional result was illustrated by Stigler (1989); he used this result to derive an explicit expression for bivariate normal orthant probabilities, namely $\Pr[Z_1 \le 0, Z_2 \le 0]$, where $(Z_1, Z_2)^T$ has a standard bivariate normal distribution with correlation coefficient ρ, as follows. Define $V = (Z_2 - \rho Z_1)/\sqrt{1 - \rho^2}$; then V is standard normal and independent of Z_1 (since the covariance is 0). Noting now that $Z_2 \le 0$ is equivalent to $V \le -cZ_1$, where $c = \rho/\sqrt{1 - \rho^2}$, we have

$$\Pr[Z_1 \le 0, Z_2 \le 0] = \Pr[Z_1 \le 0, V \le -cZ_1]$$

$$= \Pr[Z_1 \le 0, V \le c|Z_1|].$$

Now by the symmetry of the distribution of Z_1 (and since $\Pr[Z_1 = 0] = 0$)

$$\Pr[Z_1 \le 0, V \le c|Z_1|] = \Pr[Z_1 > 0, V \le c|Z_1|]$$

$$= \frac{1}{2}\Pr[V \le c|Z_1|]$$

$$= \frac{1}{2}\Pr\left[\frac{V}{|Z_1|} \le c\right]$$

$$= \frac{1}{2}\left\{\frac{1}{2} + \frac{1}{\pi}\tan^{-1}c\right\}$$

$$= \frac{1}{2}\left\{\frac{1}{2} + \frac{1}{\pi}\sin^{-1}\rho\right\}$$

from the fact that $V/|Z_1|$ has a standard Cauchy distribution.

However, it should be mentioned that the common distribution of U and V need not be normal. For example, the ratio U/V has a Cauchy distribution if

$$p_U(u) = \sqrt{2}\,\pi^{-1}(1 + u^4)^{-1},$$
$$p_V(v) = \sqrt{2}\,\pi^{-1}(1 + v^4)^{-1} \tag{16.57}$$

[Laha (1959b)]. Other examples have been given by Laha (1959a), Mauldon (1956), Fox (1965), Steck (1958), Kotlarski (1960), and Roslonek (1970). Note that since U and V have identical distributions, V/U and U/V must have identical distributions. Hence, if X has the probability density function (16.3), so does X^{-1}.

Since, provided b_1 and b_2 are not both zero,

$$\left(b_1\left(\frac{V}{U}\right) + b_2\right)^{-1} = b_1(b_1^2 + b_2^2)^{-1}\left[\frac{b_1 U - b_2 V}{b_2 U + b_1 V}\right] + b_2(b_1^2 + b_2^2)^{-1},$$

it follows that if X has a Cauchy distribution, so does $(b_1 X + b_2)^{-1}$. This result was obtained in this way by Savage (1966), and by direct calculation by Menon (1966). [Note that if U and V are independent standard normal variables, so are $(b_1 U - b_2 V)(b_1^2 + b_2^2)^{-1/2}$ and $(b_2 U + b_1 V)(b_1^2 + b_2^2)^{-1/2}$ (see Chapter 13, Section 3).] When X and Y are independent normal variables with means μ_1 and μ_2 and variances σ_1^2 and σ_2^2, respectively, then Kamerud (1978) derived the density function of the random variable X/Y which, for the special case when $\mu_1 = \mu_2 = 0$ and $\sigma_1 = \sigma_2 = 1$, reduces to the standard Cauchy density.

Arnold and Brockett (1992) proved by simple arguments that, when $\mathbf{U} = (U_1, \ldots, U_n)^T$ has a spherically symmetric distribution, then for $i \neq j$, U_i/U_j has a standard Cauchy distribution [see also DeSilva (1979)]. More generally, Arnold and Brockett (1992) also established that when $\mathbf{V} = (V_1, \ldots, V_n)^T$ has an elliptically symmetric distribution, then for $i \neq j$, V_i/V_j has a general Cauchy distribution.

The Cauchy distribution is obtained as the limiting distribution of $n^{-1}\sum_{j=1}^{n} X_j^{-1}$ as $n \to \infty$, where $X_1, X_2, \ldots$ are independent identically distributed random variables with common density function $p_X(x)$ satisfying the conditions [Pitman and Williams (1967)]

1. $p_X(0) > 0$,
2. $p_X(x)$ continuous at $x = 0$,
3. $p_X(x)$ possesses left-hand and right-hand derivatives at $x = 0$.

Since the reciprocal of a Cauchy variable also has a Cauchy distribution, it follows that the limiting distribution of the harmonic mean, under these conditions, is Cauchy. Shapiro (1978) discussed the rate of convergence of distribution functions of sums of reciprocals of random variables to the

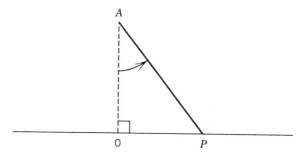

Figure 16.2 Geometric Description.

Cauchy distributions; the limit distributions in this case have also been discussed by Shapiro (1988).

The Cauchy distribution also arises in describing the distribution of the point of intersection P of a fixed straight line with another variable straight line, randomly oriented in two dimensions through a fixed point A. The distance OP of the point of intersection from the pivot (0) of the perpendicular from A to the fixed line has a Cauchy distribution with $\theta = 0$. The situation is represented diagrammatically in Figure 16.2. The angle $\angle OAP$ has a uniform (rectangular) distribution (Chapter 26) between $-\pi/2$ (corresponding to $OP = -\infty$) and $\pi/2$ (corresponding to $OP = +\infty$).

On the basis of this kind of model, the Cauchy distribution may be used to describe the distribution of points of impact of particles from a point-source A with a fixed straight line. It may be noted that if the space is of $s + 1$ dimensions instead of two dimensions, then the distance r, say, of the point of intersection with a fixed hyperplane from the foot of the perpendicular from A to the hyperplane is distributed as a multiple of central t with s degrees of freedom (Chapter 28).

Spitzer (1958) has shown that if $(R(t), \theta(t))$ are the polar coordinates of a random point under standard Brownian motion at time t (with $R(0) > 0$), then the distribution of $\{\Delta\theta(t)\}(\frac{1}{2}\log t)^{-1}$ tends to a Cauchy distribution as t tends to infinity, where $\{\Delta\theta(t)\}$ denotes the total *algebraic* angle turned through up to time t [so that $\{\Delta\theta(t)\} \equiv [\theta(t) - \theta(0)] \bmod 2\pi$]. Berman (1967) established that if τ is the first time that $R(\tau) = r_1$, then $\{\Delta\theta(\tau)\}|\log(r_1/R(0))|^{-1}$ has a Cauchy distribution with $\theta = 0$, $\lambda = |\log(r_1/R(0))|$.

Mukherjea, Nakassis, and Miyashita (1986) established an interesting property regarding the identification of Cauchy parameters by the distribution of the maximum random variable. Specifically, consider two sets of Cauchy distributions with cdf's given by

$$F_i(x) = \frac{1}{2} + \frac{1}{\pi}\tan^{-1}(a_i x), \qquad i = 1, 2, \ldots, n,$$

$$F_j^*(x) = \frac{1}{2} + \frac{1}{\pi}\tan^{-1}(b_j x), \qquad j = 1, 2, \ldots, m$$

for $-\infty < x < \infty$. Suppose that

$$F_1 F_2 \cdots F_n = F_1^* F_2^* \cdots F_m^*.$$

They then proved in this case that the a's are simply a rearrangement of the b's.

6 CHARACTERIZATIONS

Menon (1962, 1966), gave several characterizations of the Cauchy distribution. Among them is the following (from (1966)):

> If $X_1, X_2, \ldots, X_n$ are independent, identically distributed random variables, then the common distribution is a Cauchy distribution if and only if, for *any* set of real numbers $\{b_j \neq 0, a_j\}$ $(j = 1, 2, \ldots, n)$ there exist real numbers B ($\neq 0$) and A such that $\sum_{j=1}^{n}(a_j + b_j X_j)^{-1}$ has the same distribution as $A(B + X_1)^{-1}$.

Obretenov (1961) has given the following characterization:

> If X and Y are independent and have the same distribution, then a necessary and sufficient set of conditions for this to be a Cauchy distribution is (i) the characteristic function $\phi(t) = E(e^{itX}) = E(e^{itY})$ has a finite non-zero righthand derivative at $t = 0$, i.e., $\phi'_r(0) = \lambda \neq 0$, and (ii) for any pair of positive real numbers a, b there is a positive real number c (depending on a and b) such that $aX + bY$ has the same distribution as cX (or cY).

Note that the normal distribution satisfies condition (ii), but not condition (i), when a and b are chosen so that $E(aX + bY) = 0$. A further remarkable characterization, due to Williams (1969), is: "If $(1 + aX)(a - X)^{-1}$ has the same distribution as X, and $\pi^{-1} \tan^{-1} a$ is not a rational number, then X has a standard Cauchy distribution." A related characterization, also due to Williams (1969), is "If $(\theta^2 + \lambda^2)(2\theta - X)^{-1}$ has the same distribution as X and $\pi^{-1} \tan^{-1}(\theta/\lambda)$ is not a rational number, then X has a Cauchy distribution with parameters θ, λ."

As mentioned earlier in Section 5, it is known that if X and Y are independent standard normal variables, then X/Y has a standard Cauchy distribution. But the converse is not true [refer to Letac (1981) for some additional insight]. In an attempt to provide an additional condition that will guarantee the normality of X and Y, Ahsanullah and Hamedani (1988) prove the result that "if $\{\text{Min}(X, Y)\}^2$ and X/Y are distributed as chi-square with 1 degree of freedom and standard Cauchy, respectively, where X and Y are i.i.d. random variables from an absolutely continuous distribution, then X and Y are distributed as standard normal."

Knight (1976) proved the characterization result that a random variable X is of Cauchy type iff $(aX + b)/(cX + d)$ has a distribution with the same shape as X for every a, b, c, d, with $ad - bc \neq 0$. This result was extended by Knight and Meyer (1976) to the standard Cauchy distribution in the

Euclidean space $\mathbf{R}^n$. Dunau and Senateur (1987) provided a simple proof for this characterization result.

Kotlarski (1979) showed that if X has a standard Cauchy distribution, then so does $2X/(1 - X^2)$. By making use of some characterization results for the uniform distribution and the fact that $X = \tan U$, where U is a uniform random variable on the interval $[0, \pi)$, has a standard Cauchy distribution, Arnold (1979) established some characterization results for the Cauchy distribution. Specifically, he proved that if X and Y are independent absolutely continuous random variables, then any one of

1. Y and $(X + Y)/(1 - XY)$ are independent,
2. X and $(X + Y)/(1 - XY)$ are identically distributed,
3. X and $2X/(1 - X^2)$ are identically distributed,

imply that X has a standard Cauchy distribution. Norton (1983) then proved the following characterization result: For X and Y i.i.d. symmetric random variables, X has a standard Cauchy distribution iff

4. X and $X^2/(1 + X^2)$ are distributed as $2X/(1 - X^2)$ and $1/(1 + X^2)$, respectively, or
5. X and $X^2/(1 + X^2)$ are distributed as $(X + Y)/(1 - XY)$ and $1/(1 + X^2)$, respectively.

It should be noted that Norton's characterization based on condition 4 or 5 requires one further condition than Arnold's characterization based on condition 3 or 2, respectively. But the latter is under the stronger assumption that the random variables X and Y are both absolutely continuous.

Kamps (1991) showed that a recurrence relation satisfied by the single moments of order statistics established by Barnett (1966b) given by

$$E\left[(X'_{r:n})^2\right] = \frac{n}{\pi}\{E[X'_{r:n-1}] - E[X'_{r-1:n-1}]\} - 1, \qquad 3 \le r \le n - 2,$$

is in fact a characterization of the Cauchy distribution. He also proved that a generalization of Barnett's relation to the case of a doubly truncated Cauchy distribution with density function

$$p_{X_T}(x) = \frac{1}{(\tan^{-1} b - \tan^{-1} a)(1 + x^2)}, \qquad a < x < b, \quad (16.58)$$

established by Khan, Yaqub, and Parvez (1983) characterizes this doubly truncated Cauchy distribution.

Bell and Sarma (1985) and Glänzel (1987) have given some additional characterizations, with the last author basing one on truncated moments. Further characterization results are presented by Kagan, Linnik, and Rao (1973).

7   GENERATION ALGORITHMS

A convenient method of sampling from a Cauchy distribution is based on the inversion of the distribution function $F(x)$; that is, if U is a uniform $(0, 1)$ variate, then $\tan(\pi(U - \tfrac{1}{2}))$ is distributed as standard Cauchy. This method of course uses the tan function and needs one uniform $(0, 1)$ deviate per observation from the Cauchy distribution. Even though this method can be programmed easily in high-level languages, it may be relatively slow. Some efficient generation algorithms have been proposed for the Cauchy case; three of these are discussed below.

7.1   Monahan's (1979) Algorithm

Let $Y_1, Y_2, \ldots$ be i.i.d. random variables from an arbitrary distribution function F (usually chosen to be standard uniform). Let $Z_1 \equiv 1$ and $Z_2, Z_3, \ldots$ be independent Bernoulli variates with $\Pr[Z_i = 1] = a_i/a_{i-1} = p_i$ for $i > 1$ and given constants $1 = a_1 \geq a_2 \geq a_3 \geq \cdots \geq 0$. Further let E_n be the event that $\max(Y_1, \ldots, Y_n) = Y_1$ and $Z_1 = Z_2 = \cdots = Z_n = 1$. Then the algorithm in its generality involves the following steps:

1. Generate X with distribution function P (chosen to be a simple distribution, for convenience, like uniform), and let $Y_0 = G(X)$. G is an appropriately chosen function, with the only condition on it being that it maps the support of P onto the support of F.
2. Continue to generate $Y_1, Y_2, \ldots, Z_1, Z_2, \ldots$ if $\max(Y_1, Y_2, \ldots, Y_n) \leq Y_0$ and $Z_1 = Z_2 = \cdots = Z_n = 1$.
3. If $Y_{n+1} > Y_0$ or $Z_{n+1} = 0$, stop.
4. If n is even, deliver X, else go to step 1.

Then

$$\Pr[X \leq x, Y_1 \leq G(x), E_n] = \int_{-\infty}^{x} \Pr[Y_1 \leq G(y), E_n] \, dP(y), \quad (16.59)$$

$$\Pr[X \leq x, Y_1 \leq G(x), E_n, E_{n+1}^c]$$
$$= \int_{-\infty}^{x} \left[a_n F(G(y))^n - a_{n+1} F(G(y))^{n+1} \right] dP(y). \quad (16.60)$$

Let $h(w) = \sum_{n=0}^{\infty} a_n w^n$ with $1 = a_0 \geq a_1 \geq \cdots \geq 0$. Convergence is implied for $w \in [-1, 1)$ by $a_n \to 0$. Summing (16.60) over even n beginning with 0, we get the distribution function of X as

$$\frac{\int_{-\infty}^{x} h(-F(G(y))) \, dP(y)}{\int_{-\infty}^{\infty} h(-F(G(y))) \, dP(y).} \quad (16.61)$$

The denominator of (16.61) is the probability that X will be "delivered," p_D, say; the expected number of Y's needed to produce a single X, EN_F, is

$$E[N_F] = \frac{\int_{-\infty}^{x} h(F(G(y)))\, dP(y)}{\int_{-\infty}^{\infty} h(-F(G(y)))\, dP(y);} \qquad (16.62)$$

the expected number of samples from the distribution P, EN_P, is the reciprocal of the probability that X will be delivered.

For the Cauchy case, we have

$$h(w) = (1 - w/2)^{-2} = 1 + w + \frac{3}{4}w^2 + \cdots + \frac{n+1}{2^n}w^n + \cdots,$$
$$F(u) = u,\ u \in (0, 1),$$
$$P(x) = (1 + x)/2,$$
$$G(x) = 2[\sqrt{1 + x^2} - 1].$$

Algorithm

0. Generate u.
1. $X \leftarrow 2u - 1$; $n \leftarrow 0$; $y_0 \leftarrow G(X)$.
2. Generate u_1, $y_{n+1} \leftarrow u_1$.
3. If $y_{n+1} > p_{n+1}y_0$; go to 5.
4. $n \leftarrow n + 1$; go to 2.
5. $u \leftarrow (y_{n+1} - p_{n+1}y_0)/(1 - p_{n+1}y_0)$.
6. If n is odd, go to 1.
7. $u \leftarrow 2u$; if $u < 1$, deliver X.
8. Otherwise, $u \leftarrow u - 1$, and deliver $X \leftarrow 1/X$.

Monahan (1979) determined that $p_D = \pi/4$ and $EN_F \approx 1.88$. This method hence requires $\frac{1}{4}$ fewer uniform deviates than the synthetic tangent algorithm's [Kinderman, Monahan, and Ramage (1977)] $8/\pi \approx 2.56$.

7.2 Kronmal and Peterson's (1981) Acceptance-Complement Method

This method is a variant of the commonly used acceptance-rejection method for generating random variables [Knuth (1968)] that avoids rejection. In general, to generate random variables from the density $p(x)$ defined on $-\infty \le a \le x \le b \le \infty$, the method is as follows:

1. Choose a decomposition of the density $f(\cdot)$ into subdensities $g_1(\cdot)$ and $g_2(\cdot)$ such that $p(\cdot) = g_1(\cdot) + g_2(\cdot)$, with the mass of the subdensity $g_2(\cdot)$ being $q = \int_a^b g_2(x)\, dx$.
2. Choose a dominating density $h^*(\cdot) \ge g_1(\cdot)$.

Then:

3. **a.** Generate $X \sim h^*(\cdot)$.
   b. Generate $U \sim$ uniform $(0, 1)$, independent of X.
4. **a.** If $U \leq g_1(X)/h^*(X)$, then accept X.
   b. Otherwise, generate Y, independent of (X, U) from the density $g_2(\cdot)/q$, and return Y.

It may then be easily verified that

$$\Pr[X \leq t, \text{acceptance in step 4a}] = \int_a^t g_1(x)\, dx,$$

$$\Pr[Y \leq t, \text{nonacceptance in step 4a}] = \int_a^t g_2(y)\, dy$$

with the consequence that

$$\Pr[\text{random variable returned} \leq t] = \int_a^t p(x)\, dx,$$

as desired. Now, for generating standard Cauchy random variables, Kronmal and Peterson (1981) observe that if X is distributed as a truncated Cauchy with density

$$p(x) = \frac{2}{\pi(1 + x^2)} I[|x| < 1], \tag{16.63}$$

then the random variable Z defined by

$$Z \equiv X \quad \text{with probability } \frac{1}{2}$$

$$\equiv \frac{1}{X} \quad \text{with probability } \frac{1}{2} \tag{16.64}$$

will have a standard Cauchy distribution. Hence generation of X from (16.63) can be used in conjunction with the transformation in (16.64) to generate Cauchy random variables. Kronmal and Peterson use the decomposition

$$g_1(x) = \left[\frac{2}{\pi(1 + x^2)} - \left(\frac{2}{\pi} - \frac{1}{2} \right) \right] I[|x| < 1],$$

$$g_2(x) = \left(\frac{2}{\pi} - \frac{1}{2} \right) I[|x| < 1], \tag{16.65}$$

with the dominating density

$$h^*(x) = \tfrac{1}{2}I[|x| < 1]. \tag{16.66}$$

With this choice (as well as some other choices), they found this algorithm to be much more time efficient than the inverse distribution function method as well as the synthetic tangent method.

7.3 Ahrens and Dieter's (1988) Algorithm

This simple and efficient table-free method is based on the "exact approximation" method by Marsaglia (1984). The idea of an exact approximation is as follows: Let $x(t)$ be a monotonic differentiable function in $t \in [0, 1]$. Then the transformed random variable $X \leftarrow x(T)$ has the desired density $p(x)$ if the density of the T-variates is $g(t) = p(x(t))|x'(t)|$. As mentioned in the beginning of this section, the inversion method is based on the transformation $x(t) = F^{-1}(t)$ resulting in $g(t) = 1 \; \forall t \in [0, 1]$. Now, if $x(t)$ is only an easy-to-calculate approximation to $F^{-1}(t)$, $g(t)$ may still be close to 1, and then the approximation can be made exact in a statistical sense in the following way. Supposing that $\forall t \in [0, 1]$, $g(t)$ is such that $p < g(t) < r$, where $p < 1 < r$, and that the function $x(t)$ is chosen so that $h = r - p$ is small. Then generate a $U \rightarrow$ uniform $(0, 1)$ deviate, and if $U \leq p$ return $X \leftarrow x(U/p)$. Otherwise, use a new pair U, U' of uniform $(0, 1)$ deviates for an acceptance-rejection procedure: Return $X \leftarrow x(U)$ if $U' \leq (g(U) - p)/h$ or else try again with another (U, U').

For the standard Cauchy distribution, Ahrens and Dieter (1988) suggested

$$x(t) = t\left(\frac{a}{\frac{1}{4} - t^2} + b\right),$$

$$x'(t) = a\,\frac{\frac{1}{4} + t^2}{\left(\frac{1}{4} - t^2\right)^2} + b, \tag{16.67}$$

and found that the choice of $a = 0.638063$ and $b = 0.595949$ brought the function

$$g(t) = p(x)x'(t) = \frac{1}{\pi(1 + x^2)}\left(a\,\frac{\frac{1}{4} + t^2}{\left(\frac{1}{4} - t^2\right)^2} + b\right) \tag{16.68}$$

even closer to 1 in $(-\tfrac{1}{2}, \tfrac{1}{2})$.

Then the exact approximate algorithm is as follows:

0. Constants:

$$a = 0.6380631366077803$$
$$b = 0.5959486060529070$$
$$c = 0.9339962957603656$$
$$W = 0.2488702280083841$$
$$A = 0.6366197723675813$$
$$B = 0.5972997593539963$$
$$H = 0.0214949004570452$$
$$P = 4.9125013953033204$$

1. Generate U. Set $T \leftarrow U - \frac{1}{2}$ and $S \leftarrow W - T^2$. If $S \leq 0$, go to 3.
2. Return $X \leftarrow T(A/S + B)$.
3. Generate U. Set $T \leftarrow U - \frac{1}{2}$, $s \leftarrow \frac{1}{4} - T^2$, and $X \leftarrow T(a/s + b)$.
4. Generate U'. If $s^2((1 + X^2)(HU' + P) - q) + s > \frac{1}{2}$, go to 3.
5. Return X.

Ahrens and Dieter (1988) have shown that $p = 0.997738$ and $h = 0.004366$, and that the overall consumption of uniform variables is $N = p + 2h = 1.006429$ per pseudorandom observation from the standard Cauchy distribution.

8 RELATED DISTRIBUTIONS

We have already noted, in Section 2, that the Cauchy distribution is a central t distribution (Chapter 28) with one degree of freedom. It is related to other distributions in the same way as the t-distributions. Rider (1957) has studied the properties of a system that he has named the *generalized Cauchy distributions*. The probability density function of a random variable X with such a distribution is of the form:

$$p_X(x) = \frac{k\Gamma(h)}{2\lambda\Gamma(k^{-1})\Gamma(h - k^{-1})} \frac{1}{\left[1 + |(x - \theta)/\lambda|^k\right]^h}, \qquad \lambda, k, h > 0,$$

$$hk > 1. \quad (16.69)$$

For $k = 2$ and $\theta = 0$, X is distributed as $\lambda(2h - 1)^{-1/2}$ times a central t variable with $(2h - 1)$ degrees of freedom, and for $k = 2$, $h = 1$, X is distributed according to (16.1).

Symmetrically truncated Cauchy distributions, with density function (in standard form)

$$\left[2(1 + x^2)\tan^{-1}\lambda\right]^{-1}, \qquad -\lambda \leq x \leq \lambda, \qquad (16.70)$$

have been discussed in Derman (1964). The distribution is symmetrical about zero and has variance

$$\frac{\lambda - \tan^{-1} \lambda}{\tan^{-1} \lambda},$$

so the variance of the arithmetic mean of n independent variables having this distribution is

$$\frac{n^{-1}(\lambda - \tan^{-1}\lambda)}{\tan^{-1}\lambda}, \qquad (16.71)$$

while for large n the variance of the median is approximately

$$n^{-1}(\tan^{-1}\lambda)^2. \qquad (16.72)$$

Formula (16.71) is larger than (16.72) for $\lambda > 3.41$, so the median is asymptotically more efficient, as an estimator of the center of a symmetrically truncated Cauchy distribution, provided not more than $\frac{1}{2} - \pi^{-1}\tan^{-1}(3.41) = 0.0908$ of the distribution is truncated at either end.

The standard *half-Cauchy* distribution has density function

$$2\pi^{-1}(1 + x^2)^{-1}, \qquad 0 < x. \qquad (16.73)$$

The analogy with the half-normal distribution (Chapter 13, Section 7.1) is clear. Mijnheer (1968) has carried out sampling experiments to investigate methods of estimating parameters of the general half-Cauchy distribution in which extreme observations are rejected.

The *folded-Cauchy* distribution [obtained by folding the distribution (16.1) about $x = 0$] has density function

$$(\pi\lambda)^{-1}\left[\left\{1 + \left[\frac{x - \theta}{\lambda}\right]^2\right\}^{-1} + \left\{1 + \left[\frac{x + \theta}{\lambda}\right]^2\right\}^{-1}\right]$$

$$= 2(\pi\lambda)^{-1}\left[\left\{1 + \left[\frac{x - \theta}{\lambda}\right]^2\right\}\left\{1 + \left[\frac{x + \theta}{\lambda}\right]^2\right\}\right]^{-1} \qquad (16.74)$$

$$\times \left[1 + \frac{x^2 + \theta^2}{\lambda^2}\right], \qquad x > 0.$$

For $\theta \le \lambda/\sqrt{3}$, the mode is at $x = 0$; for $\theta > \lambda/\sqrt{3}$ it is at

$$x = \theta\left[1 + \left(\frac{\lambda}{\theta}\right)^2\right]^{1/2}\left[2\left\{1 + \left(\frac{\lambda}{\theta}\right)^2\right\}^{-1/2} - 1\right]^{1/2}. \qquad (16.75)$$

The *wrapped-up Cauchy distribution* is obtained by wrapping a Cauchy distribution around a circle and adding up the probability densities coinciding at each point. With mean angle θ and mean vector length ρ, the density function of a wrapped-up Cauchy random variable T is given by

$$p(t) = \frac{1}{2\pi} \frac{1 - \rho^2}{1 + \rho^2 - 2\rho \cos(t - \theta)}, \qquad 0 \le t < 2\pi, \quad (16.76)$$

where θ represents a location parameter and ρ a scale parameter with $0 \le \rho < 1$. Thus this is a location-scale family of distributiosn. When $\rho = 0$, (16.76) is simply the uniform distribution on the circle, and as $\rho \to 1$ (16.76) approaches a point mass distribution at θ. In general, the wrapped-up Cauchy distribution is unimodal and symmetric about θ. This distribution serves as an alternative to the Fisher–von Mises distribution for modeling symmetric data on the circle. For further discussion, one may refer to Mardia (1972) and Batschelet (1981). Kent and Tyler (1988) discussed the maximum likelihood estimation of the parameters θ and ρ. They showed that for $n \ge 3$, the MLEs exist and are unique and can be computed from the likelihood equations by means of a simple algorithm. Thus they noted that the two-parameter wrapped-up Cauchy distribution behaves just like the two-parameter Cauchy distribution as far as the maximum likelihood estimation of parameters is concerned. Reference should also be made to Best and Fisher (1979) who have used the wrapped-up Cauchy density as an envelope to present an acceptance-rejection method of simulation from the von Mises distribution.

BIBLIOGRAPHY

Abramowitz, M., and Stegun, I. A. (eds.). (1965). *Handbook of Mathematical Functions with Formulas, Graphs and Mathematical Tables*, New York: Dover.

Ahrens, J. H., and Dieter, U. (1988). Efficient table-free sampling methods for the exponential, Cauchy, and normal distributions, *Communications of the ACM*, **31**, 1330–1337.

Ahsanullah, M., and Hamedani, G. G. (1988). Some characterizations of normal distribution, *Calcutta Statistical Association Bulletin*, **37**, 95–99.

Arnold, B. C. (1979). Some characterizations of the Cauchy distribution, *Australian Journal of Statistics*, **21**, 166–169.

Arnold, B. C., and Brockett, P. L. (1992). On distributions whose component ratios are Cauchy, *The American Statistician*, **46**, 25–26.

Bai, Z. D., and Fu, J. C. (1987). On the maximum-likelihood estimator for the location parameter of a Cauchy distribution, *Canadian Journal of Statistics*, **15**, 137–146.

Bailey, B. J. R. (1973). Estimation from first principles, *Mathematical Gazette*, **57**, 169–174.

Balanda, K. P. (1987). Kurtosis comparisons of the Cauchy and double exponential distributions, *Communications in Statistics—Theory and Methods*, **16**, 579–592.

Balmer, D. W., Boulton, M., and Sack, R. A. (1974). Optimal solutions in parameter estimation problems for the Cauchy distribution, *Journal of the American Statistical Association*, **69**, 238–242.

Barnard, G. A. (1974). The two aspects of statistical estimation, *Mathematical Gazette*, **58**, 116–123.

Barnett, V. D. (1966a). Evaluation of the maximum likelihood estimator when the likelihood equation has multiple roots, *Biometrika*, **53**, 151–165.

Barnett, V. D. (1966b). Order statistics estimators of the location of the Cauchy distribution, *Journal of the American Statistical Association*, **61**, 1205–1217. Correction, *Ibid.*, **63**, 383–385.

Batschelet, E. (1981). *Circular Statistics in Biology*, San Diego: Academic Press.

Beckman, R. J., and Johnson, M. E. (1987). Fitting the Student-*t* distribution to grouped data, with application to a particle scattering experiment, *Technometrics*, **29**, 17–22.

Bell, C. B., and Sarma, Y. R. (1985). A characterization of the Cauchy distribution (Spanish), *Trabajos de Estadistica y de Investigacion Operativa*, **36**, 3–7.

Berman, S. M. (1967). An occupation time theorem for the angular component of plane Brownian motion, *Annals of Mathematical Statistics*, **38**, 25–31.

Best, D. J., and Fisher, N. I. (1979). Efficient simulation of the von Mises distribution, *Applied Statistics*, **28**, 152–157.

Bienaymé, I. J. (1853). Remarques sur les differences qui distinguent l'interpolation de M. Cauchy de la méthode des moindres carrés et qui assurent la supériorité de cette méthode, *Comptes Rendus Hebdomadaires des Séances de l'Academie des Sciences Paris*, **37**, 5–13.

Bloch, D. (1966). A note on the estimation of the location parameter of the Cauchy distribution, *Journal of the American Statistical Association*, **61**, 852–855.

Blyth, C. R. (1986). Convolutions of Cauchy distributions, *American Mathematical Monthly*, **93**, 645–647.

Bondesson, L. (1987). On the infinite divisibility of the half-Cauchy and other decreasing densities and probability functions on the non-negative line, *Scandinavian Actuarial Journal*, 225–249.

Bose, A. (1987). Some remarks on the paper "Sums of independent-squared Cauchy variables grow quadratically: Applications" by F. Eicker, *Sankhyā, Series A*, **49**, 138–140.

Brent, R. P. (1973). *Algorithms for Minimization without Derivatives*, Englewood Cliffs, NJ: Prentice-Hall.

Cane, G. J. (1974). Linear estimation of parameters of the Cauchy distribution based on sample quantiles, *Journal of the American Statistical Association*, **69**, 243–245.

Cauchy, A. L. (1853). Sur les résultats moyens d'observations de même nature, et sur les résultats les plus probables, *Comptes Rendus de l'Académie des Sciences, Paris*, **37**, 198–206.

Chan, L. K. (1970). Linear estimation of location and scale parameters of Cauchy distribution based on sample quantiles, *Journal of the American Statistical Association*, **65**, 851–859.

Chan, L. K., Chan, N. N., and Mead, E. R. (1973). Linear estimation of the parameters of the Cauchy distribution using selected order statistics, *Utilitas Mathematica*, **3**, 311–318.

Chan, L. K., and Cheng, S. W. H. (1971). On the Student's test based on sample percentiles from the normal, logistic and Cauchy distributions, *Technometrics*, **13**, 127–137.

Chernoff, H., Gastwirth, J. L., and Johns, M. V. (1967). Asymptotic distribution of linear combinations of functions of order statistics with applications to estimation, *Annals of Mathematical Statistics*, **38**, 52–72.

Copas, J. B. (1975). On the unimodality of the likelihood for the Cauchy distribution, *Biometrika*, **62**, 701–704.

David, F. N., and Johnson, N. L. (1954). Statistical treatment of censored data. I. Fundamental formulae, *Biometrika*, **41**, 228–240.

Derman, C. (1964). Some notes on the Cauchy distribution, National Bureau of Standards Technical Note, Nos. 3–6, Washington, DC.

DeSilva, B. M. (1979). The quotient of certain stable random variables, *Sankhyā*, Series B, **40**, 279–281.

Dugue, M. D. (1941). Sur certaines composantes des lois de Cauchy, *Comptes Rendus de l'Academie des Sciences, Paris*, 718–719.

Dunau, J.-L. and Senateur, H. (1987). An elementary proof of the Knight-Meyer characterization of the Cauchy distribution, *Journal of Multivariate Analysis*, **22**, 74–78.

Dwass, M. (1985). On the convolution of Cauchy distributions, *American Mathematical Monthly*, **92**, 55–57.

Edwards, A. W. F. (1972). *Likelihood*, Cambridge: Cambridge University Press.

Efron, B. (1975). Defining the curvature of a statistical problem (with application to second-order efficiency), with discussion, *Annals of Statistics*, **3**, 1189–1242.

Efron, B., and Hinkley, D. V. (1978). Assessing the accuracy of the maximum likelihood estimator: Observed versus expected Fisher information, *Biometrika*, **65**, 457–487.

Eicker, F. (1985). Sums of independent-squared Cauchy variables grow quadratically: Applications, *Sankhyā*, Series A, **47**, 133–140.

Eisenhart, C. (1971). The development of the concept of the best mean of a set of measurements from antiquity to the present day, *1971 A.S.A. Presidential Address*, Fort Collins, Colorado.

Ferguson, T. S. (1978). Maximum likelihood estimates of the parameters of the Cauchy distribution for samples of size 3 and 4, *Journal of the American Statistical Association*, **73**, 211–213.

Feuerverger, A., and McDunnough, P. (1981a). On the efficiency of empirical characteristic function procedures, *Journal of the Royal Statistical Society*, Series B, **43**, 20–27.

Feuerverger, A., and McDunnough, P. (1981b). On efficient inference in symmetric stable laws and processes, *Proceedings of the International Symposium on Statistics and Related Topics*, M. Csörgö, D. A. Dawson, J. N. K. Rao, and A. K. Md. E. Saleh (editors), Amsterdam: North-Holland.

Fisher, R. A., and Tippett, L. H. C. (1928). Limiting forms of the frequency distributions of the smallest and the largest member of a sample, *Proceedings of the Cambridge Philosophical Society*, **24**, 180–190.

Fox, C. (1965). A family of distributions with the same ratio property as normal distribution, *Canadian Mathematical Bulletin*, **8**, 631–636.

Franck, W. E. (1981). The most powerful invariant test of normal vs. Cauchy with applications to stable alternatives, *Journal of the American Statistical Association*, **76**, 1002–1005.

Fréchet, M. (1927). Sur la loi de probabilité de l'écart maximum, *Annales de la Société Polonaise de Mathématique, Cracow*, **6**, 93–116.

Fulvio, A. (1965). Sulla decomposizione de una variabile casuale seguent e la legge di Cauchy, *Bolletino della Unione Matematica Italiana*, **20**, 177–180.

Gabrielsen, G. (1982). On the unimodality of the Cauchy distributions: Some comments, *Biometrika*, **69**, 677–678.

Galambos, J. (1987). *The Asymptotic Theory of Extreme Order Statistics* (Second edition), Melbourne, FL: Kreiger.

Glänzel, W. (1987). A characterization theorem based on truncated moments and its application to some distribution families, *Mathematical Statistics and Probability Theory* **B**, P. Bauer, et al. (editors), 75–84, Dordrecht: Reidel.

Goria, M. (1978). Fractional absolute moments of the Cauchy distribution, *Quaderni di Statistica e Matematica Applicata alle Scienze Economico-Sociali, University of Trento, Trento (Italy)*, **1/2**, 89–96.

Gradshteyn, I. S., and Ryzhik, I. M. (1965). *Tables of Integrals, Series, and Products*, San Diego: Academic Press.

Grandi, G. (1718). Note al trattato del Galileo del moto naturale accellerato, *Opere Di Gallileo Galilei* Tomo Terzo, p. 393. Florence.

Gumbel, E. J., and Keeney, R. D. (1950). The extremal quotient, *Annals of Mathematical Statistics*, **21**, 523–537.

Gupta, A. K., and Govindarajulu, Z. (1975). Distribution of the quotient of two independent Hotelling's T^2-variates, *Communications in Statistics*, **4**, 449–453.

Haas, G., Bain, L. J., and Antle, C. (1970). Inferences for the Cauchy distribution based on maximum likelihood estimators, *Biometrika*, **57**, 403–408.

Hall, P. (1978). Some asymptotic expansions of moments of order statistics, *Stochastic Processes and their Applications*, **7**, 265–275.

Haq, M. (1971). Estimation of the scalar parameter in a Cauchy population, *Journal of Sciences—Physical Section*, **1**, 67–68.

Higgins, J. J., and Tichenor, D. M. (1977). Window estimates of location and scale with applications to the Cauchy distribution, *Applied Mathematics and Computation*, **3**, 113–126.

Higgins, J. J., and Tichenor, D. M. (1978). Efficiencies for window estimates of the parameters of the Cauchy distribution, *Applied Mathematics and Computation*, **4**, 157–166.

Hinkley, D. V. (1978). Likelihood inference about location and scale parameters, *Biometrika*, **65**, 253–261.

Horn, P. S. (1983). A measure for peakedness, *The American Statistician*, **37**, 55–56.

Howlader, H. A., and Weiss, G. (1987). Considerations on the approximation of Bayesian ratios of integrals, *Computer Science and Statistics: Proceedings of the Nineteenth Symposium on the Interface*, 406–409.

Howlader, H. A., and Weiss, G. (1988a). On Bayesian estimation of the Cauchy parameters, *Sankhyā, Series B*, **50**, 350–361.

Howlader, H. A., and Weiss, G. (1988b). Bayesian reliability estimation of a two-parameter Cauchy distribution, *Biometrical Journal*, **30**, 329–337.

Joshi, P. C. (1969). Bounds and approximations for the moments of order statistics, *Journal of the American Statistical Association*, **64**, 1617–1624.

Kadiyala, K. R., and Murthy, K. S. R. (1977). Estimation of regression equation with Cauchy disturbances, *Canadian Journal of Statistics*, **5**, 111–120.

Kagan, A. M., Linnik, Yu. V., and Rao, C. R. (1973). *Characterization Problems of Mathematical Statistics*, New York: Wiley.

Kamerud, D. B. (1978). Solution to Problem 6104: The random variable $X/Y, X, Y$ Normal, *American Mathematical Monthly*, **85**, 206.

Kamps, U. (1991). A characterizing property of Cauchy, doubly truncated Cauchy and related distributions, Technical Report, Institut für Statistik und Wirtschaftsmathematik, RWTH, Aachen, Germany: Aachen University of Technology.

Kariya, T., Sinha, B. K., and Subramanyam, K. (1981). Nearly efficient estimators based on order statistics, Technical Report No. 81–05, Pittsburgh: University of Pittsburgh.

Kendall, M. G., and Stuart, A. (1961). *The Advanced Theory of Statistics*, **2**. London: Griffin.

Kent, J. T., and Tyler, D. E. (1988). Maximum likelihood estimation for the wrapped Cauchy distribution, *Journal of Applied Statistics*, **15**, 247–254.

Khan, A. H., Yaqub, M., and Parvez, S. (1983). Recurrence relations between moments of order statistics, *Naval Research Logistics Quarterly*, **30**, 419–441. Corrigendum, *Ibid.*, **32**, 693.

Kinderman, A. J., Monahan, J. F., and Ramage, J. G. (1977). Computer methods for sampling from Student's t distribution, *Mathematics of Computation*, **31**, 1009–1018.

Knight, F. B. (1976). A characterization of the Cauchy type, *Proceedings of the American Mathematical Society*, **55**, 130–135.

Knight, F. B., and Meyer, P. A. (1976). Une Caracterisation de la loi de Cauchy, *Zeitschrift fuer Wahrscheinlichkeitstheorie und Verwandte Gebiete*, **34**, 129–134.

Knuth, D. E. (1968). *The Art of Computer Programming: Fundamental Algorithms*, **1**, Reading, MA: Addision-Wesley.

Kotlarski, I. I. (1960). On random variables whose quotient follows the Cauchy law, *Colloquium Mathematicum*, **7**, 277–284.

Kotlarski, I. I. (1977). Problem 6164, *American Mathematical Monthly*, **84**, 575.

Kotlarski, I. I. (1979). An exercise involving Cauchy random variables, *American Mathematical Monthly*, **86**, 229.

Koutrouvelis, I. A. (1982). Estimation of location and scale in Cauchy distributions using the empirical characteristic function, *Biometrika*, **69**, 205–213.

Kronmal, R. A., and Peterson, A. V., Jr. (1981). A variant of the acceptance-rejection method for computer generation of random variables, *Journal of the American Statistical Association*, **76**, 446–451.

Laha, R. G. (1959a). On the law of Cauchy and Gauss, *Annals of Mathematical Statistics*, **30**, 1165–1174.

Laha, R. G. (1959b). On a class of distribution functions where the quotient follows the Cauchy law, *Transactions of the American Mathematical Society*, **93**, 205–215.

Lawless, J. F. (1972). Conditional confidence interval procedures for the location and scale parameters of the Cauchy and logistic distributions, *Biometrika*, **59**, 377–386.

Letac, G. (1981). Isotropy and sphericity: some characterizations of the normal distribution, *Annals of Statistics*, **9**, 408–417.

Lindley, D. V. (1980). Approximate Bayesian methods (with discussion), *Trabajos de Estadistica y de Investigacion Operativa*, **31**, 232–245.

Mardia, K. V. (1972). *Statistics of Directional Data*, San Diego: Academic Press.

Marsaglia, G. (1984). The exact-approximation method for generating random variables in a computer, *Journal of the American Statistical Association*, **79**, 218–221.

Mauldon, J. H. (1956). Characterizing properties of statistical distributions, *Quarterly Journal of Mathematics, Oxford*, **7**, 155–160.

McCullagh, P. (1992a). Conditional inference and Cauchy models, *Biometrika*, **79**, 247–259.

McCullagh, P. (1992b). On the choice of ancillary in the Cauchy location-scale problem, *Bahadur Festschrift*, S. K. Mitra (editor), To appear.

McCullagh, P. (1992c). Some examples of harmonic measures in statistics, Preprint.

McCullagh, P. (1993). On the distribution of the Cauchy maximum-likelihood estimator, *Proceedings of the Royal Society of London, Series A*, **440**, 475–479.

Menon, M. V. (1962). A characterization of the Cauchy distribution, *Annals of Mathematical Statistics*, **33**, 1267–1271.

Menon, M. V. (1966). Another characteristic property of the Cauchy distribution, *Annals of Mathematical Statistics*, **37**, 289–294.

Mijnheer, J. L. (1968). Steekproeven uit de halve Cauchy verdeling, *Statistica Neerlandica*, **22**, 97–101.

Mises, R. von (1936). La Distribution de la plus grande de *n* valeurs, *Revue Mathématique de l'Union Interbalkanique*, **1**, 1–20.

Monahan, J. F. (1979). Extensions of von Neumann's methods for generating random variables, *Mathematics of Computation*, **33**, 1065–1069.

Mukherjea, A., Nakassis, A., and Miyashita, J. (1986). The problem of identification of parameters by the distribution of the maximum random variable, *Journal of Multivariate Analysis*, **18**, 178–186.

Nelsen, R. B. (1985). Letter to the editor, *American Mathematical Monthly*, **92**, 679.

Norton, R. M. (1983). A characterization of the Cauchy distribution, *Sankhyā, Series A*, **45**, 247–252.

Obretenov, A. (1961). A property characterizing the Cauchy distribution, *Fiziko-Mathematichesko Spisaniye, Bülgarska Akademiya na Naukite*, **4**, No. 37, 40–43. (In Bulgarian)

Ogawa, J. (1951). Contributions to the theory of systematic statistics, I, *Osaka Mathematical Journal*, **3**, 175–213.

Ogawa, J. (1960). Determination of optimum spacings for the estimation of the scale parameter of an exponential distribution based on sample quantiles, *Annals of the Institute of Statistical Mathematics*, **12**, 135–141.

Philippou, A. N. (1988). Asymptotically optimal tests for the logarithmic, logistic and Cauchy distributions based on the concept of contiguity, *Asymptotic Statistics 2: Proceedings of the Third Prague Symposium of Asymptotic Statistics*, P. Mandl and M. Huskova (editors), 379–386, Amsterdam: North-Holland.

Pitman, E. J. G., and Williams, E. J. (1967). Cauchy-distributed functions of Cauchy variates, *Annals of Mathematical Statistics*, **38**, 916–918.

Poisson, S. D. (1824). Sur la probabilité des résultats moyens des observations, *Connaissance des Tems pour l'an 1827*, 273–302.

Puglisi, A. (1966). Sulla decomposizione della legge di probabilità di Cauchy, *Bolletino della Unione Matematica Italiana*, **21**, 12–18.

Raghunandanan, K., and Srinivasan, R. (1972). Simplified estimation of parameters in a Cauchy distribution, *Statistica Neerlandica*, **26**, 25–27.

Reeds, J. A. (1985). Asymptotic number of roots of Cauchy likelihood equations, *Annals of Statistics*, **13**, 775–784.

Rider, P. R. (1957). Generalized Cauchy distributions, *Annals of the Institute of Statistical Mathematics*, **9**, 215–223.

Rider, P. R. (1960). Variance of the median of samples from a Cauchy distributions, *Journal of the American Statistical Association*, **55**, 322–323.

Rohatgi, V. K., Steutel, F. W., and Székely, G. J. (1990). Infinite divisibility of products and quotients of i.i.d. random variables, *Mathematical Scientist*, **15**, 53–59.

Rosenberger, J. L., and Gasko, M. (1983). Comparing location estimators: Trimmed means, medians and trimean, *Understanding Robust and Exploratory Data Analysis*, D. C. Hoaglin, F. Mosteller, and J. W. Tukey (editors), 297–338, New York: Wiley.

Roslonek, E. (1970). On pairs of random variables whose sum follows the Cauchy distribution, *Demonstratio Mathematica*, **2**, 223–230.

Rothenberg, T. J., Fisher, F. M., and Tilanus, C. B. (1964). A note on estimation from a Cauchy sample, *Journal of the American Statistical Association*, **59**, 460–463.

Salama, I. A. (1983). On some distributional properties of Cauchy random variables, *Technical Report*.

Saleh, A. K. Md. E., Hassanein, K. M., and Brown, E. F. (1985). Optimum spacings for the joint estimation and tests of hypothesis of location and scale parameters of the Cauchy distribution, *Communications in Statistics—Theory and Methods*, **14**, 247–254.

Sarhan, A. E., and Greenberg, B. G. (eds.). (1962). *Contributions to Order Statistics*, New York: Wiley.

Savage, L. J. (1966). *A geometrical approach to the special stable distributions*, Technical Report No. 1, Department of Statistics, New Haven, Connecticut: Yale University.

Sen, P. K. (1959). On the moments of the sample quantiles, *Calcutta Statistical Association Bulletin*, **9**, 1–19.

Shapiro, J. M. (1978). On the rate of convergence of distribution functions of sums of reciprocals of random variables to the Cauchy distribution, *Houston Journal of*

Mathematics, **4**, 439–445.

Shapiro, J. M. (1988). Limit distributions for sums of reciprocals of independent random variables, *Houston Journal of Mathematics*, **14**, 281–290.

Spiegelhalter, D. J. (1985). Exact Bayesian inference on the parameters of a Cauchy distribution with vague prior information, *Bayesian Statistics 2: Proceedings of the Second Valencia International Meeting on Bayesian Statistics*, J. M. Bernardo, M. H. DeGroot, D. V. Lindley, and A. F. M. Smith (editors), 743–749, Amsterdam: Elsevier.

Spitzer, F. (1958). Some theorems concerning 2-dimensional Brownian motion, *Transactions of the American Mathematical Society*, **87**, 187–197.

Springer, M. D., and Thompson, W. E. (1966). The distribution of products of independent random variables, *SIAM Journal of Applied Mathematics*, **14**, 511–526.

Steck, G. P. (1958). A uniqueness property not enjoyed by the normal distribution, *Annals of Mathematical Statistics*, **29**, 604–606.

Stigler, S. M. (1974). Studies in the history of probability and statistics. XXXIII. Cauchy and the witch of Agnesi: An historical note on the Cauchy distribution, *Biometrika*, **61**, 375–379.

Stigler, S. M. (1989). Letter to the Editor: Normal orthant probabilities, *The American Statistician*, **43**, 291.

Stoops, G., and Barr, D. (1971). Moments of certain Cauchy order statistics, *The American Statistician*, **25**, 51.

Sugiura, N. (1962). On the orthogonal inverse expansion with an application to the moments of order statistics, *Osaka Mathematical Journal*, **14**, 253–263.

Tierney, L., and Kadane, J. (1986). Accurate approximations for posterior moments and marginals, *Journal of the American Statistical Association*, **81**, 82–86.

van Zwet, W. R. (1964). *Convex Transformations of Random Variables*, Mathematical Centre Tracts 7, Amsterdam: Mathematisch Centrum.

Vaughan, D. C. (1993). The exact values of the expected values, variances and covariances of the order statistics from the Cauchy distribution, Preprint.

Weiss, G., and Howlader, H. (1988). Linear scale estimation, *Journal of Statistical Computation and Simulation*, **29**, 117–126.

Weiss, L. (1966). The relative maxima of the likelihood function. II, *Skandinavisk Aktuarietidskrift*, 119–121.

Williams, E. J. (1969). Cauchy-distributed functions and a characterization of the Cauchy distribution, *Annals of Mathematical Statistics*, **40**, 1083–1085.

Wingo, D. R. (1983). Estimating the location of the Cauchy distribution by numerical global optimization, *Communications in Statistics—Simulation and Computation*, **12**, 201–212.

Zhang, H. -C. (1990). Distribution of the squared sum of the independent Cauchy variables, *Statistical Inference in Elliptically Contoured and Related Distributions*, K. -T. Fang and T. W. Anderson (editors), 193–200, New York: Allerton Press.

CHAPTER 17

Gamma Distributions

1   DEFINITION

A random variable X has a gamma distribution if its probability density function is of form

$$p_X(x) = \frac{(x - \gamma)^{\alpha - 1} \exp[-(x - \gamma)/\beta]}{\beta^\alpha \Gamma(\alpha)}, \qquad \alpha > 0, \beta > 0; x > \gamma. \tag{17.1}$$

This distribution–denoted gamma (α, β, γ)—is Type III of Pearson's system (Chapter 12, Section 4). It depends on three parameters α, β, and γ. If $\gamma = 0$, the distribution is termed a *two-parameter gamma distribution*, de-noted gamma (α, β); see equation (17.23).

The *standard form* of distribution is obtained by setting $\beta = 1$ and $\gamma = 0$. This gives

$$p_X(x) = \frac{x^{\alpha - 1}e^{-x}}{\Gamma(\alpha)}, \qquad x \geq 0. \tag{17.2}$$

If $\alpha = 1$, we have an *exponential distribution* (see Chapter 19). If α is a positive integer, we have an *Erlang distribution*.

The distributions of $Y = -X$, namely

$$p_Y(y) = \frac{(-y - \gamma)^{\alpha - 1} \exp[(y + \gamma)/\beta]}{\beta^\alpha \Gamma(\alpha)}, \qquad y \leq -\gamma, \tag{17.1'}$$

and

$$p_Y(y) = \frac{(-y)^{\alpha - 1}e^y}{\Gamma(\alpha)}, \qquad y \leq 0, \tag{17.2'}$$

337

are also gamma distributions. But such distributions rarely need to be considered, and we will not discuss them further here.

The probability integral of distribution (17.2) is

$$\Pr[X \le x] = [\Gamma(\alpha)]^{-1} \int_0^x t^{\alpha-1} e^{-t} \, dt. \tag{17.3}$$

This is an *incomplete gamma function ratio*. The quantity

$$\Gamma_x(\alpha) = \int_0^x t^{\alpha-1} e^{-t} \, dt \tag{17.4}$$

is sometimes called an *incomplete gamma function*, but this name is also quite commonly applied to the ratio (17.3) (Chapter 1, Section 3).

This ratio depends on x and α, and it would be natural to use a notation representing it as a function of these variables. However, Pearson (1922) found it more convenient to use $u = x\alpha^{-1/2}$ in place of x for tabulation purposes, and he *defined* the incomplete gamma function as

$$I(u, \alpha - 1) = \frac{1}{\Gamma(\alpha)} \int_0^{u\sqrt{\alpha}} t^{\alpha-1} e^{-t} \, dt. \tag{17.5}$$

The main importance of the (standard) gamma distribution in statistical theory is the fact that if $U_1, U_2, \ldots, U_\nu$ are independent unit normal variables, the distribution of $\sum_{j=1}^\nu U_j^2$ is of form (17.1) with $\alpha = \nu/2$, $\beta = 2$, and $\gamma = 0$. This particular form of gamma distribution is called a *chi-square distribution with ν degrees of freedom*. The corresponding random variable is often denoted by χ_ν^2, and we will follow this practice. It is clear that $\frac{1}{2}\sum_{j=1}^\nu U_j^2$ has a standard gamma distribution with $\alpha = \nu/2$. Expressed symbolically:

$$p_{\chi_\nu^2}(x^2) = \left\{2^{\nu/2}\Gamma(\tfrac{1}{2}\nu)\right\}^{-1}(x^2)^{(\nu/2)-1} \exp(-\tfrac{1}{2}x^2), \qquad x^2 \ge 0. \tag{17.6}$$

Although in the definition above ν must be an integer, the distribution (17.6) is also called a "χ^2 distribution with ν degrees of freedom" if ν is *any* positive number. This distribution is discussed in detail in Chapter 18.

2 MOMENTS AND OTHER PROPERTIES

The moment generating function of the standard gamma distribution (17.2) is

$$E[e^{tX}] = \{\Gamma(\alpha)\}^{-1} \int_0^\infty x^{\alpha-1} \exp[-(1-t)x] \, dx = (1-t)^{-\alpha}, \qquad t < 1. \tag{17.7}$$

The characteristic function is $(1 - it)^{-\alpha}$.

Since distributions of form (17.1) can be obtained from those of form (17.2) by the linear transformation $X = (X' - \gamma)/\beta$, there is no difficulty in deriving formulas for moments, generating functions, and so on, for (17.1) from those for (17.2).

The formula for the rth moment about zero of distribution (17.2) is

$$\mu_r' = \{\Gamma(\alpha)\}^{-1} \int_0^\infty x^{\alpha+r-1} e^{-x} \, dx = \frac{\Gamma(\alpha + r)}{\Gamma(\alpha)}. \qquad (17.8)$$

From (17.8) cumulants can be obtained. These are very simple:

$$\kappa_r = (r - 1)! \alpha. \qquad (17.9)$$

Hence for distribution (17.2)

$$E[X] = \text{Var}(X) = \alpha,$$
$$\mu_3 = 2\alpha, \qquad (17.10)$$
$$\mu_4 = 3\alpha^2 + 6\alpha,$$

so

$$\alpha_3 = \sqrt{\beta_1} = 2\alpha^{-1/2},$$
$$\alpha_4 = \beta_2 = 3 + 6\alpha^{-1}. \qquad (17.11)$$

The mean deviation of distribution (17.2) is

$$\frac{2\alpha^\alpha e^{-\alpha}}{\Gamma(\alpha)}. \qquad (17.12)$$

The standard distribution (17.2) has a single mode at $x = \alpha - 1$ if $\alpha \geq 1$. [Distribution (17.1) has a mode at $x = \gamma + \beta(\alpha - 1)$.] If $\alpha < 1$, $p_X(x)$ tends to infinity as x tends to zero; if $\alpha = 1$ (the standard exponential distribution), $\lim_{x \to 0} p_X(x) = 1$.

There are points of inflexion, equidistant from the mode, at

$$x = \alpha - 1 \pm \sqrt{\alpha - 1} \qquad (17.13)$$

(provided that the values are real and positive). The standardized variable

$$W = \frac{X - \alpha}{\sqrt{\alpha}} \qquad (17.14)$$

is referred to as *the frequency factor* by hydrologists in flood frequency analysis [see, e.g., Phien (1991) or Chow (1969)].

Vodă (1974) studies a reparametrized version of the gamma distribution (17.1), with $\gamma = 0$, whose pdf is

$$p_X(x; \alpha, \theta) = \left(\frac{\alpha}{\theta}\right)^\alpha \times \frac{1}{\Gamma(\alpha)} x^{\alpha-1} \exp\left(\frac{-\alpha x}{\theta}\right), \qquad x > 0; \alpha, \theta > 0.$$

(17.15)

Some typical probability density functions are shown in Figures 17.1 and 17.2. Figure 17.2 shows three different gamma distributions (17.1), each having the same expected value (zero) and standard deviation (unity).

$$\alpha = 1; \quad p(x) = \exp[-(x + 1)] \qquad\qquad (x > -1) \quad \text{Mode at } -1$$

$$\alpha = 4; \quad p(x) = \frac{8}{3}(x + 2)^3 \exp[-2(x + 2)] \qquad (x > -2) \quad \text{Mode at } -\tfrac{1}{2}$$

$$\alpha = 9; \quad p(x) = \frac{2187}{4480}(x + 3)^8 \exp[-3(x + 3)] \quad (x > -3) \quad \text{Mode at } -\tfrac{1}{3}$$

It can be seen from Figure 17.1 that, as α increases, the shape of the curve becomes similar to the normal probability density curve. In fact the standardized gamma distribution tends to the unit normal distribution as the value of the parameter α tends to infinity:

$$\lim_{\alpha \to \infty} \Pr\left[(X - \alpha)\alpha^{-1/2} \leq u\right] = \Phi(u) \tag{17.16}$$

for all real values of u, where $\Phi(u) = (2\pi)^{-1/2}\int_{-\infty}^{u} \exp(-\tfrac{1}{2}t^2)\,dt$.
A similar result holds for the general distribution (17.1), namely

$$\lim_{\alpha \to \infty} \Pr\left[\left\{\frac{X - \gamma}{\beta} - \alpha\right\}\alpha^{-1/2} \leq u\right] = \Phi(u). \tag{17.17}$$

For χ_ν^2,

$$\lim_{\nu \to \infty} \Pr\left[(\chi_\nu^2 - \nu)(2\nu)^{-1/2} \leq u\right] = \Phi(u). \tag{17.18}$$

It can be checked from (17.10) and (17.11) that $\alpha_3 \to 0$, $\alpha_4 \to 3$ (the values for the normal distribution) as α, ν, respectively, tend to infinity.

One of the most important properties of the distribution is the *reproductive property*: If X_1, X_2 are independent random variables each having a distribution of form (17.1), with possibly different values α', α'' of α, but with *common values* of β and γ, then $(X_1 + X_2)$ also has a distribution of this form, with the same values of β and γ, and with $\alpha = \alpha' + \alpha''$.

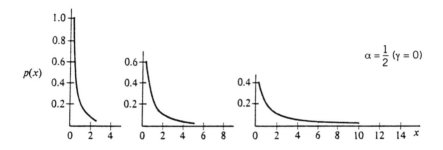

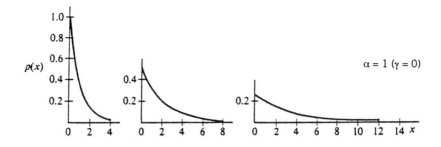

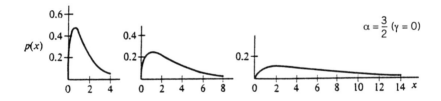

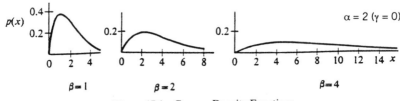

Figure 17.1 Gamma Density Functions

For distribution (17.2) Gini's concentration ratio is

$$G = \frac{\Gamma\left(\alpha + \frac{1}{2}\right)}{\sqrt{\pi}\,\Gamma(\alpha + 1)}.$$ (17.19a)

The Lorenz concentration ratio is

$$L = 2B_{0.5}(\alpha, \alpha + 1),$$ (17.19b)

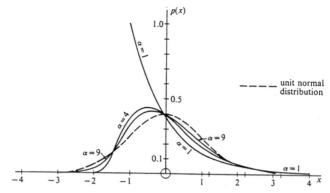

Figure 17.2 Standardized Type III Density Functions

$\alpha = 1;$ $p(x) = \exp[-(x + 1)]$ $(x > -1)$ Mode at -1
$\alpha = 4;$ $p(x) = (8/3)(x + 2)^3 \exp[-2(x + 2)]$ $(x > -2)$ Mode at $-1/2$
$\alpha = 9;$ $p(x) = (2187/4480)(x + 3)^8 \exp[-3(x + 3)]$ $(x > -3)$ Mode at $-1/3$

where $B_p(a, b) = \int_0^p y^{a-1}(1 - y)^{b-1}\, dy$ is the incomplete beta function. The Pietra ratio is

$$\rho = \left\{ \frac{\alpha^\alpha e^{-\alpha}}{\Gamma(\alpha + 2)} \right\} {}_1F_1(2, \alpha + 2; \alpha). \tag{17.20}$$

The Theil entropy measure of inequality is

$$T = \frac{1}{\alpha} + \Psi(\alpha) - \log \alpha \tag{17.21}$$

$[\Psi(\alpha) = d \log \Gamma(\alpha)/d\alpha]$ [Salem and Mount (1974); McDonald and Jensen (1979)]. Saunders and Moran (1978) show that the ε-quantile of the gamma distribution (17.1), denoted by $y_{\varepsilon|\alpha}$, where

$$\int_0^{y_{\varepsilon|\alpha}} e^{-y} y^{\alpha-1}\, dy / \Gamma(\alpha) = \varepsilon \tag{17.22}$$

has the property that when $1 > \varepsilon_2 > \varepsilon_1 > 0$, the ratio $\gamma_\alpha = y_{\varepsilon_2|\alpha}/y_{\varepsilon_1|\alpha}$ is a decreasing function of α and thus the equation

$$\gamma_\alpha = b$$

has a unique solution, $\phi(b)$ for any $b \in (1, \infty)$. Moreover $\gamma_\alpha \to 1$ as $\alpha \to \infty$.

The reproductive property is utilized, among other things, for determination of gamma priors in Bayesian reliability analysis [see Waller and Waterman (1978)]. Also $y_{\varepsilon_2|\alpha} - y_{\varepsilon_1|\alpha}$ increases with α, implying that the gamma distributions are "ordered in dispersion" (see Chapter 33).

3 GENESIS AND APPLICATIONS

Lancaster (1966) quotes from Laplace (1836) in which the latter obtains a gamma distribution as the posterior distribution of the "precision constant" ($h = \frac{1}{2}\sigma^{-2}$), (Chapter 13, Section 1) given the values of n independent normal variables with zero mean and standard deviation σ (assuming a "uniform" prior distribution for h). Lancaster (1966) also states that Bienaymé (1838) obtained the (continuous) χ^2 distribution as the limiting distribution of the (discrete) random variable $\sum_{i=1}^{k}(N_i - np_i)^2(np_i)^{-1}$, where $(N_1, \ldots, N_k)$ have a joint multinomial distribution with parameters $n, p_1, p_2, \ldots, p_k$.

The gamma distribution appears naturally in the theory associated with normally distributed random variables, as the distribution of the sum of squares of independent unit normal variables. (See Chapter 18.) The use of the gamma distribution to approximate the distribution of quadratic forms (particularly positive definite quadratic forms) in multinormally distributed variables is well established and widespread. One of the earliest examples was its use, in 1938, to approximate the distribution of the denominator in a test criterion for difference between expected values of two normal populations with possibly different variances [Welch (1938)]. It has been used in this way many times since. The use of gamma distributions to represent distributions of range and quasi-ranges in random samples for a normal population has been discussed in Chapter 13. In most applications the two-parameter form ($\gamma = 0$),

$$p_X(x) = \frac{x^{\alpha-1}e^{-x/\beta}}{\beta^\alpha \Gamma(\alpha)}, \qquad x > 0; \alpha > 0, \beta > 0, \qquad (17.23)$$

is used (this is equivalent to approximating by the distribution of $\frac{1}{2}\beta\chi^2_{2\alpha}$). However, the three-parameter form has also been used with good effect [e.g., see Pearson (1963)].

The gamma distribution may be used in place of the normal distribution as "parent" distribution in expansions of Gram-Charlier type (Chapter 12, Section 4.2). Series with Laguerre polynomial multipliers rather than Hermite polynomial multipliers are obtained in this situation. Formulas for use with such expansions and their properties have been described by Khamis (1960). These *Laguerre series* have been used by Barton (1953) and Tiku (1964a, b) to approximate the distributions of "smooth test" (for goodness-of-fit) statistics and noncentral F (Chapter 30).

In applied work, gamma distributions give useful representations of many physical situations. They have been used to make realistic adjustments to exponential distributions in representing lifetimes. The "reproductive" property (mentioned in Section 1) leads to the appearance of gamma distributions in the theory of random counters and other topics associated with random processes in time, in particular in meteorological precipitation processes [Kotz and Neumann (1963); Das (1955)]. Some other applications from diverse fields are described in papers cited in references. Among the latter papers Salem and Mount (1974) provide a comparison of the gamma and lognormal distributions and demonstrate that the gamma distributions provides a better fit for personal income data in the United States for the years 1960 to 1969.

Dennis and Patil (1984) discuss applications of gamma distribution in statistical ecology (standard model for sampling dispersion). Among more recent applications we note Costantino and Desharnais's (1981) empirical fit of the steady-state abundances of laboratory flour beetle (Tribolium) populations. Dennis and Patil (1984) generalize this result and show that a gamma distribution is an approximate stationary distribution for the abundance of a population fluctuating around a stable equilibrium. Starting from the stochastic model of population growth

$$\frac{dn}{dt} = n[g(u) + h(n)z(t)],$$

where n is population density at time t, $g(u)$ is the specific growth rate, $z(t)$ is a Gaussian process (noise) with variability σ^2, and $h(n)$ is a function specifying the density dependence on the effects of the noise, Dennis and Patil (1984) approximate Wright's formula for the equilibrium pdf by a gamma distribution emphasizing its right-skewness in the distribution of single-species abundances at equilibrium and its positive range. Then, modifying their deterministic model from $du/dt = ng(u)$ to $du/dt = n[g(u) - p(u)]$, where $p(u)$ is a specific rate describing the effects of predation, harvesting and other forces, Dennis and Patil (1984) arrive at a weighted gamma distribution for the equilibrium pdf.

Gamma distributions share with lognormal distributions (Chapter 14) the ability to mimic closely a normal distribution (by choosing α large enough) while representing an essentially positive random variable (by choosing $\gamma \geq 0$).

4 TABLES AND COMPUTATIONAL ALGORITHMS

In 1922 there appeared a comprehensive *Tables of the Incomplete Γ-Function*, edited by Pearson (1922). This contains values of $I(u, p)$ [see (17.5) with $\alpha = p + 1$] to seven decimal places for $p = -1(0.05)0(0.1)5(0.2)50$ and u at intervals of 0.1. These are supplemented by a table of values of

$$\log I(u, p) - (p + 1)\log u$$

for $p = -1(0.05)0(0.1)10$ and $u = 0.1(0.1)1.5$. This function was chosen to make interpolation easier, particularly for low values of p (Section 5).

Harter (1964) published tables of $I(u, p)$ to nine decimal places for $p = -0.5(0.5)74(1)164$ and u at intervals of 0.1. In this work he covers a greater range of values of p and has two extra decimal places, although Pearson (1922) has p at finer intervals. Harter (1969) published further tables of Type III distributions, giving the 0.01, 0.05, 0.1, 0.5, 1, 2, 2.5, 4, 5, 10(10)90, 95, 96, 97.5, 98, 99, 99.5, 99.9, 99.95, and 99.99 percentage points to 5 decimal places for $\sqrt{\beta_1} = 0.0(0.1)4.8(0.2)9.0$. Harter (1971) extends his 1969 tables and provides percentage points of the one-parameter gamma distribution (Pearson Type III) corresponding to cumulative probabilities 0.002 and 0.998 as well as 0.429624 and 0.570376. The first pair is used for determination of the magnitude of an event (flood) corresponding to 500-year return period, while the second corresponds to a return period of 32,762 years (the so-called mean annual flood according to the guidelines of the U.S. Department of Housing and Urban Development). These are the most important direct tables of $I(u, p)$.

Pearson (1922) gave the more general formula for distribution (17.1):

$$\Pr[X \le x] = e^{-y} \sum_{j=0}^{\infty} \left\{ \frac{y^{\alpha+j}}{\Gamma(\alpha + j + 1)} \right\} \quad \text{for } y = (x - \gamma)/\beta > 0. \quad (17.24)$$

Salvosa (1929, 1930) published tables of the probability integral, probability density function and its first six derivatives for distribution (17.1), with β and γ so chosen that X is standardized (i.e., $\beta = \alpha_3/2$; $\gamma = -2/\alpha_3$). Values are given to six decimal places for α_3 $(= 2\alpha^{-1/2}) = 0.1(0.1)1.1$ at intervals of 0.01 for x. Cohen, Helm, and Sugg (1969) have calculated tables of the probability integral to nine decimal places for $\alpha_3 = 0.1(0.1)2.0(0.2)3.0(0.5)6.0$ at intervals of 0.01 for x. Bobée and Morin (1973) provide tables of percentage points of order statistics for gamma distributions.

Thom (1968) has given tables to four decimal places of the distribution function $\Gamma_x(\alpha)/\Gamma(\alpha)$:

1. for $\alpha = 0.5(0.5)15.0(1)36$ and $x = 0.0001, 0.001,$ $0.004(0.002)0.020(0.02)0.80(0.1)2.0(0.2)3.0(0.5)$—the tabulation is continued for increasing x until the value of the tabulated function exceeds 0.9900;

2. values of x satisfying the equation

$$\frac{\Gamma_x(\alpha)}{\Gamma(\alpha)} = \varepsilon$$

for $\alpha = 0.5(0.5)15.0(1)36$ and $\varepsilon = 0.01, 0.05(0.05)0.95, 0.99$. Burgin (1975) provides some interesting numerical computations of the gamma and associated functions. He cites the following representation for the incomplete gamma function ratio:

$$I(u, p) = \frac{e^{-u}}{\Gamma(p + 1)} u^p \sum_{n=0}^{\infty} \left(\frac{u^{n+1}}{\prod_{i=1}^{n}(p + i + 1)} \right). \quad (17.25)$$

Lau (1980) uses a series expansion similar to Wilk, Gnanadesikan, and Huyett (1962a) to calculate the incomplete gamma function ratio. Moore (1982) finds this series expansion not always satisfactory. Bhattacharjee (1970) provides an alternative algorithm. Phien (1991) presents an algorithm for computing the quantile $y_{\varepsilon|\alpha}$. Newton's method seems to be appropriate to solving the equation

$$I(y_{\varepsilon|\alpha}, \alpha + 1) = \varepsilon.$$

Efficient computation requires a powerful algorithm for computing the incomplete gamma function ratio and a good initial value y_0. Phien used Moore's algorithm (1982) for calculation of the incomplete gamma function ratio and the approximate value provided by Hoshi and Burges (1981) is used for the initial value y_0.

Moore's algorithm is highly accurate. In the experiment conducted by Phien (1991) Moore algorithm's value agreed with those values tabulated by E. S. Pearson (1963) up to the sixth decimal place.

5 APPROXIMATION AND GENERATION OF GAMMA RANDOM VARIABLES

The best-known approximations for probability integrals of gamma distributions have been developed in connection with the χ^2 distribution. Modifications to apply to general gamma distributions are direct, using the linear transformation $y = 2(x - \gamma)/\beta$. The reader should consult Section 5 of Chapter 18.

From (17.25) it may be observed that for u small, $u^{-(p+1)}I(u, p)$ is very approximately, a linear function of u. It is for this reason that the values

$$\log I(u, p) - (p + 1)\log u$$

tabulated by Pearson (1922) lead to relatively easy interpolation. Gray,

Thompson, and McWilliams (1969) have obtained the relatively simple approximation:

$$\frac{1}{x^{\alpha-1}e^{-x}}\int_x^\infty t^{\alpha-1}e^{-t}\,dt \doteq \frac{x}{x-\alpha+1}\left[1 - \frac{\alpha-1}{(x-\alpha+1)^2 + 2x}\right], \quad (17.26)$$

which gives good results when x is sufficiently large.

If Y has the standard uniform distribution (Chapter 26, Section 1)

$$p_Y(y) = 1, \qquad 0 < y < 1, \qquad (17.27)$$

then $(-2\log Y)$ is distributed as χ^2 with 2 degrees of freedom. If $Y_1, Y_2, \ldots, Y_s$ each have distribution (17.27) and are independent then $\sum_{j=1}^s(-2\log Y_j)$ is distributed as χ^2_{2s}; that is, it has a gamma distribution with $\alpha = s$, $\beta = 2$, and $\gamma = 0$. Using this relation, it is possible to generate gamma distributed variables from tables of random numbers. Extension to cases when α is not an integer can be effected by methods of the kind described by Bánkövi (1964).

If X has distribution (17.2), then the moment generating function of $\log X$ is

$$E[e^{t\log X}] = E[X^t] = \frac{\Gamma(\alpha+t)}{\Gamma(\alpha)}. \qquad (17.28)$$

Hence the rth cumulant of $\log X$ is

$$\kappa_r(\log X) = \psi^{(r-1)}(\alpha). \qquad (17.29)$$

Note that for α large

$$\begin{aligned}\beta_1(\log X) &\doteq \alpha^{-1}, \\ \beta_2(\log X) &\doteq 3 + 2\alpha^{-1},\end{aligned} \qquad (17.30)$$

which may be compared with

$$\begin{aligned}\beta_1(X) &= 4\alpha^{-1}, \\ \beta_2(X) &= 3 + 6\alpha^{-1}.\end{aligned}$$

The distribution of $\log X$ is more nearly normal than the distribution of X. Although this approximation is not generally used directly, it is often very useful when approximating the distributions of functions of independent gamma variables.

For example, suppose that $X_1, X_2, \ldots, X_k$ are independent variables, each distributed as χ^2 with ν degrees of freedom. Then the distribution of

$$R_k = \frac{\max(X_1, X_2, \ldots, X_k)}{\min(X_1, X_2, \ldots, X_k)} \qquad (17.31)$$

may be approximated by noting that

$$\log R_k = \max(\log X_1, \ldots, \log X_k) - \min(\log X_1, \ldots, \log X_k)$$

is approximately distributed as the range of k independent normal variables each having the same expected value, and standard deviation

$$\sqrt{\psi^{(1)}\left(\frac{\nu}{2}\right)} \doteqdot \sqrt{\frac{2}{\nu - 1}}. \tag{17.32}$$

[See also (17.107).]

Gray and Schucany (1968) and Gray and Lewis (1971) apply the method of H- and B_n-transforms (see Chapter 12, Section 4) to approximate the tail probabilities of chi-squared (and hence of gamma) distributions. Alfers and Dinges (1984) sought an approximate normalizing transformation. They obtained a polynomial expression.

A recent survey of approximations for gamma distribution quantiles is contained in a paper by Phien (1991). He notes that for a variable X with pdf given by (17.2), the standardized variable is $W = (X - \alpha)/\sqrt{\alpha}$ (for which $E[W] = 0$, $\text{Var}(W) = 1$, and $\sqrt{\beta_1(W)} = 2/\sqrt{\alpha}$). The Wilson-Hilferty chi-squared approximation (see Chapter 18, Section 5) gives

$$W_\varepsilon \doteqdot \sqrt{\alpha}\left[\left\{1 - \tfrac{1}{9}\alpha^{-1} + \tfrac{1}{3}\alpha^{-1/2}z_\varepsilon\right\}^3 - 1\right], \tag{17.33}$$

where $\Phi(z_\varepsilon) = \varepsilon$. When α is small (skewness is large), the values of W_ε given by (17.33) are too low.

Kirby (1972) modifies (17.33) and obtains the following approximation:

$$W_\varepsilon \doteqdot A(U - B), \tag{17.34}$$

where $A = \max(\sqrt{\alpha}, 0.40)$, $B = 1 + 0.0144\,[\max(0, 2\alpha^{-1/2} - 2.25)]^2$, and

$$U = \max\left[B - \sqrt{\alpha}/A, \left\{1 - \left(\tfrac{1}{6}D\right)^2 + \left(\tfrac{1}{6}D\right)z_\varepsilon\right\}^3\right]$$

with

$$D = 2\alpha^{-1/2} - 0.063\{\max(0, 2\alpha^{-1/2} - 1)\}^{1.85}.$$

Kirby (1972) provides tables to assist in the calculation of A, B, and D. Kirby's approximation was in turn modified by Hoshi and Burges (1981) who express A, B, $B - A\sqrt{\alpha}$, and D as polynomials of degree 5 in $\alpha^{-1/2}$. Phien (1991) reproduces the values given by Hoshi and Burges (1981). Harter (1969) provided tables of exact values of W_ε, to which Bobée (1979) fitted polynomials in $\alpha^{-1/2}$ of degree four. For $\sqrt{\beta_1} = 2/\sqrt{\alpha} < 4$ (i.e., $\alpha > \tfrac{1}{4}$) Bobée's approximation is superior to those of Hoshi and Burges (1981) and Kirby (1979), but it is not satisfactory for smaller values of α.

Tadikamalla and Ramberg (1975), Wheeler (1975), and Tadikamalla (1977) approximate the gamma distribution (17.2) by a four-parameter Burr distribution (see Chapter 12, Section 4.5) by equating the first four moments—expected value, variance, skewness, and kurtosis.

For the Burr distribution with cdf,

$$F_X(x) = 1 - \left\{ 1 + \left(\frac{x - a}{b} \right)^c \right\}^{-k},$$

the $100p\%$ points of the largest and smallest order statistics from a random sample of size n are

$$a + b\left\{ (1 - p)^{-1/(nk)} - 1 \right\}^{1/c}, \tag{17.35a}$$

$$a + b\left\{ (1 - p^{1/n})^{-1/k} - 1 \right\}^{1/c}, \tag{17.35b}$$

respectively. These equations give good approximations to the corresponding value for the appropriate gamma distributions. Tables are available from Tadikamalla (1977) to facilitate calculation of appropriate values for a, b, k, and c.

Values of c and k for given values of α are given in Tadikamalla and Ramberg (1975) and Wheeler (1975). These two papers were published in the *same* issue of the same journal, and yet they do not cross-reference each other. (Fortunately the values of c and k given in the two papers agree.)

Many papers on generation of gamma random variables have been written in the years 1964 to 1990. It is impossible to survey them in detail. We draw attention to Ahrens and Dieter (1974, "GO algorithm"), Fishman (1976), Jöhnk (1964), Odell and Newman (1972), Wallace (1974), and Whittaker (1974)—all are based on the general von Neumann rejection method. Cheng and Feast (1979, 1980) use the ratio of uniform random variables on the lines suggested by Kinderman and Monahan (1977).

Bowman and Beauchamp (1975) warn of pitfalls with some gamma distribution simulation routines. They note that an algorithm given by Phillips and Beightler (1972) does not actually generate random variables with gamma distributions, but rather with Weibull distributions (see Chapter 21).

6 CHARACTERIZATIONS

If X_1 and X_2 are independent standard gamma random variables [i.e., having distributions of form (17.2), possibly with different values of α; α_1, α_2, say], then the random variables

$$X_1 + X_2 \quad \text{and} \quad \frac{X_1}{X_1 + X_2}$$

are mutually independent. [Their distributions are, respectively, a standard gamma with $\alpha = \alpha_1 + \alpha_2$ and a standard beta (Chapter 25) with parameters α_1, α_2.]

Lukacs (1965) showed that this property characterizes the gamma distribution in that, if X_1 and X_2 are independent positive random variables, and $X_1 + X_2$ and $X_1/(X_1 + X_2)$ are mutually independent, then X_1 and X_2 must each have gamma distributions of form (17.1) with $\gamma = 0$, common β, but possibly different values of α. If it be assumed that X_1 and X_2 have finite second moments and identical distributions, it is sufficient to require that the regression function

$$E\left[\frac{a_{11}X_1^2 + 2a_{12}X_1X_2 + a_{22}X_2^2}{(X_1 + X_2)^2} \,\middle|\, X_1 + X_2\right], \qquad a_{11} + a_{22} \neq 2a_{12},$$

be independent of $X_1 + X_2$ to ensure that the common distribution is a gamma distribution [Laha (1964)].

Marsaglia (1974) extended Lukacs's result by removing the condition that X_1 and X_2 should be positive. He shows that "If X_1 and X_2 are independent nondegenerate random variables, then $X_1 + X_2$ is independent of X_1/X_2 if and only if there is a constant c such that cX_1 and cX_2 have standard gamma distributions." Marsaglia (1989) provides a simpler proof of this result. He uses a method of deriving Lukacs's (1965) result, which was developed by Findeisen (1978), without use of characteristic functions (although a disclaimer, suggested by the referees of the Findeison paper, claims that characteristic functions are implicit in the argument). Marsaglia (1989) also remarks that the "$X_1 + X_2, X_1/X_2$" characterization has been used in developing computer methods for generating random points on surfaces by projections of points with independent (not necessarily positive) components.

Earlier Marsaglia (1974) had obtained the following result "Let $X_1, X_2, \ldots, X_n$ $(n \geq 2)$ be independent random variables. Then the vector

$$\left(\frac{X_1}{S_n}, \frac{X_2}{S_n}, \ldots, \frac{X_n}{S_n}\right),$$

where $S_n = \sum_{j=1}^{n} X_j$, is independent of S_n if and only if there is constant c such that $cX_1, cX_2, \ldots, cX_n$ are gamma." Wang and Chang (1977) used this result to develop several sensitive nonparametric tests of exponentiality. Many multivariate generalizations of these results are surveyed in Wang (1981).

On the one hand, the distribution of X_1/X_2 is not sufficient to establish that each $|X_j|$ has a gamma distribution. If X_j is distributed as standard gamma (α_j), $(j = 1, 2)$ and X_1 and X_2 are mutually independent then the probability density function of $G = X_1/X_2$ is

$$p_G(g) = [B(\alpha_1, \alpha_2)]^{-1} g^{\alpha_1 - 1} (1 + g)^{-(\alpha_1 + \alpha_2)}, \qquad g > 0, \quad (17.36)$$

which is a Pearson Type VI distribution (see Chapter 12 and also Chapter 25). However, it is possible for X_1 and X_2 to be independent, identically distributed positive random variables, and for $G = X_1/X_2$ to have distribution (17.36), without each X_j having a gamma distribution [Laha (1954); Mauldon (1956); Kotlarski (1962, 1965)]. However, Kotlarski (1967) showed that the joint distribution of ratios X_1/X_3, X_2/X_3 (in a similar situation) does characterize the distribution (up to a constant multiplier).

It follows that any result depending *only* on the distribution (17.36) of the ratio X_1/X_2 cannot characterize the distribution of each X_j. In particular, it can be shown that if X_1 and X_2 are independent and identically distributed as in (17.2), then

$$\sqrt{\frac{a}{2}} \left(\sqrt{\frac{X_1}{X_2}} - \sqrt{\frac{X_2}{X_1}} \right)$$

has a t_{2a} distribution (as defined in Chapter 28), although this property is not sufficient to establish the form of the common distribution of X_1 and X_2 (given they are positive, independent, and identically distributed).

However, if X_3 is a third random variable (with the same properties relative to X_1 and X_2), then the joint distribution of

$$\sqrt{\frac{a}{2}} \left(\sqrt{\frac{X_1}{X_2}} - \sqrt{\frac{X_2}{X_1}} \right) \quad \text{and} \quad \sqrt{\frac{a}{2}} \left(\sqrt{\frac{X_1}{X_3}} - \sqrt{\frac{X_3}{X_1}} \right)$$

is sufficient to establish that common distribution is a gamma distribution with $\gamma = 0$. [Kotlarski (1967).]

Khatri and Rao (1968) have obtained the following characterizations of the gamma distribution, based on constancy of various regression functions:

1. If $X_1, X_2, \ldots, X_n$ $(n \geq 3)$ are independent positive random variables and

$$Y_1 = \sum_{i=1}^{n} b_{1i} X_i, \qquad b_{1i} \neq 0, i = 1, 2, \ldots, n,$$

$$Y_j = \prod_{i=1}^{n} X_i^{b_{ji}}, \qquad j = 2, \ldots, n,$$

with the $(n - 1) \times n$ matrix (b_{ji}) $(j = 2, \ldots, n, i = 1, 2, \ldots, n)$ nonsingular, then the constancy of

$$E[Y_1 | Y_2, \ldots, Y_n]$$

ensures that the X's must have a common gamma distribution (unless they have zero variances).

Setting $b_{11} = b_{12} = \cdots = b_{1n} = 1$ and $b_{j,j-1} = -1$, $b_{j,j} = 1$, with all other b's zero, the condition becomes the constancy of

$$E\left[\sum_{j=1}^{n} X_j \Bigg| \frac{X_2}{X_1}, \frac{X_3}{X_1}, \ldots, \frac{X_n}{X_1} \right].$$

2. In the conditions of **1**, if $E[X_j^{-1}] \neq 0$ $(j = 1, 2, \ldots, n)$ and

$$Z_1 = \sum_{i=1}^{n} b_{1i} X_i^{-1},$$

$$Z_j = \sum_{i=1}^{n} b_{ji} X_i, \qquad j = 2, \ldots, n,$$

with the b's satisfying the same conditions as in **1**, then the constancy of

$$E[Z_1 | Z_2, Z_3, \ldots, Z_n]$$

ensures that each X_j has a gamma distribution (not necessarily the *same* for all j), unless they have zero variances. Choosing special values of b's as in **1**, we obtain the condition that $E[\sum_{j=1}^{n} X_j^{-1} | X_2 - X_1, \ldots, X_n - X_1]$ should be constant.

3. Under the same conditions as in **1**, if $E[X_1 \log X_1]$ is finite, then the constancy of

$$E\left[\sum_{j=1}^{n} a_j X_j \Bigg| \prod_{i=1}^{n} X_i^{b_i} \right],$$

with $\sum_{j=1}^{n} a_j b_j = 0$, $|b_n| > \max(|b_1|, |b_2|, \ldots, |b_{n-1}|)$, and $a_j b_j / a_n b_n < 0$ for all $j = 1, 2, \ldots, n - 1$ ensures that X_1 has a gamma distribution (unless it has zero variance).

As a special case, setting $a_1 = a_2 = \cdots = a_n = 1$, $b_n = n - 1$, $b_1 = b_2 = \cdots = b_{n-1} = -1$, we obtain the condition as the constancy of

$$E\left[\sum_{j=1}^{n} X_j \Bigg| X_n^{n-1} \left\{ \prod_{j=1}^{n-1} X_j \right\}^{-1} \right].$$

4. If $X_1, \ldots, X_n$ are independent, positive, *and identically distributed* random variables, and if $E[X_i^{-1}] \neq 0$ $(i = 1, 2, \ldots, n)$ and

$$E\left[\sum_{j=1}^{n} a_j X_j^{-1} \,\middle|\, \sum_{j=1}^{n} b_j X_j\right]$$

is constant with the same conditions on the a's and b's as in **3**, the common distribution of the X's is a gamma distribution (unless it has a zero variance).

Giving the a's and b's the same special values as in **3**, we obtain the condition of constancy of

$$E\left[\sum_{j=1}^{n} X_j^{-1} \,\middle|\, X_n - \overline{X}\right],$$

where $\overline{X} = n^{-1} \sum_{j=1}^{n} X_j$.

Khatri and Rao (1968) have also obtained a number of further conditions characterizing gamma distributions. Hall and Simons (1969) have shown that if X and Y are mutually independent and nondegenerate, and if

$$E\left[X^2(X+Y)^{-2}\,\middle|\,X+Y\right] \quad \text{and} \quad E\left[Y^2(X+Y)^{-2}\,\middle|\,X+Y\right]$$

do not depend on $X + Y$, then either X and Y or $-X$ and $-Y$ have two-parameter gamma distributions with a common value of the scale parameter β.

The following characterization based on conditional expectation was suggested by Wang (1972): "If $\beta = (b_{jk})$ be a $n \times n$ real matrix satisfying

$$\sum_{j=1}^{n} b_{jj} = 1 \quad \text{and} \quad \sum_{j,k} b_{jk} = c \neq 0,$$

$\mathbf{X} = (X_1, X_2, \ldots, X_n)'$ is a $n \times 1$ random vector, $\mathbf{Q}^* = \mathbf{X}'\boldsymbol{\beta}\mathbf{X}$, and $L = \mathbf{C}'\mathbf{X}$, where $\mathbf{C} = (c_1, c_2, \ldots, c_n)'$ is an $n \times 1$ real vector such that $\sum_{j=1}^{n} c_j = 1$ and $c_j = c_k \neq 0$ for some j, k, then provided the distribution F of X_i $(i = 1, 2, \ldots, n)$ is nondegenerate, the conditional expectation $E[\mathbf{Q}^*|L]$ is zero almost everywhere if and only if each X_i has a gamma distribution." If N is a random variable defined by

$$N = 0 \quad \text{if } X_1 > x$$

and

$$N = n \quad \text{if } X_1 + \cdots + X_n \leq x < X_1 + \cdots + X_{n+1},$$

where $X_1, X_2, \ldots, X_n$ are independent random variables each having distribution (17.1) with α an integer and with $\gamma = 0$ (Erlang distribution), then N has the generalized Poisson distribution (Chapter 9, Section 3)

$$\Pr[N \leq n] = e^{-x/\beta} \sum_{j=0}^{(n+1)\alpha-1} \left\{ \frac{(x/\beta)^j}{j!} \right\}.$$

Nabeya (1950) showed that for $\alpha = 1$, the converse is true. That is to say, it is a characterization of the common distribution of $X_1, \ldots, X_n$ (*exponential* in this case) given that they are positive, independent, identically distributed, and continuous. Goodman (1952) extended the result to apply to any positive integer value of $\alpha \leq 2$, thus providing a characterization of a gamma distribution.

Under conditions (4), Linnik, Rukhin, and Stzelic (1970) characterize the gamma distribution by the property

$$E\left\{ P_k\left(\frac{X_1}{S_n}, \ldots, \frac{X_n}{S_n} \right) \middle| S_n \right\} \quad \text{does not depend on} \quad S_n = \sum_{j=1}^{n} X_j,$$

where P_k is a polynomial of degree k, $n \geq 2$, and $n > k$. Some conditions on the behavior of the cdf of the positive i.i.d. X_i and its derivative in an interval $[0, \varepsilon]$ were required to prove the validity of this result.

If $X_1, X_2, \ldots, X_n$ are independent gamma variables with the same scale parameter β_j, and $S(X_1, X_2, \ldots, X_n)$ is a statistic invariant under the scale transformation $X \to cX$ (for all $c \neq 0$), then $U = \sum_{j=1}^{n} X_j$ and $S = S(X_1, X_2, \ldots, X_n)$, are independent [Wang (1981)]. However, whether the gamma distribution can be characterized by the independence of U and S seems still to be an open problem.

The following characterization has been proved by Wang (1981): Let $X_1, \ldots, X_n$ ($n > 2$) be nondegenerate i.i.d. positive random variables and I_1 and I_2 be arbitrary nonempty subsets of $(1, 2, \ldots, n)$ of size $k \leq n/2$. Define $T_1 = \prod_{j \in I_1} X_j$ and $T_2 = \prod_{j \in I_2} X_j$. If

$$E[T_1^2 | U] = \theta E(T_1 + T_2)^2,$$

then the X_i's have a (two-parameter) gamma distribution (17.23).

Characterization of gamma distributions by the negative binomial has been provided by Engel and Zijlstra (1980):

Let events A occur independently and at random so that the number of events occurring in a given interval is a Poisson random variable with rate θ. The waiting time between events has a negative exponential distribution with mean θ^{-1}, and the total time T between $r + 1$ events has a gamma (r, θ^{-1}) distribution.

Consider now a second process independent of the first in which events B occur at an average rate β. If we start from a specified instant and count the number N_B of B events occurring before the rth A event, the distribution of N_B is negative binomial with parameters r and $p = \alpha/(\alpha + \beta)$.

Engel and Zijlstra (1980) have shown that T has a gamma (r, α) distribution if and only if N_B has a negative binomial distribution.

Letac (1985) characterizes the gamma distributions as follows: Given two positive independent random variables X and Y, if the distribution of Y is defined by its moments

$$E[Y^s] = \left(1 + \frac{s}{a}\right)^{\alpha+s} \quad \text{for } s > 0$$

and a fixed given positive α, then $X \exp\{(-X/\alpha)Y\}$ and X have the same distribution if and only if the distribution of X is gamma with shape parameter α [as in (17.2)]. Motivation for this characterization is that $(UV)^{uv}$ and U^u have the same distribution provided U and V are independent uniform $[0, 1]$ variables.

She (1988) has proved several theorems dealing with characterizations of the gamma distribution based on regression properties; see also Wesolowski (1990) for a characterization result based on constant regression, and Yeo and Milne (1991) for a characterization based on a mixture-type distribution.

7   ESTIMATION

Estimation of parameters of gamma distribution has also received extensive attention in the literature in the last two decades. The contributions of Bowman and Shenton and of A. C. Cohen and his coworkers should be particularly mentioned. Bowman and Shenton's (1988) monograph provides detailed analysis of maximum likelihood estimators for two-parameter gamma distributions with emphasis on the shape parameter and presents valuable information on distributions and moments of these estimators, including their joint distributions. Careful discussion of estimation problems associated with the three-parameter gamma density is also presented. The major emphasis of the monograph is on the distribution of sample standard deviation, skewness and kurtosis in random samples from a Gaussian density. The authors also deal with the moments of the moment estimators. The list of references in the monograph covers the development of the authors' work on this area from 1968 onwards.

A. C. Cohen's contributions to estimation of parameters of the gamma distribution are covered in the monographs by Cohen and Whitten (1988) and Balakrishnan and Cohen (1991) with an emphasis on modified moment estimators and censored samples. As in Chapters 14 and 15 we will concentrate on results not available in monographic literature and especially those discussed in less easily available sources.

7.1 Three Parameters Unknown

We will first consider estimation for the three parameter distribution (17.1), although in many cases it is possible to assume γ is zero, and estimate only α and β in (17.1). Given values of n independent random variables $X_1, X_2, \ldots, X_n$, each distributed as in (17.1), the equations satisfied by the maximum likelihood estimators $\hat{\alpha}$, $\hat{\beta}$, and $\hat{\gamma}$ of α, β, and γ, respectively, are

$$\sum_{j=1}^{n} \log(X_j - \hat{\gamma}) - n \log \hat{\beta} - n\psi(\hat{\alpha}) = 0, \qquad (17.37a)$$

$$\sum_{j=1}^{n} (X_j - \hat{\gamma}) - n\hat{\alpha}\hat{\beta} = 0, \qquad (17.37b)$$

$$- \sum_{j=1}^{n} (X_j - \hat{\gamma})^{-1} + n\{\hat{\beta}(\hat{\alpha} - 1)\}^{-1} = 0. \qquad (17.37c)$$

From (17.37c) it can be seen that if $\hat{\alpha}$ is less than 1, then some X_j's must be less than $\hat{\gamma}$. This is anomalous, since for $x < \gamma$ the probability density function (17.1) is zero. It is also clear that equations (17.37) will give rather unstable results if $\hat{\alpha}$ is near to 1, even though it exceeds 1. It is best, therefore, not to use these equations unless it is expected that $\hat{\alpha}$ is at least 2.5, say.

It is possible to solve equations (17.37) by iterative methods. A convenient (but not the only) method is to use (17.37a) to determine a new value for $\hat{\beta}$, given $\hat{\alpha}$ and $\hat{\gamma}$. Then (17.37b) for a new $\hat{\gamma}$, given $\hat{\alpha}$ and $\hat{\beta}$, and (17.37c) for a new $\hat{\alpha}$, given $\hat{\beta}$ and $\hat{\gamma}$.

The asymptotic variance-covariance matrix of $\sqrt{n}\,\hat{\alpha}$, $\sqrt{n}\,\hat{\beta}$, and $\sqrt{n}\,\hat{\gamma}$ is the inverse of the matrix

$$\begin{pmatrix} \psi'(\alpha) & \beta^{-1} & \beta^{-1}(\alpha - 1)^{-1} \\ \beta^{-1} & \alpha\beta^{-2} & \beta^{-2} \\ \beta^{-1}(\alpha - 1)^{-1} & \beta^{-2} & \beta^{-2}(\alpha - 2)^{-1} \end{pmatrix}.$$

The determinant of this matrix is

$$\beta^4 \left[\frac{2\psi'(\alpha)}{\alpha - 2} - \frac{2\alpha - 3}{(\alpha - 1)^2(\alpha - 2)} \right].$$

Hence

$$\text{Var}(\hat{\alpha}) \doteq 2n^{-1}\left[2\psi'(\alpha) - (2\alpha - 3)(\alpha - 1)^{-2}\right]^{-1}, \qquad (17.38a)$$

$$\text{Var}(\hat{\beta}) \doteq n^{-1}\beta^2\left[(\alpha - 1)^2\psi'(\alpha) - \alpha + 2\right]\left[2(\alpha - 1)^2\psi'(\alpha) - 2\alpha + 3\right]^{-1}, \qquad (17.38b)$$

$$\text{Var}(\hat{\gamma}) \doteq n^{-1}\beta^2(\alpha - 2)\{\alpha\psi'(\alpha) - 1\}\left[2\psi'(\alpha) - (2\alpha - 3)(\alpha - 1)^{-2}\right]^{-1}. \qquad (17.38c)$$

Using the approximation

$$\psi'(\alpha) \doteq \alpha^{-1} + \tfrac{1}{2}\alpha^{-2} + \tfrac{1}{6}\alpha^{-3}, \qquad (17.39)$$

we obtain the simple formulas·

$$\text{Var}(\hat{\alpha}) \doteq 6n^{-1}\alpha^3, \qquad (17.38a)'$$

$$\text{Var}(\hat{\beta}) \doteq 3n^{-1}\beta^2\alpha, \qquad (17.38b)'$$

$$\text{Var}(\hat{\gamma}) \doteq \tfrac{3}{2}n^{-1}\beta^2\alpha^3, \qquad (17.38c)'$$

giving the orders of magnitude of the variances when α is large. Fisher (1922) obtained the more precise approximation:

$$\text{Var}(\hat{\alpha}) \doteq 6n^{-1}\left[(\alpha - 1)^3 + \tfrac{1}{5}(\alpha - 1)\right] \qquad (17.40)$$

by using more terms in the expansion (17.39).

If the method of moments is used to estimate α, β, and γ, the following simple formulas are obtained:

$$\tilde{\gamma} + \tilde{\alpha}\tilde{\beta} = \overline{X}, \qquad (17.41a)$$

$$\tilde{\alpha}\tilde{\beta}^2 = m_2, \qquad (17.41b)$$

$$2\tilde{\alpha}\tilde{\beta}^3 = m_3, \qquad (17.41c)$$

where

$$\overline{X} = n^{-1}\sum_{j=1}^{n} X_j,$$

$$m_2 = n^{-1}\sum_{j=1}^{n}\left(X_j - \overline{X}\right)^2,$$

$$m_3 = n^{-1}\sum_{j=1}^{n}\left(X_j - \overline{X}\right)^3$$

are the sample mean, second, and third central moments. (Since this method would be used only when n is rather large, there is no need to attempt to make the estimators unbiased; it is also not clear whether this would improve the accuracy of estimation.) Note that (17.41a) and (17.37b)' are identical. From equations (17.41) the following formulas for the *moment estimators* $\tilde{\alpha}$, $\tilde{\beta}$, and $\tilde{\gamma}$ are obtained:

$$\tilde{\alpha} = \frac{4m_2^3}{m_3^2} = \frac{4}{b_1} \qquad \left(\text{where } \sqrt{b_1} = \frac{m_3}{m_2^{3/2}}\right) \qquad (17.42a)$$

$$\tilde{\beta} = \frac{\frac{1}{2}m_3}{m_2}, \qquad (17.42b)$$

$$\tilde{\gamma} = \overline{X} - \frac{2m_2^2}{m_3}. \qquad (17.42c)$$

Although these are simple formulas, the estimators are often, unfortunately, considerably less accurate than the maximum likelihood estimators $\hat{\alpha}$, $\hat{\beta}$, and $\hat{\gamma}$.

It can be shown that if n and α are large

$$\text{Var}(\tilde{\alpha}) \doteq 6\alpha(\alpha + 1)(\alpha + 5)n^{-1} \qquad (17.43)$$

[Fisher (1922)]. Comparing (17.40) and (17.43), it can be seen that the ratio of approximate values $\text{Var}(\hat{\alpha})/\text{Var}(\tilde{\alpha})$ is substantially less than 1 unless α is rather large. The ratio

$$\frac{(\alpha - 1)^3 + \frac{1}{5}(\alpha - 1)}{\alpha(\alpha + 1)(\alpha + 5)}$$

increases with α and reaches the value 0.8 at $\alpha = 39.1$.

On the other hand, we have already noted that when α is less than 2.5, the maximum likelihood estimators are of doubtful utility. It then becomes necessary to consider yet other methods of estimation. When α is less than 1, the distribution is shaped like a reversed J, with the probability density function tending to infinity as x tends to γ (see Figure 17.1). If n is large (as it usually is if a three-parameter distribution is being fitted), it is reasonable to estimate γ as the smallest observed value among $X_1, X_2, \ldots, X_n$, or a value slightly smaller than this. Estimation of α and β then proceeds as for the two-parameter case, to be described later. Using the value of α so estimated, a new value for γ can be estimated, and so on.

As in the case of the lognormal and inverse Gaussian distributions (Chapters 14 and 15), Cohen and Whitten (1988) advocate the use of modified moment estimators

$$E[X] = \bar{X} = \tilde{\gamma} + \tilde{\alpha}\tilde{\beta}, \tag{17.44a}$$

$$\text{Var}(X) = s^2 = \tilde{\alpha}\tilde{\beta}^2, \tag{17.44b}$$

$$E[F(X'_{1:n})] = \frac{1}{n+1} = F\left(\frac{X'_{1:n} - \bar{X}}{s}; 0, 1, \tilde{\alpha}\right). \tag{17.44c}$$

Tables and graphs to facilitate solution of (17.44c) are given by Cohen and Whitten (1986, 1988), Bai, Jakeman, and Taylor (1990). Note that (17.37b) and (17.37c) can be written as

$$\hat{\alpha} = \sum_{i=1}^{n}\left(\frac{X_i - \hat{\gamma}}{n\hat{\beta}}\right), \tag{17.45a}$$

$$\hat{\beta} = \sum_{i=1}^{n}\left(\frac{X_i - \hat{\gamma}}{n}\right) - \frac{n}{\sum_{i=1}^{n}(X_i - \hat{\gamma})^{-1}}, \tag{17.45b}$$

respectively. Assuming a value for $\hat{\gamma}$, $\hat{\alpha}(\hat{\gamma})$, and $\hat{\beta}(\hat{\gamma})$ can be computed, the corresponding likelihood $L(\hat{\gamma})$ calculated. The value of $\hat{\gamma}$ maximizing $L(\hat{\gamma})$ is then found numerically. [See Bai, Jakeman, and Taylor (1990).]

Cheng and Amin (1983) applied their maximum product of spacings (MPS) estimator method (see Chapters 12, 14, and 15) to provide consistent estimators of α, β, and γ. This method yields the following first-order equations:

$$\frac{\partial \log G}{\partial \gamma} \equiv \sum_{i=1}^{n+1} \frac{\int_{X'_{i-1}}^{X'_i}[-(\alpha - 1) + \beta^{-1}(x - \gamma)](x - \gamma)^{\alpha-2}e^{-(x-\gamma)/\beta}\, dx}{(n+1)\int_{X'_{i-1}}^{X'_i}(x - \gamma)^{\alpha-1}e^{-(x-\gamma)/\beta}\, dx}, \tag{17.46a}$$

$$\frac{\partial \log G}{\partial \beta} \equiv -\frac{\alpha}{\beta} + \frac{1}{n+1}\sum_{i=1}^{n+1} \frac{\int_{X'_{i-1}}^{X'_i}(x - \gamma)^{\alpha}\beta^{-2}e^{-(x-\gamma)/\beta}\, dx}{\int_{X'_{i-1}}^{X'_i}(x - \gamma)^{\alpha-1}e^{-(x-\gamma)/\beta}\, dx} = 0, \tag{17.46b}$$

$$\frac{\partial \log G}{\partial \alpha} \equiv -\log \beta - \log \psi(\alpha)$$

$$+ \sum_{i=1}^{n+1} \frac{\int_{X'_{i-1}}^{X'_i}(x - \gamma)^{\alpha-1}e^{-(x-\gamma)/\beta}\log(x - \gamma)\, dx}{(n+1)\int_{X'_{i-1}}^{X'_i}(x - \gamma)^{\alpha-1}e^{-(x-\gamma)/\beta}\, dx} = 0, \tag{17.46c}$$

where $X'_1 \leq X'_2 \leq \cdots \leq X'_n$ are the order statistics corresponding to $X_1, \ldots, X_n$.

7.2 Some Parameters Unknown

We now consider estimation when the value of one of the three parameters α, β, and γ is known. The commonest situation is when the value of γ is known (usually it is zero). Occasionally α is known (at least approximately) but not β or γ. Inadmissibility of standard estimators of gamma parameters in the case when $\gamma = 0$ has received attention in the literature.

Let $X_1, X_2, \ldots, X_k$ be independent random variables with two-parameter gamma distributions with parameters α_i, β_i $(i = 1, \ldots, k)$ where the values of the α_i's are known but the β_i's (> 0) are unknown. Berger (1980) considered weighted quadratic losses $\sum_{i=1}^{k} \beta_i^{-m}(\delta_i \beta_i^{-1} - 1)^2$ for $m = 0, 2, 1, -1$, and showed that the standard estimator of $(\beta_1, \beta_2, \ldots, \beta_k)$, namely $(X_1/(\alpha_1 + 1), \ldots, X_k/(\alpha_k + 1))$ is inadmissible for $k \geq 2$ except when $m = 0$, in which case it is inadmissible for $k \geq 3$. Ghosh and Parsian (1980) also discussed this problem for the same weighted quadratic losses. Gupta (1984) has shown that inadmissibility also holds for the loss function

$$L(\underline{\beta}', \underline{\delta}) = \sum_{i=1}^{k} \delta_i \beta_i - \sum_{i=1}^{k} \log \delta_i \beta_i - k.$$

The vector of natural estimators $(X_1/\alpha_1, \ldots, X_k/\alpha_k)$ is an inadmissible estimator of $(\beta_1, \ldots, \beta_k)$ for $k \geq 3$. It would seem, however, that the critical dimension for inadmissibility is typically 2; three dimensions are required only in special cases. The problem whether the natural estimator is admissible for $k = 2$ is an open question. Zubrzycki (1966) has considered the case when β is known to exceed some positive number β_0. He has shown that with a loss function $(\beta^* - \beta)^2/\beta^2$, where β^* denotes an estimator of β, and given a single observed value of X, estimators

$$\beta^* = (\alpha + 1)^{-1}X + b, \tag{17.47}$$

with

$$\beta_0(\alpha + 1)^{-1} \leq b \leq 2\beta_0(\alpha + 1)^{-1},$$

have minimax risk [equal to $(\alpha + 1)^{-1}$] and are admissible in the class of estimators linear in X.

If γ is known, the maximum likelihood estimators of α and β might be denoted $\hat{\alpha}(\gamma), \hat{\beta}(\gamma)$ to indicate their dependence on γ. We will, however, simply use $\hat{\alpha}$ and $\hat{\beta}$; no confusion between this use and that in Section 7.1 should arise.

If γ is known to be zero, the probability density function is of form (17.23). If $X_1, X_2, \ldots, X_n$ are independent random variables each having distribution (17.23), then equations for the maximum likelihood estimators $\hat{\alpha}, \hat{\beta}$ are

$$n^{-1} \sum_{j=1}^{n} \log X_j = \log \hat{\beta} + \psi(\hat{\alpha}), \qquad (17.48a)$$

$$\bar{X} = \hat{\alpha}\hat{\beta}. \qquad (17.48b)$$

From (17.48b), $\hat{\beta} = \bar{X}/\hat{\alpha}$. Inserting this in (17.48a), we obtain the following equation for $\hat{\alpha}$:

$$n^{-1} \sum_{j=1}^{n} \log X_j - \log \bar{X} = \psi(\hat{\alpha}) - \log \hat{\alpha}; \qquad (17.48c)$$

that is,

$$R_n = \log\left[\frac{\text{Arithmetic mean }(X_1, X_2, \ldots, X_n)}{\text{Geometric mean }(X_1, X_2, \ldots, X_n)}\right] = \log \hat{\alpha} - \psi(\hat{\alpha}).$$

(Note that $R_n \geq 0$.)

It is readily seen that the estimator $\hat{\alpha}$, of α the shape parameter, and the ratio $\hat{\beta}/\beta$ are distributed independently of β. In particular, the variance of $\hat{\beta}/\beta$ does not depend on the parent population value of β. The value of $\hat{\alpha}$ can be determined by inverse interpolation in a table of the function $[\log \alpha - \psi(\alpha)]$. Such a table has been published by Masuyama and Kuroiwa (1952). Chapman (1956) has published a table giving the results of such inverse interpolation (i.e., values of $\hat{\alpha}$) corresponding to a few values of the ratio of arithmetic to geometric mean. (He reported that a more complete table was available from the Laboratory of Statistical Research, University of Washington.)

Greenwood and Durand (1960) pointed out that the function $\alpha[\log \alpha - \psi(\alpha)]$ progresses much more smoothly than does $[\log \alpha - \psi(\alpha)]$ and so is more convenient for interpolation. They gave a table of values of $\alpha[\log \alpha - \psi(\alpha)]$ as a *function of* α to eight decimal places for argument values $0.00(0.01)1.40$, and to seven decimal places for argument values $1.4(0.2)18.0$. This method eliminates the necessity of inverse interpolation and assures high accuracy using linear interpolation. Bain and Engelhardt (1975) show that $2n\alpha \log R_n$ is approximately distributed as $c\chi_\nu^2$ for appropriate values of c and ν (depending on n and α). For $\alpha \geq 2$ we have $2n\alpha \log R_n$ approximately distributed as χ_{n-1}^2. [See also (17.107).]

Wilk, Gnanadesikan, and Huyett (1962a, b) noted that the solution of equation (17.48c) is very nearly a linear function of

$$H = \frac{\text{Arithmetic mean}}{\text{Arithmetic mean} - \text{Geometric mean}},$$

except when $\hat{\alpha}$ is less than about 2. They give a table of solutions of equation (17.48c) to five decimal places for

$$H = 1.000(0.001)1.010(0.002)1.030(0.005)1.080(0.01)1.16(0.02)1.40(0.05)$$

$$2.00(0.1)3.0(0.2)5.0(0.5)7.0(1)10(2)20(10)50.$$

For $H > 1.001$, linear interpolation gives four decimal place accuracy for $\hat{\alpha}$.

If $\hat{\alpha}$ is large enough, the approximation $\psi(\alpha) \doteqdot \log(\alpha - \frac{1}{2})$ may be used. Then from (17.48c) we have

$$\frac{\text{Arithmetic mean}}{\text{Geometric mean}} \doteqdot \frac{\hat{\alpha}}{\hat{\alpha} - \frac{1}{2}};$$

that is,

$$\hat{\alpha} \doteqdot \frac{\text{Arithmetic mean}}{2(\text{Arithmetic mean} - \text{Geometric mean})} = \tfrac{1}{2}H. \qquad (17.49)$$

For a better approximation $1/12 (= 0.083)$ should be subtracted from the right-hand side.

Thom (1968) suggests the approximation

$$\hat{\alpha}_T \doteqdot \tfrac{1}{4}R_n^{-1}\left(1 + \sqrt{1 + \tfrac{4}{3}R_n}\right). \qquad (17.50)$$

Thom further suggests adding the correction $[(\hat{\alpha}_T - 1)(24 - 96\hat{\alpha}_T)^{-1} + 0.0092]$ if $\hat{\alpha}_T > 0.9$, and gives a table of corrections for $\hat{\alpha}_T < 0.9$. It is stated that with these corrections the value of $\hat{\alpha}_T$ should be correct to three decimal places.

Asymptotic formulas (as $n \to \infty$) for the variances of $\sqrt{n}\,\hat{\alpha}$ and $\sqrt{n}\,\hat{\beta}$, and the correlation between these statistics, are

$$\begin{cases} \text{Var}(\sqrt{n}\,\hat{\alpha}) \doteqdot \alpha\{\alpha\psi'(\alpha) - 1\}^{-1}, \\ \text{Var}(\sqrt{n}\,\hat{\beta}) \doteqdot \beta^2\psi'(\alpha)\{\alpha\psi'(\alpha) - 1\}^{-1}, \\ \text{Corr}(\hat{\alpha}, \hat{\beta}) \doteqdot -\{\alpha\psi'(\alpha)\}^{-1/2}. \end{cases} \qquad (17.51)$$

Masuyama and Kuroiwa (1952) give tables with values of $\alpha\{\alpha\psi'(\alpha) - 1\}^{-1}$ and $\psi'(\alpha)\{\alpha\psi'(\alpha) - 1\}^{-1}$. If the approximation $\psi'(\alpha) \doteq (\alpha - \frac{1}{2})^{-1}$, useful for α large, is used, we have

$$\begin{cases} \mathrm{Var}(\sqrt{n}\,\hat{\alpha}) \doteq 2\alpha(\alpha - \frac{1}{2}), \\ \mathrm{Var}(\sqrt{n}\,\hat{\beta}) \doteq \beta^2\alpha, \\ \mathrm{Corr}(\hat{\alpha}, \hat{\beta}) \doteq -\sqrt{1 - \frac{1}{2}\alpha^{-1}}\,. \end{cases} \tag{17.52}$$

Bowman and Shenton (1968) investigated the approximate solutions [due to Greenwood and Durand (1960)]

$$\hat{\alpha} \doteq R_n^{-1}(0.5\,00876 + 0.1648852\,R_n - 0.0544274R_n^2), \qquad 0 < R_n \le 0.5772, \tag{17.53a}$$

$$\hat{\alpha} \doteq R_n^{-1}(17.79728 + 11.968477R_n + R_n^2)^{-1}(8.898919 + 9.059950R_n$$
$$+ 0.9775373R_n^2), \qquad 0.5772 \le R_n \le 17. \tag{17.53b}$$

The error of (17.53a) does not exceed 0.0088% and that of (17.53b) does not exceed 0.0054%.

If α is known but not β or γ, maximum likelihood estimators $\hat{\beta} = \hat{\beta}(\alpha)$, $\hat{\gamma} = \hat{\gamma}(\alpha)$ satisfy equations (17.37b)′ and (17.37c)′ with $\hat{\alpha}$ replaced by α. From (17.37b)′,

$$\hat{\gamma} = \overline{X} - \alpha\hat{\beta},$$

and hence (17.37b)′ can be written as an equation for β,

$$(\alpha - 1)\hat{\beta} = \left[n^{-1} \sum_{j=1}^{n} \left(X_j - \overline{X} + \alpha\hat{\beta} \right)^{-1} \right]^{-1}. \tag{17.37c}''$$

Alternatively, using the first two sample moments, we have for moment estimators $\tilde{\beta} = \tilde{\beta}(\alpha)$ and $\tilde{\gamma} = \tilde{\gamma}(\alpha)$,

$$\tilde{\gamma} = \overline{X} - \alpha\tilde{\beta},$$

$$\alpha\tilde{\beta}^2 = m_2 \qquad [\text{cf. (17.41a) and (17.41b)}],$$

whence

$$\tilde{\beta} = \sqrt{\frac{m_2}{\alpha}}\,, \tag{17.54}$$

$$\tilde{\gamma} = \overline{X} - \sqrt{\alpha m_2}\,.$$

In this case (α known) for n large,

$$\begin{cases} \mathrm{Var}(\hat{\beta}) \doteq \beta^2 n^{-1}, \\ \mathrm{Var}(\tilde{\gamma}) \doteq \tfrac{1}{2}\beta^2\alpha(\alpha - 2)n^{-1}, \\ \mathrm{Corr}(\hat{\beta}, \hat{\gamma}) \doteq -\dfrac{(\alpha + 1)}{\alpha + 3} \end{cases} \qquad (17.55)$$

while

$$\begin{cases} \mathrm{Var}(\tilde{\beta}) \doteq \tfrac{1}{2}\beta^2(1 + 3\alpha^{-1})n^{-1}, \\ \mathrm{Var}(\tilde{\gamma}) \doteq \tfrac{1}{2}\beta^2\alpha(\alpha + 3)n^{-1}, \\ \mathrm{Corr}(\tilde{\beta}, \hat{\gamma}) \doteq -\dfrac{(\alpha + 1)}{\alpha + 3}. \end{cases} \qquad (17.56)$$

The advantage of the maximum likelihood estimators is not so great in this case as when all three parameters have to be estimated.

Glaser (1976a) observes that the distribution of R_n is the same as that of $\prod_{i=1}^{n-1} V_i$, where the V_i's are independently distributed beta random variables with parameters α and i/n ($i = 1, \ldots, n - 1$). Various methods for calculation of the distribution of R_n and its lower critical values are available in the literature.

Provost (1988) provides expression for the jth moment of R_n and provides an expression for its probability density function by inverting the Mellin transform. In their seminal paper and book Bowman and Shenton (1983, 1988) provide an approximation to the distribution of R_n, along with a new approximation to the inverse function $\hat{\alpha} = \phi^{-1}(R_n)$, namely

$$\hat{\alpha} = \frac{1}{2R_n} + \frac{1}{6} - \frac{R_n}{18} - \frac{4R_n^2}{135} + \frac{47R_n^3}{810} \cdots . \qquad (17.57a)$$

This can be used when R_n is not too small; if R_n is small, the formula

$$\hat{\alpha} \sim (R_n + \log R_n)^{-1} \qquad (17.57b)$$

is suggested. They also suggest using Thom's formula (17.50) as a starting value for an iterative procedure, calculating the mth iterate, $\hat{\alpha}_m$ from the formula

$$\hat{\alpha}_m = \frac{\hat{\alpha}_{m-1}\{\log \hat{\alpha}_{m-1} - \psi(\hat{\alpha}_{m-1})\}}{R_n}. \qquad (17.58)$$

They observe that about ten iterations suffices for reasonable accuracy.

Bowman and Shenton (1983) also obtain the formula

$$\kappa_s(R_n) = (-1)^s\{n^{1-s}\psi^{(s-1)}(\alpha) - \psi^{(s-1)}(n\alpha)\} \qquad (17.59)$$

for the sth cumulant of R_n. As $n \to \infty$,

$$\mu_1'(R_n) \sim \frac{n-1}{2n\alpha}, \qquad (17.60a)$$

$$\mu_2(R_n) \sim \frac{n-1}{2n^2\alpha^2}, \qquad (17.60b)$$

$$\mu_3(R_n) \sim -2\sqrt{\frac{2}{n-1}}, \qquad (17.60c)$$

$$\mu_4(R_n) \sim 3 + \frac{12}{n-1}. \qquad (17.60d)$$

These values suggest that for n large R_n is approximately distributed as $\chi_{n-1}^2/(2n\alpha)$. For n large

$$E[\hat{\alpha}] \sim \frac{n\alpha}{n-3} - \frac{2}{3(n-3)} + \frac{n-1}{9(n-3)n\alpha} + \frac{7(n^2-9)}{54(n^2-9)n^2\alpha^2}$$

$$+ \frac{(n^2-1)(26n^3 + 33n^2 - 324n - 707)}{810(n^2-9)(n+5)n^3\alpha^3} + \cdots, \qquad (17.61a)$$

$$\text{Var}(\hat{\alpha}) \sim \frac{2n^2\alpha^2}{(n-3)^2(n-5)} - \frac{2n(n+1)\alpha}{3(n-3)^2(n-5)}$$

$$+ \frac{2(n^2 - 3n + 8)}{9(n-3)^2(n-5)} + \cdots \qquad (17.61b)$$

$$\sqrt{\beta_1(\hat{\alpha})} \sim \frac{4\sqrt{(2n-10)}}{n-7}\left\{1 - \frac{(n-3)^2}{24n^2\alpha^2} + \cdots\right\}, \qquad (17.61c)$$

$$\beta_2(\hat{\alpha}) \sim \frac{3(n+9)(n-5)}{(n-7)(n-9)}\left\{1 - \frac{4(n-3)^2}{3(n+9)n^2\alpha^2} - \cdots\right\}$$

$$\text{[Bowman and Shenton (1988)].} \quad (17.61d)$$

Bowman and Shenton (1982) give tables of coefficients of series expansions for the expected value, variance and third and fourth central moments of $\hat{\alpha}$ and $\hat{\beta}$ (terms up to n^{-6}) for $\alpha = 0.2(0.1)3(0.2)5(0.5)15$ and $n = 6(1)50(5)100(10)150$.

Choi and Wette (1969) present results of sampling investigations. As an example, with $\alpha = 2$ and $\beta = 1$ they obtain the average values (arithmetic means of 100 repetitions) shown below:

n	$\hat{\alpha}$	$\hat{\beta}$
40	2.10	1.03
120	2.04	1.02
200	2.03	1.01

The positive bias in $\hat{\alpha}$ is to be expected from the expansion (17.61a). Note that the next term in the expansion is

$$+ \frac{(n^2 - 1)\pi(n)}{17010(n^2 - 9)(n + 5)(n + 7)n^4\alpha^4},$$

where

$$\pi(n) = 1004n^5 + 9363n^4 + 13358n^3 - 82019n^2 - 296760n - 288472.$$

The expansion is asymptotic in nature, so only limited accuracy can be expected from it's use. It is more accurate than an expansion simply in descending powers of n.

Anderson and Ray (1975), noting that the bias in $\hat{\alpha}$ can be considerable when n is small, suggest using the following, less biased estimator based on (17.61a):

$$\hat{\alpha}^* = \frac{n - 3}{n}\hat{\alpha} + \frac{2}{3n}. \qquad (17.62)$$

For estimating $\theta = \beta^{-1}$, these authors suggest $\hat{\beta}^{-1}f(\hat{\alpha}^*)$, where

$$f(a) = \left\{1 + \frac{3na}{(n - 3)(na - 1)}\left(1 + \frac{1}{9a} - \frac{1}{na} + \frac{n - 1}{27na^2}\right)\right\}. \qquad (17.63)$$

Shenton and Bowman (1973) introduce the "almost unbiased" estimators

$$\tilde{\alpha} = \bar{X}\left\{\frac{2nR_n}{n - 1} - \frac{2nR_n^2}{2(n - 1)} + \frac{4n(n + 1)R_n^3}{9(n - 1)(n + 3)}\right.$$
$$\left. - \frac{2n(7n^2 + 60n + 7)R_n^4}{135(n - 1)(n + 3)(n + 5)}\right\}, \qquad (17.64a)$$

$$\tilde{\beta} = \frac{n - 3}{2nR_n} + \frac{n + 1}{6n} - \frac{(n + 1)R_n}{18n} - \frac{(4n^2 - 10n + 4)R_n^2}{135n(n + 3)}. \qquad (17.64b)$$

Dahiya and Gurland (1978) develop generalized minimum chi-squared estimators of α and β. Stacy (1973) proposed "quasimaximum likelihood" estimators for $1/\alpha$ and β,

$$\left(\frac{1}{\alpha}\right)' = n(n-1)^{-1} \sum (Z_i - n^{-1})\log Z_i, \qquad (17.65a)$$

$$\beta' = \bar{X}\left(\frac{1}{\alpha}\right)', \qquad (17.65b)$$

where $Z_i = X_i/n\bar{X}$ $(i = 1,\ldots,n)$.

These estimators arise from the maximum likelihood estimators for Stacy and Mihram's (1965) generalized gamma distribution (see Section 8). The estimators are unbiased and

$$\mathrm{Var}\left(\left(\frac{1}{\alpha}\right)'\right) = \frac{1}{(n-1)\alpha^2}\left\{1 + \frac{n\alpha^2 + \psi'(\alpha+1)}{n\alpha+1}\right\}, \qquad (17.66a)$$

$$\mathrm{Var}(\tilde{\beta}) = \frac{\beta^2}{(n-1)\alpha}\{1 + \alpha + \alpha^2\psi'(\alpha+1)\}. \qquad (17.66b)$$

The asymptotic efficiencies are for $(1/\alpha)'$

$$\frac{n-1}{n}\left[\{1 + \alpha^2\psi'(\alpha+1)\}\left\{1 + \frac{n\alpha^2\psi'(\alpha+1)}{n\alpha+1}\right\}\right]^{-1}, \qquad (17.67a)$$

and for $\tilde{\beta}$

$$\frac{n-1}{n}\{1 + \alpha + \alpha^2\psi'(\alpha+1)\}^{-1}. \qquad (17.67b)$$

Grouped data can be tested by standard methods. Rosaiah, Kantam, and Narasimham (1991) give a detailed analysis for fitting a two-parameter gamma distribution. They provide tables for optimum group length d, in the sense of (1) minimum asymptotic generalized variance or (2) minimum sum of asymptotic variances, for $\alpha = 2(1)5$ and number of groups, $k = 2(1)10$ for the cases when neither α nor β are known or when only α or only β is known. The group boundaries are $0, d\beta, 2d\beta, \ldots, (k-1)d\beta, \infty$. Since it is necessary to know both α and β to use the tables, they can be regarded as only providing a basis for intelligent selection of group length. Tables are also provided for use when group lengths may vary.

7.3   Estimation of Shape Parameter (β and γ Known)

Bowman and Shenton (1970) carried out an extensive study of the distributional properties of two different estimators of the shape parameters, α, of the gamma distribution (17.2). These were as follows:

1. The maximum likelihood estimator (MLE) $\hat{\alpha}$.
2. Thom's (1968) estimator $\hat{\alpha}_T$ [see (17.50)].

We recall that $\hat{\alpha}$ is the solution of the equation

$$R_n = \log \hat{\alpha} - \psi(\hat{\alpha}), \qquad (17.48c)'$$

while

$$\hat{\alpha}_T = \tfrac{1}{4}R_n^{-1}\left(1 + \sqrt{1 + \tfrac{4}{3}R_n}\right), \qquad (17.50)$$

which is applicable when $\tilde{\alpha}_T$ is not too small.

3. The moment estimator

$$\tilde{\alpha} = \frac{m_1'^2}{m_2} = \frac{\overline{X}^2}{S^2}. \qquad (17.68)$$

The estimator (17.68) is much easier to compute than either $\hat{\alpha}$ or $\hat{\alpha}_T$. All three estimators are unaffected by the value of the scale parameter (β). Dusenberry and Bowman (1977) compare these three estimators. In regard to $\tilde{\alpha}$, they apply the techniques of David and Johnson (1951) to calculate the cdf of $\tilde{\alpha}$ by noting that

$$\Pr[\tilde{\alpha} < a] = \Pr\left[\frac{\overline{X}^2}{S^2} < a\right] = \Pr[\overline{X}^2 - aS^2 < 0].$$

It is not difficult to determine the moments of ($\overline{X}^2 - aS^2$). These are then used to approximate the required probability. Dusenberry and Bowman (1977) use the Cornish-Fisher expansion to evaluate percentage points of the distribution of $\tilde{\alpha}$. They plot values of $\Pr[\tilde{\alpha} < a]$ for

$$\alpha = 0.5(0.5)3.0(1)5(5)20,$$

$$n = 50, 75, 100, 150, 200, 300, 750, 1000.$$

Figure 17.3 is the plot for $\alpha = 2.0$.

The estimator $\hat{\alpha}_T$ is quite close to $\hat{\alpha}$. Although $\hat{\alpha}_T$ has a slight systematic bias, this is offset by a somewhat lower variance and greater ease of calculation. The moment estimator $\tilde{\alpha}$ is much easier to compute than either

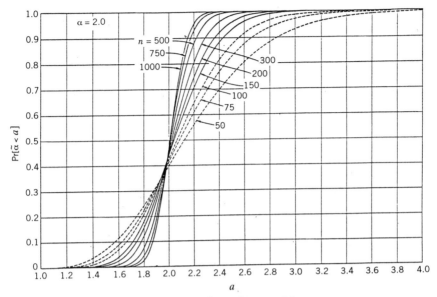

Figure 17.3 $\Pr[\tilde{\alpha} < a]$ for $\alpha = 2.0$.

α or $\hat{\alpha}_T$ but has a greater bias and variance than either of these two estimators. As measured by $\sqrt{\beta_1}$ and β_2, the distribution of $\tilde{\alpha}$ is closer to normality than that of either $\hat{\alpha}$ or $\hat{\alpha}_T$ for nearly all the values of α and n considered by Dusenberry and Bowman (1977).

Blischke (1971) constructed a BAN (best asymptotically normal) estimator of α, when β is known, as

$$\tilde{\alpha}' = \tilde{\alpha} + \frac{n^{-1}\sum_{i=1}^{n} \log(X_i/\beta) - \psi(\tilde{\alpha})}{\psi'(\tilde{\alpha})}. \tag{17.69}$$

Since $\tilde{\alpha}$ is a consistent estimator of α and

$$E\left[\log\left(\frac{X}{\beta}\right)\right] = \psi(\alpha),$$

the estimator $\hat{\alpha}'$ is asymptotically unbiased.

Results from 50 simulated samples of sizes $n = 11$, 31, and 51 are set out in Table 17.1, comparing values of $\hat{\alpha}$ and $\tilde{\alpha}'$. Note that $\hat{\alpha}$ has a positive bias, except for the case where $\alpha = 2.00$ and $n = 31$. Huque and Katti (1976) and Bar-Lev and Reiser (1983) discuss maximum *conditional* likelihood estimation of α. There is little to choose between the two estimators. Note the generally positive bias of the maximum likelihood estimator.

Table 17.1   Results of Simulation for $\tilde{\alpha}$ and $\hat{\alpha}$

		BAN Estimator ($\hat{\alpha}'$)		ML Estimator ($\hat{\alpha}$)	
α	n	Arithmetic Mean	Sample Variance	Arithmetic Mean	Sample Variance
0.25	11	0.234[a]	0.00565[a]	0.269	0.0067
	31	0.243	0.00214	0.258	0.0020
	51	0.238	0.00088	0.252	0.0010
0.50	11	0.530	0.0271	0.543	0.0219
	31	0.476	0.0059	0.498	0.0052
	51	0.494	0.0025	0.506	0.0045
0.75	11	0.760	0.0480	0.773	0.0339
	31	0.743	0.0101	0.765	0.0135
	51	0.745	0.0063	0.747	0.0069
1.25	11	1.228	0.0678	1.301	0.0879
	31	1.195	0.0204	1.257	0.0221
	51	1.253	0.0171	1.260	0.0172
1.50	11	1.533	0.0958	1.537	0.1010
	31	1.487	0.0362	1.508	0.0355
	51	1.529	0.0233	1.525	0.0199
2.00	11	2.063	0.121	2.036	0.1270
	31	1.965	0.067	1.996	0.0514
	51	1.979	0.031	2.004	0.0321

[a]Based on 48 values; in two cases negative estimates were obtained.

7.4   Order Statistics and Estimators Based on Order Statistics

A considerable amount of work has been done in evaluating the lower moments of order statistics $X_1' \leq X_2' \leq \cdots \leq X_n'$ corresponding to sets of independent random variables $X_1, \ldots, X_n$ having a common standard gamma distribution of form (17.2). Since γ and β are purely location and scale parameters, the results are easily extended to the general form (17.1). [Moments of order statistics of the exponential distribution ($\alpha = 1$) are discussed in some detail in Chapter 19, Section 6.] Tables of moments of order statistics from random samples of size n from the standard gamma distribution (17.2) are summarized in Table 17.2.

We next note that Kabe (1966) has obtained a convenient formula for the characteristic function of any linear function $\sum_{j=1}^{n} a_j X_j'$ of the order statistics. The characteristic function is

$$E\left[\exp\left\{it \sum_{j=1}^{n} a_j X_j'\right\}\right] = n! [\Gamma(\alpha)]^{-n} \int_0^\infty \cdots \int_0^{x_3} \int_0^{x_2} \left\{\prod_{j=1}^{n} x_j\right\}^{\alpha-1}$$

$$\times \exp\left\{-\sum_{j=1}^{n} (1 - ia_j t) x_j\right\} dx_1 \, dx_2 \ldots dx_n. \quad (17.70)$$

Applying the transformation

$$x_r = \prod_{j=r}^{n} w_j$$

so that $0 < w_j < 1$ for $j = 1, 2, \ldots, (n-1)$, and $w_n > 0$, we obtain the formula

$$n![\Gamma(\alpha)]^{-n} \int_0^1 \int_0^1 \cdots \int_0^1 \int_1^{\infty} \left\{ \prod_{j=1}^{n} w_j^{j\alpha-1} \right\} \exp\{-w_n D(w)\} \, dw_n \, dw_{n-1} \ldots dw_1$$

$$= \frac{n!\Gamma(n\alpha)}{[\Gamma(\alpha)]^n} \int_0^1 \int_0^1 \cdots \int_0^1 \left\{ \prod_{j=1}^{n-1} w_j^{j\alpha-1} \right\} [D(w)]^{-n\alpha} \, dw_{n-1} \ldots dw_1,$$

$$(17.71)$$

where

$$D(w) = (1 - ia_n t) + w_{n-1}(1 - ia_{n-1}t) + w_{n-2}w_{n-1}(1 - ia_{n-2}t) + \cdots$$
$$+ w_1 w_2 \ldots w_{n-1}(1 - ia_1 t).$$

Table 17.2   Details of Available Tables on Moments of Gamma Order Statistics

Reference	Values of n	Shape Parameter α	Order of Moments	Serial Number (j) of Order Statistics	Number Figures[c]
Gupta (1961)[a]	1(1)10 11(1)15	1(1)5 1(1)5	1(1)4 7(1)4	1(1)n 1	6 s.f.
Breiter and Krishnaiah (1968)	1(1)16	0.5(1)5.5 (0.5)10.5	1(1)4	1(1)n	5 s.f.
Harter (1970)	1(1)40	0.5(0.5)4.0	1	1(1)n	5 d.p.
Prescott (1974)	2(1)10	2(1)5	1	1(1)n	4 d.p.
Walter and Stitt (1988)	1(1)25 1(1)5 10(5)25	1(1)10(5)20 1, 5(5)20 1, 5(5)20	1 1 1	1, n 1(1)n 1, 5(5)n	4 d.p.
Balasooriya and Hapuarachchi (1991)[b]	1(1)10 (5)40	5(1)8	1(1)5	1(1)2	5 d.p.

[a]The values of BS (1992) occasionally differ from those of Gupta (1961) by more than 0.00001.
[b]The last of these references also includes covariances for $n = 15(5)25$, $\alpha = 2(1)5$, extending Prescott (1974) which includes covariances for $n = 2(1)10$ and the same values of α. More extensive tables of expected values of order statistics for $n = 15(1)40$, $\alpha = 5(1)8$, and covariances for $n = 2(1)25$ and $\alpha = 2(1)8$ are available on request from Balasooriya and Hapuarachchi (Memorial University of Newfoundland, St. John's).
[c]s.f. = significant figures; d.p. = decimal places.

The multiple integral can be expanded as a series of beta functions. Although we will not use it here directly, equation (17.71) is very convenient as a starting point for studying the distributions of linear functions of order statistics from gamma distributions.

The distribution of X'_r, the rth smallest among n independent random variables each having distribution (17.2), has the probability density function

$$p_{X'_r}(x) = \frac{n!}{(r-1)!(n-r)!}\left[\frac{\Gamma_x(\alpha)}{\Gamma(\alpha)}\right]^{r-1}\left[1 - \frac{\Gamma_x(\alpha)}{\Gamma(\alpha)}\right]^{n-r}\frac{x^{\alpha-1}e^{-x}}{\Gamma(\alpha)},$$

$$x > 0. \quad (17.72)$$

In general, this expression does not lend itself to simple analytic treatment. However, if α is a positive integer,

$$\frac{\Gamma_y(\alpha)}{\Gamma(\alpha)} = \frac{1}{\Gamma(\alpha)}\int_0^y t^{\alpha-1}e^{-t}\,dt = 1 - e^{-y}\sum_{j=0}^{\alpha-1}\left\{\frac{y^j}{j!}\right\} \qquad \text{[cf. (17.24a)]}$$

and so (17.72) becomes

$$p_{X'_r}(x) = \frac{n!}{(r-1)!(n-r)!}\left[1 - e^{-x}\sum_{j=0}^{\alpha-1}\left\{\frac{x^j}{j!}\right\}\right]^{r-1}$$

$$\times\left[\sum_{j=0}^{\alpha-1}\left\{\frac{x^j}{j!}\right\}\right]^{n-r}\frac{x^{\alpha-1}e^{-(n-r+1)x}}{(\alpha-1)!}, \qquad x > 0. \quad (17.73)$$

In this case it is possible to express all moments (of integer order) of X'_r and all product moments (of integer orders) of order statistics as finite sums of terms involving factorials, although these expression will usually be cumbersome.

Johnson (1952) has obtained an approximation to the distribution of range $(X'_n - X'_1)$ for random samples from (17.2). Tiku and Malik (1972) provide β_1 and β_2 values of the rth-order statistic X'_r for the gamma distribution and compare them with approximate values obtained by equating the first three moments of χ_f^2 with those of $(X'_r + a)/g$. Numerical comparisons are presented in Table 17.3. The agreement is excellent except for $\alpha = 0.01$. It is, however, doubtful whether any method of approximation based on moments can provide accurate values for the extreme lower tail, as the classical paper of E. S. Pearson (1963) indicated.

Lingappaiah (1974) obtained the joint density of X'_r and X'_s ($s > r$), from the standard gamma distribution (17.1), and he derives from it that of the difference $U_{r,s} = X'_s - X'_r$. More recently Lingappaiah (1990) has obtained the distribution of $U_{r,s}$ in the presence of a single outlier—when $n - 1$ observations are from a standard gamma α distribution but one observation is from a gamma $\alpha + \delta$ distribution. Typical values of $\Pr[U_{r,s} > u]$ under these conditions are shown in Table 17.4.

Table 17.3 Comparison of the Percentage Points of rth-Order Statistics from Standard Gamma Distribution with values from Tiku and Malik's (1972) Approximation (with $\alpha = 2$, sample size $n = 6$)

		Upper Percentage Points				
		50	75	90	95	99
$r = 2$	Exact	1.135	1.536	1.963	2.250	2.854
	Approximate	1.135	1.535	1.962	2.249	2.854
$r = 4$	Exact	2.376	3.036	3.744	4.225	5.254
	Approximate	2.370	3.040	3.754	4.234	5.245
		Lower Percentage Points				
		25	10	05	01	
$r = 2$	Exact	0.807	0.570	0.453	0.280	
	Approximate	0.808	0.571	0.453	0.276	
$r = 4$	Exact	1.829	1.423	1.214	0.885	
	Approximate	1.821	1.423	1.224	0.924	

As is to be expected, in Table 17.4 $\Pr[U_{r,s} > u]$ increases with δ. Note that the values depend only on αu, so a single table (with $\alpha = 1$) might suffice. Lingappaiah (1991) also provides tables (to 5 d.p.) with negative moments $E[X_r'^{-i}]$ of the standard gamma α distribution for $n = 2(1)5$ with $i = -(\alpha - 1), -(\alpha - 2), \ldots, -1$. As far as we know, the most general recurrence relationship in the literature for moments of order statistics from gamma α samples are those in Thomas and Moothathu (1991).

These formulas relate values of (descending) factorial moments

$$\mu_{n:n}^{(i)} = E\left[X_n'^{(i)}\right] = E\left[X_n'(X_n' - 1) \cdots (X_n' - i + 1)\right]$$

of the greatest X in a random sample of size n from the standard gamma distribution (17.1). The recurrence formula is

$$\frac{\Gamma(k + \alpha n - n + 1)}{\{\Gamma(\alpha)\}^n} \times \frac{1}{n^{k + n(\alpha - 1)}} = \sum_{j=1}^{n} A_{n-1,j} \mu_{n,n}^{(k-n+j)},$$

$$n \geq 2; \, k \geq \max[1 - \alpha, n(1 - \alpha)], \quad (17.74)$$

Table 17.4 Values of $\Pr[U_{r,s} > u]$ for $n = 5$, $r = 1$, $s = 2$

	$\alpha = 0.5$				$\alpha = 1.0$			
u	0.25	0.5	0.75	1.0	0.25	0.5	0.75	1.0
δ								
0	0.6065	0.3679	0.2231	0.1353	0.3679	0.1353	0.0498	0.0183
1	0.6672	0.4414	0.2901	0.1895	0.4414	0.1895	0.0796	0.0329
2	0.6831	0.4654	0.3161	0.2138	0.4654	0.2138	0.0968	0.0432

Note: Included are four observations from gamma α, one from gamma $\alpha + \delta$.

where

$$A_{0,1} = 1,$$

$$A_{r,1} = -r^{-1}\{k - r + 1 + (n - r)(\alpha - 1)\}A_{r-1,1},$$

$$A_{r,j} = r^{-1}(n - r)A_{r-1,j-1} - r^{-1}\{k - r + j + (n - r)(\alpha - 1)\}A_{r-1,j}$$

$$(j = 2, 3, \ldots, r - 1),$$

$$A_{r,r} = r^{-1}(n - r)A_{r-1,r}.$$

If α is an integer, expected values and moments of the other order statistics can be evaluated using Joshi's (1979) formula:

$$\mu_{k:n}^{\prime(i)} = \mu_{k-1:n-1}^{\prime(i)} + in^{-1}\Gamma(\alpha) \sum_{j=0}^{\alpha-1} \left\{ \frac{\mu_{k:n}^{\prime(i+j-\alpha)}}{j!} \right\}, \tag{17.75}$$

with $\mu_{k:n}^{\prime(i)} = E[X_{k:n}^{\prime i}]$; $\mu_{k:n}^{\prime(0)} = 1$; $\mu_{0:n}^{\prime(i)} = 0$ for $i \geq 1$. Prescott (1974) shows that

$$E[X_r' X_s'] = \frac{C}{\{\Gamma(\alpha)\}^2} \sum_b \sum_c \sum_u \sum_v \sum_j (-1)^{b+c} \binom{r-1}{b}\binom{s-r-1}{c}$$

$$\times a_u(\alpha, t - 1)a_v(\alpha, q - 1) \frac{\Gamma(u + \alpha + 1)\Gamma(u + \alpha + j + 1)}{t^{u+\alpha-j+1}(t + q)^{v+\alpha+j+1} j!},$$

$$\tag{17.76}$$

where $t = n - s + c + 1$, $q = b + s - r - c$, and summation is over $0 \leq b \leq r - 1$, $0 \leq c \leq r - s - 1$, $0 \leq u \leq (\alpha - 1)(t - 1)$, $0 \leq v \leq (\alpha - 1)(q - 1)$, $0 \leq j \leq u + \alpha$, and

$$C = \frac{n!}{(r - 1)!(s - r - 1)!(n - s)!},$$

and where $a_g(h, i)$ is the coefficient of z^g in the expansion of $\{\sum_{j=1}^{h-1}(z^j/j!)\}^i$.

Gupta (1962) obtained recurrence formulas for the crude moments $\mu_{k:n}^{\prime(i)} = E[X_{k:n}^{\prime i}]$ of the kth-order statistic $(X_{k:n}')$. These are

$$\mu_{1:n}^{\prime(i)} = \frac{n}{\Gamma(\alpha)} \sum_{j=0}^{(n-1)(\alpha-1)} a_j(\alpha, n - 1) \frac{\Gamma(\alpha + i + j)}{n^{\alpha+i+j}}, \tag{17.77a}$$

$$\mu_{k:n}^{\prime(i)} = \frac{n!}{(k - 1)!(n - k)!} \sum_{j=0}^{k-1} (-1)^j \binom{k-1}{j} \frac{\mu_{1:n-k+j+1}^{\prime(i)}}{n - k + j + 1}, \tag{17.77b}$$

where $a_j(\alpha, p)$ is as defined in (17.76); see Balakrishnan and Cohen (1991).

Prescott (1974) provided a table of variances and covariances of all order statistics for $n = 2(1)10$ with $\alpha = 2(1)5$ to four decimal places. (He used a log-gamma computer routine, accurate to about ten decimal places, to avoid errors occurring in tables prepared by Gupta (1962)—which have been ascribed to the way in which ratios of gamma functions were calculated.)

Using Gupta's tables (1962), it is possible to construct best linear unbiased estimators of the parameter β if α and γ are known. Coefficients of such estimators have been given by Musson (1965). Coefficients for best linear unbiased estimators, not using all of the sample values have been given by

1. Karns (1963) using only one-order statistic,
2. Bruce (1964) using the least M values out of n,
3. Hill (1965) using only the least number of order statistics (from a complete or censored sample) to give a specified efficiency relative to the best linear unbiased estimator using all available order statistics,
4. Särndal (1964) using the best k-order statistics. (Särndal also considers estimation of β and γ, α being known.)

Returning now to situations where it is necessary to estimate all three parameters α, β, and γ, we consider maximum likelihood estimation when the least r_1 and greatest r_2 of the X's have been censored. The maximum likelihood equations are [introducing $\hat{Z}_j = (X'_j - \hat{\gamma})/\hat{\beta}$ for convenience]

$$
\sum_{j=r_1+1}^{n-r_2} \log \hat{Z}_j - n\psi'(\hat{\alpha}) + \frac{\Gamma'(\hat{\alpha}) - \Gamma'_{\hat{Z}_{n-r_2}}(\hat{\alpha})}{\Gamma(\hat{\alpha}) - \Gamma_{\hat{Z}_{n-r_2}}(\hat{\alpha})} r_2 + \frac{\Gamma'_{\hat{Z}_{r_1+1}}(\hat{\alpha})}{\Gamma_{\hat{Z}_{r_1+1}}(\hat{\alpha})} r_1 = 0,
$$

$$(17.78a)$$

$$
-(n - r_1 - r_2)\hat{\alpha} + \sum_{j=r_1+1}^{n-r_2} \hat{Z}_j + \frac{\hat{Z}_{n-r_2}^{\hat{\alpha}} e^{-\hat{Z}_{n-r_2}}}{\Gamma(\hat{\alpha}) - \Gamma_{\hat{Z}_{n-r_2}}(\hat{\alpha})} r_2
$$

$$
- \frac{\hat{Z}_{r_1+1}^{\hat{\alpha}} e^{-\hat{Z}_{r_1+1}}}{\Gamma_{\hat{Z}_{r_1+1}}(\hat{\alpha})} r_1 = 0, \tag{17.78b}
$$

$$
-(\hat{\alpha} - 1) \sum_{j=r_1+1}^{n-r_2} \hat{Z}_j^{-1} + (n - r_1 - r_2) + \frac{\hat{Z}_{n-r_2}^{\hat{\alpha}-1} e^{-\hat{Z}_{n-r_2}}}{\Gamma(\hat{\alpha}) - \Gamma_{\hat{Z}_{n-r_2}}(\hat{\alpha})} r_2
$$

$$
- \frac{\hat{Z}_{r_1+1}^{\hat{\alpha}-1} e^{-\hat{Z}_{r_1+1}}}{\Gamma_{\hat{Z}_{r_1+1}}(\hat{\alpha})} r_1 = 0. \tag{17.78c}
$$

The equations simplify if either $r_1 = 0$ or $r_2 = 0$. For the case $r_1 = 0$ (censoring from above) a method of solving the equations is given by Harter and Moore (1965); also see Balakrishnan and Cohen (1991).

Estimation is simplified if the value of γ is known. Without loss of generality it may be arranged (if γ is known) to make $\gamma = 0$ (by adding, if necessary, a suitable constant to each observed value). For this case, with data censored from above ($r_1 = 0$), Wilk, Gnanadesikan, and Huyett (1962a, b) have provided tables which considerably facilitate solution of the maximum likelihood equations. They express these equations in terms of

$$P = \frac{\left(\Pi_{j=1}^{n-r_2} X_j'\right)^{1/(n-r_2)}}{X_{n-r_2}'}, \tag{17.79}$$

$$S = \frac{\left(\Sigma_{j=1}^{n-r_2} X_j'\right)}{(n-r_2) X_{n-r_2}'}; \tag{17.80}$$

that is, the ratios of the geometric and arithmetic means of the available observed values to their maximum. The maximum likelihood equations for $\hat{\alpha}$ and $\hat{\beta}$ are

$$(n-r_2)\log P = n\left[\psi'(\hat{\alpha}) - \log\left(\frac{X_{n-r_2}'}{\hat{\beta}}\right)\right] - r_2\frac{\partial \log J(\hat{\alpha})}{\partial \hat{\alpha}}, \tag{17.81a}$$

$$\frac{SX_{n-r_2}'}{\hat{\beta}} = \hat{\alpha} - \left(\frac{r_2}{n-r_2}\right)\frac{e^{-X_{n-r_2}'/\hat{\beta}}}{J(\hat{\alpha})}, \tag{17.81b}$$

where

$$J(\hat{\alpha}) = \int_1^\infty t^{\hat{\alpha}-1} e^{-X_{n-r_2}'t/\hat{\beta}}\, dt. \tag{17.82}$$

Note that r_2 and n enter the equations only in terms of the ratio r_2/n, and X_{n-r_2}' and $\hat{\beta}$ only as the ratio $X_{n-r_2}'/\hat{\beta}$. Wilk, Gnanadesikan, and Huyett (1962b) provide tables, giving $\hat{\alpha}$ and $\hat{\alpha}\hat{\beta}/X_{n-r_2}'$ to three decimal places for

$$\frac{n}{r_2} = 1.0, 1.1, 1.2(0.2)2.0, 2.3, 2.6, 3.0,$$

$$P = 0.04(0.04)1.00, \quad \text{and} \quad S = 0.08(0.04)1.00.$$

The values for $n/r_2 = 1$ of course correspond to uncensored samples. A special table, which we have already mentioned in Section 7.2, is provided for this case. Wilk, Gnanadesikan, and Lauh (1966) discuss generalizations and modifications of these techniques for estimation of an unknown common scale parameter based on order statistics from a sample of gamma random variables with known shape parameters not necessarily all equal.

If α is known, it is possible to use "gamma probability paper," as described by Wilk, Gnanadesikan, and Huyett (1962a) to estimate β and γ graphically. This entails plotting the observed order statistics against the corresponding expected values for the standard distribution (17.2) (which of course depends on α) or, if these are not available, the values ξ_j satisfying the equations

$$\frac{j}{n+1} = [\Gamma(\alpha)]^{-1} \int_0^{\xi_j} x^{\alpha-1} e^{-x}\, dx. \qquad (17.83)$$

In the case of *progressively censored sampling*, Cohen and Norgaard (1977) and Cohen and Whitten (1988) suggest the following procedure for solving the maximum likelihood equations (in the case when $\alpha > 1$): Let n denote the total sample size and D, the number of failing items, for which there are completely determined life spans. Suppose that censoring occurs in k stages at times $T_1 < T_2 < \cdots < T_k$ and that C_j surviving items are selected randomly and withdrawn (censored) from further observation at time T_j. Then

$$n = D + \sum_{j=1}^{k} C_j. \qquad (17.84)$$

The sample data consist of the ordered life span observations $\{X_i\}$ ($i = 1, 2, \ldots, D$), the censoring times $\{T_j\}$ and the numbers of censored items $\{C_j\}$ ($j = 1, 2, \ldots, k$). The likelihood function is

$$L = K \prod_{i=1}^{D} p_X(X_i) \prod_{j=1}^{k} \{1 - F_X(T_j)\}^{C_j}, \qquad (17.85)$$

where K is a constant, and $p_X(\cdot)$ and $F_X(\cdot)$ are the pdf and cdf of the lifetime distribution, respectively.

For the three-parameter gamma lifetime distribution (17.1), we have

$$\log L = -D \log \Gamma(\alpha) - n\alpha \log \beta - \beta^{-1} \sum_{i=1}^{D} (X_i - \gamma)$$

$$+ (\alpha - 1) \sum_{i=1}^{D} \log(X_i - \gamma) + \sum_{j=1}^{k} C_j \log(1 - F_j) + \log K, \qquad (17.86)$$

where

$$F_j = \{\beta^{\alpha} \Gamma(\alpha)\}^{-1} \int_0^{T_j - \gamma} y^{\alpha-1} \exp\left(\frac{-y}{\beta}\right) dy. \qquad (17.87)$$

The corresponding maximum likelihood equations are

$$\frac{\partial \log L}{\partial \alpha} = -D\psi(\alpha) - D \log \beta + \sum_{i=1}^{D} \log(X_i - \gamma) - \sum_{j=1}^{k} \frac{C_j}{1 - F_j} \frac{\partial F_j}{\partial \alpha} = 0,$$

$$(17.88a)$$

$$\frac{\partial \log L}{\partial \beta} = -\frac{D\alpha}{\beta} + \frac{1}{\beta^2} \sum_{i=1}^{D} (X_i - \gamma) - \sum_{j=1}^{k} \frac{C_j}{1 - F_j} \frac{\partial F_j}{\partial \beta} = 0, \qquad (17.88b)$$

$$\frac{\partial \log L}{\partial \gamma} = \frac{D}{\beta} - (\alpha - 1) \sum_{i=1}^{D} (X_i - \gamma)^{-1} - \sum_{j=1}^{k} \frac{C_j}{1 - F_j} \frac{\partial F_j}{\partial \gamma} = 0 \qquad (17.88c)$$

[cf. (17.37a)–(17.37c) for the case of uncensored samples].

Evaluation of the partial derivatives of F_j with respect to α, β, and γ and computational details are given in Cohen and Whitten (1988). They caution that convergence problems may arise in iterative solution of equations (17.88) unless $\alpha \gg 1$ (in our opinion, $\alpha \geq 2.5$). Cohen and Norgaard (1977) assert that for $\alpha \geq 4$, the formulas can be used "without any hesitation."

When α is less than 1, the likelihood function tends to infinity as $\gamma \to X_1'$. Cohen and Norgaard (1977) suggest setting an initial value $\hat{\gamma} = X_1' - \frac{1}{2}\eta$, where η is "the precision with which observations are made," and then proceeding iteratively. They also provide computational details for calculation of the asymptotic variance-covariance matrix of $\hat{\alpha}$, $\hat{\beta}$, and $\hat{\gamma}$.

The maximum likelihood estimator $\hat{\alpha}$ is the solution of

$$n^{-1} \sum_{i=1}^{n} \log\left(\frac{X_i}{\beta}\right) = \psi(\hat{\alpha}). \qquad (17.90)$$

A median estimator α^* is the solution of

$$\log\left(\frac{\text{Median}(X)}{\beta}\right) = \psi(\alpha^*). \qquad (17.91)$$

Harner, Shyu, and Trutzer (1991) carried out simulation studies of robustness of these estimators with respect to contamination of a gamma $(\alpha, 1)$ distribution by a gamma $(\alpha_0, 1)$ distribution. They took sample sizes $n = 25, 75$, with $\alpha = 1, 2, 5$ and $\alpha_0 = 0.1, 1, 5, 10$; the proportions of contaminant [gamma $(\alpha_0, 1)$] were $p = 0, 0.01, 0.05$, and 0.1. They concluded the following:

1. The moment estimator $(\tilde{\alpha})$ "greatly overestimates" the value of α.
2. The maximum likelihood estimator $(\hat{\alpha})$ "is competitive except when $\alpha_0 = 0.1$."
3. The median estimator (α^*) is "fairly stable over all combinations of the simulation parameters," with positive bias for small α, decreasing to "negligible amounts" as α or n increase.

Harner, Shyu, and Trutzer (1991) also consider trimmed mean estimators $(\hat{\alpha}_{\theta_1,\theta_2})$ satisfying the equation

$$\frac{\sum_{i=k_1}^{k_2} \log(X_i'/\beta)}{n - [\theta_1 n + \frac{1}{2}] - [\theta_2 n] - 1} = \psi(\alpha), \qquad (17.92)$$

where X_i' is the ith order statistic among $X_1, \ldots, X_n$, $k_1 = [\theta_1 n + \frac{1}{2}] + 1$ and $k_2 = n - [\theta_2 n] - 1$. Simulations were carried out for the same values of the parameters as for $\tilde{\alpha}$, $\hat{\alpha}$, and α^*, with

$$(\theta_1, \theta_2) = (0.025, 0.075) \quad \text{and} \quad (0.0, 0.1).$$

The trimmed means gave good results with $(\theta_1, \theta_2) = (0.025, 0.075)$, and it was suggested that a trimmed mean, omitting the first few order statistics, might be used in place of the maximum likelihood estimator if α is thought to be small (giving rise to observations near zero).

8   RELATED DISTRIBUTIONS

If Y has the standard uniform (rectangular) distribution

$$p_Y(y) = 1, \qquad 0 \le y \le 1, \qquad (17.93)$$

then $Z = -\log Y$ has the exponential distribution

$$p_Z(z) = e^{-z}, \qquad 0 \le z, \qquad (17.94)$$

which is a special form of gamma distribution. If $Y_1, Y_2, \ldots, Y_k$ are independent random variables, each distributed as Y, and $Z_j = -\log Y_j$ ($j = 1, \ldots, k$), then $Z_{(k)} = \sum_{j=1}^{k} Z_j$ has a gamma distribution with parameters $\alpha = k$, $\beta = 1$, $\gamma = 0$. [$2Z_{(k)}$ is distributed as χ_{2k}^2; see Chapter 18.] Relationships between gamma and beta distributions are described in Chapter 25 (see also Section 6 of this chapter).

Apart from noting these interesting relationships, we will devote this section to an account of classes of distributions that are related to gamma distributions, in particular

1. truncated gamma distributions,
2. compound gamma distributions,
3. transformed gamma distributions especially the generalized gamma distributions (which are assigned a special section of their own),
4. distributions of mixtures, sums, and products of gamma variables.

8.1   Truncated Gamma Distributions

The most common form of truncation of gamma distributions, when used in life-testing situations, is truncation from above. This is omission of values exceeding a fixed number τ, which is usually (though not always) known. If τ is not known, and the distribution before truncation is of the general form (17.1), there are four parameters $(\alpha, \beta, \gamma, \tau)$ to estimate, and technical problems become formidable. However, it is not difficult to construct fairly simple (but quite likely not very accurate) formulas for estimating these parameters.

Fortunately it is often possible to assume that γ is zero in these situations [see Parr and Webster (1965) for examples], and we will restrict ourselves to this case. We will suppose that we have observations that can be regarded as observed values of independent random variables $X_1, X_2, \ldots, X_n$, each having the probability density function

$$\frac{x^{\alpha-1}e^{-x/\beta}}{\int_0^\tau t^{\alpha-1}e^{-t/\beta}\,dt}, \qquad 0 \le x \le \tau. \tag{17.95}$$

This may be denoted as a gamma $(\alpha, \beta | \tau)$ distribution. Estimation of the parameters α and β has been discussed by Chapman (1956), Cohen (1950, 1951), Das (1955), Des Raj (1953), and Iyer and Singh (1963).

The moments of distribution (17.95) are conveniently expressed in terms of incomplete gamma functions:

$$\mu_r'(X) = \beta^r \Gamma_{\tau/\beta}(\alpha + r)/\Gamma_{\tau/\beta}(\alpha). \tag{17.96}$$

Gross (1971) notes that this is an increasing function of both α and β and, further, that μ_{r+1}'/μ_r' is an increasing function of α for $r > 0$.

The preceding results imply that

$$0 \le \mu_1' \le \frac{\beta\tau}{\beta+1} \qquad \text{for all } \alpha > 0, \tag{17.97}$$

$$\frac{\mu_{r+s}'}{\mu_r'} \le (\beta+r)\frac{\tau^s}{\beta+r+s} \qquad \text{for all } \alpha > 0. \tag{17.98}$$

Nath (1975) obtained the minimum variance unbiased estimator of the reliability function $(R(t) = \Pr[X > t])$ for the gamma $(\alpha, \beta | \tau)$ distribution *with integer* α. In his analysis he showed that the sum of n independent gamma$(\alpha, \beta | \tau)$ variables, $Y_n = X_1 + \cdots + X_n$ has cdf

$$p_{Y_n}(y | \alpha, \beta; \tau) = \frac{C^n \{\Gamma(\alpha)\}^n}{\beta^{n\alpha} \Gamma(n\alpha)} e^{-y/\beta}$$

$$\times \sum_{r=0}^{k_y} (-1)^r \binom{n}{r} (y - r\tau)^{n\alpha - 1} \theta\left(n\alpha - 1, \frac{\tau}{y - r\tau}\right)$$

for $k_y \tau < y < (k_y + 1)\tau$; $k = 0, 1, \ldots, n - 1$, (17.99)

with

$$\theta\left(n\alpha - 1, \frac{\tau}{y - r\tau}\right) = \sum_{\mathbf{r}}^* \frac{r! s(\mathbf{r})!}{\prod_{j=0}^{\alpha-1}(r_j)! \prod_{j=0}^{\alpha-1}(j!)^{r_j}} \binom{n\alpha - 1}{s(\mathbf{r})} \left(\frac{\tau}{y - r\tau}\right)^{s(\mathbf{r})},$$

(17.100)

where $s(\mathbf{r}) = \sum_{j=0}^{\alpha-1} j r_j$; the multinomial summation $\sum^*$ is over all nonnegative $\mathbf{r} = (r_1, \ldots, r_{\alpha-1})$ satisfying $\sum_{j=0}^{\alpha-1} r_j = r$.

The formula for the MVU estimator of $R(t)$ appears to be extremely cumbersome, although Nath (1975) claims that "it is not so in practical application, particularly when the sample size is small."

As $\tau \to \infty$, the distribution of Y_n tends to gamma $(n\alpha, \beta)$, as is to be expected. The MVU estimator of $R(t)$ tends to the incomplete beta function ratio $I_{t/Y_n}(\alpha, (n - 1)\alpha)$, corresponding to Basu's (1964) MVU estimator of $R(t)$, with corrected factorial term. [See also Wani and Kabe (1971), and for exponential distributions, Pugh (1963).]

8.2 Compound Gamma Distributions

Starting from (17.1), compound gamma distributions can be constructed by assigning joint distributions to α, β, and γ. The great majority of such distributions used in applied work start from (17.2) (i.e., with $\gamma = 0$) and assign a distribution to one of α and β (usually β).

If β^{-1} itself be supposed to have a gamma (δ, b^{-1}) distribution with

$$p_{\beta^{-1}}(x) = \frac{b^\delta x^{\delta - 1} e^{-xb}}{\Gamma(\delta)}, \qquad 0 \leq x, \qquad (17.101)$$

the resulting compound distribution has probability density function

$$p_X(x) = \frac{\Gamma(\alpha + \delta) \cdot b^\delta}{\Gamma(\alpha) \Gamma(\delta)} x^{\alpha - 1} (x + b)^{-(\alpha + \delta)}, \qquad 0 \leq x. \quad (17.102)$$

This belongs to Type VI of Pearson's system (Chapter 12, Section 4). It can be expressed formally as

$$\text{Gamma } (\alpha, \beta, 0) \bigwedge_{\beta^{-1}} \text{Gamma } (\delta, b^{-1}, 0) \equiv \text{Type VI.}$$

See Dubey (1970) for further details and extensions.

The noncentral χ^2 distribution (Chapter 29) is a mixture of gamma distributions, each with β equal to 2 and α distributed as $\nu + 2j$, where j is a Poisson $\frac{1}{2}\lambda$ variable. Formally

$$\text{Gamma } (\alpha, 2, 0) \bigwedge_{(\alpha-\nu)/2} \text{Poisson}\left(\tfrac{1}{2}\lambda\right) \equiv \chi_\nu'^2(\lambda),$$

or equivalently

$$\chi_\alpha^2 \bigwedge_{(\alpha-\nu)/2} \text{Poisson } \left(\tfrac{1}{2}\lambda\right) \equiv \chi_\nu'^2(\lambda).$$

Bhattacharya (1966) has considered the distribution

$$\text{Gamma } (\alpha, \beta, 0) \bigwedge_{\beta} \text{Gamma } (\alpha', \beta', 0).$$

This distribution is considerably more complicated than when β^{-1} has a gamma distribution. The probability density function is

$$2[\beta' \Gamma(\alpha) \Gamma(\alpha')]^{-1} \left(\frac{x}{\beta'}\right)^{\frac{1}{2}(\alpha+\alpha')-1} K_{\alpha'-\alpha}\left(2\sqrt{\frac{x}{\beta'}}\right), \qquad 0 < x, \quad (17.103)$$

where $K_{\alpha'-\alpha}(2\sqrt{x/\beta'})$ is the modified Bessel function of the third kind, of order $\alpha' - \alpha$ [see Abramowitz and Stegun (1965)].

8.3 Transformed Gamma Distributions

Olshen (1937) published an account of a systematic investigation into the distribution of $\log X$ when X has a standard gamma distribution (17.2). The moment-generating function of $\log X$ is

$$E[e^{t \log X}] = E[X^t] = \frac{\Gamma(\alpha + t)}{\Gamma(\alpha)}, \qquad (17.104)$$

and the cumulant generating function is

$$\log \Gamma(\alpha + t) - \log \Gamma(\alpha).$$

Hence

$$\kappa_r(\log X) = \psi^{(r-1)}(\alpha). \tag{17.105}$$

Introducing the approximation

$$\psi^{(s)}(\alpha) \doteq (-1)^{s-1}(s-1)!\left(\alpha - \tfrac{1}{2}\right)^{-s}, \qquad s \ge 1,$$

we see that the shape factors of $\log X$ are approximately

$$\alpha_3(\log X) \doteq -\left(\alpha - \tfrac{1}{2}\right)^{-1/2},$$

$$\alpha_4(\log X) \doteq 3 + 2\left(\alpha - \tfrac{1}{2}\right)^{-1}. \tag{17.106}$$

By comparison with $\alpha_3(X)$ and $\alpha_4(X)$ [see (17.11)], it can be seen that these are closer to the "normal values" 0 and 3 than are the moment ratios of the original distribution of X. The approximation "$\log X$ is normally distributed with expected value $\psi(\alpha)$ and variance $\psi'(\alpha)$" is likely to be fairly accurate for α sufficiently large. The accuracy of this approximation has been studied by Bartlett and Kendall (1946).

If the approximation is accepted, it provides an approximation to the distribution of

$$R_n = \frac{\max(X_1, \ldots, X_n)}{\min(X_1, \ldots, X_n)}, \tag{17.107}$$

where the X_j's are independent chi squared random variables with ν degrees of freedom. For

$$\log R_n = \max(\log X_1, \ldots, \log X_n) - \min(\log X_1, \ldots, \log X_n)$$

$$= \text{range}(\log X_1, \ldots, \log X_n)$$

is approximately distributed as

$$\sqrt{\psi'(\nu/2)} \times (\text{range of } n \text{ independent unit normal variables}).$$

The log-gamma distribution is defined as the distribution of a random variable Y when $-\log Y$ has a gamma (α, β) distribution. The pdf is

$$p_Y(y|\alpha, \beta) = \frac{1}{\beta^\alpha \Gamma(\alpha)} \times \frac{(-\log y)^{\alpha-1}}{y^{1+1/\beta}}, \qquad 0 < y < 1. \tag{17.108}$$

[See Consul and Jain (1971).] The rth moment about zero is

$$E[Y^r] = E[e^{rX}]$$

$$= \{\beta^\alpha \Gamma(\alpha)\}^{-1} \int_0^\infty x^{\alpha-1} \exp\{-(\beta^{-1}+r)x\}\, dx = \left(\frac{\beta}{\beta+r}\right)^\alpha.$$

$$(17.109a)$$

In particular,

$$E[Y] = \left(\frac{\beta}{\beta+1}\right)^\alpha, \tag{17.109b}$$

$$\mathrm{Var}(Y) = \left(\frac{\beta}{\beta+2}\right)^\alpha - \left(\frac{\beta}{\beta+1}\right)^{2\alpha}, \tag{17.109c}$$

$$\alpha_3(Y) = \left(\frac{\beta}{\beta+3}\right)^\alpha - 3\left(\frac{\beta}{\beta+2}\right)^\alpha\left(\frac{\beta}{\beta+1}\right)^\alpha + 2\left(\frac{\beta}{\beta+1}\right)^{3\alpha}. \tag{17.109d}$$

Generalized gamma distributions with Z^c distributed as gamma (α, β, γ) will be described in Section 8.7.

8.4 Convolutions of Gamma Distributions

The distribution of $Y = \sum_{i=1}^n X_i$ with $X_i \sim$ gamma (α_i, β_i) has received special attention in recent literature. The most recent result is due to Sim (1992) who shows that Y has pdf

$$p_Y(y|\boldsymbol{\alpha}, \boldsymbol{\beta}) = \frac{1}{\Gamma(\alpha)}\left(\prod_{i=1}^n \beta_i^{-\alpha_i}\right) y^{\alpha-1} \exp\left(-\frac{y}{\beta_n}\right)$$

$$\times \sum_{r=0}^\infty \frac{b_n(r)(\alpha_{n-1}^{(r)})}{(\alpha_n^{(r)})r!}\{(\beta_n^{-1} - \beta_{n-1}^{-1})y\}^r, \tag{17.110}$$

where $\alpha = \sum_{i=1}^n \alpha_i$,

$$b_i(r) = \begin{cases} 1, & i = 2, \\[2mm] \displaystyle\sum_{j=0}^r \frac{b_{i-1}(j)(\alpha_{i-2}^{(j)})(-r)^{(j)}}{\alpha_{i-1}^{(j)} r!}c_i^j, & i = 3, 4, \ldots, j^n, \end{cases}$$

for $r = 0, 1, 2$ and

$$c_i = \frac{\beta_{i-2}^{-1} - \beta_{i-1}^{-1}}{\beta_i^{-1} - \beta_{i-1}^{-1}}.$$

Sim (1992) claims that this result is easily obtained by mathematical induction. Moschopoulos (1985) gives the alternative expression

$$p_Y(y|\alpha, \beta) = C \sum_{i=0}^{\infty} \left[\frac{\delta_i y^{\alpha+i-1} e^{-y/\beta_i}}{\beta_i^{\alpha+i} \Gamma(\alpha + i)} \right], \qquad (17.110)'$$

where

$$\delta_{i+1} = (i + 1)^{-1} \sum_{i=1}^{k+1} \left[\sum_{j=1}^{n} \alpha_j \left(1 - \frac{\beta_i}{\beta_j} \right)^i \right] \delta_{k+1-i}.$$

His derivation is based on inversion of the moment generating function $\prod_{i=1}^{n}(1 - \beta_i t)^{-\alpha_i}$. Further expressions have been given by Mathai (1982), for cases when (1) all α_i's are equal and (2) all α_i's are integers.

Ratios of independent gamma (α_i, β_i) variables $(i = 1, 2)$ are distributed as Pearson Type VI [see also F distributions (Chapter 27)]. Distributions of products (and ratios) of gamma variables have been discussed in Malik (1967, 1968a), Lee, Holland, and Flueck (1979), and Springer (1979).

As an example, here is a derivation of the distribution of $Y = X_1 X_2$, where X_1 and X_2 are mutually independent random variables and X_j has a gamma $(\alpha_j, 1)$ distribution $(j = 1, 2)$. The joint density of X_1 and X_2 is

$$p_{X_1, X_2}(x_1, x_2) = \frac{1}{\Gamma(\alpha_1)\Gamma(\alpha_2)} x_1^{\alpha_1-1} x_2^{\alpha_2-1} e^{-x_1-x_2}, \qquad x_1, x_2 > 0.$$
$$(17.110a)$$

Making the one-to-one transformation

$$y = x_1 x_2, \ z = x_1 \text{ or inversely } x_1 = z, \ x_2 = y/z$$

with Jacobian

$$\frac{\partial(x_1, x_2)}{\partial(y, z)} = \begin{vmatrix} 0 & 1 \\ 1/z & -y/z^2 \end{vmatrix} = -1/z,$$

the joint density of $Y = X_1 X_2$ and $Z = X_1$ is

$$
p_{Y,Z}(y, z) = \frac{1}{\Gamma(\alpha_1)\Gamma(\alpha_2)} z^{\alpha_1 - 1} (y/z)^{\alpha_2 - 1} e^{-z - (y/z)} \cdot \frac{1}{z}
$$

$$
= \frac{1}{\Gamma(\alpha_1)\Gamma(\alpha_2)} y^{\alpha_2 - 1} z^{\alpha_1 - \alpha_2 - 1} e^{-z - (y/z)}, \qquad y, z > 0.
$$

Hence, the density function of $Y = X_1 X_2$ is

$$
p_Y(y) = \frac{1}{\Gamma(\alpha_1)\Gamma(\alpha_2)} y^{\alpha_2 - 1} \int_0^\infty z^{\alpha_1 - \alpha_2 - 1} e^{-z - (y/z)} \, dz
$$

$$
= \frac{1}{\Gamma(\alpha_1)\Gamma(\alpha_2)} y^{\frac{1}{2}(\alpha_1 + \alpha_2) - 1} \int_0^\infty t^{\alpha_1 - \alpha_2 - 1} e^{-(t + t^{-1})\sqrt{y}} \, dt \qquad \left(z = t\sqrt{y} \right)
$$

$$
= \frac{2}{\Gamma(\alpha_1)\Gamma(\alpha_2)} y^{\frac{1}{2}(\alpha_1 + \alpha_2) - 1} K_{\alpha_1 - \alpha_2}(2\sqrt{y}), \qquad y > 0, \qquad (17.110b)
$$

where $K_g(h)$ is the modified Bessel function of the third kind [see Abramowitz and Stegun (1965)].

Kotz and Srinivasan (1969) have obtained this result by the use of Mellin transforms.

The distribution of the ratio X_1/X_2 will be discussed in Chapter 27. The book by Springer (1979) contains many examples of distributions of products and sums of gamma (and beta) variables.

8.5 Mixtures of Gamma Distributions

A k-component mixture, in proportions $p_1 : p_2 : \cdots : p_k \left(\sum_{i=1}^k p_i = 1 \right)$, of gamma (α, β) distributions has pdf,

$$
p_X(x \mid \boldsymbol{\alpha}, \boldsymbol{\beta}; \mathbf{p}) = \sum_{i=1}^k p_i \frac{x^{\alpha_i - 1}}{\beta_i^{\alpha_i} \Gamma(\alpha_i)} \exp\left(-\frac{x}{\beta_i} \right). \qquad (17.111)
$$

Kanno (1982) has described an iterative procedure for fitting a two-component distribution of type (17.111). He emphasizes that the choice of initial values for $(\alpha_1, \alpha_2; \beta_1, \beta_2; p_1)$ substantially affects the convergence of the procedure.

8.6 Reflected Gamma Distributions

Borghi (1965) introduced a "reflected version of the gamma distribution" derived by reflecting the three-parameter distribution (17.1) about γ. It has pdf,

$$p_X(x) = \frac{1}{2} \frac{|x - \gamma|^{\alpha - 1}}{\beta^{\alpha} \Gamma(\alpha)} \exp\left(-\frac{|x - \gamma|}{\beta}\right), \qquad \alpha, \beta > 0. \quad (17.112)$$

The standard form has $\gamma = 0$, $\beta = 1$ [cf. (17.2)] with pdf,

$$p_X(x) = \frac{1}{2} \frac{|x|^{\alpha - 1}}{\Gamma(\alpha)} \exp(-|x|). \qquad (17.113)$$

Kantam and Narasimham (1991) have given tables of the BLUEs of β and γ (when α is known) for $\alpha = 2(1)5$ and sample size $n = 2(1)10$ for complete and censored (symmetric- and right-) samples. They also studied the properties of the quasi-midrange $[\frac{1}{2}(X_i' + X_{n-i+1}')]$, the sample median, trimmed mean, and Winsorized mean as alternative estimators for γ, and the mean deviation from the median, and the quasi-range $(X_{n-i+1}' - X_i')$ as bases for alternative estimators for β. Among their conclusions they note that although the median is a good alternative to the BLUE of γ when $\alpha = 1$ [the double exponential, or Laplace, distribution (see Chapter 24)], this is not the case for larger values of α.

Harvey (1967) has considered a more general form of reflected distribution, with pdf

$$K\left\{a + \frac{|x - \gamma|}{b}\right\}^{\alpha - 1} \exp\left\{-\frac{|x - \gamma|}{b}\right\}, \qquad a, b, \alpha > 0, \quad (17.114)$$

where

$$K = \left\{2\int_0^{\infty}\left(a + \frac{t}{b}\right)e^{-t/b}\,dt\right\}^{-1}$$

$$= \frac{1}{2}[be^a\{\Gamma(\alpha) - \Gamma_a(\alpha)\}].$$

An attractive feature of this class of distributions, as compared with that of Borghi, is that the density is not, in general, zero at the point of symmetry (γ).

If a, b, and α are known, the maximum likelihood estimator $\hat{\gamma}$ satisfies the equation

$$(\alpha - 1)\sum_{i=1}^{n}\left\{\left(a + \frac{|X_i - \hat{\gamma}|}{b}\right)\mathrm{sgn}(X_i - \hat{\gamma})\right\} = \sum_{i=1}^{n}\mathrm{sgn}(X_i - \hat{\gamma}). \quad (17.115)$$

Wesolowski (1993) presented a characterization for the reflected generalized gamma distribution. His result is that, for two i.i.d. random variables X and Y, symmetric distribution of X/Y is independent of $|X|^\gamma + |Y|^\gamma$ ($\gamma > 0$) if and only if the random variables have the reflected generalized gamma distribution.

8.7 Generalized Gamma Distributions

Generalized gamma distributions were discussed by Amoroso (1925), who fitted such a distribution to an observed distribution of income rates. Between 1925 and 1962, however, there appeared to be little interest in this family of distributions. An interesting physical model generating generalized gamma distributions was described by Lienhard and Meyer (1967).

If it be supposed that $\{(Z - \gamma)/\beta\}^c = X$ (with $c > 0$) has the standard gamma distribution (17.2), then the probability density function of Z is

$$p_Z(z|\alpha, \beta, \gamma, c) = \frac{c(z - \gamma)^{c\alpha - 1}}{\beta^{c\alpha}\Gamma(\alpha)} \exp\left[-\left(\frac{z - \gamma}{\beta}\right)^c\right], \qquad z \geq \gamma. \quad (17.116)$$

This was defined (with $\gamma = 0$) by Stacy (1962) as the family of generalized gamma distributions [see also Cohen (1969)]. It includes Weibull distributions ($\alpha = 1$), half-normal distributions ($\alpha = \frac{1}{2}$, $c = 2$, $\gamma = 0$), and of course ordinary gamma distributions ($c = 1$). In addition the lognormal distribution is a limiting special case when $\alpha \to \infty$. If $\gamma = 0$, the cdf can be expressed as

$$F_Z(z; \beta, \alpha, c) = G\left[\left(\frac{z}{\beta}\right)^c; \alpha\right], \qquad (17.117)$$

where $G(a, \alpha)$ is the incomplete gamma function ratio $\Gamma_a(\alpha)/\Gamma(\alpha)$.

Since $\{(Z - \gamma)/\beta\}^c$ has a standard gamma distribution, it is clear that [from (17.8)]

$$E\left[\left\{\frac{Z - \gamma}{\beta}\right\}^r\right] = E\left[\left\{\frac{Z - \gamma}{\beta}\right\}^{c(r/c)}\right]$$

$$= \frac{\Gamma(\alpha + r/c)}{\Gamma(\alpha)}. \qquad (17.118)$$

The moments and shape factors of Z can be deduced from (17.118). We may note, in particular, that if $c = 2$, then $(Z - \gamma)^2$ has a gamma distribution, and that if $c = 2$, $\gamma = 0$, $\beta = 2$, and $\alpha = \frac{1}{2}\nu$, then Z is distributed as χ_ν.

If $\gamma = 0$, the incomplete rth moment (up to ζ) of Z is

$$\frac{c}{\beta^{c\alpha}\Gamma(\alpha)} \int_0^\zeta z^{c\alpha+r-1} \exp\left\{-\left(\frac{z}{\beta}\right)^c\right\} dz$$

$$= \frac{\beta^h \exp\{-(\zeta/\beta)^c\}}{(\alpha+r/c)\Gamma(\alpha)} \left(\frac{\zeta}{\beta}\right)^{c\alpha+r} {}_1F_1\left(\left.\begin{matrix} 1 \\ \alpha+r+r/c \end{matrix}\right| \left(\frac{\zeta}{\beta}\right)^c\right), \quad (17.119)$$

where ${}_1F_1\left(\left.\begin{matrix} a \\ b \end{matrix}\right| y\right)$ is the confluent hypergeometric function, which is a special case of the generalized hypergeometric function ${}_pF_q\left(\left.\begin{matrix} a_1,\ldots,a_p \\ b_1,\ldots,b_q \end{matrix}\right| x\right)$ (see Chapter 1).

The generalized gamma distribution has two shape parameters, α and c. Taguchi (1980) suggested an alternative parametrization, introducing parameters $h = \alpha^{-1}$, $d = c\alpha$ in place of α and c. He refers to d and h as skewness and "sharpness" parameters, respectively, and suggests that they be estimated using a generalized concentration curve (closely related to the Gini concentration coefficient).

If $\gamma = 0$, we have

$$\log Z = c^{-1}(\log X + \log \beta), \quad (17.120)$$

with X having a standard gamma (α) distribution. Hence the shape factors $(\alpha_3$ and $\alpha_4)$ of $\log Z$ are the same as those of $\log X$, and they depend only on α, not on c. The distribution is bell-shaped for $c\alpha > 1$ and reverse J-shaped otherwise. If the values of c and γ are known, problems of estimation can be reduced to similar problems for ordinary gamma distributions by using the transformed variable $(Z - \gamma)^c$.

The two parameters c and α define the shape of the distribution (17.116). These distributions cover an area in the (β_1, β_2) plane. Note that, as for the Weibull distribution, there is a value $c = c(\alpha)$ for which $\beta_1 = 0$. For $c < c(\alpha)$, $\sqrt{\beta_1} > 0$; for $c > c(\alpha)$, $\sqrt{\beta_1} < 0$.

Roy (1984) presented a characterization of the generalized gamma distribution based on a conditional distribution along the lines of Patil and Seshadri (1964). Mees and Gerard (1984) applied the generalized gamma distribution to model the distributional shape of seed germination curves, and also discussed the estimation of the three parameters as well as the germination capacity through the maximum likelihood estimation method.

Johnson and Kotz (1972) studied power transformations of gamma (α, β) variables, which generate the (Stacy) generalized gamma distributions. They provide a diagram (Figure 17.4) showing regions of the $(\sqrt{\beta_1}, \beta_2)$ plane corresponding to distributions of [gamma $(\alpha)]^\lambda$ variables [i.e., (Stacy) generalized gamma distributions with $c = \lambda^{-1}$]. (As $\lambda \to 0$, and so $c \to \infty$, the distribution tends to that of the logarithm of a gamma variable.)

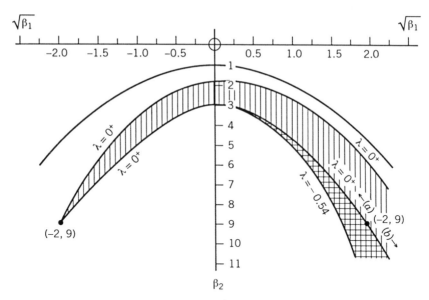

Figure 17.4 Region of Existence of (Gamma)$^\lambda$ Distributions, Shaded; Region Where Two Distributions Exist, Cross-hatched.

In the figure the region is bounded by the lines $\lambda = 0^+$, the power-function line $[p_Z(z) = \theta z^{\theta-1} \ (0 < z < 1, \ \theta > 0)]$ the log Type III line, and $\lambda = -0.54$. There are two distributions for each point between the $\lambda = -0.54$ and $\lambda = 0$ lines. This area is cross-hatched. Cammilleri (1971) calls the generalized gamma with $\gamma = 0$, $x > 0$ the *generalized seminormal density*. He studies the distribution of the product XY, where X and Y are two independent random variables having generalized seminormal distributions with distinct values of the parameters α and β but the same parameter c.

Pakes and Khattree (1992) show that the distribution of Z is determined by the moments only if $c > \frac{1}{2}$. (Compare with a related result for the lognormal distribution in Section 3 of Chapter 14.) Indeed, if $c < \frac{1}{2}$, any distribution with pdf

$$p_Z(z|\alpha, \beta, 0, c)\{1 - \varepsilon \sin(-2\pi c\alpha + z^c \tan 2\pi c)\} \qquad (17.121)$$

has the same moments for $-1 \le \varepsilon < 1$, including $\varepsilon = 0$.

Roberts (1971) shows that a necessary and sufficient condition for $|W|^{1/c}$ ($c \ne 0$) to have a gamma (α, β) distribution is that the pdf is of form

$$p_W(w) = h(w)|w|^{c\alpha - 1} \exp\left(-\frac{|w|^c}{\beta}\right), \qquad (17.122)$$

with

$$h(w) + h(-w) = |c|\{\beta^{\alpha}\Gamma(\alpha)\}^{-1} \qquad \text{for all } w.$$

[Roberts and Geisser (1966) had previously obtained this result for $c = 2$.] Jakuszenkowa (1973) further extended this result, obtaining a characterization in terms of the joint distribution of variables

$$\sqrt{Z_i Z_0^{-1}} - \sqrt{Z_i^{-1} Z_0}, \qquad i = 1, \ldots, n,$$

where $Z_0, Z_1, \ldots, Z_n$ are mutually independent positive random variables.

Lefante and Turner (1985) calculated an "average likelihood" for distribution (17.116) with $\gamma = 0$ by integrating the likelihood (based on sample values $Z_1, Z_2, \ldots, Z_n$):

$$\frac{c^n (\prod_{i=1}^{n} Z_i^c)^{\alpha - 1/c}}{\{\Gamma(\alpha)\}^n} (\beta^{-c})^{n\alpha} \exp\left\{-(\beta^{-c}) \sum_{i=1}^{n} Z_i^c\right\}$$

with respect to β^{-c} (from zero to infinity), obtaining the value

$$\frac{|c|^n \Gamma(n\alpha + 1)}{\{\Gamma(\alpha)\}^n} \left(\prod_{i=1}^{n} Z_i^c\right)^{\alpha - 1/c} \left(\sum_{i=1}^{n} Z_i^c\right)^{-(n\alpha + 1)}. \qquad (17.123)$$

They describe the use of this statistic in a form of Bayesian estimation.

The essential difficulty in estimation for generalized gamma distributions with $\gamma = 0$ is estimation of c. In fact, if c is known, one can apply the transformation $X = Z^c$ and use the methods appropriate to gamma distributions.

Maximum likelihood estimation of c, α, and β (assuming γ to be known) was described by Parr and Webster (1965). Taking $\gamma = 0$, the maximum likelihood equations are [Hager and Bain (1970)]:

$$-n\hat{\alpha} + \sum_{i=1}^{n} \left(\frac{Z_i}{\hat{\beta}}\right)^{\hat{c}} = 0, \qquad (17.124a)$$

$$\frac{n}{\hat{c}} + \hat{\alpha} \sum_{i=1}^{n} \log\left(\frac{Z_i}{\hat{\beta}}\right) - \sum_{i=1}^{n} \left(\frac{Z_i}{\hat{\beta}}\right)^{\hat{c}} \log\left(\frac{Z_i}{\hat{\beta}}\right) = 0, \qquad (17.124b)$$

$$-n\psi(\hat{\alpha}) + \hat{c} \sum_{i=1}^{n} \log\left(\frac{Z_i}{\hat{\beta}}\right) = 0. \qquad (17.124c)$$

Equation (17.124a) yields $\hat{\beta} = \hat{\beta}(\hat{\alpha}, \hat{c}) = \left\{\sum_{i=1}^{n} Z_i^{\hat{c}} / (n\hat{\alpha})\right\}^{1/\hat{c}}$. Substituting this expression for $\hat{\beta}$ into (17.124b) we have

$$\hat{\alpha} = \hat{\alpha}(\hat{c}) = \frac{1}{\hat{c}\left[\left(\sum_{i=1}^{n} \log Z_i / n\right) - \left(\sum_{i=1}^{n} Z_i^{\hat{c}} \log Z_i\right) / \left(\sum_{i=1}^{n} Z_i^{\hat{c}}\right)\right]}. \qquad (17.125)$$

Substituting into (17.124c) yields an equation in the single unknown $\hat{c}$, and then the problem is to solve $H(\hat{c}) = 0$ for $\hat{c}$, where

$$H(\hat{c}) = -\psi(\hat{\alpha}) + \hat{c}\frac{\sum_{i=1}^{n} \log Z_i}{n} - \log\left(\sum_{i=1}^{n} Z_i^{\hat{c}}\right) + \log n\hat{\alpha}, \quad (17.126)$$

and $\hat{\alpha}$ is given by (17.125).

Extensive investigation of $H(\hat{c})$ conducted by Hager (1970) in his doctoral dissertation indicated that it is not always possible to determine if $H(\hat{c}) = 0$ has a root. The complication is that as $\hat{c} \rightarrow 0$, $\hat{\alpha}$ becomes quite large. There are conflicting reports in the literature as to the possibility of solving the equation $H(\hat{c}) = 0$. Hager and Bain (1970) and Lawless (1980) report persistent divergence of their iterative numerical methods. Their experience led them to conclude that MLE may not exist unless $n > 400$, say. Stacy (1973), on the other hand, reports on multiple distinct solutions to $H(\hat{c}) = 0$ for sample sizes as small as 20. It has been found that the Newton-Raphson method does not work well and that the existence of solutions to the log-likelihood equations is sometimes in doubt. When maximum likelihood estimates could be found, the asymptotic normal distribution of $\hat{\alpha}$ was not in effect even for samples of size 400 when sampling with α values of 1 and 2.

Lawless's (1980) method maximizes the likelihood function over a subset of the parameters with the remaining parameters temporarily held fixed at some initial values. This is then followed by a heuristic interpolation scheme that refines further values of the fixed parameter. This approach guarantees only a local maximum of the likelihood function and is not very efficient computationally.

Wingo (1987b) recommends the root isolation method developed by Jones, Waller, and Feldman (1978) which assures globally optimal maximum likelihood estimators of generalized gamma distribution parameters. See also Wingo (1987a) for additional justification of the method.

Cohen and Whitten (1988) recommend the reparametrization $\theta = \beta^c$, which results in the three equations:

$$\hat{\theta} = \sum_{i=1}^{n} \frac{Z_i^{\hat{c}}}{n\hat{\alpha}}, \quad (17.127a)$$

$$\sum_{i=1}^{n} Z_i^{\hat{c}} - \frac{n\hat{\alpha}\sum_{i=1}^{n} Z_i^{\hat{c}} \log Z_i}{(n/\hat{c}) + \hat{\alpha}\sum_{i=1}^{n} \log Z_i} = 0, \quad (17.127b)$$

$$\hat{c} \sum_{i=1}^{n} \log Z_i - n \log \hat{\theta} - \psi(\hat{\alpha}) \equiv D(\hat{\theta}, \hat{\alpha}, \hat{c}) = 0. \quad (17.127c)$$

They suggest, for example, starting with a first approximation c_1, for $\hat{c}$, substituting it into (17.127b) and solving this equation for $\hat{\alpha} - \alpha_1$, substituting

α_1 and c_1 into (17.127a) and solving for $\hat{\theta} - \theta_1$, and then substituting the three approximation into (17.127c). The cycle is to be repeated until a pair α_i and c_i of successive approximations are sufficiently close to each other, and then interpolating for the final estimates. They do not address directly the convergence problem.

We note two possible parametrizations of the three-parameter generalized gamma distribution (with $\gamma = 0$): the form

$$p_Z(z; a, d, b) = \frac{b}{a^d \Gamma(d/b)} z^{d-1} e^{-(z/a)^b}, \qquad a, d, b > 0; \ z \geq 0,$$

$$(17.128a)$$

originally suggested by Stacy (1962), and another obtained by introducing $k = d/b$ allowing the power parameter to be negative

$$p_Z(z; a, k, b) = \frac{|b|}{\Gamma(k) a^{bk}} z^{bk-1} e^{-(z/a)^b}, \qquad a, k > 0, b \neq 0; \ z \geq 0.$$

$$(17.128b)$$

The second form was suggested by Stacy and Mihram (1965) and is commonly used in practice. Straightforward moment estimation of parameters based on moments of $\log Z$ was proposed by Stacy and Mihram (1965) (see pages 395–396).

Estimation in Four-parameter Generalized Gamma Distribution

Harter (1966) studied maximum likelihood estimation of all four parameters α, β, γ, and c. He gives tables of asymptotic variances and covariances and includes an extract from these tables in Harter (1967), which also contains the results of a number of applications of the method of maximum likelihood to complete samples of size 40, and also to such samples censored by exclusion of the greatest 10, 20, or 30 observed values [also see Harter (1970)]. From the evidence of some sampling experiments, it appears that

1. maximum likelihood estimates have a definite bias in samples of this size,
2. estimators of α and c have high negative correlations.

The generalized gamma distribution is not recommended as a model for analysis of sample data, unless the sample size is large enough to group the data in a frequency table. Cohen and Whitten (1988) suggest a minimum sample size of 100 observations—even more if a four-parameter model is employed.

They recommend a modified moment estimation procedure, equating sample and population values for the first three moments but equating the cdf at the first-order statistic to $(n + 1)^{-1}$, in place of equating fourth sample and population moments (which can involve large sampling errors). First, assuming a trial value c_1 for c, solve the fourth equation for the moment-ratio α_3 $(= \sqrt{\beta_1})$ and proceed iteratively until a final value $\tilde{c}$, say, is obtained producing close agreement with the sample value of α_3. Using this value for c, the estimators for α, β, and γ are calculated from the remaining equations.

Cohen and Whitten (1988) provide extensive tables of α_3 as a function of c and α for $c = 0.2, 0.5(0.5)3.5$, and also of the corresponding cdf for $c = 0.5(0.5)3.5$. Rao, Kantam, and Narasimham (1991) has derived Lloyd-type (1952) BLUE of the location γ and scale β parameters of the four-parameter generalized gamma distribution. They give coefficients of BLUEs based on order statistics $(Z'_{r_1+1}, \ldots, Z'_{n-r_2})$—that is, excluding the r_1 least and the r_2 greatest—for $c = 0.25, 0.5, 2, 4$, $\alpha = 2(1)4$, and $n = 3(1)10$, together with their variances. This paper also includes analyses (for three-parameter generalized gamma) for (1) Gupta-type BLUEs, γ_G^* and β_G^*, of γ and β, respectively, in which the order statistics are treated as if they were uncorrelated and had a common variance (see Chapter 13), (2) "trimmed-type" estimators

$$\beta_T^* = \frac{\Sigma^*(Z'_i - Z'_{r_1+1})}{\Sigma^* E^*\left[Z'_i - Z'_{r_1+1}|\beta = 1\right]},$$ (17.129)

(3) Winsorized-type estimators

$$\beta_W^* = \frac{r_1 Z'_{r_1+1} + \Sigma^* Z'_i + r_2 Z'_{n-r_2} - n Z'_{r_1+1}}{E^*\left[r_1 Z'_{r_1+1} + \Sigma^* Z'_i + r_2 Z'_{n-r_2} - n Z'_{r_1+1}|\beta = 1\right]},$$ (17.130)

and (4) median-type estimators

$$\beta_M^* = \begin{cases} \dfrac{Z'_{(n+1)/2} - Z'_{r_1+1}}{E^*\left[Z'_{(n+1)/2} - Z'_{r_1+1}|\beta = 1\right]} & \text{if } n \text{ is odd,} \\[4mm] \dfrac{Z'_{(n+2)/2} + Z'_{n/2} - 2Z'_{r_1+1}}{E^*\left[Z'_{(n+2)/2} + Z'_{n/2} - 2Z'_{r_1+1}|\beta = 1\right]} & \text{if } n \text{ is even.} \end{cases}$$ (17.131)

In each of equations (17.129)–(17.131) the expected values (E^*) are calculated on the basis $\beta = 1$; in (17.29) and (17.130), Σ^* denotes summation from $i = r_1 + 1$ to $i = n - r_2$. Rao, Kantam, and Narasimham (1991) report that in all patterns of censoring Gupta-type BLUEs (γ_G^* and β_G^*) perform well, with minimum relative efficiencies of 84% for $c = 2$ and 87% for $c = 4$. In decreasing order of preference for the other estimators of β we have $\beta_W^*, \beta_T^*, \beta_M^*$, but for $0.25 \le c \le 1$, β_T^* is preferable to β_W^*.

Some investigations have been aimed at estimation of reliability—that is, the survival function $\Pr[Z > z] = 1 - F_Z(z)$. Since this is just a particular function of the parameters, one would expect difficulties to arise when there are difficulties in estimation of α, β, and γ. So it is not surprising that Hager, Bain, and Antle (1971) and Hager and Bain (1970) found that no MLE of reliability may exist for small sample sizes or when c is small.

Log (Generalized) Gamma Distribution Estimation

Stacy and Mihram (1965) were first to propose a method estimation based on the moments of $\log Z$. [Of course if the known value of γ is not zero, $\log(Z - \gamma)$ would be used.] The moment-generating function of $T = \log(Z/\beta)$ is

$$E[e^{Tt}] = E\left[\left(\frac{Z}{\beta}\right)^t\right] = \frac{\Gamma\left(\alpha + \dfrac{t}{c}\right)}{\Gamma(\alpha)} \qquad \text{(cf. Section 2).}$$

Hence the cumulant generating function of T is

$$\log \Gamma\left(\alpha + \frac{t}{c}\right) - \log \Gamma(\alpha),$$

$$\kappa_r(T) = c^{-r}\psi^{(r-1)}(\alpha),$$

so that

$$\begin{cases} \kappa_1(\log Z) = c^{-1}\psi(\alpha) + \log \beta \\ \kappa_r(\log Z) = c^{-r}\psi^{(r-1)}(\alpha), \qquad r \geq 2. \end{cases} \tag{17.132}$$

The equations

$$\begin{cases} \mu'_1(\log Z) = c^{-1}\psi(\alpha) + \log \beta \\ \mu_2(\log Z) = c^{-2}\psi'(\alpha) \\ \mu_3(\log Z) = c^{-3}\psi''(\alpha) \end{cases} \tag{17.133}$$

can be written in the form

$$\frac{\mu_3}{\mu_2^{3/2}} = \frac{\psi''(\alpha)}{[\psi'(\alpha)]^{3/2}} \tag{17.133a}$$

$$c = \frac{\mu_2 \psi''(\alpha)}{\mu_3 \psi'(\alpha)} \tag{17.133b}$$

$$\beta = \exp[\mu'_1 - c^{-1}\psi(\alpha)]. \tag{17.133c}$$

Replacing population moments in (17.133) by sample moments and solving successively for α, c, and β gives the required moment estimators.

Lawless (1980) has illustrated the usefulness of the log-gamma distribution as a life-test model. The maximum likelihood estimation of the parameters has been discussed by Lawless (1980) and Prentice (1974). More recently, Balakrishnan and Chan (1994a) studied order statistics from the log-gamma distribution and developed best linear unbiased estimators for the parameters. The asymptotic best linear unbiased estimation of the location and scale parameters based on general Type II censored samples is discussed by Balakrishnan and Chan (1994b). By using the maximum likelihood estimates, Balakrishnan and Chan (1994c) determined tolerance limits (both upper and lower) based on Type II censored samples. The maximum likelihood estimation for the three-parameter case is considered by Balakrishnan and Chan (1994d). Young and Bakir (1987) discuss the log-gamma regression model and associated issues.

Generalized Gamma Convolutions and Related Families

The concept of a *generalized gamma convolution* (GGC) was introduced by Thorin (1977a) and used by him to establish infinite divisibility of Pareto distributions, and later [Thorin (1977b)] of lognormal distributions. It was, however, Bondesson (1979) who refined the notion; he provided a deep analysis of this family of distributions and its relationships to other families. His recent book [Bondesson (1992)] contains a consolidated account of current work in the field, including an extensive although "not exhaustive" bibliography. As yet GGC distributions have not found much application in statistical work. Here we give only a brief summary of GGC theory; we refer the reader to Bondesson's book for a more extensive discussion.

The concept of GGC arose in the context of establishing infinite divisibility, but we will be mainly interested in the distributions themselves— in particular, in the facts that several well-known distributions are included in the class of GGC. The moment-generating function (mgf) of the gamma(α, β) distribution is $(1 - \beta t)^{-\alpha}$ and the convolution of k gamma(α_i, β_i) variables ($i = 1, \ldots, k$) has mgf

$$\prod_{i=1}^{k} (1 - \beta_i t)^{-\alpha_i} = \exp\left\{ -\sum_{i=1}^{k} \alpha_i \log(1 - \beta_i t) \right\}. \qquad (17.134)$$

The class of GGC distributions is defined to include limits of convolutions of generalized gamma distributions as $k \to \infty$, with the β_i's following a mixed nonnegative distribution with possible probability greater than zero of taking a specific value.

A formal definition is that the class of GGC distributions has mgf of form

$$\varphi(t) = \exp\left\{ at - \int \log(1 - \beta t) U(d\beta) \right\}, \qquad (17.135)$$

where $U(d\beta)$ is a nonnegative measure over $\beta > 0$, satisfying

$$\int_0^1 |\log \beta| U(d\beta) < \infty,$$

$$\int_1^\infty \beta^{-1} U(\alpha\beta) < \infty.$$

All GGCs enjoy the following properties:

1. They are unimodal.
2. They are infinitely divisible.
3. They are (of course) closed with respect to convolution.
4. If X and Y are independent, X is a GGC variable, and Y has a gamma distribution, then X/Y is a GGC variable.
5. All pdfs of form

$$Kx^{\beta-1} \prod_{j=1}^{M} (1 + c_j x)^{\gamma_j}, \qquad \beta, c_j, \gamma_j > 0,$$

are GGC.
6. If $f(x)$ is the pdf of a GGC distribution, so is $Kf(x)(1 + cx)^{-\gamma}$ for any $\gamma \geq 0$. (In particular $Kx^{-1}f(x)$ is the pdf of a GGC.)

The class of GGC distributions includes Pareto, lognormal, inverse Gaussian and Type VI Pearson (beta of second kind) distributions.

An important tool in identifying a distribution as a GGC distribution is the result that if its pdf $f(x)$ satisfies the condition "$f(xy)f(x/y)$ is a strictly monotonic function of $y + y^{-1}$ for every x" [termed "hyperbolically completely monotone" by Bondesson (1992)], then it must be a GGC distribution. This is a sufficient, but not a necessary, condition for a distribution to belong to the GGC class.

BIBLIOGRAPHY

Abramowitz, M., and Stegun, I. A. (editors) (1965). *Handbook of Mathematical Functions with Formulas, Graphs and Mathematical Tables*, New York: Dover.

Abu Salih, M. S. (1987). Bayesian estimation of a generalized gamma distribution, *Arabian Journal for Science and Engineering (Saudi Arabia)*, 225–230.

Achcar, J. A., and Bolfarine, H. (1986). The log-linear model with a generalized gamma distribution for the error: A Bayesian approach, *Statistics & Probability Letters*, **4**, 325–332.

Ahrens, J. H., and Dieter, U. (1974). Computer methods for sampling from gamma, beta, Poisson and binomial distributions, *Computing*, **12**, 223–246.

Alexander, G. N. (1962). The use of the gamma distribution in estimating regulated output from storages, *Transactions in Civil Engineering, Institute of Engineers, Australia*, **4**, 29–34.

Alfers, D., and Dinges, H. (1984). A normal approximation for beta and gamma tail probabilities, *Zeitschrift fuer Wahrscheinlichkeitstheorie*, **65**, 399–419.

Amoroso, L. (1925). Ricerche intorno alla curva dei redditi, *Annali di Mathematica*, Series IV, **2**, 123–159.

Anderson, C. W., and Ray, W. D. (1975). Improved maximum likelihood estimators for the gamma distribution, *Communications in Statistics*, **4**, 437–448.

Atkinson, A. C., and Pearce, M. C. (1976). The computer generation of beta, gamma and normal random variables, *Journal of the Royal Statistical Society, Series A*, **139**, 431–461.

Bai, J., Jakeman, A. J., and Taylor, J. A. (1990). Percentile estimation of the three-parameter gamma and lognormal distributions: methods of moments versus maximum likelihood, *Mathematics and Computers in Simulation*, **32**, 167–192.

Bain, L. J., and Engelhardt, M. (1975). A two-moment chi-square approximation for the statistic $\log(\bar{x}/\tilde{x})$, *Journal of the American Statistical Association*, **70**, 948–950.

Balakrishnan, N., and Chan, P. S. (1994a). Log-gamma order statistics and linear estimation of parameters, *Computational Statistics & Data Analysis* (to appear).

Balakrishnan, N., and Chan, P. S. (1994b). Asymptotic best linear unbiased estimation for log-gamma distribution, *Sankhyā, Series B* (to appear).

Balakrishnan, N., and Chan, P. S. (1994c). Maximum likelihood estimation for the log-gamma distribution under Type-II censored samples and associated inference, In *Recent Advances in Life-testing and Reliability*, N. Balakrishnan (editor), Boca Raton, FL: CRC Press (to appear).

Balakrishnan, N., and Chan, P. S. (1994d). Maximum likelihood estimation for the three-parameter log-gamma distribution, In *Recent Advances in Life-testing and Reliability*, N. Balakrishnan (editor), Boca Raton, FL: CRC Press (to appear).

Balakrishnan, N., and Cohen, A. C. (1991). *Order Statistics and Inference: Estimation Methods*, San Diego: Academic Press.

Balasooriya, U., and Hapuarachchi, K. P. (1992). Extended tables for the moments of gamma distribution order statistics, *IEEE Transactions on Reliability*, **41**, 256–264.

Bánkövi, G. (1964). A note on the generation of beta distributed and gamma distributed random variables, *Mathematical Proceedings of the Hungarian Academy of Science, Series A*, **9**, 555–562.

Bar-Lev, S. K., and Reiser, B. (1963). A note on maximum conditional likelihood estimation for the gamma distribution, *Sankhyā, Series B*, **45**, 300–302.

Bartlett, M. S., and Kendall, D. G. (1946). The statistical analysis of variance heterogeneity and the logarithmic transformation, *Journal of the Royal Statistical Society, Series B*, **8**, 128–138.

Barton, D. E. (1953). The probability distribution function of a sum of squares, *Trabajos de Estadística*, **4**, 199–207.

Basu, A. P. (1964). Estimates of reliability for some distributions useful in life testing, *Technometrics*, **6**, 215–219.

Basu, A. P. (1981). The estimation of $P(X < Y)$ for distributions useful in life testing, *Naval Research Logistics Quarterly*, **28**, 383–392.

Beard, R. E. (1948). Some experiments in the use of the incomplete gamma function for the approximate calculation of actuarial functions, *Proceedings of the Centennial Assembly, Institute of Actuaries*, **2**, 89–107.

Berger, J. (1980). Improving on inadmissible estimators in continuous exponential families with applications to simultaneous estimation of gamma scale parameters, *Annals of Statistics*, **8**, 545–571.

Berger, R. L., and Casella, G. (1992). Deriving generalized means as least squares and maximum likelihood estimates, *The American Statistician*, **46**, 279–282.

Berman, M. (1981). The maximum likelihood estimators of the parameters of the gamma distribution are always positively biased, *Communications in Statistics—Theory and Methods*, **10**, 693–697.

Berndt, G. D. (1958). Power functions of the gamma distribution, *Annals of Mathematical Statistics*, **29**, 302–306.

Bhattacharya, S. K. (1966). A modified Bessel function model in life testing, *Metrika*, **11**, 133–144.

Bhattacharjee, G. P. (1970). Algorithm AS 32. The incomplete gamma integral, *Applied Statistics*, **9**, 285–287.

Bienaymé, I. J. (1838). Mémoire sur la probabilité des résultats moyens des observations; demonstration directe de la régle de Laplace, *Mémoires de l'Académie de Sciences de l'Institute de France, Paris, Series Étrangers*, **5**, 513–558.

Bienaymé, I. J. (1852). Mémoire sur la probabilité des erreurs d'aprés la méthode de moindres carrés, *Liouville's Journal de Mathématiques Pures et Appliquées*, **17**, 33–78.

Birnbaum, Z. W., and Saunders, S. C. (1958). A statistical model for life-length of materials, *Journal of the American Statistical Association*, **53**, 151–160.

Blischke, W. R. (1971). Further results on estimation of the parameters of the Pearson Type III distribution, Report ARL 71-0063, Wright-Patterson Air Force Base, OH.

Blischke, W. R., Glinski, A. M., Johns, M. V., Mundle, P. B., and Truelove, A. J. (1965). On non-regular estimation, minimum variance bounds and the Pearson Type III distribution, Report ARL 65-177, Wright-Patterson Air Force Base, OH.

Block, H. W., and Rao, B. Raja (1974). Some generalized distributions based on Stacy's generalized gamma distribution, *Scandinavian Actuarial Journal*, 185–189.

Bobée, B., and Morin, G. (1973). Statistique d'ordre de la loi Pearson III et de sa forme dérivée à asymetrie négative, *Revue de Statistique Appliquée*, **21**, 4.

Bondesson, L. (1979). A general result on infinite divisibility, *Annals of Probability*, **7**, 965–979.

Bondesson, L. (1992). *Generalized Gamma Convolutions and Related Classes of Distributions and Densities*, New York: Springer-Verlag.

Borghi, O. (1965). Sobre una distribución de frecuencias, *Trabajos de Estadística*, **16**, 171–192.

Bowman, K. O., and Beauchamp, J. J. (1975). Pitfalls with some gamma variate simulation routines, *Journal of Statistical Computation and Simulation*, **4**, 141–154.

Bowman, K. O., and Shenton, L. R. (1968). Properties of estimators for the gamma distribution, Report CTC-1, Union Carbide Corp., Oak Ridge, Tennessee.

Bowman, K. O., and Shenton, L. R. (1970). Small sample properties of estimators for the gamma distribution, Report CTC-28, UCCND, Oak Ridge, Tennessee.

Bowman, K. O., and Shenton, L. R. (1978). Coefficient of variation on sampling from a gamma universe, ICQC'78, Tokyo, D1-19-D1-24.

Bowman, K. O., and Shenton, L. R. (1982). Properties of estimators for the gamma distribution, *Communications in Statistics—Simulation and Computation*, **11**, 377-519.

Bowman, K. O., and Shenton, L. R. (1983). Maximum likelihood estimators for the gamma distribution revisited, *Communications in Statistics—Simulation and Computation*, **12**, 697-710.

Bowman, K. O., and Shenton, L. R. (1988). *Properties of Estimators for the Gamma Distribution*, New York: Marcel Dekker.

Box, G. E. P., and Tiao, G. C. (1962). A further look at robustness via Bayes's theorem, *Biometrika*, **49**, 419-432.

Breiter, M. C., and Krishnaiah, P. R. (1967). Tables for the Moments of Gamma Order Statistics, Report ARL 67-1066, Wright-Patterson Air Force Base, OH. (Also *Sankhyā, Series B*, **30**, 59-72.)

Broeder, G. G. den (1955). On parameter estimation for truncated Pearson type III distributions, *Annals of Mathematical Statistics*, **26**, 659-663.

Bruce, R. A. (1964). Estimation of the scale parameter of the gamma distribution by the use of M order statistics, Unpublished thesis, Air Force Institute of Technology, Wright-Patterson Air Force Base, OH.

Burgin, T. A. (1975). The gamma distribution and inventory control, *Operations Research Quarterly*, **26**, 507-525.

Cammilleri, G. (1972). Di una distribuzione del prodotto di variabili stocastiche, *Università di Palermo, Annali della Facoltà di Economia e Commercio*, **25**, No. 4, 87-97.

Cantor, A. K., and Knapp, R. G. (1985). A test of the equality of survival distributions, based on paired observations from conditionally independent exponential distributions, *IEEE Transactions on Reliability*, **R-34**, 342-346.

Chapman, D. G. (1956). Estimating the parameters of a truncated gamma distribution, *Annals of Mathematical Statistics*, **27**, 498-506.

Cheng, R. C. H. (1977). The generation of gamma variables with non-integral shape parameters, *Applied Statistics*, **26**, 71-75.

Cheng, R. C. H., and Amin, N. A. K. (1983). Estimating parameters in continuous univariate distributions with a shifted origin, *Journal of the Royal Statistical Society, Series B*, **45**, 394-403.

Cheng, R. C. H., and Feast, G. M. (1979). Some simple gamma variate generators, *Applied Statistics*, **28**, 290-295.

Cheng, R. C. H., and Feast, G. M. (1980). Gamma variate generators with increased shape parameter range, *Communications of the ACM*, **23**, 389-394.

Choi, S. C., and Wette, R. (1969). Maximum likelihood estimation of the parameters of the gamma distribution and their bias, *Technometrics*, **11**, 683-690.

Chow, V. T. (1969). *Handbook of Applied Hydrology*, New York: McGraw-Hill.

Clark, C. (1951). Urban population densities, *Journal of the Royal Statistical Society, Series A*, **114**, 490–496.

Cobb, L. (1981). The multimodal exponential families of statistical catastrophe theory, *Statistical Distributions in Scientific Work*, **4**, C. Taillie, G. P. Patil, and B. Baldessari (editors), 67–90, Dordrecht: Reidel.

Cohen, A. C. (1950). Estimating parameters of Pearson type III populations from truncated samples, *Journal of the American Statistical Association*, **45**, 411–423.

Cohen, A. C. (1951). Estimation of parameters in truncated Pearson frequency distributions, *Annals of Mathematical Statistics*, **22**, 256–265.

Cohen, A. C. (1969). A Generalization of the Weibull Distribution, NASA Contractor Report, No. 61293, Cont. NAS 8-11175, Marshall Space Flight Center, Alabama.

Cohen, A. C., Helm, F. R., and Sugg, M. (1969). Tables of Areas of the Standardized Pearson Type III Density Function, Report NASA CR-61266, NASA, Marshall Space Flight Center, Alabama.

Cohen, A. C., and Norgaard, M. J. (1977). Progressively censored sampling in the three-parameter gamma distribution, *Technometrics*, **9**, 333–340.

Cohen, A. C., and Whitten, B. J. (1986). Modified moment estimation for the three-parameter gamma distribution, *Journal of Quality Technology*, **17**, 147–154.

Cohen, A. C., and Whitten, B. J. (1988). *Parameter Estimation in Reliability and Life Span Models*, New York: Marcel Dekker.

Constantine, K., Karson, M., and Tse, S. K. (1986). Estimation of $P(Y < X)$ in the gamma case, *Communications in Statistics—Simulation and Computation*, **15**, 365–388.

Constantine, K., Karson, M., and Tse, S. K. (1989). Bootstrapping estimators of $P(X < Y)$ in the gamma case, *Journal of Statistical Computation and Simulation*, **33**, 217–231.

Constantine, K., Karson, J., and Tse, S. K. (1990). Confidence interval estimation of $P(Y < X)$ in the gamma case, *Communications in Statistics—Simulation and Computation*, **19**, 225–244.

Consul, P. C., and Jain, G. C. (1971). On the log-gamma distribution and its properties, *Statistische Hefte*, **12**, 100–106.

Costantino, R. F., and Desharnais, R. A. (1981). Gamma distributions of adult numbers for Tribolium populations in the regions of their steady states, *Journal of Animal Ecology*, **50**, 667–681.

Czuber, E. (1981). *Theorie der Beobachtungsfehler*, Leipzig: Teubner.

Dahiya, R. C., and Gurland, J. (1978). Estimating the parameters of a gamma distribution, *Trabajos de Estadistica*, **29/2**, 81–87.

Damsleth, E. (1975). Conjugate classes for gamma distributions, *Scandinavian Journal of Statistics*, **2**, 80–84.

Das, S. C. (1955). The fitting of truncated type III curves to daily rainfall data, *Australian Journal of Physics*, **7**, 298–304.

Das Gupta, A. (1984). Admissibility in the gamma distribution: Two examples, *Sankhyā, Series A*, 395–407.

David, F. N., and Johnson, N. L. (1951). The effect of non-normality on the power function of the *F*-test in the analysis of variance, *Biometrika*, **38**, 43–57.

Dennis, B., and Patil, G. P. (1984). The gamma distribution and the weighted multimodal gamma distributions as models of population abundance, *Mathematical Biosciences*, **68**, 187–212.

Des Raj, S. (1953). Estimation of the parameters of type III populations from truncated samples, *Journal of the American Statistical Association*, **48**, 336–349.

DiCiccio, T. J. (1987). Approximate inference for the generalized gamma distribution, *Technometrics*, **29**, 33–40.

Dubey, S. D. (1970). Compound gamma, beta and *F* distributions, *Metrika*, **16**, 27–31.

Dusenberry, W. E., and Bowman, K. O. (1977). The moment estimator for the shape parameter of the gamma distribution, *Communications in Statistics—Simulation and Computation*, **6**, 1–19.

Elderton, W. P. (1902). Tables for testing the goodness of fit of theory to observation, *Biometrika*, **1**, 155–163.

Ellis, R. L. (1844). On a question in the theory of probabilities, *Cambridge Mathematical Journal*, **4**, 127–132.

Engel, J., and Zijlstra, M. (1980). A characterisation of the gamma distribution by the negative binomial distribution, *Journal of Applied Probability*, **17**, 1138–1144.

Esteban, J. (1981). Income-share elasticity, density functions and the size distribution of income, Manuscript, *University of Barcelona*.

Ferreri, C. (1984). On the hypergeometric birth process and some implications about the Gamma distribution representation, *Journal of the Royal Statistical Society, Series B*, **46**, 52–57.

Findeisen, P. (1978). A simple proof of a classical theorem which characterizes the gamma distribution, *Annals of Statistics*, **6**, 1165–1167.

Fisher, R. A. (1921). On the mathematical foundations of theoretical statistics, *Philosophical Transactions of the Royal Society of London, Series A*, **222**, 309–368.

Fisher, R. A., and Cornish, E. A. (1960). The percentile points of distributions having known cumulants, *Technometrics*, **2**, 209–225.

Fishman, G. S. (1976). Sampling from the gamma distribution on a computer, *Communications of the Association for Computing Machinery*, **7**, 407–409.

Fraser, D. A. S. (1990). Tail probabilities from observed likelihood, *Biometrika*, **77**, 65–76.

Galambos, J., and Kotz, S. (1978). Characterizations of probability distributions, *Lecture Notes in Mathematics*, **675**, New York: Springer-Verlag.

Garti, Y., and Consoli, T. (1954). Sur la densité de probabilité du produit de variables aléatoires de Pearson du Type III, *Studies in Mathematics and Mechanics*, 301–309, Presented to R. von Mises, San Diego: Academic Press.

Geldston, S. (1962). *Probability Distribution at the Output of a Logarithmic Receiver*, Research Report PIBMRI-1087-62, Microwave Research Institute, Polytechnic Institute of Brooklyn, NY.

Ghosh, J. K., and Singh, R. (1970). Estimation of the reciprocal of the scale parameters of a gamma density, *Annals of the Institute of Statistical Mathematics*, **22**, 51–55.

Ghosh, M., and Parsian, A. (1980). Admissible and minimax multiparameter estimation in exponential families, *Journal of Multivariate Analysis*, **10**, 551–564.

Glaser, R. E. (1973). Inferences for a gamma distributed random variable with both parameters unknown with applications to reliability, Technical Report 154, Department of Statistics, Palo Alto: Stanford University.

Glaser, R. E. (1976a). The ratio of the geometric mean to the arithmetic mean for a random sample from a gamma distribution, *Journal of the American Statistical Association*, **71**, 480–487.

Glaser, R. E. (1976b). Exact critical values for Bartlett's test for homogeneity of variances, *Journal of the American Statistical Association*, **71**, 488–490.

Goodhardt, G. J., and Chatfield, C. (1973). Gamma distribution in consumer purchasing, *Nature*, **244**, 316.

Goodman, L. A. (1952). On the Poisson-gamma distribution problem, *Annals of the Institute of Statistical Mathematics*, **3**, 123–125.

Gray, H. L., and Lewis, T. O. (1971). Approximation of tail probabilities by means of the B_n-transformation, *Journal of the American Statistical Association*, **66**, 897–899.

Gray, H. L., and Schucany, W. R. (1968). On the evaluation of distribution functions, *Journal of the American Statistical Association*, **63**, 715–720.

Gray, H. L., Thompson, R. W., and McWilliams, G. V. (1969). A new approximation for the chi-square integral, *Mathematics of Computation*, **23**, 85–89.

Greenwood, J. A., and Durand, D. (1960). Aids for fitting the gamma distribution by maximum likelihood, *Technometrics*, **2**, 55–65.

Grice, J. V., and Bain, L. J. (1980). Inferences concerning the mean of the gamma distribution, *Journal of the American Statistical Association*, **75**, 929–933.

Gross, A. J. (1971). Monotonicity properties of the moments of truncated gamma and Weibull density functions, *Technometrics*, **13**, 851–857.

Gupta, A. D. (1984). Admissibility in the gamma distribution, *Sankhyā, Series A*, **46**, 395–407.

Gupta, S. S. (1960). Order statistics from the gamma distribution, *Technometrics*, **2**, 243–262.

Gupta, S. S. (1962). Gamma distribution, In *Contributions to Order Statistics*, A. E. Sarhan and B. G. Greenberg (editors), 431–450, New York: Wiley.

Gupta, S. S., and Groll, P. A. (1961). Gamma distribution in acceptance sampling based on life tests, *Journal of the American Statistical Association*, **56**, 942–970.

Hager, H. W. (1970). Statistical Inference for the Generalized Gamma Distribution, Ph.D. dissertation, Department of Statistics, Rolla: University of Missouri.

Hager, H. W., and Bain, L. J. (1970). Inferential procedures for the generalized gamma distribution, *Journal of the American Statistical Association*, **65**, 1601–1609.

Hager, H. W., Bain, L. J., and Antle, C. E. (1971). Reliability estimation for the generalized gamma distribution and robustness of the Weibull model, *Technometrics*, **13**, 547–557.

Hall, W. J., and Simons, G. (1969). Characterizations of the gamma distribution, *Sankhyā, Series A*, **31**, 385–390.

Harner, E. J., Shyu, H. -R., and Trutzer, V. (1991). A simulation study of the robustness of estimators for the shape parameter of the gamma distribution, *The Frontiers of Statistical Computation, Simulation, and Modeling* (Vol. 1 of the Proceedings of ICOSCO-1), 317–322.

Harter, H. L. (1964). *New Tables of the Incomplete gamma Function Ratio and of Percentage Points of the Chi-square and Beta Distributions*, Washington, DC: Government Printing Office.

Harter, H. L. (1966). Asymptotic variances and covariances of maximum-likelihood estimators, from censored samples, of the parameters of a four-parameter generalized gamma population, ARL Report 66-0158, Wright-Patterson Air Force Base, OH.

Harter, H. L. (1967). Maximum likelihood estimation of the parameters of a four-parameter generalized gamma population from complete and censored samples, *Technometrics*, **9**, 159–165.

Harter, H. L. (1969). A new table of percentage points of the Pearson type III distribution, *Technometrics*, **11**, 177–187.

Harter, H. L. (1970). *Order Statistics and their Use in Testing and Estimation*, **1** and **2**, Washington, DC: Government Printing Office.

Harter, H. L. (1971). More percentage points of the Pearson type III distribution, *Technometrics*, **13**, 203–204.

Harter, H. L., and Moore, A. H. (1965). Maximum likelihood estimation of the parameters of gamma and Weibull populations from complete and from censored samples, *Technometrics*, **7**, 639–643.

Harvey, H. (1967). *A Family of Averages in Statistics*, Morrisville, PA: Annals Press.

Henery, R. J. (1983). Permutation probabilities for gamma random variables, *Journal of Applied Probability*, **20**, 822–834.

Hill, T. D. (1965). Estimation of the scale parameter of the gamma distribution by the use of L order statistics, Unpublished thesis, Air Force Institute of Technology, Wright-Patterson Air Force Base, OH.

Hobbs, J. R., Moore, A. H., and James, W. (1984). Minimum distance estimation of the three parameters of the gamma distribution, *IEEE Transactions on Reliability*, **R33**, 237–240.

Hoshi, K., and Burges, S. J. (1981). Approximate estimation of the derivative of a standard gamma quantile for use in confidence interval estimates, *Journal of Hydrology*, **53**, 317–325.

Hsu, Y. S., and Amunrud, L. R. (1981). A generalization of the Poisson and general-gamma distributions with application, *American Journal of Mathematical and Management Sciences*, **1**, 167–188.

Huang, W. J., and Cheng, L. -S. (1989). Note on a characterization of gamma distributions, *Statistics & Probability Letters*, **8**, 485–487.

Huque, F., and Katti, S. K. (1976). A note on maximum conditional likelihood estimators, *Sankhyā, Series B*, **38**, 1–13.

Ismail, R., Jeyaratnam, S., and Panchapakesan, S. (1986). Estimation of $\Pr[X > Y]$ for gamma distributions, *Journal of Statistical Computation and Simulation*, **26**, 253–267.

Iyer, P. V. K., and Singh, N. (1963). Estimation of the mean and the standard deviation of a type III population from censored samples, *Journal of the Indian Statistical Association*, **1**, 161–166.

Jackson, O. A. Y. (1969). Fitting a gamma or log-normal distribution to fibre-diameter measurements of wool tops, *Applied Statistics*, **18**, 70–75.

Jakuszenkowa, H. (1973). On some property of the generalized gamma-distribution, *Prace Matematyczne*, **17**, 237–243.

Jensen, J. L. (1986). Inference for the mean of a gamma distribution with unknown shape parameter, *Scandinavian Journal of Statistics*, **13**, 135–151.

Jensen, J. L., and Kristensen, L. B. (1991). Saddlepoint approximations to exact test and improved likelihood ratio tests for the gamma distribution, *Communications in Statistics—Theory and Methods*, **20**, 1515–1532.

Johnson, N. L. (1952). Approximations to the probability integral of the distribution of range, *Biometrika*, **39**, 417–419.

Johnson, N. L., and Kotz, S. (1972). Power transformation of gamma variables, *Biometrika*, **59**, 226–229.

Jöhnk, M. D. (1964). Erzeugung von betaverteilten und gammaverteilten Zufallszahlen, *Metrika*, **8**, 5–15.

Jones, B., Waller, W. G., and Feldman, A. D. (1978). Root isolation using function values; *BIT: Nordisk Tidskrift for Informationsbehandlung*, **18**, 311–319.

Joshi, P. C. (1979). On the moments of gamma order statistics, *Naval Research Logistics Quarterly*, **26**, 675–679.

Kabe, D. G. (1966). Dirichlet's transformation and distributions of linear functions of ordered gamma variates, *Annals of the Institute of Statistical Mathematics*, **18**, 367–374.

Kagan, A. M., and Rukhin, A. L. (1967). On the estimation theory of the scale parameter, *Teoriya Veroyatnostei i ee Primeneniya*, **12**, 735–741. (In Russian)

Kaluszka, M. (1986). Admissible and minimax estimators of λ^r in the gamma distribution with truncated parameter space, *Metrika*, **33**, 363–375.

Kanno, R. (1982). Maximum likelihood estimation of parameters for a mixture of two gamma distributions, *Reports of Statistical Application Research, JUSE*, **29**, No. 3, 14–24.

Kantam, R. R. L., and Narasimham, V. L. (1991). Linear estimation in reflected gamma distribution, *Sankhyā, Series B*, **53**, 25–47.

Karns, R. C. (1963). Scale parameter estimation of the gamma probability function based on one order statistic, Unpublished thesis, Air Force Institute of Technology, Wright-Patterson Air Force Base, OH.

Keating, J. P., Glaser, R. E., and Ketchum, N. S. (1990). Testing hypotheses about the shape parameter of a gamma distribution, *Technometrics*, **32**, 67–82.

Khamis, S. H. (1960). Incomplete gamma function expansions of statistical distribution functions, *Bulletin of the International Statistical Institute*, **37**, 385–396.

Khan, A. H., and Khan, R. U. (1983). Recurrence relation between the moments of order statistics from generalized gamma distribution, *Journal of Statistical Research (Bangladesh)*, **17**, 75–82.

Khatri, C. G., and Rao, C. R. (1968). Some characterizations of the gamma distribution, *Sankhyā, Series A*, **30**, 157–166.

Kinderman, A. J., and Monahan, J. F. (1977). Computer generation of random variables using the ratio of uniform deviates, *Association for Computing Machinery, Monograph Transactions, Mathematical Software*, **3**, 257–260.

Kirby, W. (1972). Computer-oriented Wilson-Hilferty transformation that preserves the first three moments and the lower bound of the Pearson type 3 distribution, *Water Resources Research*, **8**, 1251–1254.

Klebaner, F. C. (1989). Stochastic difference equations and generalized gamma distribution, *Annals of Probability*, **17**, 178–188.

Klinken, J. van (1961). A method for inquiring whether the Γ-distribution represents the frequency distribution of industrial accident costs, *Actuariële Studiën*, **3**, 83–92.

Kotlarski, I. (1962). On pairs of independent variables whose quotients follow some known distribution, *Colloquium Mathematicum*, **9**, 151–162.

Kotlarski, I. (1965). A certain problem connected with the gamma distribution, *Zeszyty Naukowe Politechniki Warszawskiej*, **6**, 21–28. (In Polish)

Kotlarski, I. (1967). On characterizing the gamma and the normal distribution, *Pacific Journal of Mathematics*, **20**, 69–76.

Kotz, S., and Neumann, J. (1963). On distribution of precipitation amounts for the periods of increasing length, *Journal of Geophysical Research*, **68**, 3635–3641. (Addendum, **69**, 800–801.)

Kotz, S., and Srinivasan, R. (1969). Distribution of product and quotient of Bessel function variates, *Annals of the Institute of Statistical Mathematics*, **21**, 201–210.

Kotz, S., and Steutel, F. W. (1988). Note on a characterization of exponential distribution, *Statistics & Probability Letters*, **6**, 201–203.

Kruskal, W. H. (1946). Helmert's distribution, *American Mathematical Monthly*, **53**, 435–438.

Kübler, H. (1979). On the fitting of the three parameter distributions: Lognormal, gamma, Weibull, *Statistische Hefte*, **20**, 68–123.

Laha, R. G. (1954). On a characterization of the gamma distribution, *Annals of Mathematical Statistics*, **25**, 784–787.

Laha, R. G. (1964). On a problem connected with beta and gamma distributions, *Transactions of the American Mathematical Society*, **113**, 287–298.

Lancaster, H. O. (1966). Forerunners of the Pearson χ^2, *Australian Journal of Statistics*, **8**, 117–126. (See also W. Kruskal's review in *Mathematical Reviews*, **36**, 916.)

Laplace, P. S. (1836). *Théorie Analytique des Probabilités* (Supplement to third edition).

Lau, C. L. (1980). A simple series for the incomplete gamma integral, *Applied Statistics*, **29**, 113–114.

Lavender, D. E. (1967). On the Distribution of the Sum of Independent Doubly Truncated Gamma Variables, Report NASA CR-61184.

Lawless, J. F. (1980). Inference in the generalized gamma and log gamma distributions, *Technometrics*, **22**, 409–419.

Lee, M. -L. T., and Gross, A. J. (1989). Properties of conditionally independent generalized gamma distributions, *Probability in the Engineering and Informational Sciences*, **3**, 289–297.

Lee, R. Y., Holland, B. S., and Flueck, J. A. (1979). Distribution of a ratio of correlated Gamma random variables, *SIAM Journal Applied Mathematics*, **36**, 304–320.

Lefante, J. J., Jr., and Turner, M. E., Jr. (1985). The average likelihood and a fiducial approximation: One parameter members of the generalized gamma distributions, *Communications in Statistics—Theory and Methods*, **14**, 419–436.

Leslie, R. T., and Khalique, A. (1980). A note on bias reduction in moment estimators of the gamma distribution, *Pakistan Journal Statistics, Series A*, **2**, 33–40.

Letac, G. (1985). A characterization of the gamma distribution, *Advances in Applied Probability*, **17**, 911–912.

Lienhard, H. J., and Meyer, P. L. (1967). A physical basis for the generalized gamma distribution, *Quarterly of Applied Mathematics*, **25**, 330–334.

Lingappaiah, G. S. (1974). Prediction in samples from the gamma distribution as applied to life testing, *Australian Journal of Statistics*, **16**, 30–32.

Lingappaiah, G. S. (1976). Effect of outliers on the estimation of parameters, *Metrika*, **23**, 27–30.

Lingappaiah, G. S. (1991). Prediction in samples from a gamma population in the presence of an outlier, *Bulletin of the Malaysian Mathematical Society (Second Series)*, **14**, 1–14.

Linnik, Yu. V., Rukhin, A. L., and Stzelic, S. I. (1970). Gamma distribution and partial sufficiency of polynomials, *Proceedings of the Steklov Institute of Mathematics*, **111**, 45–58.

Lloyd, E. H. (1952). Least-squares estimation of location and scale parameters using order statistics, *Biometrika*, **39**, 88–95.

Lukacs, E. (1965). A characterization of the gamma distribution, *Annals of Mathematical Statistics*, **26**, 319–324.

Lwin, T., and Singh, N. (1974). Bayesian analysis of the gamma distribution model in reliability estimation, *IEEE Transactions in Reliability*, **R-23**, 314–319.

Malik, H. J. (1967). Exact distribution of the quotient of independent generalized gamma variables, *Canadian Mathematical Bulletin*, **10**, 463–465.

Malik, H. J. (1968a). Exact distribution of the product of independent generalized gamma variables with the same shape parameter, *Annals of Mathematical Statistics*, **39**, 1751–1752.

Malik, H. J. (1968b). A characterization of the generalized gamma distribution, *Metron*, **27**(3–4), 41–45.

Malik, H. J. (1976). Distribution of a linear function and the ratio of two independent linear functions of independent generalized gamma variables, *Naval Research Logistics Quarterly*, **23**, 339–343.

Marsaglia, G. (1974). Extension and application of Lukacs' characterization of the gamma distribution, *Proc. Symposium Statistics & Related Topics*, Carleton University, Ottawa.

Marsaglia, G. (1989). The $X + Y$, X/Y characterization of the gamma distribution, In *Contributions to Probability and Statistics*, L. J. Bleier, M. D. Perlman, S. J. Press, and A. R. Sampson (editors), 91–98, New York: Springer-Verlag.

Masuyama, M., and Kuroiwa, Y. (1952). Table for the likelihood solutions of gamma distribution and its medical applications, *Reports of Statistical Application Research*, (*JUSE*), **1**, 18–23.

Mathai, A. M. (1972). Products and ratios of generalized gamma variates, *Skandinavisk Aktuarietidskrift*, 192–198.

Mathai, A. M. (1982). Storage capacity of a dam with gamma type inputs, *Annals of the Institute of Statistical Mathematics*, **34**, 591–597.

Mauldon, J. G. (1956). Characterizing properties of statistical distributions, *Quarterly Journal of Mathematics* (*Oxford, 2nd Series*) **27**, 155–160.

McDonald, J. B. (1984). Some generalized functions for the size distribution of income, *Econometrica*, **52**, 647–663.

McDonald, J. B., and Jensen, B. C. (1979). An analysis of estimators of alternative measures of income inequality associated with the gamma distribution function, *Journal of American Statistical Association*, **74**, 856–860.

Mees, E., and Gerard, G. (1984). The generalized gamma distribution: A model for seed germination curves, *Biometrie-Praximetrie*, **24**, 101–115.

Menon, M. V. (1966). Characterization theorems for some unimodal distributions, *Journal of the Royal Statistical Society*, Series B, **28**, 143–145.

Miller, R. B. (1980). Bayesian analysis of the two-parameter gamma distribution, *Technometrics*, **22**, 65–69.

Moore, R. J. (1982). Algorithm AS 187: Derivatives of the incomplete gamma integral, *Applied Statistics*, **31**, 330–335.

Moschopoulos, P. G. (1985). The distribution of the sum of independent gamma random variables, *Annals of the Institute of Statistical Mathematics*, **37**, 541–544.

Mosleh, A., and Apostolakis, G. (1982). Some properties of distributions useful in the study of rare events, *IEEE Transactions on Reliability*, **R-31**, 87–94.

Musson, T. A. (1965). Linear estimation of the location and scale parameters of the Weibull and gamma probability distributions by the use of order statistics, Unpublished thesis, Air Force Institute of Technology, Wright-Patterson Air Force Base, OH.

Nabeya, S. (1950). On a relation between exponential law and Poisson's law, *Annals of the Institute of Statistical Mathematics*, **2**, 13–16.

Nath, G. B. (1975). Unbiased estimates of reliability for the truncated gamma distribution, *Scandinavian Actuarial Journal*, 181–186.

Norris, N. (1966). Maximum-likelihood estimators of the relative scale parameters of Type III and Type IV populations, Presented at Institute of Mathematical Statistics Meeting, August, 1966, New Brunswick, NJ.

Odell, P. L., and Newman, T. G. (1972). *The Generation of Random Variates*, London: Charles Griffin.

Oginski, L. (1991). Estimation of parameters of a mixture of an arbitrary number of generalized gamma distributions, *Zeszyti Naukowe Matematiyczne, Lódz*, **22**, 5–12.

Olshen, A. C. (1937). Transformation of the Pearson type III distributions, *Annals of Mathematical Statistics*, **8**, 176–200.

Osaki, S., and Li, X. (1988). Characterization of gamma and negative binomial distributions, *IEEE Transactions on Reliability*, **R-37**, 379–382.

Pakes, A. G. (1992). On characterizations through mixed sums, *Australian Journal of Statistics*, **34**, 323–339.

Pakes, A. G., and Khattree, R. (1992). Length-biasing characterizations of laws and the moment problem, *Australian Journal of Statistics*, **34**, 307–326.

Parr, Van B., and Webster, J. T. (1965). A method for discriminating between failure density functions used in reliability predictions, *Technometrics*, **7**, 1–10.

Patil, G. P., and Seshadri, V. (1964). Characterization theorems for some univariate probability distributions, *Journal of the Royal Statistical Society, Series B*, **26**, 286–292.

Paul, S. R., and Thiagarajah, K. (1992). Multi-sample test of equal gamma distribution scale parameters in presence of unknown common shape parameter, *Communications in Statistics—Theory and Methods*, **21**, 1633–1650.

Pearson, E. S. (1963). Some problems arising in approximating to probability distributions using moments, *Biometrika*, **50**, 95–111.

Pearson, E. S., and Hartley, H. O. (1954). *Biometrika Tables for Statisticians*, **1**, Cambridge: Cambridge University Press.

Pearson, K. (1900). On a criterion that a given system of deviations from the probable in the case of a correlated system of variables is such that it can be reasonably supposed to have arisen from random sampling, *Philosophical Magazine, 5th Series*, **50**, 157–175.

Pearson, K. (ed.). (1922). *Tables of the Incomplete Γ-Function*, H. M. Stationery Office, London. (Reprint 1934, Cambridge University Press.)

Pearson, K. (1931). Historical note on the distribution of the standard deviation of samples of any size drawn from an indefinitely large normal parent population, *Biometrika*, **23**, 416–418.

Phien, H. N. (1991). On the computation of gamma quantiles, In *The Frontiers of Statistical Computation, Simulation and Modeling* (Vol. 1 of the Proceedings of ICOSCO-I), 253–269.

Phillips, D. T., and Beightler, C. S. (1972). Procedures for generating gamma variates with non-integer parameter sets, *Journal of Statistical Computation and Simulation*, **1**, 197–208.

Pitman, E. J. G. (1939). Tests of hypotheses concerning location and scale parameters, *Biometrika*, **31**, 200–215.

Plucinska, A. (1965). On certain problems connected with a division of a normal population into parts, *Zastosowania Matematyki*, **8**, 117–125. (In Polish)

Plucinska, A. (1966). On a general form of the probability density function and its application to the investigation of the distribution of rheostat resistance, *Zastosowania Matematyki*, **9**, 9–19. (In Polish)

Plucinska, A. (1967). The reliability of a compound system under consideration of the system elements prices, *Zastosowania Matematyki*, **9**, 123–134. (In Polish)

Podolski, H. (1972). The distribution of a product of n independent random variables with generalized gamma distribution, *Demonstratio Mathematica*, **4**, 119–123.

Pogurova, V. I. (1965). On the calculation of quantiles of the Γ-distribution, *Teoriya Veroyatnostei i ee Primeneniya*, **10**, 746–749. (In Russian. English translation, 677–680).

Prentice, R. L. (1974). A log-gamma model and its maximum likelihood estimation, *Biometrika*, **61**, 539–544.

Prescott, P. (1974). Variances and covariances of order statistics from the gamma distribution, *Biometrika*, **61**, 607–613.

Provost, S. B. (1988). The exact density of a statistic related to the shape parameter of a gamma variate, *Metrika*, **35**, 191–196.

Pugh, E. L. (1963). The best estimate of reliability in the exponential case, *Operations Research*, **11**, 57–61.

Pusz, J., and Wesolowski, J. (1992). A non-Lukacsian regressional characterization of the gamma distribution, *Applied Mathematics Letters*, **5**, 81–84.

Radhakrishna, C., Rao, A. V. D., and Anjaneyulu, G. V. S. R. (1992). Estimation of parameters in a two component mixture generalized gamma distribution, *Communications in Statistics—Theory and Methods*, **21**, 1799–1805.

Raja Rao, B., and Garg, M. L. (1969). A note on the generalized (positive) Cauchy distribution, *Canadian Mathematical Bulletin*, **12**, 865–868.

Rao, A. V., Kantam, R. R., and Narasimham, V. L. (1991). Linear estimation of location and scale parameters in the generalized gamma distribution, *Communications in Statistics—Theory and Methods*, **20**, 3823–3848.

Reiser, B., and Rocke, D. M. (1991). Inference for stress-strength problems under the gamma distribution, *preprint*.

Revfeim, K. J. A. (1991). Approximation for the cumulative and inverse gamma distribution, *Statistica Neerlandica*, **45**, 328–331.

Roberts, C. D. (1971). On the distribution of random variables whose mth absolute power is gamma, *Sankhyā, Series A*, **33**, 229–232.

Roberts, C., and Geisser, S. (1966). A necessary and sufficient condition for the square of a random variable to be gamma, *Biometrika*, **53**, 275–277.

Rosaiah, K., Kantam, R. R. L., and Narasimham, V. L. (1991). Optimum class limits for ML estimation in two-parameter gamma distribution from a grouped data, *Communications in Statistics—Simulation and Computation*, **20**, 1173–1189.

Roy, D. (1984). A characterization of the generalized gamma distribution, *Calcutta Statistical Association Bulletin*, **33**, 137–141.

Roy, J. (1972). Evaluation of gamma score, *Sankhyā, Series B*, **34**, 27–32.

Salem, A. B., and Mount, T. D. (1974). A convenient descriptive model of income distribution: the gamma density, *Econometrica*, **42**, 1115–1127.

Salvosa, L. R. (1929). *Generalizations of the Normal Curve of Error*, Ann Arbor: Edwards Brothers.

Salvosa, L. R. (1930). Tables of Pearson's type III function, *Annals of Mathematical Statistics*, **1**, 191–198.

Sarma, P. V. S., Srinivasa Rao, K. S., and Prabhakara Rao, R. (1990). On a family of bimodal distributions, *Sankhyā, Series B*, **52**, 287–292.

Särndal, C. -E. (1964). Estimation of the parameters of the gamma distribution by sample quantiles, *Technometrics*, **6**, 405–414.

Saunders, I. W., and Moran, P. A. P. (1978). On the quantiles of the gamma and *F* distributions, *Journal of Applied Probability*, **15**, 426–432.

She, S. Y. (1988). A class of characterization theorems for determining whether a population has a Γ-distribution, *Huaihua Shizhuan Xuebao*, **7**, 30–35.

Shenton, L. R., and Bowman, K. O. (1972). Further remarks on m.l.e. for the gamma distribution, *Technometrics*, **14**, 725–733.

Shenton, L. R., and Bowman, K. O. (1973). Comments on the gamma distribution and uses in rainfall data, *The Third Conference on Probability and Statistics in Atmospheric Science*, 8 pp.

Shiue, W. K., Bain, L. J., and Engelhardt, M. (1988). Test of equal gamma distribution means with unknown and unequal shape parameters, *Technometrics*, **30**, 169–174.

Siddiqui, M. M., and Weiss, G. H. (1963). Families of distributions for hourly median power and instantaneous power of received radio signals, *Journal of Research; National Bureau of Standards*, **67D**, 753–762.

Sim, C. H. (1992). Point process with correlated gamma interarrival times, *Statistics & Probability Letters*, **15**, 135–141.

Simpson, J. (1972). Use of the gamma distribution in single-cloud rainfall analysis, *Monthly Weather Review (100)*, **4**, 309–312.

Singh, R. (1972). Admissible estimators of λ^r in gamma distribution with quadratic loss, *Trabajos de Estadística*, **23**, 129–134.

Singh, S. K., and Maddala, G. S. (1976). A function for the size distribution of incomes, *Econometrica*, **44**, 963–970.

Slutskii, E. E. (1950). *Tablitsi dlya Vichisleniya Nepolnoi Γ-funktsii i Veroyatnosti* χ^2, (Tables for Computing the Incomplete Gamma Function and χ^2 Probabilities), A. N. Kolomogorov (editor), Moscow: Akademia Nauk SSSR.

Springer, M. D. (1979). *Algebra of Random Variables*, New York: Wiley.

Stacy, E. W. (1962). A generalization of the gamma distribution, *Annals of Mathematical Statistics*, **33**, 1187–1192.

Stacy, E. W. (1973). Quasimaximum likelihood estimators for two-parameter gamma distributions, *IBM Journal of Research and Development*, **17**, 115–124.

Stacy, E. W., and Mihram, G. A. (1965). Parameter estimation for a generalized gamma distribution, *Technometrics*, **7**, 349–358.

Stammberger, A. (1967). Über einige Nomogramme zur Statistik, *Wissenshaftliche Zeitschrift der Humboldt-Universität Berlin, Mathematisch-Naturwissenschaftliche Reihe*, **16**, 1, 86–93.

Tadikamalla, P. R. (1977). An approximation to the moments and the percentiles of gamma order statistics, *Sankhyā, Series B*, **39**, 372–381.

Tadikamalla, P. R. (1978). Computer generation of gamma random variables, *Communications of the Association for Computing Machinery*, **21**, 419–422.

Tadikamalla, P. R., and Ramberg, J. S. (1975). An approximate method for generating gamma and other variates, *Journal of Statistical Computation and Simulation*, **3**, 275–282.

Taguchi, T. (1980). On an interpretation and an estimation of shape parameters of the generalized gamma distribution, *Metron*, **38**, 27–40.

Taillie, C. (1981). Lorenz ordering within the generalized gamma family of income distributions, In *Statistical Distributions in Scientific Work*, **6**, C. Taillie, G. P. Patil, and B. Balderssari (editors), 181–192, Dordrecht: Reidel.

Thom, H. C. S. (1968). *Direct and Inverse Tables of the Gamma Distribution*, Silver Spring, MD; Environmental Data Service.

Thomas, P. Y., and Moothathu, T. S. K. (1991). Recurrence relations for moments of different orders of extremes from gamma distribution, *Communications in Statistics—Theory and Methods*, **20**, 945–950.

Thorin, O. (1977a). On the infinite divisibility of the ratio distribution, *Scandinavian Actuarial Journal*, 31–40.

Thorin, O. (1977b). On the infinite divisibility of the lognormal distribution, *Scandinavian Actuarial Journal*, 121–148.

Thorin, O. (1978). An extension of the notion of a generalized Γ-convolution, *Scandinavian Actuarial Journal*, 141–149.

Tiku, M. L. (1964). Approximating the general non-normal variance ratio sampling distribution, *Biometrika*, **51**, 83–95.

Tiku, M. L., and Malik, H. J. (1972). On the distribution of order statistics, *Australian Journal of Statistics*, **14**, 103–108.

Tricomi, F. G. (1950). Sulle funzione gamma incomplete, *Annali di Matematica Pura ed Applicata*, **31**, 263–279.

Vodă, V. G. (1974). The study of a certain gamma-type distribution, *Studii si Cenetări Matematice*, **26**, 311–322.

Wallace, N. D. (1974). Computer generation of gamma random variates with non-integral shape parameters, *Communications of the Association for Computing Machinery*, **7**, 691–695.

Waller, R. A., and Waterman, M. S. (1978). Percentiles for the gamma distribution, *SIAM Review*, **20**, 856.

Walter, S. D., and Stitt, L. W. (1988). Extended tables for moments of gamma distribution order statistics, *Communications in Statistics—Simulation and Computation*, **17**, 471–487.

Wang, Y. H. (1972). On characterization of certain probability distribution, *Proceedings of the Cambridge Philosophical Society*, **71**, 347–352.

Wang, Y. H. (1981). Extension of Lukacs' characterization of the gamma distribution, *Analytical Methods in Probability Theory*, D. Dugué, E. Lukacs, and V. K. Rohatgi (editors), New York: Springer-Verlag.

Wang, Y. H., and Chang, S. A. (1977). A new approach to the nonparametric tests of exponential distribution with unknown parameters, In *The Theory and Applications of Reliability II*, 235–258, San Diego: Academic Press.

Wani, J. K., and Kabe, D. G. (1971). Point estimation of reliability of a system comprised of *K* elements from the same gamma model, *Technometrics*, **13**, 859–864.

Wasilewski, M. J. (1967). Sur certaines propriétés de la distribution gamma généralisée, *Revue de Statistique Appliquée*, **15**, No. 1, 95–105.

Welch, B. L. (1938). The significance of the difference between two means when the population variances are unequal, *Biometrika*, **29**, 350–361.

Wesolowski, J. (1990). A constant regression characterization of the gamma law, *Advances in Applied Probability*, **22**, 488–490.

Wesolowski, J. (1993). Some characterizations connected with properties of the quotient of independent random variables (preprint).

Wheeler, D. J. (1975). An approximation for simulation of gamma distributions, *Journal of Statistical Computation and Simulation*, **3**, 225–232.

Whittaker, J. (1974). Generating gamma and beta random variables with non-integral shape parameters, *Applied Statistics*, **23**, 210–214.

Wilk, M. B., and Gnanadesikan, R. (1964). Graphical methods for internal comparisons in multiresponse experiments, *Annals of Mathematical Statistics*, **35**, 613–631.

Wilk, M. B., Gnanadesikan, R., and Huyett, M. J. (1962a). Probability plots for the gamma distribution, *Technometrics*, **4**, 1–20.

Wilk, M. B., Gnanadesikan, R., and Huyett, M. J. (1962b). Estimation of parameters of the gamma distribution using order statistics, *Biometrika*, **49**, 525–545.

Wilk, M. B., Gnanadesikan, R., and Lauh, E. (1966). Scale parameter estimation from the order statistics of unequal gamma components, *Annals of Mathematical Statistics*, **37**, 152–176.

Wilson, E. B., and Hilferty, M. M. (1931). The distribution of chi-square, *Proceedings of the National Academy of Sciences, Washington*, **17**, 684–688.

Wingo, D. R. (1987a). Computing globally optimal maximum likelihood estimates of generalized gamma distribution parameters—Some new numerical approaches and analytical results, In *Proceedings of the Nineteenth Symposium on Interface of Computer Science and Statistics*.

Wingo, D. R. (1987b). Computing maximum-likelihood parameter estimates of the generalized gamma distribution by numerical root isolation, *IEEE Transactions on Reliability*, **R-36**, 586–590.

Wishart, J. (1927). On the approximate quadrature of certain skew curves, with an account of the researches of Thomas Bayes, *Biometrika*, **19**, 1–38. (Correction, *Ibid.*, **19**, 442.)

Wong, A. C. M. (1992). Inferences on the shape parameter of a gamma distribution: A conditional approach, *Technometrics*, **34**, 348–351.

Wong, A. C. M. (1993). A note on inference for the mean parameter of the gamma distribution, *Statistics & Probability Letters*, **17**, 61–66.

Yehia, A. Y., and Ahmed, A. N. (1987). A characterization of the generalized gamma distribution using residual moments, Manuscript, Department of Mathematics, Cairo University.

Yeo, G. F., and Milne, R. K. (1991). On characterizations of beta and gamma distributions, *Statistics & Probability Letters*, **11**, 239–242.

Young, D. H. (1971). Moment relations for order statistics of the standardized gamma distribution and the inverse multinomial distribution, *Biometrika*, **58**, 637–640.

Young, D. H., and Bakir, S. T. (1987). Bias correction for a generalized log-gamma regression model, *Technometrics*, **29**, 183–191.

Zubrzycki, S. (1966). Explicit formulas for minimax admissible estimators in some cases of restrictions imposed on the parameter, *Zastosowania Matematyki*, **9**, 31–52.

CHAPTER 18

Chi-Square Distributions, Including Chi and Rayleigh

1 HISTORICAL REMARKS

As we noted in Chapter 17, the *standard chi-square distribution with ν degrees of freedom* is in fact a gamma distribution with shape parameter $\nu/2$ and scale parameter 2. As a result many properties and results discussed for the gamma distribution in Chapter 17 will continue to hold (with suitable changes) for the chi-square distribution. In this chapter, however, we present specific developments that have taken place on the chi-square distribution and related issues. It is important to mention here the book by Lancaster (1969) which, even though it is more than 20 years old, gives a fine account of this distribution and will still serve as a good basic reference source.

Lancaster (1966) noted that Bienaymé (1838) obtained the chi-square distribution as the limiting distribution of the random variable $\Sigma_{i=1}^{k}(N_i - np_i)^2/(np_i)$, where $N_1, N_2, \ldots, N_k$ have a joint multinomial distribution with parameters $n, p_1, p_2, \ldots, p_k$ (see Chapter 11 of the first edition). It is also well-known that if $U_1, U_2, \ldots, U_\nu$ are independent standard normal variables, then $\Sigma_{i=1}^{\nu}U_i^2$ has a chi-square distribution with ν degrees of freedom (here ν has to be an integer by definition, but the distribution is defined for any real $\nu > 0$ as can be seen in the next section). Lancaster (1966) also pointed out that the result, that if $V_1, V_2, \ldots, V_k$ are independent chi-square random variables each with two degrees of freedom (i.e., they are exponential random variables), then $\Sigma_{i=1}^{k}V_i$ is distributed as chi-square with $2k$ degrees of freedom, was demonstrated by Ellis (1844). The general reproductive property of chi-square distributions was proved by Bienaymé (1852) and also by Helmert (1875) using a different method. Helmert (1876a, b) also established the result that if $X_1, X_2, \ldots, X_n$ are independent normal $N(\mu, \sigma^2)$ variables then $\Sigma_{i=1}^{n}(X_i - \bar{X})^2/\sigma^2$ (with $\bar{X}$ being the sample mean) is distributed as a chi-square variable with $n - 1$ degrees of freedom and that this variable is statistically independent of $\bar{X}$. Because of this historical fact

415

Kruskal (1946) recommended calling the joint distribution of the two random variables *Helmert's distribution*; this recommendation was also supported by Lancaster (1966).

Kendall (1971), through his investigations, concluded that Abbé (1863) was the first to deduce the chi-square law; interested readers may also refer to Sheynin (1966, 1971, 1988), Matsunawa (1981), and Plackett (1983) for additional illuminating details on the history of the chi-square distribution. In 1860 Maxwell derived the chi-square distribution for the case $\nu = 3$; during the years 1878 to 1881, Boltzmann used the chi-square distribution first with $\nu = 2$ and then with $\nu = 3$ (in 1878) and published a general formula later in 1881. The formula provided by Boltzmann (1878) for the case $\nu = 3$, however, contained an error. Quite interestingly, Sheynin (1988) (following a clue from Bol'shev) discovered that Gauss (1816), in one of his papers in *geodetics* (the branch of applied mathematics that determines the shape and area of large tracts of land, the exact position of geographical points, and the curvature, shape, and dimensions of the earth), presented a derivation of the asymptotic distribution of the chi-square (as $\nu \to \infty$) which (in modern day notation) is the well-known normal distribution with mean ν and standard deviation $\sqrt{2\nu}$. Sheynin (1971, 1988) also pointed out that Herschel came close to deriving chi-square for $\nu = 2$ in the year 1869 [as was discovered earlier by Kruskal (1946)].

The chi-square distribution also appeared in Pearson (1900) as the approximate distribution for the *chi-square statistics* used for various tests in contingency tables (of course the exact distribution of this statistic is discrete). The use of chi-square distribution to approximate the distribution of quadratic forms (particularly positive definite ones) in multinormally distributed variables is well established and widespread. One of the earliest ones in this direction is the work of Welch (1938) who used a chi-square approximation for the distribution of the denominator in a test criterion for difference in means of two normal populations with possibly different variances. Interested readers may refer to the commentaries of Lancaster (1982), Koch and Bhapkar (1982), and Koch and Stokes (1982) to see many other situations where the chi-square distribution plays a vital role.

2 DEFINITION

Let us denote a chi-square random variable with ν degrees of freedom by χ_ν^2. Then the probability density function of χ_ν^2 is

$$p_{\chi_\nu^2}(x) = \frac{1}{2^{\nu/2}\Gamma(\nu/2)} e^{-x/2} x^{(\nu/2)-1}, \qquad x \geq 0, \qquad (18.1)$$

where $\Gamma(\cdot)$ is the complete gamma function. As mentioned in the last section, this is precisely the density function of $\sum_{i=1}^{\nu} U_i^2$ when U_i's are independent standard normal variables. Although in this derivation of the density, ν must be an integer, the distribution in (18.1) is referred to as a χ^2 *distribution with ν degrees of freedom for any positive ν.*

From (18.1) we find

$$\frac{d}{dx} p_{\chi_\nu^2}(x) = \frac{e^{-x/2}}{2^{\nu/2}\Gamma(\nu/2)} \left\{ -\frac{1}{2} x^{(\nu/2)-1} + \left(\frac{\nu}{2} - 1\right) x^{(\nu/2)-2} \right\} \quad (18.2)$$

which immediately reveals that for $\nu \leq 2$ the mode of the distribution is at 0 while for $\nu > 2$ the mode is at $\nu - 2$. For any $\nu > 0$, we can also find the cumulative distribution function from (1) to be

$$F_{\chi_\nu^2}(x) = \Pr[\chi_\nu^2 \leq x] = \frac{\Gamma_{x/2}(\nu/2)}{\Gamma(\nu/2)}, \qquad x > 0, \quad (18.3)$$

where $\Gamma_x(\alpha)$ denotes the *incomplete gamma function* defined by

$$\Gamma_x(\alpha) = \int_0^x e^{-t} t^{\alpha-1} dt, \qquad x > 0. \quad (18.4)$$

Of course, for even values of ν, by repeated integration it can be shown from (18.3) that $1 - F_{\chi_\nu^2}(x)$ is equal to the probability that a Poisson random variable with mean $x/2$ does not exceed $(\nu/2) - 1$. For the purpose of illustration, plots of $p_{\chi_\nu^2}(x)$ are presented in Figure 18.1 for $\nu = 1(1)8$. Gulli (1989) recently made an analogy between the χ_ν^2 density function in (18.1) and the impulse response of a fractional system.

The distribution of the positive square root of a variable having a chi-square distribution with ν degrees of freedom is called a *chi-distribution with ν degrees of freedom*, and the corresponding variable denoted by χ_ν. From (18.1) we then obtain the pdf of χ_ν as

$$p_{\chi_\nu}(y) = \frac{1}{2^{(\nu/2)-1}\Gamma(\nu/2)} e^{-y^2/2} y^{\nu-1}, \qquad y > 0, \nu > 0. \quad (18.5)$$

This includes as special cases the half-normal ($\nu = 1$), Rayleigh ($\nu = 2$), and Maxwell-Boltzmann ($\nu = 3$) density functions. From (18.5), we immediately find that for $\nu \leq 1$ the mode of the distribution is at 0, while for $\nu > 1$ the mode is at $\sqrt{\nu - 1}$. One can also find the cumulative distribution of χ_ν easily from (18.3) to be

$$F_{\chi_\nu}(y) = \Pr[\chi_\nu \leq y] = \frac{\Gamma_{y^2/2}(\nu/2)}{\Gamma(\nu/2)}, \qquad y > 0, \nu > 0. \quad (18.6)$$

For the purpose of illustration, plots of $p_{\chi_\nu}(y)$ are presented in Figure 18.2 for $\nu = 1(1)8$.

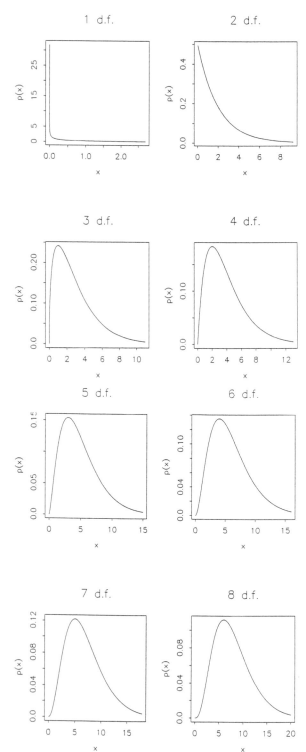

Figure 18.1 Plots of Chi-square Density in (18.1)

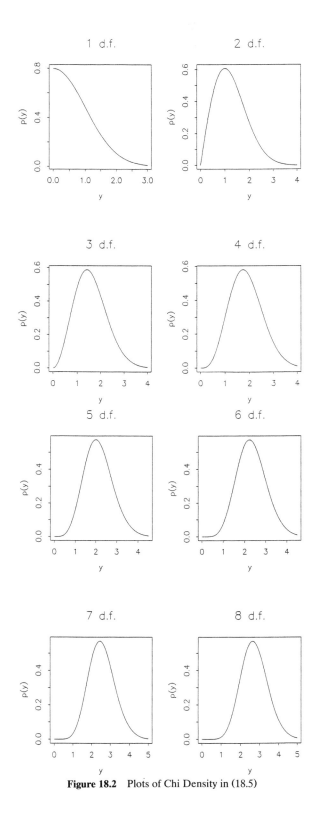

Figure 18.2   Plots of Chi Density in (18.5)

3 MOMENTS AND OTHER PROPERTIES

From (18.1) we find the moment-generating function of χ_ν^2 to be

$$E\left[e^{t\chi_\nu^2}\right] = \frac{1}{2^{\nu/2}\Gamma(\nu/2)}\int_0^\infty e^{-x(1-2t)/2}x^{(\nu/2)-1}\,dx$$

$$= (1-2t)^{-\nu/2}, \qquad t < \frac{1}{2}, \tag{18.7}$$

and, similarly, the characteristic function to be $(1 - 2it)^{-\nu/2}$. From (18.7), one may obtain the raw moments of χ_ν^2. Instead, one can derive the rth raw moment of χ_ν^2 directly from (18.1) to be

$$E\left[\chi_\nu^2\right]^r = \mu_r'(\chi_\nu^2) = \frac{1}{2^{\nu/2}\Gamma(\nu/2)}\int_0^\infty e^{-(x/2)}x^{(\nu/2)+r-1}\,dx$$

$$= \frac{2^r\Gamma[(\nu/2)+r]}{\Gamma(\nu/2)}$$

$$= \nu(\nu+2)\cdots\{\nu+2(r-1)\}. \tag{18.8}$$

From (18.7) we also find the cumulant generating function of χ_ν^2 to be $-(\nu/2)\log(1 - 2t)$, from which we find the rth cumulant to be

$$\kappa_r(\chi_\nu^2) = \nu 2^{r-1}(r-1)!, \qquad r = 1,2,3,\ldots. \tag{18.9}$$

Thus for the χ_ν^2 distribution we find that

$$E\left[\chi_\nu^2\right] = \nu, \qquad \mathrm{Var}(\chi_\nu^2) = 2\nu,$$

$$\sqrt{\beta_1(\chi_\nu^2)} = \sqrt{\frac{8}{\nu}}, \qquad \beta_2(\chi_\nu^2) = 3 + \frac{12}{\nu} \tag{18.10}$$

and the mean deviation to be

$$\frac{e^{-\nu/2}\nu^{\nu/2}}{2^{(\nu/2)-1}\Gamma(\nu/2)}. \tag{18.11}$$

It can be seen from the plots of the χ_ν^2 density function in Figure 18.1 that as ν increases, the shape of the curve becomes similar to that of the normal density curve. As a matter of fact, the χ_ν^2 distribution (after standardization) tends to the unit normal distribution; that is,

$$\lim_{\nu \to \infty} \Pr\left[\frac{\chi_\nu^2 - \nu}{\sqrt{2\nu}} \le x\right] = \Phi(x), \tag{18.12}$$

where $\Phi(\cdot)$ denotes the cumulative distribution function of a unit normal variable. It can also be checked from (18.10) that $\sqrt{\beta_1(\chi_\nu^2)} \to 0$ and $\beta_2(\chi_\nu^2) \to 3$ (the values for the unit normal distribution) as $\nu \to \infty$.

One of the important properties of the chi-square distribution is its *reproductive property* (described for the gamma distribution in Chapter 17): If X_1 and X_2 are independently distributed as $\chi_{\nu_1}^2$ and $\chi_{\nu_2}^2$, respectively, then $X_1 + X_2$ is distributed as $\chi_{\nu_1 + \nu_2}^2$.

Causey (1986) derived a formula for he expected absolute departure of χ_ν^2 variable from its median, that is, $\text{Disp}(\chi_\nu^2) = E[|\chi_\nu^2 - \text{Median} (\chi_\nu^2)|]$. Since for any continuous random variable X, the measure $E[|X - c|]$ (as a function of c) is minimum when $c = \text{Median} (X)$, this measure can be regarded as a measure of dispersion of the variable X. For the χ_ν^2 distribution, for example, Causey reported the values of the median and this measure of dispersion as

ν	1	2	3	4
Median(χ_ν^2)	0.4549	1.386	2.366	3.357
Disp(χ_ν^2)	0.8573	1.386	1.779	2.103

The moments of a χ_ν random variable can be derived easily from (18.8). For example, the rth raw moment is given by

$$E[\chi_\nu]^r = \mu'_r(\chi_\nu) = \mu'_r\left(\left[\chi_\nu^2\right]^{r/2}\right)$$
$$= \frac{2^{r/2}\Gamma[(\nu + r)/2]}{\Gamma(\nu/2)}, \qquad r = 1, 2, \dots. \tag{18.13}$$

Specifically, we have

$$E[\chi_\nu] = \sqrt{2}\,\frac{\Gamma[(\nu + 1)/2]}{\Gamma(\nu/2)} \quad \text{and} \quad \text{Var}(\chi_\nu) = \nu - 2\left\{\frac{\Gamma[(\nu + 1)/2]}{\Gamma(\nu/2)}\right\}^2. \tag{18.14}$$

Johnson and Welch (1939) gave formulas for computing the first six cumulants of χ_ν; they also presented a series expansion for $E[\chi_\nu]$ in (18.14) as:

$$E[\chi_\nu] = \sqrt{\nu}\left\{1 - \frac{1}{4\nu} + \frac{1}{32\nu^2} + \frac{5}{128\nu^3} - \frac{21}{2048\nu^4}\right.$$
$$\left. - \frac{399}{8192\nu^5} + \frac{869}{65536\nu^6} + \cdots\right\}. \tag{18.15}$$

They also presented the following recurrence relations for the cumulants $\kappa_r(\chi_\nu)$ (with $\kappa_1(\chi_\nu) = E[\chi_\nu]$):

$$\kappa_2(\chi_\nu) = \nu - \{\kappa_1(\chi_\nu)\}^2,$$

$$\kappa_3(\chi_\nu) = \kappa_1(\chi_\nu)\{1 - 2\kappa_2(\chi_\nu)\},$$

$$\kappa_4(\chi_\nu) = \tfrac{1}{2} - (2\nu - 1)\{1 - 2\kappa_2(\chi_\nu)\} - \tfrac{3}{2}\{1 - 2\kappa_2(\chi_\nu)\}^2,$$

$$\kappa_5(\chi_\nu) = \kappa_1(\chi_\nu)\Big[-2\kappa_4(\chi_\nu) + 3\{1 - 2\kappa_2(\chi_\nu)\}^2\Big],$$

and

$$\kappa_6(\chi_\nu) = 2(2\nu - 1)\kappa_4(\chi_\nu) + 3\{1 - 2\kappa_2(\chi_\nu)\}$$
$$- 12(2\nu - 1)\{1 - 2\kappa_2(\chi_\nu)\}^2 - 15\{1 - 2\kappa_2(\chi_\nu)\}^3.$$

Deutler (1984) pointed out correctly that the terms in the formula for $\kappa_r(\chi_\nu)$ are each of the same order and hence a dangerous numerical cancellation results for large ν, since the χ_ν-distribution tends to normal for large ν and as a result $\kappa_r(\chi_\nu/\sqrt{\nu}) \to 0$ for $r \geq 3$ as $\nu \to \infty$. Therefore by making use of Luke's (1969) series expression for the ratio of two gamma functions, Deutler (1984) derived series expansions for the cumulants $\kappa_r(\chi_\nu)$ similar to the one in (18.15) for $E[\chi_\nu]$. For example,

$$\nu^{-3/2}\kappa_3(\chi_\nu) = \frac{1}{2^2\nu^2} + \frac{1}{2^4\nu^3} - \frac{13}{2^7\nu^4} - \frac{75}{2^9\nu^5} + \frac{1215}{2^{13}\nu^6}$$
$$+ \frac{17403}{2^{15}\nu^7} - \frac{122101}{2^{18}\nu^8} + \cdots,$$

$$\nu^{-2}\kappa_4(\chi_\nu) = \frac{3}{2^4\nu^4} + \frac{3}{2^4\nu^5} - \frac{45}{2^7\nu^6} - \frac{57}{2^6\nu^7} + \frac{4875}{2^{12}\nu^8} + \cdots.$$

These series expressions demonstrate how quickly $\kappa_r(\chi_\nu/\sqrt{\nu}) \to 0$ for $r \geq 3$ as ν increases, giving a rough criterion for the convergence rate of $(\chi_\nu/\sqrt{\nu}) - 1$ to standard normality. Keiding, Jensen, and Ranek (1972), among others, studied properties of the χ_ν distribution and observed that it quickly approaches normal distribution as ν gets large.

4   TABLES AND NOMOGRAMS

One of the earliest tables of probability integrals of chi-square distributions was published in 1902. It contained values of $\Pr[\chi_\nu^2 > x]$ to six decimal places for $\nu = 2(1)29$ and $x = 1(1)30(10)70$, and was prepared by Elderton (1902). In 1922 Pearson edited a comprehensive *Tables of the Incomplete Γ-Function*.

These tables contained values of the function $I(u, p)$ defined by (also see Chapter 17)

$$I(u, p) = \frac{1}{\Gamma(p+1)} \int_0^{u\sqrt{p+1}} e^{-t} t^p \, dt = \frac{\Gamma_{u\sqrt{p+1}}(p+1)}{\Gamma(p+1)} \quad (18.16)$$

to seven decimal places for $p = -1(0.05)0(0.1)5(0.2)50$ and u at intervals of 0.1. These values were also supplemented by a table of values of

$$\log I(u, p) - (p+1)\log u \quad (18.17)$$

for $p = -1(0.05)0(0.1)10$ and $u = 0.1(0.1)1.5$. This function was chosen to make interpolation easier, particularly for low values of p. Harter (1964a, b) published tables of $I(u, p)$ to nine decimal places for $p = -0.5(0.5)74(1)164$ and u at intervals of 0.1 (thus covering a greater range of values of p than that considered by Pearson (1922), although not at such fine intervals, and also with a higher accuracy) and a table of percentage points of the χ_ν^2 distribution. As mentioned earlier in Section 2, special values of the survival function (or the cumulative distribution function) of a chi-square distribution can be obtained from tables of the cumulative Poisson distribution using the formulas (for integral values of ν)

$$\Pr\left[\chi_{2\nu}^2 \geq x\right] = e^{-x/2} \sum_{i=0}^{\nu-1} (x/2)^i / i! \quad (18.18)$$

and

$$\Pr\left[\chi_{2\nu-1}^2 \geq x\right] = e^{-x/2} \sum_{i=0}^{\nu-2} \left\{(x/2)^{i+\frac{1}{2}} / \Gamma(i + \tfrac{3}{2})\right\} + 2\{1 - \Phi(\sqrt{x})\}. \quad (18.19)$$

Khamis and Rudert (1965) published extensive tables of $\Pr[\chi_{2\nu}^2 \geq x]$ to ten decimal places for

$$\nu = 0.05(0.05)10(0.1)20(0.25)70,$$

$$x = 0.0001(0.0001)0.001(0.001)0.01(0.01)1.00(0.05)6.0(0.1)$$

$$16.0(0.5)66(1)166(2)250.$$

The varied intervals for x were chosen to make interpolation easier just as Pearson (1922) used equal intervals for $u = x/(2\sqrt{\nu})$ for the same reason. Thompson (1941) and Pearson and Hartley (1954) presented tables of lower ε percentage points of χ_ν^2, viz., $\chi_{\nu,\varepsilon}^2$ such that

$$\Pr\left[\chi_\nu^2 \leq \chi_{\nu,\varepsilon}^2\right] = F_{\chi_\nu^2}(\chi_{\nu,\varepsilon}^2) = \varepsilon \quad (18.20)$$

to six significant figures for $\nu = 1(1)30(10)100$ and $\varepsilon = 0.005, 0.01, 0.025,$ $0.05, 0.10, 0.25, 0.50, 0.75, 0.90, 0.95, 0.975, 0.99, 0.995$. Pearson and Hartley (1954) also presented values for $\varepsilon = 0.999$. Vanderbeck and Cooke (1961) used the same values of ε, with the addition of the value 0.80, and tabulated $\chi^2_{\nu, \varepsilon}$ to four decimal places or significant digits, whichever turned out to be more accurate (except for $\nu = 1$), for $\nu = 1(1)300$; an approximation was used by them when $\nu > 30$. The less easily accessible tables prepared by these authors give values of $1 - F_{\chi^2_\nu}(x)$ for $x = 0.1(0.1)3.2$ with $\nu = 0.05(0.05)0.2(0.1)6.0$, and for $x = 3.2(0.2)7.0(0.5)10(1)35$ with $\nu = 0.1(0.1)0.4(0.2)6.0$. The noninteger choices of ν in these tables is a valuable feature. To facilitate interpolation for small values of x, tables of $F_{\chi^2_\nu}(x)/(x/2)^{\nu/2}$ were also presented for $x = 0.05(0.05)0.2(0.1)1.0$ with $\nu = 0.05(0.05)0.2(0.1)6.0$. For large values of x, another table presented values of $1 - F_{\chi^2_\nu}(x)$ for $\sqrt{2x} - \sqrt{2\nu} = -4.0(0.1)4.8$ with $\sqrt{2/\nu} = 0.02(0.02)0.22(0.01)0.25$.

Hald and Sinkbaek (1950) presented tables of $\chi^2_{\nu, \varepsilon}$ to three decimal places or significant digits, whichever turned out to be more accurate, for $\nu = 1(1)100$ and $\varepsilon = 0.0005, 0.001, 0.005, 0.01, 0.025, 0.05, 0.1(0.1)0.9, 0.95,$ $0.975, 0.99, 0.995, 0.999, 0.9995$. Harter (1964a) presented tables of $\chi^2_{\nu, \varepsilon}$ to six significant figures for $\nu = 1(1)150(2)330$ and the above choices of ε and also $\varepsilon = 0.0001$ and 0.9999. Part of these tables corresponding to $\nu = 1(1)100$ was published by Harter (1964b). As pointed out by Bol'shev in his review of these tables, the correct values for $\chi^2_{60, 0.6}$, $\chi^2_{74, 0.8}$, and $\chi^2_{80, 0.4}$ are 62.1348, 83.9965, and 76.1879, respectively. Russell and Lal (1969) tabulated values of $1 - F_{\chi^2_\nu}(x)$ to five decimal places for $\nu = 1(1)50$ and $x = 0.001(0.001)0.01(0.01)0.1(0.1)10.0$. Slustskii (1950) compiled the first comprehensive tables of the χ^2_ν distribution in the Soviet Union.

A nomogram produced by Boyd (1965) connects values of $P = \Pr[\chi^2_\nu \geq \chi^2_0]$ with those of ν and χ^2_0 by means of a straight-edge. Boyd's nomogram is presented here in Figure 18.3. Stammberger (1967) produced another nomogram for the χ^2_ν distribution. Krauth and Steinebach (1976) produced extended tables of percentage points of the chi-square distribution with at most ten degrees of freedom. Specifically they tabulated the percentage points corresponding to

$$\varepsilon = 0.0000001, 0.0000005, 0.000001, 0.000005, 0.00001, 0.00005,$$

$$0.0001(0.0001)0.001(0.001)0.01(0.005)0.1(0.01)0.2, 0.8(0.1)0.9$$

$$(0.005)0.99(0.001)0.999(0.0001)0.9999, 0.99995, 0.99999,$$

$$0.999995, 0.999999, 0.9999995, 0.9999999.$$

Gilbert (1977) provided a simple formula for interpolating tables of χ^2_ν. He implemented an approximation of the form

$$\chi^2_{\nu, \varepsilon} \simeq a_0 + a_1 \nu + a_2 \nu^2 + a_3 \log \nu, \tag{18.21}$$

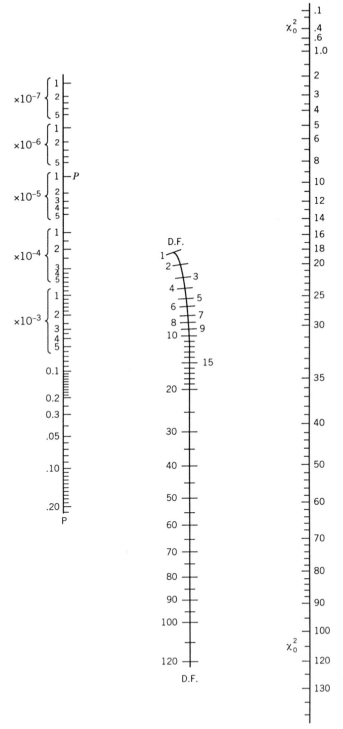

Figure 18.3 Boyd's χ^2 Nomogram

where the coefficients a_i depend only on ε (or the significance level $1 - \varepsilon$). By employing Dwyer's (1951) method of pivotal condensation to solve the normal equations obtained from (18.21) for $\nu = 1(1)30$ and the tabled values of $\chi^2_{\nu, \varepsilon}$ for $\varepsilon = 0.95, 0.99$ and 0.999, Gilbert determined the following values for the coefficients a_i:

| | SIGNIFICANCE LEVEL, $1 - \varepsilon$ | | |
	0.05	**0.01**	**0.001**
a_0	2.518232	5.174627	9.205913
a_1	1.282189	1.402766	1.542498
a_2	-0.00211427	-0.00303260	-0.00411568
a_3	1.371169	1.858993	2.314035

Gilbert has discussed the maximum error incurred while using the interpolation formula in (18.21); he has shown that this interpolation method provides quite reasonable values (often correct to two decimal places or more) for ν small and even nonintegral for the values of ε considered.

5   APPROXIMATIONS AND COMPUTATIONAL ALGORITHMS

As mentioned earlier in Section 3, the standardized χ^2_ν distribution tends to the unit normal distribution as $\nu \to \infty$ [see (18.12)]. The simple approximation obtained from (18.12) given by

$$F_{\chi^2_\nu}(x) \simeq \Phi\left(\frac{x - \nu}{\sqrt{2\nu}}\right), \tag{18.22}$$

however, is not very accurate unless ν is rather large. Better approximations may be obtained by using the asymptotic normality of various functions of χ^2_ν, even though only approximate standardization is effected. Among the best-known simple approximations are Fisher's (1922) approximation given by

$$F_{\chi^2_\nu}(x) \simeq \Phi\left(\sqrt{2x} - \sqrt{2\nu - 1}\right) \tag{18.23}$$

and the Wilson-Hilferty (1931) approximation given by

$$F_{\chi^2_\nu}(x) \simeq \Phi\left(\sqrt{\frac{9\nu}{2}}\left\{\left(\frac{x}{\nu}\right)^{1/3} - 1 + \frac{2}{9\nu}\right\}\right). \tag{18.24}$$

Of these two approximations the second one is definitely more accurate, but both the approximations are better than the one in (18.22). From the approximations in (18.23) and (18.24), one may obtain approximations to the percentage point $\chi^2_{\nu,\varepsilon}$ as

$$\chi^2_{\nu,\varepsilon} \simeq \frac{1}{2}\left(U_\varepsilon + \sqrt{2\nu - 1}\right)^2, \tag{18.25}$$

$$\chi^2_{\nu,\varepsilon} \simeq \nu\left(\sqrt{\frac{2}{9\nu}} U_\varepsilon + 1 - \frac{2}{9\nu}\right)^3 \tag{18.26}$$

respectively; here U_ε denotes $\Phi^{-1}(\varepsilon)$, the lower ε percentage point of the standard normal distribution. It should be mentioned that the addition of $(U_\varepsilon^2 - 1)/6$ to (18.25) will make it very nearly equal to the usually better approximation in (18.26), unless ν is small or ε is close to 0 or 1. This point is illustrated by the values in Table 18.1.

Table 18.1 Comparison of Approximations to χ^2 Percentile Points

| | | | Approximation | | Difference | |
ε	ν	$\chi^2_{\nu,\varepsilon}$	(18.25)	(18.26)	(18.26)–(18.25)	$\frac{1}{6}(U_\varepsilon^2 - 1)$
0.01	5	0.5543	0.2269	0.5031	0.2762	
	10	2.5582	2.0656	2.5122	0.4466	0.7353
	25	11.5240	10.9215	11.4927	0.5712	
0.05	5	1.1455	0.9182	1.1282	0.3100	
	10	3.9403	3.6830	3.9315	0.2485	0.2843
	25	14.6114	14.3388	14.6086	0.2698	
0.10	5	1.6103	1.4765	1.6098	0.1333	
	10	4.8652	4.7350	4.8695	0.1345	0.1071
	25	16.4734	16.3503	16.4788	0.1285	
0.50	5	4.3515	4.5000	4.3625	−0.1375	
	10	9.3418	9.5000	9.3480	−0.1520	−0.1667
	25	24.3366	24.5000	24.3392	−0.1608	
0.90	5	9.2364	9.1658	9.2078	0.0420	
	10	15.9872	15.9073	15.9677	0.0604	0.1071
	25	34.3816	34.2920	34.3701	0.0781	
0.95	5	11.0705	10.7873	11.0439	0.2666	
	10	18.3070	18.0225	18.2918	0.2693	0.2843
	25	37.6525	37.3667	37.6452	0.2785	
0.99	5	15.0862	14.1850	14.4599	0.2749	
	10	23.2093	22.3463	23.2393	0.8930	0.7353
	25	44.3141	43.4904	44.3375	0.8471	

Vanderbeck and Cooke (1961) used the Cornish-Fisher expansion to give

$$
\chi^2_{\nu,\varepsilon} \simeq \nu + \sqrt{2\nu}\, U_\varepsilon + \frac{2}{3}\left(U_\varepsilon^2 - 1\right) + \frac{1}{9\sqrt{2\nu}}\left(U_\varepsilon^3 - 7U_\varepsilon\right)
$$

$$
- \frac{1}{405\nu}\left(106U_\varepsilon^4 - 486U_\varepsilon^2 + 168\right)
$$

$$
+ \frac{1}{4860\nu\sqrt{2\nu}}\left(9U_\varepsilon^5 + 256U_\varepsilon^3 - 433U_\varepsilon\right) \tag{18.27}
$$

[also see Goldberg and Levine (1946); Peiser (1943)]. They also presented a table comparing exact values of $\chi^2_{\nu,\varepsilon}$ with those given by (18.27) for $\nu = 10(10)100$ and $\varepsilon = 0.005, 0.010, 0.025, 0.05, 0.10, 0.25, 0.50, 0.75, 0.90, 0.95, 0.975, 0.99, 0.995$. For $\nu \geq 60$ the difference was only occasionally greater than 0.0001. In fact, even when $\nu = 30$, the difference exceeded 0.0001 only for ε outside the range 0.1–0.9.

A power series expression for $F_{\chi^2_\nu}(x)$ can be derived in the following way:

$$
F_{\chi^2_\nu}(x) = \frac{1}{2^{\nu/2}\Gamma(\nu/2)} \int_0^x e^{-t/2} t^{(\nu/2)-1}\, dt
$$

$$
= \frac{1}{2^{\nu/2}\Gamma(\nu/2)} \int_0^x \sum_{i=0}^{\infty} (-1)^i \{ t^{(\nu/2)+i-1}/(2^i i!) \}\, dt
$$

$$
= \frac{2(x/2)^{\nu/2}}{\Gamma(\nu/2)} \sum_{i=0}^{\infty} (-1)^i \frac{x^i}{(\nu + 2i)2^i i!}. \tag{18.28}
$$

The above series converges to all $x > 0$, and the true value will lie between $\sum_{i=0}^{m}$ and $\sum_{i=0}^{m+1}$. The series expression in (18.28) may be used as a basis for evaluating $F_{\chi^2_\nu}(x)$ approximately when x is small (e.g., $x < \nu$). Pearson and Hartley (1954) suggested an iterative procedure for calculating $\chi^2_{\nu,\varepsilon}$ based on (18.28). Since $F_{\chi^2_\nu}(\chi^2_{\nu,\varepsilon}) = \varepsilon$, (18.28) can be written in the form

$$
\chi^2_{\nu,\varepsilon} = 2\left\{ \varepsilon\Gamma\left(\frac{\nu}{2}\right) \right\}^{2/\nu} \left\{ \sum_{i=0}^{\infty} (-1)^i \frac{(\chi^2_{\nu,\varepsilon})^i}{(\nu + 2i)2^i i!} \right\}^{-2/\nu}. \tag{18.29}
$$

By inserting a trial value of $\chi^2_{\nu,\varepsilon}$ on the right-hand side of (18.29), a new value can be obtained. The first term of the series could be used for the initial value, while one additional term could be included at each successive iteration. Harter (1964a, b) used this iterative method in the construction of his tables.

Gray, Thompson, and McWilliams (1969) used an approximation for the incomplete gamma integral, together with Stirling's approximation, to get

$$1 - F_{\chi_\nu^2}(x) \simeq \frac{\nu\sqrt{\nu}\,e^{\nu/2}}{(\nu + \frac{1}{6})\sqrt{\pi}} \left(\frac{x}{\nu}\right)^{\nu/2} e^{-x/2}\left\{1 - \frac{2(\nu - 2)}{(x - \nu + 2)^2 + 4\nu}\right\}.$$

$$(18.30)$$

This approximation gives an accuracy of three decimal places when $1 - F_{\chi_\nu^2}(x)$ is of order 0.1, even for ν as small as 2.

Wallace (1959) derived definite bounds for the survival function of the χ_ν^2 distribution and for the corresponding equivalent standard normal deviate. He showed that

$$d_\nu\{1 - \Phi(w(x))\} > 1 - F_{\chi_\nu^2}(x) > d_\nu e^{1/(9\nu)}\{1 - \Phi(w_2(x))\}, \quad (18.31)$$

where

$$d_\nu = \left(\frac{\nu}{2}\right)^{(\nu-1)/2} \frac{e^{-\nu/2}\sqrt{2\pi}}{\Gamma(\nu/2)},$$

$$w(x) = \left\{x - \nu - \nu \log\left(\frac{x}{\nu}\right)\right\}^{1/2},$$

$$w_2(x) = w(x) + \frac{\sqrt{2}}{3\sqrt{\nu}}.$$

From Stirling's formula it is easily noted that d_ν is very nearly equal to 1. Defining the equivalent normal deviate by the equation

$$1 - F_{\chi_\nu^2}(x) = 1 - \Phi(u(x)), \quad (18.32)$$

Wallace also proved that

$$w(x) \le u(x) \le w_2(x) + \frac{1}{w_2(x)} \max\left(0, \frac{1}{d_\nu}e^{-1/(9\nu)} - 1\right) \quad (18.33)$$

with the lower limit being valid for $\nu > 0.37$ and the max can be replaced by $0.6\sqrt{\nu}$. While the bounds in (18.33) are less precise than the Wilson-Hilferty approximation in (18.24) for most of the distribution, the upper bound in (18.33) gives useful approximation in the extreme upper tail. Hill (1969) presented an expansion of $u(x)$ in terms of $w(x)$ as

$$u(x) \simeq w(x) + \frac{\sqrt{2}}{3\sqrt{\nu}} - \frac{1}{18\nu}(w(x))^2 - \frac{\sqrt{2}}{810\nu\sqrt{\nu}}\{(w(x))^2 - 13\}$$

$$+ \frac{7}{9720\nu^2}\{6(w(x))^3 + 17w(x)\} + \cdots, \tag{18.34}$$

which gives $\Phi(u(x))$ to five decimal places for $\nu > 20$ (in fact, to eight decimal places if terms up to order ν^{-4} are retained). Cornish (1969) reported high accuracy for Hill's expansion in (18.34). Hill also presented a Cornish-Fisher type expansion for the chi-square percentiles as

$$\chi^2_{\nu,\varepsilon} \simeq \left(\nu - \frac{2}{3}\right)\exp\left\{U_\varepsilon c_\nu - \frac{1}{6}U_\varepsilon^2 + \frac{1}{36c_\nu}(U_\varepsilon^3 - U_\varepsilon)\right.$$

$$- \frac{1}{1620c_\nu^2}(6U_\varepsilon^4 - 31U_\varepsilon^2 - 32)$$

$$\left. + \frac{1}{38880c_\nu^3}(9U_\varepsilon^5 - 308U_\varepsilon^3 - 481U_\varepsilon)\right\}, \tag{18.35}$$

where $c_\nu = \sqrt{(\nu/2) - (1/3)}$ and U_ε is the lower ε percentage point of the standard normal distribution (as before). The approximation in (18.35) is much more accurate than Vanderbeck and Cooke's (1961) Cornish-Fisher expansion in (18.27) for ε values close to 0 or 1.

In his algorithm for the computation of χ^2-quantiles, Goldstein (1973) used the extended formula

$$\chi^2_{\nu,\varepsilon} \simeq \nu\left[1 - \frac{2}{9\nu} + \frac{1}{1215\nu^2}(4x^4 + 16x^2 - 28)\right.$$

$$+ \frac{1}{229635\nu^3}(8x^6 + 720x^4 + 3216x^2 + 2904)$$

$$+ \sqrt{\frac{2}{\nu}}\left\{\frac{x}{3} + \frac{1}{162\nu}(-x^3 + 3x) - \frac{1}{5832\nu^2}(3x^5 + 40x^3 + 45x)\right.$$

$$\left.\left. + \frac{1}{7873200\nu^3}(301x^7 - 1519x^5 - 32769x^3 - 79349x)\right\}\right]^3, \tag{18.36}$$

to which the following polynomial was fitted:

$$
\begin{aligned}
\chi^2_{\nu,\varepsilon} \simeq \nu \Bigg\{ &\left(1.0000886 - \frac{0.2237368}{\nu} - \frac{0.01513904}{\nu^2} \right) \\
&+ \frac{x}{\sqrt{\nu}} \left(0.4713941 + \frac{0.02607083}{\nu} - \frac{0.008986007}{\nu^2} \right) \\
&+ \frac{x^2}{\nu} \left(0.0001348028 + \frac{0.01128186}{\nu} + \frac{0.02277679}{\nu^2} \right) \\
&+ \frac{x^3}{\nu\sqrt{\nu}} \left(-0.008553069 - \frac{0.01153761}{\nu} - \frac{0.01323293}{\nu^2} \right) \\
&+ \frac{x^4}{\nu^2} \left(0.00312558 + \frac{0.005169654}{\nu} - \frac{0.006950356}{\nu^2} \right) \\
&+ \frac{x^5}{\nu^2\sqrt{\nu}} \left(-0.0008426812 + \frac{0.00253001}{\nu} + \frac{0.001060438}{\nu^2} \right) \\
&+ \frac{x^6}{\nu^3} \left(0.00009780499 - \frac{0.001450117}{\nu} + \frac{0.001565326}{\nu^2} \right) \Bigg\}^3,
\end{aligned}
$$

$$(18.37)$$

where $x \equiv U_\varepsilon$ is the lower ε percentage point of the standard normal distribution. Goldstein's approximation in (18.36) seems to be quite good even for extreme significance limits. The polynomial fit in (18.37) is slightly better than (18.36), but is unwieldy and is useful primarily for computer implementation.

Severo and Zelen (1960) considered the approximation to the normal distribution function $\Phi(x + h)$ as given by the Taylor series expansion about x, and the following terms of the Gram-Charlier series

$$
F_{\chi^2_\nu}(x) = \Phi(x) - \frac{\gamma_1}{6}\Phi^{(3)}(x) + \frac{\gamma_2}{24}\Phi^{(4)}(x),
$$

where $\Phi^{(k)}(x)$ is the kth derivative of $\Phi(x)$ and γ_1 and γ_2 are the measures of skewness and kurtosis, respectively, given by

$$
\gamma_1 = \frac{\sqrt{128}}{27\nu\sqrt{\nu}} \quad \text{and} \quad \gamma_2 = -\frac{4}{9\nu} - \frac{64}{81\nu^2},
$$

to propose a modification to the Wilson-Hilferty formula in (18.26) as

$$\chi^2_{\nu,\varepsilon} \simeq \nu \left\{ 1 - \frac{2}{9\nu} + \sqrt{\frac{2}{9\nu}} \, (U_\varepsilon - h_\nu) \right\}^3, \qquad (18.38)$$

where

$$h_\nu = -\frac{2}{27\nu} \left(\frac{2\sqrt{2}\,(U_\varepsilon^2 - 1)}{3\sqrt{\nu}} - \frac{(U_\varepsilon^3 - 3U_\varepsilon)}{4} \right). \qquad (18.39)$$

In obtaining the correction factor h_ν in (18.39), Severo and Zelen (1960) used only the first term in the above expression of γ_2. By using both terms of γ_2 and proceeding similarly, Zar (1978) proposed a modified Severo-Zelen approximation as

$$\chi^2_{\nu,\varepsilon} \simeq \nu \left\{ 1 - \frac{2}{9\nu} + \sqrt{\frac{2}{9\nu}} \, (U_\varepsilon - h'_\nu) \right\}^3, \qquad (18.40)$$

where

$$h'_\nu = \frac{1}{486\nu^2} \left\{ (9\nu + 16)(U_\varepsilon^3 - 3U_\varepsilon) - 24\sqrt{2\nu}\,(U_\varepsilon^2 - 1) \right\}. \quad (18.41)$$

Another simple approximation to the percentage points of the χ^2_ν distribution is the time-honored Haldane's (1937) approximation given by

$$\chi^2_{\nu,\varepsilon} \simeq \frac{1}{13} \left[12\nu \left\{ \frac{5U_\varepsilon}{6\left(1 - \frac{1}{18\nu}\right)\sqrt{2\nu}} + 1 - \frac{5}{18\nu}\left(1 + \frac{7}{48\nu}\right) \right\}^{13/5} + \nu \right].$$

$$(18.42)$$

Aroian (1943) used the coefficient of skewness ($= \sqrt{8/\nu}$) of the χ^2_ν distribution to determine empirically a relationship

$$\chi^2_{\nu,\varepsilon} \simeq \nu + \sqrt{2\nu} \left\{ U_\varepsilon + b_1 \sqrt{\frac{8}{\nu}} + b_2\left(\frac{8}{\nu}\right) \right\} \qquad (18.43)$$

reporting values of b_1 and b_2 for various choices of ε.

Hoaglin (1977), upon examining the tables of percentage points for $\nu = 3$ to 30, gave an approximation for the right-tail of the χ^2_ν distribution to be

$$\chi^2_{\nu,\varepsilon} \simeq \left\{ 1.00991\sqrt{\nu} + 1.95188\sqrt{-\log_{10}(1 - \varepsilon)} - 1.14485 \right\}^2, \quad (18.44)$$

which he called the "additive" fit. A better approximation, which he called the "extended" fit, is

$$\chi^2_{\nu,\varepsilon} \simeq \left\{ 1.06807\sqrt{\nu} + 2.13161\sqrt{-\log_{10}(1-\varepsilon)} \right.$$
$$\left. -0.04589\sqrt{\nu[-\log_{10}(1-\varepsilon)]} - 1.37266 \right\}^2, \quad (18.45)$$

and a "simplified" fit was given as

$$\chi^2_{\nu,\varepsilon} \simeq \left\{ \sqrt{\nu} + 2\sqrt{-\log_{10}(1-\varepsilon)} - \frac{7}{6} \right\}^2. \quad (18.46)$$

Similarly, for the left-tail, Hoaglin provided the "additive" fit to be

$$\chi^2_{\nu,\varepsilon} \simeq \left\{ 0.97657\sqrt{\nu} + 1.46049\sqrt{-\log_{10}(1-\varepsilon)} + 0.59025 \right\}^2, \quad (18.47)$$

the "extended fit" to be

$$\chi^2_{\nu,\varepsilon} \simeq \left\{ 1.14309\sqrt{\nu} - 0.94591\sqrt{-\log_{10}(1-\varepsilon)} \right.$$
$$\left. -0.13138\sqrt{\nu[-\log_{10}(1-\varepsilon)]} - 0.06198 \right\}^2, \quad (18.48)$$

and a "simplified fit" to be

$$\chi^2_{\nu,\varepsilon} \simeq \left\{ \sqrt{\nu} - 1.5\sqrt{-\log_{10}(1-\varepsilon)} - 0.6 \right\}^2. \quad (18.49)$$

A very interesting comparative study of all these approximation formulas was carried out by Zar (1978) at a number of different significance levels. He also included in this study an improved approximation derived from Severo and Zelen's approximation in (18.38) and its modification in (18.40) given by

$$\chi^2_{\nu,\varepsilon} \simeq \nu \left\{ 1 - \frac{2}{9\nu} + \sqrt{\frac{2}{9\nu}} \left(U_\varepsilon - \frac{K}{\nu} \right) \right\}^3, \quad (18.50)$$

where $K = H = ch_c \simeq \nu h_\nu$ or $K = H' = ch'_c \simeq \nu h'_\nu$, whichever provides the better approximation, where h_ν and h'_ν are as given in (18.39) and (18.41), respectively. The values of c, and H or H' that were found empirically to yield the most accurate approximation using (18.50) have been tabulated by Zar (1978).

Zar (1978) assessed the accuracy of all the above described approximation formulas by determining the minimum value of ν necessary to achieve a specified absolute relative error for various values of ε. These are reported in Table 18.2. The relative error defined by (Approximate value − Actual

Table 18.2 Accuracy of Several Chi-squared Approximation Formulas

$1-\varepsilon =$	0.999	0.995	0.990	0.975	0.950	0.900	0.750	0.500	0.250	0.100	0.050	0.025	0.010	0.005	0.001
$|RE| = 1\%$															
F	—	123	97	65	42	21	7	17	10	4	18	33	54	69	107
WH	25	17	14	9	6	4	4	3	1	2	2	2	1	1	6
EWH	6	5	4	3	3	2	1	1	2	2	2	2	2	2	2
PWH	1	3	2	2	2	2	2	2	1	2	2	1	2	2	2
SZ	12	9	8	6	5	4	3	3	2	3	4	4	4	4	4
MSZ	12	8	7	6	5	5	4	3	3	4	4	3	1	1	5
ESZ	9	7	6	6	5	4	2	1	2	2	1	1	1	1	1
A	21	15	14	10	8	5	5	4	3	1	2	3	4	4	6
H	19	14	12	8	6	4	3	3	2	2	2	1	1	2	4
CF	8	7	7	6	5	4	2	2	2	1	1	2	2	2	3
ESZ & CF		6 (17)	4 (16)	4 (14)	3 (14)	3 (9)			2 (6)					1 (9)	
$|RE| = 0.5\%$															
F	—	—	—	119	75	35	17	34	19	10	42	74	118	—	12
WH	36	24	19	12	8	4	6	4	2	4	4	3	2	5	3
EWH	7	5	5	4	3	2	2	2	2	2	2	2	2	2	3
PWH	4	3	3	3	2	2	2	2	1	2	2	1	2	2	3
SZ	17	12	10	7	6	5	4	4	2	4	5	5	6	5	5
MSZ	15	10	8	7	6	5	5	4	3	5	5	5	3	3	7
ESZ	9	8	7	6	6	5	3	1	2	2	1	1	1	3	1
A	26	19	17	13	10	6	6	5	3	1	3	4	6	7	9
H	25	18	15	11	8	5	6	4	3	2	2	1	3	4	6
CF	9	9	8	7	6	4	4	4	2	3	2	2	3	3	3
ESZ & CF		7	4	4	3	3	2	3	2					2	

$|RE| = 0.1\%$

F	—	—	—	—	—	82	93	—	89	136	—	—	—	—	—
WH	47	24	14	5	9	12	9	9	15	5	14	26	43	57	87
EWH	4	3	3	3	3	3	3	2	2	3	4	5	6	7	9
PWH	3	2	3	3	2	2	3	3	4	3	2	5	4	4	5
SZ	7	10	11	11	10	8	3	7	7	7	9	12	16	21	31
MSZ	15	4	7	10	10	9	6	7	9	9	9	11	14	18	28
ESZ	4	4	2	2	7	5	4	2	4	6	7	9	10	12	12
A	21	17	14	10	3	2	7	10	10	9	16	22	28	30	41
H	18	12	9	5	3	4	6	6	7	7	13	18	26	32	45
CF	6	5	5	3	3	3	4	5	4	7	8	10	12	13	13
ESZ & CF	—	4	—	—	—	—	2	—	—	5	6	8	9	11	—

$|RE| = 0.05\%$

F	—	—	—	—	—	20	25	12	25	18	18	37	62	83	131
WH	80	43	26	7	15	20	3	4	4	4	6	6	7	9	10
EWH	5	3	4	4	4	3	3	3	5	3	6	5	6	6	8
PWH	3	2	3	3	3	2	4	9	16	12	15	15	20	26	42
SZ	9	14	15	15	14	11	8	9	11	10	10	13	18	23	37
MSZ	21	5	10	13	14	13	4	3	4	7	10	12	13	15	12
ESZ	6	7	3	13	2	6	9	12	13	11	22	27	49	35	48
A	29	24	19	13	11	3	8	9	10	8	16	25	35	43	61
H	28	19	14	8	3	6	5	5	6	11	10	11	15	15	16
CF	7	7	6	4	5	3	3	—	—	5	9	9	10	12	—
ESZ & CF	8	5	—	—	—	—	—	—	—	—	—	—	—	—	—

Note: Shown is the minimum ν necessary to achieve the indicated accuracy (RE = relative error). F = Fisher (18.25); WH = Wilson-Hilferty (18.26); EWH = extended Wilson-Hilferty (18.37); PWH = polynomial extension of Wilson-Hilferty (18.36); SZ = Severo-Zelen (18.38); MSZ = modified Severo-Zelen (18.40); ESZ = empirically modified Severo-Zelen (18.50); A = Aroian (18.93); H = Haldane (18.42); CF = Cornish-Fisher (18.27); ESZ & CF = mean of ESZ and CF, use for ν up to that in parentheses, with CF alone used for higher ν. The — indicates that the indicated accuracy was not reached for ν up to 140.

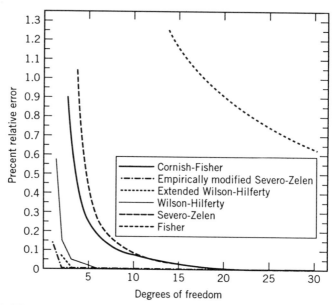

Figure 18.4   The Accuracy of Several Formulas for Approximating χ^2 at the 5% Significance Level ($\varepsilon = 0.95$). Fisher (18.25), Wilson and Hilferty (18.26); extended Wilson-Hilferty (18.36), Severo and Zelen (18.38), Cornish-Fisher expansion (18.27), empirically modified Severo-Zelen formula (18.50). On this plot, Haldane's approximation (18.42) would appear very near the Cornish-Fisher approximation, the modified Severo-Zelen (18.40) would appear very near the Severo-Zelen, and the Aroian (18.43) would lie between the Wilson-Hilferty and the Cornish-Fisher. Relative error = (Approximation − Actual value)/(Actual value)

value)/(Actual value) is presented in Figure 18.4, for example, for the various approximation formulas at the 5% significance level (i.e., $\varepsilon = 0.95$). It is clear from Table 18.2 and Figure 18.4 that Fisher's approximation in (18.25) is the poorest. The Wilson-Hilferty approximation in (18.26) is impressively good, especially for ε no more extreme than 0.05 or 0.995. The Severo-Zelen approximation in (18.38) and its modification in (18.40) do not differ much from the Wilson-Hilferty, except when ε is more extreme than 0.10 or 0.99, at which they become better. The extended Wilson-Hilferty formula in (18.36) is even better, and the polynomial fit in (18.37) improves a bit on that. Hence these two approximations are quite appropriate for computer implementation. The direct approximations of Gilbert in (18.21) and of Hoaglin in (18.44)–(18.49) do not provide asymptotically improving approximations to $\chi^2_{\nu,\varepsilon}$ as ν increases as do the other procedures. This is to be expected after all, since these approximations were developed based on tabulated values for ν up to 30. Further these approximations as well as that of Aroian in (18.43) are not decided improvements over the Wilson-Hilferty approximation in (18.26) and hence are not as good as the Cornish-Fisher approximation in (18.27) or the modified Severo-Zelen approximation in

(18.50). The Haldane approximation in (18.42), as simple as it may be, provides good results and generally just a little inferior to the approximations secured by the Cornish-Fisher formula in (18.27) for ε in the range 0.025 to 0.999, but it does become poor at extreme levels of significance. In conclusion, the Wilson-Hilferty approximation in (18.26) will perform quite satisfactorily for many routine uses, except when ν is very small. Should more accuracy be required, the Cornish-Fisher approximation in (18.27) could be used. For computer implementation, the extended Wilson-Hilferty approximations in (18.36) or (18.37) could be used. One may instead use the algorithm of Best and Roberts (1975) that is described below in order to achieve greater accuracy. Interestingly Zar (1978) also noted that the average of the Cornish-Fisher approximation in (18.27) and the formula in (18.50) provided better approximations than either of them in many cases as one tended to overestimate and the other tended to underestimate.

Similar comparative studies have also been carried out by Sahai and Thompson (1974), Ling (1977, 1978), and on a very minor scale by Liddell (1983). In the former the authors concluded that for $\nu \geq 40$ the four-term Cornish-Fisher expansion considered by Peiser (1943) [see (18.27)] is accurate enough with an error less than 0.1%; however, for small ν and values of ε quite close to 1, one may need to use additionally seventh and eighth terms in the approximation in (18.27) which are given by

$$\frac{1}{25515\nu^2}\left(12U_\varepsilon^6 - 243U_\varepsilon^4 - 923U_\varepsilon^2 + 1472\right)$$

and

$$-\frac{1}{9185400\nu^2\sqrt{2\nu}}\left(3753U_\varepsilon^7 + 4353U_\varepsilon^5 - 289517U_\varepsilon^3 - 289717U_\varepsilon\right),$$

respectively.

Ling (1977, 1978), on the other hand, compared the asymptotic—the Fisher and Wilson-Hilferty approximations of $F_{\chi_\nu^2}(x)$ in (18.22), (18.23), and (18.24), respectively—with the approximation due to Peizer and Pratt (1968) given by

$$F_{\chi_\nu^2}(x) \simeq \Phi\left\{-\frac{(1/3) + (0.08/\nu)}{\sqrt{2\nu - 2}}\right\} \qquad \text{if } x = \nu - 1,$$

$$\simeq \Phi\left\{\frac{x - \nu + (2/3) - (0.08/\nu)}{|x - (\nu - 1)|}\right.$$

$$\left. \times\left[(\nu - 1)\log\left(\frac{\nu - 1}{x}\right) + x - (\nu - 1)\right]^{1/2}\right\} \quad \text{if } x \neq \nu - 1.$$

$$(18.51)$$

Table 18.3 Maximum Absolute Error in Approximations of $F_{\chi_\nu^2}(x)$ in (18.22), (18.23), (18.24), and (18.51)

ν	(18.22)	(18.23)	(18.51)	(18.24)
5	$8.4(-2)^a$	$2.0(-2)$	$3.3(-4)$	$2.6(-3)$
6	$7.7(-2)$	$1.9(-2)$	$2.1(-4)$	$2.2(-3)$
7	$7.1(-2)$	$1.7(-2)$	$1.4(-4)$	$1.8(-3)$
8	$6.7(-2)$	$1.6(-2)$	$1.0(-4)$	$1.6(-3)$
9	$6.3(-2)$	$1.5(-2)$	$7.5(-5)$	$1.4(-3)$
10	$6.0(-2)$	$1.5(-2)$	$5.8(-5)$	$1.3(-3)$
11	$5.7(-2)$	$1.4(-2)$	$4.6(-5)$	$1.1(-3)$
12	$5.4(-2)$	$1.3(-2)$	$3.7(-5)$	$1.0(-3)$
13	$5.2(-2)$	$1.3(-2)$	$3.1(-5)$	$9.6(-4)$
14	$5.0(-2)$	$1.2(-2)$	$2.6(-5)$	$8.9(-4)$
15	$4.9(-2)$	$1.2(-2)$	$2.2(-5)$	$8.2(-4)$
20	$4.2(-2)$	$1.0(-2)$	$1.2(-5)$	$6.1(-4)$
25	$3.8(-2)$	$9.3(-3)$	$7.3(-6)$	$4.8(-4)$
30	$3.4(-2)$	$8.5(-3)$	$5.0(-6)$	$3.9(-4)$
40	$3.0(-2)$	$7.4(-3)$	$2.9(-6)$	$2.9(-4)$
50	$2.7(-2)$	$6.6(-3)$	$1.9(-6)$	$2.3(-4)$
60	$2.4(-2)$	$6.1(-3)$	$1.4(-6)$	$1.9(-4)$
100	$1.9(-2)$	$4.7(-3)$	$5.5(-7)$	$1.1(-4)$
120	$1.7(-2)$	$4.3(-3)$	$4.1(-7)$	$9.2(-5)$
150	$1.5(-2)$	$3.8(-3)$	$2.8(-7)$	$7.3(-5)$
200	$1.3(-2)$	$3.3(-3)$	$1.8(-7)$	$5.4(-5)$
240	$1.2(-2)$	$3.0(-3)$	$1.3(-7)$	$4.5(-5)$

$^a c(d) = c \times 10^d$.

Ling determined the maximum absolute error in these approximations of $F_{\chi_\nu^2}(x)$ in the interval $(0.0001, 0.9999)$ with x in increments of 0.001 for integral values of ν between 5 and 240. These values are presented in Table 18.3 from which it is quite evident that the Peizer-Pratt approximation in (18.51) is even better than the good Wilson-Hilferty approximation in (18.24) in the range studied and probably beyond.

El Lozy (1982), through a similar study, determined that Hill's modified Cornish-Fisher expansion in (18.35) performs better than the Peizer-Pratt approximation in (18.51), but that this greater accuracy is achieved at the cost of involving considerably more mathematical operations. One may also refer to Narula and Li (1977) to get an idea about the computational effort involved (in terms of number of arithmetic and special function operations) in the different methods of approximating the distribution function $F_{\chi_\nu^2}(x)$ and of approximating the percentage points $\chi_{\nu,\varepsilon}^2$.

Best and Roberts (1975) provided an algorithm (along with a Function Subprogram in Fortran) for computing quite accurately the percentage points of the χ^2_ν distribution. This algorithm is both more general and more accurate than an algorithm given earlier by Goldstein (1973). In this method, for a given value of ε, $\chi^2_{\nu, \varepsilon}$ is found from the Taylor series expansion [Hill and Davis (1968)] given by

$$\chi^2_{\nu, \varepsilon} = \chi^2_0 + \sum_r \frac{1}{r!} c_r(\chi^2_0) \left(\frac{E}{p_{\chi^2_\nu}(\chi^2_0)} \right)^r, \qquad (18.52)$$

where χ^2_0 is a suitable starting approximation,

$$c_1(u) = 1,$$

$$c_{r+1}(u) = \left\{ r \left[\frac{1}{2} - \frac{1}{u} \left(\frac{\nu}{2} - 1 \right) \right] + \frac{d}{du} \right\} c_r(u),$$

$$E = \varepsilon - F_{\chi^2_\nu}(\chi^2_0).$$

For many choices of ν and ε, the Wilson-Hilferty approximation in (18.26) is used for χ^2_0 (say, χ^2_{01}). However, better starting approximations than this one are used for the three limiting cases $\varepsilon \to 0$, $\varepsilon \to 1$, and $\nu \to 0$:

1. $\varepsilon \to 0$ (small $\chi^2_{\nu, \varepsilon}$): $\chi^2_{02} = \{\varepsilon \nu 2^{(\nu/2)-1} \Gamma(\nu/2)\}^{2/\nu}$.
2. $\varepsilon \to 1$ (large $\chi^2_{\nu, \varepsilon}$): $\chi^2_{03} = -2\{\log(1 - \varepsilon) - [(\nu/2) - 1]\log(\chi^2_{01}/2) + \log \Gamma(\nu/2)\}$.
3. $\nu \to 0$: For the special case $\nu \leq 0.32$, ε is expressed in terms of Hastings's (1955) approximation to the exponential integral, and in this case χ^2_{04} is determined by Newton-Raphson iteration.

In this vein, it should be added here that Shea (1988) has presented an algorithm (along with a Function Subprogram in Fortran 77) for computing accurately the incomplete gamma integral and hence the cumulative distribution function of χ^2_ν. This algorithm performs more effectively than the ones given earlier by Hill and Pike (1967), Bhattacharjee (1970), El Lozy (1976), Gautschi (1979a, b), and Lau (1980) [also see Rice and Gaines Das (1985)].

Further discussions have been made by Acock and Stavig (1976), Gander (1977), Fettis (1979), Watts (1979), Bukač and Burstein (1980), Cheng and Fu (1983), Chernick and Murthy (1983), and Haines (1988) regarding some other approximations for the percentage points of χ^2_ν distribution. Similarly Dunlap and Duffy (1975), Weintraub (1979), Sletten (1980), Craig (1984), Knüsel (1986), and Terrell (1987) have all discussed the computation of the distribution function of χ^2_ν distribution. Interestingly, for even integral values of ν, Terrell (1987) noted that the following continued fraction approximation

$$F_{\chi_\nu^2}(x) = 2p_{\chi_\nu^2}(x)\left[\frac{x}{\nu -}\frac{\nu x}{\nu + 2 +}\frac{2x}{\nu + 4 -}\frac{(\nu + 2)x}{\nu + 6 +}\frac{4x}{\nu + 8 -}\cdots\right]$$

$$(18.53)$$

converges significantly faster than the Poisson series expression in (18.18) for the left-tail of the distribution. Lin (1988) presented a very simple approximation to the cumulative distribution (for large ν) as

$$F_{\chi_\nu^2}(x) = \begin{cases} \frac{1}{2}e^{b_1 z + a_1 z^2} & \text{if } z \leq 0, \\ 1 - \frac{1}{2}e^{b_2 z + a_2 z^2} & \text{if } z > 0, \end{cases}$$

$$(18.54)$$

where $z = \sqrt{x} - \sqrt{\nu}$, $a_1 = -0.9911$, $b_1 = 0.8055$, $a_2 = -0.6763$, and $b_2 = -1.2451$. As Lin (1988) noted, a simple approximation for $\chi_{\nu,\varepsilon}^2$ can also be secured from (18.54) by equating the right-hand side to ε, taking logs, and then solving for z from a quadratic equation.

Tables to facilitate the construction of shortest confidence intervals for a parameter θ based on a pivotal quantity Q having χ_ν^2 distribution have been prepared by many authors including Tate and Klett (1959), Guenther (1969), and Akhlagi and Parsian (1986). These tables give the values of a and b such that $\Pr[a \leq Q \leq b] = 1 - \alpha$ which yields the shortest confidence interval for the parameter θ, for various choices of ν and α.

It must be remarked here that the Wilson-Hilferty approximation in (18.24) or (18.26) is essentially based on the fact that the cube root transformation of chi-square brings it close to normality. Hawkins and Wixley (1986) showed that the fourth root transformation transforms the chi-squared distribution to very near normality for all degrees of freedom. The fourth root transformation turns out to be also superior than the cube root transformation for small degrees of freedom like $\nu = 1$ and 2. The fourth root transformation also provides very close approximations to the tail percentage points. However, for large degrees of freedom, the cube root transformation turns out to be better, but both transformations are very accurate as measured by the moments and percentage points and even more so by the coverage probabilities. These points are well illustrated in Table 18.4 which shows, for each selected value of ν the coefficients of skewness and kurtosis, and the actual left- and right-tail areas at a nominal two-sided 90% of both the fourth root and the cube root transformations, fitted by the method of moments to the transformed variable.

Mention should also be made here to the work of Hernandez and Johnson (1980) and Taylor (1985) who used the Kullback-Leibler information to determine optimal values of λ (power of the transform) for certain choices of

Table 18.4 Comparisons of the Cube Root and the Fourth Root Transformations

	Skewness		Kurtosis		True cdf at Nominal 5% Levels			
ν	3rd Root	4th Root	3rd Root	4th Root	3rd Root	4th Root	3rd Root	4th Root
1	0.41	0.08	2.68	2.48	2.19	4.90	93.75	94.88
2	0.17	−0.08	2.73	2.75	4.52	5.52	94.52	95.40
3	0.09	−0.12	2.81	2.87	4.81	5.50	94.74	95.45
10	0.01	−0.10	2.98	2.99	4.98	5.31	94.96	95.32
50	0.00	−0.05	2.99	3.00	5.00	5.14	95.00	95.15
100	0.00	−0.04	3.00	3.00	5.00	5.10	95.00	95.10

ν. For example,

ν	1	2	3	4	5
λ	0.2084	0.2654	0.2887	0.3006	0.3124

It is clear from these values, once again, that the fourth root transformation will be more accurate than the cube root transformation for small values of ν, while the latter becomes superior for large values of ν.

Nelson (1987) in fact suggested applying the cube root transformation of the chi-square values and then using the 2σ limits on an $\bar{x}$ chart, instead of a control chart for chi-square values directly (since the latter will look peculiar, even when the process is in control, due to the fact that the χ^2 distribution is skewed to the right).

6 CHARACTERIZATIONS

First of all, it should be mentioned here that all the characterization results presented for the gamma distribution in Chapter 17 can simply be stated as characterization results for the chi-square distribution (with minor changes, of course). Hence we will not repeat those results once again; we present here some key characterization results for the chi-square that are based on quadratic forms of normal variables (as one would expect).

Let $V_1, V_2, \ldots, V_n$ be a random sample of size n from a normal $N(0, \sigma^2)$ population, and Q a quadratic form in these variables, with symmetric matrix A of rank r, $0 < r \le n$. Then the variable Q/σ^2 is distributed as χ_r^2 if and only if $A^2 = A$. Further suppose that Q^* is another quadratic form in V_i's with real symmetric matrix B. Then the variables Q and Q^* are stochastically independent if and only if $AB = O$.

Next, we note the celebrated Cochran's (1934) theorem which characterizes a χ^2 distribution based on its decomposition: "Let $\sum_{i=1}^{n} V_i^2 = \sum_{j=1}^{k} Q_j$, where Q_j is a quadratic form in $V_1, V_2, \ldots, V_n$ (a random sample from

$N(0, \sigma^2))$ with associated matrix A_j of rank r_j, $j = 1, 2, \ldots, k$. Then, the random variables Q_j/σ^2, $j = 1, 2, \ldots, k$, are mutually independent with Q_j/σ^2 distributed as $\chi^2_{r_j}$ if and only if $\sum_{j=1}^{k} r_j = n$."

Dykstra and Hewett (1972) presented two interesting examples of absolutely continuous bivariate distributions which shed further light in this regard. One example simply illustrates the fact that the sum of two random variables can be distributed as χ^2 with one of the variables also distributed as χ^2 and the other variable positive but not necessarily distributed as χ^2. Their other example illustrates the fact that the sum of two variables can be distributed as χ^2, with each variable also distributed as χ^2 and their degrees of freedom adding up properly, but still the two variables need not be independent.

Ruben (1974, 1975) characterized the normal distribution through the sample variance; the relevance of these results to this chapter comes from the chi-square distribution of the sample variance. Ruben's (1975) characterization is as follows: "Let $X_1, X_2, \ldots,$ be i.i.d. random variables. Define $Q_n = \sum_{i=1}^{n} (X_i - \bar{X}_n)^2$, where $\bar{X}_n = \sum_{i=1}^{n} X_i/n$. Let p and q be distinct integers ≥ 2. Then constants λ_p and λ_q exist such that Q_p and Q_q are distributed as $\lambda_p \chi^2_{p-1}$ and $\lambda_q \chi^2_{q-1}$, respectively, if and only if X_1 is normally distributed." A simpler proof for this characterization result was provided by Bondesson (1977).

Khatri (1963, 1978), Shanbhag (1968), and Good (1969) all presented necessary and sufficient conditions for a quadratic form in multinormal random variables to be distributed as chi-square. Let $X \sim MVN(Q, \Sigma)$, where Σ is nonnegative definite and can be written as $\Sigma = BB^T$, B being an $n \times s$ matrix of rank $s = \text{rank}(\Sigma)$. Then the quadratic form $Q = X^T A X$ is distributed as chi-square if and only if (1) $\Sigma A \Sigma A \Sigma = \Sigma A \Sigma$ [Khatri (1963)], (2) $(A\Sigma)^3 = (A\Sigma)^2$ or $\text{trace}(A\Sigma) = \text{trace}((A\Sigma)^2) = \text{rank}(\Sigma A \Sigma)$ [Shanbhag (1968)], (3) the eigenvalues of $A\Sigma$ are zeros and ones only or $\text{trace}(A\Sigma) = \text{trace}((A\Sigma)^2) = \cdots = \text{trace}((A\Sigma)^n)$ [Good (1969)]. [Refer to Khatri (1978) for some comments on all these results.]

In trying to interpret what is so unique about the sum of squares in the definition of "a chi-square distribution with n degrees of freedom as the probability distribution of the sum of squares of n independent standard normal variates," Mitra (1974) presented an interesting uniqueness result concerning the sum of squares. Let $V_1, V_2, \ldots, V_n$ be i.i.d. normal variables with mean 0 and an unknown variance $\theta \in \Omega$, where the parameter space Ω is sufficiently rich so that $\sum_{i=1}^{n} V_i^2$ is a complete sufficient statistic for θ. Then, if $g(\mathbf{V})$ is a statistic such that $g(\mathbf{V})/\theta$ is distributed as χ^2_n for every $\theta \in \Omega$, Mitra proved that $g(\mathbf{V}) = \sum_{i=1}^{n} V_i^2$ a.e. It should be mentioned here that the degrees of freedom being n is very crucial in this result, for there is no dearth of examples of statistics distributed as χ^2 with less than n degrees of freedom that are not even second degree functions in V_i's. Mitra (1974) also conjectured that the result may be true when the parameter space Ω is a set of finite cardinality.

Hussain (1986) has shown that a χ^2 random variable with even degrees of freedom can be decomposed and expressed as a sum of two identically distributed random variables each with a mixed chi-square distribution. More specifically, if $X \sim \chi^2_{2n}$, then there exists a representation $X = U + V$ with U and V being identically distributed, each as a mixture of χ^2 distributions with degrees of freedom $2j - 1$, $j = 1, 2, \ldots, n$; further the random variables U and V are negatively correlated with the coefficient of correlation $\rho_{UV} = -(n-1)/(n+1)$. For the special case $n = 1$, we have $X \sim \chi^2_2$ and $\rho_{UV} = 0$ which is also evident from the fact that U and V are independent random variables each having χ^2_1 distribution in this case (due to Cochran's theorem stated in the beginning of this section).

7 SIMULATIONAL ALGORITHMS

First of all, it should be mentioned that a pseudorandom observation from a chi-square distribution with integer ν can be generated as the sum of squares of ν standard normal observations (generated by one of the efficient algorithms discussed in Chapter 13). This naturally can be quite time-consuming even for moderately large ν. One could also use an accurate approximation of $F_{\chi^2_\nu}^{-1}(\cdot)$ (discussed earlier in Section 5) along with an efficient standard uniform random generator to produce an observation from χ^2_ν distribution via the usual inverse cumulative distribution function method.

On the other hand, one could use one of the many algorithms for generating gamma observations discussed earlier in Chapter 17 to produce a χ^2_ν observation by appropriately scaling the simulated gamma deviate (with shape parameter $\alpha = \nu/2$ and $\beta = 2$). Of course this requires the gamma algorithm to be able to cover an extended range ($\alpha \geq 1/2$) instead of the usual lower bound of 1 (since $\nu \geq 1$). For example, the gamma simulational algorithms due to Best (1978) and Kinderman and Monahan (1980) are valid only for $\alpha > 1$. But the algorithm given by Cheng and Feast (1980) meets the minimum requirement of $\alpha \geq 1/2$ and hence may be utilized for the purpose of simulating pseudo-random observations from χ^2_ν distribution.

If one requires to simulate random observations from χ_ν distribution, one could do this by first simulating random observations from χ^2_ν distribution and then taking the positive square roots of these observations. One could also simulate random observations from χ_ν distribution directly by using the algorithm of Monahan (1987), which is essentially a ratio of uniforms algorithm [Kinderman and Monahan (1977)]. Of course Monahan's algorithm can be used to generate observations from χ^2_ν distribution (by taking squares of the simulated χ_ν observations) or to generate random observations from the gamma distribution with shape parameter $\alpha = \nu/2$ (by taking squares of the simulated χ_ν observations and then dividing by 2). Monahan (1987) also carried out a time-comparative study of his algorithm against that of Cheng and Feast (1980). The time-comparisons of these two algorithms, as reported

Table 18.5   A Time-Comparative Study of Two Algorithms

	$\nu = 1$	$\nu = 5$	$\nu = 9$	Varied #1[a]	Varied #2[b]
		Gamma Distribution			
Monahan	49.2	53.4	53.4	325.4	324.6
Cheng-Feast	54.2	53.6	53.8	335.8	338.0
		Chi-square Distribution			
Monahan	47.8	52.0	51.6	322.4	317.8
Cheng-Feast	55.0	54.6	55.0	342.2	348.2
		Chi Distribution			
Monahan	46.8	51.6	51.4	320.2	316.2
Cheng-Feast	60.6	60.0	60.6	370.4	376.2

Note: Time is given in milliseconds for the average of five replications of 10,000 samples, except 50,000 for variable ν.
[a]Varied #1 covered the range $(1, 11)$.
[b]Varied #2 covered $(1, 101)$.

by Monahan (1987), are presented in Table 18.5. These comparisons reveal clearly that Monahan's algorithm is the faster of the two, although the time improvements are not dramatic.

8   DISTRIBUTIONS OF LINEAR COMBINATIONS

In this section, we present some important developments regarding the derivation (or computations) of distributions of linear combinations of independent χ^2 variables. More detailed treatment of this topic may be found in general sources on distributions of quadratic forms in normal variables. Let us consider specifically the distribution

$$Q_k = \sum_{i=1}^{k} c_i \chi_{\nu_i}^2, \qquad (18.55)$$

where c_i's are known constants. (Without loss of generality we can take $c_1 > c_2 > \cdots > c_k$.) Two efficient methods of determining percentage points of Q_k are those given by Imhof (1961) and Johnson and Kotz (1968); the former inverts numerically the characteristic function of the distribution of Q_k to get $\Pr[Q_k < t]$ for specified t, while the latter gets the distribution based on Laguerre series. Exact significance points of Q_k, for selected nonnegative values of c_i and with all $\nu_i = 1$, were first published by Grad and Solomon (1955) and Solomon (1960) for $k = 2$ and 3, by Johnson and Kotz

(1968) for $k = 4$ and 5, and by Solomon and Stephens (1977) for $k = 6, 8$, and 10.

Davis (1977) showed that the distribution of Q_k in (18.55), with c_i nonnegative and any specified values of v_i, is derivable from a kth-order linear differential equation. The computation of percentage points of Q_k by analytic continuation of the series solution at the origin becomes feasible by this method for a useful range of parameter values. This method thus avoids the slow convergence of the power series for large values of t (since the power series is about $t = 0$ and is effective only for small values of t).

For the case when all v_i's are equal to 1, Jensen and Solomon (1972) proposed a Wilson-Hilferty type approximation by taking $Z = (Q_k/E[Q_k])^h$ to be approximately a normal distribution with mean and variance depending on the first three cumulants of Q_k. Proceeding similarly, Solomon and Stephens (1977) proposed two more approximations, one by fitting a Pearson curve with the same first four moments as Q_k, and the other by fitting $Q_k = A(\chi_p^2)^r$ where A, p, and r are determined by the first three moments of Q_k. To evaluate the performance of these approximations, they carried out a study comparing the exact (E) values taken from Solomon's (1960) and Johnson and Kotz's (1968) tables, Imhof's (I) approximate value, the Jensen-Solomon (J) approximate value, the four-moment Pearson (P) approximate value, and the three-moment χ^2 (S) approximate value, of the α percentage points for different choices of c_i's. These values are presented in Table 18.6.

From Table 18.6 it is clear that the Imhof method gives quite accurate values (to be expected since the method is essentially exact) at all levels of α. All three approximations P, J, and S give excellent accuracy in the upper tail, while S performs better than the other two in the lower tail. However, all (except Imhof's) approximations become relatively less accurate in the lower tails as the coefficients of skewness and kurtosis of Q_k increase.

For the positive definite case ($c_k > 0$), Oman and Zacks (1981) used Robbins and Pitman's (1949) method of mixtures to derive a very good approximation for the distribution function of Q_k for any specified values of c_i's and v_i's. For example, for $\alpha > 1$ using this mixture method, it is well-known that [Neuts and Zacks (1968)] the distribution of $a\chi_v^2$ can be represented as a negative binomial mixture of central chi-squared distributions:

$$a\chi_v^2 \sim \chi_{v+2J}^2, \qquad (18.56)$$

where $J \sim NB(\psi, \mu)$ with its probability mass function

$$g(j|\psi, \mu) = \frac{\Gamma(\mu + j)}{\Gamma(\mu)\Gamma(j + 1)} \psi^j (1 - \psi)^\mu, \qquad j = 0, 1, \ldots, \quad (18.57)$$

Table 18.6 Comparison of Exact Percentage Points and Approximations for Q_k

k	c-Values	$[\sqrt{\beta_1}, \beta_2]$	Method	0.01	0.025	0.05	0.10	0.90	0.95	0.975	0.99
								α			
3	0.5, 0.4, 0.1	[1.974, 9.000]	E	—	—	0.098	0.164	2.188	2.820	—	—
			I	0.032	0.060	0.097	0.164	2.187	2.818	3.450	4.290
			P	0.061	0.079	0.109	0.166	2.191	2.818	3.447	4.284
			S	0.027	0.053	0.091	0.158	2.184	2.809	3.441	4.275
3	0.5, 0.3, 0.2	[1.932, 9.000]	E	—	—	0.110	0.183	2.122	2.708	—	—
			I	0.036	0.068	0.110	0.183	2.122	2.708	3.302	4.102
			P	0.056	0.082	0.119	0.187	2.127	2.712	3.302	4.093
			S	0.040	0.071	0.114	0.186	2.125	2.713	3.308	4.103
			J	0.029	0.060	0.103	0.175	2.132	2.740	3.365	4.217
4	1.5, 1.5, 0.5, 0.5	[1.771, 7.920]	E	0.261	0.425	0.627	0.947	8.120	10.203	12.283	15.033
			I	0.259	0.424	0.627	0.947	8.120	10.202	12.282	15.032
			P	0.387	0.510	0.673	0.953	8.141	10.205	12.261	14.984
			S	0.252	0.415	0.614	0.933	8.115	10.177	12.247	14.990
			J	0.212	0.373	0.374	0.894	8.152	10.287	12.456	15.385

4	2.5, 0.5, 0.5, 0.5	[2.444, 12.612]	E	0.226	0.369	0.545	0.826	8.540	11.343	14.281	18.297
			I	0.225	0.368	0.545	0.826	8.540	11.342	14.279	18.294
			P	0.723	—	0.807	0.939	8.661	11.403	—	18.133
			S	0.202	0.331	0.495	0.765	8.564	11.232	14.051	17.980
			J	0.180	0.295	0.440	0.683	8.707	11.766	15.141	20.094
5	1.8, 1.8, 0.6, 0.4, 0.4	[1.773, 7.957]	E	0.448	0.677	0.943	1.347	9.921	12.419	14.914	18.214
			I	0.449	0.677	0.943	1.347	9.921	12.419	14.913	18.213
			P	0.667	0.813	1.011	1.351	9.951	12.419	14.879	18.143
			S	0.415	0.640	0.904	1.372	9.902	12.358	14.837	18.130
			J	0.370	0.589	0.850	1.253	9.983	12.567	15.208	18.801
5	3.0, 0.5, 0.5, 0.5, 0.5	[2.460, 12.750]	E	0.403	0.609	0.848	1.211	10.407	13.778	17.307	22.259
			I	0.404	0.609	0.848	1.211	10.407	13.778	17.308	22.127
			P	1.104	—	1.201	1.355	10.562	13.847	—	21.919
			S	0.344	0.528	0.750	1.103	10.428	13.591	16.958	21.735
			J	0.324	0.485	0.680	0.992	10.645	14.285	18.568	24.813

with $\psi = (a - 1)/a$ and $\mu = \nu/2$. As a result one can write

$$\Pr[a\chi_\nu^2 \leq x] = \sum_{j=0}^{\infty} g(j|\psi, \mu) F_{\chi_{\nu+2j}^2}(x), \qquad 0 \leq x < \infty. \quad (18.58)$$

Using this method, Oman and Zacks (1981) obtained the mixture representation for the distribution function of Q_k as

$$\Pr[Q_k \leq x] = \sum_{j=0}^{\infty} F_{\chi_{\nu+2j}^2}(x) \Pr[J = j], \quad (18.59)$$

where $\nu = \nu_1 + \nu_2 + \cdots + \nu_k$ and $J = J_1 + J_2 + \cdots + J_k$, and the probability mass function of J can be obtained as a k-fold convolution of the corresponding mass functions of $NB(\psi_i, \mu_i)$, which incidentally is not negative binomial since the ψ_i's are not equal. These authors have also suggested an approximation for the distribution of J in order to reduce the amount of computation required for (18.59). The formula in (18.59) provides quite an accurate approximation for the percentage points of Q_k both in lower and upper tails, as is illustrated in Table 18.7 for some selected choices of k, c_i's, and ν_i's.

It is of interest to add here that for the special case when all ν_i's are 1, Buckley and Eagleson (1988) [based on earlier work of Hall (1983)] derived an approximation for the distribution function of Q_k, and they also derived an explicit upper bound for the error involved in this approximation. Wood (1989) proposed a three-parameter F approximation to the distribution of Q_k and showed that this approximation has clear superiority in the lower tail over both the Buckley and Eagleson approximation as well as the classical Satterthwaite (1946) approximation; but, the S approximation of Solomon and Stephens (mentioned earlier) seems to have an overall edge over Wood's F-approximation.

Generalizing the method of mixtures due to Robbins and Pitman (1949), Ruben (1962) considered a linear combination of noncentral chi-squares (with noncentrality parameters δ_i^2) as

$$Q_k^* = \sum_{i=1}^{k} c_i \chi_{\nu_i}^2(\delta_i^2), \quad (18.60)$$

and presented an infinite series expression for its distribution function as

$$\Pr[Q_k^* < x] = \sum_{j=0}^{\infty} a_j F_{\chi_{\nu+2j}^2}\left(\frac{x}{\beta}\right), \quad (18.61)$$

where $\nu = \nu_1 + \nu_2 + \cdots + \nu_k$ and β is an arbitrary positive constant; Ruben has also shown that (18.61) converges uniformly on every finite interval of

Table 18.7 Percentage Points Using Imhof's Method (I), Solomon and Stephens's Method (S), and Mixture Approximation (M)

$c_1(\nu_1),\ldots,$ $c_k(\nu_k)$	Method	N	α							
			0.01	0.025	0.05	0.10	0.90	0.95	0.975	0.99
0.5(1), 0.4(1), 0.1(1)	I		0.032	0.060	0.097	0.164	2.187	1.818	3.450	4.290
	S		0.027	0.053	0.091	0.158	2.184	2.809	3.441	4.275
	M	29	0.031	0.059	0.098	0.164	2.187	2.818	3.451	4.292
0.5(1), 0.3(1), 0.2(1)	I		0.036	0.068	0.110	0.183	2.122	2.708	3.302	4.102
	S		0.040	0.071	0.114	0.186	2.125	2.713	3.308	4.103
	M	12	0.036	0.067	0.110	0.183	2.121	2.708	3.303	4.105
1.5(2), 0.5(2)	I		0.259	0.424	0.627	0.947	8.120	10.202	12.282	15.032
	S		0.252	0.415	0.614	0.933	8.115	10.177	12.247	14.990
	M	18	0.259	0.425	0.627	0.947	8.120	10.203	12.283	15.040
2.5(1), 0.5(3)	I		0.225	0.368	0.545	0.826	8.540	11.342	14.279	18.294
	S		0.202	0.331	0.495	0.765	8.564	11.232	14.051	17.980
	M	26	0.224	0.368	0.545	0.826	8.540	11.343	14.280	18.296
1.8(2), 0.6(1), 0.4(2)	I		0.449	0.677	0.943	1.347	9.921	12.419	14.913	18.213
	S		0.415	0.640	0.904	1.372	9.902	12.358	14.837	18.130
	M	29	0.447	0.677	0.943	1.347	9.921	12.419	14.915	18.216
3.0(1), 0.5(4)	I		0.404	0.609	0.848	1.211	10.407	13.778	17.308	22.127
	S		0.344	0.528	0.750	1.103	10.428	13.591	16.958	21.735
	M	31	0.403	0.609	0.848	1.211	10.407	13.778	17.307	22.129
0.2(4), 0.1(2)	I		0.135	0.197	0.263	0.356	1.800	2.147	2.483	2.914
	S		0.139	0.198	0.263	0.355	1.801	2.148	2.484	2.912
	M	14	0.139	0.198	0.263	0.355	1.801	2.148	2.484	2.916
0.4(1), 0.2(1), 0.1(4)	I		0.122	0.178	0.238	0.324	1.879	2.332	2.799	3.429
	S		0.135	0.186	0.242	0.324	1.885	2.335	2.794	3.420
	M	20	0.125	0.178	0.237	0.323	1.879	2.333	2.798	3.434
0.3(1), 0.2(1), 0.1(2), 0.05(6)	I		0.208	0.268	0.331	0.416	1.741	2.101	2.463	2.950
	S		0.213	0.270	0.330	0.413	1.741	2.096	2.454	2.937
	M	33	0.209	0.268	0.330	0.414	1.746	2.105	2.464	2.940
0.4(1), 0.1(3), 0.05(6)	I		0.199	0.257	0.317	0.398	1.786	2.228	2.630	3.336
	S		0.204	0.256	0.312	0.389	1.794	2.214	2.655	3.274
	M	40	0.201	0.258	0.317	0.398	1.787	2.229	2.694	3.334

$x > 0$. If (18.61) is actually a mixture representation, that is, $a_j \geq 0$ and $\sum_{j=0}^{\infty} a_j = 1$, then the truncated series (up to $J - 1$) will necessarily lie between 0 and 1, and the truncation error will be bounded above by

$$\left| \sum_{j=J}^{\infty} a_j F_{\chi_{\nu+2j}^2} \left(\frac{x}{\beta} \right) \right| \leq \left| 1 - \sum_{j=0}^{J-1} a_j F_{\chi_{\nu+2J}^2} \left(\frac{x}{\beta} \right) \right|. \qquad (18.62)$$

If some a_j's are negative, the truncation error will not necessarily satisfy (18.62).

Based on Ruben's formulas, Sheil and O'Muircheartaigh (1977) and Farebrother (1984a, b, c) have provided algorithms for determining the distribution of a linear combination of chi-square (not necessarily central) random variables. Following the lines of Imhof, Davies (1980) presented an algorithm for computing the distribution of a variable

$$Q = \sum_{i=1}^{k} c_i \chi^2_{\nu_i}(\delta^2_i) + \delta X_0, \qquad (18.63)$$

where X_0 is an independent standard normal variable, by employing the method of numerical inversion of the characteristic function developed earlier by Davies (1973).

Field (1993) has recently used a uniform saddlepoint approximation to obtain extreme tail areas for a linear combination of independent chi-square variables. This approximation uses a transformation of the tail area coming out of the Fourier inversion of the characteristic function which yields a quadratic in the exponent. The Jacobian of the transformation, when expanded in a Taylor series, leads to this approximation. Field has displayed that the approximation is quite accurate in extreme tails.

Finally, mention should be made of the works of Charnet and Rathie (1985), Shirahata (1988), and Provost (1986), with the last author deriving the exact distribution of the ratio of a linear combination of χ^2 variables over the root of a product of χ^2 variables. Gordon and Ramig (1983) discussed the distribution of the sum of correlated χ^2 variables (components of a multivariate chi-square distribution), while Hanumara and Barry (1978) discussed the distribution of a difference between independent χ^2 variables.

9 RELATED DISTRIBUTIONS

As mentioned earlier in Section 1 and also in Chapter 17, χ^2_ν distributions are special gamma distributions. In fact χ^2_ν is distributed as gamma with shape parameter $\alpha = \nu/2$ and scale parameter $\beta = 2$. As pointed out in Sections 2 and 4, the chi-square distribution is also related to the Poisson distribution. Specifically, for even values of ν, the survival function of a χ^2_ν distribution at x is same as the cumulative distribution function of a Poisson random variable (with mean $x/2$) at $(\nu/2) - 1$. It has also been seen earlier in Chapter 15 that chi-square distributions arise naturally in the decomposition of inverse Gaussian variates.

Just as the sum of squares of ν independent standard normal variables has a χ^2_ν distribution, the sum of squares of ν independent $N(\mu_j, 1)$, $j =$

$1, 2, \ldots, \nu$, variables has a *noncentral chi-square distribution with ν degrees of freedom and noncentrality parameter* $\delta^2 = \sum_{j=1}^{\nu} \mu_j^2$. Han (1975) has expressed the cdf of a noncentral chi-square distribution with odd degrees of freedom as functions of the cdf and pdf of the standard normal distribution. For more details and a discussion on this noncentral χ_ν^2 distribution, refer to Chapter 29. One may also similarly define noncentral chi distributions.

Dion and Fridshal (1982) published an interesting conjecture that relates the quantiles of two independent χ^2 random variables with the quantiles of an appropriate F distribution. Specifically, let $\chi^2_{\nu,(1+\gamma)/2}$ and $\chi^2_{\nu,(1-\gamma)/2}$ denote the lower $(1 + \gamma)/2$ and $(1 - \gamma)/2$ percentage points of χ_ν^2 distribution, that is, $\Pr[\chi_\nu^2 \leq \chi^2_{\nu,(1+\gamma)/2}] = (1 + \gamma)/2$ and $\Pr[\chi_\nu^2 \leq \chi^2_{\nu,(1-\gamma)/2}] = (1 - \gamma)/2$, for any $\gamma \in [0, 1)$. Similarly, let $\chi^2_{\nu',(1+\gamma)/2}$ and $\chi^2_{\nu',(1-\gamma)/2}$ denote the lower $(1 + \gamma)/2$ and $(1 - \gamma)/2$ percentage points of $\chi_{\nu'}^2$ distribution. Finally, let $F_{\nu',\nu,(1+\gamma)/2}$ and $F_{\nu',\nu,(1-\gamma)/2}$ denote the lower $(1 + \gamma)/2$ and $(1 - \gamma)/2$ percentage points of the F-distribution with (ν', ν) degrees of freedom. Then Dion and Fridshal (1982) conjectured the quantile inequalities

$$F_{\nu',\nu,(1-\gamma)/2} \geq \frac{\chi^2_{\nu',(1-\gamma)/2}/\nu'}{\chi^2_{\nu,(1+\gamma)/2}/\nu},$$

$$F_{\nu',\nu,(1+\gamma)/2} \leq \frac{\chi^2_{\nu',(1+\gamma)/2}/\nu'}{\chi^2_{\nu,(1-\gamma)/2}/\nu}. \qquad (18.63)'$$

Burk et al. (1984) proved the above conjecture for the case of equal degrees of freedom (i.e., $\nu = \nu'$). In this case they also proved that (18.63)$'$ holds for a broader class of distributions than χ^2 and gave sufficient conditions on the density functions in order for (18.63)$'$ to hold. It is also of interest to mention here that Burk et al., through numerical studies, showed that the conjecture in (18.63)$'$ is false in some small interval $0 \leq \gamma < \gamma_0$ but holds in the interval $\gamma_0 \leq \gamma < 1$.

Ghosh (1973) proved some monotonicity theorems for χ^2 and related F-distributions. He then remarked that the adequacy of interpolation formulae for $\chi^2_{\nu,\varepsilon}$ (discussed in Section 5) and $F_{\nu',\nu,\varepsilon}$ can be judged by using his theorems regarding increasing or decreasing property of these quantities. Following this work, Sarkar (1983) proved that $\chi^2_{\nu,\varepsilon}$ have the *log-concave* property in $\nu > 0$ for fixed $\varepsilon \in (0, 1)$. He also proved that $F_{\nu',\nu,\varepsilon}$ have the log-concave (*log-convex*) property in $\nu' > 0$ $(\nu > 0)$ for fixed $\nu > 0$ $(\nu' > 0)$ and $\varepsilon \in (0, 1)$. Incidentally he also established that $\chi^2_{\nu,\varepsilon}$ possess the *total positivity of order* 2 property in (ν, ε), and similarly the related $F_{\nu',\nu,\varepsilon}$ possess the same property in both (ν, ε) and (ν', ε). Sarkar (1980) has also derived similar results for the distribution functions of χ_ν^2 (central as well as noncentral) and $F_{\nu',\nu}$.

The distribution of the product (W) of two independent variables, distributed as $\chi^2_{\nu_1}$ and $\chi^2_{\nu_2}$ respectively, can be obtained from (17.110.b) as the distribution of $4Y$ with $a_1 = \nu_1/2$ and $a_2 = \nu_2/2$. The resulting density of $W \stackrel{d}{=} \chi^2_{\nu_1}\chi^2_{\nu_2}$ is given by

$$p_W(w) = \frac{1}{2^{\frac{1}{2}(\nu_1+\nu_2)-1}\Gamma(\nu_1/2)\Gamma(\nu_2/2)} y^{\frac{1}{4}(\nu_1+\nu_2)-1} K_{\frac{1}{2}(\nu_1-\nu_2)}(\sqrt{w}),$$

$$w > 0, \quad (18.64)$$

where $K_g(h)$ is the modified Bessel function of the third kind.

The probability distribution of the distance from the origin to a point $(Y_1, Y_2, \ldots, Y_N)$ in N-dimensional Euclidean space, where the Y_i's are normal variables gives rise to the *generalized Rayleigh distribution*. Specifically, when Y_i's are i.i.d. $N(0, \sigma^2)$ variables, the probability density function of $X = \sqrt{\sum_{i=1}^N Y_i^2}$ is given by

$$p_X(x; N, \sigma) = \frac{2}{(2\sigma^2)^{N/2}\Gamma(N/2)} x^{N-1} e^{-x^2/(2\sigma^2)}, \qquad x > 0, \sigma > 0.$$

$$(18.65)$$

This generalized form of the Rayleigh distribution is also referred in litera-ture as the "chi distribution with N degrees of freedom and scale parameter σ." The density function in (18.5) thus corresponds to the standard chi distribution ($\sigma = 1$) with ν degrees of freedom. The density function in (18.65) is unimodal and the mode is easily seen to be at $\sqrt{N-1}\,\sigma$. The kth moment of X is given by

$$E[X^k] = \frac{2}{(2\sigma^2)^{N/2}\Gamma(N/2)} \int_0^\infty e^{-x^2/(2\sigma^2)} x^{N+k-1}\, dx$$

$$= \frac{2^{k/2}\sigma^k \Gamma[(N+k)/2]}{\Gamma(N/2)}. \qquad (18.66)$$

From the expression of the raw moment in (18.66), one may easily obtain the mean, variance, and the coefficients of skewness and kurtosis of the distribu-tion in (18.65) for any specified value of N, the degrees of freedom. It is also clear from the form of the density function in (18.65) that for a fixed value of N, the distribution belongs to the one-parameter exponential family of distributions; further, based on a sample of size n, namely $X_1, X_2, \ldots, X_n$, a sufficient statistic for σ is $\sum_{i=1}^n X_i^2$. The likelihood function based on this sample is

$$L = \frac{2^n}{2^{Nn/2}\{\Gamma(N/2)\}^n \sigma^{Nn}} \left\{ \prod_{i=1}^{n} X_i^{N-1} \right\} e^{-\sum_{i=1}^{n} X_i^2/(2\sigma^2)} \qquad (18.67)$$

which immediately yields the maximum likelihood estimator of σ to be

$$\hat{\sigma} = \left\{ \frac{1}{Nn} \sum_{i=1}^{n} X_i^2 \right\}^{1/2}. \qquad (18.68)$$

The special case of $N = 2$ in (18.65) is the Rayleigh density function, which has been studied quite extensively in the statistical and engineering literature. Due to its importance and also the great deal of attention it has received over the years, specific developments with regard to the Rayleigh distribution are described in some length in Section 10 of this chapter.

It should also be mentioned here that the function in (18.65), for the special case of $N = 1$, reduces to

$$\frac{1}{\sigma}\sqrt{\frac{2}{\pi}} e^{-x^2/(2\sigma^2)}, \qquad x > 0, \sigma > 0,$$

which is the density function of a half-normal or folded normal random variable. This distribution is discussed in Chapter 13. Furthermore, for the special case of $N = 3$, (18.65) reduces to

$$\frac{2^{3/2}}{\sigma^3\sqrt{\pi}} x^2 e^{-x^2/(2\sigma^2)}, \qquad x > 0, \sigma > 0,$$

which is the density function known in the statistics and physics literatures as "Maxwell-Boltzmann density function." This distribution arises as the distribution of the magnitude of the velocity of a randomly chosen molecule of a gas in a closed container under the assumption that the gas is not flowing and that the pressure in the gas is the same in all directions. Classical estimation and tests of hypotheses for the parameter σ have been discussed by Iliescu and Vodă (1974). Tyagi and Bhattacharya (1989) have developed a Bayesian estimator of the average velocity and of the distribution function, under the squared error loss function.

It should be mentioned here that for the chi distribution in (18.65), with N being assumed to be known, estimation of the parameter σ has been considered by several authors. For example, Stacy and Mihram (1965) discussed the estimation based on a complete sample while Dyer (1973) considered the BLU estimation of σ based on k optimally selected quantiles for various choices of N. By considering the two-parameter form of the chi distribution with probability density function

$$p_X(x; N, \mu, \sigma) = \frac{2}{(2\sigma^2)^{N/2} \Gamma(N/2)} (x - \mu)^{N-1} e^{-(x-\mu)^2/(2\sigma^2)},$$

$$x > \mu, \sigma > 0, \quad (18.69)$$

Harter (1970) dealt with the maximum likelihood estimation of the parameters μ and σ based on complete and Type II censored samples. For this two-parameter chi distribution with N degrees of freedom, Hassanein, Saleh, and Brown (1989) worked out the asymptotic best linear unbiased estimators of μ and σ based on k optimally selected sample quantiles and presented the necessary tables of optimum spacings and the corresponding coefficients and the variances and covariance of these estimates for $N = 3(1)30$ and $k = 2(1)6$. Ali, Umbach, and Saleh (1992) recently carried out a similar study on the asymptotic linear estimation of life functions such as the quantile, survival and hazard functions.

Azzalini (1985) considered a random variable Z with density function

$$\phi(z; \lambda) = 2\Phi(\lambda z)\phi(z), \quad -\infty < z < \infty, \quad (18.70)$$

where $\phi(\cdot)$ and $\Phi(\cdot)$ are standard normal density and distribution functions, respectively. He called Z as a *skew-normal random variable with parameter λ*. This density is related to the chi density in (18.5) as it can be shown that (18.70), in the limiting case of $\lambda \to \infty$, tends to the half-normal density function which, as mentioned earlier, is a chi distribution with $\nu = 1$ degree of freedom.

Chi-bar-square distributions are simply mixtures of chi-square distributions, mixed over their degrees of freedom. They often arise when testing hypotheses that involve inequality constraints [Barlow et al. (1972)]. Suppose that $\{P_n\}$ is a sequence of probability distributions, each of which has support contained in the nonnegative integers and has nonzero finite mean μ_n and finite variance σ_n^2. Then Y_n, the random variable having the chi-bar-square distribution associated with P_n, has its survival function to be

$$\Pr[Y_n > x] = \sum_{j=0}^{\infty} p_{n,j} \Pr[\chi_j^2 > x], \quad (18.71)$$

where $p_{n,j} = P_n(j)$. It may then be easily verified that

$$E[Y_n] = \sum_{j=0}^{\infty} p_{n,j} E[X_j^2] = \sum_{j=0}^{\infty} j p_{n,j} = \mu_n, \qquad (18.72)$$

$$E[Y_n^2] = \sum_{j=0}^{\infty} p_{n,j} E\left[(X_j^2)^2\right] = \sum_{j=0}^{\infty} (j^2 + 2j) p_{n,j}$$

$$= \sigma_n^2 + \mu_n^2 + 2\mu_n, \qquad (18.73)$$

and consequently

$$\text{Var}(Y_n) = \sigma_n^2 + 2\mu_n. \qquad (18.74)$$

Recently Dykstra (1991) has determined necessary and sufficient conditions on the distributions $\{P_n\}_{n=1}^{\infty}$ that will ensure that $Z_n = (Y_n - \mu_n)/(\sigma_n^2 + 2\mu_n)^{1/2}$ converges in distribution to the standard normal. Specifically, he has shown that the asymptotic normality occurs for the chi-bar-square distribution if either the ratio of the mean to the variance of the mixing distribution goes to ∞ or the weighting distribution itself is asymptotically normal; these are the only ways for asymptotic normality to hold, other than a combination of the two phenomena. As an example, if we consider the earlier described (see also Chapter 29) noncentral chi-square random variable $\chi_\nu^2(\delta^2)$, with ν degrees of freedom and δ^2 as noncentrality parameter, then we can write

$$\Pr\left[\chi_\nu^2(\delta^2) > x\right] = \sum_{i=0}^{\infty} p_i\left(\frac{\delta^2}{2}\right) \Pr\left[\chi_{2i+\nu}^2 > x\right], \qquad (18.75)$$

where $p_i(\lambda)$ is the probability that a Poisson (λ) random variable takes on the value i (see Chapter 4). The weight distribution in (18.71) for this distribution will be that of $2X + \nu$ when X has a Poisson $(\delta^2/2)$ distribution. Of course this weighting distribution has mean $\delta^2 + \nu$ and variance $2\delta^2$; also $2X + \nu$ is asymptotically normal as $\delta^2 \to \infty$. Hence we have the asymptotic normality of a noncentral chi-square distribution in the case when ν is fixed and $\delta^2 \to \infty$. This result will continue to hold even when ν is allowed to vary.

By considering the one-parameter exponential family of distributions with pdf [Barndorff-Nielsen (1978)]

$$f(x; \theta) = \exp[a(x)b(\theta) + c(\theta) + h(x)],$$

Rahman and Gupta (1993) called the subfamily, having $-2a(X)b(\theta)$ to be distributed as chi-square with appropriate degrees of freedom, as the family of *transformed chi-square distributions*. They show that a necessary and sufficient condition for belonging to this subfamily is

$$\frac{2c'(\theta)b(\theta)}{b'(\theta)} = k,$$

where k is positive and free from θ. It may be easily verified that distributions like the normal, log-normal, gamma, exponential, Rayleigh, Pareto, Weibull, Erlang, inverse Gaussian, and Maxwell all (discussed in this volume) belong to this transformed chi-square family. Rahman and Gupta (1993) have also discussed some inference problems with regard to this family of distributions.

10 SPECIFIC DEVELOPMENTS IN THE RAYLEIGH DISTRIBUTION

10.1 Historical Remarks

The Rayleigh distribution was originally derived by Lord Rayleigh (J. W. Strutt) (1880, 1919) in connection with a problem in the field of acoustics. Miller (1964) derived the Rayleigh distribution as the probability distribution of the distance from the origin to a point $(Y_1, Y_2, \ldots, Y_N)$ in N-dimensional Euclidean space, where the Y_i's are independent and identically distributed $N(0, \sigma^2)$ variables. Siddiqui (1962) showed that the Rayleigh amplitude distribution (the distribution of the power or amplitude of electronic waves received through a scattering medium) is the asymptotic distribution of a two-dimensional random walk. Polovko (1968) noted that some types of electrovacuum devices have the feature that they age rapidly with time even though they may have no manufacturing defects. The Rayleigh distribution is quite appropriate for modeling the lifetimes of such units as it possesses a linearly increasing hazard rate [see (18.79)]. Quoting the early works of Hertz (1909) and Skellam (1952), Cliff and Ord (1975) point out the remarkable property that the Rayleigh distribution arises as the distribution of the distance between an individual and its nearest neighbor when the spatial pattern is generated by a Poisson process. Hirano (1986) has presented a brief account of the history and properties of this distribution.

10.2 Basic Properties

A Rayleigh random variable X has probability density function [with $N = 2$ from (18.65)]

$$p_X(x) = \frac{x}{\sigma^2}e^{-x^2/(2\sigma^2)}, \qquad 0 \le x < \infty, \sigma > 0, \tag{18.76}$$

cumulative distribution function

$$F_X(x) = 1 - e^{-x^2/(2\sigma^2)}, \qquad 0 \le x < \infty, \sigma > 0, \qquad (18.77)$$

survival or reliability function

$$R_X(x) = 1 - F_X(x) = e^{-x^2/(2\sigma^2)}, \qquad 0 \le x < \infty, \sigma > 0, \quad (18.78)$$

and hazard function

$$h_X(x) = \frac{p_X(x)}{R_X(x)} = \frac{x}{\sigma^2}, \qquad 0 \le x < \infty, \sigma > 0. \qquad (18.79)$$

The mean residual life or the life expectancy of a Rayleigh component is obtained from (18.78) as

$$e_x = \int_x^\infty \frac{R_X(u)}{R_X(x)} \, du = e^{x^2/(2\sigma^2)} \int_x^\infty e^{-u^2/(2\sigma^2)} \, du.$$

Raja Rao and Talwalker (1989) derived lower and upper bounds for this function e_x.

As mentioned earlier, it is the linearly increasing hazard or failure rate in (18.79) that makes the Rayleigh distribution a possible model for the lifetimes of components that age rapidly with time. From (18.78) we observe that for small enough values of x the reliability of a component having the Rayleigh distribution decreases with time more slowly than the reliability of a component having an exponential distribution (whose hazard rate is constant); see Chapter 19. However, for larger values of x the reliability of the Rayleigh component decreases with time more rapidly than in the case of an exponential component [see Kodlin (1967)].

It should be mentioned here that the Rayleigh survival function in (18.78) has been generalized by Bradley, Bradley, and Naftel (1984) who have studied a survival function of the form

$$R(t) = \exp\left\{ -\mu(1 - e^{-\beta t})^v - \alpha_0 t - \alpha_1 t^2 \right\}$$

which is appropriately referred to as the generalized Gompertz-Rayleigh model. This general model includes the Makeham-Gompertz model studied earlier by Bailey, Homer, and Summe (1977) and Bailey (1978) as a special case.

Plots of the density function $p_X(x)$ are presented for $\sigma = 0.5(0.5)3.0$ in Figure 18.5 and for $\sigma = 4.0(1.0)8.0$ in Figure 18.6. The case when $\sigma = 1$ is referred to as the standard Rayleigh density function.

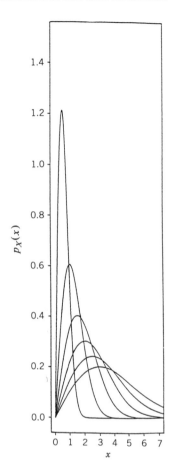

Figure 18.5   Rayleigh Densities for $\sigma = 0.5(0.5)3.0$

From (18.76) and (18.77), it is readily noted that the mode of the distribution is at σ [also see the comment after (18.65)] and that the pth percentile of the distribution is

$$x = F_X^{-1}(p) = \sigma\{-2\log(1-p)\}^{1/2}, \qquad 0 < p < 1. \qquad (18.80)$$

Further, the coefficient of variation of this distribution is $\sqrt{4\pi - 1}$ and the median of the distribution is $\sigma\sqrt{\log 4} \doteq 1.17741\sigma$. Furthermore we obtain immediately from (18.66) the rth raw moment of X to be

$$\mu_r' = E[X^r] = \sigma^r 2^{r/2}\Gamma\left(\frac{r}{2} + 1\right), \qquad (18.81)$$

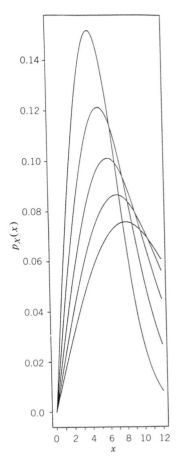

Figure 18.6 Rayleigh Densities for $\sigma = 4.0(1.0)8.0$

Equation (18.81) yields specifically the mean, variance and the coefficients of skewness and kurtosis of X to be

Mean $= E[X] = \sigma\sqrt{\pi/2} \doteq 1.25331\sigma,$

Variance $= \text{Var}(X) = \sigma^2(4 - \pi)/2 \doteq 0.42920\sigma^2,$

Skewness $= \sqrt{\beta_1(X)} = 2(\pi - 3)\sqrt{\pi}/(4 - \pi)^{3/2} \doteq 0.63111,$

Kurtosis $= \beta_2(X) = (32 - 3\pi^2)/(4 - \pi)^2 \doteq 3.24509.$

10.3 Order Statistics and Properties

Let $X_1, X_2, \ldots, X_n$ be random variables each having the Rayleigh distribution in (18.76). Let $X'_1 \le X'_2 \le \cdots \le X'_n$ denote the order statistics obtained by arranging the above sample in increasing order of magnitude. Then Lieblein (1955) has shown that for $1 \le r \le n$,

$$E[X'_r] = \frac{n!}{(r-1)!(n-r)!} \int_0^\infty \{1 - e^{-x^2/(2\sigma^2)}\}^{r-1} \{e^{-x^2/(2\sigma^2)}\}^{n-r+1} \frac{x^2}{\sigma^2} dx$$

$$= \sigma \sqrt{\frac{\pi}{2}} \frac{n!}{(r-1)!(n-r)!} \sum_{i=0}^{r-1} \frac{(-1)^{r-1-i} \binom{r-1}{i}}{(n-i)^{3/2}} \tag{18.82}$$

and

$$E[X'^2_r] = \frac{n!}{(r-1)!(n-r)!} \int_0^\infty \{1 - e^{-x^2/(2\sigma^2)}\}^{r-1} \{e^{-x^2/(2\sigma^2)}\}^{n-r+1} \frac{x^3}{\sigma^2} dx$$

$$= 2\sigma^2 \frac{n!}{(r-1)!(n-r)!} \sum_{i=0}^{r-1} \frac{(-1)^{r-1-i} \binom{r-1}{i}}{(n-i)^2}. \tag{18.83}$$

Similarly Dyer and Whisenand (1973a) have shown that for $1 \le r < s \le n$,

$$E[X'_r X'_s] = \frac{n!}{(r-1)!(s-r-1)!(n-s)!} \int_0^\infty \int_0^y \{1 - e^{-x^2/(2\sigma^2)}\}^{r-1}$$

$$\times \{e^{-x^2/(2\sigma^2)} - e^{-y^2/(2\sigma^2)}\}^{s-r-1} \{e^{-y^2/(2\sigma^2)}\}^{n-s+1} \frac{x^2 y^2}{\sigma^4} e^{-x^2/(2\sigma^2)} dx\,dy$$

$$= \sigma^2 \frac{n!}{(r-1)!(s-r-1)!(n-s)!} \sum_{i=0}^{s-r-1} \sum_{j=0}^{r-1} (-1)^{i+j} \binom{s-r-1}{i} \binom{r-1}{j}$$

$$\times H(s - r - i + j, n - s + i + 1), \tag{18.84}$$

where the function $H(a, b)$ is defined as

$$H(a, b) = \int_0^\infty \int_0^y x^2 y^2 e^{-ax^2/2} e^{-by^2/2} dx\,dy, \qquad a, b > 0. \tag{18.85}$$

Integration by parts immediately yields

$$H(a,b) = \int_0^\infty y^2 e^{-by^2/2} \left\{ -\frac{y}{a} e^{-ay^2/2} + \sqrt{\frac{\pi}{2a^3}} \, \text{Erf}\left(\sqrt{\frac{a}{2}}\, y\right) \right\} dy,$$

where $\text{Erf}(\cdot)$ denotes the error function (see Chapter 13). Now upon using the integral formulas

$$\int_0^\infty y^3 e^{-(a+b)y^2/2} \, dy = \frac{2}{(a+b)^2},$$

$$\int_0^\infty \text{Erf}\left(\sqrt{\frac{a}{2}}\, y\right) e^{-by^2/2} y^2 \, dy = \sqrt{\frac{2}{\pi b^3}} \left\{ \frac{\pi}{2} - \tan^{-1}\sqrt{\frac{b}{a}} + \frac{\sqrt{ab}}{a+b} \right\}$$

[see Ng and Geller (1969)], we obtain

$$H(a,b) = \frac{1}{(ab)^{3/2}} \left\{ \frac{\pi}{2} - \tan^{-1}\sqrt{\frac{b}{a}} + \frac{\sqrt{ab}\,(a-b)}{(a+b)^2} \right\}. \quad (18.86)$$

As pointed out by Dyer and Whisenand (1973a), formula (18.84) can effectively be used along with formula (18.86) in computing the product moments of order statistics. It should be mentioned here that an alternate formula for the product moments in terms of the incomplete beta function was derived by Lieblein (1955); reference may also be made to Shoukri (1982).

10.4 Inference

Let $X_1, X_2, \ldots, X_n$ be a random sample from the Rayleigh population with probability density function as in (18.76). Let $X_1' \le \cdots \le X_n'$ be the corresponding order statistics. In this section we present the various results on inference for the parameter σ.

Best Linear Unbiased Estimation
Suppose that $X_{r+1}' \le X_{r+2}' \le \cdots \le X_{n-s}'$ is the available Type II censored sample, where the smallest r and the largest s observations were not observed due to experimental restrictions. Then, by using the generalized least-squares approach originally due to Aitken (1934) and Lloyd (1952), one may obtain the best linear unbiased estimator (BLUE) of the scale parameter

σ of the Rayleigh distribution in (18.77) as

$$\sigma^* = \frac{\boldsymbol{\alpha}^T \boldsymbol{\Sigma}^{-1}}{\boldsymbol{\alpha}^T \boldsymbol{\Sigma}^{-1} \boldsymbol{\alpha}} \mathbf{X} = \sum_{i=r+1}^{n-s} a_i X_i', \tag{18.87}$$

where

$$\mathbf{X} = (X_{r+1}', X_{r+2}', \ldots, X_{n-s}')^T,$$

$$\boldsymbol{\alpha} = \frac{1}{\sigma} E[\mathbf{X}] \quad \text{and} \quad \boldsymbol{\Sigma} = \frac{1}{\sigma^2} \text{Var}(\mathbf{X}).$$

Further the variance of the BLUE in (18.87) is

$$\text{Var}(\sigma^*) = \frac{\sigma^2}{\boldsymbol{\alpha}^T \boldsymbol{\Sigma}^{-1} \boldsymbol{\alpha}}. \tag{18.88}$$

For details, one may refer to David (1981) and Balakrishnan and Cohen (1991). Dyer and Whisenand (1973a) have computed the coefficients a_i and Var(σ^*) for n up to 15 and all possible choices of r and s (i.e., $r + s \leq n - 1$). Balakrishnan and Ambagaspitiya (1993) have recently tabulated the coefficients a_i and values of Var(σ^*)/σ^2 for n up to 30 for the case of right-censored samples ($r = 0$). These values are presented in Table 18.8 for sample sizes up to 10 with $r = 0$; the efficiency of σ^* based on censored samples relative to the complete sample are also reported. More elaborate tables have also been presented by Adatia (1994).

It should be mentioned here that Hirai (1972) has used Downton's (1966) general approach to derive quadratic coefficients estimators for the two-parameter Rayleigh distribution. This method, applicable only for complete samples, is shown to yield highly efficient estimators even in the case of small sample sizes [also see Hirai (1976)].

Asymptotic Best Linear Unbiased Estimation
The asymptotic best linear unbiased estimate (ABLUE) of σ based on k optimally selected order statistics has been discussed by Dyer and Whisenand (1973b), Dyer (1973), and Cheng (1980). By following the steps of Ogawa (1951) [also see Balakrishnan and Cohen (1991)], these authors have shown that the ABLUE of σ based on the spacing $\{\lambda_1, \lambda_2, \ldots, \lambda_k\}$ is

$$\sigma^{**} = \sum_{i=1}^{k} b_i X_{[n\lambda_i]+1}', \tag{18.89}$$

where

$$b_i = \frac{f_i}{K_2} \left\{ \frac{f_i \xi_i - f_{i-1} \xi_{i-1}}{\lambda_i - \lambda_{i-1}} - \frac{f_{i+1} \xi_{i+1} - f_i \xi_i}{\lambda_{i+1} - \lambda_i} \right\}, \qquad i = 1, \ldots, k,$$

$$K_2 = \sum_{i=1}^{k+1} \frac{(f_i \xi_i - f_{i-1} \xi_{i-1})^2}{\lambda_i - \lambda_{i-1}},$$

$$\xi_i = \{-2\log(1 - \lambda_i)\}^{1/2} \quad \text{and} \quad f_i = \xi_i(1 - \lambda_i), \qquad i = 1, 2, \ldots, k,$$

with

$$\lambda_0 = 0, \; \lambda_{k+1} = 1, \quad \text{and} \quad f_0 \xi_0 = 0 = f_{k+1} \xi_{k+1}.$$

(Note that the λ_i's do not depend on n.) Further the asymptotic variance of the ABLUE in (18.89) is

$$A \operatorname{Var}(\sigma^{**}) = \frac{\sigma^2}{nK_2}. \tag{18.90}$$

From (18.90) it is readily seen that the spacing which maximizes K_2 is the optimal spacing. These optimal spacings have been numerically determined by Dyer and Whisenand (1973b), Dyer (1973), and Cheng (1980). It should be mentioned here that Cheng has shown that the optimal spacing $\{\lambda_i\}$ satisfies the relation

$$\lambda_i = 1 - \lambda_{k-i+1}^*, \qquad i = 1, 2, \ldots, k,$$

where $\{\lambda_i^*\}$ is the optimal spacing for the ABLUE μ^{**} (when σ is known) for the extreme-value distribution of the largest values determined earlier by Chan and Kabir (1969). An optimal t-test [based on σ^{**} in (18.89)] has also been proposed by Cheng (1980) for testing hypotheses regarding σ. In Table 18.9, the values of the optimal λ_i, the corresponding coefficients b_i, and K_2 are presented for k up to 10; Adatia (1994) has recently prepared tables of optimal BLUE based on k selected order statistics for sample sizes $n = 20(1)40$ and $k = 2(1)4$.

It is also of interest to mention here that D'Agostino and Lee (1975) have used the technique of Chernoff, Gastwirth and Johns (1967) to derive the ABLUE of σ based on Type II censored samples. They show that this estimator, in addition to having a simple closed form, possesses good efficiency even for samples as small as 5.

Table 18.8 Coefficients a_i of the BLUE of σ and Efficiency Relative to the BLUE Based on Complete Sample for $n = 2(1)10$, $r = 0$, and $s = 0(1)n - 2$

n	s	a_i	n	s	a_i	n	s	a_i	n	s	a_i
2	0	0.24097	5	2	0.09957	8	0	0.02996	10	0	0.02149
		0.48534			0.15151			0.04464			0.03171
		100.0%			0.67477			0.05739			0.04034
3	0	0.13004			59.3%			0.06977			0.04841
		0.21305	5	3	0.14600			0.08274			0.05641
		0.35498			0.35498			0.09746			0.06475
		100.0%			100.0%			0.11634			0.07388
		79.7%			1.03016			0.15894			0.08457
3	1	0.19180	6	0	0.04599			100.0%			0.09865
		0.71088			0.06969	8	1	0.03422			0.13149
		65.8%			0.09142			0.05097			100.0%
4	0	0.08434			0.11434			0.06549	10	1	0.02387
		0.13250			0.14236			0.07953			0.03522
		0.18370			0.20229			100.0%			0.04479
		0.28212			100.0%	8	4	0.05933			0.05373
		100.0%	7	0	0.03655			0.08793	10	4	0.05739
					0.05485			0.11216			0.06872
					0.07110			0.74509			0.07982
					0.08739			49.4%			0.09114
					0.10537	8	5	0.07820			0.44466
					0.12799			0.11540			69.8%
					0.17785			0.99106	10	5	0.03570
					100.0%			36.8%			0.05256
			7	1	0.04261	8	6	0.11440			0.06668
					0.06390			0.15894			0.07971
					0.08272			100.0%			0.09237
					0.10145			1.27394			0.56855
					0.12177			27.7%			59.7%
					0.31934	9	0	0.05933	10	—	0.04268
					32.3%			0.08793			0.06276
								0.11216			
								0.74509			

(continued — coefficients for $n = 9$:)

n	s	a_i
9	3	0.04781, 0.06103, 0.07351, 0.08604, 0.09924, 0.36983 — 77.6%
9	4	0.03762, 0.05565, 0.07094, 0.08529, 0.09952, 0.49590 — 66.4%
9	—	0.07820, 0.11540, 0.99106 — 36.8%; 0.11440, 1.38007 — 24.1%
9	0	0.02514, 0.03725, 0.04761, 0.04500 — 100.0%

n	r	coefficients	efficiency
4	1	0.11181, 0.17429, 0.53470	74.5%
4	2	0.16426, 0.36696, 0.88430	83.1%
5	0	0.06038, 0.09279, 0.12404, 0.16048, 0.23520	100.0%
5	1	0.07530, 0.11534, 0.15324, 0.43358	49.0%
6	1	0.05512, 0.08341, 0.10912, 0.13579	85.6%
6	2	0.06864, 0.10354, 0.13477, 0.55338	71.1%
6	3	0.09063, 0.13594, 0.79308	66.2%
6	4	0.13274, 1.15833	49.3%
7	2	0.05103, 0.07640, 0.09867, 0.12049, 0.47254	87.4%
7	3	0.06348, 0.09478, 0.12187, 0.65522	56.6%
7	4	0.08373, 0.12440, 0.89716	42.1%
7	5	0.12255	
8	2	0.09412, 0.11041, 0.28344	87.4%
8	2	0.03988, 0.05935, 0.07615, 0.09228, 0.10882, 0.41420	74.8%
8	3	0.04773, 0.07092, 0.09079, 0.10964, 0.56289	62.1%
9	1	0.05745, 0.06743, 0.07816, 0.09057, 0.10672, 0.14384	100.0%
9	2	0.02827, 0.04189, 0.05351, 0.06453, 0.07566, 0.08755, 0.10106, 0.25529, 0.03229	88.8%
9	5	0.06647, 0.08456, 0.10137, 0.64294	55.1%
9	6	0.05590, 0.08238, 0.10448, 0.82629	43.9%
9	7	0.07364, 0.10812, 1.07720	32.6%
9	7	0.10768, 1.47869	21.4%
10	2	0.06258, 0.07176, 0.08174, 0.09325, 0.23258	90.0%
10	3	0.02685, 0.03959, 0.05033, 0.06033, 0.07019, 0.08036, 0.09127, 0.33479, 0.03066, 0.04518	79.9%
10	6	0.07948, 0.09477, 0.71543	49.5%
10	6	0.05301, 0.07777, 0.09821, 0.90085	
10	7	0.06980, 0.10207, 1.15721	29.5%
10	8	0.10203, 1.57120	19.3%

Table 18.9 Optimum Spacings $\{\lambda_i\}$, Corresponding Coefficients b_i, and the Value of K_2 for the ABLUE σ^{**} (Complete Samples)

$k =$	2	3	4	5	6	7	8	9	10
λ_1	0.63853	0.52953	0.45143	0.39306	0.34790	0.31198	0.28274	0.25848	0.23805
λ_2	0.92655	0.82994	0.74192	0.66705	0.60421	0.55134	0.50651	0.46814	0.43500
λ_3		0.96545	0.90671	0.84336	0.78288	0.72769	0.67820	0.63407	0.59475
λ_4			0.98105	0.94338	0.89785	0.85062	0.80469	0.76138	0.72118
λ_5				0.98850	0.96308	0.92972	0.89286	0.85517	0.81818
λ_6					0.99250	0.97460	0.94959	0.92055	0.88965
λ_7						0.99484	0.98178	0.96262	0.93946
λ_8							0.99630	0.98649	0.97152
λ_9								0.99726	0.98971
λ_{10}									0.99791
b_1	0.37319	0.27489	0.21407	0.17305	0.14374	0.12191	0.10512	0.09187	0.08120
b_2	0.20462	0.21326	0.19429	0.17201	0.15167	0.13419	0.11942	0.10696	0.09640
b_3		0.10061	0.13013	0.13495	0.13040	0.12251	0.11376	0.10517	0.09714
b_4			0.05759	0.08501	0.09630	0.09920	0.09784	0.09440	0.09002
b_5				0.03628	0.05861	0.07084	0.07664	0.07856	0.07819
b_6					0.02443	0.04217	0.05356	0.06024	0.06371
b_7						0.01728	0.03139	0.04147	0.04814
b_8							0.01270	0.02402	0.03276
b_9								0.00962	0.01881
b_{10}									0.00747
K_2	3.28105	3.56419	3.70764	3.79029	3.84224	3.87702	3.90144	3.91924	3.93263

Maximum Likelihood Estimation

Based on a complete sample $X_1, X_2, \ldots, X_n$ of size n from the Rayleigh population with pdf as in (18.76), the likelihood function is [see (18.67)]

$$L(X_1, \ldots, X_n; \sigma) = \frac{1}{\sigma^{2n}} \left\{ \prod_{i=1}^{n} X_i \right\} e^{-\sum_{i=1}^{n} X_i^2/(2\sigma^2)} \qquad (18.91)$$

from which we immediately obtain the maximum likelihood estimator (MLE) of σ as [see (18.68)]

$$\hat{\sigma} = \left\{ \frac{1}{2n} \sum_{i=1}^{n} X_i^2 \right\}^{1/2}. \qquad (18.92)$$

Of course from (18.81) we can also propose a simple moment estimator for σ as $\bar{X}\sqrt{2/\pi}$. As Cohen and Whitten (1988) and Balakrishnan and Cohen (1991) point out, since $X_i^2 \sim \sigma^2 \chi_2^2$ $(i = 1, \ldots, n)$, $2n\hat{\sigma}^2/\sigma^2$ has a chi-square distribution with $2n$ degrees of freedom. By making use of this exact distributional result, we can construct a $100(1 - \alpha)\%$ confidence interval for σ as

$$\left(\sqrt{\frac{2n\hat{\sigma}^2}{\chi^2_{2n, 1-\frac{\alpha}{2}}}} , \sqrt{\frac{2n\hat{\sigma}^2}{\chi^2_{2n, \frac{\alpha}{2}}}} \right), \tag{18.93}$$

where $\chi^2_{2n, \frac{\alpha}{2}}$ and $\chi^2_{2n, 1-(\frac{\alpha}{2})}$ are respectively the lower and upper $\frac{\alpha}{2}$ percentage points of the chi-square distribution with $2n$ degrees of freedom. Further, since the Rayleigh density in (18.76) belongs to the general exponential family of distributions, it immediately follows from (18.91) that $\sum_{i=1}^{n} X_i^2$ is both complete and sufficient, and that the MLE of σ^2 given by $\sum_{i=1}^{n} X_i^2/(2n)$ is unbiased and also the uniformly minimum variance unbiased estimator (UMVUE) of σ^2. Now upon using the fact that $2n\hat{\sigma}^2/\sigma^2 \sim \chi^2_{2n}$ and the asymptotic expansion $\Gamma(x + a)/\Gamma(x) \approx x^a [1 + (a^2 + a)/(2x)]$, as done by Sinha and Howlader (1983), the MLE $\hat{\sigma}$ in (18.92) can be shown to be asymptotically unbiased for σ. This fact can alternatively be verified by using Kendall, Stuart, and Ord's (1991) series expression for the mean of the chi distribution; for example, refer to Balakrishnan and Cohen (1991, p. 143).

Furthermore it is easily observed that the MLE of the reliability function $R_X(t)$ in (18.78) is given by

$$\hat{R}_X(t) = \exp\left\{ \frac{-nt^2}{\sum_{i=1}^{n} X_i^2} \right\}.$$

Also, following Basu (1964), we have the UMVUE of $R_X(t)$ as

$$\tilde{R}_X(t) = \begin{cases} \left(1 - \dfrac{t^2}{\sum_{i=1}^{n} X_i^2} \right)^{n-1} & \text{for } 0 \le t < \sum_{i=1}^{n} X_i^2, \\ 0 & \text{for } t \ge \sum_{i=1}^{n} X_i^2. \end{cases}$$

It is of interest to note here that

$$\lim_{n \to \infty} \tilde{R}_X(t) = \lim_{n \to \infty} \left(1 - \frac{t^2}{\sum_{i=1}^{n} X_i^2} \right)^{n-1} = \exp\left\{ \frac{-nt^2}{\sum_{i=1}^{n} X_i^2} \right\} = \hat{R}_X(t).$$

Hence the UMVUE of the reliability function $R_X(t)$ tends asymptotically to the MLE.

Suppose that the available sample is singly right censored at time T. That is, out of a total of N items placed on test n are completely observed ($\le T$) and c are censored ($> T$) so that $N = n + c$. Note that T is a fixed constant in the case of Type I censoring and that $T = X'_n$ in the case of Type II censoring. For samples of this type the MLE of σ may be explicitly derived to be [Cohen and Whitten (1988)]

$$\hat{\sigma} = \left\{ \frac{\sum_{i=1}^{n} X_i'^2 + cT^2}{2n} \right\}^{1/2}. \tag{18.94}$$

Suppose that the independent sample X_i, $i = 1, 2, \ldots, n$, arises from a singly right-truncated Rayleigh distribution (truncated at $x = T$). That is, the sample consists of n random observations each of which is in the time interval $[0, T]$. Then, by introducing the unknown parameter $T_0 = T/\sigma$, Cohen and Whitten (1988) have shown that the MLE of T_0 satisfies the equation

$$J_2(\hat{T}_0) = \frac{1}{nT^2} \sum_{i=1}^{n} X_i^2, \tag{18.95}$$

where

$$J_2(z) = \frac{2}{z^2} - \frac{\phi(z)}{\phi(0) - \phi(z)}; \tag{18.96}$$

$\phi(\cdot)$ is the pdf of the standard normal distribution. To facilitate the calculation of $\hat{T}_0$ and hence of $\hat{\sigma} \ (= T/\hat{T}_0)$, Cohen and Whitten (1988) presented tables and a graph of the function $J_2(z)$ [see also Balakrishnan and Cohen (1991, p. 142)]. A graph of the function $J_2(z)$ is presented here in Figure 18.7. To use this method of estimation in a practical situation, we first need

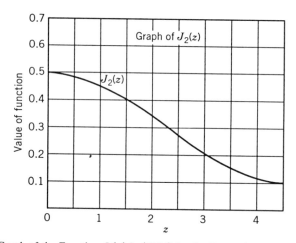

Figure 18.7 Graph of the Function $J_2(z)$ in (18.96) for the Determination of the MLE of σ in the Singly Right-Truncated Rayleigh Distribution

to find the value of $J_2(\hat{T}_0)$ in (18.95) from the observed data, read the corresponding value of $\hat{T}_0$ from Figure 18.7, and then determine the MLE of σ as $\hat{\sigma} = T/\hat{T}_0$.

It should be mentioned here that by considering a doubly Type II censored sample $X'_{r+1} \le X'_{r+2} \le \cdots \le X'_{n-s}$ from (18.76) (with r smallest and s largest observations censored from a sample of size n), Lee, Kapadia, and Brock (1980) derived the modified maximum likelihood estimator of the parameter σ. Specifically, by considering the likelihood function

$$L = \frac{n!}{r! s! \sigma^{n-r-s}} \{F(z_{r+1})\}^r \{1 - F(z_{n-s})\}^s \prod_{i=r+1}^{n-s} p(z_i) \qquad (18.97)$$

and the likelihood equation for σ given by

$$\frac{d \log L}{d\sigma} = -\frac{1}{\sigma} \left[2(n - r - s) + rz_{r+1} \frac{p(z_{r+1})}{F(z_{r+1})} \right.$$

$$\left. - sz_{n-s} \frac{p(z_{n-s})}{1 - F(z_{n-s})} - \sum_{i=r+1}^{n-s} z_i^2 \right] = 0, \quad (18.98)$$

where $z_i = X'_i/\sigma$ are the standardized order statistics and $p(z)$ and $F(z)$ are the standard pdf and cdf given by

$$p(z) = ze^{-z^2/2} \qquad \text{for } z > 0,$$

$$F(z) = 1 - e^{-z^2/2} \qquad \text{for } z > 0,$$

Lee, Kapadia, and Brock (1980) derived a modified maximum likelihood estimator for σ by approximating the likelihood equation in (18.98). Upon using the linear approximation

$$g(z_{r+1}) = \frac{p(z_{r+1})}{F(z_{r+1})} = \frac{z_{r+1}e^{-z_{r+1}^2/2}}{1 - e^{-z_{r+1}^2/2}} \simeq \alpha + \beta z_{r+1}, \qquad (18.99)$$

where α and β are such that

$$\beta = \frac{g(h_2) - g(h_1)}{h_2 - h_1} \qquad \text{and} \qquad \alpha = g(h_1) - h_1\beta, \qquad (18.100)$$

with h_1 and h_2 being chosen so that

$$h_1 = F^{-1} \left\{ \frac{r}{n} - \sqrt{\frac{1}{n}\left(\frac{r}{n}\right)\left(\frac{n-r}{n}\right)} \right\},$$

$$h_2 = F^{-1} \left\{ \frac{r}{n} + \sqrt{\frac{1}{n}\left(\frac{r}{n}\right)\left(\frac{n-r}{n}\right)} \right\},$$

Lee, Kapadia, and Brock (1980) derived the modified maximum likelihood estimator for σ to be

$$\hat{\sigma} = \frac{1}{2G}\{-D + \sqrt{D^2 + 4GB}\}, \tag{18.101}$$

where

$$G = 2(n - r - s),$$
$$D = r\alpha X'_{r+1},$$
$$B = r\beta X'^2_{r+1} + X'^2_{n-s} + \sum_{i=r+1}^{n-s} X'^2_i.$$

These results have been reported by Tiku, Tan, and Balakrishnan (1986) (with an algebraic error). It is of interest to mention here that Lee, Kapadia, and Brock (1980) also suggested an estimator based on linear approximation twice in order to achieve greater accuracy in the approximation in case of small sample sizes.

Based on a complete sample $X_1, X_2, \ldots, X_n$ from a two-parameter Rayleigh population with pdf,

$$p_X(x) = \frac{x - \mu}{\sigma^2}e^{-(x-\mu)^2/(2\sigma^2)}, \qquad x \geq \mu, \sigma > 0, \tag{18.102}$$

we have the likelihood function as

$$L(X_1, \ldots, X_n; \mu, \sigma) = \frac{1}{\sigma^{2n}}\prod_{i=1}^{n}(X_i - \mu)e^{-\sum_{i=1}^{n}(X_i-\mu)^2/(2\sigma^2)}, \tag{18.103}$$

from which the likelihood equations for σ and μ are immediately obtained as

$$\frac{\partial \log L}{\partial \sigma} = -\frac{2n}{\sigma} + \frac{1}{\sigma^3}\sum_{i=1}^{n}(X_i - \mu)^2 = 0 \tag{18.104}$$

and

$$\frac{\partial \log L}{\partial \mu} = -\sum_{i=1}^{n}\left\{\frac{1}{X_i - \mu}\right\} + \frac{1}{\sigma^2}\sum_{i=1}^{n}(X_i - \mu) = 0. \tag{18.105}$$

Upon writing (18.104) as

$$\hat{\sigma}^2 = \frac{1}{2n}\sum_{i=1}^{n}(X_i - \hat{\mu})^2 \tag{18.106}$$

and then eliminating σ^2 from (18.105), we obtain the likelihood equation for μ as

$$\frac{n(\overline{X} - \hat{\mu})}{\sum_{i=1}^{n}[1/(X_i - \hat{\mu})]} - \frac{1}{2n}\sum_{i=1}^{n}(X_i - \hat{\mu})^2 = 0. \qquad (18.107)$$

As mentioned by Balakrishnan and Cohen (1991, pp. 140–141), one may solve for $\hat{\mu}$ iteratively from (18.107) and then determine $\hat{\sigma}$ from (18.106).

From (18.81) we can also employ the estimating equations $E[X] = \overline{X}$ (the sample mean) and $\text{Var}(X) = s^2$ (the sample variance) and derive simple moment estimators for σ and μ as $\tilde{\sigma} = s\sqrt{2/(4 - \pi)}$ and $\tilde{\mu} = \overline{X} - \tilde{\sigma}\sqrt{\pi/2}$. Alternatively, we can employ the estimating equations $E[X_1'] = X_1'$ and $E[X] = \overline{X}$ and derive the following modified moment estimators [see Balakrishnan and Cohen (1991) for more details] for σ and μ as

$$\tilde{\sigma} = \frac{\overline{X} - X_1'}{\sqrt{\pi/2} - \sqrt{\pi/2n}} \quad \text{and} \quad \tilde{\mu} = \overline{X} - \tilde{\sigma}\sqrt{\frac{\pi}{2}}. \qquad (18.108)$$

Bayesian Estimation

Let us consider the one-parameter Rayleigh distribution with pdf as in (18.76). In situations where very little is known about the parameter σ of interest, it is customary to use Jeffreys's (1961) noninformative (or vague) prior

$$g(\sigma) \, \alpha \, |I(\sigma)|^{1/2} = \frac{1}{\sigma} \qquad (18.109)$$

[since $I(\sigma) = E\{-(d^2/d\sigma^2)\log p_X(x|\sigma)\} = 1/\sigma^2$ is the Fisher's information based on a single observation] due to its invariance under parametric transformations [see Box and Tiao (1973)]. Combining the likelihood function in (18.91) with the vague prior in (18.109), we obtain the posterior density of σ as

$$\Pi(\sigma|\mathbf{X}) = \frac{2\left(\frac{1}{2}\sum_{i=1}^{n}X_i^2\right)^n}{\Gamma(n)\sigma^{2n+1}}e^{-\sum_{i=1}^{n}X_i^2/(2\sigma^2)}. \qquad (18.110)$$

In estimating a parameter, the squared-error loss function can be used when decisions become gradually more damaging for larger error. Under this squared-error loss function, as shown by Sinha and Howlader (1983), the Bayesian estimator of σ is the mean of the posterior density in (18.110) and is given by

$$\hat{\sigma}_B = E[\sigma|\mathbf{X}] = \frac{\Gamma(n - \frac{1}{2})}{\sqrt{2}\,\Gamma(n)} \left\{ \sum_{i=1}^{n} X_i^2 \right\}^{1/2}. \tag{18.111}$$

Note that the Bayesian estimator $\hat{\sigma}_B$ in (18.111) tends to the MLE $\hat{\sigma}$ in (18.92) as $n \to \infty$.

Similarly the posterior density of σ^2 is

$$\Pi(\sigma^2|\mathbf{X}) = \frac{\left(\frac{1}{2}\sum_{i=1}^{n} X_i^2\right)^n}{\Gamma(n)(\sigma^2)^{n+1}} e^{-\sum_{i=1}^{n} X_i^2 /(2\sigma^2)} \tag{18.112}$$

from which the Bayesian estimator of σ^2, under the squared-error loss function, may be obtained as

$$(\hat{\sigma}^2)_B = \frac{1}{2n - 2} \sum_{i=1}^{n} X_i^2. \tag{18.113}$$

For large n, the Bayesian estimator of σ^2 in (18.113) becomes $\sum_{i=1}^{n} X_i^2/(2n)$, which is the MLE of σ^2.

Next, upon substituting $\sigma^2 = t^2/\{-2\log(R_X(t))\}$ in (18.112), we obtain the posterior density of the reliability function $R_X(t)$ as

$$\Pi(R_X(t)|\mathbf{X}) = \frac{\left[(1/t^2)\sum_{i=1}^{n} X_i^2\right]^n}{\Gamma(n)} (R_X(t))^{(1/t^2)\sum_{i=1}^{n} X_i^2 - 1}$$

$$\times \{-\log(R_X(t))\}^{(n-1)/2}, \qquad 0 < R_X(t) < 1, \tag{18.114}$$

from which the Bayesian estimator of $R_X(t)$, under the squared-error loss function, is obtained as

$$\left(\widehat{R_X(t)}\right)_B = E[R_X(t)|\mathbf{X}] = \left(1 + \frac{t^2}{\sum_{i=1}^{n} X_i^2}\right)^{-n/2}. \tag{18.115}$$

It is easy to note that the Bayesian estimator $\left(\widehat{R_X(t)}\right)$ in (18.115) tends to the MLE $\hat{R}_X(t) = \exp\{-nt^2/\sum_{i=1}^{n} X_i^2\}$.

Proceeding in a similar manner, Sinha and Howlader (1983) obtained the posterior density of the hazard function $h_X(t)$ in (18.79) to be

$$\Pi(h_X(t)|\mathbf{X}) = \frac{\left(\frac{1}{2}\Sigma_{i=1}^n X_i^2\right)^n}{\Gamma(n)t^n}(h_X(t))^{n-1}e^{-(1/2t)h_X(t)\Sigma_{i=1}^n X_i^2}, \qquad 0 < h_X(t) < \infty,$$

(18.116)

and the Bayesian estimator of $h_X(t)$, under the squared-error loss function, to be

$$\left(\widehat{h_X(t)}\right)_B = E[h_X(t)|\mathbf{X}] = \frac{2nt}{\Sigma_{i=1}^n X_i^2} \tag{18.117}$$

which is exactly the same as the MLE of $h_X(t)$.

Sinha and Howlader (1983) also developed Bayesian intervals for the parameter σ. Since $\Sigma_{i=1}^n X_i^2/\sigma^2$ has a chi-square distribution with $2n$ degrees of freedom, the interval in (18.93) simply becomes the $100(1 - \alpha)\%$ equal-tail credible interval for σ [Edwards, Lindman, and Savage (1963)]. After observing that the posterior density of σ in (18.110) is unimodal, the highest posterior density interval (H_L, H_U) of σ satisfies the following two equations simultaneously:

$$\Pr\left[\frac{\Sigma_{i=1}^n X_i^2}{H_U^2} \leq \chi_{2n}^2 \leq \frac{\Sigma_{i=1}^n X_i^2}{H_L^2}\right] = 1 - \alpha, \tag{18.118}$$

$$e^{-\Sigma_{i=1}^n X_i^2[(1/H_L^2)-(1/H_U^2)]/2} = \left(\frac{H_L}{H_U}\right)^{2n+1}. \tag{18.119}$$

Similarly, by noting from the posterior density of $R_X(t)$ in (18.114) that $-2\Sigma_{i=1}^n X_i^2 \log(R_X(t))/t^2$ has a chi-square distribution with $2n$ degrees of freedom, the highest posterior density interval $(H_L(t), H_U(t))$ of the reliability $R_X(t)$ satisfies the following two equations simultaneously:

$$\Pr\left[-\frac{2}{t^2}\sum_{i=1}^n X_i^2 \log(H_U(t)) \leq \chi_{2n}^2 \leq -\frac{2}{t^2}\sum_{i=1}^n X_i^2 \log(H_L(t))\right] = 1 - \alpha,$$

(18.120)

$$\left\{\frac{\log(H_U(t))}{\log(H_L(t))}\right\}^{n-1} = \left\{\frac{H_L(t)}{H_U(t)}\right\}^{(1/t^2)\Sigma_{i=1}^n X_i^2-1}. \tag{18.121}$$

Mention has to be made of the work of Ariyawansa and Templeton (1984) who considered the problem of statistical inference on the parameter σ

based on a doubly Type II censored sample using Fraser's (1968) structural inference approach. Mention should also be made to Bühler (1983), Corotis, Sigl, and Klein (1978), and Barros and Estevan (1983) who have implemented methods of estimation and fitted Rayleigh distributions in some practical situations. While the first paper introduced the Rayleigh distribution in the analysis of blurred photographs due to camera shake, the last two papers make use of the Rayleigh distribution in the evaluation of wind power from short wind records.

10.5 Prediction

When X has a one-parameter Rayleigh distribution with density as in (18.76), then it is easy to see that X^2 has a one-parameter exponential distribution (since $X^2 \sim \sigma^2 \chi_2^2$). Due to this fact, some of the known results on the prediction limits and prediction intervals in the one-parameter exponential distribution can be suitably adopted for the Rayleigh distribution. Specifically the prediction problems that are of interest are of the following types [Aitchison and Dunsmore (1975)]:

1. After observing the first r failure times out of n items placed on a lifetest, one wants to construct a prediction interval for X_s', the sth failure time, for $r < s \leq n$.
2. After observing a sample size n (or even a Type II censored sample), one may be interested in constructing a prediction interval for failure times from a "future sample" of items to be placed on a life test.

Balakrishnan and Ambagaspitiya (1993) and Raqab (1992) discussed the prediction problems for the Rayleigh distribution in great detail. While the former discusses prediction based on the BLUE discussed in Section 10.4.1, the latter discusses predictors based on maximum likelihood method, conditional median approach, and linear method; it also compares these various predictors through Monte Carlo simulations.

   Howlader (1985) constructed prediction intervals for problem (2) from the Bayesian viewpoint. Given the data $\mathbf{X}$, with the posterior density of σ as in (18.110), we have the conditional joint density of a future independent observation y and σ as

$$h(y, \sigma | \mathbf{X}) = f(y | \sigma, \mathbf{X}) \Pi(\sigma | \mathbf{X})$$

$$= f(y | \sigma) \Pi(\sigma | \mathbf{X}) \qquad \text{since } y \text{ is independent of } \mathbf{X}$$

$$= \frac{y}{\sigma^2} e^{-y^2/(2\sigma^2)} \frac{2\left(\frac{1}{2}\sum_{i=1}^{n} X_i^2\right)^n}{\Gamma(n)\sigma^{2n+1}} e^{-\sum_{i=1}^{n} X_i^2/(2\sigma^2)}. \qquad (18.122)$$

Integrating out σ in (18.122), we obtain the predictive density of y to be

$$\Pi(y|\mathbf{X}) = \frac{2ny\left(\sum_{i=1}^{n} X_i^2\right)^n}{\left(y^2 + \sum_{i=1}^{n} X_i^2\right)^{n+1}},\qquad(18.123)$$

from which it is easy to see that y^2 has a beta distribution of the second kind. Under the squared-error loss function, the Bayesian predictive estimator of y is simply

$$\hat{y}_B = E[y|\mathbf{X}] = \frac{\sqrt{\pi}\,\Gamma\left(n - \frac{1}{2}\right)}{2\Gamma(n)}\left(\sum_{i=1}^{n} X_i^2\right)^{1/2}.\qquad(18.124)$$

From (18.93), Howlader (1985) also derived the $100(1 - \alpha)\%$ highest posterior density prediction interval (c_1, c_2) for y to be the simultaneous solution of

$$\left(\frac{\sum_{i=1}^{n} X_i^2}{c_1^2 + \sum_{i=1}^{n} X_i^2}\right)^n - \left(\frac{\sum_{i=1}^{n} X_i^2}{c_2^2 + \sum_{i=1}^{n} X_i^2}\right)^n = 1 - \alpha,\qquad(18.125)$$

$$\left(\frac{c_2^2 + \sum_{i=1}^{n} X_i^2}{c_1^2 + \sum_{i=1}^{n} X_i^2}\right)^{n+1} = \frac{c_2}{c_1}.\qquad(18.126)$$

It should also be added here that Howlader (1985) further discussed the prediction of the kth order statistic y_k' from a future sample of m observations. Sinha (1990) developed Bayesian prediction limits for a future independent observation y under the asymptotically locally invariant prior due to Hartigan (1964) which assigns a prior distribution for σ as $g(\sigma)$ and satisfies the equation

$$\frac{d}{d\sigma}\log g(\sigma) = -\frac{E(I_1 I_2)}{E(I_2)},\qquad(18.127)$$

where $I_i = (d^i/d\sigma^i)\log f(x|\sigma)$ if $E(I_1) = 0$ and $E(I_1^2) + E(I_2) = 0$; for the case of the Rayleigh distribution in (18.76), it yields the prior $g(\sigma) \propto 1/\sigma^3$.

10.6   Record Values and Related Issues

Distributions and Moments
Let $X_{U(1)}, X_{U(2)}, \dots$ be the upper record values arising from a sequence $\{X_i\}$ of independent and identically distributed Rayleigh random variables with pdf and cdf as in (18.76) and (18.77), respectively. Balakrishnan and Chan (1993) have studied the properties of record values and also discussed some related issues. Let us denote $E[X_{U(n)}^k]$ by $\alpha_n^{(k)}$, $E[X_{U(m)} X_{U(n)}]$ by $\alpha_{m,n}$,

$\text{Var}(X_{U(n)})$ by $\Sigma_{n,n}$, and $\text{Cov}(X_{U(m)}, X_{U(n)})$ by $\Sigma_{m,n}$. Then for $n \geq 1$ and $k \geq 0$ we obtain

$$
\alpha_n^{(k)} = \frac{1}{\Gamma(n)} \int_0^\infty x^k \{-\log(1 - F_X(x))\}^{n-1} p_X(x) \, dx
$$

$$
= \frac{1}{\Gamma(n)} \int_0^\infty x^k \left(\frac{x^2}{2\sigma^2}\right)^{n-1} \times \frac{x}{\sigma^2} e^{-x^2/(2\sigma^2)} \, dx
$$

$$
= \frac{\sigma^k 2^{k/2} \Gamma\left[n + \dfrac{k}{2}\right]}{\Gamma(n)}. \tag{18.128}
$$

Similarly we obtain for $1 \leq m < n$,

$$
\alpha_{m,n} = \frac{1}{\Gamma(m)\Gamma(n-m)} \int_0^\infty \int_0^y xy \left(\frac{x^2}{2\sigma^2}\right)^{m-1} \frac{x}{\sigma^2} \left(\frac{y^2}{2\sigma^2} - \frac{x^2}{2\sigma^2}\right)^{n-m-1}
$$

$$
\times \frac{y}{\sigma^2} e^{-y^2/(2\sigma^2)} \, dx \, dy
$$

$$
= 2\sigma^2 \frac{\Gamma(m + \frac{1}{2})\Gamma(n+1)}{\Gamma(m)\Gamma(n + \frac{1}{2})}. \tag{18.129}
$$

From (18.128) and (18.129), for example, we get

$$
\alpha_n = E[X_{U(n)}] = \frac{\sigma\sqrt{2}\,\Gamma(n + \frac{1}{2})}{\Gamma(n)}, \tag{18.130}
$$

$$
\Sigma_{n,n} = \text{Var}(X_{U(n)}) = 2\sigma^2 \left\{ n - \left(\frac{\Gamma(n + \frac{1}{2})}{\Gamma(n)}\right)^2 \right\}, \tag{18.131}
$$

$$
\Sigma_{m,n} = \text{Cov}(X_{U(m)}, X_{U(n)}) = 2\sigma^2 \frac{\Gamma(m + \frac{1}{2})}{\Gamma(m)} \left\{ \frac{\Gamma(n+1)}{\Gamma(n + \frac{1}{2})} - \frac{\Gamma(n + \frac{1}{2})}{\Gamma(n)} \right\}. \tag{18.132}
$$

These results can be used to develop some inference procedures based on record values, as demonstrated by Balakrishnan and Chan (1993).

Estimation

Suppose the first n upper record values from the Rayleigh population are available. Then, by proceeding along the lines in Section 10.4 and denoting

$$\mathbf{X} = \left(X_{U(1)}, X_{U(2)}, \ldots, X_{U(n)} \right)^T,$$

$$\boldsymbol{\alpha} = \frac{1}{\sigma}(\alpha_1, \alpha_2, \ldots, \alpha_n)^T \quad \text{and} \quad \boldsymbol{\Sigma} = \frac{1}{\sigma^2}\left((\Sigma_{i,j}) \right)^n_{i,j=1},$$

we derive the best linear unbiased estimator (BLUE) of the scale parameter σ of the Rayleigh distribution as

$$\sigma^* = \frac{\boldsymbol{\alpha}^T \boldsymbol{\Sigma}^{-1}}{\boldsymbol{\alpha}^T \boldsymbol{\Sigma}^{-1} \boldsymbol{\alpha}} \mathbf{X} = \sum_{i=1}^{n} c_i X_{U(i)} \tag{18.133}$$

and its variance as

$$\text{Var}(\sigma^*) = \frac{\sigma^2}{\boldsymbol{\alpha}^T \boldsymbol{\Sigma}^{-1} \boldsymbol{\alpha}}. \tag{18.134}$$

By noting that $\Sigma_{i,j}$ is of the specific form $p_i q_j$ (for $i \leq j$), and that therefore the matrix $\boldsymbol{\Sigma}$ is invertible and its inverse a simple tridiagonal matrix, Balakrishnan and Chan (1993) have simplified the expression in (18.133) in order to derive the BLUE as

$$\sigma^* = \frac{\Gamma(n)}{\sqrt{2}\,\Gamma(n + \frac{1}{2})} X_{U(n)} \tag{18.135}$$

and its variance to be

$$\text{Var}(\sigma^*) = \sigma^2 \left\{ \frac{n\Gamma^2(n)}{\Gamma^2(n + \frac{1}{2})} - 1 \right\}. \tag{18.136}$$

Based on the first n upper record values, the likelihood function is then given by

$$L = \frac{p_X(x_1)}{1 - F_X(x_1)} \frac{p_X(x_2)}{1 - F_X(x_2)} \cdots \frac{p_X(x_{n-1})}{1 - F_X(x_{n-1})} p_X(x_n)$$

$$= \left(\prod_{i=1}^{n} x_i \right) \sigma^{-2n} e^{-x_n^2/(2\sigma^2)}, \qquad 0 \leq x_1 < x_2 < \cdots < x_n < \infty. \tag{18.137}$$

Equation (18.137) yields the maximum likelihood estimator of σ as

$$\hat{\sigma} = \frac{1}{\sqrt{2n}} X_{U(n)} \tag{18.138}$$

and its variance is given by

$$\text{Var}(\hat{\sigma}) = \sigma^2 \left\{ 1 - \frac{\Gamma^2(n + \frac{1}{2})}{n \Gamma^2(n)} \right\}. \tag{18.139}$$

Since $\lim_{n \to \infty} n^{-1/2} [\Gamma(n + \frac{1}{2})/\Gamma(n)] = 1$ [e.g., Abramowitz and Stegun (1973, p. 257)], it is quite clear from (18.135) and (18.138) that the BLUE and the MLE are asymptotically identical.

Prediction

Suppose the first m upper records $X_{U(1)}, \ldots, X_{U(m)}$ have been observed. Then the BLU predicted value of the nth record $X_{U(n)}$, $n \geq m + 1$, is given by

$$X^*_{U(n)} = \frac{\Gamma(m)}{\Gamma(m + \frac{1}{2})} \frac{\Gamma(n + \frac{1}{2})}{\Gamma(n)} X_{U(m)}. \tag{18.140}$$

We can also construct a prediction interval for $X_{U(n)}$ by considering the pivotal quantity

$$T_{m,n} = \sqrt{2} \frac{\Gamma(m + \frac{1}{2})}{\Gamma(m)} \left\{ \frac{X_{U(n)}}{X_{U(m)}} - 1 \right\}. \tag{18.141}$$

Since $X^2/(2\sigma^2) \sim \chi_2^2/2$ is distributed as a standard exponential random variable, we can show that $X_{U(m)}^2 / X_{U(n)}^2$ is distributed as a beta$(m, n - m)$ variate [see Chapter 25 and Dunsmore (1983)]. Using this property, the $100(1 - \alpha)\%$ prediction interval for $X_{U(n)}$ has been derived by Balakrishnan and Chan (1993) to be

$$\left[X_{U(m)}, \frac{X_{U(m)}}{\sqrt{b_\alpha(m, n - m)}} \right], \tag{18.142}$$

where $b_\alpha(m, n - m)$ is the lower α percentage point of the beta$(m, n - m)$ distribution. Balakrishnan and Chan (1993) have extended all these results to the case where the upper record values arise from a sequence of i.i.d. variables from a two-parameter Rayleigh distribution with density function as in (18.102). For more details on related work, interested readers should refer to Ahsanullah (1988), Arnold and Balakrishnan (1989), and Arnold, Balakrishnan, and Nagaraja (1992).

10.7 Related Distributions

As indicated earlier in Section 9, the Rayleigh distribution is simply the chi distribution with two degrees of freedom and scale parameter σ. The inverse Rayleigh distribution with probability density function

$$\frac{2\theta}{x^3}e^{-\theta/x^2}, \qquad x > 0, \theta > 0,$$

has been considered by Iliescu and Vodă (1973) and Vodă (1972). Similarly the log-Rayleigh distribution has been discussed by Shepherd (1975).

It is of interest to point out here that Vodă (1976a) considered a distribution with probability density function

$$p_X(x; k, \theta) = \frac{2\theta^{k+1}}{\Gamma(k+1)}x^{2k+1}e^{-\theta x^2}, \qquad x > 0, \theta > 0, k \geq 0, \quad (18.143)$$

and termed it the "generalized Rayleigh distribution." The Rayleigh density in (18.76) is obtained as a special case of (18.143) when $k = 0$. One should bear in mind that while the chi distribution in (18.65) is defined for positive integral values of the parameter N, the generalized Rayleigh distribution in (18.143) is defined for nonnegative real values of the parameter k. Vodă (1976a) has discussed some properties of this generalized distribution and also studied the maximum likelihood estimate and the best linear estimate of the parameter θ under complete and censored samples [also see Vodă (1977)]. Further Vodă (1975, 1976a) considered the left-truncated form of the distribution in (18.143) given by

$$p_X(x; k, \theta) = \frac{p_{X_T}(x; k, \theta)}{1 - F_X(x_T; k, \theta)}, \qquad x \geq x_T, \theta > 0, k \geq 0,$$

where x_T is the truncation point, and discussed the maximum likelihood estimation of the parameter θ. Vodă (1976b) also considered a two-component mixture of generalized Rayleigh distributions of the form $\pi p_X(x; k, \theta_1) + (1 - \pi)p_X(x; k, \theta_2)$ and discussed the method of moments estimation of the parameters θ_1, θ_2, and π. This work is an extension of the work by Krysicki (1963) who had earlier discussed the problem of estimation for a two-component mixture of Rayleigh distributions.

The Rayleigh density function in (18.76) is also a special case of the Weibull distribution with density function

$$p_X(x) = \frac{c}{\sigma^c}x^{c-1}e^{-x^c/\sigma^c}, \qquad 0 \leq x < \infty, c > 0, \sigma > 0.$$

This distribution is discussed at great length in Chapter 21.

Many suggestions have been made regarding the use of a distribution with its hazard function being a lower-order polynomial in the fields of life testing and reliability; for example, see Bain (1974), Gross and Clark (1975), and Lawless (1982). The linear-exponential distribution with density

$$p_X(x) = (\lambda + vx)e^{-(\lambda x + vx^2/2)}, \qquad x \geq 0, \qquad (18.144)$$

and its hazard function varying as a linear function is one such distribution. This distribution and its applications have been discussed in detail by Broadbent (1958), Carbone, Kellerhouse and Gehan (1967), Gehan and Siddiqui (1973), and Balakrishnan and Malik (1986).

Balakrishnan and Kocherlakota (1985) considered the double Rayleigh distribution with probability density function

$$p_X(x) = \frac{|x - \mu|}{2\sigma^2} e^{-(x-\mu)^2/2\sigma^2}, \qquad -\infty < x < \infty, \ -\infty < \mu < \infty, \ \sigma > 0$$

$$(18.145)$$

[the two-parameter Rayleigh distribution in (18.102) is simply the density function in (18.145) folded at $x = \mu$, and hence the distribution in (18.145) is appropriately termed as the double Rayleigh]. They studied the order statistics from this distribution and constructed the best linear unbiased estimators of the parameters μ and σ. Balakrishnan and Kocherlakota also carried out a similar work on a more general double Weibull distribution.

Another interesting distribution to which Rayleigh is related is the Rice distribution. In the data transmission theory the Rice distribution function plays a fundamental role in topics concerned with normally derived processes as the first-order statistics of the envelope of additive narrowband signal and normal noise processes [e.g., see Rice (1944, 1945, 1948); Middleton (1960); Levin (1973); and Dozio, Taute, and Hein (1991)]. The Rice density function of kind (μ, σ) is the two-parameter density function given by

$$\frac{x}{\sigma^2} e^{-(x^2 + \mu^2)/(2\sigma^2)} I_0\left(\frac{\mu x}{\sigma^2}\right) H(x), \qquad -\infty < \mu < \infty, \ \sigma > 0, \quad (18.146)$$

where H is the Heaviside unit step and I_0 is the modified Bessel function of the first kind and order zero [Watson (1944)]. It is clear that (18.146) reduces to the Rayleigh density in (18.76) in the particular case when $\mu = 0$. In the context of statistical properties of additive narrowband signal and normal noise processes, σ^2 in (18.146) represents the mean-squared value of the cosine and sine components of the real stationary normal narrowband noise process, while μ in (18.146) represents the amplitude of the narrowband signal ensemble. When μ/σ is small, the Rice density function in (18.146) is very close to the Rayleigh density function, and when x is far out on the tail of the curve or when μ is large, the Rice density function in (18.146) behaves like a normal density with mean μ and variance σ^2 [see Dozio, Taute and Hein (1991)].

Barnett, Clough, and Kedem (1993) have considered the fitting of Rayleigh and left-truncated Rayleigh distributions to model high peak distributions while studying the phenomenon of acoustic emission. It was Ono (1976) who earlier studied the peak distribution of continuous and burst-type acoustic emission signals and displayed that the Rayleigh distribution serves as an adequate model for the peak magnitude observed in continuous-type acoustic emission regardless of bandwidth.

BIBLIOGRAPHY

Abbé, E. (1863). Über die Gesetzmässigkeit der Vertheilung der Fehler bei Beobachtungsreihen, *Gas. Abh.*, **2**, 55–81.

Abramowitz, M., and Stegun, I. A. (eds.) (1973). *Handbook of Mathematical Functions with Formulas, Graphs and Mathematical Tables* (Second edition), New York: Dover Publications.

Acock, A. C., and Stavig, G. R. (1976). Normal deviate approximations of χ^2, *Perceptual and Motor Skills*, **42**, 220–221.

Adatia, A. (1994). Best linear unbiased estimator of the Rayleigh scale parameter based on fairly large censored samples, *IEEE Transactions on Reliability* (to appear).

Ahsanullah, M. (1988). *Introduction to Record Statistics*, Needham Heights, MA: Ginn Press.

Aitchison, J., and Dunsmore, I. R. (1975). *Statistical Prediction Analysis*, Cambridge: Cambridge University Press.

Aitken, A. C. (1934). On least squares and linear combination of observations, *Proceedings of the Royal Society of Edinburgh*, **55**, 42–48.

Akhalgi, M. R. A., and Parsian, A. (1986). A note on shortest confidence intervals, *Communications in Statistics—Simulation and Computation*, **15**, 425–433.

Ali, M. M., Umbach, D., and Saleh, A. K. Md. E. (1992). Estimating life functions of chi distribution using selected order statistics, *IEEE Transactions on Industrial Engineering Research and Development*, **24**, 88–98.

Ariyawansa, K. A., and Templeton, J. G. C. (1984). Structural inference on the parameter of the Rayleigh distribution from doubly censored samples, *Statistische Hefte*, **25**, 181–199.

Arnold, B. C., and Balakrishnan, N. (1989). *Relations, Bounds and Approximations for Order Statistics*, Lecture Notes in Statistics, **53**, New York: Springer-Verlag.

Arnold, B. C., Balakrishnan, N., and Nagaraja, H. N. (1992). *A First Course in Order Statistics*, New York: Wiley.

Aroian, L. A. (1943). A new approximation to the levels of significance of the chi-square distribution, *Annals of Mathematical Statistics*, **14**, 93–95.

Azzalini, A. (1985). A class of distributions which includes the normal ones, *Scandinavian Journal of Statistics*, **12**, 171–178.

Bailey, R. C. (1978). Limiting forms of the Makeham model and their use for survival analysis of transplant studies, *Biometrics*, **34**, 725–726.

Bailey, R. C., Homer, L. D., and Summe, J. P. (1977). A proposal for the analysis of kidney graft survival, *Transplantation*, **24**, 309–315.

Bain, L. J. (1974). Analysis for the linear failure rate distribution, *Technometrics*, **16**, 551–559.

Balakrishnan, N., and Ambagaspitiya, R. S. (1993). Best linear unbiased estimate and associated inference for the Rayleigh distribution (submitted for publication).

Balakrishnan, N., and Chan, P. S. (1993). Record values from Rayleigh and Weibull distributions and associated inference, Presented at the International Conference on Extremes and Applications held at Gaithersburg, Maryland, May 1993.

Balakrishnan, N., and Cohen, A. C. (1991). *Order Statistics and Inference: Estimation Methods*, San Diego: Academic Press.

Balakrishnan, N., and Kocherlakota, S. (1985). On the double Weibull distribution: Order statistics and estimation, *Sankhyā, Series B*, **47**, 161–178.

Balakrishnan, N., and Malik, H. J. (1986). Order statistics from the linear-exponential distribution, Part I: Increasing hazard rate case, *Communications in Statistics—Theory and Methods*, **15**, 179–203.

Barlow, R. E., Bartholomew, D. J., Bremner, J. M., and Brunk, H. D. (1972). *Statistical Inference under Order Restrictions*, New York: Wiley.

Barndorff-Nielsen, O. (1978). *Information and Exponential Families in Statistical Theory*, New York: Wiley.

Barnett, J. T., Clough, R. B., and Kedem, B. (1993). Power considerations in acoustic emission (Preprint).

Barros, V. R., and Estevan, E. A. (1983). On the evaluation of wind power from short wind records, *Journal of Climate and Applied Meteorology*, **22**, 1116–1123.

Basu, A. P. (1964). Estimates of reliability for some distributions useful in life tests, *Technometrics*, **6**, 215–219.

Best, D. J. (1978). Letter to the editor, *Applied Statistics*, **27**, 181.

Best, D. J., and Roberts, D. E. (1975). Algorithm AS 91. The percentage points of the χ^2 distribution, *Applied Statistics*, **24**, 385–388.

Bhattacharjee, G. P. (1970). The incomplete gamma integral, *Applied Statistics*, **19**, 285–287.

Bienaymé, I. J. (1838). Mémoire sur la probabilité des résultats moyens des observations; demonstration directe de la règle de Laplace, *Mémoires de l'Académie de Sciences de l'Institut de France, Paris, Series Étrangers*, **5**, 513–558.

Bienaymé, I. J. (1852). Mémoire sur la probabilité des erreurs d'aprés la méthode de moindres carrés, *Liouville's Journal de Mathématiques Pures et Appliquées*, **17**, 33–78.

Boltzmann, L. (1878). Weitere Bemerkungen über einige Probleme der mechanischen Wärmetheorie, *Wiss. Abh.*, **2**, 250–288.

Boltzmann, L. (1881). Über einige das Wärmegleichgewicht betreffende Sätze, *Wiss. Abh.*, **2**, 572–581.

Bondesson, L. (1977). The sample variance, properly normalized, is χ^2-distributed for the normal law only, *Sankhyā, Series A*, **39**, 303–304.

Box, G. E. P., and Tiao, G. C. (1973). *Bayesian Inference in Statistical Analysis*, Reading, MA: Addison-Wesley.

Boyd, W. C. (1965). A nomogram for chi-square, *Journal of the American Statistical Association*, **60**, 344–346.

Bradley, D. H., Bradley, E. L., and Naftel, D. C. (1984). A generalized Gompertz-Rayleigh model as a survival distribution, *Mathematical Biosciences*, **70**, 195–202.

Broadbent, S. (1958). Simple mortality rates, *Applied Statistics*, **7**, 86–95.

Buckley, M. J., and Eagleson, G. K. (1988). An approximation to the distribution of quadratic forms in normal random variables, *Australian Journal of Statistics*, **30**, 150–159.

Bühler, E. (1983). Eine Anwendung der Rayleigh-Verteilung auf ein Problem der Photographie, *Statistische Hefte*, **24**, 155–159.

Bukač, J., and Burstein, H. (1980). Approximations of Student's t and chi-square percentage points, *Communications in Statistics—Simulation and Computation*, **9**, 665–672.

Burk, F., Dion, L., Fridshal, D., Langford, E., O'Cinneide, C., and Parsons, T. (1984). On a conjecture relating χ^2 and F quantiles, *Communications in Statistics—Theory and Methods*, **13**, 661–670.

Carbone, P. O., Kellerhouse, L. E., and Gehan, E. A. (1967). Plasmacytic myeloma: A study of the relationship of survival to various clinical manifestations and anomolous protein type in 112 patients, *American Journal of Medicine*, **42**, 937–948.

Causey, B. D. (1986). Expected absolute departure of chi-square from its median, *Communications in Statistics—Simulation and Computation*, **15**, 181–183.

Chan, L. K., and Kabir, A. B. M. L. (1969). Optimum quantiles for the linear estimation of the parameters of the extreme value distribution in complete and censored samples, *Naval Research Logistics Quarterly*, **16**, 381–404.

Charnet, R., and Rathie, P. N. (1985). The distribution of linear combination of independent random variables, *Gujarat Statistical Review*, **12**, 17–26.

Cheng, R. C. H., and Feast, G. M. (1980). Gamma variate generators with increased shape parameter range, *Communications of the Association of Computing Machinery*, **23**, 389–394.

Cheng, S. W. (1980). On the ABLUE and optimum t-test of the parameter of Rayleigh distribution, *Tamkang Journal of Mathematics*, **11**, 11–17.

Cheng, S. W., and Fu, J. C. (1983). An algorithm to obtain the critical values of the t, χ^2 and F distributions, *Statistics & Probability Letters*, **1**, 223–227.

Chernick, M. R., and Murthy, V. K. (1973). Chi-square percentiles: Old and new approximations with applications to sample size determination, *American Journal of Mathematical and Management Sciences*, **3**, 145–161.

Chernoff, H., Gastwirth, J. L., and Johns, M. V., Jr. (1967). Asymptotic distribution of linear combinations of functions of order statistics with applications to estimation, *Annals of Mathematical Statistics*, **38**, 52–72.

Cliff, A. D., and Ord, J. K. (1975). Model building and the analysis of spatial pattern in human geography (with discussion), *Journal of the Royal Statistical Society, Series B*, **37**, 297–348.

Cochran, W. G. (1934). The distribution of quadratic forms in a normal system, with applications to the analysis of covariance, *Proceedings of the Cambridge Philosophical Society*, **30**, 178–191.

Cohen, A. C., and Whitten, B. J. (1988). *Parameter Estimation in Reliability and Life Span Models*, New York, Marcel Dekker.

Cornish, E. A. (1969). Fisher Memorial Lecture, *37th International Statistical Institute Session*, London, England.

Corotis, R. B., Sigl, A. B., and Klein, J. (1978). Probability models of wind velocity magnitude and persistence, *Solar Energy*, **20**, 483–493.

Craig, R. J. (1984). Normal family distribution functions: FORTRAN and BASIC programs, *Journal of Quality Technology*, **16**, 232–236.

D'Agostino, R. B., and Lee, A. F. S. (1975). Asymptotically best linear unbiased estimation of the Rayleigh parameter for complete and tail-censored samples, *IEEE Transactions on Reliability*, **24**, 156–157.

David, H. A. (1981). *Order Statistics* (Second edition), New York: Wiley.

Davies, R. B. (1973). Numerical inversion of a characteristic function, *Biometrika*, **60**, 415–417.

Davies, R. B. (1980). Algorithm AS 155. The distribution of a linear combination of χ^2 random variables, *Applied Statistics*, **29**, 323–333.

Davis, A. W. (1977). A differential equation approach to linear combinations of independent chi-squares, *Journal of the American Statistical Association*, **72**, 212–214.

Deutler, T. (1984). A series expansion for the cumulants of the χ-distribution and a Cornish-Fisher expansion for the noncentrality parameter of the noncentral *t*-distribution, *Communications in Statistics—Simulation and Computation*, **13**, 507–513.

Dion, L., and Fridshal, D. (1982). A relationship between chi-square and *F* critical values, *Communications in Statistics—Simulation and Computation*, **11**, 233–235.

Downton, F. (1966). Linear estimates with polynomial coefficients, *Biometrika*, **53**, 129–141.

Dozio, M., Taute, R., and Hein, H. -J. (1991). Topics of Rice-statistics: The problem of decomposition of superpositions, *Wissenschaft. Zeitschrift, Martin-Luther Universität Halle-Wittenberg, Mathematik-Naturwissenschaft*, **40**, 73–86.

Dunlap, W. P., and Duffy, J. A. (1975). FORTRAN IV functions for calculating exact probabilities associated with z, χ^2, t and F values, *Behavior Research Methods and Instrumentation*, **7**, 59–60.

Dunsmore, I. R. (1983). The future occurrence of records, *Annals of Institute of Statistical Mathematics*, **35**, 267–277.

Dwyer, P. S. (1951). *Linear Computations*, New York: Wiley.

Dyer, D. D. (1973). Estimation of the scale parameter of the chi distribution based on sample quantiles, *Technometrics*, **15**, 489–496.

Dyer, D. D., and Whisenand, C. W. (1973a). Best linear unbiased estimator of the parameter of the Rayleigh distribution. Part I: Small sample theory for censored order statistics, *IEEE Transactions on Reliability*, **22**, 27–34.

Dyer, D. D., and Whisenand, C. W. (1973b). Best linear unbiased estimator of the parameter of the Rayleigh distribution. Part II: Optimum theory for selected order statistics, *IEEE Transactions on Reliability*, **22**, 229–231.

Dykstra, R. (1991). Asymptotic normality for chi-bar-square distributions, *Canadian Journal of Statistics*, **19**, 297–306.

Dykstra, R. L., and Hewett, J. E. (1972). Examples of decompositions of chi-squared variables, *The American Statistician*, **26**, 42–43.

Edwards, W., Lindman, H., and Savage, L. J. (1963). Bayesian statistical inference for psychological research, *Psychology Reviews*, **70**, 193–242.

Elderton, W. P. (1902). Tables for testing the goodness of fit of theory to observation, *Biometrika*, **1**, 155–163.

Ellis, R. L. (1844). On a question in the theory of probabilities, *Cambridge Mathematical Journal*, **4**, 127–132.

El Lozy, M. (1976). Remark on Algorithm 299. Chi-squared integral, *ACM Transactions on Mathematical Software*, **2**, 393–395.

El Lozy, M. (1982). Efficient computation of the distribution functions of Student's t, chi-squared and F to moderate accuracy, *Journal of Statistical Computation and Simulation*, **14**, 179–189.

Farebrother, R. W. (1984a). Algorithm AS 204. The distribution of a positive linear combination of χ^2 random variables, *Applied Statistics*, **33**, 332–339.

Farebrother, R. W. (1984b). Remark AS R52. The distribution of a linear combination of central χ^2 random variables. A remark on AS 153: Pan's procedure for the tail probabilities of the Durbin-Watson statistic, *Applied Statistics*, **33**, 363–366.

Farebrother, R. W. (1984c). A remark on Algorithms AS 106, AS 153 and AS 155. The distribution of a linear combination of χ^2 random variables, *Applied Statistics*, **33**, 366–369.

Fettis, H. E. (1979). An asymptotic expansion for the upper percentage points of the χ^2-distribution, *Mathematics of Computation*, **33**, 1059–1064.

Field, C. (1993). Tail areas of linear combinations of chi-squares and non-central chi-squares, *Journal of Statistical Computation and Simulation*, **45**, 243–248.

Fisher, R. A. (1922). On the interpretation of χ^2 from contingency tables and calculation of P, *Journal of the Royal Statistical Society, Series A*, **85**, 87–94.

Fraser, D. A. S. (1968). *The Structure of Inference*, New York: Wiley.

Gander, W. (1977). A machine independent algorithm for computing percentage points of the χ^2-distribution, *Zeitschrift für angewandte Mathematik und Physik*, **28**, 1133–1136.

Gauss, C. F. (1816) (1880). Bestimmung der Genauigkeit der Beobachtungen, *Werke 4, Göttingen*, 109–117.

Gautschi, W. (1979a). A computational procedure for incomplete gamma functions, *ACM Transactions on Mathematical Software*, **5**, 466–481.

Gautschi, W. (1979b). Incomplete gamma functions, *ACM Transactions on Mathematical Software*, **5**, 482–489.

Gehan, E. A., and Siddiqui, M. M. (1973). Simple regression methods for survival time studies, *Journal of the American Statistical Association*, **68**, 848–856.

Ghosh, B. K. (1973). Some monotonicity theorems for χ^2, F and t distributions with applications, *Journal of the Royal Statistical Society, Series B*, **35**, 480–492.

Gilbert, R. I. (1977). A simple formula for interpolating tables of χ^2, *Biometrics*, **33**, 383–385.

Goldberg, H., and Levine, H. (1946). Approximate formulas for the percentage points and normalization of t and χ^2, *Annals of Mathematical Statistics*, **17**, 216–225.

Goldstein, R. B. (1973). Algorithm 451: Chi-square quantiles, *Communications of the Association of Computing Machinery*, **16**, 483–485.

Good, I. J. (1969). Conditions for a quadratic form to have a chi-square distribution, *Biometrika*, **56**, 215–216. Correction, **57**, 225.

Gordon, N. H., and Ramig, P. F. (1983). Cumulative distribution function of the sum of correlated chi-squared random variables, *Journal of Statistical Computation and Simulation*, **17**, 1–9.

Grad, A., and Solomon, H. (1955). Distribution of quadratic forms and some applications, *Annals of Mathematical Statistics*, **26**, 464–477.

Gray, H. L., Thompson, R. W., and McWilliams, G. V. (1969). A new approximation for the chi-square integral, *Mathematics of Computation*, **23**, 85–89.

Gross, A. J., and Clark, V. A. (1975). *Survival Distribution: Reliability Applications in the Biomedical Sciences*, New York: Wiley.

Guenther, W. C. (1969). Shortest confidence intervals, *The American Statistician*, **23**, 22–25.

Gulli, C. (1989). Analogy between the chi-squared density function and the impulse response of a fractional system, *Signal Processing*, **17**, 129–140.

Haines, P. D. (1988). A closed form approximation for calculating the percentage points of the F and t distributions, *Applied Statistics*, **37**, 95–100.

Hald, A., and Sinkbaek, S. A. (1950). A table of percentage points of the χ^2 distribution, *Skandinavisk Aktuarietidskrift*, **33**, 168–175.

Haldane, J. B. S. (1937). The approximate normalization of a class of frequency distributions, *Biometrika*, **29**, 392–404.

Hall, P. (1983). Chi-squared approximations to the distribution of a sum of independent random variables, *Annals of Probability*, **11**, 1028–1036.

Han, C.-P. (1975). Some relationships between noncentral chi-squared and normal distributions, *Biometrika*, **62**, 213–214.

Hanumara, R. C., and Barry, E. P. (1978). A table of some percentage points of the distribution of a difference between independent chi-square variables, *Journal of Statistical Computation and Simulation*, **6**, 169–181.

Harter, H. L. (1964a). *New Tables of the Incomplete Gamma Function Ratio and of Percentage Points of the Chi-square and Beta Distributions*, Washington, DC: Government Printing Office.

Harter, H. L. (1964b). A new table of percentage points of the chi-square distribution, *Biometrika*, **51**, 231–239.

Harter, H. L. (1970). *Order Statistics and Their Use in Testing and Estimation*, Vol. 2, Washington, DC: Government Printing Office.

Hartigan, J. A. (1964). Invariant prior distributions, *Annals of Mathematical Statistics*, **35**, 836–845.

Hassanein, K. M., Saleh, A. K. Md. E., and Brown, E. F. (1989). Simultaneous estimation of the location and scale parameters of the chi-distribution by selected order statistics, *Soochow Journal of Mathematics*, **15**, 55–64.

Hastings, C., Jr. (1955). *Approximations for Digital Computers*, Princeton: Princeton University Press.

Hawkins, D. M., and Wixley, R. A. J. (1986). A note on the transformation of chi-square variables to normality, *The American Statistician*, **40**, 296–298.

Helmert, F. R. (1875). Über die Berechnung der wahrscheinlichen Fehlers aus einer endlichen Anzahl wahrer Beobachtungsfehler, *Zeitschrift für angewandte Mathematik und Physik*, **20**, 300–303.

Helmert, F. R. (1876a). Die Genauigkeit der Formel von Peters zue Berechnung des wahrscheinlichen Beobachtungsfehlers directer Beobachtungen gleicher Genauigkeit, *Astronomische Nachrichten*, **88**, columns 113–120.

Helmert, F. R. (1876b). Über die Wahrscheinlichkeit der Potenzsummen der Beobachtungsfehler und über einige damit in Zusammenhänge stehende Fragen, *Zeitschrift für angewandte Mathematik und Physik*, **21**, 192–218.

Hernandez, F., and Johnson, R. A. (1980). The large-sample behavior of transformations to normality, *Journal of the American Statistical Association*, **75**, 855–861.

Herschel, J. F. W. (1869). *Familiar Lectures on Scientific Subjects*, 496–507, New York: Routledge.

Hertz, P. (1909). Über die gegenseitigen durchshnittlichen Abstand von Punkten, die mit bekannter mittlerer Dichte im Raum angeordnet sind, *Mathematische Annalen*, **67**, 387–398.

Hill, G. W. (1969). Progress results on asymptotic approximations for Student's t and chi-squared, *Personal communication*.

Hill, G. W., and Davis, A. W. (1968). Generalized asymptotic expansions of Cornish-Fisher type, *Annals of Mathematical Statistics*, **39**, 1264–1273.

Hill, I. D., and Pike, M. C. (1967). Algorithm 299. Chi-squared integral, *Communications of the Association of Computing Machinery*, **10**, 243–244.

Hirai, A. S. (1972). Estimation of scale and location parameters, using quadratic coefficients, *Australian Journal of Statistics*, **14**, 222–226.

Hirai, A. S. (1976). Estimation of the scale-parameter from the Type II censored Rayleigh distribution, *Punjab University Journal of Mathematics*, **9**, 61–67.

Hirano, K. (1986). Rayleigh distribution, In *Encyclopedia of Statistical Sciences*, **7**, S. Kotz, N. L. Johnson, and C. B. Read (editors), 647–649, New York: Wiley.

Hoaglin, D. C. (1977). Direct approximation for chi-squared percentage points, *Journal of the American Statistical Association*, **72**, 508–515.

Howlader, H. A. (1985). HPD prediction intervals for Rayleigh distribution, *IEEE Transactions on Reliability*, **34**, 121–123.

Hussain, A. (1986). A note on a decomposition of χ^2-distribution, *Pakistan Journal of Statistics*, **2**, 41–42.

Iliescu, D. V., and Vodă, V. Gh. (1973). Studiul variabilei aleatoare repartizate invers Rayleigh, *Studii si Cercetari Matematica, Academia Republicii Socialiste Romania, Bucharest*, **25**, 1507–1521.

Iliescu, D. V., and Vodă, V. Gh. (1974). Asupra variabilei aleatoare repartizate după legea Lui Maxwell, *Studii si Cercetari Matematica, Academia Republicii Socialiste Romania, Bucharest*, **26**, 169–178.

Imhof, J. P. (1961). Computing the distribution of quadratic forms in normal variables, *Biometrika*, **48**, 417–426.

Jeffreys, H. (1961). *Theory of Probability*, Oxford: Clarendon Press.

Jensen, D. R., and Solomon, H. (1972). A Gaussian approximation to the distribution of a definite quadratic form, *Journal of the American Statistical Association*, **67**, 898–902.

Johnson, N. L., and Kotz, S. (1968). Tables of distributions of positive definite quadratic forms in central normal variables, *Sankhyā, Series B*, **30**, 303–314.

Johnson, N. L., and Welch, B. L. (1939). On the calculation of the cumulants of the χ-distribution, *Biometrika*, **31**, 216–218.

Keiding, N., Jensen, S. T., and Ranek, L. (1972). Maximum likelihood estimation of the size distribution of liver cell nuclei from the observed distribution in a plane section, *Biometrics*, **28**, 813–829.

Kendall, M. G. (1971). The work of E. Abbé, *Biometrika*, **58**, 369–373.

Kendall, M. G., Stuart, A., and Ord, J. K. (1991). *Advanced Theory of Statistics* (Sixth edition), **1**, New York: Oxford University Press.

Khamis, S. H., and Rudert, W. (1965). *Tables of the incomplete gamma function ratio: chi-square integral, Poisson distribution*, Darmstadt: Justus von Liebig.

Khatri, C. G. (1963). Further contribution to Wishartness and independence of second degree polynomials in normal vectors, *Journal of the Indian Statistical Association*, **1**, 61–70.

Khatri, C. G. (1978). A remark on the necessary and sufficient conditions for a quadratic form to be distributed as chi-squared, *Biometrika*, **65**, 239–240.

Kinderman, A. J., and Monahan, J. F. (1977). Computer generation of random variables using the ratio of uniform deviates, *ACM Transactions on Mathematical Software*, **3**, 257–260.

Kinderman, A. J., and Monahan, J. F. (1980). New methods for generating Student's t and gamma variables, *Computing*, **25**, 369–377.

Knüsel, L. (1986). Computation of the chi-square and Poisson distribution, *SIAM Journal of Scientific and Statistical Computation*, **7**, 1022–1036.

Koch, G. G., and Bhapkar, V. P. (1982). Chi-square tests, In *Encyclopedia of Statistical Sciences*, **1**, S. Kotz and N. L. Johnson (editors), 442–457, New York: Wiley.

Koch, G. G., and Stokes, M. E. (1982). Chi-square tests: Numerical examples, In *Encyclopedia of Statistical Sciences*, **1**, S. Kotz and N. L. Johnson (editors), 457–472, New York: Wiley.

Kodlin, D. (1967). A new response time distribution, *Biometrics*, **23**, 227–239.

Krauth, J., and Steinebach, J. (1976). Extended tables of the percentage points of the chi-square distribution for at most ten degrees of freedom, *Biometrische Zeitschift*, **18**, 13–22.

Kruskal, W. H. (1946). Helmert's distribution, *American Mathematical Monthly*, **53**, 435–438.

Krysicki, W. (1963). Application de la méthode des moments a l'estimation des paramètres d'un melange de deux distributions de Rayleigh, *Revue de Statistique Appliquée*, **11**, 25–45.

Lancaster, H. O. (1966). Forerunners of the Pearson χ^2, *Australian Journal of Statistics*, **8**, 117–126.

Lancaster, H. O. (1969). *The Chi-Squared Distribution*, New York: Wiley.

Lancaster, H. O. (1982). Chi-square distribution, In *Encyclopedia of Statistical Sciences*, **1**, S. Kotz and N. L. Johnson (editors), 439–442, New York: Wiley.

Lau, C. L. (1980). A simple series for the incomplete gamma integral, *Applied Statistics*, **29**, 113–114.

Lawless, J. F. (1982). *Statistical Models and Methods for Lifetime Data*, New York: Wiley.

Lee, K. R., Kapadia, C. H., and Brock, D. B. (1980). On estimating the scale parameter of the Rayleigh distribution from doubly censored samples, *Statistische Hefte*, **21**, 14–29.

Levin, B. R. (1973). *Fondaments Théoriques de la Radiotechnique Statistique*, Moscow: Editions mir.

Liddell, D. (1983). On the distributions of χ^2 and F, *Bulletin in Applied Statistics*, **10**, 240–243. Correction, *Journal of Applied Statistics*, **11**, 119.

Lieblein, J. (1955). On moments of order statistics from the Weibull distribution, *Annals of Mathematical Statistics*, **26**, 330–333.

Lin, J. -T. (1988). Approximating the cumulative chi-square distribution and its inverse, *The Statistician*, **37**, 3–5.

Ling, R. F. (1977). On the maximum absolute errors of some approximations for t, chi-squared, and F tail probabilities, *ASA Proceedings on Statistical Computing*, 299–304.

Ling, R. F. (1978). A study of the accuracy of some approximations for t, χ^2, and F tail probabilities, *Journal of the American Statistical Association*, **73**, 274–283.

Lloyd, E. H. (1952). Least-squares estimation of location and scale parameters using order statistics, *Biometrika*, **39**, 88–95.

Luke, Y. L. (1969). *The Special Functions and Their Approximations*, **1**, San Diego: Academic Press.

Matsunawa, T. (1981). Some reflexions and historical reviews on the chi-square distribution, *Proceedings of the Institute of Statistical Mathematics*, **29**, 109–127. (In Japanese)

Maxwell, J. C. (1860, 1927). Illustrations of the dynamical theory of gases, *Scientific Papers*, **1**, 377–410. Paris: Librairies Scientifiques Hermann.

Middleton, D. (1960). *An Introduction to Statistical Communication Theory*, New York: McGraw-Hill.

Miller, K. S. (1964). *Multidimensional Gaussian Distributions*, New York: Wiley.

Mitra, S. K. (1974). On a definition of the chi-square distribution, *Gujarat Statistical Review*, **1**, 33–36.

Monahan, J. F. (1987). An algorithm for generating chi random variables, *ACM Transactions on Mathematical Software*, **13**, 168–172.

Narula, S. C., and Li, F. S. (1977). Approximations to the chi-square distribution, *Journal of Statistical Computation and Simulation*, **5**, 267–277.

Nelson, L. S. (1987). A chi-square control chart for several proportions, *Journal of Quality Technology*, **19**, 229–231.

Neuts, M. F., and Zacks, S. (1967). On mixtures of χ^2 and F-distributions which yield distributions of the same family, *Annals of the Institute of Statistical Mathematics*, **19**, 527–536.

Ng, E. W., and Geller, M. (1969). A table of integrals of the error function, *Journal of Research of National Bureau of Standards, B. Mathematical Sciences*, **73**, 1–20.

Ogawa, J. (1951). Contributions to the theory of systematic statistics, I, *Osaka Mathematical Journal*, **3**, 175–213.

Oman, S. D., and Zacks, S. (1981). A mixture approximation to the distribution of a weighted sum of chi-squared variables, *Journal of Statistical Computation and Simulation*, **13**, 215–224.

Ono, K. (1976). Amplitude distribution analysis of acoustic emission signals, *Materials Evaluation*, **34**, 177–184.

Pearson, E. S., and Hartley, H. O. (1954). *Biometrika Tables for Statisticians*, **1**, Cambridge: Cambridge University Press.

Pearson, K. (1900). On a criterion that a given system of deviations from the probable in the case of a correlated system of variables is such that it can be reasonably supposed to have arisen from random sampling, *Philosophical Magazine, 5th Series*, **50**, 157–175.

Pearson, K. (ed.) (1922). *Tables of the Incomplete Γ-Function*, H. M. Stationery Office, London, England (Reprint 1934, Cambridge University Press).

Peiser, A. M. (1943). Asymptotic formulas for significance levels of certain distributions, *Annals of Mathematical Statistics*, **14**, 56–62.

Peizer, D. B., and Pratt, J. W. (1968). A normal approximation for binomial, F, beta and other common, related tail probabilities I, *Journal of the American Statistical Association*, **63**, 1416–1456.

Plackett, R. L. (1983). K. Pearson and the chi-squared test, *International Statistical Review*, **51**, 59–72.

Polovko, A. M. (1968). *Fundamentals of Reliability Theory*, San Diego: Academic Press.

Provost, S. B. (1986). The exact distribution of the ratio of a linear combination of chi-square variables over the root of a product of chi-square variables, *Canadian Journal of Statistics*, **14**, 61–67.

Rahman, M. S., and Gupta, R. P. (1993). Family of transformed chi-square distributions, *Communications in Statistics—Theory and Methods*, **22**, 135–146.

Raja Rao, B., and Talwalker, S. (1989). Bounds on life expectancy for the Rayleigh and Weibull distributions, *Mathematical Biosciences*, **96**, 95–115.

Raqab, M. Z. (1992). *Predictors of Future Order Statistics from Type-II Censored Samples*, Unpublished Ph.D. thesis, Columbus: Ohio State University.

Rayleigh, J. W. S. (1880). On the resultant of a large number of vibrations of the same pitch and of arbitrary phase, *Philosophical Magazine*, 5th Series, **10**, 73–78.

Rayleigh, J. W. S. (1919). *Philosophical Magazine*, 6th Series, **37**, 321–347.

Rice, L. R., and Gaines Das, R. E. (1985). A remark on algorithms AS 32 and AS 147: The incomplete gamma integral, *Applied Statistics*, **34**, 326.

Rice, S. O. (1944). Mathematical analysis of random noise, I, *Bell System Technical Journal*, **23**, 282–332.

Rice, S. O. (1945). Mathematical analysis of random noise, II, *Bell System Technical Journal*, **24**, 46–156.

Rice, S. O. (1948). Statistical properties of a sine wave plus random noise, *Bell System Technical Journal*, **27**, 109–157.

Robbins, H. E., and Pitman, E. J. G. (1949). Applications of the method of mixtures to quadratic forms in normal variates, *Annals of Mathematical Statistics*, **20**, 552–560.

Ruben, H. (1962). Probability content of regions under spherical normal distributions. IV. The distribution of homogeneous and non-homogeneous quadratic functions of normal variables, *Annals of Mathematical Statistics*, **33**, 542–570.

Ruben, H. (1974). A new characterization of the normal distribution through the sample variance, *Sankhyā*, *Series A*, **36**, 379–388.

Ruben, H. (1975). A further characterization of normality through the sample variance, *Sankhyā*, *Series A*, **37**, 72–81.

Russell, W., and Lal, M. (1969). *Tables of Chi-square Probability Function*, Department of Mathematics, St. Johns: Memorial University of Newfoundland. (Reviewed in *Mathematics of Computation*, **23**, 211–212.)

Sahai, H., and Thompson, W. O. (1974). Comparisons of approximations to the percentiles of the t, χ^2, and F distributions, *Journal of Statistical Computation and Simulation*, **3**, 81–93.

Sarkar, S. K. (1980). Inference from incomplete multivariate samples—Some testing and related problems, *Doctoral dissertation*, Pittsburgh: University of Pittsburgh.

Sarkar, S. K. (1983). Some results on χ^2 and a related distribution, *Sankhyā*, *Series A*, **45**, 253–255.

Satterthwaite, F. E. (1946). An approximate distribution of estimates of variance components, *Biometrika*, **33**, 110–114.

Severo, N. C., and Zelen, M. (1960). Normal approximation to the chi-square and non-central F probability function, *Biometrika*, **47**, 411–416.

Shanbhag, D. N. (1968). Some remarks concerning Khatri's result on quadratic forms, *Biometrika*, **55**, 593–596.

Shea, B. L. (1988). Algorithm AS 239. Chi-squared and incomplete gamma integral, *Applied Statistics*, **37**, 466–473.

Sheil, J., and O'Muircheartaigh, I. (1977). Algorithm AS 106: The distribution of non-negative quadratic forms in normal variables, *Applied Statistics*, **26**, 92–98.

Shepherd, W. L. (1975). Computation of moments of a log-Rayleigh distributed random variable, *Proceedings of Twentieth Conference on Design of Experiments in Army Research Development and Testing*, Parts 1 and 2, ARO Report No. 75-2, 441–448.

Sheynin, O. B. (1966). Origin of the theory of errors, *Nature*, **211**, 1003–1004.

Sheynin, O. B. (1971). On the history of some statistical laws of distribution, *Biometrika*, **58**, 234–236.

Sheynin, O. B. (1988). C. F. Gauss and the χ-square distribution, *NTM-Schriftenr. Gesch. Naturwiss., Technik, Med.-Leipzig*, **25**, 21–22.

Shirahata, S. (1988). Computation of the exact distribution of a weighted sum of chi-square variables, *Bulletin of the Computational Statistics of Japan*, **1**, 37–44. (In Japanese)

Shoukri, M. M. (1982). A note on estimation and test for the Rayleigh amplitude distribution, *Communications in Statistics—Theory and Methods*, **11**, 1919–1934.

Siddiqui, M. M. (1962). Some problems connected with Rayleigh distributions, *Journal of Research of National Bureau of Standards, Series D*, **66**, 167–174.

Sinha, S. K. (1990). On the prediction limits for Rayleigh life distribution, *Calcutta Statistical Association Bulletin*, **39**, 104–109.

Sinha, S. K., and Howlader, H. A. (1983). Credible and HPD intervals of the parameter and reliability of Rayleigh distribution, *IEEE Transactions on Reliability*, **32**, 217–220.

Skellam, J. G. (1952). Studies in statistical ecology: I, Spatial pattern, *Biometrika*, **39**, 346–362.

Sletten, O. (1980). Algorithms for hand calculators to approximate Gaussian and chi-square probabilities, *Educational and Psychological Measurement*, **40**, 899–910.

Slutskii, E. E. (1950). *Tablitsi dlya Vichisleniya Nepolnoi Γ-funktsii i Veroyatnosti χ^2*, (Tables for Computing the Incomplete Gamma function and χ^2 probabilities), A. N. Kolmogorov (editor), Moscow: Akademia Nauk SSSR.

Solomon, H. (1960). Distribution of quadratic forms—Tables and applications, *Applied Mathematics and Statistics Laboratories, Technical Report 45*, Stanford, CA: Stanford University.

Solomon, H., and Stephens, M. A. (1977). Distribution of a sum of weighted chi-square variables, *Journal of the American Statistical Association*, **72**, 881–885.

Stacy, E. W., and Mihram, G. A. (1965). Parameter estimation for a generalized gamma distribution, *Technometrics*, **7**, 349–358.

Stammberger, A. (1967). Über einige Nomogramme zur Statistik, *Wissenshaftliche Zeitschrift der Humboldt-Universität Berlin, Mathematisch-Naturwissenschaftliche Reihi*, **16**, 186–193.

Tate, R. F., and Klett, G. W. (1959). Optimal confidence intervals for the variance of a normal distribution, *Journal of the American Statistical Association*, **54**, 674–682.

Taylor, J. M. G. (1985). Power transformations to symmetry, *Biometrika*, **72**, 145–152.

Terrell, G. R. (1987). Chi-squared left-tail probabilities, *Journal of Statistical Computation and Simulation*, **28**, 264–266.

Thompson, C. M. (1941). Tables of percentage points of the chi-square distribution, *Biometrika*, **28**, 187–191.

Tiku, M. L., Tan, W. Y., and Balakrishnan, N. (1986). *Robust Inference*, New York: Marcel Dekker.

Tyagi, R. K., and Bhattacharya, S. K. (1989). Bayes estimator of the Maxwell's velocity distribution function, *Statistica*, **49**, 563–566.

Vanderbeck, J. P., and Cooke, J. R. (1961). *Extended Table of Percentage Points of the Chi-Square Distribution*, Nauweps Report 7770, U.S. Naval Ordnance Test Station, China Lake, CA.

Vodă, V. Gh. (1972). On the "inverse Rayleigh" distributed random variable, *Reports in Statistical Applied Research JUSE*, **19**, 13–21.

Vodă, V. Gh. (1975). Note on the truncated Rayleigh variate, *Revista Colombiana de Matematicas*, **9**, 1–7.

Vodă, V. Gh. (1976a). Inferential procedures on a generalized Rayleigh variate, I, *Aplikace Matematiky*, **21**, 395–412.

Vodă, V. Gh. (1976b). Inferential procedures on a generalized Rayleigh variate, II, *Aplikace Matematiky*, **21**, 413–419.

Vodă, V. Gh. (1977). Some inferences on Rayleigh distribution, *ASQC Technical Conference Transactions*, 588–589.

Wallace, D. L. (1959). Bounds on normal approximations to Student's t and the chi-square distributions, *Annals of Mathematical Statistics*, **30**, 1121–1130.

Watson, G. N. (1944). *A Treatise on the Theory of Bessel Functions* (Second edition), Cambridge: Cambridge University Press.

Watts, D. G. (1979). Simple revealing approximations to F and χ^2 percentage points, *The Statistician*, **28**, 141–142.

Weintraub, S. (1979). Non-linear least squares fit of the chi-square and t distributions, *Journal of Statistical Computation and Simulation*, **10**, 79–84.

Welch, B. L. (1938). The significance of the difference between two means when the population variances are unequal, *Biometrika*, **29**, 350–361.

Wilson, E. B., and Hilferty, M. M. (1931). The distribution of chi-square, *Proceedings of the National Academy of Sciences*, **17**, 684–688.

Wood, A. T. A. (1989). An F approximation to the distribution of a linear combination of chi-squared variables, *Communications in Statistics—Simulation and Computation*, **18**, 1439–1456.

Zar, J. H. (1978). Approximations for the percentage points of the chi-squared distribution, *Applied Statistics*, **27**, 280–290.

CHAPTER 19

Exponential Distributions

1 DEFINITION

The random variable X has an *exponential* (or *negative exponential*) *distribution* if it has a probability density function of form

$$p_X(x) = \sigma^{-1} \exp\left[-\frac{(x - \theta)}{\sigma} \right], \qquad x > \theta; \sigma > 0. \qquad (19.1)$$

Figure 19.1 gives a graphical representation of this function, with $\theta > 0$. This is a special case of the gamma distribution, the subject of Chapter 17. The exponential distribution has a separate chapter because of its considerable importance and widespread use in statistical procedures.

Very often it is reasonable to take $\theta = 0$. The special case of (19.1) so obtained is called the *one-parameter* exponential distribution. If $\theta = 0$ and $\sigma = 1$, the distribution is called the *standard* exponential distribution. The pdf is

$$p_X(x) = \exp(-x), \qquad x > 0. \qquad (19.2)$$

The mathematics associated with the exponential distribution is often of a simple nature, and so it is possible to obtain explicit formulas in terms of elementary functions, without troublesome quadratures. For this reason models constructed from exponential variables are sometimes used as an approximate representation of other models that are more appropriate for a particular application.

2 GENESIS

There are many situations in which one would expect an exponential distribution to give a useful description of observed variation. One of the most widely quoted is that of events recurring "at random in time." In particular, suppose that the future lifetime of an individual has the same distribution, no matter

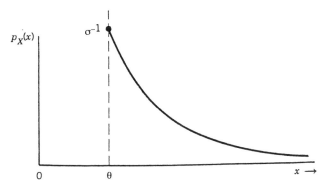

Figure 19.1 Exponential Density Function

how old it is at present. This can be written formally (X representing lifetime)

$$\Pr[X \le x_0 + x | X > x_0] = \Pr[X \le x] \qquad \text{for all } x_0 > 0, x > 0.$$

X must be a continuous positive random variable. If it has a probability density function $p_X(x)$, then the conditional probability density function, given that X is greater than x_0, is

$$\frac{p_X(x)}{1 - F_X(x_0)}, \qquad x > x_0 > 0.$$

Since the conditional distribution of the future lifetime $(X - x_0)$ is the same as the (unconditional) distribution of X, we have, say,

$$\frac{p_X(x_0)}{1 - F_X(x_0)} = p_X(0) = p_0.$$

It follows that if $F_X(x_0) \ne 1$, $p_0 > 0$ and $F_X(x)$ satisfy the differential equation

$$\frac{dF_X(x)}{dx} = p_0[1 - F_X(x)],$$

whence $1 - F_X(x) \propto e^{-p_0 x}$. Introducing the condition $\lim_{x \to 0} F_X(x) = 0$, we find that

$$1 - F_X(x) = e^{-p_0 x}, \tag{19.3}$$

that is,

$$F_X(x) = 1 - e^{-p_0 x} = p_0 \int_0^x e^{-p_0 t} \, dt.$$

This shows that the probability density function of X is of form (19.1) with $\theta = 0$, $\sigma = p_0^{-1}$.

There are other situations in which exponential distributions appear to be the most natural. Many of these do, however, have as an essential feature the random recurrence (often in time) of an event.

In applying the Monte Carlo method it is often required to transform random variables from a standard rectangular distribution to exponential random variables. An ingenious method was suggested at an early date by von Neumann (1951). Let $\{X_i; \ i = 0, 1, \ldots\}$ be a sequence of independent random variables from the standard rectangular distribution, and define a random variable N taking positive integer values through $\{X_i\}$ by the inequalities

$$X_1 < X_0, \quad \sum_{j=1}^{2} X_j < X_0, \ldots, \quad \sum_{j=1}^{N-1} X_j < X_0, \quad \sum_{j=1}^{N} X_j > X_0.$$

We "accept" the sequence $\{X_i\}$ if N is odd, otherwise we "reject" it and repeat the process until N turns out odd. Let T be the number of sequences rejected before an odd N appears ($T = 0, 1, \ldots$) and X_0 be the value of the first variable in the accepted sequence. Then $Y = T + X_0$ is an exponential random variable with the standard density e^{-x}.

A rather more convenient method was suggested by Marsaglia (1961). Let N be a nonnegative random integer with the geometric distribution [Chapter 5, Section 2, equation (5.8)]

$$\Pr[N = n] = (1 - e^{-\lambda})e^{-n\lambda}, \qquad n = 0, 1, \ldots,$$

and let M be a positive random integer with the zero-truncated Poisson distribution [Chapter 4, Section 10, equation (4.73)]

$$\Pr[M = m] = (1 - e^{-\lambda})^{-1} \frac{e^{-\lambda}\lambda^m}{m!}, \qquad m = 1, 2, \ldots.$$

Finally let $\{X_i; \ i = 1, 2, \ldots\}$ be a sequence of independent random variables each having a standard rectangular distribution (Chapter 26). Then

$$Y = \lambda\{N + \min(X_1, \ldots, X_M)\}$$

has the standard exponential distribution.

Sibuya (1962) gave a statistical interpretation of the procedure, recommended that the value of the parameter λ be taken as 0.5 or log 2, and extended the technique to the chi-square distribution. Bánkövi (1964) investigated a similar technique. A table of 10,000 exponential random numbers is given by Barnett (1965).

3 SOME REMARKS ON HISTORY

Over the last 40 years the study of estimators based on samples from an exponential population has been closely associated with the study of order statistics. Lloyd (1952) described a method for obtaining the best linear unbiased estimators (BLUEs) of the parameters of a distribution, using order statistics. Epstein and Sobel (1953) presented the maximum likelihood estimator of the scale parameter σ, of the one-parameter exponential distribution in the case of censoring from the right. Epstein and Sobel (1954) extended the foregoing analysis to the two-parameter exponential distribution. Sarhan (1954) employed the method derived by Lloyd to obtain the BLUEs of σ and θ for the two-parameter exponential distribution in the case of no censoring. Sarhan (1955) extended his results to censoring. Sarhan noted that in the special case of the one-parameter exponential distribution, his results agreed with those of Epstein and Sobel, and therefore his estimator of σ was not only the best linear unbiased estimator but also the maximum likelihood estimator of σ. Epstein (1960) extended his own results to estimators of σ and θ for the one- and two-parameter exponential distributions in the cases of censoring from the right and/or left. For the two-parameter exponential distribution his maximum likelihood estimators coincided with the BLUEs of Sarhan (1960), but for the one-parameter exponential distribution there was agreement only in the case of censoring from the right. Many other contributions by Epstein and Sobel are included in the references.

In the light of the applicability of order statistics to the exponential distribution it became quite natural to attempt estimation of the parameter by use of the sample quasi-ranges. Rider (1959) derived the probability density function and the cumulants of the quasi-range of the standardized exponential distribution and Fukuta (1960) derived "best" linear estimators of σ and θ by two sample quasi-ranges. The next step would quite reasonably be that of determining the two order statistics that would supply the best linear unbiased estimator of σ and θ for the two-parameter distribution; this was in fact done numerically by Sarhan, Greenberg, and Ogawa (1963). They employed the method of Lloyd to obtain the best linear estimators of σ and θ based on the two-order statistics $X'_{l:n}$ and $X'_{m:n}$, and then compared numerically the relative efficiencies of the estimators for various pairs of values (l, m). Harter (1961), using a similar approach to that of Sarhan and his coworkers, presented the best linear estimators of σ for the one-parameter distribution based on one- and two-order statistics. Harter mentioned in this paper that he was not aware of any analytical process by which the optimum pair of order statistics $X'_{l:n}$ and $X'_{m:n}$ can be determined. Siddiqui (1963) presented an analytical method based on the Euler-Maclaurin formula for obtaining the optimum pair of BLUE order statistics. Since 1963 a considerable number of additional, more refined, results have been obtained. Some of these results are presented in Section 7.

4   MOMENTS AND GENERATING FUNCTIONS

The moment generating function of a random variable X with probability density function (19.1) is

$$E[e^{tX}] = (1 - \sigma t)^{-1} e^{t\theta} \quad \left(= (1 - \sigma t)^{-1} \text{ if } \theta = 0 \right). \quad (19.4)$$

The characteristic function is $(1 - i\sigma t)^{-1} e^{it\theta}$.

The central moment generating function is

$$E[e^{t(X-\theta-\sigma)}] = (1 - \sigma t)^{-1} e^{-t\sigma}.$$

The cumulant generating function is $\log E[e^{tX}] = t\theta - \log(1 - \sigma t)$. Hence the cumulants are

$$\kappa_1 = \theta + \sigma \quad (= E[X])$$

$$\kappa_r = (r - 1)! \sigma^r, \quad r > 1. \quad (19.5)$$

Setting $r = 2, 3, 4$, we find that

$$\text{Var}(X) = \mu_2 = \sigma^2$$

$$\mu_3 = 2\sigma^3$$

$$\mu_4 = 9\sigma^4.$$

Note that if $\theta = 0$ and $\sigma = 1$, then $E[X] = 1 = \text{Var}(X)$.

The first two moment ratios are

$$\sqrt{\beta_1} = 2, \quad \beta_2 = 9.$$

The mean deviation is

$$2\sigma \int_1^\infty (x - 1) e^{-x} \, dx = 2e^{-1}\sigma. \quad (19.6)$$

Note that

$$\frac{\text{Mean deviation}}{\text{Standard deviation}} = \frac{2}{e} = 0.736. \tag{19.7}$$

The median of the distribution is $\theta + \sigma \log_e 2$. The mode of this distribution is at the lowest value θ of the range of variation.

The information generating function [$(u - 1)$-th frequency moment] is $\sigma^{1-u} u^{-1}$. The entropy is $1 + \log \sigma$. The hazard rate (σ^{-1}) is constant. This is an important property of exponential distributions.

5 APPLICATIONS

As has already been mentioned in Section 1, the exponential distribution is applied in a very wide variety of statistical procedures. Currently among the most prominent applications are in the field of life-testing. The *lifetime* (or *life characteristic*, as it is often called) can be usefully represented by an exponential random variable, with (usually) a relatively simple associated theory. Sometimes the representation is not adequate; in such cases a modification of the exponential distribution [often a Weibull distribution (Chapter 21) is used].

Another application is producing usable approximate solutions to difficult distributional problems. An ingenious application of the exponential distribution to approximate a sequential procedure is due to Ray (1957). He wished to calculate the distribution of the smallest n for which $\sum_{i=1}^{n} U_i^2 < K_n$, where $U_1, U_2, \ldots$ are independent unit normal variables and $K_1, K_2, \ldots$ are specified positive constants. By replacing this by the distribution of the smallest even n, he obtained a problem in which the sums $\sum_{i=1}^{n} U_i^2$ are replaced by sums of independent exponential variables (actually χ^2's with two degrees of freedom each).

Vardeman and Ray (1985) investigated the average sum lengths for CUSUM schemes when observations are exponentially distributed. They show that in this case the Page (1954) integral equation whose solution gives average sum lengths for one-sided CUSUM schemes can be solved *without* resorting to approximation. They provide tables of average run lengths for the exponential case.

6 ORDER STATISTICS

Let $X_1' \leq X_2' \leq \cdots \leq X_n'$ be the order statistics obtained from a sample of size n from the standard exponential distribution in (19.2). Then, the joint density of all n order statistics is

$$p_{X'_1,\ldots,X'_n}(x_1,\ldots,x_n) = n!\,e^{-\Sigma_{i=1}^n x_i}, \qquad 0 \le x_1 \le \cdots \le x_n < \infty. \quad (19.8)$$

By making the transformation

$$Y_i = (n - i + 1)(X'_i - X'_{i-1}), \qquad i = 1, 2, \ldots, n \text{ (with } X'_0 \equiv 0), \quad (19.9)$$

we obtain from (19.8) the joint density function of $Y_1, Y_2, \ldots, Y_n$ to be

$$p_{Y_1,\ldots,Y_n}(y_1,\ldots,y_n) = e^{-\Sigma_{i=1}^n y_i}, \qquad 0 \le y_1,\ldots, y_n < \infty. \quad (19.10)$$

That is, the Y_i's (termed *normalized spacings*) are independent and identically distributed standard exponential random variables. This result is originally due to Sukhatme (1937). Also from (19.9),

$$X'_i \overset{d}{=} \sum_{j=1}^{i} \left\{ \frac{Y_j}{n - j + 1} \right\}, \qquad i = 1, 2, \ldots, n. \quad (19.11)$$

From (19.11) it is clear that the exponential order statistics form an *additive Markov chain* as shown originally by Rényi (1953).

The additive Markov chain representation in (19.11) makes it possible to write down explicit expressions for the single and product moments of X'_i. For example, we have

$$E[X'_i] = \sum_{j=1}^{i} \frac{E[Y_j]}{n - j + 1} = \sum_{j=1}^{i} \frac{1}{n - j + 1},$$
$$i = 1, 2, \ldots, n, \quad (19.12)$$

$$\mathrm{Var}(X'_i) = \sum_{j=1}^{i} \frac{\mathrm{Var}(Y_j)}{(n - j + 1)^2} = \sum_{j=1}^{i} \frac{1}{(n - j + 1)^2},$$
$$i = 1, 2, \ldots, n, \quad (19.13)$$

$$\mathrm{Cov}(X'_i, X'_k) = \sum_{j=1}^{i} \frac{\mathrm{Var}(Y_j)}{(n - j + 1)^2} = \sum_{j=1}^{i} \frac{1}{(n - j + 1)^2}$$
$$= \mathrm{Var}(X'_i) \qquad \text{for } 1 \le i < k \le n. \quad (19.14)$$

This special structure of the variance-covariance matrix of exponential order statistics makes it possible to derive the best linear unbiased estimators of the parameters in an explicit form (see Section 7).

Interestingly the result that the normalized spacings Y_i's defined in (19.9) are i.i.d. standard exponential random variables has been generalized by

Viveros and Balakrishnan (1994) to the case of a Type II progressively censored sample. To be precise, let m be the number of failures observed before termination of a life test of n individuals and $X'_1 \leq X'_2 \leq \cdots \leq X'_m$ be the observed ordered life lengths. Let R_i designate the number of units removed at the time of the ith failure (Type II censoring),

$$0 \leq R_i \leq n - \sum_{j=1}^{i-1} (R_j + 1) - 1, \qquad i = 2, 3, \ldots, m - 1,$$

with

$$0 \leq R_1 \leq n - 1,$$

$$R_n = n - \sum_{j=1}^{m-1} (R_j + 1) - 1.$$

The resulting data are referred to as a *Type II progressively censored sample* [e.g., see Nelson (1982); Lawless (1982); Cohen and Whitten (1988); Balakrishnan and Cohen (1991); Cohen (1991)]. Defining the ith normalized spacing between $X'_1, X'_2, \ldots, X'_m$ as

$$Y_i = \left\{ n - \sum_{j=1}^{i-1} (R_j + 1) \right\} (X'_i - X'_{i-1}), \qquad i = 1, 2, \ldots, m, \text{ (with } X'_0 \equiv 0),$$

$$(19.15)$$

Viveros and Balakrishnan (1994) have proved that Y_i's are i.i.d. exponential random variables. Sukhatme's result, presented earlier, is a particular case of this general result corresponding to $R_1 = R_2 = \cdots = R_m = 0$. In this general case an additive Markov chain representation for X'_i similar to the one in (19.11) is possible; it can then be used in writing down explicit expressions for the means, variances, and covariances of X'_i that are similar to those in (19.12)–(19.14). The special structure of the variance-covariance matrix of exponential order statistics observed earlier in (19.13)–(19.14) occurs in this general case of Type II progressively censored sample and enables the best linear unbiased estimators of the parameters to be derived in an explicit form (as described in Section 7c).

Due to the close relationship between the geometric and the exponential distributions, there also exists a close relationship between the dependence structure of order statistics from the geometric distribution and those from the exponential distribution. Steutel and Thiemann (1989) showed some of these similarities. For example, by introducing a second subscript to denote sample size, (19.11) can be written as

$$X'_{i:n} \stackrel{d}{=} \sum_{j=1}^{i} \frac{Y_j}{(n-j+1)} \stackrel{d}{=} \sum_{j=1}^{i} X'_{1:n-j+1}, \qquad (19.16)$$

where the $X'_{1:n-j+1}$'s are independent. Steutel and Thiemann (1989) established the following parallel relationship for the geometric order statistics:

$$Z'_{i:n} \stackrel{d}{=} \sum_{j=1}^{i} Z'_{1:n-j+1} + \left[\sum_{j=1}^{i} \left\langle \frac{Y_j}{n-j+1} \right\rangle \right], \qquad (19.17)$$

where $Z'_{i:n}$ denotes the ith order statistic from a random sample of size n from a geometric (p) distribution, the Y_j's are independent exponential (σ) random variables with $\sigma = 1/[-\log(1-p)]$, $[Y]$ denotes the integer part of Y, and $\langle Y \rangle$ denotes the fractional part of Y. Further, all the random variables on the right-hand side of (19.17) are independent.

Arnold and Villaseñor (1989) and Arnold and Nagaraja (1991) discussed the Lorenz order relationships among order statistics from exponential samples. The Lorenz curve associated with X is

$$L_X(u) = \frac{\int_0^u F_X^{-1}(t)\,dt}{\int_0^1 F_X^{-1}(t)\,dt}, \qquad 0 \le u \le 1 \qquad (19.18)$$

(see Chapter 12, Section 1). Given two nonnegative random variables X and Y (with finite positive mean), we say X exhibits less inequality (or variability) than Y *in the Lorenz sense*, and write $X <_L Y$, if $L_X(u) \ge L_Y(u)$ for all $u \in [0,1]$; if the inequality is an equality, we write $X =_L Y$. If $L_X(u)$ and $L_Y(u)$ cross, X and Y are not comparable in the Lorenz sense. Arnold and Nagaraja (1991) proved that for $i \le j$,

$$X'_{j:m} <_L X'_{i:n} \quad \text{iff } (n-i+1)E[X'_{i:n}] \le (m-j+1)E[X'_{j:m}]. \quad (19.19)$$

As direct consequences of this result, they established for the exponential order statistics that

1. $X'_{i:n+1} <_L X'_{i:n}$,
2. $X'_{i+1:n+1} <_L X'_{i:n}$,
3. $X'_{i+1:n} <_L X'_{i:n}$ iff $E[X'_{i:n}] \le 1$; otherwise, $X'_{i:n}$ and $X'_{i+1:n}$ are not Lorenz ordered.

These authors have also discussed the Lorenz ordering of linear functions of exponential order statistics.

By making use of the underlying differential equation of a standard exponential distribution, given by $p_X(x) = 1 - F_X(x)$, Joshi (1978, 1982) derived the following recurrence relations:

$$E[X_{1:n}^{'m}] = \frac{m}{n}E[X_{1:n}^{'m-1}], \qquad n \ge 1, \quad m = 1, 2, \ldots, \qquad (19.20a)$$

$$E[X_{i:n}^{'m}] = E[X_{i-1:n-1}^{'m}] + \frac{m}{n}E[X_{i:n}^{'m-1}],$$

and

$$2 \le i \le n, \quad m = 1, 2, \ldots, \qquad (19.20b)$$

$$E[X_{i:n}' X_{i+1:n}'] = E[X_{i:n}'^{2}] + \frac{1}{n-i}E[X_{i:n}'], \qquad 1 \le i \le n-1, \quad (19.20c)$$

and

$$E[X_{i:n}' X_{j:n}'] = E[X_{i:n}' X_{j-1:n}'] + \frac{1}{n-j+1}E[X_{i:n}'],$$

$$1 \le i < j \le n, \quad j - i \ge 2. \quad (19.20d)$$

These recurrence relations can be used in a simple recursive manner in order to compute all the single and product moments (in particular, the means, variances, and covariances) of all order statistics. Balakrishnan and Gupta (1992) extended these results and derived relations that will enable one to find the moments and cross-moments (of order up to 4) of order statistics. They then used these results to determine the mean, variance, and the coefficients of skewness and kurtosis of a general linear function of exponential order statistics and approximate its distribution. Through this approach Balakrishnan and Gupta (1992) justify a chi-square approximation for the distribution of the best linear unbiased estimator of the mean lifetime based on doubly Type II censored samples (see Section 7). Interestingly the relations in (19.20a)–(19.20d), under certain conditions, can also be shown to be characterizations of the exponential distribution [Lin (1988, 1989)] (see Section 8). Balakrishnan and Malik (1986) derived similar recurrence relations for the single and product moments of order statistics from a linear-exponential distribution with increasing hazard rate. Sen and Bhattacharyya (1994) have discussed inferential issues relating to this distribution.

By making use of the facts that $X = -\log U$ has a standard exponential distribution, when U has a uniform $(0, 1)$ distribution (see Chapter 26), and that $-\log U$ is a monotonic decreasing function of U, we have

$$X_{i:n}' \overset{d}{=} -\log U_{n-i+1:n}', \qquad 1 \le i \le n,$$

and hence

$$V_i = \left(\frac{U'_{i:n}}{U'_{i+1,n}}\right)^i \overset{d}{=} \exp\{-i(X'_{n-i+1:n} - X'_{n-i:n})\} \overset{d}{=} e^{-Y'_{n-i+1}}, \quad (19.21)$$

where Y_i's are the normalized exponential spacings defined in (19.9). Since Y_i's are i.i.d. exponential, as seen earlier, it immediately follows from (19.21) that

$$V_1 = \frac{U'_{1:n}}{U'_{2:n}}, \quad V_2 = \left(\frac{U'_{2:n}}{U'_{3:n}}\right)^2, \dots, V_{n-1} = \left(\frac{U'_{n-1:n}}{U'_{n:n}}\right)^{n-1}, \quad V_n = U'^{n}_{n:n}$$

$$(19.22)$$

are i.i.d. uniform $(0, 1)$ random variables. This result was derived originally by Malmquist (1950) and is now used effectively to simulate order statistics from uniform distribution without resorting to any ordering.

Joshi (1978, 1982) also considered order statistics from a right-truncated exponential population with density

$$p_X(x) = \begin{cases} \dfrac{1}{P} e^{-x}, & 0 \le x \le P_1, \\ 0, & \text{otherwise}, \end{cases} \quad (19.23)$$

where $P_1 = -\log(1 - P)$ and $1 - P$ $(0 < P < 1)$ is the proportion of truncation on the right of the standard exponential distribution. By making use of the underlying differential equation given by

$$p_X(x) = [1 - F_X(x)] + \left(\frac{1 - P}{P}\right)$$

and proceeding along the lines used to prove relations (19.20a)–(19.20d), Joshi (1978, 1982) established the following recurrence relations:

$$E[X'^{m}_{1:n}] = \frac{m}{n} E[X'^{m-1}_{1:n}]$$
$$- \left(\frac{1 - P}{P}\right) E[X'^{m}_{1:n-1}], \quad n \ge 2, m = 1, 2, \dots,$$

$$(19.24a)$$

$$E[X'^{m}_{i:n}] = \frac{1}{P} E[X'^{m}_{i-1:n-1}] + \frac{m}{n} E[X'^{m-1}_{i:n}]$$
$$- \left(\frac{1 - P}{P}\right) E[X'^{m}_{i:n-1}], \quad 2 \le i \le n - 1, m = 1, 2, \dots,$$

$$(19.24b)$$

$$E[X_{n:n}'^m] = \frac{1}{P}E[X_{n-1:n-1}'^m] + \frac{m}{n}E[X_{n:n}'^{m-1}]$$

$$- \left(\frac{1-P}{P}\right)P_1^m, \qquad n \ge 2, \, m = 1, 2, \ldots, \qquad (19.24c)$$

$$E[X_{n-1:n}'X_{n:n}'] = E[X_{n-1:n}'^2] + E[X_{n-1:n}']$$

$$- n\left(\frac{1-P}{P}\right)\left\{P_1 E[X_{n-1:n-1}'] - E[X_{n-1:n-1}'^2]\right\},$$

$$n \ge 2, \quad (19.24d)$$

$$E[X_{i:n}'X_{i+1:n}'] = E[X_{i:n}'^2] + \frac{1}{n-i}$$

$$\times \left\{E[X_{i:n}'] - n\left(\frac{1-P}{P}\right)\left\{E[X_{i:n-1}'X_{i+1:n-1}']\right.\right.$$

$$\left.\left. - E[X_{i:n-1}'^2]\right\}\right\}, \qquad 1 \le i \le n-2, \quad (19.24e)$$

$$E[X_{i:n}'X_{j:n}'] = E[X_{i:n}'X_{j-1:n}'] + \frac{1}{n-j+1}$$

$$\times \left\{E[X_{i:n}'] - n\left(\frac{1-P}{P}\right)\left\{E[X_{i:n-1}'X_{j:n-1}']\right.\right.$$

$$\left.\left. - E[X_{i:n-1}'X_{j-1:n-1}']\right\}\right\},$$

$$1 \le i < j \le n-1; \, j - i \ge 2, \quad (19.24f)$$

and

$$E[X_{i:n}'X_{n:n}'] = E[X_{i:n}'X_{n-1:n}'] + E[X_{i:n}']$$

$$- n\left(\frac{1-P}{P}\right)\left\{P_1 E[X_{i:n-1}'] - E[X_{i:n-1}'X_{n-1:n-1}']\right\},$$

$$1 \le i \le n-2. \quad (19.24g)$$

Saleh, Scott, and Junkins (1975) derived exact (but somewhat cumbersome) explicit expressions for the first two single moments and the product moments of order statistics in this right-truncated exponential case. Balakrishnan and Gupta (1992) extended the results of Joshi and established recurrence relations that will enable one to find the moments and cross-moments (of order up to four) of order statistics.

By considering the doubly truncated exponential distribution with density

$$p_X(x) = \begin{cases} \dfrac{1}{P-Q}e^{-x}, & Q_1 \leq x \leq P_1, \\ 0, & \text{otherwise,} \end{cases} \tag{19.25}$$

where Q and $1 - P$ $(0 < Q < P < 1)$ are the proportions of truncation on the left and right of the standard exponential distribution, respectively, and $Q_1 = -\log(1 - Q)$ and $P_1 = -\log(1 - P)$, Joshi (1979) and Balakrishnan and Joshi (1984) derived several recurrence relations satisfied by the single and the product moments of order statistics. Khan, Yaqub, and Parvez (1983) tabulated these quantities for some value of P, Q, and n. Distributions of some systematic statistics like the sample range and quasi-range were derived in this case by Joshi and Balakrishnan (1984).

7 ESTIMATION

Before 1959 a considerable amount of work had been done on inference procedures for the exponential distribution with both censored and uncensored data. (See numerous references at the end of this chapter.) It was realized, in the 1960s and 1970s, that although the exponential distribution can be handled rather easily, the consequent analysis is often poorly robust [e.g., see Zelen and Dannemiller (1961)]. Nevertheless, the study of properties of this distribution, and especially construction of estimation and testing procedures has continued steadily, during the last 30 years, with some emphasis on Bayesian analysis and order statistics methodology, and an explosion of results on characterizations. To keep this chapter within reasonable bounds, it was necessary to be very selective in our citations and descriptions of results. We first discuss classical estimators.

7.1 Classical Estimation

If $X_1, X_2, \ldots, X_n$ are independent random variables each having the probability density function (19.1), then the maximum likelihood estimators of θ and σ are

$$\hat{\theta} = \min(X_1, X_2, \ldots, X_n),$$

$$\hat{\sigma} = n^{-1} \sum_{i=1}^{n} (X_i - \hat{\theta}) = \bar{X} - \hat{\theta}. \tag{19.26}$$

If θ is known, the maximum likelihood estimator of σ is $(\bar{X} - \theta)$. Even with σ known, $\hat{\theta}$ above is still the maximum likelihood estimator of θ.

The probability density function of $\hat{\theta}$ is

$$p_{\hat{\theta}}(t) = \left(\frac{n}{\sigma}\right)\exp\left[-\frac{n(t-\theta)}{\sigma}\right], \qquad t > \theta, \qquad (19.27)$$

which is of the same form as (19.1) but with σ replaced by σ/n. The variance of $\hat{\theta}$ is therefore σ^2/n^2, and its expected value is $\theta + \sigma/n$. It is interesting to note that the variance is proportional to n^{-2} and not to n^{-1}.

The expected value of $\hat{\sigma}$ $[= \bar{X} - \hat{\theta}]$ is $\sigma(1 - n^{-1})$, and its variance is

$$\sigma^2[n^{-1} + n^{-2} - 2n^{-3}].$$

The expected value of $(\bar{X} - \theta)$ is σ and its variance is $\sigma^2 n^{-1}$.

A function of special interest in some applications is the value of the probability that X exceeds a value x_0; the *reliability function* $R(x_0)$. If $\theta = 0$ so that (19.1) becomes

$$p_X(x) = \sigma^{-1}\exp\left(-\frac{x}{\sigma}\right), \qquad x > 0, \sigma > 0, \qquad (19.28)$$

then

$$R(x_0) = e^{-x_0/\sigma}. \qquad (19.29)$$

Inserting the maximum likelihood estimator, $\hat{\sigma} = n^{-1}\sum_{i=1}^{n}X_i$ in place of σ, would give the estimator $\exp(-x_0 n/\sum_{i=1}^{n}X_i)$. This is the maximum likelihood estimator of the reliability $R(x_0)$. It is biased, but a minimum variance unbiased estimator can be obtained by using the Blackwell-Rao theorem.

The statistic

$$T = \begin{cases} 1 & \text{if } X_1 > x_0, \\ 0 & \text{if } X_1 \le x_0, \end{cases}$$

is an unbiased estimator of $\exp(-x_0/\sigma)$. Since $\sum_{i=1}^{n}X_i$ is a complete sufficient statistic for σ [and so also for $\exp(-x_0/\sigma)$], the required minimum variance unbiased estimator is

$$E\left[T\left|\sum_{i=1}^{n}X_i\right.\right] = \Pr\left[X_1 > x_0\left|\sum_{i=1}^{n}X_i\right.\right].$$

The ratio $X_1/\sum_{i=1}^{n}X_i$ has a beta distribution with parameters 1, $n - 1$, *and is independent of* $\sum_{i=1}^{n} X_i$ (Chapter 17, Section 6). Hence

$$E\left[T \middle| \sum_{i=1}^{n} X_i\right] = (n-1)\int_{x_0/\sum_{i=1}^{n}X_i}^{1} (1-z)^{n-2}\,dz$$

$$= \begin{cases} \left(1 - \dfrac{x_0}{\sum_{i=1}^{n}X_i}\right)^{n-1}, & x_0 < \sum_{i=1}^{n} X_i \\ 0, & x_0 \geq \sum_{i=1}^{n} X_i \end{cases} \tag{19.30}$$

which is the required minimum variance unbiased estimator. This formula was obtained by Pugh (1963). (Pugh claims this is the "best" estimator but does not compare its mean square error with that of competing estimators.) The sampling distribution of the maximum likelihood estimator of parameter σ in (19.28), based on a "time-censored" sample, was derived by Bartholomew (1963).

Moment estimators $(\tilde{\theta}, \tilde{\sigma})$ of (θ, σ) can be obtained by equating sample and population values of the mean and variance. They are

$$\tilde{\theta} = \overline{X} - \tilde{\sigma} \tag{19.31a}$$

$$\tilde{\sigma} = \text{sample standard deviation.} \tag{19.31b}$$

Cohen and Helm (1973) discuss modified moment estimators obtained by replacing (19.31b) by an equation that puts the first-order statistic X_1' equal to its expected value. This gives

$$\tilde{\theta}^* + n^{-1}\tilde{\sigma}^* = X_1', \tag{19.31c}$$

which leads to

$$\tilde{\theta}^* = \frac{nX_1' - \overline{X}}{n-1}, \tag{19.31d}$$

and

$$\tilde{\sigma}^* = \frac{n(\overline{X} - X_1')}{n-1}. \tag{19.31e}$$

They show that these are minimum variance unbiased estimators (and a fortiori BLUEs). Also

$$\text{Var}(\tilde{\theta}^*) = \frac{\sigma^2}{n(n-1)}, \tag{19.32a}$$

$$\text{Var}(\tilde{\sigma}^*) = \frac{\sigma^2}{n-1}, \tag{19.32b}$$

$$\text{Cov}(\tilde{\theta}^*, \tilde{\sigma}^*) = \frac{\sigma^2}{n(n-1)}, \tag{19.32c}$$

so that $\text{Corr}(\tilde{\theta}^*, \tilde{\sigma}^*) = 1/\sqrt{n}$. Further, since $\tilde{\sigma}^*$ is distributed as $\frac{1}{2}(n-1)^{-1}\chi^2_{2(n-1)}$, a $100(1-\alpha)\%$ confidence interval for σ is (in the obvious notation)

$$\left(\frac{2(n-1)}{\chi^2_{2(n-1),\,1-\frac{\alpha}{2}}} \tilde{\sigma}^*, \; \frac{2(n-1)}{\chi^2_{2(n-1),\,\frac{\alpha}{2}}} \tilde{\sigma}^* \right). \tag{19.33}$$

7.2 Grouped Data

In a monograph Kulldorff (1961) discussed a general theory of estimation based on grouped or partially grouped samples. By *grouped* we mean that in disjoint intervals of the distribution range, only the numbers of observed values that have fallen in the intervals are available, and not the individual sample values. The distribution of the observed numbers is a multinomial distribution with probabilities that are functions of the parameter. If individual observations are available in *some* intervals, the sample is *partially grouped*.

Kulldorff devoted a large part of his book to the estimation of the exponential distribution because of its simplicity. The cases studied included completely or partially grouped data, θ unknown, σ unknown, both θ and σ unknown, a finite number of intervals, an infinite number of intervals, intervals of equal length, and intervals of unequal length. Here we describe only the maximum likelihood estimator of σ when θ is known, the intervals are not of equal length, and the number of intervals is finite.

Let $0 = x_0 < x_1 < \cdots < x_{k-1} < \infty$ be the dividing points and $N_1, \ldots, N_k$, $(\sum_{i=1}^{k} N_i = n)$ the numbers of observed values in the respective intervals. Then the maximum likelihood estimator $\hat{\sigma}$ is the unique solution of

$$\sum_{i=1}^{k-1} \frac{N_i(x_i - x_{i-1})}{e^{(x_i - x_{i-1})/\hat{\sigma}} - 1} - \sum_{i=2}^{k} N_i x_{i-1} = 0 \tag{19.34}$$

which exists if and only if $N_1 < n$ and $N_k < n$. For large n,

$$n \, \mathrm{Var}(\hat{\sigma}^{-1}) \doteq \left(\sum_{i=1}^{k-1} \frac{(x_i - x_{i-1})^2}{e^{x_i/\sigma} - e^{x_{i-1}/\sigma}} \right)^{-1}. \tag{19.35}$$

For a given k, the dividing points that minimize the asymptotic variance are

$$\frac{x_i}{\sigma} = \sum_{j=k-i}^{k-1} \delta_j, \tag{19.36}$$

where

$$\delta_1 = g^{-1}(2), \quad \delta_i = g^{-1}\{2 + \delta_{i-1} - g(\delta_{i-1})\},$$

$$g(x) = x(1 - e^{-x})^{-1}.$$

For example,

$$k = 2, \quad \frac{x_1}{\sigma} = 1.5936;$$

$$k = 3, \quad \frac{x_1}{\sigma} = 1.0176, \quad \frac{x_2}{\sigma} = 2.6112.$$

The simplicity of mathematical analysis for the exponential distribution permits us to construct convenient Bayesian estimators of parameters of (19.1). Some initial results in this area (for censored samples) are presented in Varde (1969), who also compared their performance with more natural (at least in this case) and efficient maximum likelihood and minimum variance unbiased estimators.

7.3 Estimators Using Selected Quantiles

In life-test analysis it is often supported that lifetime can be represented by a random variable with probability density function

$$p_X(x) = \sigma^{-1} \exp\left(-\frac{x}{\sigma} \right), \quad x > 0, \sigma > 0. \tag{19.28$'$}$$

If a number n of items are observed with lifetimes commencing simultaneously, then, as each life concludes, observations of lifetime become available sequentially, starting with the shortest lifetime X_1' of the n items, followed by the 2nd, 3rd... shortest lifetimes $X_2', X_3', \ldots$, respectively. Clearly it will be advantageous if useful inferences can be made at a relatively early stage, without waiting for completion of the longer lifetimes. This means that inference must be based on observed values of the k, say, shortest lifetimes, or in more general terms, the first k-*order statistics* (see Section 6). On account of the practical importance of these analyses, statistical techniques have been worked out in considerable detail. Here we will describe only methods of estimation, but a considerable range of test procedures is also available.

From (19.8) we find that if

$$V_2 = X_2' - X_1', \quad V_3 = X_3' - X_2', \ldots, V_n = X_n' - X_{n-1}',$$

then

1. $X_1', V_2, V_3, \ldots, V_n$ are mutually independent,
2. the distribution of X_1' is exponential (19.1) with $\theta = 0$, $\sigma = n^{-1}$,
3. the distribution of V_j is exponential (19.1) with $\theta = 0$,
   $\sigma = (n - j + 1)^{-1}, j = 2, \ldots, n.$

Since $X_j' = X_1' + V_2 + \cdots + V_j$ ($j \geq 2$), it follows that all linear functions of the order statistics can be expressed as linear functions of the independent random variables $X_1', V_2, \ldots, V_n$. This form of representation [suggested by Sukhatme (1937) and Rényi (1953); see also Epstein and Sobel (1954) for a similar result] is very helpful in solving distribution problems associated with methods described in the remainder of this section. A similar kind of representation can be applied to gamma distributions, though the results are not so simple.

It is necessary to distinguish between *censoring* (often refered as *Type II censoring*), in which the order statistics that will be used (e.g., the r smallest values) are decided in advance, and *truncation* (or *Type I censoring*) in which the range of values that will be used (e.g., all observations less than T_0) is decided in advance (regardless of how many observations fall within the specific limits). Truncation (or Type I censoring) by omission of all observations less than a fixed value T_0 (> 0) has the effect that observed values may be represented by a random variable with probability density function

$$\sigma^{-1} \exp\left[-\frac{x - T_0}{\sigma}\right], \qquad x > T_0, \sigma > 0, \tag{19.37}$$

which is again of form (19.1) [with T_0 (known) replacing θ], and so presents no special difficulties. However, if (as is more commonly the case) truncation is by omission of all values greater than T_0 (> 0), then the corresponding probability density function is

$$p_X(x) = \left[1 - \exp\left(-\frac{T_0}{\sigma}\right)\right]^{-1} \sigma^{-1} \exp\left(-\frac{x}{\sigma}\right), \qquad 0 < x < T_0, \sigma > 0.$$

$$\tag{19.38}$$

If m observations are obtained, they can be represented by independent random variables $X_1, X_2, \ldots, X_m$, each with distribution (19.38). The maximum likelihood equation for an estimator $\hat{\sigma}_{T_0}$ of σ is

$$\hat{\sigma}_{T_0} = m^{-1} \sum_{j=1}^{m} x_j + T_0 \left[\exp\left(\frac{T_0}{\hat{\sigma}_{T_0}}\right) - 1 \right]^{-1}. \qquad (19.39)$$

This equation may be solved by an iterative process. In this work the table of Barton, David, and Merrington (1963) is useful.

Wright, Engelhardt, and Bain (1978) and Piegorsch (1987) studied inference on θ and σ for distribution (19.1) under Type I censoring. Wright, Engelhardt, and Bain (1978) presented procedures based on the conditional distribution of "failure" times, given the number D of failures occurring before the censoring time τ. They distinguished between testing with and without replacement. In the first case X_1' and D are sufficient statistics, and in the second D and $S = \Sigma_{j=1}^{D} X_j'$ are sufficient.

Assuming testing without replacement, they utilized the facts that given $D = d$, the conditional distribution of $[(X - X_1')/(X - \theta)]^d$ is uniform on $(0, 1)$, and that for fixed θ, D is sufficient and complete for σ, while for fixed σ, X_1' is sufficient and complete for θ.

Assuming testing without replacement leads to a rather complicated conditional density for X_1', given D and S. However, Wright, Engelhardt, and Bain (1978) provide a table of exact percentage points of X_1' for small and moderate D and an approximation for large D. [In the case where $\theta = 0$, Bartholomew (1963) developed confidence intervals based on the maximum likelihood estimator

$$\hat{\sigma} = \frac{\Sigma_{i=1}^{D} X_i' + (n - D)\tau}{D} = \frac{T_{D:n}}{D} \qquad (19.40)$$

where τ is the termination time, (provided that $D > 0$)].

Piegorsch (1987) uses somewhat simpler methods, based on the approximate distribution of the likelihood ratio (LR) test statistic and some other approximations reviewed in Lawless (1982, Sec. 3.5.2). He discusses small sample performance of these procedures, based on Monte Carlo evaluation. He introduces the sets

$$\Delta = \{i: \min(t_i, \tau) = t_i, i = 1, \ldots, n\}$$

and

$$\Gamma = \{i: \min(t_i, \tau) = \tau, i = 1, \ldots, n\}$$

and proposes estimators

$$\tilde{\theta} = X'_1 \tag{19.41a}$$

and

$$\tilde{\sigma} = \left[\Sigma_\Delta(X_i - \tilde{\theta}) + \Sigma_\Gamma \max(\tau - \tilde{\theta}, 0)\right], \qquad D > 0, \tag{19.41b}$$

and observed lifetimes t_i and fixed censoring time at τ.

The LR statistic for testing $\sigma = \sigma_0$ is

$$2D\{\tilde{\sigma}\sigma_0^{-1} - \log(\tilde{\sigma}\sigma_0^{-1}) - 1\}. \tag{19.42}$$

The asymptotic distribution of this statistic is χ^2 with one degree of freedom if $\theta = \theta_0$. To construct approximate $100(1 - \alpha)\%$ confidence limits for θ, Piegorsch (1987) suggests solving the equations

$$\tilde{\sigma}\sigma^{-1} - \log(\tilde{\sigma}\sigma^{-1}) = 1 + \chi^2_{1,1-\alpha}(2D)^{-1}.$$

This equation always has two solutions, one for $\tilde{\sigma}/\sigma < 1$ and the other for $\tilde{\sigma}/\sigma > 1$. These solutions L_σ and U_σ are then used to construct an asymptotic $100(1 - \alpha)\%$ confidence interval on σ of the form

$$\frac{\tilde{\sigma}}{U_\sigma} \le \sigma \le \frac{\tilde{\sigma}}{L_\sigma}. \tag{19.43}$$

A similar approach yields asymptotic $100(1 - \alpha)\%$ confidence limits for θ. For smaller sample sizes ($n \le 10$), the conditional inference on σ given by Wright, Engelhardt, and Bain (1978) may be preferred. However, Piegorsch's method for θ performs well even for $n = 5$ with an F-based approximation providing errors closer to nominal than the LR-based values particularly at $\alpha = 0.05$.

Joint regions for (σ, θ) can be constructed using Bonferroni's approach, with a rectangular region corresponding to the cross-product of univariate $1 - \frac{\alpha}{2}$ confidence intervals on θ and σ (the elliptic procedure using the asymptotic normality of the MLEs breaks down, yielding a hyperbolic rather than an elliptic region). Ranking and subset selection procedures for exponential population with Type I (and Type II) censored data are discussed in Berger and Kim (1985).

We will now restrict ourselves to Type II censored samples only. We will give details only for the case where $\theta = 0$ [so that the probability density function is as in (19.28)] and where censoring results in omission of the largest $n - k$ values (i.e., observation of the k smallest values, where k is specified prior to obtaining the observations). The joint probability density function of the k (> 1) smallest observations in a random sample of size n is

$$p_{X'_1,\ldots,X'_k}(x_1,\ldots,x_k) = \frac{n!}{(n-k)!\sigma^k} \exp\left(-\frac{\sum_{j=1}^{k-1}x_j + (n-k+1)x_k}{\sigma}\right),$$

$$0 \le x_1 \le x_2 \le \cdots \le x_k. \quad (19.44)$$

The maximum likelihood estimator of σ is

$$Y'_k = k^{-1}\left\{\sum_{j=1}^{k-1} X'_j + (n-k+1)X'_k\right\} = k^{-1}T_{k:n}. \quad (19.45)$$

This statistic is distributed as $(\frac{1}{2}\sigma/k) \times (\chi^2$ with $2k$ degrees of freedom). The expected value of Y'_k is therefore σ, and its variance is σ^2/k. The limits

$$\frac{2kY'_k}{\chi^2_{2k,1-\frac{1}{2}\alpha}} \quad \text{and} \quad \frac{2kY'_k}{\chi^2_{2k,\frac{1}{2}\alpha}}$$

define a confidence interval for σ with confidence coefficient $100(1-\alpha)\%$.

A wide variety of estimators of σ and θ based on order statistics is available. Many references at the end of this chapter contain discussions of such estimators. Among the problems discussed are estimation

1. by linear functions of a limited number (usually not more than about five) of order statistics—this includes both the choice of coefficients in the linear function, and of the order statistics to be used;
2. when only the k *largest* values are observed;
3. when predetermined numbers of both the smallest and largest values are omitted;
4. conversely, when only these values are known—that is, the middle group of observations is omitted.

In all cases formulas have been obtained appropriate to estimation of σ, knowing θ; of θ, knowing σ; and of both θ and σ, neither being known. We now discuss some of the more useful of these formulas.

The variance-covariance matrix of the order statistics $X'_1 \le X'_2 \le \cdots \le X'_n$ has elements

$$\text{Var}(X'_r) = \sigma^2 \sum_{j=1}^{r} (n-j+1)^{-2} = \text{Cov}(X'_r, X'_s), \quad r < s. \quad (19.46)$$

Also

$$E[X'_r] = \theta + \sigma \sum_{j=1}^{r} (n - j + 1)^{-1}. \tag{19.47}$$

From these relationships it is straightforward to construct best linear unbiased estimators based on k-selected order statistics $X'_{n_1}, X'_{n_2}, \dots, X'_{n_k}$ with

$$n_1 < n_2 < \cdots < n_k.$$

It will be convenient to use the notation

$$w_{mi} = \sum_{j=n_{i-1}}^{n_i - 1} (n - j)^{-m}, \tag{19.48}$$

with $n_0 = 0$, and w_{10}/w_{20}, $w_{1,k+1}/w_{2,k+1}$ each defined to be zero.
   If θ is known, the best linear unbiased estimator of σ is

$$\tilde{\sigma} = \left[\sum_{i=1}^{k} \left(\frac{w_{1i}}{w_{2i}} - \frac{w_{1,i+1}}{w_{2,i+1}} \right) X'_{n_i} - \frac{w_{11}}{w_{21}} \theta \right] \left[\sum_{i=1}^{k} \frac{w_{1i}^2}{w_{2i}} \right]^{-1}. \tag{19.49}$$

The variance of $\tilde{\sigma}$ is $\sigma^2 (\sum_{i=1}^{k} w_{1i}^2 w_{2i}^{-1})^{-1}$. Some special cases are

1. $n_i = i$, $k = n$ (complete sample):

$$\tilde{\sigma} = n^{-1} \sum_{i=1}^{n} X_i,$$

$$\mathrm{Var}(\tilde{\sigma}) = \frac{\sigma^2}{n}.$$

2. $n_i = r_1 + i$, $k = n - r_1 - r_2$ (censoring r_1 smallest and r_2 largest values):

$$\tilde{\sigma} = K^{-1} \left[\frac{\sum_{i=1}^{r_1+1} (n - i + 1)^{-1}}{\sum_{i=1}^{r_1+1} (n - i + 1)^{-2}} (X'_{r_1+1} - \theta) - (n - r_1) X'_{r_1+1} \right.$$

$$\left. + (n - r_1 - k) X'_{r_1+k} + \sum_{i=r_1+1}^{r_1+k} X'_i \right] \tag{19.50}$$

with

$$K = \left\{ \sum_{i=1}^{r_1+1} (n - i + 1)^{-1} \right\}^2 \left\{ \sum_{i=1}^{r_1+1} (n - i + 1)^{-2} \right\}^{-1} + k - 1.$$

(If $r_1 = 0$, this is the case of omission of the r_2 largest values only, already discussed. If $r_2 = 0$, it is the case of omission of the r_1 smallest values only.) Balakrishnan (1990) and Balasubramanian and Balakrishnan (1992) have derived similar linear estimators for the case when the available sample is multiply Type-II censored. Fei and Kong (1994) have discussed the interval estimation of parameters in this case, while Kong and Fei (1994) have established limit theorems for the maximum likelihood estimators based on multiply censored samples.

3. $n_i = 1$ for $i = 1, 2, \ldots, r_1$, $n_j = n - r_2 + j - r_1$ for $j = r_1 + 1, \ldots,$ $r_1 + r_2$, $k = r_1 + r_2$. This corresponds to censoring by omission of a central group of values, retaining only the r_1 smallest and r_2 largest values. In this case

$$\tilde{\sigma} = \left[r_1 + r_2 - 1 + \frac{w'^2}{w} \right]^{-1} \times \left\{ \sum_{i=1}^{r_1-1} X_i' + \left(n - r_1 + 1 - \frac{w'}{w} \right) X_{r_1}' \right.$$

$$\left. + \left(\frac{w'}{w} - r_2 + 1 \right) X_{r_2+1}' + \sum_{i=r+2}^{n} X_i' - n\theta \right\}, \tag{19.51}$$

where

$$w' = \sum_{i=r_1+1}^{n-r_2+1} (n - i + 1)^{-1},$$

$$w = \sum_{i=r_1+1}^{n-r_2+1} (n - i + 1)^{-2}.$$

4. $k = 1$ (estimation from a single order statistic X_{n_1}'). In this case

$$\tilde{\sigma} = \left[\sum_{i=1}^{n_1} (n - i + 1)^{-1} \right]^{-1} (X_{n_1}' - \theta), \tag{19.52a}$$

$$\mathrm{Var}(\tilde{\sigma}) = \sigma^2 \left[\sum_{i=1}^{n_1} (n - i + 1)^{-2} \right] \left[\sum_{i=1}^{n_1} (n - i + 1)^{-1} \right]^{-2}. \tag{19.52b}$$

The variance is a minimum (with respect to n_1), when n_1 is the nearest integer to $0.79681(n + 1) - 0.39841 + 1.16312(n + 1)^{-1}$ [Siddiqui

(1963)]. Epstein (1962) has noted that the efficiency of the unbiased estimator based on $X'_r - \theta$ is never less than 96% if $r/n \leq \frac{1}{2}$, or 90% if $r/n \leq \frac{2}{3}$.

5. $k = 2$ (estimation from two order statistics X'_{n_1}, X'_{n_2}). The variance of $\tilde{\sigma}$ is a minimum when n_1 is the nearest integer to $0.6386(n + \frac{1}{2})$ and n_2 is the nearest integer to $0.9266(n + \frac{1}{2})$ [Siddiqui (1963)].

For small samples Sarhan and Greenberg (1958) give the following optimal choices:

Sample Size (n)	2–4	5–7	8–11	12–15	16–18	19–21
n_1	$n - 1$	$n - 2$	$n - 3$	$n - 4$	$n - 6$	$n - 7$
n_2	n	n	n	n	$n - 1$	$n - 1$

If θ is not known, the optimal choices are different. Sarhan and Greenberg (1958) give the following optimal choices for n_1 and n_2 (with $k = 2$):

Sample Size (n)	2–6	7–10	11–15	16–20	21
n_1	1	1	1	1	1
n_2	n	$n - 1$	$n - 2$	$n - 3$	$n - 4$

The best (in fact only) linear unbiased estimator of σ, using only X'_{n_1} and X'_{n_2} is

$$\left(X'_{n_2} - X'_{n_1} \right) \left[\sum_{j=n_1}^{n_2-1} (n - j)^{-1} \right]^{-1} \qquad (19.53)$$

and its variance is

$$\left[\sum_{j=n_1}^{n_2-1} (n - j)^{-2} \right] \left[\sum_{j=n_1}^{n_2-1} (n - j)^{-1} \right]^{-2} \sigma^2. \qquad (19.54)$$

The optimal value of n_1 is always 1, whatever the values of n and n_2.

Kulldorff (1962) considered the problem of choosing $n_1, n_2, \ldots, n_k$ to minimize $\mathrm{Var}(\tilde{\sigma})$ or $\mathrm{Var}(\tilde{\theta})$ for k fixed. He tabulated the optimal n_i's and coefficients for small values of k and n. Earlier, Sarhan and Greenberg (1958) treated the asymptotic (large n) case (for θ known) giving the optimal percentiles $n_1/n, n_2/n, \ldots, n_k/n$, for k fixed. These tables are reproduced by Ogawa (1962). Saleh and Ali (1966) and Saleh (1966) proved the uniqueness of the optimal selection and extended the results to censored cases.

The table of Zabransky, Sibuya, and Saleh (1966) is most exhaustive and covers uncensored and censored samples, finite and asymptotic cases for a wide range. Sibuya (1969) gave the algorithms for computing their tables, and unified previous results in simpler form.

Epstein (1962) gave a number of useful results. He pointed out that $X'_r - \theta$ is approximately distributed as $\frac{1}{2}(\sigma/r) \times (\chi^2$ with $2r$ degrees of freedom). Approximate confidence intervals for σ can be constructed on this basis. Epstein also gave formulas for the minimum variance unbiased estimators of θ and σ for the two-parameter distribution (19.1), based on $X'_1, X'_2, \ldots, X'_k$. These are

$$\sigma^* = (k-1)^{-1} \sum_{j=1}^{k-1} (n-j)(X'_{j+1} - X'_j), \qquad (19.55\text{a})$$

$$\theta^* = X'_1 - n^{-1}\sigma^*; \qquad (19.55\text{b})$$

σ^* is distributed as $\frac{1}{2}\sigma(k-1)^{-1} \times [\chi^2$ with $2(k-1)$ degrees of freedom]. $100(1-\alpha)\%$ confidence limits for σ are

$$\frac{2(k-1)\sigma^*}{\chi^2_{2(k-1), 1-\frac{\alpha}{2}}} \quad \text{and} \quad \frac{2(k-1)\sigma^*}{\chi^2_{2(k-1), \frac{\alpha}{2}}}.$$

$100(1-\alpha)\%$ confidence limits for θ are

$$X'_1 - F_{2, 2(k-1), 1-\alpha}\sigma^* n^{-1} \quad \text{and} \quad X'_1 \qquad (19.56)$$

(using the notation of Chapter 27, Section 1).

If only $X'_{k_1+1}, \ldots, X'_{n-k_2}$ are to be used, the minimum variance unbiased estimators are

$$\sigma^* = (n - k_1 - k_2 - 1)^{-1} \sum_{j=k_1+1}^{n-k_2-1} (n-j+1)(X'_{j+1} - X'_j), \quad (19.57\text{a})$$

$$\theta^* = X'_{k_1+1} - \sigma^* \sum_{j=0}^{k_1} (n-j)^{-1}; \qquad (19.57\text{b})$$

σ^* is distributed as $\frac{1}{2}\sigma(n - k_1 - k_2 - 1)^{-1} \times [\chi^2$ with $2(n - k_1 - k_2 - 1)$ degrees of freedom] and

$$\text{Var}(\theta^*) = \sigma^2 \left[\sum_{j=0}^{k_1} (n-j)^{-2} + (n - k_1 - k_2 - 1)^{-1} \left\{ \sum_{j=0}^{k_1} (n-j)^{-1} \right\}^2 \right].$$

$$(19.58)$$

The minimum variance unbiased estimator (MVUE) of the reliability function based on $X'_1, \ldots, X'_k$ is

$$
\hat{R}(t) = \begin{cases} \left(1 - \dfrac{t}{T_{k:n}}\right)^{k-1} & \text{if } t \le T_{k:n}, \\ 0 & \text{if } t \ge T_{k:n}. \end{cases} \tag{19.59}
$$

$[T_{k:n}$ is defined in (19.45)].

The mean square error of $\hat{R}(t)$ is

$$
MSE(\hat{R}(t)) = \int_c^\infty \left(1 - \frac{c}{u}\right)^{2r-2} \gamma(u; r)\, du
$$

where

$$
\gamma(u; r) = \frac{1}{\Gamma(r)} u^{r-1} e^{-u},
$$

$$
c = \frac{t}{\sigma}
$$

[Basu (1964)]. Based on past data, one may have a prior (guessed) value of σ, σ_0, say, that can be utilized for statistical inference. The so-called *shrinkage estimators* arising in this situation perform better than the MVUE if the value σ_0 is close to σ. A preliminary test can be conducted to check the closeness of σ_0 to σ. Chiou (1987) proposes incorporating a preliminary test on σ, using an estimator of type

$$
R_{PT}(t) = \begin{cases} \exp\left(-\dfrac{t}{\sigma_0}\right) & \text{if } H_0: \sigma = \sigma_0 \text{ is not rejected,} \\ \hat{R}(t) & \text{otherwise.} \end{cases} \tag{19.60}
$$

Since $2T_{k:n}/\sigma$ has a χ^2 distribution with $2k$ degrees of freedom, H_0 is not rejected if

$$
C_1 \le \frac{2T_{k:n}}{\sigma_0} \le C_2,
$$

where C_1 (C_2) are the $\frac{\alpha}{2}$ lower (upper) percentage points of χ^2_{2k}. Thus

$$
\hat{R}_{PT}(t) = \begin{cases} \exp\left(-\dfrac{t}{\sigma_0}\right) & \text{if } C_1 \le \dfrac{2T_{k:n}}{\sigma_0} \le C_2, \\ \hat{R}(t) & \text{otherwise.} \end{cases} \tag{19.61}
$$

Chiou (1987) provides optimal critical values for the preliminary test and their corresponding level of significance, based on the minimax regret criterion.

Alternatively, Chiou (1992) proposes the "shrinkage estimator"

$$\hat{R}_2(t) = \begin{cases} \omega \exp\left(-\dfrac{t}{\sigma_0}\right) + (1 - \omega)\hat{R}(t) & \text{if } C_1 \le \dfrac{2T_{k:n}}{\sigma_0} \le C_2 \\ \hat{R}(t) & \text{otherwise,} \end{cases} \quad (19.62)$$

where ω is a shrinkage coefficient suitably specified. Chiou provides a table of optimum values of shrinkage coefficient ω for $t/\sigma_0 = 0.25(0.25)2.0$ and $k = 4(2)10$ as well as critical values C_1 and C_2 for the preliminary test.

Zacks and Even (1966) compared the performance of MVUE and the maximum likelihood estimator in terms of mean square error for small samples. The MLE is more efficient than MVUE over the interval $0.5 < t/\sigma < 3.5$. Over the "effective intervals where σ_0/σ is in the vicinity of 1 [(0.7, 1.4) for $r = 4$ and $t/\sigma_0 = 2$ as an example], Chiou's (1992) shrinkage estimators are more efficient, but none of the estimators for $R(t)$ investigated so far is uniformly better than others over the whole possible range of σ/σ_0."

Cohen and Whitten (1988) discuss estimation of parameters in the case of *progressively censored samples*. Censoring is progressively carried out in k stages at times $\tau_1 < \tau_2 < \cdots < \tau_j < \cdots < \tau_k$. At jth stage of censoring, C_j sample specimens, selected randomly from the survivors at time τ_j, are removed. In addition we have n full-term observations and our sample consists of $(X_1, X_2, \ldots, X_n)$ plus k partial term observations $\{C_j\tau_j\}$ ($j \ge 1, \ldots, k$). Thus $N = n + r$, where $r = \sum_1^k C_j$ ($C_j \ge 1$ corresponds to censoring and $C_j = 0$ to noncensoring), and the *sum total* (ST) term of all N observations in the sample is

$$ST = \sum_{i=1}^{n} X_i + \sum_{j=1}^{k} C_j\tau_j.$$

The modified maximum likelihood estimators (MMLEs) of σ and θ in the case of progressive censoring, obtained by solving

$$\frac{\partial \ln L}{\partial \sigma} = 0 \quad \text{and} \quad E[X_1'] = X_1'$$

are given by

$$\tilde{\sigma} = \frac{ST - N\tilde{\theta}}{n}$$

$$\tilde{\theta} = \frac{ST - NX_1'}{n - 1}. \quad (19.63a)$$

Explicitly

$$\tilde{\sigma} = \frac{ST - NX_1'}{n - 1},$$

$$\tilde{\theta} = \frac{nX_1' - N^{-1}ST}{n - 1}. \tag{19.63b}$$

(For the uncensored case $C_j = 0$, $n = N$, and $ST = n\overline{X}$.)

Estimation of parameters from general location-scale families (of which the two-parameter exponential distribution is a member) under progressive censored sampling was studied by Viveros and Balakrishnan (1994). They follow the classical arguments for conditional inference for location and scale parameters as expounded, for example, in Lawless (1982).

A modified hybrid censoring model was investigated by Zacks (1986). Let τ_0 be the fixed time at which the (Type I) censoring occurs. Let $X_{k:n}'$ denote the kth-order statistic of a sample of n i.i.d. random variables from the one-parameter exponential distribution (19.28).

Let $X_k^* = \min(X_{k:n}', \tau_0)$. *The recorded* random variable is the time-censored kth-order statistic. Let $X_{k1}^*, \ldots, X_{km}^*$ denote m i.i.d. random variables distributed as X_k^*. Zacks (1986) shows the existence and uniqueness of a MLE of σ based on the sample $\mathbf{X}_{k,m}^* = (X_{k1}^*, \ldots, X_{km}^*)$.

He also investigates the properties of the moment estimator (ME) that is the root of the equation in σ

$$\sigma \sum_{j=0}^{k-1} \frac{1}{n-j} \left[1 - B(j; n; e^{-\tau_0/\sigma}) \right] = \overline{X}_k^* = m^{-1} \sum_{j=1}^{m} X_{kj}^*, \tag{19.64}$$

where $B(j; n; p)$ is the cdf of the binomial distribution with parameters (n, p).

Numerical comparisons show that in the case of MLE "censoring has a dramatic effect on the possibility of estimating σ efficiently. The efficiency of the MLE drops to almost zero when σ is in the neighborhood of τ_0." The asymptotic relative efficiency of ME relative to MLE is an increasing function of $\tau_0/\sigma \equiv \eta$. When $\eta = 1$, the ARE of the ME is about 62%, whereas at $\eta = 3$ it is 99%. On the other hand, under censoring, the MLE is considerably more efficient than the ME when η is close to 1.

7.4   Estimation of Quantiles

Robertson (1977), in an important paper, develops estimation procedures for quantiles of the exponential distribution (19.28) such that mean square errors in the *predicted* distribution function is minimized. For complete random samples of size n, a particular quantile, represented as $K\sigma$, has estimators of the form $K^*\overline{X}$. The optimal estimator of $K\sigma$ (with squared error loss) is in fact $K_0\overline{X}$, where

$$K_0 = n\left[\frac{\exp\{K/(n+1)\} - 1}{2 - \exp\{K/(n+1)\}}\right]. \tag{19.65}$$

Its mean square error is

$$e^{-2K} - e^{-K}\left[2 - \exp\left(\frac{K}{n+1}\right)\right]^{n+1}. \tag{19.66}$$

For a linear estimator $\sum_{i=1}^{n} a_i X_i$, the mean square error is

$$\int_0^\infty \cdots \int_0^\infty \{e^{-K} - e^{\Sigma a_i x_i}\}^2 e^{-\Sigma x_i} \, dx_1 \, dx_2 \cdots dx_n$$

$$= e^{-2K} - 2e^{-K} \prod_{i=1}^{n} (1 + a_i)^{-1} + \prod_{i=1}^{n} (1 + 2a_i)^{-1} \tag{19.67}$$

where $\sigma = 1$ for convenience, yielding minimizing values

$$a_i = \frac{[\exp\{K/(n+1)\} - 1]}{2 - \exp\{K(n+1)\}} \qquad \text{for all } i. \tag{19.68}$$

For $n < (K/\log 2) - 1$, this estimator will break down, and the optimum choice will be $K_0 = \infty$. [Indeed one cannot reasonably expect to estimate the 99% point ($K = 4.605$) using a sample of size 5 [$< (K/\log 2) - 1$).] Comparison of $K_0\overline{X}$ with $K\overline{X}$ (the natural K) and $K_2\overline{X}$ [chosen to make $\exp(-K_2\overline{X})$ an unbiased estimator of $\exp(-K\sigma)$] has been carried out by Robertson (1977). For small n or large K (extreme percentiles) there are considerable differences among the three estimators.

To estimate $K\sigma$ by means of a *single* order statistic the estimator is of the form $K_3 X_r'$, where the optimal choice for large n turns out to be $r \doteq \alpha n$ with $\alpha = 0.79681213$ (the positive root of equation $e^{-2\alpha} = 1 - \alpha$) and

$$K_3 = \frac{K}{2\alpha} + \frac{3K^2 - 2(1 - \alpha)K}{16\alpha^2(1 - \alpha)n} + O(n^{-2}). \tag{19.69}$$

For large n, $K_3 X_r'$ has efficiency approximately 65% [$\doteq 4\alpha(1 - \alpha)\%$] relative to $K_0\overline{X}$, the optimal estimator based on the complete sample.

7.5 Bayesian Estimation

One-Parameter Exponential Distributions
In Bayesian estimation prior distributions are often assigned to the hazard rate ($\lambda = \sigma^{-1}$) rather than to the expected value (σ). We therefore use the

pdf

$$p_X(x|\lambda) = \lambda e^{-\lambda x} \qquad (x > 0; \lambda > 0). \qquad (19.28)''$$

If the prior on λ is uniform over $(0, M)$, then the posterior density of λ, given a random sample $X_1, X_2, \ldots, X_n$ is

$$\frac{\lambda^n \exp(-\lambda \Sigma_{i=1}^n X_i)}{\int_0^M \lambda^n \exp(-\lambda \Sigma_{i=1}^n X_i)\, d\lambda}, \qquad 0 < \lambda < M. \qquad (19.70)$$

For large M this is approximately a gamma $(n + 1, \{\Sigma_{i=1}^n X_i\}^{-1})$ distribution with pdf

$$\frac{(\Sigma_{i=1}^n X_i)^{n+1} \lambda^n}{\Gamma(n+1)} \exp\left(-\lambda \sum_{i=1}^n X_i\right). \qquad (19.71)$$

If this posterior pdf is used as a new prior, and a random sample $Y_1, Y_2, \ldots, Y_m$ obtained, the new posterior is of the same form as (19.71), with n increased to $(n + m)$ and $\Sigma_{i=1}^n X_i$ increased by $\Sigma_{j=1}^m Y_j$. Thus the gamma prior is a "natural conjugate" prior for λ [see Barlow and Proschan (1979)].

Waller et al. (1977) develop an interesting and potentially fruitful approach to determination of the parameters of a gamma (α, β) prior on λ in the pdf (19.28)$''$, based on values of

$$\Pr[\lambda < \lambda_j] = p_j, \qquad j = 1, 2; \lambda_1 < \lambda_2,$$

decided upon by a researcher. Extensive tables and graphs are provided giving values of α and β for selected values λ_0 and p_0 for which

$$\Pr[\lambda < \lambda_0] = p_0. \qquad (19.72)$$

By overlapping transparencies of graphs for (λ_1, p_1) and (λ_2, p_2), appropriate values of α and β can be determined. If (19.72) is satisfied then λ_0/β is constant for given α. For small values of α, Waller et al. (1977) recommended the approximation $\lambda \sim \beta p^{1/\alpha}$.

For situations where only the first k "failures" are observed, with k specified *in advance* (Type II censoring), the MLE of λ is

$$\hat{\lambda} = kT_k^{-1}, \qquad (19.73)$$

with $T_k = \Sigma_{i=1}^k X_i' + (n - k)X_k'$. With a gamma (a, b^{-1}) prior on λ, we arrive at a posterior gamma $(a + k, (b + T_k)^{-1})$ distribution of λ. Similar results are obtained for truncated sampling (Type I censoring), and also for inverse binomial sampling (when a failed or truncated unit is replaced).

If λ^{-1} $(= \sigma)$ is ascribed a "uniform" distribution, the posterior distribution for σ would have pdf

$$\frac{(\sum_{i=1}^{n} X_i)^{n-1}}{\Gamma(n-1)} \left(\frac{1}{\sigma}\right)^n \exp\left(-\frac{1}{\sigma} \sum_{i=1}^{n} X_i\right), \qquad \sigma > 0, \qquad (19.74)$$

which is called the *inverted gamma* distribution.

Since an inverted gamma prior on σ also results in an inverted gamma posterior distribution for σ, it is a natural conjugate in this situation. This is not surprising since, if λ has a gamma distribution, σ $(= \lambda^{-1})$ has an inverted gamma distribution.

Bayesian estimation of the reliability function

$$R(t) = \Pr[X > t] = \exp\left(-\frac{t}{\sigma}\right) \qquad (19.29)'$$

has been studied by Sinha and Guttman (1976). Assuming a so-called vague Jeffreys's prior density (proportional to σ^{-1}) on σ, and using only the first k order statistics, they show that the posterior density of $R(t)$ is

$$\frac{S_k^k}{\Gamma(k)} \{-\log R(t)\}\{R(t)\}^{S_k-1}, \qquad R(t) \geq 0, \qquad (19.75)$$

where $S_k = T_k/t$, which can be regarded as the "cumulative life per unit time" up to time t. The posterior density for σ in this situation is

$$\frac{T_k^k}{\Gamma(k)} \sigma^{-(k+1)} \exp\left(-\frac{T_k}{\sigma}\right). \qquad (19.76)$$

See also Shoukri (1983) for further discussions.

Villén-Altamizano (1990) used a gamma (p, a^{-1}) prior for λ $(= \sigma^{-1})$, obtaining the posterior density of $R(1) = \exp(-\sigma^{-1})$ given the results (W_j, I_j) $(j = 1, \ldots, n)$ from a randomly censored sample of size n with lifetimes $\{X_j\}$ and censoring times $\{Y_j\}$, where

$$W_j = \min(X_j, Y_j)$$

and

$$I_j = I(X_j \le Y_j) = \begin{cases} 1 & \text{if } X_j \le Y_j, \\ 0 & \text{if } X_j > Y_j, \end{cases}$$

in the form

$$p_R(r|W, I) = \frac{(a + W)^{p+I}}{\Gamma(p + 1)} r^{a+W-1}(-\log r)^{p+I-1}, \qquad 0 < r < 1, \quad (19.77)$$

where $W = \sum_{j=1}^{n} W_j$ and $I = \sum_{j=1}^{n} I_j$.
   The posterior expected value is

$$\tilde{R}_2 = E[R|W, I] = \left(\frac{a + W}{a + W + 1}\right)^{p+I}. \qquad (19.78)$$

This is the Bayesian estimator (optimal for quadratic loss). The mode of the posterior distribution is

$$\tilde{R}_3 = \begin{cases} \exp\left(-\dfrac{p + I - 1}{a + W - 1}\right) & \text{if } W > 1 - a, \\ 0 & \text{otherwise.} \end{cases} \qquad (19.79)$$

Two-Parameter Exponential Distributions
For the two-parameter distribution

$$p_X(x) = \sigma^{-1} \exp\left(-\frac{x - \theta}{\sigma}\right), \qquad x \ge \theta > 0; \sigma > 0, \qquad (19.1)'$$

Sinha and Guttman (1976) ascribed a joint prior to θ and σ that is proportional to σ^{-a} ($a > 0$), and obtained the following results for Type I censoring using the first k order statistics in a random sample of size n. The posterior expected values (Bayes estimators) of θ and σ are

$$E[\theta|\mathbf{X}] = \frac{C_k\{n(k + a - 3)X_1' - C_{k-1}^{-1}S_k'\}}{n(k + a - 3)}, \qquad (19.80)$$

$$E[\sigma|\mathbf{X}] = \frac{C_k S_k'}{C_{k-1}(k + a - 3)}, \qquad (19.81)$$

where

$$S'_k = \sum_{i=1}^{k} (X'_i - X'_1) + (n - k)(X'_k - X'_1),$$

$$C_k = \left\{ 1 - \left(1 + \frac{nX'_1}{S'_k} \right)^{-(k+a-2)} \right\}^{-1}.$$

For the reliability function

$$R(t) = \exp\left\{ -\frac{t - \theta}{\sigma} \right\},$$

we have

$$E[R(t)|\mathbf{X}] = \frac{n}{n+1} C_k \left\{ \left(1 + \frac{t - X'_1}{S_k} \right)^{-(k+a-2)} - \left(1 + \frac{t + nX'_1}{S_k} \right)^{-(k+a-2)} \right\}, \qquad t > X'_1. \quad (19.82)$$

This estimator was found to be reasonably robust with respect to a. The posterior distribution of $R(t)$ is quite complicated [see also Pierce (1973)]. Trader (1985) used a truncated normal prior.

7.6 Miscellaneous

Maximum probability estimation (MPE) is a method of estimation proposed by Weiss and Wolfowitz (1967) [see also Weiss (1985)]. The essential idea is to seek an interval of predetermined length l, say, that maximizes the integral of the likelihood function over the interval. In the case of the two-parameter exponential distribution (19.1), the analysis for estimation of θ takes an especially simple form. Given a random sample of size n with values $X_1, \ldots, X_n$, the likelihood function is

$$L(\mathbf{X}|\theta, \sigma) = \begin{cases} \sigma^{-n} \exp\left\{ -\dfrac{\sum_{i=1}^{n}(X_i - \theta)}{\sigma} \right\}, & \sigma > 0, \theta \le \min X_i = X'_1, \\ 0, & \theta > X'_1. \end{cases}$$

$$(19.83)$$

Since $L = 0$ for $\theta > X_1'$ and L is an increasing function of θ for $\theta \leq X_1'$, the MPE interval of length l is

$$(X_1' - l, X_1') \tag{19.84}$$

(whatever the value of σ). Blyth (1982) followed this analysis by considering choice of attractive values for l, if $(X_1' - \frac{1}{2}l)$ is used as a point estimator of θ. He found that

1. for minimum mean square error, $\frac{1}{2}l = n^{-1}\sigma$,
2. for minimum absolute error, $\frac{1}{2}l = n^{-1}\sigma \log 2$.

Introducing a quadratic loss function, Arnold (1970), and independently Zidek (1973) and Brewster (1974), derived optimal estimators of σ for cases where θ is unknown. The MLE of σ, $\hat{\sigma} = \bar{X} - X_1'$, is also the best affine equivariant estimator (BAEE) of σ. However, the estimator is inadmissible for a wide class of loss functions satisfying conditions of differentiability and bowl-shapedness. Modified estimators, aimed at reducing this drawback, were suggested by the above authors.

Estimators of the ratio $(\delta = \sigma_2/\sigma_1)$ of scale parameters of two distributions π_1, π_2 of form (19.1) with values (σ_j, θ_j) $(j = 1, 2)$ for the parameters (σ, θ), respectively, were investigated by Madi and Tsui (1990). Suppose that there are two independent ordered random sample values

$$X_{11}' \leq X_{12}' \leq \cdots \leq X_{1n_1}' \quad \text{from } \pi_1$$

and

$$X_{21}' \leq X_{22}' \leq \cdots \leq X_{2n_2}' \quad \text{from } \pi_2.$$

With loss function $W(\delta, \delta^*)$ whence δ^* is an estimator of δ, the statistics

$$\left(\bar{X}_1 - X_{11}', \bar{X}_2 - X_{21}'\right)\left(\frac{X_{11}'}{\bar{X}_1 - X_{11}'}, \frac{X_{21}'}{\bar{X}_2 - X_{21}'}\right)$$

are sufficient for the four parameters σ_1, σ_2, θ_1, and θ_2. Assuming that $W(\delta, \delta^*) = W(\delta^*/\delta)$, the BAEE estimator is

$$\delta^* = \begin{cases} \dfrac{\bar{X}_1 - X_{11}'}{\bar{X}_2 - X_{21}'} \min\left[c_0, c_1 \dfrac{\bar{X}_1}{\bar{X}_1 - X_{11}'}\right] & \text{if } X_{11}' > 0, \\[4mm] \dfrac{\bar{X}_1 - X_{11}'}{\bar{X}_2 - X_{21}'} c_0 & \text{if } X_{11}' \leq 0, \end{cases} \tag{19.85}$$

where c_0 minimizes $\int_0^\infty t^{n_1-2} W(ct)(1 + n_1 t/n_2)^{-(n_1+n_2-2)} dt$ and c_1 minimizes $\int_0^\infty t^{n_1-1} W(ct)(1 + n_1 t/n_2)^{-(n_1+n_2-1)} dt$ provided that

1. W is differentiable,
2. $W(y)$ is bowl-shaped [i.e., $W(y)$ decreases with y for $y \le y_0$, increases for $y > y_0$], attaining its minimum value at $y = 1$,
3. if $\sigma_1 = \sigma_2 = 1$, then

$$E\left[\left| W'\left(c \frac{\overline{X}_1 - X'_{11}}{\overline{X}_2 - X'_{21}} \right) \right|\right] \quad \text{is finite for all } c > 1.$$

A "smooth" version of δ^* can be obtained by replacing the multiplier of $(\overline{X}_1 - X'_{11})/(\overline{X}_2 - X'_{21})$ by a more refined function. The resulting estimator has

$$c_0 = 1 - 3n_2^{-1},$$

$$c_1 = n_2^{-1}(n_1 + 1)^{-1} n_1(n_2 - 3) \quad \text{for } n_2 > 3 \quad \text{and } W\left(\frac{\delta^*}{\delta}\right) = 1 - \left(\frac{\delta^*}{\delta}\right)^2;$$

$$(19.86a)$$

$$c_0 = n_2^{-1}(n_1 - 1)^{-1} n_1(n_2 - 2),$$

$$c_1 = 1 - 2n_2^{-1} \quad \text{for } W\left(\frac{\delta^*}{\delta}\right) = \left(\frac{\delta^*}{\delta}\right) - \log\left(\frac{\delta^*}{\delta}\right) - 1. \quad (19.86b)$$

Effects of Outliers

The effects of outliers on the estimation of σ in the one-parameter case (19.28) has received considerable attention in the literature since the early 1970s. Among the first studies were those of Kale and Sinha (1971) and Joshi (1972). They initially studied situations where $n - 1$ independent random variables each had distribution (19.28), while one further random variable (the "outlier") has a pdf of the same form, but with σ replaced by σ/α.

For situations where the identity of the outlier is not known (and the probability that X_j is the outlier is n^{-1} for $j = 1, \ldots, n$), Kale and Sinha (1971) suggest consideration of the class of estimators (for σ):

$$\tilde{\sigma}_k = \frac{T_k}{k + 1}, \quad (19.87)$$

where T_k is defined as in (19.73). Joshi (1972) tabulates optimum values for k, minimizing the mean square error. If $k = n$ [with $\tilde{\sigma}_n = \sum_{i=1}^{n} X_i/(n + 1)$], we have

$$\text{MSE}(\tilde{\sigma}_n/\sigma) = (n + 1)^{-1} + 2(\alpha^{-1} - 1)^2(n + 1)^{-2}, \quad (19.88)$$

where σ/α is the mean of the outlier.

Generally, the optimum k has to be found numerically. Joshi (1972) found that for $0.55 \leq \alpha \leq 1$ the optimal value of k is n. Later [Joshi (1988)] he found that the optimal value of k is fairly stable over wide ranges of values of α and suggested, as a rule of thumb,

Take $k = n$ for $0.5 \leq \alpha < 1.0$
Take $k = n - 1$ for $0.25 \leq \alpha < 0.5$
Take $k = n - 2$ for $0.05 \leq \alpha \leq 0.25$.

Chikkagoudar and Kunchur (1980) suggest the estimator

$$V = \sum_{j=1}^{n} \left(\frac{1}{n} - \frac{2j}{n^2(n + 1)} \right) X_j'. \tag{19.89}$$

Comparison between $\tilde{\sigma}_k$ and V shows that neither estimator dominates the other (in terms of mean square error loss). [T_k (for optimal k) is superior for values of α near 0 or 1.] Balakrishnan and Barnett (1994) have recently proposed some generalized estimators of this form and discussed their properties.

From (1991) has studied robust estimators that are general linear functions of order statistics, aiming to obtain optimal (or near-optimal) values for the coefficients c_j in $\sum_{j=1}^{n} c_j X_j'$, for various values of α and n. Optimal choice calls for rather complicated calculations, so From advocates use of simplified estimators with very nearly optimal mean square error, which is always less than those of $\tilde{\sigma}_k$ and V. This estimator uses

$$c_j = \begin{cases} d_1 & \text{for } 1 \leq j \leq m, \\ d_2 & \text{for } m + 1 \leq j \leq n, \end{cases}$$

with $d_1 > d_2$ and appropriate integer m. Tables are provided giving optimal values of (d_1, d_2, m) for $n = 2(1)15(5)30(10)50$ and $\alpha = 0.05, 0.15$.

If the value of α is known, Joshi (1972) suggests using the estimator

$$(n - 1 + \alpha^{-1})\tilde{\sigma}_{n-1}/n. \tag{19.90}$$

He also suggests estimating α (if it is unknown) by an iterative procedure, solving the equation

$$n\tilde{\sigma}_n = (n - 1 + \alpha^{-1})\tilde{\sigma}_{n-1} \tag{19.91}$$

to estimate α, and then using optimal k to get a new estimator $\tilde{\sigma}_n$ of σ, and so on; see also Jevanand and Nair (1993).

Through a systematic study of order statistics from independent and non-identically distributed exponential random variables, Balakrishnan (1994) has established several recurrence relations for the single and the product moments of order statistics. These results will enable the computation of all the single and product moments of order statistics arising from a sample containing p outliers (recursively in p by starting with $p = 0$ or the i.i.d. case). Through this recursive computational process, Balakrishnan (1994) has extended the work of Joshi (1972) to the p-outlier case and determined the optimal trimmed and Winsorized estimators. The robustness properties of various linear estimators, including the Chikkagoudar–Kunchur estimator in (19.89), have also been examined by Balakrishnan (1994).

Veale (1981) has investigated cases where the identity of the outlier is known. This of course leads to much simpler analysis.

Returning to uncontaminated (no-outliers) data, the estimation of σ in the standard (one-parameter) case (19.28), subject to the condition that σ is no less than σ_0, was studied by Gupta and Basu (1980). [Related cases $\sigma < \sigma_0$, or σ in (σ_0, σ_1) can be analyzed similarly by using appropriate transformations.]

Natural estimators of σ (given $\sigma > \sigma_0$) are

$$\hat{\sigma} = \max(\overline{X}, \sigma_0) \qquad \text{(the MLE)} \qquad (19.92)$$

or

$$\sigma^* = n^{-1} \sum_{j=1}^{n} \max(X_j, \sigma_0). \qquad (19.93)$$

Numerical studies show that the mean square error (MSE) of σ^* is less than that of $\hat{\sigma}$ for small n, when σ_0/σ is small. Indeed, even for $n = 30$, σ^* has the smaller MSE if $\sigma_0/\sigma \leq 0.3$.

Estimation of the probability $P = \Pr[Y < X]$, where X and Y are independent exponential variables, has received prominent attention in the literature. The common interpretations of this probability is a measure of the reliability or performance of an item of strength Y subject to a stress X, or probability that one component fails prior to another component of some device.

Tong (1974, 1975) obtained the uniformly minimum variance unbiased estimator (UMVUE) of P when X and Y are independent one-parameter exponential (19.28) variables. Kelley, Kelley, and Schucany (1976) derived the variance of the UMVUE of P. Beg (1980) obtained the UMVUE of P when X and Y have two-parameter exponential (19.1) distributions with unequal scale and location parameters. Gupta and Gupta (1988) obtained the MLE, the UMVUE, and a Bayesian estimator of P when the location parameters are unequal but there is a common scale parameter. Bartoszewicz (1977) tackled the problem in the exponential case for different types of censoring. Reiser, Faraggi, and Guttman (1993) discuss the choice of sample sizes for the experiments dealing with inference on $\Pr[Y < X]$ in an acceptance sampling theory framework with exponential variables. Most recently Bai and Hong (1992) revisited the problem, obtaining the UMVUE of P with unequal sample sizes when X and Y are independent two-parameter exponential random variables with an unknown common location parameter. Kocherlakota and Balakrishnan (1986) have discussed one-sided as well as two-sided acceptance sampling plans based on Type-II censored samples.

If X and Y are independent one-parameter exponential (19.28) random variables with hazard rates (σ^{-1}) λ and μ, respectively, then

$$P = \Pr[Y < X] = \mu(\lambda + \mu)^{-1}. \tag{19.94}$$

Given two independent random samples $X_1, \ldots, X_m$ and $Y_1, \ldots, Y_n$, the MVUE of P is

$$P^* = \begin{cases} \displaystyle\sum_{j=0}^{m-1} (-1)^j \frac{(m-1)^{(j)}}{n^{[j]}} \left(\frac{n\overline{Y}}{m\overline{X}} \right)^j & \text{if } n\overline{Y} \le m\overline{X}, \\[4mm] 1 - \displaystyle\sum_{j=0}^{n-1} (-1)^n \frac{(n-1)^{(j)}}{m^{[j]}} \left(\frac{m\overline{X}}{n\overline{Y}} \right)^j & \text{if } n\overline{Y} > m\overline{X}. \end{cases} \tag{19.95}$$

[Tong (1974)] where $a^{(b)} = a(a-1)\cdots(a-b+1)$, $a^{[b]} = a(a+1)\cdots$ $(a+b-1)$.

If μ is known so that there is no need to sample for Y values, the MVUE is

$$P^{**} = \sum_{j=0}^{m-1} (-1)^j (m-1)^{(j)} (\mu\mu\overline{X})^{-j}$$

$$+ (-1)^m (m\mu\overline{X})^{-m+1} \exp(-m\mu\overline{X}). \tag{19.96}$$

If there is Type II censoring, with only the first g order statistics $X'_1, \ldots, X'_g$ of X and the first h order statistics $Y'_1, \ldots, Y'_h$ of Y being available, the UMVUE of P is

$$\hat{P}^* = \begin{cases} \displaystyle\sum_{j=0}^{g-1} (-1)^j \frac{(g-1)^{(j)}}{h^{[j]}} \left(\frac{H}{G} \right)^j & \text{if } H \le G, \\[4mm] 1 - \displaystyle\sum_{j=0}^{h-1} (-1)^j \frac{(h-1)^{(j)}}{g^{[j]}} \left(\frac{G}{H} \right)^j & \text{if } H > G, \end{cases} \tag{19.97}$$

where $G = \sum_{j=1}^{g} X'_j + (m-g)X'_g$, $H = \sum_{j=1}^{h} Y'_j + (n-h)Y'_h$. For $g = m$ and $h = n$, (19.97) reduces to (19.95). If each failed item is replaced immediately by a new item having the same lifetime distribution, the UMVUE of P is obtained from (19.97) by replacing G by mX'_g and H by nY'_h.

The maximum likelihood estimator of P (given complete sample values) is

$$\hat{P} = \frac{\overline{X}}{\overline{X} + \overline{Y}}. \qquad (19.98)$$

If $m = n$ the MLE is unbiased ($E[\hat{P}] = \mu(\lambda + \mu)^{-1}$), and hence it is also the UMVUE of P, as noted by Chious and Cohen (1984). Expressions for the UMVUE of P for the general case of two-parameter exponential random variables are complex, even in the case of common (unknown) location parameter.

However, the MLE of P is given by

$$\hat{P} = \frac{\overline{X} - T}{\overline{X} + \overline{Y} - 2T}, \qquad (19.99)$$

where $T = \min(\mathbf{X}, \mathbf{Y}) = \min(X'_1, Y'_1)$ is the maximum likelihood estimator of the common location parameter. As $m, n \to \infty$, with $m/n \to \gamma$, asymptotically

$$\sqrt{n}\,(\hat{P} - P) \xrightarrow{d} N(0, \sigma^2), \qquad (19.100)$$

where $\sigma^2 = P^2(1 - P)^2\{\gamma(1 - \gamma)\}$ and the $\hat{P}$ and the UMVUE are asymptotically equivalent in this case.

In view of (19.94), the problem of choice of sample sizes for estimating P with specified accuracy is equivalent to the solution to the sample size problem provided by Reiser, Faraggi, and Guttman (1993) for the ratio of two exponential scale parameters. Given P_1, α, P_2, β with $0 < P_2 < P_1 < 0$, $0 < \alpha < 1$, $0 < \beta < 1$, it is required to find an acceptance rule of the form $P > P_c$, with m and n such that (1) if $P = P_1$, the probability of acceptance is $1 - \alpha$, and (2) if $P = P_2$, the probability of acceptance is β. In this case m and n must satisfy

$$\frac{F_{2n, 2m; \alpha}}{F_{2n, 2m; (1-\beta)}} = \frac{P_2}{P_1}, \qquad (19.101)$$

where $F_{2n, 2m; \alpha}$ is the α percentage point of the $F_{2n, 2m}$ distribution.

When a random sample of size n has been censored by omission of the greatest $n - k$ observed values, it may be of interest to estimate (predict) the value of a specific one of the omitted values X_r', say $(r > k)$. Lawless (1971) used the pivotal variable

$$G_1 = \frac{X_r' - X_k'}{T_k} \qquad [T_k \text{ as defined in } (19.73)] \qquad (19.102)$$

to obtain the upper $100(1 - \alpha)\%$ prediction limit (UPL),

$$\text{UPL}(X_r') = X_k' + g_{1,1-\alpha}T_k \qquad (19.103)$$

where $\Pr[G_1 \le g_{1,1-\alpha}] = 1 - \alpha$. The cdf of G_1 is

$$\Pr[G_1 \le g] = \frac{k}{B(r - k, n - r + 1)}$$

$$\times \sum_{i=1}^{r-k-1} \binom{r - k - 1}{i}(-1)^i(n - r + i + 1)^{-1}$$

$$\times \{1 + (n - r + i + 1)g\}^{-k}. \qquad (19.104)$$

Lingappaiah (1973) suggested using

$$G_2 = \frac{X_r' - X_k'}{X_k'} \qquad (19.105)$$

in place of G_1, noting that these would lead to simpler calculations and that X_k' contains "all the prior information" about X_r'. However, Kaminsky (1977) pointed out that this is only true if σ is *known*. He also noted that the probability that the UPL based on G_1 exceeds that based on G_2 is close to (but below) 0.5 in a number of representative cases. Furthermore he showed that the asymptotic probability (as $n \to \infty$) that the length of predictor interval based on G_2 exceeds that based on G_1 is zero.

For the two-parameter distribution (19.1) similar procedures can be followed, noting that the variables $X_i' - X_1'$ are distributed as order statistics from a random sample of size $n - 1$ from the distribution (19.28). Likeš (1974) extends Lawless's (1971) treatment to this case, using the statistic

$$G_1' = \frac{X_r' - X_k'}{T_k'} \qquad (19.106)$$

with $T_k' = \sum_{i=2}^k X_i' + (n - k)X_k' - (n - 1)X_1'$. Tables of quantiles of G_1' were constructed by Likeš and Nedělka (1973). Further results on these lines were obtained by Lawless (1977), and later incorporated in his important textbook,

Lawless (1982). Prediction intervals under "hybrid censoring"—termination when either k "failures" have occurred, or at a fixed time T, whichever comes first—have been studied by Ebrahimi (1992).

Sometimes, it is desired to "predict" statistics based on a *future* random sample from the *same* exponential population. If all observed values in the original sample $(X'_1 \le \cdots \le X'_{n_1})$ are available the (minimum variance unbiased) estimator of the sth order statistic (Y'_s) in a future sample of size n_2 is in fact the best *linear* unbiased estimator.

A predicted interval for the *range* of a future sample has been constructed by Colangelo and Patel (1992), using the cdf of the ratio (W_2/W_1) of ranges of independent random samples of sizes n_1 and n_2. We have

$$\Pr\left[\frac{W_2}{W_1} \le y\right]$$

$$= (n_1 - 1) \int_0^\infty e^{-x}(1 - e^{-x})^{n_1-2}(1 - e^{-yx})^{n_2-1}\, dx$$

$$= (n_1 - 1) \sum_{i=0}^{n_1-2} \sum_{j=0}^{n_2-1} \binom{n_1-2}{i}\binom{n_2-1}{j}(-1)^{i+j}(1 + i + jy)^{-1}.$$

$$(19.107)$$

The authors provide tables of value K_γ such that $\Pr[(W_2/W_1) \le K_\gamma] = \gamma$ for $n_1, n_2 = 5(5)25$ and $\gamma = 0.005, 0.001, 0.025, 0.05, 0.1, 0.9, 0.95, 0.975, 0.99, 0.995$. From these tables $100(1 - \alpha_1 - \alpha_2)\%$ prediction intervals $(K_{\alpha_1}W, K_{1-\alpha_2}W)$ can be constructed.

Evans and Nigm (1980) [see also Dunsmore (1974)] use Bayesian methods to obtain a predicted interval for X'_s, based on a Type I censored data $(X'_1, \ldots, X'_k)$ $(k < s)$. They point out that with the Bayesian approach it is easy to take into account prior information on the value of θ [in (19.1)] such as $\theta > 0$.

Dunsmore's (1974) result, which does not allow for the condition $\theta > 0$, may serve as a reasonably good approximation even when the true underlying distribution is the left-truncated exponential, since in typical life-testing situations the amount of probability given to negative X's using Dunsmore's approach is small. Dunsmore's (1974) results, which do not impose positivity constraint, can be used as starting values in an iterative solution and in many cases may provide an adequate approximation.

8 CHARACTERIZATIONS

In the preparation of this section, some 200 works published on this subject during the last 35 years [from Fisz (1958) onward] have been studied. These include three monographs—Galambos and Kotz (1978), Azlarov and Volodin

(1986) and Kakosyan, Klebanov, and Melamed (1984)—and the survey of Kotz (1974). The stream of publications shows no signs of drying up. We are highly impressed by the ingenuity displayed in much of the work on this topic, though we must admit to having some doubts on adequate justification for publishing some of it. It is of course impossible to mention all the published results, and the inevitable choices presented considerable difficulties. We would like to make it clear that lack of citation of a result is not necessarily indicative of assessment of lesser value, on our part, and certainly not in any objective sense.

The main criteria for inclusion were chronological novelty (post 1978) and evidence of development of the four basic types of characterization delineated in the Galambos and Kotz (1978) monograph. We believe the latter is the clearest (most "transparent") available system of classification, enabling us to distinguish, in many cases, among overlapping, albeit independent, results. Our sincere apologies go to researchers whose names are not mentioned or whose contributions are included only in part.

We have also attempted to include new types of characterizations, developed since 1978, although, in our opinion, these constitute only a minority in the flood of recently published results. For readers' convenience we produce here—with minor modifications—the four distinct types of characterizations presented in Galambos and Kotz (1978), ignoring some mathematical niceties. See Table 19.1.

Table 19.1   A Dictionary of Characterizations of the Exponential Distribution

	Mathematical Properties	Statistical Interpretation
I.	$\bar{F}(x_1 + x_2) = \bar{F}(x_1)\bar{F}(x_2)$ for $x_1, x_2 > 0$	Lack of memory property
II.	$\bar{F}(nx) = \{\bar{F}(x)\}^n$ for $n \geq 1$, $x \geq 0$	$nX_1' \stackrel{d}{=} X$ (characterization in terms of the minimum value in random sample of size n)
III.	$\displaystyle\int_z^\infty \bar{F}(x)\,dx = \bar{F}(z)$ for $z \geq 0$ and successive iterations of this integral equation	Characterizations in terms of truncated distributions, the constancy of conditional expectations, conditional variances, and conditional higher moments
IV.	$\bar{F}'(x) = c\bar{F}(x)$ for $x \geq 0$	Constancy of the hazard rate

Remark

In a substantial number of papers, characterizations in terms of independence of order statistics and some other relations between them are directly reduced to property I in Table 19.1 via the property

$$\left\{ \frac{\overline{F}(x + u)}{\overline{F}(x)} \right\}^{n} \qquad \text{is independent of } x \text{ for all } n \geq 1, u \geq 0.$$

The underlying tool (in many cases) is the logarithmic version of the celebrated Cauchy functional equation

$$f(x + y) = f(x)f(y), \qquad \forall \, x,y \geq 0 \qquad (19.108)$$

or its more abstract equivalent, the Choquet-Deny equation. This is especially true in type I characterizations.

The major part of the edifice of characterizations of exponential distributions is rooted, in the final analysis, in this equation, or in its integral form, although very often in a hardly recognizable guise. Once again we note that what follows is a condensed account of a selective catalog of characterizations, motivated by two (occasionally conflicting) aims—emphasizing the more recent results and assessing the (potential) applicability of these results in practical situations.

There are quite severe restraints on size, since this is just one section of a volume containing ten chapters. We have also kept in mind the aim of reflecting, and reporting objectively on, the "state of the art" in statistical distribution theory in the last decade of the twentieth century. The results are presented, more or less, in chronological order, and we have tried to hint at mathematical tools used in their derivation. The close relationship between exponential and Poisson distributions is glossed over in this section. Some relevant characterizations are sited in Chapter 4 (on Poisson distributions), and other results relating to exponential distributions appear in Chapter 6 (on negative binomial, including geometric, distributions).

8.1 Characterizations Based on Lack of Memory and on Distributions of Order Statistics

The lack of memory property is identical with the constancy of hazard rate property (group IV) provided that the existence of a density [absolute continuity of $F_X(x)$] is assumed, so we will not treat the latter separately. We will, however, include in this subsection characterizations based on distributions of order statistics. We now consider extensions of the simple lack of memory property as characterizations of exponential distributions.

We first discuss conditions in terms of *relevation* functions. The earliest concept of this kind introduced by Krakowski (1973) in order to represent the distribution of the time T to failure of the *second* component in a system, where each item is replaced on failure by an item of the *same age*. If the lifetime cdf of each item is $F_X(x)$, with $F_X(0) = 0$, and the density $p_X(x) = F_X'(x)$ exists, the relevation cdf is

$$F_T(t) = \Pr[T \le t] = \int_0^t p_X(x)\left\{1 - \frac{\bar{F}_X(t)}{\bar{F}_X(x)}\right\} dx. \qquad (19.109)$$

The similarity is apparent to the cdf of the convolution of two X's (corresponding to replacement by a *new* item), which is

$$\int_0^t p_X(x)\{1 - \bar{F}_X(t - x)\} dx. \qquad (19.110)$$

The two cdf's are in fact identical if $F_X(x)$ is of form (19.28). Under certain mild assumptions on $F_X(x)$, the converse is also true.

In fact, if (19.109) and (19.110) are identical, or equivalently if

$$\bar{F}_X(t) \int_0^t p_X(x)\{\bar{F}_X(x)\}^{-1} dx = \int_0^t p_X(x) \bar{F}_X(t - x) \, dx, \qquad (19.111)$$

then provided that $F_X(x)$ is absolutely continuous $F_X(x) = 1 - e^{-x/\sigma}$.

For the case where the replacement item has lifetimes X_2 with a cdf $F_{X_2}(x)$ differing from $F_{X_1}(x)$, Grosswald, Kotz, and Johnson (1980) and Westcott (1981), among others, extend (19.111) to

$$\bar{F}_{X_1}(t) \int_0^t p_X(x)\{\bar{F}_{X_2}(x)\}^{-1} dx = \int_0^t p_X(x) \bar{F}_{X_2}(t - x) \, dx. \qquad (19.112)$$

If $F_{X_2}(x)$ is continuous and $F_{X_1}(x)$ is absolutely continuous, then $F_{X_2}(x) = 1 - e^{-x/\sigma}$ whatever $F_{X_1}(x)$ may be. Rao and Shanbhag (1986) show that this result holds even under less stringent conditions than continuity of $F_{X_2}(x)$.

Ahsanullah (1976, 1977, 1978a, b) considers the following equation:

$$\int_0^\infty p_X(x)\left[\left\{\frac{\bar{F}_X(x + z)}{\bar{F}_X(x)}\right\} - \bar{F}(z)\right] dx = 0. \qquad (19.113)$$

This can be regarded as an "integrated" lack of memory condition. It is clearly satisfied for exponential distributions of type (19.28) since $\bar{F}_X(x) = e^{-x/\sigma}$ results in $\bar{F}_X(x + z)/\bar{F}_X(x) = e^{-z/\sigma} = \bar{F}_X(z)$. Conditions under which the converse also holds are more difficult to establish than those sufficing for (19.111) and (19.112). Grosswald and Kotz (1981), and also Shimizu (1978), found that satisfaction of the condition

$$\lim_{x \to \infty} e^{-bx}\bar{F}''(x) = 0 \qquad \text{for some } b > 0$$

is sufficient. Further refinements were obtained by Yanushkyavichyus (1989).

A theorem of Lau and Rao (1982, 1984) reveals that *integrated* lack of memory properties generally lead to solutions of type

$$\bar{F}_X(x) = g(x)e^{-\lambda x}, \qquad (19.114)$$

where

$$g(x + w) = g(x) \qquad \text{for some } w.$$

[This generally does reduce to $e^{-\lambda x}$ provided the support of w is not a lattice.]

In this subsection it will be convenient, in our study of characterizations using order statistics, to denote the rth order statistic in a random sample of size n by the symbol $X'_{r:n}$ because we will need to refer to different sample sizes in the same equation. Puri and Rubin (1970) showed that the condition

$$X'_{2:2} - X'_{1:2} \stackrel{d}{=} X, \qquad (19.115)$$

which is equivalent to

$$2\int_0^\infty \bar{F}(x + y)p_X(y)\,dy = \bar{F}(x), \qquad (19.115)'$$

characterizes a (nonlattice) exponential distribution. Ahsanullah (1976) claimed that the conditions

$$X'_{s_i:n} - X'_{r:n} \stackrel{d}{=} X'_{s_i-r:n-r}, \qquad i = 1, 2, \quad 1 < r < s_1 < s_2 < n \quad (19.116)$$

characterize exponential distribution (provided the distribution is absolutely continuous). [Gather (1989) points out that Ahsanullah's proof involves the assumption that $\bar{F}(x + u)/\{\bar{F}(u)\bar{F}(x)\} - 1$ has the same sign (always positive or always negative) for all $u > 0$ and any fixed $x > 0$.]

Another result of Ahsanullah's (1977)—that

$$(n - 1)(X'_{i+1:n} - X'_{i:n}) \overset{d}{=} X \qquad \text{for some } i \text{ and } n \qquad (19.117)$$

characterizes an exponential distribution—is effectively based on the functional equation

$$\bar{F}(mx) = \{\bar{F}(x)\}^m. \qquad (19.118)$$

In fact property II of Table 19.2 can be derived from I. Since

$$\bar{F}(s_1 + s_2 + s_3) = \bar{F}(s_1 + s_2)\bar{F}(s_3) = \bar{F}(s_1)\bar{F}(s_2)\bar{F}(s_3)$$

and generally

$$\bar{F}\left(\sum_{j=1}^{m} s_j\right) = \prod_{j=1}^{m} \bar{F}(s_j), \qquad (19.119a)$$

by taking $s_1 = s_2 = \cdots s_m = x$, we have

$$\bar{F}(mx) = \{\bar{F}(x)\}^m. \qquad (19.119b)$$

We therefore do not assign a separate subsection to property II, and proceed to property III.

Recently, Liang and Balakrishnan (1992) proved a characterization result for exponential distributions with possibly different scale parameters. Their result is as follows: "Let $X_1, X_2, \ldots, X_n$ be mutually independent nonnegative random variables having absolutely continuous distribution functions $F_i(x)$ over its support $[0, \infty)$ and the corresponding density function $p_i(x) > 0$ for $x > 0$. Let A denote the event that $X_i - X_{i+1} > 0$ for all $i = 1, 2, \ldots,$ $n - 1$. Then, conditional on the event A, $X_i - X_{i+1}$ and X_{i+1} are independent for all $i = 1, 2, \ldots, k$ if and only if X_i $(i = 1, 2, \ldots, k)$ are exponentially distributed random variables (possibly with different scale parameters), where $1 \leq k \leq n - 1$."

A similar characterization of geometric distributions has also been established by Liang and Balakrishnan (1993). Fosam and Shanbhag (1994) have shown that these characterization results can be proved through Lau and Rao's (1982, 1984) theorem mentioned above under weaker assumptions.

8.2 Characterizations Based on Conditional Expectation (Regression)

If X is a nonnegative random variable, and $h(z)$ is a real-valued measurable function for $z > 0$, such that the expected value of $|h(X)|$ is finite, we have the following characterizations based on the condition

$$E[h(X - y)|X \geq y] = E[h(X)] \qquad (19.120)$$

for all nonnegative values of y. If $h(z)$ is linear, the corresponding condition

$$E[X|X \geq y] = y + E[X] \qquad (19.121)$$

characterizes the exponential distributions [Shanbhag (1970); Dallas (1973)].

With $h(z)$ a polynomial in z the same result was derived by Sahobov and Geshev (1974). An even more general result was obtained by Klebanov (1980). His characterization of exponential distributions is set out formally below.

The cdf $F_X(x)$ is of form (19.28) if and only if for some absolutely continuous $h(z)$ and for some value of σ,

$$g(z) = \begin{cases} h'(z)\exp(-\sigma z) & \text{for } z \geq 0, \\ 0 & \text{for } z < 0, \end{cases} \qquad (19.122)$$

is integrable in Lebesgue measure, and also

1. $\lim_{x \to \infty}\{\overline{F}_X(x)\exp(\sigma x)\} = 0$,

2. $-\left(\dfrac{1}{2\pi}\right)\log\left\{1 - \dfrac{1}{E[h(X)] - h(0)}\right\}\int_{-\infty}^{\infty} g(-z)e^{itz}\, dz \leq 1$.

We have the following characterizations as corollaries:

1. $E[e^{-X}|X \geq y] = e^{-y}E[X]$ for all $y \geq 0$.
2. $E[\exp\{-\frac{1}{2}(X - y)^2\}|X \geq y] = E[\exp(-\frac{1}{2}X^2)]$ for all $y \geq 0$.
3. $E[(X - y)^\alpha|X \geq y] = E[X^\alpha]$ for all $y \geq 0$ and for $0 < \alpha \leq 4$ provided that $\lim_{x \to \infty}e^{\delta x}\{1 - F_X(x)\} = 0$ for some $\delta > 0$.

Introducing the further condition on $F_X(x)$ [in addition to $F_X(0) = 0$] that $E[X^2]$ is finite, we have

$$\text{Var}(X|X > y) = c \qquad \text{for all } y \geq 0 \qquad (19.123)$$

which characterizes distribution (19.28), as does

$$E[(X - y)^2|X > y] = c \qquad \text{for all } y \geq 0. \qquad (19.124)$$

[Azlarov, Dzamirzaev, and Sultanova (1972); Laurent (1974)].

Dallas (1979) showed that if $F_X(0) = 0$ and $E[X^r]$ is finite for some positive integer r, then

$$E\left[(X - y)^r | X > y\right] = c \qquad \text{for all } y \geq 0 \qquad (19.125)$$

characterizes (19.28). Nagaraja's (1975) characterization avoids introducing constancy of conditional expectation directly. The condition is

$$\{E[X - y | X > y]\}^2 = \text{Var}(X | X > y) \qquad \text{for all } y > 0. \quad (19.126)$$

The condition

$$E\left[\{h(X) - h(y)\}^m | X > y\right] = c, \qquad (19.127)$$

where $h(\cdot)$ is strictly increasing and differentiable characterizes

$$F_X(x) = 1 - \exp\left(-\frac{h(x)}{c}\right),$$

as does the condition

$$E[h(X) | X > y] = h(y) + c. \qquad (19.128)$$

Kotlarski (1972) obtained a generalization of this case.

Characterizations due to Guerrieri (1965), Reinhardt (1968), Hoang (1968), Shanbhag (1970), Obretenov (1970), Azlarov, Dzamirzaev, and Sultanova (1972), Laurent (1974), Vartak (1974), Dallas (1974), Deryagin and Polesitskaya (1975), and Gupta (1975) are all equivalent to, or included in Sahobov and Geshev's (1974) result (see the following paragraph).

As an indication of the mathematical tools employed in Sahobov and Geshev's (1974) work, note that their equation [see also Dallas (1974)]

$$E\left[(X - y)^r | X > y\right] = E[X^r] \qquad (19.129)$$

for all $y \geq 0$ and some integer $r > 0$ is mathematically equivalent to the functional equation

$$\int_y^\infty (u - y)^{r-1} \overline{F}(u) \, du = c_r \overline{F}(y), \qquad (19.129)'$$

which leads to an equation of form

$$\frac{d^r \overline{F}(y)}{dy^r} = c_r^* \overline{F}(y). \qquad (19.130)$$

The general solution of this equation is

$$\bar{F}(y) = c_0 e^{-y/\sigma_n} + \sum_{i=1}^{n-1} c_i \exp(z_i y) \tag{19.131}$$

where $\sigma_n > 0$ and the z_i's are (possibly) complex numbers.

The conditions that $\bar{F}(y)$ is bounded, nonincreasing, and nonnegative ensure that $c_i \equiv 0$ for $i = 1, \ldots, n - 1$ so that

$$\bar{F}(y) = c_0 \exp(-y/\sigma_n)$$

—that is, it is the survivor function of an exponential distribution. Similarly the condition $\text{Var}(X - y | X \geq y) = c$ for all $y \geq 0$ is equivalent to

$$\frac{2}{\bar{F}(y)} \int_y^\infty (x - y)^2 \bar{F}(x) \, dx - \left\{ \frac{1}{\bar{F}(y)} \int_y^\infty \bar{F}(x) \, dx \right\}^2 = c,$$

leading to

$$\frac{2\psi(y)}{\psi''(y)} = \left[\frac{\psi'(y)}{\psi''(y)} \right]^2, \tag{19.132}$$

where $\psi(y) = \int_y^\infty (x - y) \bar{F}(x) \, dx$, and $\psi''(y) = \bar{F}(y)$. Solutions of (19.132) are

1. $\psi(y) = c_0 \exp(\pm \sqrt{y} / K)$,
2. $\psi(y) \equiv$ quadratic function of y.

However only $\psi(y) = \psi_0 \exp(-\sqrt{y} / K)$ satisfies the boundary conditions.

Wang and Srivastava (1980) obtained a characterization in terms of order statistics. If X has a continuous cdf and has finite expected value, then

$$E\left[(n - k)^{-1} \sum_{i=k+1}^n (X'_{i:n} - X'_{k:n}) | X'_{k:n} = y \right] = c \qquad \text{for all } y \geq 0$$

characterizes a exponential distribution. Mukherjee and Roy (1986) noted that since

$$E\left[(n - k)^{-1} \sum_{i=k+1}^n (X'_{i:n} - X'_{k:n}) | X'_{k:n} = y \right] = E[X - y | X > y],$$

this result is equivalent to the results of Reinhardt (1968) and Shanbhag (1970). They also showed that the exponential distribution is characterized by constancy of the conditional coefficient of variation of $(n - k)^{-1}\sum_{i=k+1}^{n}(X'_{i:n} - X'_{k:n})$, given $X'_{k:n}$.

In a later paper Roy and Mukherjee (1991) show that among distributions with finite variance and monotone $E[\sum_{j=2}^{n}(X'_{j:n} - X'_{1:n})|X'_{1:n} = x] = r(x)$, the condition $\text{Cov}(X'_{1:n}, \sum_{j=2}^{n}(X'_{j:n} - X'_{1:n})) = 0$ characterizes exponential (19.1) distributions [see also Swanepoel (1991)]. Kotz and Shanbhag (1980) pointed out that any distribution with $F_X(0) = 0$—as is typical for lifetime distributions—is uniquely determined by its mean residual life function $E[X - u|X > u]$, and many of the results cited above can be derived from this fact.

The following example of a stable characterization is appealing. Van Beek and Braat (1973) showed that if $E[X] = 1$ and the residual lifetime at a randomly chosen "age" is distributed as X/q, then if $q = 1$,

$$F_X(x) = 1 - e^{-x},$$

while if $0 < q < 1$,

$$F_X(x) = 1 - \frac{\int_{-\infty}^{\infty} q^{u^2/2} \exp(-xq^{u-(1/2)})\nu(du)}{\int_{-\infty}^{\infty} q^{u^2/2}\nu(du)}$$

for any periodic measure ν of period 1. [It is straightforward to allow for $E[X] \neq 1$.]

All distributions of the above class have the same moments

$$\mu'_r = rq^{-r(r-1)/2}.$$

Although the moments do not determine ("characterize") the distributions, Vardi, Shepp, and Logan (1981) have shown that they differ little unless q is very small. In fact for $q > e^{-2\pi} \doteq 0.001867$, the maximum absolute difference between the cdfs of any two distributions in the class is less than

$$\frac{\sqrt{2\pi\alpha^{-1}} \exp\left(\frac{1}{8}\pi^2\alpha^{-1}\right)}{(\sqrt{2\pi\alpha^{-1}})\sinh(\pi^2\alpha^{-1})} \sim 8\exp\left(-\frac{7}{8}\pi^2\alpha^{-1}\right),$$

where $\alpha = -\log q$.

Vardi, Shepp, and Logan (1981) note that for $q > 0.8$, the absolute difference cannot exceed 10^{-15}. This may be regarded as a remarkable example of "stability" in a characterization. Vardi and his coworkers also showed that it is not possible to have $q > 1$.

Suppose X_i, $i = 1, 2, 3$, are i.i.d., nonnegative continuous random variables that are square-integrable such that $E[X_i] = \text{Var}(X_i) = 1$ for $i = 1, 2, 3$. Let

$$U = \frac{1}{b}X_2 - \frac{1}{a}X_1 \quad \text{and} \quad W = \frac{1}{b}X_3 - \frac{1}{a}X_1.$$

Then, Pusz (1993) has proved that X_i's are exponentially distributed if and only if

$$E[U|W] = \begin{cases} W + \dfrac{a}{(a+b)b} & \text{if } W \le 0 \\[2ex] \dfrac{a}{(a+b)b} & \text{if } W > 0 \end{cases} \quad \text{and} \quad \text{Var}(U|W) = \frac{(a+b)^2 + b^2}{(a+b)^2 b^2}.$$

8.3 Record Values

Characterization based on *record values* has been a fashionable topic during the last 15 years. A record value in a sequence $X_1, X_2, \ldots$ is defined as any X_j that is greater than the preceding values $(X_1, \ldots, X_{j-1})$. [Lower record values (e.g., time to run 100 m.) may similarly be defined as any X_j that is smaller than the preceding values $(X_1, \ldots, X_{j-1})$.]

The record *times* are the values of j for which X_j is a record value. In terms of order statistics $(X'_{1:n} \le X'_{2:n} \le \cdots \le X'_{n:n})$, if the uth record time is $n_{(u)}$, then

$$n_{(1)} = 1, \quad n_{(u)} = \min\left(m > n_{(u-1)}; X_m > X'_{n_{(u-1)}:n_{(u-1)}}\right), \quad (19.133)$$

and the uth record value is

$$R_{(u)} = X'_{n_{(u)}:n_{(u)}}. \quad (19.134)$$

The starting point of much research in this field was Tata's (1969) theorem, which states that if $F_X(x)$ is absolutely continuous, it is an exponential cdf if and only if $R_{(1)}$ and $R_{(2)} - R_{(1)}$ are mutually independent. [Galambos and Kotz (1978) provide a detailed discussion.] A flood of generalizations, extensions, and variants followed the original work. These are well summarized in Deheuvels (1984).

8.4 Miscellaneous

Another line of development started from results obtained by Ferguson (1964) and Crawford (1966), stating that quite generally if $X'_{1:2}$ and $X'_{2:2} - X'_{1:2}$ are mutually independent, then either $F_X(x)$ is a geometric cdf (see Chapter 5) or

$$F_X(x) = 1 - \exp\{-b(x - a)\}, \quad x \ge a, b > 0.$$

An independent flood of generalizations followed, the most notable being due to Rossberg (1972a, b). Among his results, we quote the following:

1. If $\bar{F}_X(x)$ is neither degenerate nor a lattice distribution, the condition

$$\Pr[X'_{k+1:n} - X'_{k:n} > x] = \{\bar{F}(x)\}^{n-k} \qquad (19.135)$$

characterizes distributions of form (19.28).

2. If $F(x)$ and $G(x)$ are cdfs, such that $\lim_{x \to 0+} F(x) = 0$ and $\lim_{x \to 0+} G(x) = 1$, $F(x)$ not being a lattice distribution, then the condition

$$\alpha F(x) + \beta G(x) = F(x) * G(x) \qquad (19.136)$$

[the convolution of independent variables with cdfs $F(x), G(x)$, respectively] is equivalent to $F(x)$ being of form (19.28) and $G(x)$ being the cdf of the negative of a variable of the same form (19.28), with the ratio of the parameters (σ_F/σ_G) being β/α.

8.5 Stability

The "stability" of certain characterizations of the exponential distribution—namely the closeness of a distribution to an exponential distribution consequent upon near (but not exact) satisfaction of a characterization condition—has been studied by Shimizu (1980, 1981). He showed, among other things, that if $X_1, \ldots, X_n$ are i.i.d. with survival function $\bar{F}(x)$, and

$$\Pr[\min(a_1 X_1, \ldots, a_n X_n) < x] = [1 - R(x)]\bar{F}(x) \qquad (19.137)$$

for $0 \le x \le x_0$ with $x_0 > 0$, $a_j > 0$, $\sum_{j=1}^{n} a_j^{-1} < \infty$ where

$$R(x) = \begin{cases} \varepsilon\zeta x^2, & x \le 1, \\ \delta\zeta, & x > 1, \end{cases} \quad |\zeta| < 1, 0 < \varepsilon, \delta < \tfrac{1}{2},$$

then

$$\bar{F}(x) = \begin{cases} e^{-x/\sigma}(1 + \varepsilon_1 \zeta x^2), & 0 \le x \le 1, \\ e^{-x/\sigma}\{1 + \phi\zeta x \exp(\phi x)\}, & 1 < x < x_0, \end{cases} \qquad (19.138)$$

where $\lim_{x \to 0+}\{F(x)/x\} = \sigma^{-1}$, $\varepsilon_1 = 2\varepsilon(1 - a)^{-1}$, $\phi = 2(\varepsilon + \delta)(1 - a)^{-1}$, and $a^{-1} = \max(a_1^{-1}, \ldots, a_n^{-1})$. [See also Yanushkyevichene (1983).]

From Yanushkyevichene (1983) and Yanushkyavichyus (1989) we have another stability theorem: If $F_X(x)$ is continuous and strictly monotone over the domain $0 < F_X(x) < 1$ and has a finite first moment, and if for some $1 \le j \le K$ there exists a positive constant γ such that for F-almost all $x > 0$,

$$\left| E[X_{j+1:n} - X_{j:n} | X_{j:n} = x] - \gamma \right| \le \varepsilon,$$

then for all $x \geq 0$,

$$\left| \overline{F}_X(x) - \exp(-\lambda_* x) \right| \leq C\varepsilon^{1/(n-j)} \exp(-\lambda x),$$

where λ_*, λ, and C depend only on γ and $n - j$.

The following result gives stability theorem for record value characterizations. Let $\{R_j\}$ be record values of a sequence of i.i.d. nonnegative random variables with a continuous distribution function $F(x)$, possessing a finite first moment and also strictly monotonic over $0 < F(x) < 1$. If for some fixed j there exists a positive constant γ such that for all $x \geq 0$,

$$\left| E\left[R_{j+1} - R_j | R_j = x \right] - \gamma \right| \leq \varepsilon,$$

then

$$\left| \overline{F}(x) - \exp(-\lambda_0 x) \right| \leq C\varepsilon \exp(\lambda x), \qquad x \geq 0,$$

where $\lambda = \gamma^{-1} - \delta > 0$, δ a fixed positive constant, and C and λ_0 depend only on γ and δ.

Before leaving the topic of characterizations we note the following simple transformations [for distribution (19.2)] that relate the exponential distribution to other commonly employed distributions.

Transformation	$\overline{F}_Y(y) = 1 - F_Y(y)$	Distribution
$Y = \log X$	$\exp(-e^y)$	Extreme value (Chapter 22)
$Y = \exp(X/\tau)$	$(1/y)^\tau \ (y \geq 1)$	Pareto (Chapter 20)
$Y = 1 - e^{-X}$	$1 - y \ (0 \leq y \leq 1)$	Uniform (Chapter 26)
$Y = X^{1/c} \ (c > 0)$	$\exp(-y^c)\,(y > 0)$	Weibull (Chapter 21)

Characterizations of these distributions can be used (with appropriate transformation) to characterize exponential distributions, and conversely.

9   MIXTURES OF EXPONENTIAL DISTRIBUTIONS

Mixtures of exponential distributions have important life-testing applications. A batch of electronic tubes may consist of two subpopulations, with each subpopulation having its own characteristic mean life and the two subpopulations being in proportions ω_1 and $\omega_2 = 1 - \omega_1$. (Each unit of the population may be regarded as if it contains a "tag" that indicates the subpopulation to which it belongs and hence defines the way in which that particular item will fail.) Finally, the failure times for the ith subpopulation are assumed to be independently distributed with probability density function (19.28) with $\sigma = \sigma_i$ $(i = 1, 2)$.

If ω_1 is the proportion of units belonging to subpopulation $i = 1$, then the cumulative distribution function of T, the time to failure, is

$$\omega_1\left[1 - \exp\left(\frac{-t}{\sigma_1}\right)\right] + (1 - \omega_1)\left[1 - \exp\left(\frac{-t}{\sigma_2}\right)\right], \qquad t \geq 0, \quad (19.139a)$$

and the density function is

$$\omega_1\sigma_1^{-1}e^{-t/\sigma_1} + (1 - \omega_1)\sigma_2^{-1}e^{-t/\sigma_2}, \qquad t \geq 0. \qquad (19.139b)$$

This distribution has also been applied (under the name *Schuhl distribution*) to the distances between elements in traffic [Petigny (1966)]. Mixtures of exponential distributions have been found to represent some demographic distributions [Susarla and Pathala (1965)].

Gumbel (1939) gave a method for fitting a mixture of two exponential distributions, using the first two sample moments. Cases where σ has a continuous distribution have been studied by Bhattacharya and Holla (1965) and Bhattacharya (1966). In the latter paper it is shown that if σ has an exponential distribution (with parameters $0, \beta$) the probability density function of T is $2\beta^{-1}K_0(2(t/\beta)^{1/2})$ $(0 < t)$, where $K_0(\cdot)$ is a modified Bessel function of the third kind of zero order (see Chapter 1, Section A12). If σ has a beta distribution (Chapter 25) over the range $(0, \theta)$ the probability density function can be expressed in terms of a Whittaker function. In the earlier paper the special case where σ has a uniform distribution is considered.

We now consider methods of estimating the parameters ω_1, σ_1, and σ_2 in (19.139a) and (19.139b).

Case A: Relative Magnitude of Subpopulation Parameters Not Known
Mendenhall and Hader (1958) derived the following equations for the maximum likelihood estimators, expressed in terms of

$$\hat{\omega}_1 = \frac{R_1}{n} + \hat{k}\frac{(n - R)}{n}, \qquad (19.140a)$$

$$\hat{\beta}_1 = \overline{X}_1 + \hat{k}\frac{(n - R)}{R_1}, \qquad (19.140b)$$

$$\hat{\beta}_2 = \overline{X}_2 + \frac{(1 - \hat{k})(n - R)}{R_2}, \qquad (19.140c)$$

where

$$\overline{X}_i = \frac{\sum_{j=1}^{R_i} X_{ij}}{R_i}, \qquad i = 1, 2,$$

and

$$\hat{k} = \cfrac{1}{1 + \left(\cfrac{1 - \hat{\omega}_1}{\hat{\omega}_1}\right)\exp\left[(1/\hat{\beta}_1) - (1/\hat{\beta}_2)\right]}.$$

Here

n = the number of units on test,
t_0 = the time at which test is terminated,
R = the total number of failures during test,
R_1 = the number of failures from subpopulation 1,
R_2 = the number of failures from subpopulation 2 (note that $R_1 + R_2 = R$),
$X_{ij} = T_{ij}/t_0$, T_{ij} being the time at which the jth failure from the ith population occurs,
$\hat{\beta}_i$ is the estimator of $\beta_i = \sigma_i/t_0$ ($i = 1, 2$).

The estimators of β_1, β_2, and ω_1 must be obtained by solving the simultaneous equations (19.140). A method of solving these approximately was given by Mendenhall and Hader (1958).

Case B: Relative Magnitude of Subpopulation Parameters Known
In the special case $\beta_1 = \beta_2 = \beta$ the maximum likelihood estimators become

$$\hat{\omega}_1 = \frac{r_1}{r}. \tag{19.140a$'$}$$

$$\hat{\beta} = \frac{r_1\overline{X}_1 + r_2\overline{X}_2 + (n - r)}{r}. \tag{19.140b$'$}$$

In general, we have to use the method in case A to obtain estimators, but if we have crossover, that is,

$$\hat{\beta}_1 > \hat{\beta}_2 \quad \text{while} \quad \beta_1 \leq \beta_2$$

or

$$\hat{\beta}_1 < \hat{\beta}_2 \quad \text{while} \quad \beta_1 \geq \beta_2,$$

we use case B estimators corresponding to the special case $\beta_1 = \beta_2 = \beta$.

Rider (1961) used the "method of moments" to obtain rather involved estimators of mixture exponential parameters. Proschan (1963) also discussed the problem of mixture of several exponential distributions and derived some interesting results concerning the resulting mixed population. For the case of mixture distributions with pdf

$$(1 - \omega)\sigma^{-1} \exp\left(-\frac{x}{\sigma}\right) + \omega\left(\frac{\sigma}{\alpha}\right)^{-1} \exp\left(-\frac{\alpha x}{\sigma}\right),$$

$$x > 0, \sigma > 0, \alpha > 0, 0 \leq \omega \leq 1$$

(the *number* of "outliers" in the sample not being specified), Willemain et al. (1992) carried out extensive Monte Carlo comparisons of seven estimators of σ, using the relative closeness (difference of absolute values of errors of estimation) as a criterion. More details on mixtures of exponential distributions can be found in books by Everitt and Hand (1981) and Titterington, Smith, and Makov (1985).

Mixed distributions with

$$F_X(x) = \omega + (1 - \omega)(1 - e^{-t/\sigma}), \qquad t \geq 0, 0 < \omega < 1, \sigma > 0, \quad (19.141)$$

can arise in clinical trial situations where there is a probability ω of immediate failure but survivors have a lifetime distribution as in (19.28). Kleyle and Dahiya (1975) have considered estimation of the parameters ω and σ.

McClean (1986) studied estimation for the mixed exponential distribution where the failure time data is grouped. The problem originated from the data of job status of university graduates in the Northern Ireland collected quarterly for various grades over $5\frac{1}{4}$ year period on whether or not they were still in service. Bartholomew (1982) has shown that a mixed exponential distribution provides a good fit to these data.

Mixtures of generalized exponential distributions have also found application in "mine dust" analysis [Tallis and Light (1968)]. In particular, a problem that arises is that of estimating the parameters α, β, and γ of the distribution with density function

$$p_X(x) = \gamma A e^{-\alpha\sqrt{x}} + (1 - \gamma)B e^{-\beta\sqrt{x}}, \qquad x > a,$$

where A, B, and a are known constants.

The method of maximum likelihood is often regarded as the standard method of estimating parameters. In this case, however, we would obtain rather complicated equations, and methods using *half-moment* equations have been developed [Joffe (1964)]. Half-moment equations are less complicated. Although they must be solved iteratively, they can be handled on a pocket calculator. The three half-moment equations are

$$m'_{\frac{r}{2}} = \tilde{\gamma}\mu'_{\frac{r}{2},\tilde{\alpha}} + (1 - \tilde{\gamma})\mu'_{\frac{r}{2},\tilde{\beta}}, \qquad r = 1, 2, 3, \quad (19.142)$$

where

$$m'_{\frac{r}{2}} = \frac{1}{n} \sum_{i=1}^{n} X_i^{r/2} \qquad (n \text{ is the number of elements})$$

and

$$\mu'_{\frac{r}{2}, \tilde{\alpha}} = \frac{\tilde{\alpha}^2 e^{\tilde{\alpha}\sqrt{a}}}{2(1 + \tilde{\alpha}\sqrt{a})} \int_a^{\infty} x^{r/2} e^{-\alpha\sqrt{x}} \, dx.$$

Use may be made of the recurrence relationship

$$\int_a^{\infty} x^{r/2} e^{-\alpha\sqrt{x}} \, dx = \frac{2a^{(r+1)/2} e^{-\alpha\sqrt{a}}}{\alpha} + \frac{r+1}{\alpha} \int_a^{\infty} x^{(r-1)/2} e^{-\alpha\sqrt{x}} \, dx.$$

One can express $\tilde{\beta}$ as a function of $\tilde{\alpha}$ and $\tilde{\gamma}$, and the three half-moments as follows:

$$\tilde{\beta} = \frac{P - \tilde{\gamma}Q}{R - (\tilde{\gamma}/\tilde{\alpha})Q}, \qquad (19.143a)$$

where

$$P = 4m'_1 - 3\sqrt{am'_{\frac{1}{2}}},$$

$$Q = 4\mu'_{1,\tilde{\alpha}} - 3\sqrt{a\mu'_{\frac{1}{2},\tilde{\alpha}}},$$

$$R = m'_{\frac{3}{2}} - \sqrt{am'_1}.$$

$\tilde{\gamma}$ may be written as a function of $\tilde{\alpha}$, and the three half-moments as

$$\tilde{\gamma} = \frac{PU - SR}{Q[U - (S/\tilde{\alpha})] - \tau[R - (P/\tilde{\alpha})]}, \qquad (19.143b)$$

where

$$U = m'_1 - \sqrt{am'_{\frac{1}{2}}}$$

$$S = 3m'_{\frac{3}{2}} - 2\sqrt{a}$$

$$\tau = 3\mu'_{\frac{1}{2},\tilde{\alpha}} - 2\sqrt{a}.$$

A solution of these equations can be obtained iteratively by choosing a value of $\tilde{\alpha}$ and consequent values $\tilde{\beta}$ and $\tilde{\gamma}$ from (19.143a) and (19.143b).

Considerable work has also been done on the generalized mixed exponential form (often denoted by GH) in the last decade; for example, see Botta and Harris (1986), Botta, Marchal, and Harris (1987), Harris and Sykes (1987), Harris and Botta (1988), and Harris, Marchal, and Botta (1992). These generalized mixed exponential forms are linear but not necessarily convex combinations of negative exponential distribution functions (unlike the mixture form). The cumulative distribution function is given by

$$F(x) = 1 - \sum_{i=1}^{k} p_i e^{-\theta_i x},$$

with $\theta_i > 0$, $\sum_{i=1}^{k} p_i = 1$, $-\infty < p_i < \infty$, and $\theta_1 < \theta_2 < \cdots < \theta_k$ (without loss of any generality). Botta and Harris (1986) showed that the above class is dense in the set of all distribution functions relative to an appropriate metric. Harris, Marchal, and Botta (1992) have recently reviewed the relationship of these generalized hyperexponential distributions to other types of distribution functions in the Coxian family, particularly the phase types. Interestingly, by assuming the claim amount distribution to be of the generalized hyperexponential form given above, Gerber, Goovaerts, and Kaas (1987) have considered the probability and the severity of ruin in terms of the usual model of the collective risk theory in insurance.

10   RELATED DISTRIBUTIONS

There are many important distributions closely related to the exponential distribution. (See also the end of Section 8.)

If a variable X $(> \theta)$ is such that $Y = (X - \theta)^c$ has the exponential distribution (19.28), then X is said to have a *Weibull distribution* with shape parameter c. It is necessary that c be greater than -1, for otherwise the integral of

$$p_X(x) = \sigma^{-1} c (x - \theta)^{c-1} \exp\left[\frac{-(x - \theta)^c}{\sigma} \right], \qquad x > \theta, \quad (19.144)$$

between $x = \theta$ and $x = \theta' > \theta$ will be infinite. (See Chapter 21 for a detailed discussion of this distribution.) If c and θ are known (and often θ is known to be zero) the transformed variable $Y = (X - \theta)^c$ may be used, and the well-developed techniques associated with the exponential distribution become applicable. If c is not known, special techniques are needed; these techniques are discussed in Chapter 21.

If $Y = e^{-X}$ has an exponential distribution of form (19.28), then X has a distribution of *extreme value* form (Chapter 22).

Another distribution of theoretical, and some practical, importance is the *double* (or *bilateral*) exponential distribution. This is the distribution of a random variable X with probability density function

$$p_X(x) = (2\sigma)^{-1} \exp\left(\frac{-|x - \theta|}{\sigma}\right), \qquad \sigma > 0. \qquad (19.145)$$

It is also known as *Laplace's first law of error.* (*Laplace's second law of error* is the normal distribution.) The exponential distribution might be regarded as a double exponential "folded" about $x = \theta$. The double exponential distribution is the subject of Chapter 24.

The fact that the exponential distribution belongs to the class of gamma distributions has been mentioned in Section 1. Taking $\theta = 0$ and $\sigma = 2$ in (19.1), we obtain the distribution of χ^2 with two degrees of freedom. From this it can be seen that if $X_1, X_2, \ldots, X_n$ are independent random variables each with probability density function (19.1), then their arithmetic mean is distributed as $\theta + \frac{1}{2}n^{-1}\sigma \times (\chi^2$ with $2n$ degrees of freedom). We further note that if the quadratic form $\sum_{i,j=1}^{n} a_{ij}U_iU_j$ in independent unit normal variables $U_1, U_2, \ldots, U_n$ has a matrix $\mathbf{A}$ with eigenvalues $\lambda_1, \lambda_2, \ldots, \lambda_n$, then applying an appropriate linear transformation of the U's,

$$\sum_{i,j=1}^{n} a_{ij}U_iU_j = \sum_{j=1}^{n} \lambda_j U_j^{*2}, \qquad (19.146)$$

where $U_1^*, U_2^*, \ldots, U_n^*$ are independent unit normal variables. If the nonzero λ's are equal in pairs, then $\sum_{j=1}^{n}\lambda_j U_j^{*2}$ can be expressed as a linear function of independent variables each distributed as χ^2 with two degrees of freedom, and so exponentially. The distribution theory of such a variable is much simpler than that of a general quadratic form in normal variables.

In fact, if $X_1, \ldots, X_n$ are independent standard exponential variables, then the linear form

$$Y = \sum_{j=1}^{n} \lambda_j X_j, \qquad \lambda_j \neq \lambda_k,$$

has the probability density function

$$p_Y(y) = \sum_{j=1}^{n} \left(\prod_{k \neq j}(\lambda_j - \lambda_k)^{-1}\right)\lambda_j^{n-2}e^{-y/\lambda_j}, \qquad y > 0. \qquad (19.147)$$

This is a special type of mixture of exponential distributions and called the *general gamma* or the *general Erlang* distribution. It is used in queueing theory, reliability theory, and psychology. Some special patterns of λ's have been discussed by Likes (1967, 1968), McGill and Gibbon (1965), and others.

Kaminsky, Luks, and Nelson (1984) derived a recursive formula that is easily programmable for the evaluation of the distribution function of ratios of linear combinations of independent exponential random variables. The result is based on a generalization of the lack of memory property for the exponential distribution.

Let $\mathbf{a} = (a_1, \ldots, a_m)'$, $\mathbf{b} = (b_1, \ldots, b_n)'$ be *vectors of positive constants*, and let $\mathbf{X} = (X_1, X_2, \ldots, X_m)'$, $\mathbf{Y} = (Y_1, Y_2, \ldots, Y_n)'$ be *independent vectors of independent exponential random variables*, each with density $p_X(x) = e^{-x}$, $x > 0$. For any vector $\mathbf{U} = (U_1, \ldots, U_m)'$, let $\mathbf{U}^* = (U_2, \ldots, U_m)'$. Then the functions $G_{m,n} \colon ((R^+)^m \times (R^+)^n) \to [0, 1]$ given by

$$G_{m,n}(\mathbf{a}, \mathbf{b}) = \Pr[\mathbf{a}'\mathbf{X} \geq \mathbf{b}'\mathbf{Y}]$$

satisfy

$$G_{m,n}(\mathbf{a}, \mathbf{b}) = \left(\frac{b_1}{a_1 + b_1}\right) G_{m-1,n}(\mathbf{a}^*, \mathbf{b}) + \left(\frac{a_1}{a_1 + b_1}\right) G_{m,n-1}(\mathbf{a}, \mathbf{b}^*),$$

$$\text{(19.148)}$$

with boundary conditions,

$$G_{0,n}(\mathbf{0}, \mathbf{b}) = 0,$$

$$G_{m,0}(\mathbf{a}, \mathbf{0}) = 1.$$

The relationship between the exponential and rectangular distributions is discussed in more detail in Chapter 26. Here we note only the following property: If $X_1, X_2, \ldots, X_n$ each have probability density function (19.28), then $Y_j = \sum_{i=1}^{j} X_i / [\sum_{i=1}^{n} X_i]$ for $j = 1, \ldots, (n-1)$ are distributed as the order statistics of a random sample of size $n - 1$ from a rectangular distribution over the range 0 to 1.

The following relation between Poisson and exponential distributions is of importance in applications (see also Chapter 4, Sections 2 and 9): If $T_1, T_2, \ldots$ are a succession of independent random variables each having the same density function (19.28) and the random variable N is defined by the inequalities

$$T_1 + T_2 + \cdots + T_N \leq \tau < T_1 + T_2 + \cdots + T_{N+1},$$

then N has a Poisson distribution with expected value τ/σ.

An exponential distribution truncated from below is still an exponential distribution with the same scale parameter.

An exponential distribution truncated by exclusion of values exceeding x_0 has the density function

$$p_X(x) = \sigma^{-1}\left\{1 - \exp\left[-\frac{x_0 - \theta}{\sigma}\right]\right\}^{-1} \exp\left[-\frac{x - \theta}{\sigma}\right], \qquad x_0 > x > \theta.$$

$$(19.149)$$

The expected value of this distribution is

$$\theta + \sigma\left[1 - \frac{x_0 - \theta}{\sigma}\left\{\exp\left[\frac{x_0 - \theta}{\sigma}\right] - 1\right\}^{-1}\right].$$

If $X_1, X_2, \ldots, X_n$ are independent random variables, each having the distribution (19.149), then the statistic $\sum_{j=1}^{n} X_j$ is sufficient for σ, if θ be known. Bain and Weeks (1964) have shown that the density function of $Y = \sum_{j=1}^{n} X_j - n\theta$ is

$$p_Y(y) = \left[\sigma\left\{1 - \exp\left[-\frac{x_0 - \theta}{\sigma}\right]\right\}\right]^{-n} e^{-y/\sigma}[(n-1)!]^{-1}$$

$$\times \sum_{j=0}^{j_0} (-1)^j \binom{n}{j}[y - j(x_0 - \theta)]^{n-1} \qquad (19.150a)$$

for $j_0(x_0 - \theta) < y < (j_0 + 1)(x_0 - \theta)$, $j_0 = 0, 1, 2, \ldots, (n-1)$. The cumulative distribution function is

$$F_Y(y) = \left\{1 - \exp\left[-\frac{x_0 - \theta}{\sigma}\right]\right\}^{-n} \sum_{j=0}^{j_0} (-1)^j \binom{n}{j} e^{-j(x_0 - \theta)/\sigma}$$

$$\times \Pr\left[\chi_{2n}^2 < \frac{2(j_0 + d - j)(x_0 - \theta)}{\sigma}\right], \qquad (19.150b)$$

where $y = (x_0 - \theta)(j_0 + d)$, j_0 being an integer and $0 \le d \le 1$.

Deemer and Votaw (1955) showed that if θ be known to be zero, then the maximum likelihood estimator $\hat{\sigma}$ of σ is the solution of the equation

$$\bar{X}x_0^{-1} = \hat{\sigma}x_0^{-1} - (e^{x_0/\hat{\sigma}} - 1)^{-1} \qquad (19.151)$$

provided that $\bar{X} < \frac{1}{2}x_0$. If $\bar{X} > \frac{1}{2}x_0$, then $\hat{\sigma}$ is infinite; this result may be taken to mean that a truncated exponential distribution is inappropriate. [Formally, it indicates that a rectangular distribution (Chapter 26) over the range 0 to x_0 should be used.] In Deemer and Votaw (1955) there is a table of $\bar{X}/x_0$ to four decimal places as a function of $\hat{\sigma}^{-1}$ for $\hat{\sigma}^{-1} = 0.01(0.01)0.89$. For n large,

$$n\,\mathrm{Var}(\hat{\sigma}^{-1}) \doteq \left[\sigma^2 - x_0^2 e^{-x_0/\sigma}(1 - e^{-x_0/\sigma})^{-2}\right]^{-1}. \qquad (19.152)$$

Estimation of θ for the cumulative distribution function

$$1 - e^{-x/\theta}\{1 - F(x)\},$$

[where $F(x)$ is the cumulative distribution function of a non-negative random variable] was treated by Gerčbah (1967). He suggested an estimator based on a random sample of size $n = km$. Divide the sample into m groups at random, with $X'_{1(j)}$ the smallest value in the jth group. Then under certain monotonicity conditions on $F(x)$

$$\hat{\theta} = km^{-1} \sum_{j=1}^{m} X'_{1(j)}$$

is a consistent estimator of θ as k and m tend to infinity.

In extending the Marshall–Olkin bivariate exponential distribution to an absolutely continuous and not necessarily possessing memoryless property model, Ryu (1993) has recently presented a generalization of the exponential distribution derived as a marginal distribution of this generalized bivariate model. The new generalized distribution with density function

$$p_X(x) = \{\lambda_1 + \lambda_{12}(1 - e^{-sx})\}e^{-\lambda_1 x - \lambda_{12}x + \frac{\lambda_{12}}{s}(1 - e^{-sx})} \qquad (19.153)$$

has an increasing failure rate in general. Note that when $s \to \infty$, the density in (19.153) becomes the exponential density function with constant hazard rate of $\lambda_1 + \lambda_{12}$. In Figure 19.2, plots of the exponential density function and the generalized exponential density function are presented. While the exponential density function is monotonically decreasing, the density in (19.153) is initially increasing and then decreasing.

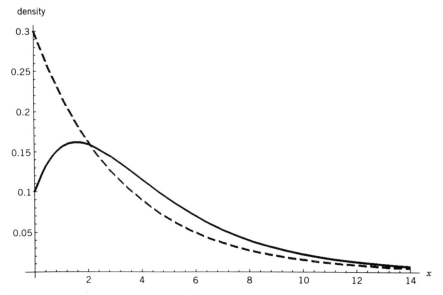

Figure 19.2 Plots of the Generalized Exponential Density in (19.153) and the Exponential Density for the Case $\lambda_1 = 0.1$, $\lambda_{12} = 0.2$ and $s = 0.5$, and $\lambda_1 = 0.1$, $\lambda_{12} = 0.2$ and $s = \infty$, Respectively. The Dashed Line Is the Exponential Density and the Solid Line Is the Generalized Exponential Density.

BIBLIOGRAPHY

Ahsanullah, M. (1976). On a characterization of the exponential distribution by order statistics, *Journal of Applied Probability*, **13**, 818–822.

Ahsanullah, M. (1977). A characteristic property of the exponential distribution, *Annals of Statistics*, **5**, 580–582.

Ahsanullah, M. (1978a). Record values and the exponential distribution, *Annals of the Institute of Statistical Mathematics*, **30**, 429–433.

Ahsanullah, M. (1978b). A characterization of the exponential distribution by spacings, *Journal of Applied Probability*, **15**, 650–653. (Similar paper in *Annals of the Institute of Statistical Mathematics*, **30**, 163–166.)

Ahsanullah, M. (1979). Characterization of the exponential distribution by record values, *Sankhyā*, *Series B*, **41**, 116–121.

Ahsanullah, M. (1981a). Record values of exponentially distributed random variables, *Statistische Hefte*, **22**, 121–127.

Ahsanullah, M. (1981b). On characterizations of the exponential distribution by spacings, *Statistische Hefte*, **22**, 316–320.

Ahsanullah, M. (1988). On a conjecture of Kakosyan, Klebanov, and Melamed, *Statistische Hefte*, **29**, 151–157.

Alzaid, A. A., Rao, C. R., and Shanbhag, D. N. (1987). Solution of the integrated Cauchy functional equation on a half line using exchangeability, *Sankhyā, Series A*, **49**, 189–194.

Arnold, B. C. (1970). Inadmissibility of the usual scale estimate for a shifted exponential distribution, *Journal of the American Statistical Association*, **65**, 1260–1264.

Arnold, B. C., and Nagaraja, H. N. (1991). Lorenz ordering of exponential order statistics, *Statistics & Probability Letters*, **11**, 485–490.

Arnold, B. C., and Villaseñor, J. A. (1989). Lorenz ordering of order statistics, Presented at the *Workshop on Stochastic Orders and Decision Under Risk*, Hamburg, Germany.

Azlarov, T. A. (1972). Stability of characteristic properties of the exponential distribution, *Lietuvos Matematikos Rinkinys*, **12**, No. 2, 5–9. [In Russian. English version in *Translations in Mathematical Statistics*, **14**, 33–38 (1978).]

Azlarov, T. A., Dzamirzaev, A. A., and Sultanova, M. M. (1972). Characterization properties of the exponential distribution, and their stability, *Random Processes and Statistical Inference, No. II, Izdat. "Fan" Uzbek. SSR, Tashkent*, 10–19. (In Russian)

Azlarov, T. A., and Volodin, N. A. (1986). *Characterization Problems Associated with the Exponential Distribution*, Berlin: Springer-Verlag.

Bai, D. S., and Hong, Y. W. (1992). Estimation of $\Pr(X < Y)$ in the exponential case with common location parameter, *Communications in Statistics—Theory and Methods*, **21**, 269–282.

Bain, L. J., and Weeks, D. L. (1964). A note on the truncated exponential distribution, *Annals of Mathematical Statistics*, **35**, 1366–1367.

Balakrishnan, N. (1990). On the maximum likelihood estimation of the location and scale parameters of exponential distribution based on multiply Type II censored samples, *Journal of Applied Statistics*, **17**, 55–61.

Balakrishnan, N. (1994). Order statistics from non-identical exponential random variables and some applications (with discussion), *Computational Statistics & Data Analysis* (to appear).

Balakrishnan, N., and Barnett, V. (1994). Outlier-robust estimation of the mean of an exponential distribution (submitted for publication).

Balakrishnan, N., and Cohen, A. C. (1991). *Order Statistics and Inference: Estimation Methods*, San Diego: Academic Press.

Balakrishnan, N., and Gupta, S. S. (1992). Higher order moments of order statistics from exponential and right-truncated exponential distributions and applications to life-testing problems, *Technical Report No. 92-07C*, Department of Statistics, Purdue University, West Lafayette, IN.

Balakrishnan, N., and Joshi, P.C. (1984). Product moments of order statistics from doubly truncated exponential distribution, *Naval Research Logistics Quarterly*, **31**, 27–31.

Balakrishnan, N., and Malik, H. J. (1986). Order statistics from the linear-exponential distribution, Part I: Increasing hazard rate case, *Communications in Statistics—Theory and Methods*, **15**, 179–203.

Balasubramanian, K., and Balakrishnan, N., (1992). Estimation for one- and two-parameter exponential distributions under multiple type-II censoring, *Statistische Hefte*, **33**, 203–216.

Bánkövi, G. (1964). A decomposition-rejection technique for generating exponential random variables, *Mathematical Proceedings of the Hungarian Academy of Science, Series A*, **9**, 573–581.

Barlow, R. E., and Proschan, F. (1979). Inference for the Exponential Life Distribution, *Report ORC 79-16*, Operations Research Center, University of California, Berkeley.

Barnett, V. D. (1965). *Random Negative Exponential Deviates*, Tracts for Computers, **27**, Cambridge: Cambridge University Press.

Bartholomew, D. J. (1963). The sampling distribution of an estimate arising in life testing, *Technometrics*, **5**, 361–374.

Bartholomew, D. J. (1982). *Stochastic Models for Social Processes* (Third edition), New York: Wiley.

Barton, D. E., David F. N., and Merrington, M. (1963). Table for the solution of the exponential equation, $\exp(b) - b/(1 - p) = 1$, *Biometrika*, **50**, 169–176.

Bartoszewicz, J. (1977). Estimation of $P(Y < X)$ in the exponential case, *Zastosowania Matematyki*, **16**, No. 1, 1–8.

Basu, A. P. (1964). Estimates of reliability for some distributions useful in life testing, *Technometrics* **6**, 215–219.

Basu, A. P. (1965a). On some tests of hypotheses relating to the exponential distribution when some outliers are present, *Journal of the American Statistical Association*, **60**, 548–559.

Basu, A. P. (1965b). On characterizing the exponential distribution by order statistics, *Annals of the Institute of Statistical Mathematics*, **17**, 93–96.

Beg, M. A. (1980). Estimation of $P(Y < X)$ for truncation parameter distributions, *Communications in Statistics—Theory and Methods*, **9**, 327–345.

Beg, M. I., and Balasubramanian, K. (1990). Distributions determined by conditioning on a single order statistic, *Metrika*, **37**, 37–43.

Beg, M. I., and Kirmani, S. N. U. A. (1974). On a characterization of exponential and related distribution, *Australian Journal of Statistics*, **16**, 163–166 (Correction, *Ibid.*, 1976, **18**, 85).

Bell, C. B., and Sarma, Y. R. K. (1980). A characterization of exponential distributions based on order statistics, *Metrika*, **27**, 263–269.

Berger, R. L., and Kim, J. S. (1985). Ranking and subset selection procedures for exponential populations with Type-I and Type-II censored data, In *Frontiers of Modern Statistical Inference Procedures*, E. J. Dudewicz (editor), 425–448, Syracuse, New York: American Sciences Press.

Bhattacharya, S. K. (1966). A modified Bessel function model in life testing, *Metrika*, **11**, 133–144.

Bhattacharya, S. K., and Holla, M. S. (1965). On a life test distribution with stochastic deviations in the mean, *Annals of the Institute of Statistical Mathematics*, **17**, 97–104.

Blyth, C. R. (1982). Maximum probability estimation in small samples, In *A Festschrift for Erich L. Lehmann*, P. Bickel et al. (editors), 83–96, Belmont, CA: Wadsworth.

Botta, R. F., and Harris, C. M. (1986). Approximation with generalized hyperexponential distributions: Weak convergence results, *Queueing Systems*, **2**, 169–190.

Botta, R. F., Marchal, W. G., and Harris, C. M. (1987). Characterizations of generalized hyperexponential distribution functions, *Communications in Statistics-Stochastic Models*, **3**, 115–148.

Brewster, J. F. (1974). Alternative estimators for the scale parameter of the exponential distribution with unknown location, *Annals of Statistics*, **2**, 553–557.

Brewster, J. F., and Zidek, J. V. (1974). Improving on equivariant estimators, *Annals of Statistics*, **2**, 21–38.

Bulgren, W. G., and Hewett, J. E. (1974). Double sample tests for hypotheses about the mean of an exponential distribution, *Technometrics*, **15**, 187–192.

Carlson, P. G. (1958). Tests of hypothesis on the exponential lower limit, *Skandinavisk Aktuarietidskrift*, **41**, 47–54.

Chikkagoudar, M. S., and Kunchur, S. H. (1980). Estimation of the mean of an exponential distribution in the presence of an outlier, *Canadian Journal of Statistics*, **8**, 59–63.

Chiou, P. (1987). A preliminary test estimator of reliability in a life-testing model, *IEEE Transactions on Reliability*, **R-36**, 408–410.

Chious, W. J., and Cohen, A. (1984). Estimating the common location parameter of exponential distribution with censored samples, *Naval Research Logistics Quarterly*, **31**, 475–482.

Choquet, D., and Deny, J. (1960). Sur l'equation de convolution $\mu = \mu * \sigma$, *Comptes Rendus de l'Académie des Sciences, Paris*, **250**, 799–801.

Cohen, A. C. (1963). Progressively censored samples in life testing, *Technometrics*, **5**, 327–339.

Cohen, A. C. (1991). *Truncated and Censored Samples: Theory and Applications*, New York: Marcel Dekker.

Cohen, A. C., and Helm, F. R. (1973). Estimators in the exponential distribution, *Technometrics*, **15**, 415–418.

Cohen, A. C., and Whitten, B. J. (1988). *Parameter Estimation for Reliability and Life Span Models*, New York: Marcel Dekker.

Colangelo, L., and Patel, J. K. (1992). Prediction intervals based on ranges and waiting times for an exponential distribution, *IEEE Transactions on Reliability*, **41**, 469–472.

Crawford, G. B. (1966). Characterization of geometric and exponential distributions, *Annals of Mathematical Statistics*, **37**, 1790–1795.

Csörgö, M., Seshadri, V., and Yalovsky, M. (1975). Applications of characterizations, in the area of goodness of fit, In *Statistical Distributions in Scientific Work*, **2**, G. P. Patil, S. Kotz, and J. K. Ord (editors), 70–90, Dordrecht: Reidel.

Dallas, A. C. (1974). A characterization of the exponential distribution, *Bulletin de la Sociéte Mathématique de Gréce*, **14**, 172–175.

Dallas, A. C. (1979). On the exponential law, *Metrika*, **26**, 105–106.

Dallas, A. C. (1981). Record values and the exponential distribution, *Journal of Applied Probability*, **18**, 949–951.

David F. N., and Johnson, N. L. (1948). The probability integral transformation when parameters are estimated from the sample, *Biometrika*, **35**, 182–190.

Deemer, W. L., and Votaw, D. F. (1955). Estimation of parameters of truncated or censored exponential distributions, *Annals of Mathematical Statistics*, **26**, 498–504.

Deheuvels, P. (1984). The characterization of distributions by order statistics and record values—a unified approach, *Journal of Applied Probability*, **21**, 326–334.

Deny, J. (1960). Sur l'equation de convolution $\mu = \mu * \sigma$, *Seminaire de Théórie de Potentiel Secretariat Mathématique, Faculté des Sciences, Paris*, **4** Ann.

Deryagin, Yu. V., and Polesitskaya, N. A. (1975). A characterization of the exponential distribution by the absence of after-effect property, *Trudy Moskovskogo Instituta Elektronicheskogo Mashinostroyeniya*, **44**, 192–198.

Dufour, R. (1982). Tests d'adjustement pour des echantillons tronques ou censures, Ph.D. thesis, Université de Montréal.

Dufour, R., Maag, U. R., and Eeden, C. van (1984). Correcting a proof of a characterization of the exponential distribution, *Journal of the Royal Statistical Society, Series B*, **46**, 238–241.

Dunsmore, I. R. (1974). The Bayesian predictive distribution in life testing models, *Technometrics*, **16**, 455–460.

Ebrahimi, N. (1992). Prediction intervals for future failures in the exponential distribution under hybrid censoring, *IEEE Transactions on Reliability*, **41**, 127–132.

Elewa, S. A., Nagarsenker, B. N., and Nagarsenker, P. B. (1991). On a test of equality of exponential populations based on censored samples, *Communications in Statistics—Simulation and Computation*, **21**, 909–917.

Elperin, T., and Gertsbakh, I. (1988). Estimation in a random censoring model with incomplete information: Exponential lifetime distribution, *IEEE Transactions on Reliability*, **37**, 223–229.

El-Sayyad, G. M. (1967). Estimation of the parameter of an exponential distribution, *Journal of the Royal Statistical Society, Series B*, **29**, 525–532.

Epstein, B. (1954). Truncated life tests in the exponential case, *Annals of Mathematical Statistics*, **25**, 555–564.

Epstein, B. (1957). Simple estimators of the parameters of exponential distributions when samples are censored, *Annals of the Institute of Statistical Mathematics*, **8**, 15–26.

Epstein, B. (1958). Exponential distribution and its role in life testing, *Industrial Quality Control*, **15**, 4–9.

Epstein, B. (1960a, b). Tests for the validity of the assumption that the underlying distribution of life is exponential, Parts I and II, *Technometrics*, **2**, 83–101 and 167–183.

Epstein, B. (1960c). Estimation of the parameters of two parameter exponential distributions from censored samples, *Technometrics*, **2**, 403–406.

Epstein, B. (1960d). Statistical life test acceptance procedures, *Technometrics*, **2**, 435–446.

Epstein, B (1960e). Estimation from life test data, *Technometrics*, **2**, 447–454.

Epstein, B. (1961). Estimates of bounded relative error for the mean life of an exponential distribution, *Technometrics*, **3**, 107–109.

Epstein, B. (1962). Simple estimates of the parameters of exponential distributions, In *Contributions to Order Statistics*, A. E. Sarhan and B. G. Greenberg (editors), 361–371, New York: Wiley.

Epstein, B., and Sobel, M. (1953). Life testing, *Journal of the American Statistical Association*, **48**, 486–502.

Epstein, B., and Sobel, M. (1954). Some theorems relevant to life testing from an exponential distribution, *Annals of Mathematical Statistics*, **25**, 373–381.

Epstein, B., and Sobel, M. (1955). Sequential life tests in the exponential case, *Annals of Mathematical Statistics*, **26**, 82–93.

Epstein, B., and Tsao, C. K. (1953). Some tests based on ordered observations from two exponential distributions, *Annals of Mathematical Statistics*, **24**, 458–466.

Evans, I. G., and Nigm, A. H. M. (1980). Bayesian prediction for the left truncated exponential distribution, *Technometrics*, **22**, 201–204.

Everitt, B. S., and Hand, D. J. (1981). *Finite Mixture Distributions*, London: Chapman and Hall.

Fairbanks, K. (1988). A two-stage life test for the exponential parameter, *Technometrics*, **30**, 175–180.

Fei, H., and Kong, F. (1994). Interval estimations for one- and two-parameter exponential distributions under multiply Type II censored samples, *Communications in Statistics—Theory and Methods* (to appear).

Ferguson, T. S. (1964). A characterization of the exponential distribution, *Annals of Mathematical Statistics*, **35**, 1199–1207.

Ferguson, T. S. (1967). On characterizing distributions by properties of order statistics, *Sankhyā, Series A*, **29**, 265–278.

Fisz, M. (1958). Characterization of some probability distributions, *Skandinavisk Aktuarietidskrift*, **41**, 65–70.

Fosam, E. B., and Shanbhag, D. N. (1994). Certain characterizations of exponential and geometric distributions, *Journal of the Royal Statistical Society, Series B*, **56**, 157–160.

From, S. G. (1991). Mean square error efficient estimation of an exponential mean under an exchangeable single outlier model, *Communications in Statistics—Simulation and Computation*, **20**, 1073–1084.

Fukuta, J. (1961). The use of sample quasi-ranges from the exponential population, *Research Reports of the Faculty of Engineering, Gifu University*, No. **11**, 40–45.

Gajek, L., and Gather, U. (1989). Characterizations of the exponential distribution by failure rate and moment properties of order statistics, In *Lecture Notes in Statistics*, **51**, J. Hülser and R.-D. Reiss (editors), 114–124, Berlin: Springer-Verlag.

Gajek, L., and Gather, U. (1991). Moment inequalities for order statistics with application to characterizations of distributions, *Metrika*, **38**, 357–367.

Galambos, J. (1972). Characterization of certain populations by independence of order statistics, *Journal of Applied Probability*, **9**, 224–230.

Galambos, J. (1975a). Characterizations of probability distributions by properties of order statistics, I (continuous case), In *Statistical Distributions in Scientific Work*, **3**, G. P. Patil, S. Kotz, and J. K. Ord (editors), 71–88, Dordrecht: Reidel.

Galambos, J. (1975b). Characterizations of probability distributions by properties of order statistics, II (discrete case), In *Statistical Distributions in Scientific Work*, **3**, G. P. Patil, S. Kotz, and J. K. Ord (editors), 89–101, Dordrecht: Reidel.

Galambos, J., and Kotz, S. (1978). *Characterizations of Probability Distributions*, Lecture Notes in Mathematics, **675**, Berlin: Springer-Verlag.

Gather, U. (1989). On a characterization of the exponential distribution by properties of the order statistics, *Statistics & Probability Letters*, **7**, 93–96.

Gerber, H. U., Goovaerts, M. J., and Kaas, R. (1987). On the probability and severity of ruin, *ASTIN Bulletin*, **17**, 151–163.

Gerčbah (Gertsbach), I. B. (1967). On estimation of parameters of a distribution with an exponential factor, *Teoriya Veroyatnostei i ee Primeneniya*, **7**, 121–123. (English translation: *Theory of Probability and Its Applications*, **12**, 110–111.)

Ghosh, M., and Razmpour, A. (1984). Estimation of common location parameters of several exponentials, *Sankhyā, Series A*, **46**, 383–394.

Gnedenko, B. V., Belyaev, Yu. K., and Solovev, A. D. (1965). *Mathematical Methods in Reliability Theory*, Moscow. (English translation, San Diego: Academic Press, 1968.)

Govindarajulu, Z. (1966). Characterization of the exponential and power distributions, *Skandinavisk Aktuarietidskrift*, **49**, 132–136. (Correction, *Scandinavian Actuarial Journal*, 1978).

Grosswald, E., and Kotz, S. (1981). An integrated lack of memory characterization of the exponential distribution, *Annals of the Institute of Statistical Mathematics*, **33**, 205–214.

Grosswald, E., Kotz, S., and Johnson, N. L. (1980). Characterizations of the exponential distribution by relevation-type equations, *Journal of Applied Probability*, **17**, 874–877.

Guenther, W. C. (1971). On the use of best tests to obtain best β-content tolerance intervals, *Statistica Neerlandica*, **25**, 191–202.

Guenther, W. C., Patil, S. A., and Uppuluri, V. R. R. (1976). One-sided β-content tolerance factors for the two parameter exponential distribution, *Technometrics*, **18**, 333–340.

Guerrieri, G. (1965). Some characteristic properties of the exponential distribution, *Giornale degli Economisti e Annali di Economica*, **24**, 427–437.

Gumbel, E. J. (1939). La dissection d'une répartition, *Annales de l'Université de Lyon*, 3-me Series, Section A, **2**, 39–51.

Gupta, A. K., and Basu, A. D. (1980). Estimation of restricted scale parameters of the exponential distribution, *Sankhyā, Series B*, **42**, 1–10.

Gupta, R. C. (1975). On characterization of distributions by conditional expectations, *Communications in Statistics*, **4**, 99–103.

Gupta, R. D., and Gupta, R. C. (1988). Estimation of $P(Y_p > \text{Max}(Y_1, Y_2, \ldots, Y_{p-1}))$ in the exponential case, *Communications in Statistics—Theory and Methods*, **17**, 911–924.

Hamdan, M. A. (1972). On a characterization by conditional expectations, *Technometrics*, **14**, 497–499.

Harris, C. M., and Botta, R. F. (1988). A note on phase-type, almost phase-type, generalized hyperexponential and Coxian distributions, *Technical Report No. GMU/22461 / 102*, Department of Operations Research and Applied Statistics, Fairfax, VA: George Mason University.

Harris, C. M., Marchal, W. G., and Botta, R. F. (1992). A note on generalized hyperexponential distributions, *Communications in Statistics—Stochastic Models*, **8**, 179–191.

Harris, C. M., and Sykes, E. A. (1987). Likelihood estimation for generalized mixed exponential distributions, *Naval Research Logistics*, **34**, 251–279.

Harter, H. L. (1961). Estimating the parameters of negative exponential populations from one or two order statistics, *Annals of Mathematical Statistics*, **32**, 1078–1090.

Harter, H. L. (1964). Exact confidence bounds, based on one order statistic, for the parameter of an exponential population, *Technometrics*, **6**, 301–317.

Hawkins, D. M., and Kotz, S. (1976). A clocking property of the exponential distribution, *Australian Journal of Statistics*, **8**, 170–172.

Hoang, H. N. (1968). Estimate of stability of a characterization of the exponential distribution, *Lietuvos Matematikos Rinkinys*, **8**, No. 1, 175–177. (In Russian)

Homan, S. M., and Lachenbruch, P. A. (1986). Robust estimation of the exponential mean parameter for small samples: Complete and censored data, *Communications in Statistics—Simulation and Computation*, **15**, 1087–1107.

Hsieh, H. K. (1981). On testing the equality of two exponential distributions, *Technometrics*, **23**, 265–269.

Hsieh, H. K. (1986). An exact test for comparing location parameters of k exponential distribution with unequal scales based on type II censored data, *Technometrics*, **28**, 157–164.

Huang, J. S. (1974a). Characterizations of the exponential distribution by order statistics, *Journal of Applied Probability*, **11**, 605–608.

Huang, J. S. (1974b). On a theorem of Ahsanullah and Rahman, *Journal of Applied Probability*, **11**, 216–218.

Huang, J. S. (1975). Characterization of distributions by the expected values of the order statistics, *Annals of the Institute of Statistical Mathematics*, **27**, 87–93.

Huzurbazar, V. S. (1955). Confidence intervals for the parameter of a distribution admitting a sufficient statistic when the range depends on the parameter, *Journal of the Royal Statistical Society, Series B*, **17**, 86–90.

Jevanand, E. S., and Nair, N. U. (1993). Estimation of exponential parameters in the presence of outliers, *Biometrical Journal*, **35**, 471–478.

Joffe, A. D. (1964). Mixed exponential estimation by the method of half moments, *Applied Statistics*, **13**, 93–98.

Joshi, P. C. (1972). Efficient estimation of the mean of an exponential distribution when an outlier is present, *Technometrics*, **14**, 137–144.

Joshi, P. C. (1978). Recurrence relations between moments of order statistics from exponential and truncated exponential distributions, *Sankhyā, Series B*, **39**, 362–371.

Joshi, P. C. (1979). A note on the moments of order statistics from doubly truncated exponential distribution, *Annals of the Institute of Statistical Mathematics*, **31**, 321–324.

Joshi, P. C. (1982). A note on the mixed moments of order statistics from exponential and truncated exponential distributions, *Journal of Statistical Planning and Inference*, **6**, 13–16.

Joshi, P. C. (1988). Estimation and testing under exchangeable exponential model with a single outlier, *Communications in Statistics—Theory and Methods*, **17**, 2315–2326.

Joshi, P. C., and Balakrishnan, N. (1984). Distribution of range and quasi-range from doubly truncated exponential distribution, *Trabajos de Estadistica y de Investigaciones Operationes*, **35**, 231–236.

Kakosyan, A. V., Klebanov, L. B., and Melamed, I. A. (1984). *Characterization of Distributions by the Method of Intensively Monotone Operators*, Lecture Notes in Mathematics *1088*, New York: Springer-Verlag.

Kale, B. K., and Sinha, S. K. (1971). Estimation of expected life in the presence of an outlier observation, *Technometrics*, **13**, 755–759.

Kambo, N. S., and Awad, A. M. (1985). Testing equality of location parameters of K exponential distributions, *Communications in Statistics—Theory and Methods*, **14**, 567–583.

Kaminsky, K. S. (1977). Comparison of prediction intervals for failure times when life is exponential, *Technometrics*, **19**, 83–86.

Kaminsky, K. S., Luks, E. M., and Nelson, P. I. (1984). Strategy, non-transitive dominance and the exponential distribution, *Australian Journal of Statistics*, **26**, 111–118.

Kelley, G. D., Kelley, J. A., and Schucany, W. R. (1976). Efficient estimation of $P(Y < X)$ in the exponential case, *Technometrics*, **18**, 359–360.

Khan, A. H., Yaqub, M., and Parvez, S. (1983). Recurrence relations between moments of order statistics, *Naval Research Logistics Quarterly*, **30**, 419–441. Corrigendum, **32**, 693 (1985).

Kleyle, R. M., and Dahiya, R. C. (1975). Estimation of parameters of mixed failure time distribution from censored samples, *Communications in Statistics*, **4**, 873–882.

Knight, W. (1965). A method of sequential estimation applicable to the hypergeometric, binomial, Poisson, and exponential distributions, *Annals of Mathematical Statistics*, **36**, 1494–1503.

Kocherlakota, S., and Balakrishnan, N. (1986). One- and two-sided sampling plans based on the exponential distribution, *Naval Research Logistics Quarterly*, **33**, 513–522.

Kong, F., and Fei, H. (1994). Limit theorems for the maximum likelihood estimators under multiple Type II censoring, *Communications in Statistics—Theory and Methods* (to appear).

Kopocinski, B. (1988). Some characterizations of the exponential distribution function, *Probability and Mathematical Statistics*, **9**, 105–111.

Kotlarski, I. I. (1972). On a characterization of some probability distributions by conditional expectations, *Sankhyā, Series A*, **34**, 461–466.

Kotz, S. (1974). Characterizations of statistical distributions, a supplement to a recent survey, *International Statistical Review*, **42**, 39–65.

Kotz, S., and Johnson, N. L. (1974). A characterization of exponential distributions by a waiting time property, *Communications in Statistics*, **3**, 257–258.

Kotz, S., and Shanbhag, D. N. (1980). Some new approaches to probability distribution, *Advances in Applied Probability*, **12**, 903–921.

Kotz, S., and Steutel, F. W. (1988). Note on a characterization of exponential distributions, *Statistics & Probability Letters*, **6**, 201–203.

Krakowski, M. (1973). The relevation transform and a generalization of the gamma distribution function, *Revue Française, Automatique et Informatique Recherche Operationelle 7, Series V*-2, 107–120.

Krishnaji, N. (1970). Characterization of the Pareto distribution through a model of unreported incomes, *Econometrica*, **38**, 251–255.

Krishnaji, N. (1971). Note on a characterizing property of the exponential distribution, *Annals of Mathematical Statistics*, **42**, 361–362.

Kulldorff, G. (1961). *Contributions to the Theory of Estimation from Grouped and Partially Grouped Samples*, Stockholm: Almqvist and Wiksell (New York: Wiley).

Kulldorff, G. (1962). On the asymptotic optimum spacings for the estimation of the scale parameter of an exponential distribution based on sample quantiles, Mimeographed report, University of Lund.

Kulldorff, G. (1963). Estimation of one or two parameters of the exponential distribution on the basis of suitably chosen order statistics, *Annals of Mathematical Statistics*, **34**, 1419–1431.

Kumar, S., and Patel, J. K. (1971). A test for comparison of two exponential distributions, *Technometrics*, **13**, 183–189.

Lau, K., and Rao, C. R. (1982). Integrated Cauchy functional equation and characterization of the exponential law, *Sankhyā, Series A*, **44**, 72–90.

Lau, K., and Rao, C. R. (1984). Solution to the integrated Cauchy functional equation on the whole line, *Sankhyā, Series A*, **46**, 311–318.

Laurent, A. G. (1974). On characterization of some distributions by truncation properties, *Journal of the American Statistical Association*, **69**, 823–827.

Laurent, A. G., and Gupta, R. C. (1969). A characterization of the exponential distribution, *Annals of Mathematical Statistics*, **40**, 1865.

Lawless, J. F. (1977). Prediction intervals for the two-parameter exponential distribution, *Technometrics*, **19**, 469–472.

Lawless, J. F. (1982). *Statistical Models and Methods for Lifetime Data*, New York: Wiley.

Leslie, J., and van Eeden, C. (1991). On a characterization of the exponential distribution based on a type 2 right censored sample, *Quebec University at Montreal Technical Report*, No. 154.

Liang, T.-C., and Balakrishnan, N. (1992). A characterization of exponential distributions through conditional independence, *Journal of the Royal Statistical Society, Series B*, **54**, 269–271.

Liang, T.-C., and Balakrishnan, N. (1993). A characterization of geometric distributions through conditional independence, *Australian Journal of Statistics*, **35**, 225–228.

Lieberman, A. (1959). Sequential life testing plans for the exponential distribution, *Industrial Quality Control*, **16**, 14–18.

Likěs, J. (1967). Distributions of some statistics in samples from exponential and power-function populations, *Journal of the American Statistical Association*, **62**, 259–271.

Likěs, J. (1968). Differences of two ordered observations divided by standard deviation in the case of negatively exponential distribution, *Metrika*, **12**, 161–172.

Likěs, J. (1974). Prediction of s-th ordered observation for the two-parameter exponential distribution, *Technometrics*, **16**, 241–244.

Likěs, J., and Nedělka, S. (1973). Note on studentized range in samples from an exponential population, *Biometrishche Zeitschrift*, **15**, 545–555.

Lin, G. D. (1987). On characterizations of distributions via moments of record values, *Probability Theory and Related Fields*, **74**, 479–483.

Lin, G. D. (1988). Characterizations of distributions via relationships between two moments of order statistics, *Journal of Statistical Planning and Inference*, **19**, 73–80.

Lin, G. D. (1988). Characterizations of uniform distributions and of exponential distributions, *Sankhyā, Series A*, **50**, 64–69.

Lin, G. D. (1989). The product moments of order statistics with applications to characterizations of distributions, *Journal of Statistical Planning and Inference*, **21**, 395–406.

Lingappaiah, G. S. (1973). Prediction in exponential life testing, *Canadian Journal of Statistics*, **1**, 113–117.

Lloyd, E. H. (1952). Least-squares estimation of location and scale parameters using order statistics, *Biometrika*, **39**, 88–95.

Madi, M., and Tsui, K.-W. (1990). Estimation of the ratio of the scale parameters of two exponential distributions with unknown location parameters, *Annals of the Institute of Statistical Mathematics* **42**, 77–87.

Malmquist, S. (1950). On a property of order statistics from a rectangular distribution, *Skandinavisk Aktuarietidskrift*, **33**, 214–222.

Marsaglia, G. (1961). Generating exponential random variables, *Annals of Mathematical Statistics*, **32**, 899–900.

Marsaglia, G., and Tubilla, A. (1975). A note on the "lack of memory" property of the exponential distribution, *Annals of Probability*, **3**, 353–354.

McClean, S. (1986). Estimation for the mixed exponential distribution using grouped follow-up data, *Applied Statistics* **15**, 31–37.

McGill, W. J., and Gibbon, J. (1965). The general-gamma distribution and reaction time, *Journal of Mathematical Psychology*, **2**, 1–18.

Mendenhall, W., and Hader, R. J. (1958). Estimation of parameters of mixed exponentially distributed failure time distributions from censored life test data, *Biometrika*, **45**, 504–520.

Menon, M. V., and Seshadri, V. (1975). A characterization theorem useful in hypothesis testing, *Contributions libre, 40 ieme Session de l'Institut International de Statistique*, 586–590.

Mukherjee, S. P., and Roy, D. (1986). Some characterizations of the exponential and related life distributions, *Calcutta Statistical Association Bulletin*, **35**, 189–197.

Nadler, J. (1960). Inverse binomial sampling plans when an exponential distribution is sampled with censoring, *Annals of Mathematical Statistics*, **31**, 1201–1204.

Nagaraja, H. N. (1977). On a characterization based on record values, *Australian Journal of Statistics*, **19**, 71–74.

Nagaraja, H. N. (1988). Record values and related statistics. A review, *Communications in Statistics—Theory and Methods*, **17**, 2223–2238.

Nelson, W. (1982). *Applied Life Data Analysis*, New York: Wiley.

Neumann, J. von (1951). Various techniques in connection with random digits, Monte Carlo methods, *National Bureau of Standards*, Applied Mathematics Series, **12**, 36–38. Washington, DC: Government Printing Office.

Obretenov, A. (1970). On a property of exponential distribution, *Fiziko Matematicesko Spisanie, Bolgarskie Akademija na Naukita, Fiziceskie Institut*, **13**, No. 1, 51–53.

Ogawa, J. (1951). Contributions to the theory of systematic statistics, I, *Osaka Mathematical Journal*, **3**, 175–213.

Ogawa, J. (1960). Determination of optimum spacings for the estimation of the scale parameters of an exponential distribution based on sample quantiles, *Annals of the Institute of Statistical Mathematics*, **12**, 135–141.

Ogawa, J. (1962). Optimum spacing and grouping for the exponential distribution, In *Contributions to Order Statistics*, A. E. Sarhan and B. G. Greenberg (editors), 371–380, New York: Wiley.

Page, E. S. (1954). Continuous inspection schemes, *Biometrika*, **41**, 100–114.

Patil, G. P. (1963). A characterization of the exponential-type distribution, *Biometrika*, **50**, 205–207.

Patil, G. P., and Seshadri, V. (1964). Characterization theorem for some univariate probability distributions, *Journal of the Royal Statistical Society, Series B*, **26**, 286–292.

Paulson, E. (1941). On certain LRT associated with exponential distribution, *Annals of Mathematical Statistics*, **12**, 301–306.

Perng, S. K. (1978). A test for equality of two exponential distributions, *Statistica Neerlandica*, **32**, 93–102.

Petigny, B. (1966). Extension de la distribution de Schuhl, *Annales des Ponts et Chaussées*, **136**, 77–84.

Pfeifer, D. (1982). Characterizations of exponential distributions by independent nonstationary record increments, *Journal of Applied Probability*, **19**, 127–135. (Corretion, *Ibid.*, **19**, 906.)

Piegorsch, W. W. (1987). Performance of likelihood based internal estimates for two-parameter exponential samples subject to type 1 censoring, *Technometrics*, **29**, 41–49.

Pierce, D. A. (1973). Fiducial, frequency, and Bayesian inference for the two parameter negative exponential distribution, *Technometrics*, **15**, 249–253.

Proschan, F. (1963). Theoretical explanation of observed decreasing failure rate, *Technometrics*, **5**, 375–383.

Pudeg, A. (1990). Charakterisierung von Wahrscheinlichkeitsverteilungen durch Verteilungseigenschaften der Ordnungsstatistiken und Rekorde, Ph.D. Dissertation, Rheinisch-Westfälischen Technischen Hochschule Aachen, Germany.

Pugh, E. L. (1963). The best estimate of reliability in the exponential case, *Operations Research*, **11**, 57–61.

Puri, P. S., and Rubin, H. (1970). A characterization based on absolute differences of two i.i.d. random variables, *Annals of Mathematical Statistics*, **41**, 2113–2122.

Pusz, J. (1992). Characterization of exponential distributions by conditional moments, *Teoriya Voryantnostei i Primeneniya*, **37**, No. 3, 599–601. [In Russian. English translation in *Theory of Probability and Its Applications*, **37**, 557–560 (1993).]

Raghavachari, M. (1965). Operating characteristic and expected sample size of a sequential probability ratio test for the simple exponential distribution, *Bulletin of the Calcutta Statistical Association*, **14**, 65–73.

Ramachandran, B. (1979). On the strong memorylessness property of the exponential and geometric probability laws, *Sankhyā, Series A*, **41**, 244–251.

Ramachandran, B. (1982). On the equation: $f(x) = \int_{[0,\infty)} f(x + y)\, d\mu(y)$, *Sankhyā, Series A*, **44**, 364–371.

Ramachandran, B., and Rao, B. L. S. P. (1984). On the equation $f(x) = \int_{-\infty}^{\infty} f(x + y)\, d\mu(y)$, *Sankhyā, Series A*, **46**, 326–338.

Rao, C. R., and Shanbhag, D. N. (1986). Recent results on characterization of probability distributions: A unified approach through extensions of Deny's theorem, *Advances in Applied Probability*, **18**, 660–678.

Ray, W. D. (1957). Sequential confidence intervals for the mean of a normal population with unknown variance, *Journal of the Royal Statistical Society, Series B*, **19**, 133–143.

Reinhardt, H. E. (1968). Characterizing the exponential distribution, *Biometrics*, **24**, 437–439.

Reiser, B., Faraggi, D., and Guttman, I. J. (1993). Choice of sample size for testing the $P(X > Y)$, *Communications in Statistics—Theory and Methods*, **21**, 559–569.

Reiser, B., and Guttman, I. J. (1986). Statistical inference for $\Pr(Y > X)$: The normal case, *Technometrics*, **28**, 253–257.

Reiser, B., and Guttman, I. J. (1989). Sample size choice for reliability verification in strength stress models, *Canadian Journal of Statistics*, **17**, 253–259.

Rényi, A. (1953). On the theory of order statistics, *Acta Mathematica Academiae Scientiarum Hungaricae*, **4**, 191–232.

Richards, D. St. P. (1981). Stability theorems for some characterizations of the exponential distribution, *Annals of the Institute of Statistical Mathematics*, **33**, 199–204.

Rider, P. R. (1959). Quasi-ranges of samples from an exponential population, *Annals of Mathemtical Statistics*, **30**, 252–254.

Rider, P. R. (1961). The method of moments applied to a mixture of two exponential distributions, *Annals of Mathematical Statistics*, **32**, 143–147.

Robertson, C. A. (1977). Estimation of quantiles of exponential distributions with minimum error in predicted distribution function, *Journal of the American Statistical Association*, **72**, 162–164.

Rossberg, H.-J. (1966). Charakterisierungsprobleme, die sich aus der von A. Rényi in die Theorie der Ranggrössen eingeführten Methode ergeben, *Monatsberichte der Deutschen Akademie der Wissenschaften zu Berlin*, **8**, 561–572.

Rossberg, H.-J. (1972a). Characterization of distribution functions by the independence of certain functions of order statistics, *Sankhyā, Series A*, **34**, 111–120.

Rossberg, H.-J. (1972b). Characterization of the exponential and the Pareto distributions by means of some properties of the distributions which the differences and quotients of order statistics are subject to, *Mathematische Operationsforschung und Statistik*, **3**, 207–216.

Roy, D., and Mukherjee, S. P. (1991). Revival of a life-time distribution, *Economic Quality Control*, **6**, 164–173.

Ryu, K. (1993). An extension of Marshall and Olkin's bivariate exponential distribution, *Journal of the American Statistical Association*, **88**, 1458–1465.

Sahobov, O. M., and Geshev, A. (1974). Characteristic property of the exponential distribution, *Natura Universita Plovdiv*, **7**, 25–28. (In Russian)

Saleh, A. K. Md. E. (1966). Estimation of the parameters of the exponential distribution based on optimum statistics in censored samples, *Annals of Mathematical Statistics*, **37**, 1717–1735.

Saleh, A. K. Md. E., and Ali, M. M. (1966). Asymptotic optimum quantiles for the estimation of the parameters of the negative exponential distribution, *Annals of Mathematical Statistics*, **37**, 143–151.

Saleh, A. K. Md. E., Scott, C., and Junkins, D. B. (1975). Exact first and second order moments of order statistics from the truncated exponential distribution, *Naval Research Logistics Quarterly*, **22**, 65–77.

Sarhan, A. E. (1954). Estimation of the mean and standard deviation by order statistics, *Annals of Mathematical Statistics*, **25**, 317–328.

Sarhan, A. E. (1955). Estimation of the mean and standard deviation by order statistics, Part III, *Annals of Mathematical Statistics*, **26**, 576–592.

Sarhan, A. E., and Greenberg, B. G. (1958). Estimation problems in the exponential distribution using order statistics, In *Proceedings in Statistics for the Technical Missile Evaluation Symposium*, 123–173, Blacksburg, VA.

Sarhan, A. E., and Greenberg, B. G. (Editors) (1962), *Contributions to Order Statistics*, New York: Wiley.

Sarhan, A. E., Greenberg, B. G., and Ogawa, J. (1963). Simplified estimates for the exponential distribution, *Annals of Mathematical Statistics*, **34**, 102–116.

Sen, A., and Bhattacharyya, G. K. (1994). Inference procedures for the linear failure rate model (preprint).

Seshadri, V., Csörgö, M., and Stephens, M. A. (1969). Tests for the exponential distribution using Kolmogorov-type statistics, *Journal of the Royal Statistical Society, Series B*, **31**, 499–509.

Shanbhag, D. N. (1970). The characterizations for exponential and geometric distributions, *Journal of the American Statistical Association*, **65**, 1256–1259.

Shanbhag, D. N. (1977). An extension of the Rao-Rubin characterization of the Poisson distribution, *Journal of Applied Probability*, **14**, 640–646.

Shimizu, R. (1978). Solution to a functional equation and its application to some characterization problems, *Sankhyā, Series A*, **40**, 319–332.

Shimizu, R. (1979). A characterization of the exponential distribution, *Annals of the Institute of Statistical Mathematics*, **31**, 367–372.

Shimizu, R. (1980). Functional equation with an error term and the stability of some characterization of the exponential distributions, *Annals of the Institute of Statistical Mathematics*, **32**, 1–16.

Shimizu, R. (1981). On the stability of characterizations of the exponential distributions, *Annals of the Institute of Statistical Mathematics*, **33**, 339–346.

Shoukri, M. (1983). Some remarks on the Bayesian inference for the exponential model, *Metron*, **41**/1, 77–85.

Sibuya, M. (1962). On exponential and other random variable generators, *Annals of the Institute of Statistical Mathematics*, **13**, 231–237.

Siddiqui, M. M. (1963). Optimum estimators of the parameters of negative exponential distributions from one or two order statistics, *Annals of Mathematical Statistics*, **34**, 117–121.

Sinha, S. K., and Guttman, I. (1976). Bayesian inference about the reliability function for the exponential distribution, *Communications in Statistics*, **5**, 471–479.

Spinelli, J. J., and Stephens, M. A. (1987). Tests for exponentiality when origin and scale parameters are unknown, *Technometrics*, **29**, 471–476.

Spurrier, J. D. (1984). An overview of tests for exponentiality, *Communications in Statistics-Theory and Methods*, **13**, 1635–1654.

Srivastava, M. S. (1967). A characterization of the exponential distribution, *American Mathematical Monthly*, **74**, 414–416.

Srivastava, R. C. (1978). Some characterizations of the exponential distribution based on record values, Abstract, *Bulletin of the Institute of Mathematical Statistics*, **7**, 283.

Srivastava, R. C. (1981). Some characterizations of the exponential distribution based on record values, In *Statistical Distributions in Scientific Work*, **4**, C. Taillie, G. P. Patil, and B. Baldessari (editors), 417–418, Dordrecht: Reidel.

Srivastava, R. C., and Wang, Y. H. (1980). A characterization of the exponential and related distributions by linear regression, *Annals of Statistics*, **8**, 217–220.

Stephens, M. A. (1986). "Tests based on EDF Statistics," and "Tests based on regression," In *Goodness-of-fit Techniques*, R. B. d'Agostino and M. A. Stephens (editors), 97–234, New York: Marcel Dekker.

Steutel, F. W., and Thiemann, J. G. F. (1989). On the independence of integer and fractional parts, *Statistica Neerlandica*, **43**, 53–59.

Sukhatme, P. V. (1936). On the analysis of k samples from exponential population with special reference to the problem of random intervals, *Statistical Research Memoirs*, **2**, 94–112.

Sukhatme, P. V. (1937). Tests of significance for samples of χ^2-population with two degrees of freedom, *Annals of Eugenics*, **8**, 52–56.

Susarla, V., and Pathala, K. S. (1965). A probability distribution for the time of first birth, *Journal of Scientific Research*, Banaras Hindu University, **16**, 59–62.

Swanepoel, J. W. H. (1991). A note on the characterization of the exponential distribution by the covariance of spacings, *South African Statistical Journal*, **25**, 73–78.

Swartz, G. B. (1975). A short proof of a characterization by conditional expectations, *IEEE Transactions on Reliability*, **R-24**, 76–77.

Tallis, G. M., and Light, R. (1968). The use of fractional moments for estimating the parameters of a mixed exponential distribution, *Technometrics*, **10**, 161–175.

Talwalker, S. (1977). A note on characterization by the conditional expectation, *Metrika*, **24**, 129–136.

Tanis, E. A. (1964). Linear forms in the order statistics from an exponential distribution, *Annals of Mathematical Statistics*, **35**, 270–276.

Tata, M. N. (1969). On outstanding values in a sequence of random variables, *Zeitschrift für Wahrscheinlichkeitstheorie und Ihre Verwandte Gebiete*, **12**, 9–20.

Tiku, M. L. (1981). Testing equality of location parameters of two exponential distribution, *Aligarh Journal of Statistics*, **1**, 1–7.

Titterington, D. M., Smith, A. F. M., and Makov, U. E. (1985). *Statistical Analysis of Finite Mixture Distributions*, New York: Wiley.

Tong, H. (1974). A note on the estimation of $P(Y < X)$ in the exponential case, *Technometrics*, **16**, 625.

Tong, H. (1975). Letter to Editor, *Technometrics*, **17**, 395.

Trader, R. L. (1985). Bayesian inference for truncated exponential distributions, *Communications in Statistics—Theory and Methods*, **14**, 585–592.

Van Beek, P., and Braat, J. (1973). The limit of sequences of iterated overshoot distribution functions, *Stochastic Processes and Their Applications*, **1**, 307–316.

Varde, S. D. (1969). Life testing and reliability estimation for the two parameter exponential distribution, *Journal of the American Statistical Association*, **64**, 621–631.

Vardeman, S., and Ray, D. (1985). Average run lengths CUSUM schemes when observations are exponentially distributed, *Technometrics*, **27**, 145–150.

Vardi, Y., Shepp, L. A., and Logan, B. F. (1981). Distribution functions invariant under residual-lifetime and length-biased sampling, *Zeitschrift für Wahrscheinlichkeitstheorie und Ihre Verwandte Gebiete*, **56**, 415–426.

Vartak, M. N. (1974). Characterization of certain classes of probability distributions, *Journal of the Indian Statistical Association*, **12**, 67–72.

Veale, J. R. (1975). Improved estimation of expected life when one identified spurious observation may be present, *Journal of the American Statistical Association*, **70**, 398–401.

Villén-Altamizano, J. (1990). Confidence intervals for reliability functions of an exponential distribution under random censorship, *Kybernetika*, **26**, 462–472.

Viveros, R., and Balakrishnan, N. (1994). Interval estimation of parameters of life from progressively censored data, *Technometrics*, **36**, 84–91.

Waller, R. A., Johnson, M. M., Waterman, M. S., and Martz, H. F., Jr. (1977). Gamma prior distribution selection for Bayesian analysis of failure rate and reliability, Los Alamos Scientific Laboratory, *Report 6A-UR-155*.

Wang, Y. H. (1971). On characterization of some probability distributions and estimation of the parameters of the Pareto distribution, Ph.D. dissertation, Columbus: Ohio State University.

Wang, Y. H., and Srivastava, R. C. (1980). A characterization of the exponential and related distributions by linear regression, *Annals of Statistics*, **8**, 217–220.

Weiss, L. (1985). Maximum Probability Estimation, In *Encyclopedia of Statistical Sciences*, S. Kotz, N. L. Johnson, and C. B. Read (editors), **5**, 351–357, New York: Wiley.

Weiss, L., and Wolfowitz, J. (1967). Maximum probability estimators, *Annals of the Institute of Statistical Mathematics*, **19**, 193–206.

Westcott, M. (1981). Letter to the Editor, *Journal of Applied Probability*, **18**, 568.

Willemain, T. R., Allahvardi, A., De Sautels, P., Eldridge, J., Gur, O., Miller, M., Panos, G., Srinivasan, A., Surithadi, J., and Topal, E. (1992). Robust estimation methods for exponential data: A Monte-Carlo comparison, *Communications in Statistics—Simulation and Computation*, **21**, 1043–1075.

Witte, H.-J. (1988). Some characterizations of distributions based on the integrated Cauchy functional equation, *Sankhyā, Series A*, **50**, 59–63.

Witte, H.-J. (1990). Characterizations of distributions of exponential or geometric type by the integrated lack of memory property and record values, *Computational Statistics & Data Analysis*, **10**, 283–288.

Witte, H.-J. (1991). Some characterizations of exponential or geometric type distributions in a nonstationary record value model, Unpublished manuscript.

Wright, F. T., Engelhardt, M., and Bain, L. J. (1978). Inferences for the two-parameter exponential distribution under type I censored sampling, *Journal of the American Statistical Association*, **73**, 650–655.

Yanushkyevichene, O. L. (1983). Estimate of stability of a characterization of the exponential law, *Theory of Probability and Its Applications*, **28**, 116–128.

Yanushkyavichyus, R. L. (1989). *Investigation of stability of characterization problems of mathematical statistics* (Abstract) Ph.D. dissertation, Vilnius University, Lithuania (appeared as a monograph in 1991).

Yeo, G. F., and Milne, R. K. (1989). On characterizations of exponential distributions, *Statistics & Probability Letters*, **7**, 303–305.

Zabransky, F., Sibuya, M., and Saleh, A. K. Md. E. (1966). Tables for the estimation of the exponential distribution (Review of an unpublished Mathematical Table, File No. 92), *Mathematics of Computation*, **20**, 621.

Zacks, S., and Even, M. (1966). The efficiencies in small samples of the maximum likelihood and best unbiased estimators of reliability functions, *Journal of the American Statistical Association*, **61**, 1033–1051.

Zelen, M., and Dannemiller, M. C. (1961). The robustness of life testing procedures derived from the exponential distribution, *Technometrics*, **3**, 29–49.

Zidek, J. V. (1973). Estimating the scale parameter of the exponential distribution with unknown location, *Annals of Statistics*, **1**, 264–278.

CHAPTER 20

Pareto Distributions

1 INTRODUCTION

Since the publication of the first edition of this book, the literature on Pareto distributions has grown, perhaps faster than for any other continuous univariate distribution. To survey comprehensively the new material would require much more space than is available for this distribution. Fortunately, the availability of an excellent text by B. C. Arnold (1983), devoted solely (except for the last chapter) to the univariate Pareto and related distributions allows us to concentrate on topics that are only briefly discussed in that text (e.g., estimation using order statistics) and also to emphasize developments since 1982. Besides a thorough discussion of characterization and measures of inequality that are closely associated with Pareto distributions, Arnold's monograph provides a comprehensive historical overview.

2 GENESIS

The Pareto distribution is named after an Italian-born Swiss professor of economics, Vilfredo Pareto (1848–1923). Pareto's law, as formulated by him (1897), dealt with the distribution of income over a population and can be stated as follows:

$$N = Ax^{-a}$$

where N is the number of persons having income $\geq x$, and A, a are parameters (a is known both as *Pareto's constant* and as a shape parameter). It was felt by Pareto that this law was universal and inevitable—regardless of taxation and social and political conditions. "Refutations" of the law have been made by several well-known economists over the past 60 years [e.g., Pigou (1932); Shirras (1935); Hayakawa (1951)]. More recently attempts have been made to explain many empirical phenomena using the Pareto distribution or some closely related form [e.g., Steindl (1965); Mandelbrot (1960, 1963, 1967); Hagstroem (1960); Ord (1975)].

Harris (1968) has pointed out that a mixture of exponential distributions, with parameter θ^{-1} having a gamma distribution, and with origin at zero, gives rise to a Pareto distribution. [See also Maguire, Pearson, and Wynn (1952).] In fact, if

$$\Pr[X \leq x|\theta] = 1 - e^{-x/\theta}$$

and $\mu = \theta^{-1}$ has a gamma distribution [(17.2) of Chapter 17], then

$$\Pr[X \leq x] = \frac{1}{\beta^{\alpha}\Gamma(\alpha)} \int_0^{\infty} t^{\alpha-1} e^{-t/\beta}(1 - e^{-tx}) \, dt$$

$$= 1 - \frac{1}{\beta^{\alpha}\Gamma(\alpha)} \int_0^{\infty} t^{\alpha-1} e^{-t(x+\beta^{-1})} \, dt$$

$$= 1 - (\beta x + 1)^{-\alpha},$$

which is of the form (20.4) below.

3 DEFINITIONS

The Pareto distribution has survived in its original form as

$$\bar{F}_X(x) = \Pr[X \geq x] = \left(\frac{k}{x}\right)^a, \qquad k > 0, a > 0, x \geq k, \qquad (20.1)$$

where $\bar{F}_X(x)$ is the probability that the income is equal to or greater than x and k represents some minimum income. As a consequence of (20.1), the cumulative distribution function of X, representing income, may be written as

$$F_X(x) = 1 - \left(\frac{k}{x}\right)^a, \qquad k > 0, a > 0, x \geq k, \qquad (20.2)$$

and the corresponding pdf is

$$p_X(x) = ak^a x^{-(a+1)}, \qquad k > 0, a > 0, x \geq k. \qquad (20.3)$$

This is a special form of Pearson Type VI distribution; we will denote it by $X \sim P(I)(k, a)$. The relation given by (20.2) is now more properly known as the *Pareto distribution of the first kind*.

For Bayesian analysis a slightly different parametrization is sometimes found to be convenient (see Section 6). In economic literature, special attention is given to determination of an appropriate value of the parameter a. For income distributions, the assumption that $a > 1$ seems to be quite realistic. In Bresciani-Turroni (1939) we find the comment:

> The actual values of "a" ... fall within comparatively narrow limits. They oscillate about 1.5, and deviations from this value appear to be caused [more] by imperfections of statistical material than by the real cause.

Some 30 years later, Cramer (1971, p. 57) remarked

> The values of "a" have increased from between 1.6 and 1.8 in the nineteenth century to between 1.9 and 2.1 in the developed countries at the present time.

Two other forms of this distribution were proposed by Pareto. One, now referred to as the *Pareto distribution of the second kind* (sometimes *Lomax distribution*), is given by

$$F_X(x) = 1 - \frac{C^a}{(x + C)^a}, \qquad x \geq 0. \tag{20.4}$$

This is also a Pearson Type VI distribution. It will be denoted by $X \sim$ P(II)(C, a). Lomax (1954) used this distribution in the analysis of business failure data.

The *standard* P(II) distribution has $C = 1$; its pdf is

$$p_X(x) = a(1 + x)^{-a-1}, \qquad x > 0, a > 0, \tag{20.5a}$$

and the survival function is

$$\bar{F}_X(x) = (1 + x)^{-a}, \qquad x > 0, a > 0. \tag{20.5b}$$

Introducing a location parameter μ, the P(II) distribution's survival function can be represented as

$$\bar{F}_X(x) = \left\{1 + \frac{x - \mu}{C}\right\}^{-a}, \qquad x > \mu; C, a > 0. \tag{20.4}'$$

In most applications μ is nonnegative. The shape parameter a is often assumed to exceed 1, so as to ensure that the expected value is finite. We will denote this distribution by P(II)(μ, C, a).

The third distribution proposed by Pareto—the *Pareto distribution of the third kind*—has the cdf

$$F_X(x) = 1 - \frac{Ce^{-bx}}{(x + C)^a}, \qquad x > 0. \tag{20.6}$$

It can be denoted P(III)($a, b; C$).

In this chapter we will concentrate mainly on the simple Pareto (I) and Pareto (II) distributions. Arnold (1983) defines a Pareto distribution by the cdf

$$F_X(x) = \left[1 + \left(\frac{x - \mu}{\sigma}\right)^{1/\gamma}\right]^{-\alpha}, \qquad x > \mu; \alpha, \gamma, \sigma > 0. \qquad (20.7)$$

We will denote this distribution as P(IV)($\mu, \sigma, \gamma, \alpha$). Note that

$$P(I)(\sigma, \alpha) \equiv P(IV)(\sigma, \sigma, 1, \alpha) \quad \text{and}$$

$$P(II)(\mu, \sigma, \alpha) \equiv P(IV)(\mu, \sigma, 1, \alpha).$$

Feller (1971) defines a Pareto distribution as the distribution of $W = Y^{-1} - 1$ when Y has a standard beta (γ_1, γ_2) distribution. The pdf of Y is

$$p_Y(y) = \{B(\gamma_1, \gamma_2)\}^{-1} y^{\gamma_1 - 1}(1 - y)^{\gamma_2 - 1}, \qquad 0 < y < 1; \gamma_1, \gamma_2 > 0. \qquad (20.8)$$

If $\gamma_2 = 1$ the distribution of W is a Lomax distribution (20.4). Arnold (1983) pointed out that the variable

$$\mu + \sigma(Y^{-1} - 1)^{\gamma}$$

is a generalization of the Pareto (IV) variable. This in turn is called a *generalized F variable* if $\mu = 0$ [Kalbfleisch and Prentice (1980)]; see Chapter 27.

A basic property of a standard (P(I)(k, a)) Pareto variable X is that

$$X = k \exp\left(\frac{V}{a}\right), \qquad (20.9)$$

where V is a standard exponential variable ($p_V(v) = \exp(-v)$, $v > 0$). [Equivalently $2a \log(X/k)$ has a χ^2 distribution with two degrees of freedom.]

This intimate relationship between P(I) and exponential distributions permits us to obtain a wide range of properties of the former—in particular, characterizations—from properties of the latter. (See Chapter 19, Section 8, for characterizations of exponential distributions.) Takano (1992) has discussed recently a relationship between the Pareto distribution and a differential equation. An expression for the U-measure of the Pareto density has also been given by Takano (1992), where it is considered as a generalized gamma convolution (see Chapter 17, Section 8.4).

4 MOMENTS AND OTHER INDICES

4.1 Moments

Provided that r is less than a, the rth moment about zero of a $P(I)(k, a)$ variable is

$$\mu'_r = ak^r(a - r)^{-1} = k^r\left(1 - \frac{r}{a}\right)^{-1}, \qquad r < a. \qquad (20.10)$$

In particular the expected value is

$$E[X] = ak(a - 1)^{-1}, \qquad a > 1, \qquad (20.11a)$$

and the variance is

$$\text{Var}(X) = ak^2(a - 1)^{-2}(a - 2)^{-1}, \qquad a > 2. \qquad (20.11b)$$

Also

$$\alpha_3(X) = \sqrt{\beta_1(X)} = 2\frac{a + 1}{a - 3}\sqrt{\frac{a - 2}{a}}, \qquad a > 3, \qquad (20.11c)$$

$$\alpha_4(X) = \beta_2(X) = \frac{3(a - 2)(3a^2 + a + 2)}{a(a - 3)(a - 4)}, \qquad a > 4. \qquad (20.11d)$$

As $a \to \infty$, $\alpha_3(x) \to 2$, and $\alpha_4(X) \to 9$. The mean deviation is $2k(a - 1)^{-1}(1 - a^{-1})^{a-1}$ $(a > 1)$ and

$$\frac{\text{Mean deviation}}{\text{Standard deviation}} = 2(1 - 2a^{-1})^{1/2}(1 - a^{-1})^{a-1}, \qquad a > 2. \qquad (20.12)$$

The value of this ratio is 0.513 when $a = 3$, and 0.597 when $a = 4$. As a tends to infinity, the ratio tends to $2e^{-1} = 0.736$. Cohen and Whitten (1988) provide tables of the skewness $[\alpha_3(X)]$ and kurtosis $[\alpha_4(X)]$ of $P(I)(k, a)$ as functions of a for $a = 4.1, 4.2(0.2)19.0, 20, 30, 50, 100$, and the expected value and variance as functions of a and k.

4.2 Alternative Measures of Location

Although the expected value does not always exist, the following alternative measures of location do exist. The population geometric mean is

$$g = k \exp\left(\frac{1}{a}\right). \qquad (20.13)$$

The population harmonic mean $(\{E[X^{-1}]\}^{-1})$ is

$$h = k(1 + a^{-1}).$$ (20.14)

The population median is

$$m = 2^{1/a}k.$$ (20.15)

One might also use the mode k, though this index does not take into account the value of a. [The mode of $P(II)(\mu, C, a)$ is at μ.]

4.3 Measures of Inequality

Arnold (1983) emphasizes that the value of the variance is of little importance in the context of modeling income distributions, as compared with some other measures of inequality. We note the Gini index, which for $P(I)(k, a)$ is

$$G = (2a - 1)^{-1}, \qquad a > 1.$$ (20.16)

Also the Lorenz curve is

$$L(u) = 1 - (1 - u)^{(a-1)/a}, \qquad a > 1.$$ (20.17)

An alternative expression for G is

$$1 - \frac{E[X'_{1:2}]}{E[X]},$$ (20.16)'

from which an extension to a generalized *nth-order Gini index*

$$G_n = 1 - \frac{E[X'_{1:n+1}]}{E[X'_{1:n}]},$$ (20.16)''

has been made. [Here $X'_{1:m}$ denotes the first order statistic (least value) in a random sample of size m.] For the $P(I)(k, a)$ distribution

$$G_n = \{n(n + 1)a - n\}^{-1}. \tag{20.18}$$

Other classical measures of inequality are the average income of poorer-than-average individuals

$$m_L = \frac{1}{F_X(E[X])} \int_k^{E[X]} x \, dF_X(x)$$

$$\left(= \left(\frac{a}{a-1}\right)\left\{1 - \left(\frac{a}{a-1}\right)^{1-a}\right\}\left\{1 - \left(\frac{a}{a-1}\right)^{-a}\right\}^{-1} \right.$$

$$\left. \text{for } P(I)(k, a) \right) \tag{20.19}$$

and the average income of richer-than-average individuals

$$m_U = \frac{1}{1 - F_X(E[X])} \int_{E[X]}^{\infty} x \, dF_X(x)$$

$$\left(= \left(\frac{a}{a-1}\right)^2 \quad \text{for } P(I)(k, a) \right). \tag{20.20}$$

The hazard rate is

$$r(x) = ax^{-1}, \tag{20.21}$$

and the cumulative hazard rate is

$$R(x) = \int_k^x r(t) \, dt = a \log\left(\frac{x}{k}\right). \tag{20.22}$$

Note that the hazard rate is a decreasing function of x. However, the *elasticity* $[d \log F(x)/d \log x]$ is constant (equal to a).

Ahmed, Haq, and Khurshid (1990) have discussed the peakedness of the Pareto distribution through a slight modification of a measure of peakedness given by Horn (1983). The modification has been made in order to take into account the skewness of the Pareto distribution.

5 ESTIMATION OF PARAMETERS

In this section we will suppose (*unless otherwise explicitly stated*) that $X_1, X_2, \ldots, X_n$ are independent random variables each distributed as in (20.2).

5.1 Least-Squares Estimators

Rearranging (20.2) and taking logarithms of both sides, we obtain

$$\log[1 - F_X(x)] = a \log k - a \log x. \qquad (20.23)$$

The parameters a and k may be estimated by least-squares from sample estimates of $F_X(x)$, using as the dependent variable the logarithm of 1 minus the cumulative distribution of the sample. The least-squares estimator of a is then

$$\tilde{a} = \frac{-n\sum_{i=1}^n \log X_i \, \log[1 - F_X(X_i)] + (\sum_{i=1}^n \log X_i)(\sum_{i=1}^n \log[1 - F_X(X_i)])}{n\sum_{i=1}^n (\log X_i)^2 - (\sum_{i=1}^n \log X_i)^2}.$$

$$(20.24)$$

The corresponding least-squares estimator of k may be obtained by substituting into (20.23) the arithmetic mean values of the dependent and independent variables along with the estimator $\tilde{a}$ and solving for k. Estimators of the parameters obtained by least-squares methods have been shown to be consistent [Quandt (1966)].

5.2 Estimators from Moments

Provided that $a > 1$, the mean of the Pareto distribution exists and is given by (20.11a). By equating this to the sample mean $\overline{X}$ and rearranging, we obtain the relation

$$a^* = \frac{\overline{X}}{\overline{X} - k^*} \qquad (20.25)$$

between estimators a^*, k^* of a and k. A formula for k^* may be found by the following argument [Quandt (1966)]: The probability that all n of the X_i's are greater than a particular value x is $(k/x)^{an}$. Let $F_{X_1'}(x)$ be the cumulative distribution function of the smallest sample value. Then

$$F_{X_1'}(x) = 1 - \left(\frac{k}{x}\right)^{an}, \qquad x \geq k. \qquad (20.26)$$

The corresponding density function is

$$p_{X_1'}(x) = \frac{ank^{an}}{x^{an+1}}. \qquad (20.27)$$

From (20.27) the expected value of X_1' (the smallest sample value) is

$$E[X_1'] = \frac{ank}{an-1}. \qquad (20.28)$$

Equating X_1' to $E[X_1']$ in (20.28), the estimator of k is found to be

$$k^* = \frac{(a^*n - 1)X_1'}{a^*n}, \qquad (20.29)$$

and substituting this in (20.25), we obtain

$$a^* = \frac{n\bar{X} - X_1'}{n(\bar{X} - X_1')}. \qquad (20.30)$$

The estimators a^* and k^* are consistent [Quandt (1966)].

5.3 Maximum Likelihood Estimation

The likelihood function for a sample $(X_1, \ldots, X_n)$ from a Pareto distribution is

$$L = \prod_{j=1}^{n} \frac{ak^a}{X_j^{a+1}}. \qquad (20.31)$$

Taking logarithms of both sides, differentiating partially with respect to the parameter a, and setting the result to zero we find the relation

$$\hat{a} = n \left[\sum_{j=1}^{n} \log\left(\frac{X_j}{\hat{k}}\right) \right]^{-1} \qquad (20.32)$$

between the maximum likelihood estimators $\hat{a}, \hat{k}$ of a, k, respectively.

A second equation (corresponding to $\partial \log L / \partial k = 0$) cannot be obtained in the usual way, since $\log L$ is unbounded with respect to k. Since k is a lower bound on the random variable X, $\log L$ must be maximized subject to the constraint

$$\hat{k} \leq \min_i X_i. \qquad (20.33)$$

By inspection, the value of $\hat{k}$ that minimizes (20.31) subject to (20.33) is

$$\hat{k} = \min_i X_i = X_1'; \qquad (20.34)$$

$\hat{a}$ may be expressed as a function of $\hat{k}$ and the geometric mean by substituting $\hat{g} = (\prod_{i=1}^{n} X_i)^{1/n}$ into (20.32) to yield

$$\hat{a} = \left[\log\left(\frac{\hat{g}}{\hat{k}} \right) \right]^{-1}. \tag{20.35}$$

We have

$$E[\hat{k}] = \frac{nka}{na - 1}, \qquad n > \frac{1}{a}, \tag{20.36a}$$

$$\mathrm{Var}(\hat{k}) = \frac{nka^2}{(na - 1)^2(na - 2)}, \qquad n > \frac{2}{a}; \tag{20.36b}$$

hence

$$\mathrm{MSE}(\hat{k}) = \frac{2k^2}{(na - 1)(na - 2)}, \qquad n > \frac{2}{a}. \tag{20.36c}$$

Also

$$E[\hat{a}] = \frac{na}{n - 2}, \qquad n > 2, \tag{20.37a}$$

$$\mathrm{Var}(\hat{a}) = \frac{n^2 a^2}{(n - 2)^2(n - 3)}, \qquad n > 3; \tag{20.37b}$$

hence

$$\mathrm{MSE}(\hat{a}) = \frac{a^2(n^2 + 4n - 12)}{(n - 2)(n - 3)}, \qquad n > 3. \tag{20.37c}$$

It has been shown by Quandt (1966) that both $\hat{a}$ and $\hat{k}$ are consistent estimators of a and k, respectively.

Note that $2na/\hat{a}$ is distributed as χ^2 with $2(n - 1)$ degrees of freedom. A $100(1 - \alpha)\%$ confidence interval for a is given by

$$\left(\hat{a}\chi^2_{2(n-1),\frac{\alpha}{2}}(2n)^{-1}, \hat{a}\chi^2_{2(n-1),1-\frac{\alpha}{2}}(2n)^{-1} \right). \tag{20.38}$$

The pdf of $\hat{k}$ is

$$p_{\hat{k}}(y) = nak^{na}y^{-na-1}, \qquad y > 0.$$

The set $(\hat{k}, \hat{a})$ is jointly sufficient for (k, a), and $\hat{k}$ and $\hat{a}$ are mutually independent [Malik (1970)]. Also $\hat{k}$ is sufficient for k when a is known and $\hat{a}$ is sufficient for a when k is known.

From (20.37a) and (20.37b),

$$a^* = (1 - 2n^{-1})\hat{a} \qquad (20.39)$$

and

$$k^* = \{1 - (n - 1)^{-1}\hat{a}^{-1}\}\hat{k} \qquad (20.40)$$

are unbiased estimators of a and k, respectively. The variance of k^* is

$$\mathrm{Var}(k^*) = a^{-1}(n - 1)^{-1}(na - 2)^{-1}k^2, \qquad n > \frac{2}{a}. \qquad (20.41)$$

Since $(\hat{a}, \hat{k})$ is a sufficient statistic for (a, k), and (a^*, k^*) is a function of $(\hat{a}, \hat{k})$, (a^*, k^*) is the uniform minimum variance estimator of (a, k). Its ellipse of concentration is contained within the ellipse of concentration of every other unbiased estimator of (a, k). From this point of view it may be called the "best" unbiased estimator of (a, k), though some biased estimators might be "better" in terms of appropriate loss functions.

If k is known the maximum likelihood estimator of a is

$$\hat{a}' = \left[\log\left(\frac{\hat{g}}{k}\right)\right]^{-1}. \qquad (20.42)$$

Note that $2na/\hat{a}'$ is distributed as χ^2 with $2n$ degrees of freedom. A $100(1 - \alpha)\%$ confidence interval for a is given by

$$\left(\hat{a}'\chi^2_{2n,\frac{\alpha}{2}}(2n)^{-1}, \hat{a}'\chi^2_{2n,1-\frac{\alpha}{2}}(2n)^{-1}\right). \qquad (20.43)$$

[Compare with (20.38) to assess the greater accuracy accruing from knowledge of the value of k.] Also

$$E[\hat{a}'] = \frac{na}{n - 1}, \qquad (20.44a)$$

$$\mathrm{Var}(\hat{a}') = \frac{n^2a^2}{(n - 1)^2(n - 2)}. \qquad (20.44b)$$

Maximum likelihood estimation from grouped data for $P(I)(k, a)$ when k is known (i.e., estimation of a when k is known) has been discussed by Fisk (1961) and Aigner and Goldberger (1970). Aigner and Goldberger apply a general method of Kulldorff (1961) for maximum likelihood estimation from grouped data, to estimation of a from $P(I)$ distributions with $k = 1$ [$P(I)(1, a)$ distributions].

Let $x_0 = 1 < x_1 < x_2 < \cdots < x_T < x_{T+1} = \infty$ be the boundaries of the $(T + 1)$ groups, and denote the number of observations, in a random sample of size n, falling into the group bounded by x_t and x_{t+1}, by N_t ($t = 0, 1, \ldots, T$; $\sum_{t=0}^{T} N_t = n$). The maximum likelihood estimator of a is the solution of the equation

$$\sum_{t=0}^{T-1} N_t \left(\frac{x_{t+1}^{-a} \log x_{t+1} - x_t^{-a} \log x_t}{x_{t+1}^{-a} - x_t^{-a}} \right) + N_T \log x_T = 0. \qquad (20.45)$$

Although the left-hand side of (20.45) is highly nonlinear in a, Aigner and Goldberger (1970) assert that it is usually amenable to iterative solution, or simply to trial-and-error search over a relevant range of values for a. It should be noted, however, that zero values of N_t pose problems for maximum likelihood estimation. It is recommended that such group intervals be disregarded (though this is likely to lead to bias in the estimator).

If the group boundaries are in equal ratios, with $x_t = cx_{t-1}$ for $t = 1, \ldots, T$ (and $x_0 = k$), then the maximum likelihood estimator of a is

$$\frac{1}{\log c} \log\left(1 + \frac{n}{\sum_{t=0}^{T} tN_t}\right). \qquad (20.46)$$

(The grouped data have a truncated log-geometric distribution; see Chapter 5, Section 11).

5.4   Estimation Based on Order Statistics

For P(I)(k, a) distribution both the moment estimation (Section 5.2) and the maximum likelihood estimation (Section 5.3) may be modified by using the equation

$$X_1' \left(= E[X_1']\right) = nak(na - 1)^{-1} \qquad (20.47)$$

together with (20.25) or (20.32), respectively. Estimation of k and a, using estimators of two quantiles, is simple to apply, in the way described below.

Select two numbers P_1 and P_2 between 0 and 1 and obtain estimators of the respective quantiles $\hat{X}_{P_1}$ and $\hat{X}_{P_2}$. [In large samples it is possible to take $\hat{X}_{P_j} = X'_{(n+1)P_j}$.] Estimators of a and k are then obtained by solving the two simultaneous equations

$$P_j = 1 - \left(\frac{k}{\hat{X}_{P_j}}\right)^a, \qquad j = 1, 2. \qquad (20.48)$$

The estimator of a is

$$\frac{\log[(1 - P_1)/(1 - P_2)]}{\log\left(\hat{X}_{P_2}/\hat{X}_{P_1}\right)}. \tag{20.49}$$

The corresponding estimator of k can be obtained from (20.48). It has been shown that the estimators of a and k obtained by this method are consistent [Quandt (1966)] (see also Section 5.8).

Koutrouvelis (1981) extends the quantile method of estimation to the use of suitably chosen order statistics $X'_{n_1}, X'_{n_2}, \ldots, X'_{n_k}$ from a random sample of (large) size n from a P(I)(k, a) distribution (see also Section 5.8). The idea is to reduce the problem to an exponential distribution by noting that $Y'_{n_i} = \log X'_{n_i}$ is an order statistic from a population with cdf $F_Y(y) = 1 - \exp(-(y - \eta)/\sigma)$, where $\eta = \log k$ and $\sigma = 1/a$. The estimators $\hat{\eta}$ and $\hat{\sigma}$ are obtained from the weighted linear regression fit to $E[\log X'_{n_i}] = \eta - \sigma \log(1 - t_i)$ $(i = 1, \ldots, k)$, where $t_i = n_i/n$. They are given by

$$\tilde{\sigma} = \sum_{i=1}^{n} b_i \log X'_{n_i}, \tag{20.50a}$$

$$\tilde{\eta} = \log X'_{n_1} - \tilde{\sigma} u_1, \tag{20.50b}$$

where

$$b_1 = -\frac{(u_2 - u_1)/\{\exp(u_2) - \exp(u_1)\}}{L},$$

$$b_i = \frac{\{(u_i - u_{i-1})/[\exp(u_i) - \exp(u_{i-1})]\} - \{(u_{i+1} - u_i)/[\exp(u_{i+1}) - \exp(u_i)]\}}{L},$$

$$i = 2, 3, \ldots, k - 1,$$

$$b_k = \frac{(u_k - u_{k-1})/\{\exp(u_k) - \exp(u_{k-1})\}}{L},$$

with

$$u_i = -\log(1 - t_i), \qquad i = 1, 2, \ldots, k,$$

$$L = \sum_{i=2}^{k} \frac{(u_i - u_{i-1})^2}{\exp(u_i) - \exp(u_{i-1})}.$$

These estimators are asymptotically BLUEs of η and σ, respectively [Ogawa (1962)], while the estimators $\tilde{k} = \exp(\tilde{\eta})$ and $\tilde{a} = 1/\tilde{\sigma}$ (with $\tilde{\eta}$ and $\tilde{\sigma}$ given above) remain asymptotically unbiased but no longer linear. Koutrouvelis

(1981) has shown that $\tilde{k}$ and $\tilde{a}$ are consistent and asymptotically normal with means k and a and variances

$$\text{Var}(\tilde{k}) \sim \frac{k^2\left(u_1^2 L^{-1} + \exp(u_1) - 1\right)}{na^2}, \tag{20.51a}$$

$$\text{Var}(\tilde{a}) \sim a^2/(nL). \tag{20.51b}$$

Since the determinant of the variance-covariance matrix of $\hat{k}$ and $\hat{a}$ is $k^2\{\exp(u_1) - 1\}/(n^2 L)$, the optimal values for $u_1, \ldots, u_k$ are obtained by maximizing the function $\{\exp(u_1) - 1\}^{-1} L$ with respect to $u_1, u_2, \ldots, u_k$ subject to the constraint $0 < u_1 < u_2 < \cdots < u_k < \infty$. The values of t_1 and of $u_1 = -\log(1 - t_1)$ must be positive. Saleh and Ali (1966) propose to first take the optimal $n_1^0 = 1$ and $t_1^0 = n_1^0/(n + \frac{1}{2})$ for fixed $t_2, \ldots, t_k$. Thus $u_1^0 = \log\{1 - 1/(n + \frac{1}{2})\}^{-1}$. The optimal choice of $u_2, \ldots, u_k$ is determined after replacing u_1 by u_1^0. Koutrouvelis (1981) provides additional details. Simulation investigations carried out by Koutrouvelis (1981) show that for parameter k, his method using two optimal sample quantiles performs almost as well as using five optimal quantiles (for $n = 25, 100,$ and 500), while for parameter a the estimator based on five sample quantiles is substantially more efficient than the two quantile procedure, and competes in efficiency with the maximum likelihood estimator. For all the cases considered $\text{RE}(\tilde{a}) \geq 0.697$ for the five quantile procedure and $\text{RE}(\tilde{a}) \geq 0.584$ for two quantile procedure ($n = 25, 100, 500$, $k = 1$, and $a = 0.5, 1.5,$ and 4.5). He also discovered that even a slight change of the definition of sample quantile has a significant effect on the values of moments of $\tilde{a}$.

For the three-parameter $P(II)(\mu, C, a)$ distribution (20.4)', the construction of BLUEs for μ and C, *when a is known* has received much attention. Kulldorff and Vannman (1973) and Vannman (1976) pioneered this work. Charek, Moore, and Coleman (1988) compared BLUEs with six minimum distance estimators of μ and C [based on Kolmogorov (D) distance, Cramér-von Mises distance and the Anderson-Darling statistic, each applied with two different minimization routines]. They found that for the values $a = 1(1)4$ and sample sizes $n = 6(3)18$, the BLUE for C had the smallest mean square error, and that the BLUE for μ also has advantageous properties.

The BLUEs developed by Kulldorff and Vannman (1973) for the case $a > 2$, are calculated from the equations

$$\tilde{\mu} = X_1' - Y[(na - 1)(na - 2) - naD]^{-1}, \tag{20.52a}$$

$$\tilde{C} = (X_1' - \tilde{\mu})(na - 1), \tag{20.52b}$$

where

$$B_0 = 1, \ B_i = \{1 - 2a^{-1}(n - i + 1)^{-1}\}B_{i-1}, \qquad i = 1, 2, \ldots, n,$$

$$D = (a + 1) \sum_{i=1}^{n} B_i + (a - 1)B_n,$$

$$Y = (a + 1) \sum_{i=1}^{n} B_i X_i' + (a - 1)B_n X_n' - DX_1'. \tag{20.52c}$$

Vannman's (1976) estimators, developed for the case $a < 2$, are given by

$$\tilde{\mu} = X_2' - \frac{\tilde{C}}{na - 1}, \tag{20.53a}$$

$$\tilde{C} = \frac{1}{U} \left[(a + 1) \sum_{i=1}^{n-2} B_i X_i' + (2a - 1)B_{n-1} X_{n-1}' \right.$$

$$\left. - \left(\frac{na - 1}{na} \right)(na - 2 - U)X_1' \right], \tag{20.53b}$$

where

$$U = \frac{(na - 2)(na - a - 2) - na(a - 2)B_{n-1}}{(na - 1)(a + 2)}$$

and the B_i's are as defined in (20.52c). For $2/n < a \le 2$, with $2/a$ an integer, Vannman's estimators simplify to

$$\tilde{C} = (na - 1)(X_1' - \hat{\mu}), \tag{20.53a}'$$

$$\tilde{\mu} = X_1' - \frac{(a + 1)(a + 2)}{(na - 2)(na - a - 2)} \left[\sum_{i=1}^{n-(2/a)} B_i X_i' + \frac{na - 2}{a + 2} X_1' \right]. \tag{20.53b}'$$

Note that in (20.53b)' only the first $n - (2/a)$ order statistics are used. Asymptotically BLUEs of C, when a is known, have been considered by Chan and Cheng (1973), who also dealt with censored cases.

For small sample sizes, calculation of BLUEs is fairly simple, but it becomes tedious as n increases. Vannman (1976) suggests using BLUEs based on the first k order statistics, with k considerably less than n. Arnold (1983) notes that the latter order statistics may be unstable.

5.5 Sequential Estimation

Hamdy and Pallotta (1987) constructed a three-stage procedure for estimating the parameters of P(I)(k, a). From a random sample $X_1, X_2, \ldots, X_n$ of size n, we have estimators

$$k_n^* = X_1' = \min(X_1, \ldots, X_n), \tag{20.54a}$$

$$a_n^* = (n - 1)\left\{ \sum_{i=1}^{n} \log\left(\frac{X_i}{X_1'}\right) \right\}^{-1}, \tag{20.54b}$$

of k and a, respectively. Hamdy and Pallotta (1987) assume that the optimal sample size needed to estimate k is of the form

$$n_0 = \beta a^{-\gamma}, \qquad \beta, \gamma > 0, \tag{20.55}$$

where β and γ are *known* constants. If a also were known, then n_0 would be determined. If a is not known, the following three-stage procedure is proposed: Start with a sample of size m (≥ 2), and take an "appropriate" constant θ. If $[\theta\beta a_m^{*-\gamma}] + 1 > m$ take an additional sample of size $[\theta\beta a_m^{*-\gamma}] + 1 - m$ so that the total sample size is $R^* = \max(m, [\theta\beta a_m^{*-\gamma}] + 1)$. In the third stage we make the final sample size to be

$$N^* = \max(R^*, [\beta a_{R^*}^{*-\gamma}] + 1)$$

by taking $N^* - R^*$ additional observations. Hamdy and Pallotta suggest taking $\theta = \frac{1}{2}$, and assert that their method is a practical alternative to one-at-a-time sequential sampling.

5.6 Minimax Estimation

Kaluszka (1986) has shown that for the P(I)(k, a) distribution, a minimax and admissible estimator (with mean square error loss function) of a', if k is known, is

$$\frac{\Gamma(n - r)}{\Gamma(n - 2r)}\left(-n \log k + \sum_{i=1}^{n} \log X_i \right)^{-r}, \qquad r < \frac{n}{2}. \tag{20.56}$$

[See also Iliescu and Vodă (1979).] The result is obtained by noting that the distribution of $\sum_{i=1}^{n} \log(X_i/k)$ is gamma (n, a^{-1}) and then applying a general method of Kaluszka (1986). [See also Chapter 17, Section 8.]

5.7 Estimation of Pareto Densities

Asrabadi (1990) discusses estimation of the $P(I)(k, a)$ pdf

$$p_X(x) = \frac{ak^a}{x^{a+1}}, \qquad x \geq k > 0, a > 0, \qquad (20.57)$$

using the popular method of finding the UMVUE of this simple function of the parameters. If k is known, then $T = \prod_{i=1}^n X_i$ is a complete sufficient statistic for the parameter a. Thus the conditional pdf of X_1, given T, is a function of T only, $p_X^*(x|T)$, say, and this is the UMVUE of $p_X(x)$. We have

$$p_X^*(x) = \frac{(n-1)\{\log T - \log x - (n-1)\log k\}^{n-2}}{x(\log T - n \log k)^{n-1}}, \qquad k \leq x \leq Tk^{-n+1}.$$

$$(20.58)$$

The corresponding estimate of the cdf

$$F_X(x) = 1 - \left(\frac{k}{x}\right)^a, \qquad x \geq k > 0, \qquad (20.2)'$$

is

$$F_X^*(x) = \begin{cases} 0 & \text{for } x < k, \\ 1 - \dfrac{\{\log T - \log x - (n-1)\log k\}^{n-1}}{(\log T - n \log k)^{n-1}} & \text{for } k \leq x \leq Tk^{-n+1}, \\ 1 & \text{for } n \geq Tk^{-n+1}. \end{cases}$$

$$(20.59)$$

This is the UMVUE of $F_X(x)$. Similarly the rth moment of $p_X^*(x)$ is the UMVUE of the rth moment of the $P(I)(k, a)$ distribution.

After successive integrations by parts we find

$$(E[X^r])^* = \frac{(n-1)!k^r r^{n-1}}{(\log T - n \log k)^{-n+1}}$$

$$\times \left\{ Tk^{-rn} - 1 - \sum_{i=1}^{r-2} \frac{(\log T - n \log k)^i r^i}{i!} \right\}. \qquad (20.60)$$

We emphasize that these simple expressions are due to the especially simple form of the pdf and cdf of the classical $P(I)(k, a)$ distribution (see analogous results for the exponential distribution in Chapter 19).

Both $p_X(x)$ and $p_X^*(x)$ are decreasing functions of x for $x \geq k$ (and equal to zero for $x < k$). The value of $p_X^*(x)$ at $x = k$ is $(n - 1)k^{-1}\{\log T - n \log k\}^{-1}$, which is the UMVUE of $k^{-1}a$, the value of $p_X(x)$ at $x = k$. Note that

$$a^* = \frac{n - 1}{n \log k} \qquad (20.61)$$

is the UMVUE of a and that the distribution of a^{*-1} is gamma $(n, a^{-1}(n - 1)^{-1})$, so $2(n - 1)a/a^*$ has a χ^2_{2n} distribution (see Chapter 18). Also

$$\mathrm{Var}(a^*) = a^2(n - 2)^{-1} \qquad (20.62)$$

and $a^{*2}(n - 1)^{-1}$ is an unbiased estimator of $\mathrm{Var}(a^*)$.

Reduction to a problem involving exponentially distributed variables is employed by Hassanein, Saleh, and Brown (1991) to estimate the survivor function

$$\bar{F}_X(x) = \left(\frac{k}{x}\right)^a, \qquad x \geq k, \qquad (20.63)$$

for P(I)(k, a) using just m ($\leq n$) of the order statistics from a random sample of size n. Choice of the appropriate order statistics is discussed in their paper.

5.8 Estimation of Pareto Quantiles

Umbach, Ali, and Hassanein (1981) consider the P(III)(μ, C, a) distribution (20.4)', with *known* shape parameter a. They wish to estimate x_ξ, the ξ-quantile of P(III)(μ, C, a) by the estimator

$$\tilde{x}_\xi(l, u) = b_l X_l' + b_u X_u', \qquad (20.64)$$

where X_l', X_u' are the lth and uth order statistics ($l < u$) in a random sample of size n, and b_l and b_u are chosen so that $\tilde{x}_\xi(l, u)$ will be asymptotically unbiased and minimum variance.

Suppose that $l = [np_l] + 1$, $u = [np_u] + 1$ ($p_l < p_u$). The asymptotic expected values of X_l' and X_u' are

$$E[X_l'] \doteq x_{p_l} = \mu + z_{p_l}k,$$
$$E[X_u'] \doteq x_{p_u} = \mu + z_{p_u}k, \qquad (20.65)$$

where z_ξ is the ξ-quantile of P(I)$(1, a)$.

Table 20.1   Values of $\begin{pmatrix} p_l, & p_u \\ b_l, & b_u \end{pmatrix}$ **for Specific Estimator of** x_ξ

of the Form (20.64)

	$\xi = 0.1$	$\xi = 0.5$	$\xi = 0.75$
$a = 0.5$	$0 \quad , 0.3596$ $0.8369, 0.1631$	$0.4374, 0.6397$ $0.8151, 0.1849$	$0 \quad , 0.3596$ $-9.4279, 10.4279$
$a = 1$	$0 \quad , 0.5000$ $0.8889, 0.1111$	$0.4000, 0.7000$ $0.8000, 0.2000$	$0 \quad , 0.5000$ $-3.0000, 4.0000$
$a = 2$	$0 \quad , 0.6180$ $0.9125, 0.0875$	$0.3572, 0.7545$ $0.7835, 0.2165$	$0.6786, 0.8772$ $0.7835, 0.2165$
$a = 3$	$0 \quad , 0.6695$ $0.9199, 0.0801$	$0.3333, 0.7796$ $0.7745, 0.2255$	$0.6667, 0.8898$ $0.7745, 0.2255$

The asymptotic variance of $\tilde{x}_\xi(l, u)$ is

$$\text{Var}\big(\tilde{x}_\xi(l,u)\big) \doteq \frac{k^2}{na^2(z_{p_u} - z_{p_l})^2} \Big\{ (z_{p_u} - z_\xi)^2 z_{p_l}^2 \, (z_{p_l}^a - 1)$$

$$+ 2 z_{p_l} z_{p_u} \, (z_{p_u} - z_\xi)(z_{p_l} - z_\xi)(z_{p_l}^a - 1)$$

$$+ (z_\xi - z_{p_l})^2 z_{p_u}^2 \, (z_{p_u}^a - 1) \Big\}. \qquad (20.66)$$

If optimal values of l and u (minimizing the asymptotic variance) include $p_l = 0$, then $p_u = 1 - k_a^{-a}$, $b_l = (k_a - z_\xi)/(k_a - 1)$, and $b_u = (z_\xi - 1)/(k_a - 1)$, where

$$ak_a^a(k_a - 1) = 2(k_a^a - 1). \qquad (20.67)$$

Expressions for the optimal values of p_l, p_u, b_l, and b_u are more complex in the general case. Optimal sets of values are shown in Table 20.1 for a few special cases. Of course one needs to have a good idea of the value of a to use these tables. Values of the ARE of $\tilde{x}_\xi$, compared to the nonparametric estimator $X'_{[n\xi]+1}$, is presented in Table 20.2 which indicates that $\tilde{x}_\xi$ performs far better than $X'_{[n\xi]+1}$ especially for values of ξ close to 0 or 1.

A different approach to estimation of quantiles was adopted by Asrabadi (1990), using formula (20.59) for the UMVUE, $F_X^*(x)$ of $F_X(x)$. The ξ-quantile of $F_X^*(x)$, viz., $X_\xi^* = F_X^{*-1}(\xi)$ is *not* a UMVUE of x_ξ, the ξ-quantile of $P(I)(k, a)$. However, from the distribution of X_ξ^* an *approximate* $100(1 - \alpha)\%$

Table 20.2   ARE for $\tilde{x}_\xi$ Compared to $X'_{[n\xi]+1}$.

	$\xi = 0.1$	$\xi = 0.5$	$\xi = 0.75$
$a = 0.5$	191%	111%	5288%
$a = 1$	277	117	133
$a = 2$	380	126	116
$a = 3$	438	130	167

confidence interval for x_ξ is given by

$$\left[1 - \left\{\frac{2\hat{a}\log(X_\xi^*/k)}{\chi_{2n,1-\frac{\alpha}{2}}^2}\right\}^{n-1}, \quad 1 - \left\{1 - \frac{2\hat{a}\log(X_\xi^*/k)}{\chi_{2n,\frac{\alpha}{2}}^2}\right\}^{n-1}\right]. \quad (20.68)$$

Saleh, Ali, and Umbach (1985) investigated *nonlinear* estimators of the quantiles of $P(I)(k, a)$ based on a few selected order statistics, using the approach developed by Koutrouvelis (1981) (see Section 5.4). The quantile

$$x_\xi = F_X^{-1}(\xi) = k(1 - \xi)^{-1/a} \quad (20.69)$$

is estimated by exploiting the relationship between Pareto and exponential distributions [see (20.9)]. If X'_r is the rth-order statistic in a random sample of size n from a $P(I)(k, a)$ distribution, then $Z'_r = \log(X'_r/k)$ is the corresponding order statistic in a sample from the exponential distribution with pdf

$$a^{-1}\exp(-a^{-1}z), \quad z \geq 0. \quad (20.70)$$

Asymptotic BLUEs $\widetilde{\log k}$ and $\tilde{a}^{-1}$, for $\log k$ and a^{-1}, respectively, for the latter situation have been thoroughly investigated [e.g., see Saleh and Ali (1966) and Koutrouvelis (1981) for a corrected version]. The corresponding estimator of x_ξ is

$$\tilde{x}_\xi = \exp\{\widetilde{\log k} - \tilde{a}^{-1}\log(1 - \xi)\}. \quad (20.71)$$

The estimator is quite efficient even for a small number (2–4) of selected order statistics and highly efficient for moderate numbers (≥ 8). It is superior to the nonparametric estimator $X'_{[n\xi]+1}$, and far superior if ξ is near 0 or 1.

5.9   Censored Data

For censored samples of size n, with only values less than a (known) value τ recorded, suppose that exact values $X_1, \ldots, X_{n-c}$ are known for only $n - c$ individuals, while for the other c individuals it is only known that the value exceeds τ. If k is known, the MLE of a is

$$\hat{a} = (n - c)\left(\sum_{i=1}^{n-c} \log X_i - n \log k + c \log \tau\right)^{-1}. \qquad (20.72)$$

If k is not known, Cohen and Whitten (1988) recommend modified maximum likelihood estimators, obtained by solving the equations

$$X_1' = E[X_1'] = \hat{k}\hat{a}n(\hat{a}n - 1)^{-1}, \qquad (20.73a)$$

$$\hat{a} = (n - c)\left(\sum_{i=1}^{n-c} \log X_i - n \log \hat{k} + c \log \tau\right)^{-1}, \qquad (20.73b)$$

iteratively, starting with $\hat{k}$ slightly less than X_1'.

Rohatgi and Saleh (1987) discuss the case where τ is not known, but only the least $n - c$ among the n sample values ($X_1' \le X_2' \le \cdots \le X_{n-c}'$) are known. All that is known of the remaining c values is that they are not less than X_{n-c}'. Extending results of Saksena and Johnson (1984), they obtain the maximum likelihood estimators

$$\hat{k} = X_1', \qquad (20.74a)$$

$$\hat{a} = (n - c)\left(\sum_{i=1}^{n-c} \log X_i' + c \log X_{n-c}' - n \log X_1'\right)^{-1}. \qquad (20.74b)$$

The UMVUEs are

$$k^* = \left(1 - \frac{n - c}{n(n - c - 1)}\frac{1}{\hat{a}}\right)\hat{k}, \qquad (20.75a)$$

$$a^* = \frac{n - c - 2}{n - c}\hat{a}. \qquad (20.75b)$$

If we have independent samples of sizes n_1, n_2 from $P(I)(k, a_1)$ and $P(I)(k, a_2)$, respectively, of which only the $n_1 - c_1, n_2 - c_2$ least values are known, UMVUE of the common value of k is

$$\hat{\hat{k}} = \left[1 - \left\{\sum_{j=1}^{2} \frac{n_j(n_j - c_j - 1)\hat{a}_j}{n_j - c_j}\right\}^{-1}\right]\hat{k}, \qquad (20.76)$$

where subscripts refer to estimators based on the jth sample values, and

$\hat{k} = \min(\hat{k}_1, \hat{k}_2)$. The variance of this UMVUE of k is estimated by

$$\hat{k}^2 \left\{ \sum_{j=1}^{2} \frac{n_j(n_j - c_j - 1)\hat{a}_j}{n_j - c_j} \right\}^2. \tag{20.77}$$

5.10 Bayesian Estimation

The following prior distributions for the parameters a and k in $P(I)(k, a)$ have been studied [* denotes natural conjugate, and $(\text{Pareto})^{-1}$ signifies distribution of the reciprocal of a $P(I)(k, a)$ variable]:

Reference	k		a
Lwin (1972)	Known		Gamma*
	Known		$\propto a^{-1}$
	Power Function		Known
	$\dfrac{m(m-1)^m(-\log n)^m}{\Gamma(m)}$		$\dfrac{k^{ma-1}a^m *}{n^a}$
Nigm and Hamdy (1987)	—		Gamma*
	$\propto k^{-1}$		$\propto a^{-1}$ (independent)
Arnold and Press	(1983)	Power function*	Gamma*
	(1989)	$(\text{Pareto})^{-1}$	Known
	(1989)	$(\text{Pareto})^{-1}$	Gamma*

Gamma distributions are natural conjugates for a; power-function distributions are natural conjugates for k.

Geisser (1984, 1985) has provided extensive analysis of the application of Bayesian methods in predicting future values of $P(I)(k, a)$ distributed random variables from observed values in a complete random sample. Nigm and Hamdy (1987) have considered a similar problem where only the r least values (first r order statistics) in a random sample of size n are available, and it is desired to predict the remaining $n - r$ values. Geisser (1985) and Arnold and Press (1989) are of the opinion that it is more natural to assess prior belief about k first and then to consider the conditional prior distribution of a, given k, than conversely. Arnold and Press (1989) also point out that Nigm and Hamdy's (1987) analysis can be performed for the case where for the $P(I)(k, a)$ distribution

$$\Pr[X'_{r+1} > x | X'_i = x_i, i = 1, \ldots, r; k, a] = \left(\frac{x}{x_r} \right)^{-(n-r)a}. \tag{20.78}$$

By considering the Pareto density function in (20.3) when the shape parameter a is known and the scale parameter k is unknown, Liang (1993) has discussed the problem of estimating the scale parameter k under a squared-error loss through the nonparametric empirical Bayes approach. He has then derived an empirical Bayes estimator of k and discussed the corresponding asymptotic optimality. Under some mild conditions, Liang (1993) has shown that this empirical Bayes estimator is asymptotically optimal and that the associated rate of convergence is of order $O(n^{-2/3})$.

6 ESTIMATION OF LORENZ CURVE AND GINI INDEX

The Lorenz curve is

$$L(p, a) = 1 - (1 - p)^{1 - a^{-1}}, \tag{20.79}$$

and the Gini index is

$$G(a) = (2a - 1)^{-1}$$

(see Section 4.3).

The maximum likelihood estimators of these two quantities are obtained by replacing a by $\hat{a}$ (if k is known) or by $\hat{a}'$ (if k is unknown) and then introducing the condition $L(p, a) > 0$. Moothathu (1985, 1990) has investigated the sampling distributions of the estimators $(\hat{L})$ of $L(p, a)$, showing them to be asymptotically consistent. The distribution is mixed, with

$$\Pr[\hat{L} = 1] = \Pr[(\text{MLE of } a) \leq 1] \tag{20.80a}$$

and density

$$p_{\hat{L}}(w) = - \frac{\eta^{n - s + 1}}{\Gamma(n - s + 1)(1 - p)^{\eta}} (1 - w)^{n - 1} (\log w)^{n - s}, \tag{20.80b}$$

with $\eta = na/\log(1 - p)$, and $s = 1\ (2)$ according as k is (is not) known.

An unbiased estimator of $L(p, a)$ is

$$\tilde{L} = 1 - (1 - p)\left[1 + \sum_{j=1}^{\infty} \frac{\{-S_1 \log(1 - p)\}^j}{n^{[j]} j!} \right]$$

$$= 1 - (1 - p) {}_0F_1(-; n; -S_1 \log(1 - p)), \tag{20.81a}$$

where $_0F_1(-; n; x)$ is defined in Chapter 1, Section A7, and a minimum variance unbiased estimator is

$$\tilde{L}^* = 1 - (1 - p)_0F_1(-; n - 1; S_2 \log(1 - p)). \qquad (20.81b)$$

In (20.81a) and (20.81b) $_0F_1(\ldots)$ stands for a confluent hypergeometric function (Chapter 1, Section A7), and

$$S_1 = n^{-1} \sum_{i=1}^{n} \log\left(\frac{X_i}{k}\right),$$

$$S_2 = n^{-1} \sum_{i=1}^{n} \log\left(\frac{X_i}{X_1'}\right). \qquad (20.81c)$$

The maximum likelihood estimator $\hat{G}$ of the Gini index has pdf

$$p_{\hat{G}}(w) = \frac{(2na)^{n-s+1}}{\Gamma(n-s+1)} \left(\frac{w}{1+w}\right)^{n-s} (1+w)^{-2} \exp\left(\frac{-2naw}{1+w}\right),$$

$$0 < w < 1, \quad (20.82)$$

where again $s = 1$ (2) according as k is (is not) known.

Uniform minimum variance unbiased estimators of G are

$$_1F_1(1; n; \tfrac{1}{2}S_1) - 1 \qquad \text{if } k \text{ is known,} \qquad (20.83a)$$

$$_1F_1(1; n - 1; \tfrac{1}{2}S_2) - 1 \qquad \text{if } k \text{ is unknown,} \qquad (20.83b)$$

where

$$_1F_1(a; b; x) = \sum_{j=0}^{\infty} \frac{a^{[r]}}{b^{[r]}} \frac{x^r}{r!}$$

with $a^{[r]} = a(a + 1) \cdots (a + r - 1)$, $b^{[r]} = b(b + 1) \cdots (b + r - 1)$; see Chapter 1, Section A7.

7 MISCELLANEOUS

The distribution of $Y = \sum_{j=1}^{n} \log X_j$ has density function

$$p_Y(y) = \frac{a^n}{\Gamma(n)} (y - n \log k)^{n-1} e^{-a(y - n \log k)}, \qquad y > n \log k \quad (20.84)$$

[Malik (1966)]. The probability density function of $Z = e^Y = X_1 X_2 \ldots X_n$ is

$$p_Z(z) = \frac{a^n k^{na}}{\Gamma(n)} \left[\log\left(\frac{z}{k^n}\right)\right]^{n-1} z^{-(a+1)}, \qquad z > k^n. \qquad (20.85)$$

The distribution of the sample geometric mean $G^* = Z^{1/n}$ has probability density function

$$p_{G^*}(g) = \frac{n^n a^n}{k\Gamma(n)} e^{-(1+an)\log(g/k)} \left\{ \log\left(\frac{g}{k}\right)\right\}^{n-1}, \qquad k < g < \infty. \quad (20.86)$$

$2an \log(G^*/k)$ is distributed as χ^2 with $2n$ degrees of freedom [Malik (1970a)]. The distributions of $\hat{h}$ and $\hat{m}$ can be directly derived from that of G^*, where $\hat{h}$ and $\hat{m}$ are the estimators of h and m in (20.14) and (20.15).

A measure of dispersion has been suggested by Muniruzzaman (1957). It is termed *geometric standard deviation*. This is

$$\lambda = \exp\left\{ \sqrt{E\left[(\log X - \log \gamma)^2\right]} \right\}. \qquad (20.87)$$

where $\gamma = k \exp(1/a)$ is the population geometric mean. The moment-generating function of $\log X$ is

$$k^t(1 - ta^{-1})^{-1}, \qquad (20.88)$$

from which

$$\mathrm{Var}(\log X) = a^{-2} = \left[\log\left(\frac{\gamma}{k}\right)\right]^2. \qquad (20.89)$$

Thus the geometric standard deviation of X is $\gamma/k \; (= \lambda)$, and the Pareto distribution can be completely specified by the location parameter γ and scale parameter λ. The maximum likelihood estimator of λ is G^*/X_1'.

The probability density function of the sample median M in samples of size $n = 2p + 1$ is

$$p_M(m) = \frac{(2p+1)! a k^{a(p+1)}}{(p!)^2} m^{-(ap+a+1)} \left[1 - \frac{m^{-a}}{k^{-a}}\right]^p, \qquad k \le m \le \infty.$$

$$(20.90)$$

The rth moment of M is

$$\mu_r'(M) = \frac{\Gamma(n+1)\Gamma[\{(n+1)/2\} - (r/a)]}{\Gamma[(n+1)/2]\Gamma[n+1-(r/a)]} k^r. \qquad (20.91)$$

Note that

$$E[M] = \frac{\Gamma(n+1)\Gamma[\frac{1}{2}(n+1) - a^{-1}]}{\Gamma[\frac{1}{2}(n+1)]\Gamma(n+1-a^{-1})} \times k. \qquad (20.92)$$

[Note also from (20.15) that the population median value is $2^{1/a}k$.]

Applying Stirling's formula to (20.92), using logarithmic expansion, and retaining terms to the order of $1/n$, we obtain

$$E[M] \doteq k2^{1/a}\left(1 + \frac{a+1}{2na^2}\right),$$

$$\text{Var}(M) \doteq k^2 2^{2/a} a^{-2} n^{-1}. \qquad (20.93)$$

It is apparent from (20.93) that

$$\lim_{n \to \infty} E[M] = 2^{1/a}k = m; \qquad [\text{cf } (20.15)]$$

Further, we see that

$$\frac{V(\hat{m})}{V(M)} \doteq 1,$$

where $\hat{m} = 2^{1/\hat{a}}\hat{k}$ is the MLE of the median (m).

On setting the parameter k in (20.2) equal to 1, we obtain

$$F_X(x) = 1 - x^{-a}, \qquad x \geq 1. \qquad (20.94)$$

This may be regarded as a *standard form* of the Pareto distribution. A location parameter ε can be introduced to yield

$$F_X(x) = 1 - (x - \varepsilon)^{-a}, \qquad x \geq 1 + \varepsilon. \qquad (20.95)$$

It can be shown [Gumbel (1958)] that if T has an exponential distribution with scale parameter θ, then $X = \exp(T) + \varepsilon$ has the Pareto distribution (20.95) with

$$a = \frac{1}{\theta}, \qquad a \geq 1,$$

$$x > 1 + \varepsilon, \qquad \varepsilon \geq 0.$$

A single-order statistic estimator using only X'_m of a, given ε, is

$$\tilde{a}|\varepsilon = \frac{1}{\tilde{\theta}} = \frac{1}{C_{mn} \log(X'_m - \varepsilon)}, \qquad (20.96)$$

where

$$C_{mn} = \frac{1}{\sum_{i=1}^{m}(n - i + 1)^{-1}}$$

[Moore and Harter (1967)]. $\tilde{a}|\varepsilon$ and $\tilde{\theta}$ are consistent estimators of the shape parameter of the Pareto distribution and the scale parameter of the exponential distribution, respectively [Moore and Harter (1967)].

It can be shown [Moore and Harter (1967)] that an exact confidence interval for a based on X'_m alone, if ε is known, is

$$\{D_{l,m,n}\log(X'_m - \varepsilon)\}^{-1} > a > \{D_{u,m,n}\log(X'_m - \varepsilon)\}^{-1}, \quad (20.97)$$

where the coefficients $D_{l,m,n}$ and $D_{u,m,n}$ have been tabulated [Harter (1964)] when $n = 1(1)20(2)40$ for m optimal. (In the sense that the value of m selected maximizes the efficiency of the confidence interval.)

8 ORDER STATISTICS AND RECORD VALUES

8.1 Order Statistics

We have already noted the distribution of the median of n $(= 2p + 1)$ independent random variables, each having the Pareto distribution. The jth smallest order statistic X'_j has density function

$$p_{X'_j}(x) = \frac{n!}{(j-1)!(n-j)!}\left(\frac{a}{x}\right)\left(\frac{k}{x}\right)^{a(n-j+1)}\left\{1 - \left(\frac{k}{x}\right)^a\right\}^{j-1}, \quad k \le x,$$

$$(20.98)$$

and its rth moment about zero is

$$\mu'_r(X'_j) = \frac{\Gamma(n+1)\Gamma(n-j+1-ra^{-1})}{\Gamma(n+1-ra^{-1})\Gamma(n-j+1)}k^r. \quad (20.99)$$

[see (20.91) which is obtained by putting $j = \frac{1}{2}(n + 1)$ in (20.99).] The joint probability density function of two order statistics X'_r, X'_s $(r < s)$ is

$$p_{X'_r, X'_s}(x_r, x_s) = \frac{n!a^2k^{2a}}{(r-1)!(s-r-1)!(n-s)!}$$

$$\times \left[1 - \left(\frac{k}{x_r}\right)^a\right]^{r-1}\left[\left(\frac{k}{x_s}\right)^a - \left(\frac{k}{x_r}\right)^a\right]^{s-r-1}$$

$$\times \left(\frac{k}{x_s}\right)^{a(n-s)}(x_r x_s)^{-(a+1)}, \quad k \le x_r \le x_s, \quad (20.100)$$

and their covariance can be calculated from

$$E[X_r' X_s'] = \frac{\Gamma(n+1)\Gamma(n-r-2a^{-1}+1)\Gamma(n-s-a^{-1}+1)}{\Gamma(n-s+1)\Gamma(n-r-a^{-1}+1)\Gamma(n-2a^{-1}+1)} k^2.$$

(20.101)

Malik (1966) showed that the characteristic function of the kth smallest order statistic X_k' in a random sample of size n is

$$\frac{\Gamma(n+1)}{\Gamma(k)} \sum_{j=0}^{\infty} \frac{\Gamma(k+ja^{-1})}{\Gamma(n+ja^{-1}+1)} \frac{(it)^j}{j!},$$

and obtained the interesting recurrence relationships

$$(k-1)E[(X_k')^r] = (k+ra^{-1}-1)E[(X_{k-1}')^r] \qquad (k > 1) \quad (20.102a)$$

$$(2a^{-1}+l-1)E[X_k' X_l'] = (a^{-1}+l-1)E[X_k' X_{l-1}'] \qquad (k < l).$$

(20.102b)

He has provided tables of:

1. $E[X_j']$ for $n \le 12$,
2. $E[X_r' X_s']$ for $n \le 12$,
3. $\text{Var}(X_j')$ and $\text{Cov}(X_r', X_s')$ for $n \le 8$

to four decimal places.

Huang (1975) showed that for P(I)(k, a)

$$E[X_i'^r] = \frac{k^{(r)} n^{(r)}}{(n-ia^{-1})^{(r)}} \qquad \text{for } a > i(n-r+1)^{-1}. \quad (20.103)$$

Balakrishnan and Joshi (1982) derived further recurrence relations among single and product moments of order statistics for different sample sizes, with $\mu_{r:n}^{(t)} = E[(X_{r:n}')^t]$ and $\mu_{r,s:n} = E[X_{r:n}' X_{s:n}']$, namely

$$(na - t)\mu_{1:n}^{(t)} = nak^t, \qquad n \ge 1, \qquad (20.104a)$$

$$(na - t)\mu_{r:n}^{(t)} = na\mu_{r-1:n-1}^{(t)}, \qquad 2 \le r \le n, \qquad (20.104b)$$

$$\{a(n-r)-1\}\mu_{r,r+1:n} = a(n-r)\mu_{r:n}^{(2)},$$

$$1 \le r \le n-1, \quad (20.104c)$$

and

$$\{a(n - s + 1) - 1\}\mu_{r,s:n} = a(n - s + 1)\mu_{r,s-1:n},$$

$$1 \le r < s \le n, s - r \ge 2. \quad (20.104d)$$

8.2 Record Values

Pareto distributions arise as limiting distributions of the waiting time W_1, the number of new observations needed to obtain a value exceeding the ath greatest among n observations already available. Formally,

$$W = \min(h; h \ge 1, X_{n+h} > X'_{n-a+1:n}). \quad (20.105)$$

It can be shown that if the parent distribution is continuous

$$\lim_{n \to \infty} \Pr[W_n^{-1} > x] = (1 + x)^{-a}, \qquad x \ge 0. \quad (20.106)$$

This is the $P(I)(1, a)$ distribution. This is a general result for any continuous parent distribution.

For cases where the parent distribution itself is Pareto, Ahsanullah and Houchens (1989) and Balakrishnan and Ahsanullah (1994a, b) have obtained several results: Let X_n $(n \ge 1)$ be a sequence of i.i.d. random variables, possessing a pdf. Define

$$Y_n = \max(X_1, \ldots, X_n), \qquad n \ge 1.$$

Then X_j is a (upper) *record value* of the sequence $\{X_n\}$ if $X_j > Y_{j-1}$ $(j > 1)$. (By definition, X_1 is a record value.)

The indices T_n at which the record values occur are the *record times*. Formally,

$$T_n = \min(j : X_j > X_{T_{n-1}}), \quad (20.107)$$

with $T_0 = 1$. A characteristic property of the Pareto distribution is that the ratios $X_{T_n}/X_{T_{n-1}}$ are independent and identically distributed as $P(I)(1, a)$. The pdf of X_{T_n} is

$$p_{X_{T_n}}(x) = \frac{\{R(x)\}^n}{n!} p_X(x), \quad (20.108)$$

where $R(x)$ is the cumulative hazard rate $a \log(x/k)$ (see Section 4.3). It follows that X_{T_n} is the product of $n - 1$ mutually independent $P(I)(k, a)$ variables.

Ahsanullah and Houchens (1989) provide tables of the variances and covariances of X_{T_m} and X_{T_n} for $a = 2.5(0.5)5.0$, $0 < m < n \leq 3$. Balakrishnan and Ahsanullah (1994a) derive recurrence relations for single and product moments of (upper) record values for standardized Lomax, P(II), distributions with cdf

$$F_X(x) = 1 - (1 + x)^{-a}, \qquad x \geq 0, \, a > 0. \qquad (20.109)$$

They show that the pdf of X_{T_n} is

$$p_{X_{T_n}}(x) = \frac{1}{n!}\left[-\log\{1 - F_X(x)\}\right]^n p_X(x). \qquad (20.110)$$

Using direct calculations, they show that

$$E\left[X_{T_{n+1}}^{r+1}\right] = \frac{(r+1)E\left[X_{T_{n+1}}^r\right] + aE\left[X_{T_n}^{r+1}\right]}{a - r - 1}, \qquad r < a - 1. \quad (20.111a)$$

In particular,

$$E\left[X_{T_{n+1}}\right] = \frac{1 + aE\left[X_{T_n}\right]}{a - 1}. \qquad (20.111b)$$

Also

$$\mathrm{Cov}\left(X_{T_m}, X_{T_n}\right) = \left(\frac{a}{a-1}\right)^{n-m} \mathrm{Var}\left(X_{T_m}\right), \qquad 0 \leq m \leq n - 1. \quad (20.111c)$$

Ahsanullah (1991) has carried out a detailed study of a *generalized Pareto (II) (Lomax)* distribution, with pdf

$$p_X(x) = \frac{1}{K}\left\{1 + \beta\frac{x - \xi}{K}\right\}^{-(1+\beta^{-1})}, \qquad \begin{pmatrix} x \geq \xi & \text{if } \beta \geq 0 \\ 0 < x < \xi - \dfrac{1}{\beta} & \text{if } \beta < 0 \end{pmatrix}. \qquad (20.112)$$

In this case X_{T_n} is distributed as

$$\xi - \frac{K}{\beta} + \frac{K}{\beta}\prod_{i=0}^{n}V_i, \qquad (20.113)$$

where the V_i's are i.i.d. as $P(I)(1, \beta^{-1})$. For this case, Balakrishnan and Ahsanullah (1994b) have established recurrence relations for single and product moments of upper record values generalizing the results given above [equations (20.111)] for the Lomax distribution.

9 CHARACTERIZATIONS

The close relationship between the exponential and Pareto distributions emphasized in Section 3 allows us to simply translate various characterizations of the exponential distribution to Pareto. These characterizations are however of limited intrinsic value for understanding the models that can be generated by and/or admit a Pareto distribution fit. Galambos and Kotz (1978) provide a detailed discussion of characterizations of the exponential and Pareto distributions; some of these characterizations are presented in Chapter 19. Most characterizations of Pareto distribution are essentially based on the mean residual life property (Hagstroem (1925)). More rigorous proofs have been developed by Arnold (1971) and Huang (1974).

A typical, more recent, result along these lines, due to Morrison (1978), states that the Pareto distribution of the second kind (with range $x > 0$) is the unique gamma mixture of exponential distributions that yields a linearly increasing mean residual lifetime $E[X|X > t]$ regardless of the mixing process. (Recall that Pareto is generated by a gamma mixing of exponentials—see Section 2). There is an extensive discussion of characterizations based on mean residual lifetime in Kotz and Shanbhag (1980). A second class of characterizations is based on the property connecting truncation to rescaling. A typical result, due to Bhattacharya (1963), is that under certain conditions on the support of the distribution, the relationship

$$\Pr[X > y | X > x_0] = \Pr\left(\frac{x_0}{k} X > y\right), \qquad \text{for all } y > x_0 \geq k, \quad (20.114)$$

characterizes P(I)(k, a) distributions with scale parameter k.

Under the assumption of the existence of moments, this characterization is equivalent to the linearity of the mean residual lifetime as a function of x_0 [Kotz and Shanbhag (1980), Dallas (1976)].

The property (20.114) can be interpreted in the following ("underreported income property") way:

If the distribution of $Y = \theta X$, where $0 < \theta \leq 1$, truncated from below at k_0, is the same as the distribution of X, then X must have a P(II) distribution with scale parameter k [Krishnaji (1970, 1971)]. Y may be regarded as reported and X as actual income.

Refinements of this interpretation have been provided by Huang (1978) and Marsaglia and Tubilla (1975). However, the assumption that underreported income has this relation to actual income is very doubtful.

Characterizations based on truncation invariance of the Gini index [Ord, Patil, and Taillie (1981)] and on a linear relation between $\log x$ and $\log\{1 - F_X(x)\}$ [Moothathu (1984)] have similar bases.

Schmittlein and Morrison (1981) claimed that the *median* residual lifetime

$$M(x_0) = \text{median}(X - x_0 | X > x_0) \qquad (20.115)$$

is of form $a + bx_0$ if and only if X has a P(II) distribution. However, Gupta and Langford (1984) showed that non-Paretian distributions with cdf of form

$$F_X(x) = 1 - \left(\frac{a}{a + bx}\right)^c g(\log(a + bx))$$

with $g(\cdot)$ a periodic function of period $\log(b + 1)$ such that

$$g(\log(a + bk)) = \left(\frac{a + bk}{a}\right)^c,$$

where $c = (\log 2)/\{\log(b + 1)\}$, also have $M(x_0)$ of this form.

Many characterizations are based on distributions of order statistics, usually involving the least observed value, X_1'.

A typical result is

Let $X_1' \le X_2' \le \cdots \le X_n'$ be the order statistics from an absolutely continuous distribution function $F_X(x)$ where the lower bound on the random variable X is given by k, that is, $F_X(k) = 0$. It can then be shown for a sample of size n from this distribution that the statistics X_1' and $(X_1' + \cdots + X_n')/X_1'$, or equivalently X_1' and $(X_2' + \cdots + X_n')/X_1'$, are independent if and only if the random variable X has the Pareto distribution. [Srivastava (1965)]

Dallas (1976) provides an essentially equivalent theorem "Under the assumption of existence of the r-th moment

$$E\left[\sum_{i=1}^{n} X_i^r | X_1' = x\right] = cx^r \text{ implies that } X \text{ is Pareto (I)."} \qquad (20.116)$$

Characterizations of the type:

$$E\left[X_{s+1}'^r | X_s' = x\right] = cx^r \text{ for } 1 \le s < n \qquad (20.117)$$

[Beg and Kirmani (1974), Khan and Khan (1987)] are simply restatements of the characterizing equation

$$\frac{p_X(x)}{1 - F_X(x)} = \frac{a}{x} \qquad (20.118)$$

of the Pareto distribution (see Section 4).

James's (1979) characterization based on the independence $(X_1/X_1', X_2/X_1', \ldots X_n/X_1')$ and X_1', falls within the general framework of shape–size analysis in Mosimann (1970).

Arnold (1983) provides an alternative, very illuminating interpretation of these results. The reader is referred to his book for further characterizations of Pareto distributions of various kinds. In our opinion most of the characterizations of Pareto (and exponential) distributions are in essence variants of the *memoryless property of the exponential* and are eventually reduced to solving a Cauchy-type functional equation or its integral variants (see Chapter 19).

10 PRODUCT AND RATIOS OF PARETO RANDOM VARIABLES

The product of n independent Pareto random variables X_i, distributed as $P(I)(k_i, a_i)$ $(i = 1, \ldots n)$ can be expressed in the form

$$\left(\prod_{i=1}^{n} k_i\right) \exp\left\{\sum_{i=1}^{n} (V_i/a_i)\right\} \qquad (20.119)$$

where the V_i's are independent standard exponential variables.

If the a_i's are all distinct but $k_i = k$ for all i, the distribution of $V = \prod_{i=1}^{n} X_i$ is given in terms of the survival function as

$$\bar{F}_Y(y) = \Pr[Y > y] = \sum_{i=1}^{n} \left\{\left(\frac{y}{k}\right)^{-a_i} \prod_{\substack{l=1 \\ l \neq i}}^{n} \left(\frac{a_i}{a_i - a_l}\right)\right\}, \qquad y > k. \quad (20.120)$$

Pederzoli and Rathie (1980), using the method of Mellin transforms, obtained an expression for the distribution of Y when the a_i's are *not* all distinct.

Castellani (1950) and Malik (1970a) have analyzed the special case when $a_1 = a_2 = \cdots = a_n = a$.

In this case,

$$\sum_{i=1}^{n} (V_i/a)$$

is distributed as a gamma (n, a^{-1}) variable, and the distribution of the product has pdf

$$p_Y(y) = \frac{\{m \log(y/m)\}^{n-1}(y/m)^{-a}(a/y)}{\Gamma(n)} \qquad (y \geq m), \quad (20.121)$$

where $m = \prod_{i=1}^{n} k_i$.

The distribution of the quotient $Z = X_1/X_2$ of two independent Pareto random variables with parameters (k_1, a_1) and (k_2, a_2) respectively, was also tackled by Pederzoli and Rathie (1980). The corresponding pdf is

$$
p_Z(z) = \begin{cases} \dfrac{a_1 a_2}{k_1 k_2^{-1}(a_1 + a_2)} \left(\dfrac{z}{k_1 k_2^{-1}} \right)^{a_2 - 1} & \text{for } z \le k_1 k_2^{-1} \\[4mm] \dfrac{a_1 a_2}{k_1 k_2^{-1}(a_1 + a_2)} \left(\dfrac{z}{k_1 k_2^{-1}} \right)^{-a_1 - 1} & \text{for } z \ge k_1 k_2^{-1} \end{cases} \qquad (20.122)
$$

The cdf is

$$
F_Z(z) = \begin{cases} \dfrac{a_1}{a_1 + a_2} \left(\dfrac{z}{k_1 k_2^{-1}} \right)^{a_2} & \text{for } z \le k_1 k_2^{-1} \\[4mm] 1 - \dfrac{a_2}{a_1 + a_2} \left(\dfrac{k_1 k_2^{-1}}{z} \right)^{a_1} & \text{for } z > k_1 k_2^{-1}. \end{cases} \qquad (20.123)
$$

Given a random sample of size n from $P(I)(k, a)$ with order statistics $X_1' \le X_2' \le \cdots \le X_n'$, the pdf of the ratio

$$
W \equiv \frac{X_i'}{X_j'} \qquad (i < j)
$$

is

$$
p_W(w) = \frac{k}{a^k B(n - j + 1, j - 1)} w^{k(n-j+1)}(1 - w^k)^{j-i-1} \qquad (0 < w < 1) \qquad (20.124)
$$

[see, e.g., Springer (1979)].

The pdf of the product $T = X_i' X_j'$ was also obtained in a closed, albeit cumbersome, form, using the Mellin transform. The exact expression involves summations with respect to r and s of terms of the type:

$$
\frac{n! a k^{2a(r+n-j+1)}}{(i - 1)!(j - i - 1)!(n - j)!(2r + n + i - 2j - s + 1)} t^{a(j-r-n-1)-1}
$$
$$
\cdot \left[1 - \left(\frac{k}{\sqrt{t}} \right)^{-a(2r+n+i-2j-s+1)} \right] = h(t, a, k; i, j, s, r), \qquad (20.125)
$$

with appropriate weights.

Explicitly:

$$p_T(t) = \sum_{r=0}^{j-i-1} \sum_{s=0}^{i-1} (-1)^{r+s-1} \binom{j-i-1}{r} \binom{i-1}{s} h(t,a,k;i,j,s,r)$$

$$(0 < t) \quad (20.126)$$

where $1 \le i < j \le n$, [see, e.g., Springer (1979)].

Distribution of products of independent Pareto (IV) variables can be derived from the distribution of products of *powers* of independent gamma variables (generalized gamma variables)—see Chapter 17, Section 8.7. Blum (1970) has discussed the distribution of the sum of independent Pareto variables.

If X has a P(I)(k, a) distribution, then $Y = X^{-1}$ has the density function

$$p_Y(y) = ak^a y^{a-1} \quad (0 < y < k^{-1}). \quad (20.127)$$

This distribution, which is a special Pearson Type I distribution, is called the *power–function distribution*. Its moments are, of course, simply the negative moments of the corresponding Pareto distribution, so that

$$\mu'_r(Y) = \frac{ak^{-r}}{a+r}. \quad (20.128)$$

Moments of order statistics for this distribution were calculated by Malik (1967), while recurrence relations for these quantities have been derived by Balakrishnan and Joshi (1981). Rider (1964) has obtained the distribution of products and quotients of maximum values of sets of independent random variables having distribution (20.127). He suggests such distributions might be used in approximate representation of the lower tail of the distribution of a random variable having a fixed lower bound (as in a gamma distribution, for example).

11 APPLICATIONS AND RELATED DISTRIBUTIONS

The discrete form of the Pareto distribution (Zipf–Estoup's law) has been discussed in Chapter 11. In many cases the Pareto distribution may be used as an approximation to the Zipf distribution. Xekalaki and Panaretos (1988) consider the Yule distribution discussed in Chapter 6, Section 10.3 to be the discrete analog of the Pareto distribution and provide a mutual characterization of these distributions. Many socioeconomic and other naturally occurring quantities are distributed according to certain statistical distributions with very long right tails. Examples of some of these empirical phenomena are distributions of city population sizes, occurrence of natural resources, stock price fluctuations, size of firms, personal incomes, and error clustering in communication circuits. Many distributions have been developed in an attempt to explain these empirical data. As Arnold (1985) points out:

> The bald fact remains that in the upper tail, income distributions are reasonably well approximated by Pareto distributions and, certainly for predictive purposes, the lack of a compelling explanatory model should not deter one from capitalizing on the appealing simplicity of Pareto's distribution.

The lognormal distribution has emerged as the main competitor for Pareto. It has been observed that while the fit of the Pareto curve may be rather good at the extremities of the income range, the fit over the whole range is often rather poor. On the other hand, the lognormal (Gibrat) distribution (Chapter 14) fits well over a large part of the income range but diverges markedly at the extremities [Fisk (1961)].

Ferguson (1962) and James (1979) discuss the Pareto distributions as limiting distributions of the generalized gamma distribution (Chapter 17) when the parameter $c \to -\infty$.

Truncated Pareto distributions, with an upper limit (k') as well as a lower limit (k), have been found by Goldberg (1967) to fit the distribution of oil fields, in a specified area, by size. A truncated $P(I)(k, a)$ distribution has density function of the form

$$p_X(x) = \frac{a}{k} \left[1 - (k/k')^a\right]^{-1} (k/x)^{a+1} \qquad (k \le x \le k') \quad (20.129)$$

and cumulative distribution function

$$F_X(x) = \begin{cases} 0 & (x < k) \\ \left[1 - (k/k')^a\right]^{-1} \left[1 - (k/x)^a\right] & (k \le x \le k') \quad (20.130a) \\ 1 & (x > k'), \end{cases}$$

or equivalently

$$F_X(x) = \begin{cases} 0 & (x < K(1 - Q)^{-1/a}) \\ (1 - Q - k^a x^{-a})/(P - Q) & (K(1 - Q)^{-1/a} \le x \le K(1 - P)^{-1/a}) \\ 1 & (x > K(1 - P)^{-1/a}) \end{cases}$$

$$(20.130b)$$

with $K = k(1 - Q)^{1/a}$, where Q and $1 - P$ $(Q < P)$ are the proportions truncated on the left and right, respectively. Moments of order statistics $E[X_{r:n}^{\prime i}] \equiv \mu_{r:n}^{(i)}$ from this distribution exist for all r, n and i because the range of the distribution is finite.

Balakrishnan and Joshi (1982) derived recurrence relations for these moments. Taking, without loss of generality, the scale parameter $k = 1$, a typical relation is:

For $n \geq 2$, $1 \leq r \leq n - 1$ and $i = 1, 2, \ldots$

$$\{(n - r + 1)a - i\}\mu_{r:n}^{(i)}$$
$$= a\left[(n - r + 1)\mu_{r-1:n}^{(i)} + nP_2\{\mu_{r:n-1}^{(i)} - \mu_{r-1:n-1}^{(i)}\}\right] \quad (20.131)$$

where

$$P_2 = \frac{P - 1}{P - Q}.$$

(Compare with similar relationship for the nontruncated case in Section 8.)

Similar relations were obtained by Khurana and Jha (1985) based on expressions for $\mu_{r:n}^{(i)}$ in terms of hypergeometric functions $_2F_1(a, b; c; t)$. A typical relation is

$$\left(\frac{i}{a} - r\right)\mu_{r:n}^{(i)} = \frac{i}{a}\frac{1 - Q}{k^a}\mu_{r:n}^{(i+a)} - r\mu_{r+1:n}^{(i)}, \quad (20.132)$$

which follows from

$$\mu_{r:n}^{(i)} = k^i(1 - Q)^{-i/a}\,_2F_1\left(i/a, r; n + 1; \frac{P - Q}{1 - Q}\right),$$

since $(a - b)\,_2F_1(a, b; c; z) = a\,_2F_1(a + 1, b; c; z) - b\,_2F_1(a, b + 1; c; z)$.

Cohen and Whitten (1988) discuss estimation of parameters for distributions truncated on the right (at point x_0). When k is known, the MLE of a is the solution (in $\hat{a}$) of the equation

$$\frac{1}{\hat{a}} + \frac{(k/x_0)^{\hat{a}}\log(k/x_0)}{1 - (k/x_0)^{\hat{a}}} = \frac{1}{n}\sum_{i=1}^{n}\log X_i - \log k. \quad (20.133)$$

Cohen and Whitten recommend the use of

$$\hat{a}_1 = n\bigg/\sum_{i=1}^{n}\log\left(\frac{X_i}{K}\right)$$

as the initial value in an iterative procedure, which would be the MLE of a if the distribution were not truncated.

A mixture of two Pareto distributions is sometimes called a *double Pareto* distribution.

Modifications of the Pareto distribution have been developed to give better representation of income distributions.

One of the best known income distributions is the *Champernowne distribution* [Champernowne (1952)]. It is supposed that the random variable X, termed "income–power," and defined by $X = \log T$, where T is the actual income, has a density function of the form

$$p_X(x) = \frac{n}{\cosh\{\alpha(x - x_0)\} + \lambda},$$

(20.134)

where n, α, x_0 and λ are parameters (n is, of course, a function of α and λ). This is included in Perks's family of distributions, described in Chapter 23. We have retained the unusual notation (n) used by Champernowne. This symbol does not have anything to do with "sample size" in the present context.

The curve given by (20.134) is symmetrical and x_0 is the median value of the income–power. If we let $\log t_0 = x_0$ then the density function of income ($T = e^X$) is

$$p_T(t) = \frac{n}{t\left[\frac{1}{2}(t/t_0)^{-\alpha} + \lambda + \frac{1}{2}(t/t_0)^{\alpha}\right]}$$

(20.135)

and t_0 is the median value of the income.

In order to find the proportion of persons with income greater than t, we integrate (20.135). The form that the integral takes will depend on the value of λ. For $-1 < \lambda < 1$ the cumulative distribution function is

$$F_T(t) = 1 - \frac{1}{\theta}\tan^{-1}\left\{\frac{\sin\theta}{\cos\theta + (t/t_0)^{\alpha}}\right\} \qquad (t > 0), \quad (20.136)$$

where

$$0 < \theta < \pi \quad \text{and} \quad \cos\theta = \lambda.$$

For $\lambda = 1$

$$F_T(t) = 1 - \frac{t_0^{\alpha}}{t_0^{\alpha} + t^{\alpha}} \qquad (t > 0).$$

(20.137)

For $\lambda > 1$

$$F_T(t) = 1 - \frac{1}{2\eta}\log\left\{\frac{t^{\alpha} + e^{\eta}t_0^{\alpha}}{t^{\alpha} + e^{-\eta}t_0^{\alpha}}\right\} \qquad (t > 0), \qquad (20.138)$$

where

$$\cosh\eta = \lambda.$$

In (20.136)–(20.138), α is equal to Pareto's constant (Section 2), and t_0 is the median value of the income. No simple interpretation exists for the parameters θ and η. The latter may be regarded as a parameter used for adjusting the kurtosis of the fitted density function (20.134) along the income power scale [Champernowne (1952)].

The three forms of the density function corresponding to (20.136)–(20.138) may be obtained by differentiation and are, respectively

$$\frac{\alpha \sin \theta}{\theta t \left\{ (t/t_0)^\alpha + 2 \cos \theta + (t_0/t)^\alpha \right\}};$$
(20.139)

$$\frac{\alpha}{t \left\{ (t/t_0)^{\alpha/2} + (t_0/t)^{\alpha/2} \right\}^2};$$
(20.140)

$$\frac{\alpha \sinh \eta}{\eta t \left\{ (t/t_0)^\alpha + 2 \cosh \eta + (t_0/t)^\alpha \right\}}.$$
(20.141)

[These are all of form (20.135), of course.] Methods of fitting are described in Champernowne (1952).

In general, it is possible to improve the fit of a distribution by the incorporation of extra parameters. To this end Champernowne (1952) proposed a five parameter model of the form:

$$F_T(t) = \begin{cases} 1 - \dfrac{1}{(1 + \sigma)\theta} \left[(\sigma - 1)\theta + 2 \tan^{-1} \left\{ \dfrac{\sin \theta}{\cos \theta + (t/t_0)^{\sigma\alpha}} \right\} \right] & \text{for } 0 \le t \le t_0, \\[4mm] 1 - \dfrac{2\sigma}{(1 + \sigma)\theta} \tan^{-1} \left\{ \dfrac{\sin \theta}{\cos \theta + (t/t_0)^\alpha} \right\} & \text{for } t \ge t_0. \end{cases}$$
(20.142)

The fifth parameter σ may be considered as a measure of skewness, in that when $\sigma > 1$ the curve exhibits positive skewness, and when $\sigma < 1$ the curve is negatively skewed. When $\sigma = 1$, the distribution (20.136) is obtained.

The limiting form of the density function of the Champernowne distribution as $\theta \to 0$ is

$$\frac{(t/t_0)^{\alpha - 1}}{t_0 \left[1 + (t/t_0)^\alpha \right]^2}.$$
(20.143)

On making the transformation $(T/t_0)^\alpha = e^\phi$, the *logistic* or *sech square density function*

$$p_\phi(t) = \frac{e^t}{\left[1 + e^t \right]^2} = \frac{e^{-t}}{\left[1 + e^{-t} \right]^2}$$
(20.144)

is obtained [see Chapter 23] [Fisk (1961)]. The shape of this distribution is quite similar to that of the normal density function. However, in the tails the sech square density is greater than the normal. For economists concerned with the upper tails of distributions, the Pareto, Champernowne, and sech square distributions are probably more useful than the lognormal, which generally gives a poor fit in the tails [Aitchison and Brown (1957)].

The cumulative distribution function corresponding to (20.143) is

$$F_T(t) = \frac{(t/t_0)^\alpha}{1 + (t/t_0)^\alpha}, \qquad (20.145)$$

that is,

$$F_T(t) = 1 - \frac{(t/t_0)^{-\alpha}}{1 + (t/t_0)^{-\alpha}}. \qquad (20.146)$$

For small t, (20.146) is close to the Pareto distribution (20.2).

A number of papers [Mandelbrot (1960, 1963, 1967)] discuss a class of distributions that has been termed "stable Paretian." In the development of the theory, two forms of the Pareto law are distinguished [Mandelbrot (1960)]:

Strong Law of Pareto. The distribution is of the form given by (20.2); that is,

$$1 - F_X(x) = \left(\frac{x}{k}\right)^{-\alpha} \qquad x \geq k;$$

$$= 1 \qquad x < k.$$

Weak or Asymptotic form of the law of Pareto. The form of this distribution is

$$1 - F_X(x) \sim \left(\frac{x}{k}\right)^{-\alpha} \qquad \text{as } x \to \infty. \qquad (20.147)$$

This implies that if $\log[1 - F_X(x)]$ is plotted against $\log x$, the resulting curve should be asymptotic to a straight line with slope equal to $-\alpha$ as x approaches infinity (see Section 9). It has been shown [Lévy (1925)] that there is a class of distributions that follow the asymptotic form of the law of Pareto and are characterized by the fact that $0 < \alpha < 2$. These are known as the *stable Paretian or stable non-Gaussian* distributions. The normal distribution ($\alpha = 2$) is also a member of the family of stable laws and has the property that it is the only one with a finite variance. A further property of the non-Gaussian stable laws when $1 < \alpha < 2$ is that the expected value is finite.

It has been shown by Lévy (1925) that the logarithm of the characteristic function of the stable Paretian distribution is

$$\log \phi(t) = i\,\delta t - \gamma |t|^\alpha [1 - i\beta(t/|t|)\tan(\alpha\pi/2)]. \qquad (20.148)$$

The parameters of (20.148) are α, β, γ and δ. The location parameter is δ; and if $\alpha > 1$, then δ is the mean of the distribution. β is an index of skewness such that when equal to zero the curve is symmetric; when $\beta > 0$ the curve is skewed to the right, and when $\beta < 0$ it is skewed to the left. The scale parameter is γ; α is intimately related to the Pareto exponent (in the sense that the limit of $x^{\alpha-1} \Pr[X > x]$ is finite and non-zero) and controls the amount of probability found in the tails of the distribution. When $0 < \alpha < 2$, the extreme tails of the stable distributions are higher than those of the normal, with the total probability in the tails increasing as α moves from 2 to 0. Explicit expressions for the density functions of stable Paretian distributions are known for only three cases; the Cauchy ($\alpha = 1$, $\beta = 0$), the normal ($\alpha = 2$), and the "coin tossing" (Bernoulli) case ($\alpha = \frac{1}{2}$, $\beta = 1$, $\delta = 0$, $\gamma = 1$) [Fama (1963)].

Putting $\beta = 0$ in (20.148), we have

$$\log \phi(t) = i \, \delta t - \gamma |t|^{\alpha}.$$

For the variable $Y = (X - \delta)\gamma^{-1/\alpha}$ we have the "standard form" of symmetric stable distribution with

$$\log \phi_Y(t) = -|t|^{\alpha}.$$

Bergström (1952) has shown that, for $\alpha > 1$,

$$p_Y(y) = \frac{1}{\pi \alpha} \sum_{j=0}^{\infty} (-1)^j \frac{\Gamma\left[(2j+1)\alpha^{-1}\right]}{(2j)!} y^{2j} \qquad (20.149)$$

so that

$$F_Y(y) = \frac{1}{2} + \sum_{j=1}^{\infty} (-1)^{j-1} \frac{\Gamma\left[(2j-1)\alpha^{-1}\right]}{(2j-1)!} y^{2j-1}. \qquad (20.150)$$

For large y, the asymptotic series

$$1 + \frac{1}{\pi} \sum_{j=1}^{\infty} (-1)^j \frac{\Gamma(j\alpha)}{j! y^{\alpha j}} \sin\left(\tfrac{1}{2} j\alpha\pi\right) \qquad (20.151)$$

is useful for calculating $F_Y(y)$, when $\alpha > 1$.

Fama and Roll (1968) give tables of values of $F_Y(y)$, to four decimal places, for

$$\alpha = 1.0(0.1)1.9(0.05)2.00$$

and

$$y = 0.05(0.05)1.00(0.1)2.0(0.2)4.0(0.4)6.0(1)8, 10, 15, 20.$$

They also give values of y_P satisfying $F_Y(y_P) = P$, to three decimal places, for the same values of α and $P = 0.52(0.02)0.94(0.01)0.97(0.005)0.995, 0.9995$. It is notable that $y_{0.72}$ varies but little with α. Its value increases from 0.827 when $\alpha = 1$ (Cauchy) to about 0.830 ($\alpha = 1.3 - 1.6$) and then decreases to 0.824 for $\alpha = 2$ (normal).

Fama and Roll (1968) suggest that this stability may be used to construct an estimator of $\gamma^{1/\alpha}$ as

$$(0.827)^{-1} \times [\text{upper 28\% order statistic–lower 28\% order statistic}],$$

which should be robust with respect to changes in α.

The Pareto distribution has been applied by astronomers to model the brightness (equivalent to magnitude) distribution of comets. Donnison (1990) used the Pareto power-law model for the distribution of the long-period comets (those that take more than 200 years to complete an orbit) and the short-period comets (those that take less than 15 years to complete an orbit). Donnison analyzed some data by assuming a Pareto distribution for the distribution of brightness. By hypothesizing that brightness indices have increased with discovery date, Donnison and Peers (1992) subsequently analyzed the same data in terms of trend analysis. Pettit (1993) has recently provided an alternative Bayesian solution to the problem by using Gibbs sampling.

12 GENERALIZED PARETO DISTRIBUTIONS

1. A class of "generalized Pareto" distributions was described by Ljubo (1965). The cumulative distribution function is

$$F_X(x) = \begin{cases} 1 - \left(\dfrac{k + \alpha}{x + \alpha}\right)^a e^{-\beta(x-k)} & (x \geq k) \\ 0 & (x < k) \end{cases} \qquad (20.152)$$

with $\beta > 0$.

2. A class more extensively studied within the framework of extreme-value distributions and used in applications is originally due to Pickands (1975) but more recently propagated by Hosking and his co-workers.

Hosking and Wallis (1987) and Dargahi-Noubary (1989), among others, investigate a generalized Pareto distribution (GPD) with the cdf

$$F_X(x; c, k) = 1 - (1 - cx/k)^{1/c}. \qquad (20.153)$$

Special cases, $c = 0$ and $c = 1$, correspond respectively to exponential distribution with mean k and the uniform distribution on $(0, k)$, while Pareto distributions correspond to negative values of c. This distribution is extensively used in the analysis of extreme events (Pickands (1975) was apparently the first to use this distribution in this context), especially in hydrology [van Montfort and Witter (1985)], as well as in reliability studies when

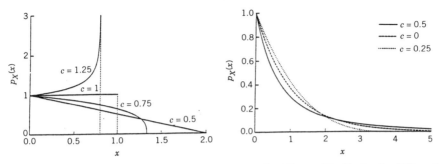

Figure 20.1 Probability Density Function of the Generalized Pareto Distribution for Different Values of the Shape Parameter c [*Source*: Hosking and Wallis (1987)]

robustness is required against heavier tailed or lighter tailed alternatives to an exponential distribution. Typical forms of these distributions are depicted in Figure 20.1.

The pdf is

$$p_X(x) = k^{-1}(1 - cx/k)^{c^{-1}-1} \qquad (c \neq 0), \qquad (20.154a)$$

or

$$p_X(x) = k^{-1}\exp(-x/k) \qquad (c = 0). \qquad (20.154b)$$

The support of the distribution is $x > 0$ for $c \leq 0$ and $0 < x < k/c$ for $c > 0$.

Dargahi-Noubary (1989) recommends GPD for use as the distribution of the excess of observed values over an arbitrarily chosen "threshold." He points out that, "the GPD arises as a class of limit distribution for the excess over a threshold, as the threshold increases toward the right-hand end of the distribution (tail)."

Moreover, "Direct connection between the GPD and extreme value distributions is as follows: If N is a Poisson random variable with mean λ and $Y_1, Y_2, \ldots Y_N$ are independent *excesses* with the distribution function [(20.153)], then

$$\Pr\left[\max(Y_1, Y_2, \ldots, Y_N) \leq x\right] = \exp\left[-\lambda(1 - cx/k)^{1/c}\right] \text{''} \qquad (20.155)$$

(an extreme value distribution; see Chapter 22).

If X has a GPD with parameters (k, c) then the conditional distribution of $X - y'$ given $X > y'$ is GPD with parameters $k - cy', c$ provided $k - cy' > 0$. (Compare with the memoryless property of the exponential distribution.)

This property of GPD is called "threshold stability." Dargahi-Noubary (1989) also advocates use of GPD (and threshold methods) for fitting a distribution of annual maxima of the wind speed (gusts) using the data provided by Cook (1982), and that of maximum floods of the Feather River for the years (1962–1980). He finds the fit based on GPD to be much better than one using classical extreme value methods.

The hazard rate $r_x(x) = p_X(x)/\{1 - F_X(x)\}$ of the GPD is $1/(k - cx)$, which is a monotonic function of x, decreasing for $c < 0$, constant for $c = 0$, and increasing for $c > 0$ [see Xekalaki (1983), who characterizes this distribution by its hazard rate].

Generalized Pareto distributions with $c > \frac{1}{2}$ have finite endpoints with $p_X(x) > 0$ at each end (see Figure 20.1). Such distributions seldom occur in statistical application. (The variance of generalized Pareto for $c \geq \frac{1}{2}$ is infinite.)

With $E[(1 - cX/k)^r] = (1 + rc)^{-1}$ for $1 + rc > 0$, we have

$$\mu = E[X] = k/(1 + c) \tag{20.156a}$$

and

$$\text{Var}(X) = \sigma^2 = k^2/\{(1 + c)^2(1 + 2c)\}. \tag{20.156b}$$

The skewness is

$$\sqrt{\beta_1} = \alpha_3 = 2(1 - c)(1 + 2c)^{1/2}/(1 + 3c), \tag{20.156c}$$

and the kurtosis is

$$\beta_2 = \alpha_4 = \frac{3(1 + 2c)(3 - c + 2c^2)}{(1 + 3c)(1 + 4c)}. \tag{20.156d}$$

In general the r-th moment exists if $c > -1/r$.

Also, since $-c^{-1} \log(1 - cX/k)$ has a standard exponential distribution,

$$E\left[\left\{-\log\left(1 - \frac{cX}{k}\right)\right\}^s\right] = c^s \Gamma(s + 1) \quad \text{for integer } s. \tag{20.156e}$$

Davison (1984) has observed that if X has a GPD (k, c) distribution, then

$$E[X - u | X > u] = \frac{k - cu}{1 + c}, \quad \text{for all } u > 0$$

if $c > -1$. Thus the plot of mean "excess" over u, versus u, should be roughly linear with the slope $-c/(1 + c)$ and intercept $k/(1 + c)$.

Hosking and Wallis (1987) provide a detailed investigation of estimation procedures for the GPD. Their general conclusion is that "maximum likelihood estimation, although asymptotically the most efficient method, does not clearly display its efficiency even in samples as large as 500, while the method of moments is generally reliable except when $c < -2$."

The maximum likelihood equations are:

$$\left[1 + n^{-1} \sum_{i=1}^{n} \log\left(1 - \hat{c}\hat{k}^{-1}X_i\right)\right]\left[n^{-1} \sum_{i=1}^{n} \left(1 + \hat{c}\hat{k}^{-1}X_i\right)^{-1}\right] = 1; \quad (20.157a)$$

$$\hat{c} = -n^{-1} \sum_{i=1}^{n} \log\left(1 - \hat{c}\hat{k}^{-1}X_i\right). \quad (20.157b)$$

Since $\log L$ approaches infinity when $c > 1$ and k/c approaches $\max(X_i)$ $= X_n'$, the MLEs are taken as the values $\hat{k}$ and $\hat{c}$ which yield a local maximum of $\log L$ (unless the Giesbrecht and Kempthorne (1976) approach of maximizing *grouped*-data likelihood function is used—as described in Chapter 14 for the lognormal distribution).

Maximum likelihood estimation of parameters works well if $c < \frac{1}{2}$, but not otherwise. For $c < \frac{1}{2}$ the asymptotic normal distribution of the maximum likelihood estimates based on a sample of size n has the mean value (c, k) and variance–covariance matrix

$$\frac{1}{n}\begin{pmatrix} (1-c)^2 & k(1-c) \\ k(1-c) & 2k^2(1-c). \end{pmatrix}. \quad (20.158)$$

The estimators are consistent, asymptotically normal, and asymptotically efficient for $c < \frac{1}{2}$. The nonregular case is discussed by Smith (1985).

Grimshaw (1993) describes an algorithm for solution of equations (20.157). He points out that $\hat{c}\hat{k}^{-1}$ can be determined from (20.157a), and this value can then be inserted in (20.157b) to obtain $\hat{c}$. If no local minimum is found in the space

$$\{c < 0, k > 0\} \cup \{0 < c \leq 1, k/c > X_n'\},$$

there is no maximum likelihood estimator. In such cases, Grimshaw (1993) recommends using "alternative estimators given by Hosking and Wallis (1987)."

The moment estimators are easily computed from the formulas

$$\tilde{k} = \frac{1}{2}\bar{X}\left(\bar{X}^2 S^{-2} + 1\right) \quad (20.159a)$$

$$\text{and} \quad \tilde{c} = \frac{1}{2}\left(\bar{X}^2/S^2 - 1\right) \quad (20.159b)$$

where $\bar{X}$ and S^2 are the sample mean and variance respectively.

Since $\text{Var}(S^2)$ is infinite for $c \leq -1/4$, these estimators are not asymptotically normal (of order n^{-1}). For $c = 0$, we have exponential distributions and the moment estimators coincide with the maximum likelihood ones. For $c > -1/4$, $(\tilde{k}, \tilde{c})$ is asymptotically bivariate normal with variance–covariance matrix

$$n \text{ var} \begin{bmatrix} \tilde{k} \\ \tilde{c} \end{bmatrix} \sim \frac{(1 + c)^2}{(1 + 2c)(1 + 3c)(1 + 4c)}$$

$$\times \begin{bmatrix} 2k^2(1 + 6c + 12c^2) & k(1 + 2c)(1 + 4c + 12c^2) \\ k(1 + 2c)(1 + 4c + 12c^2) & (1 + 2c)^2(1 + c + 6c^2) \end{bmatrix}.$$
$$(20.160)$$

Hosking and Wallis (1987) also suggest use of probability-weighted moments (PWM)

$$M_{p,r,s} = E\left[X^p \{F_X(X)\}^r \{1 - F_X(X)\}^s \right]$$

for estimation of GPD parameters. In particular,

$$a_s = M_{1,0,s} = E\left[X\{1 - F(X)\}^s \right] = \frac{k}{(s + 1)(s + 1 + k)} \quad (20.161)$$

exists for $c > -1$ and the identities

$$k = \frac{2a_0 a_1}{a_0 - 2a_1}, \qquad c = \frac{a_0}{a_0 - 2a_1} - 2, \qquad (20.162)$$

are valid. They recommend substituting a_0 and a_1 in (20.162) by the estimators

$$\tilde{a}_r = n^{-1} \sum_{j=1}^{n} \frac{(n - j)(n - j - 1) \ldots (n - j - r + 1)}{(n - 1)(n - 2) \ldots (n - r)} X'_j, \qquad r = 0, 1,$$
$$(20.163a)$$

or

$$\tilde{a}_r = n^{-1} \sum_{j=1}^{n} (1 - p_{j:n})^r X'_j, \qquad r = 0, 1, \qquad (20.163b)$$

where $p_{j:n} = (j + \gamma)/(n + \delta)$, where γ and δ are suitable constants (often $\gamma = -0.375$ and $\delta = 0.25$).

The estimators $\tilde{a}_r$ and $\tilde{a}_r$ are asymptotically equivalent.

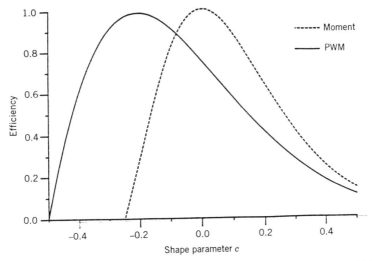

Figure 20.2 Asymptotic Efficiency, Relative to the Maximum Likelihood Estimator, of Moment and PWM Estimators of the Shape Parameter of the Generalized Pareto Distribution

Asymptotic efficiency of moment and probability-weighted moment estimators of the shape parameter of the GPD are presented in Figure 20.2.

Simulations carried out by Hosking and Wallis (1987) have shown that the estimators with the smallest mean squared error (MSE) are moment estimators when $c > 0$ and PWM estimators when $c < -0.2$, while PWM have overall lower bias. The theoretical and simulated standard deviations of the estimators differed by less than 10% provided $n \geq 50$ and $c > 0.4$.

Quantile estimators of GPD are easily obtained from the relation

$$
\begin{aligned}
x(F) &= k\{1 - (1 - F)^c\}/c, &\quad c \neq 0 \\
&= -k \log(1 - F), &\quad c = 0
\end{aligned}
\tag{20.164}
$$

by substituting k and c by their estimators $\hat{k}$ and $\hat{c}$.

Since

$$
x(F) = kF\{1 + \tfrac{1}{2}(1 - c)F + O(F^2)\} \text{ as } F \to 0,
$$

the accuracy of $\hat{x}(F)$, for small F, is essentially determined by the accuracy of $\hat{k}$.

Smith (1987) estimates the tail of a distribution that covers all three limit laws of classical extreme value theory. Given a large number of independent observations $X_1, \ldots, X_n$ with a common density f, fix a high threshold u and let N denote the number of exceedance of u and let $Y_1, \ldots, Y_N$ be the excesses. Namely $Y_i = X_j - U$ when j is the index of the i-th exceedance.

Conditionally on N the excesses are i.i.d. with the definition

$$F_u(y) = \{F(u + y) - F(u)\}/\{1 - F(u)\}, u < x_0, \quad 0 < y < x_0 - u,$$
$$(20.165)$$

where x_0 is the upper end point of F.

Smith (1987) proposes to estimate F_u by the GPD $G(\cdot; k, c)$, estimating k and c by maximum likelihood estimates. A similar approach has been taken by Hall (1982) and Joe (1987).

The justification of these approximations is based on Pickands's (1975) result stating

$$\lim_{u \to x_0} \sup_{0 < y < x_0 - u} |F_u(y) - G(u; k(u), c)| = 0 \qquad (20.166)$$

for some fixed c and function $k(u)$ if and only if F is in the domain of attraction of one of the three limit extreme value laws. (See Chapter 22.)

BIBLIOGRAPHY

Ahmed, S. A., Haq, M. -U., and Khurshid, A. (1990). Measure of peakedness in Pareto model, *The Philippine Statistician*, **39**, 61–70.

Ahsanullah, M. (1991). On record values from the generalized Pareto distribution, *Pakistan Journal of Statistics, Series A*, **7**, 129–136.

Ahsanullah, M., and Houchens, R. L. (1989). A note on record values from a Pareto distribution, *Pakistan Journal of Statistics, Series A*, **5**, 51–57.

Aigner, D. J., and Goldberger, A. S. (1970). Estimation of Pareto's law from grouped observations, *Journal of the American Statistical Association*, **65**, 712–721.

Aitchison, J., and Brown, J. A. C. (1957). *The Lognormal Distribution*, Cambridge: Cambridge University Press.

Arnold, B. C. (1971). Two characterizations of the exponential distribution, Unpublished manuscript, Ames: Iowa State University.

Arnold, B. C. (1983). *Pareto Distributions*, Fairland, MD: International Cooperative Publishing House.

Arnold, B. C. (1985). Pareto distributions, In *Encyclopedia of Statistical Sciences*, S. Kotz, N. L. Johnson and C. B. Read (editors), 568–574, New York: Wiley.

Arnold, B. C., and Press, S. J. (1983). Bayesian inference for Pareto populations, *Journal of Econometrics*, **21**, 287–306.

Arnold, B. C., and Press, S. J. (1986). Bayesian analysis of censored or grouped data from Pareto populations, In *Bayesian Inference and Decision Techniques*, P. Gores and A. Zellner (editors), 157–173, Amsterdam: North-Holland.

Arnold, B. C., and Press, S. J. (1989). Bayesian estimation and prediction for Pareto data, *Journal of the American Statistical Association*, **84**, 1079–1084.

Asrabadi, B. R. (1990). Estimation in the Pareto distribution, *Metrika*, **37**, 199–205.

Balakrishnan, N., and Ahsanullah, M. (1994a). Relations for single and product moments of record values from Lomax distribution, *Sankhya*, *Series B* (to appear).

Balakrishnan, N., and Ahsanullah, M. (1994b). Recurrence relations for single and product moments of record values from generalized Pareto distribution, *Communications in Statistics—Theory and Methods* (to appear).

Balakrishnan, N., and Joshi, P. C. (1981). Moments of order statistics from doubly truncated power function distribution, *Aligarh Journal of Statistics*, **1**, 98–105.

Balakrishnan, N., and Joshi, P. C. (1982). Moments of order statistics from doubly truncated Pareto distribution, *Journal of the Indian Statistical Association*, **20**, 109–117.

Beg, M. I., and Kirmani, S. N. U. A. (1974). On a characterization of exponential and related distributions, *Australian Journal of Statistics*, **16**, 163–166.

Benjamin, J. R., and Cornell, C. A. (1970). *Probability, Statistics and Decision for Civil Engineers*, New York: McGraw-Hill.

Berger, J. M., and Mandelbrot, B. (1963). A new model for error clustering in telephone circuits, *IBM Journal of Research and Development*, **7**, 224–236.

Bergstrom, H. (1952). On some expansions of stable distributions, *Arkiv for Matematik*, **2**, 375–378.

Bhattacharya, N. (1963). A property of the Pareto distribution, *Sankhyā*, *Series B*, **25**, 195–196.

Blum, M. (1970). On the sums of independently distributed Pareto variates, *SIAM Journal of Applied Mathematics*, **19**, 191–198.

Bresciani-Turroni, C. (1939). Annual survey of statistical data: Pareto's law and index of inequality of incomes, *Econometrica*, **7**, 107–133.

Brillinger, D. (1963). Necessary and sufficient conditions for a statistical problem to be invariant under a Lie group, *Annals of Mathematical Statistics*, **34**, 492–500.

Castellani, M. (1950). On multinomial distributions with limited freedom: A stochastic genesis of Pareto's and Pearson's curves, *Annals of Mathematical Statistics*, **21**, 289–293.

Champernowne, D. G. (1952). The graduation of income distribution, *Econometrica*, **20**, 591–615.

Chan, L. K., and Cheng, S. W. (1973). On the optimum spacing for asymptotically best linear estimate of the scale parameter of the Pareto distribution, *Tamkang Journal of Mathematics*, **4**, 9–21.

Charek, D. J., Moore, A. H., and Coleman, J. W. (1988). A comparison of estimation techniques for the three parameter Pareto distribution, *Communications in Statistics—Simulation and Computation*, **17**, 1395–1407.

Cohen, A. C., and Whitten, B. J. (1988). *Parametric Estimation in Reliability and Life Span Models*, New York: Marcel Dekker.

Cook, N. J. (1982). Towards better estimation of extreme winds, *Journal of Wind Engineering and Industrial Aerodynamics*, **9**, 295–323.

Cramer, J. S. (1971). *Empirical Econometrics*, Amsterdam: North-Holland.

Dallas, A. C. (1976). Characterizing the Pareto and power distributions, *Annals of the Institute of Statistical Mathematics*, **28**, 491–497.

Dargahi-Noubary, G. R. (1989). On tail estimation: An improved method, *Mathematical Geology*, **21**, 829–842.

Davison, A. C. (1984). Modelling excesses over high thresholds, with an application, In *Statistical Extremes and Applications*, J. Tiago de Oliveira (editor), 461–482, Dordrecht: Reidel.

Donnison, J. R. (1990). The distribution of cometary magnitudes, *Monthly Notices of the Royal Astronomical Society*, **245**, 658–663.

Donnison, J. R., and Peers, H. W. (1992). An approach to trend analysis in data with special reference to cometary magnitudes, *Monthly Notices of the Royal Astronomical Society*, **256**, 647–654.

Fama, E. F. (1963). Mandelbrot and the stable Paretian hypothesis, *Journal of Business, University of Chicago*, **36**, 420–429.

Fama, E. F., and Roll, R. (1968). Some properties of symmetric stable distributions, *Journal of the American Statistical Association*, **63**, 817–836.

Feller, W. (1971). *An Introduction to Probability Theory and its Applications*, **2** (Second edition), New York: Wiley.

Ferguson, T. S. (1962). Location and scale parameters in exponential families of distributions, *Annals of Mathematical Statistics*, **33**, 986–1001 (Correction, *Ibid.*, **34**, 1603).

Fisk, P. R. (1961). The graduation of income distributions, *Econometrics*, **29**, 171–185.

Freiling, E. C. (1966). *A Comparison of the Fallout Mass-Size Distributions Calculated by Lognormal and Power-Law Models*, San Francisco: U.S. Naval Radiological Defense Laboratory.

Galambos, J., and Kotz, S. (1978). *Characterization of Probability Distributions*, Lecture Notes in Mathematics, **675**, Berlin: Springer-Verlag.

Geisser, S. (1984). Predicting Pareto and exponential observables, *Canadian Journal of Statistics*, **12**, 143–152.

Geisser, S. (1985). Interval prediction for Pareto and exponential observables, *Journal of Econometrics*, **29**, 173–185.

Giesbrecht, F., and Kempthorne, O. (1976). Maximum likelihood estimation of the 3-parameter log-normal distribution, *Journal of the Royal Statistical Society, Series B*, **38**, 257–264.

Goldberg, G. (1967). *The Pareto Law and the Pyramid Distribution*, Publication No. 505, Houston, TX: Shell Development.

Grimshaw, S. D. (1993). Computing maximum likelihood estimates for the generalized Pareto distributions, *Technometrics*, **35**, 185–191.

Gumbel, E. J. (1958). *Statistics of Extremes*, New York: Columbia University Press.

Gupta, R. C., and Langford, E. S. (1984). On the determination of a distribution by its median residual life function: a functional equation, *Journal of Applied Probability*, **21**, 120–128.

Hagstroem, K. -G. (1925). La loi de Pareto et la reassurance, *Skandinavisk Aktuarietidskrift*, **8**, 65–88.

Hagstroem, K. -G. (1960). Remarks on Pareto distributions, *Skandinavisk Aktuarietid-skrift*, **43**, 59–71.

Hall, P. (1982). On estimating the end point of a distribution, *Annals of Statistics*, **10**, 556–568.

Hamdy, H. I., and Pallotta, W. J. (1987). Triple sampling procedure for estimating the scale parameters of Pareto distributions, *Communications in Statistics—Theory and Methods*, **16**, 2155–2164.

Harris, C. M. (1968). The Pareto distribution as a queue service discipline, *Operations Research*, **16**, 307–313.

Harter, H. L. (1964). Exact confidence bounds based on one order statistic, for the parameter of an exponential population, *Technometrics*, **6**, 301–317.

Hassanein, K. M., Saleh, A. K. Md. E., and Brown, E. F. (1991). Estimating survivor function using optimally selected order statistics, *Metrika*, **38**, 99–115.

Hayakawa, M. (1951). The application of Pareto's law of income to Japanese data, *Econometrica*, **19**, 174–183.

Hogg, R. V., and Klugman, S. A. (1984). *Loss Distributions*, New York: Wiley.

Horn, P. S. (1983). A measure of peakedness, *The American Statistician*, **37**, 55–56.

Hosking, J. R. M., and Wallis, J. R. (1987). Parameter and quantile estimation for the generalized Pareto distribution, *Technometrics*, **29**, 339–349.

Houchens, R. L. (1984). *Record value theory and inference*, Ph.D. dissertation, Riverside: University of California.

Huang, J. S. (1974). On a theorem of Ahsanullah and Rahman, *Journal of Applied Probability*, **11**, 216–218.

Huang, J. S. (1975). A note on order statistics from the Pareto distribution, *Scandinavian Actuarial Journal*, **2**, 187–190.

Huang, J. S. (1978). On a "lack of memory" property, Technical Report No. 84, Statistical Series, Ontario, Canada: University of Guelph.

Iliescu, D., and Vodă, K. (1979). Some notes on Pareto distribution, *Revue Roumaine des Mathemetiques Pures et Appliquees*, **4**, 328–337.

James, I. R. (1979). Characterization of a family of distributions by the independence of size and shape variables, *Annals of Statistics*, **7**, 869–881.

Joe, H. (1987). Estimation of quantiles of the maximum of N observations, *Biometrika*, **74**, 347–354.

Kabe, D. G. (1972). On moments of order statistics from the Pareto distribution, *Skandinavisk Aktuarietidskrift*, **55**, 179–181.

Kalbfleisch, J. D., and Prentice, R. L. (1980). *The Statistical Analysis of Failure Time Data*, New York: Wiley.

Kaluszka, M. (1986). Admissible and minimax estimators of a^r in the gamma distribution with truncated space, *Metrika*, **33**, 363–375.

Kern, D. M. (1983). Minimum variance estimation in the Pareto distribution, *Metrika*, **30**, 15–19.

Khan, A. H., and Ali, M. M. (1987). Characterizations of probability distributions through higher order gap, *Communications in Statistics—Theory and Methods*, **16**, 1281–1287.

Khan, A. H., and Khan, S. A. (1987). Moments of order statistics from Burr distribution and its characterizations, *Metron*, **45**, 21–29.

Khurana, A. P., and Jha, V. D. (1985). Order statistics of a sample and of an extended sample from discrete distributions, *Journal of the Indian Society for Agricultural Statistics*, **37**, 63–70.

Kotz, S., and Shanbhag, D. N. (1980). Some new approaches to probability distributions, *Advances in Applied Probability*, **12**, 903–921.

Koutrouvelis, I. A. (1981). Large-sample quantile estimation in Pareto laws, *Communications in Statistics—Theory and Methods*, **10**, 189–201.

Krishnaji, N. (1970). Characterization of the Pareto distribution through a model of under-reported incomes, *Econometrica*, **38**, 251–255.

Krishnaji, N. (1971). Note on a characterizing property of the exponential distribution, *Annals of Mathematical Statistics*, **42**, 361–362.

Kulldorff, G. (1961). *Contributions to the Theory of Estimation from Grouped and Partially Grouped Samples*, Uppsala, Sweden: Almqvist and Wiksall.

Kulldorff, G., and Vannman, K. (1973). Estimation of the location and scale parameters of a Pareto distribution by linear function of order statistics, *Journal of the American Statistical Association*, **68**, 218–227.

Lévy, P. (1925). *Calcul des Probabilites*, Paris: Gauthier-Villars.

Liang, T. -C. (1993). Convergence rates for empirical Bayes estimation of the scale parameter in a Pareto distribution, *Computational Statistics & Data Analysis*, **16**, 35–45.

Ljubo, M. (1965). Curves and concentration indices for certain generalized Pareto distributions, *Statistical Review*, **15**, 257–260. (In Serbo-Croatian, English summary.)

Lomax, K. S. (1954). Business failures. Another example of the analysis of failure data, *Journal of the American Statistical Association*, **49**, 847–852.

Lwin, T. (1972). Estimation of the tail of the Paretian law, *Skandinavisk Aktuarietidskrift*, **55**, 170–178.

Maguire, B. A., Pearson, E. S., and Wynn, H. A. (1952). The time intervals between industrial accidents, *Biometrika*, **39**, 168–180.

Malik, H. J. (1966). Exact moments of order statistics from the Pareto distribution, *Skandinavisk Aktuarietidskrift*, **49**, 144–157.

Malik, H. J. (1967). Exact moments of order statistics for a power-function population, *Skandinavisk Aktuarietidskrift*, **50**, 64–69.

Malik, H. J. (1968). Estimation of the parameters of the power-function population, *Metron*, **27**, 196–203.

Malik, H. J. (1970a). Distribution of product statistics from a Pareto population, *Metrika*, **15**, 19–22.

Malik, H. J. (1970b). Estimation of the parameter of the Pareto distribution, *Metrika*, **15**, 126–132.

Malik, H. J., and Trudel, R. (1976). Distribution of Product and Quotient of Order Statistics, Technical Report No. 30, Ontario, Canada: University of Guelph.

Mandelbrot, B. (1960). The Pareto-Levy law and the distribution of income, *International Economic Review*, **1**, 79–106.

Mandelbrot, B. (1963). The variation of certain speculative prices, *Journal of Business, University of Chicago*, **36**, 394–419.

Mandelbrot, B. (1967). The variation of some other speculative prices, *Journal of Business, University of Chicago*, **40**, 393–413.

Marsaglia, G., and Tubilla, A. (1975). A note on the "lack of memory" property of the exponential distribution, *Annals of Probability*, **3**, 353–354.

Moore, A. H., and Harter, H. L. (1967). One-order statistics conditional estimators of the shape parameter of the limited and Pareto distributions and the scale parameters of Type II asymptotic distributions of smallest and largest values, *IEEE Transactions on Reliability*, **16**, 100–103.

Moothathu, T. S. K. (1981). On Lorenz curves of lognormal and Pareto distributions, *Journal of the Indian Statistical Association*, **19**, 103–108.

Moothathu, T. S. K. (1984). Characterizations of power function and Pareto distributions in terms of their Lorenz curves, *Research Review*, **3**, 118–123.

Moothathu, T. S. K. (1985). Sampling distributions of Lorenz curve and Gini index of the Pareto distribution, *Sankhyā, Series B*, **47**, 247–258.

Moothathu, T. S. K. (1990). The best estimator and a strongly consistent, asymptotically normal, unbiased estimator of Lorenz curve, Gini index and Theil entropy index of Pareto distribution, *Sankhyā, Series B*, **52**, 115–127.

Morrison, D. G. (1978). On linearly increasing mean residual lifetimes, *Journal of Applied Probability*, **15**, 617–620.

Mosimann, J. E. (1970). Size allometry: Size and shape variables with characterizations of the lognormal and generalized gamma distributions, *Journal of the American Statistical Association*, **65**, 930–945.

Muniruzzaman, A. N. M. (1950). On some distributions in connection with Pareto's law, *Proceedings of the First Pakistan Statistical Conference*, 90–93.

Muniruzzaman, A. N. M. (1957). On measures of location and dispersion and tests of hypotheses on a Pareto population, *Bulletin of the Calcutta Statistical Association*, **7**, 115–123.

Nigm, A. M., and Hamdy, H. I. (1987). Bayesian prediction bands for the Pareto lifetime model, *Communications in Statistics—Theory and Methods*, **16**, 1761–1772.

Ogawa, J. (1962). Optimum spacing and grouping for the exponential distribution, In *Contributions to Order Statistics*, A. E. Sarhan and B. G. Greenberg (editors), 371–380, New York: Wiley.

Ord, J. K. (1975). Statistical models for personal income distributions, In *Statistical Distributions in Scientific Work*, **2**, G. P. Patil, S. Kotz, and J. K. Ord (editors), 151–158, Dordrecht: Reidel.

Ord, J. K., Patil, G. P., and Taillie, C. (1981). Relationships between income distributions for individuals and for households, In *Statistical Distributions in Scientific Work*, **6**, C. Taillie, G. P. Patil and B. A. Baldessari (editors), 203–210, Dordrecht: Reidel.

Pareto, V. (1987). *Cours d'Economie Politique*, Paris: Rouge et Cie.

Pederzoli, G., and Rathie, P. N. (1980). Distribution of product and quotient of Pareto variables, *Metron*, **27**, 165–169.

Pettit, L. I. (1993). Inferences about ordered parameters—an astronomical problem, *The Statistician*, **42**, 491–496.

Pickands, J. (1975). Statistical inference using extreme order statistics, *Annals of Statistics*, **3**, 119–131.

Pigou, A. C. (1932). *The Economics of Welfare*, London: Macmillan.

Porter, J. E., Coleman, J. W., and Moore, A. H. (1992). Modified KS, AD and C-VM tests for the Pareto distribution with unknown location and scale parameters, *IEEE Transactions on Reliability*, **41**, 112–117.

Quandt, R. E. (1964). Statistical discrimination among alternative hypotheses and some economic regularities, *Journal of Regional Science*, **5**, 1–23.

Quandt, R. E. (1966). Old and new methods of estimation and the Pareto distribution, *Metrika*, **10**, 55–82.

Rider, P. R. (1964). Distribution of product and quotient of maximum values in samples from a power-function population, *Journal of the American Statistical Association*, **59**, 877–880.

Rohatgi, V. K., and Saleh, A. K. Md. E. (1987). Estimation of the common scale parameter of two Pareto distributions in censored samples, *Naval Research Logistics*, **34**, 235–238.

Rytgaard, M. (1990). Estimation in Pareto distribution, *ASTIN Bulletin*, **20**, 201–216.

Saksena, S. K., and Johnson, A. M. (1984). Best unbiased estimators for the parameters of a two-parameter Pareto distribution, *Metrika*, **31**, 77–83.

Saleh, A. K. Md. E., and Ali, M. M. (1966). Asymptotic optimum quantiles for the estimation of the parameter of the negative exponential distribution, *Annals of Mathematical Statistics*, **37**, 143–152.

Saleh, A. K. Md. E., Ali, M. M., and Umbach, D. (1985). Large sample estimation of Pareto quantiles using selected order statistics, *Metrika*, **32**, 49–56.

Schmittlein, D., and Morrison, D. (1981). The median residual time: A characterization theorem and an application, *Operations Research*, **29**, 392–399.

Shirras, G. F. (1935). The Pareto law and the distribution of income, *Economic Journal*, **45**, 663–681.

Smith, R. L. (1984). Threshold methods for sample extremes, In *Statistical Extremes and Applications*, J. Tiago de Oliveira (editor), 621–638, Dordrecht: Reidel.

Smith, R. L. (1985). Maximum likelihood estimation in a class of non-regular cases, *Biometrika*, **72**, 67–90.

Smith, R. L. (1987). Estimating tails of probability distributions, *Annals of Statistics*, **15**, 1174–1207.

Springer, M. D. (1979). *The Algebra of Random Variables*, New York: Wiley.

Srivastava, M. S. (1965). A characterization of Pareto's distribution and $(k + 1)x^k/\theta^{(k+1)}$, *Annals of Mathematical Statistics*, **36**, 361–362.

Steindl, J. (1965). *Random Processes and the Growth of Firms*, New York: Hafner.

Takano, K. (1992). On relations between the Pareto distribution and a differential equation, *Bulletin of the Faculty of Science of Ibaraki University, Series A*, **24**, 15–30.

Umbach, D., Ali, M. M., and Hassanein, K. H. (1981). Estimation of Pareto quantiles using two order statistics, *Communications in Statistics—Theory and Methods*, **10**, 1933–1941.

Van Montfort, M. A. J., and Witter, J. V. (1985). Testing exponentiality against generalized Pareto distribution, *Journal of Hydrology*, **78**, 305–315.

Vännman, K. (1976). Estimators based on order statistics from a Pareto distribution, *Journal of the American Statistical Association*, **71**, 704–708.

Xekalaki, E. (1983). Hazard functions and life distributions in discrete time, *Communications in Statistics—Theory and Methods*, **12**, 2503–2509.

Xekalaki, E., and Panaretos, J. (1988). On the association of the Pareto and the Yule distribution, *Theory of Probability and Its Applications*, **33**, 191–195.

CHAPTER 21

Weibull Distributions

1 HISTORICAL REMARKS

The Weibull distribution is named after the Swedish physicist, Waloddi Weibull, who (1939a, b) used it to represent the distribution of the breaking strength of materials and in 1951 for a wide variety of other applications. The close agreement that Weibull demonstrated between his observed data and those predicted with the fitted Weibull models was quite impressive. He used the distribution to model data secured from problems dealing with yield strength of Bofors' steel, fiber strength of Indian cotton, length of syrtoideas, fatigue life of an ST-37 steel, statures of adult males born in the British Isles, and breadth of beans of *Phaseolus vulgaris*. Interestingly Rosen and Rammler had used this distribution earlier in 1933 to describe the "laws governing the fineness of powdered coal." In the Russian statistical literature this distribution is often referred to as the *Weibull-Gnedenko distribution*, since it is one of the three types of limit distributions for the sample maximum established rigorously by Gnedenko (1943) [e.g., see Gnedenko, Belyaev, and Solov'ev (1965)]. The name *Frechét distribution* is also used sometimes due to the fact that it was Frechét (1927) who first identified this distribution to be an extremal distribution [later on shown to be one of three possible solutions by Fisher and Tippett (1928)].

The Weibull distribution includes the exponential and the Rayleigh distributions as special cases (see Chapters 19 and 18). Since the hazard function (or the failure rate) of this distribution is a decreasing function when the shape parameter c is less than 1, a constant when c equals 1 (the exponential case), and an increasing function when c is greater than 1, the use of the distribution in reliability and quality control work was advocated by many authors following Weibull (1951), Kao (1958, 1959), and Berrettoni (1964). Due to the nature of the hazard function described above, the distribution often becomes suitable where the conditions for "strict randomness" of the exponential distribution are not satisfied, with the shape parameter c having a characteristic or predictable value depending upon the fundamental nature of the problem being considered. Unlike in the case of the exponential distribution, probabilistic bases for the Weibull distribution are not commonly encountered in situations where the distribution is actually employed. However, Malik (1975) and Franck (1988) have assigned some simple physical meanings and interpretations for the Weibull distribution, thus providing natural applications of this distribution in reliability problems particularly dealing with wearing styles [see also Harter (1986)]. The distribution, being a power transformation of the exponential, presents a convenient way of introducing some flexibility in the model through the power (or the shape parameter) c. Some situations where the Weibull distribution will likely arise have been mentioned by Gittus (1967). Hallinan (1993) has recently provided an excellent review of the Weibull distribution by presenting historical facts, and the many different forms of this distribution as used by practitioners and possible confusions and errors that arise due to this non-uniqueness.

The Weibull distribution is undeniably the distribution that has received maximum attention during the past 23 years since the first edition of this volume was published. This is clearly evident from the large number of references (most of which have been published since 1970) at the end of this chapter.

2 DEFINITION

A random variable X has a *Weibull distribution* if there are values of the parameters c (> 0), α (> 0), and ξ_0 such that

$$Y = \left(\frac{X - \xi_0}{\alpha}\right)^c \tag{21.1}$$

has the standard exponential distribution with probability density function

$$p_Y(y) = e^{-y}, \qquad y > 0. \tag{21.2}$$

The probability density function of the Weibull random variable X is then

$$p_X(x) = \frac{c}{\alpha}\left(\frac{x - \xi_0}{\alpha}\right)^{c-1} e^{-\{(x-\xi_0)/\alpha\}^c}, \qquad x > \xi_0. \tag{21.3}$$

The cumulative distribution function is

$$F_X(x) = 1 - e^{-\{(x-\xi_0)/\alpha\}^c}, \qquad x > \xi_0, \tag{21.4}$$

and the survival (or the reliability) function is

$$R_X(x) = 1 - F_X(x) = e^{-\{(x-\xi_0)/\alpha\}^c}, \qquad x > \xi_0. \tag{21.5}$$

From (21.3) and (21.5) we get the hazard function

$$h_X(x) = \frac{p_X(x)}{R_X(x)} = \frac{c}{\alpha}\left(\frac{x - \xi_0}{\alpha}\right)^{c-1}, \qquad x > \xi_0. \qquad (21.6)$$

As mentioned earlier, it is clear that the hazard function in (21.6) is decreasing for $c < 1$, constant for $c = 1$, and increasing for $c > 1$. The mean residual life or the life expectancy of a Weibull component is obtained from (21.5) as

$$e_x = \int_x^\infty \frac{R_X(u)}{R_X(x)}\,du = e^{[(x-\xi_0)/\alpha]^c}\int_x^\infty e^{-[(u-\xi_0)/\alpha]^c}\,du.$$

Raja Rao and Talwalker (1989) derived lower and upper bounds for this function.

For $c > 1$ the Weibull density function in (21.3) tends to 0 as $x \to \xi_0$, and there is a single mode at

$$x = \alpha\left(\frac{c - 1}{c}\right)^{1/c} + \xi_0. \qquad (21.7)$$

This value tends to $\alpha + \xi_0$ very rapidly as $c \to \infty$. For $0 < c \le 1$ the mode is at ξ_0 and the density is a decreasing function of x for all $x > \xi_0$.

From (21.4) we observe that the median of the distribution is

$$\alpha(\log 2)^{1/c} + \xi_0. \qquad (21.8)$$

It is also of interest to note that for all values of c,

$$F_X(\xi_0 + \alpha) = 1 - e^{-1} \doteq 0.63. \qquad (21.9)$$

The standard form of the distribution will have $\xi_0 = 0$ and $\alpha = 1$ so that the standard density function is

$$p_X(x) = cx^{c-1}e^{-x^c}, \qquad x > 0, c > 0 \qquad (21.10)$$

and the corresponding cumulative distribution function is

$$F_X(x) = 1 - e^{-x^c}, \qquad x > 0, c > 0. \qquad (21.11)$$

The distribution of X now depends on the shape parameter c alone. The plots of the standard density function in (21.10) for $c = 0.25, 0.5, 1, 1.5, 2, 3, 4, 5$ are presented in Figure 21.1. The moment ratios, coefficient of variation, and standardized cumulants $\kappa_r / \kappa_2^{r/2}$ of the standard distribution are of course the same as those of the distribution in (21.3), which depend only on c and not on ξ_0 or α (as they are location and scale parameters). Moments corresponding to the three-parameter distribution in (21.3) are easily derived from those corresponding to the standard distribution in (21.10) by using the transformation $X' = \xi_0 + \alpha X$.

Since X^c has the standard exponential distribution in (21.2), the rth raw moment of X is also the (r/c)th moment about zero of the standard

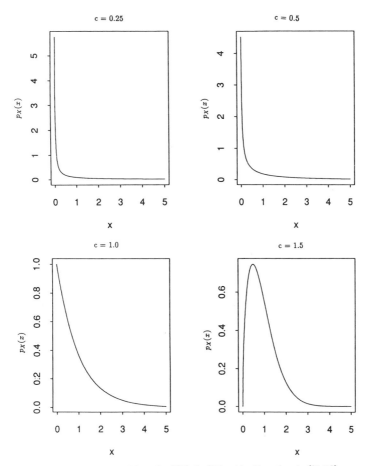

Figure 21.1   Plots of Standard Weibull Density Function in (21.10)

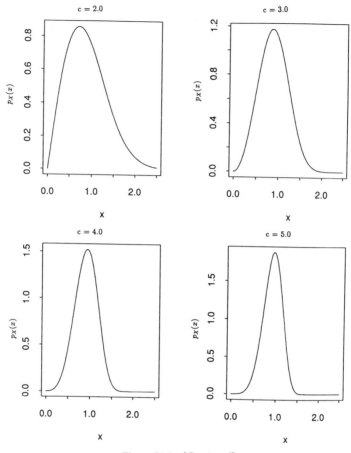

Figure 21.1 (*Continued*)

exponential random variable (see Chapter 19). Hence

$$\mu'_r = E[X^r] = \Gamma\left(\frac{r}{c} + 1\right) \tag{21.12}$$

from which we get

$$E[X] = \Gamma\left(\frac{1}{c} + 1\right)$$

$$\text{Var}(X) = \Gamma\left(\frac{2}{c} + 1\right) - \left\{\Gamma\left(\frac{1}{c} + 1\right)\right\}^2. \tag{21.13}$$

Similar expressions for the moment ratios can easily be written down explicitly. When c is large, we obtain from (21.13) that

$$E[X] \simeq 1 - \frac{\gamma}{c} + \frac{1}{2c^2}\left(\frac{\pi^2}{6} + \gamma^2\right) \simeq 1 - \frac{0.57722}{c} + \frac{0.98905}{c^2},$$

$$\mathrm{Var}(X) \simeq \frac{\pi^2}{6c^2} \simeq \frac{1.64493}{c^2}, \tag{21.14}$$

where γ is Euler's constant.

Table 21.1 gives some numerical values of the mean, standard deviation, and the coefficients of skewness and kurtosis for various values of c. It may be noted from this table that for $c = 3.6$ (approx.) the coefficient of skewness $\sqrt{\beta_1}$ is zero. In fact, for values of c in the neighborhood of 3.6, the Weibull distribution is similar in shape to a normal distribution. Dubey (1967a) made a detailed study of the closeness of agreement between the two distributions. It is of interest to note from the table that the coefficient of kurtosis β_2 has a minimum value of about 2.71 when $c = 3.35$ (approx.).

Cohen (1973) and Rousu (1973) both examined skewness and kurtosis as a function of the shape parameter c. With $c_0 = 3.6023494257197$, β_1 is observed to be a decreasing function of c in the interval $c < c_0$ and an increasing function in the interval $c > c_0$ ($\beta_1 = 0$ for $c = c_0$). With $c_1 = 3.35$, β_2 is also observed to be a decreasing function of c in the interval $c < c_1$ and an increasing function in the interval $c > c_1$. As mentioned above,

Table 21.1 Moments of Weibull Distributions

c	Mean	Standard Deviation	$\sqrt{\beta_1}$	β_2
1.2	0.9407	0.7872	1.52	6.24
1.4	0.9114	0.6596	1.20	4.84
1.6	0.8966	0.5737	0.96	4.04
1.8	0.8893	0.5112	0.78	3.56
2.0	0.8862	0.4633	0.63	3.25
2.2	0.8856	0.4249	0.51	3.04
2.4	0.8865	0.3935	0.40	2.91
2.6	0.8882	0.3670	0.32	2.82
2.8	0.8905	0.3443	0.24	2.76
3.0	0.8930	0.3245	0.17	2.73
3.2	0.8957	0.3072	0.11	2.71
3.4	0.8984	0.2918	0.05	2.71
3.6	0.9011	0.2780	0.00	2.72
3.8	0.9038	0.2656	−0.04	2.73
4.0	0.9064	0.2543	−0.09	2.75

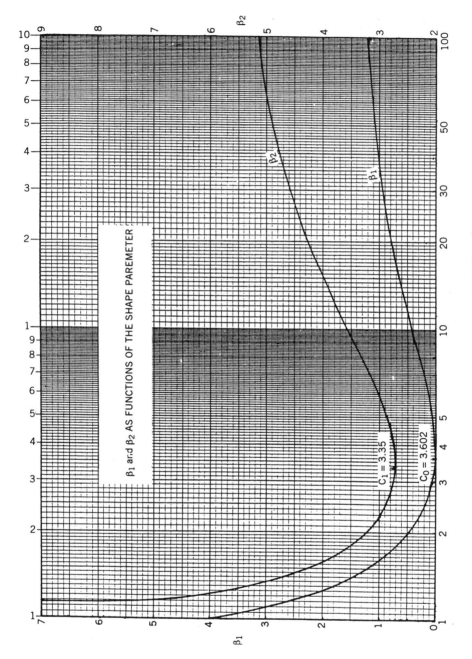

Figure 21.2 β_1 and β_2 as Functions of the Shape Parameter

Dubey (1967a) has studied the behavior of the Weibull distribution in considerable detail for values of c in the vicinity of c_0 and c_1. Graphs of both β_1 and β_2 as functions of c are presented in Figure 21.2.

When β_2 is plotted as a function of β_1, two distinct lines or branches are observed intersecting at $(\beta_1 = 0, \beta_2 = 2.72)$. The point of intersection corresponds to $c = c_0$, while the two separate branches correspond to the intervals $c < c_0$ and $c > c_0$. These two (β_1, β_2) lines for the Weibull distribution along with the corresponding lines for the log-normal distribution and for the Pearson Type III (gamma) and Type V distributions (see Chapter 12) have been plotted in Figure 21.3. As shown in Figure 21.3, the (β_1, β_2) line for the Weibull distribution with $c < c_0$ lies in the Type I region (region below Type III) and extends approximately parallel to the Type III line until the two lines intersect off the graph at $(\beta_1 = 4.0, \beta_2 = 9.0)$. The Weibull line for $c > c_0$ originates in the Type I region and extends approximately parallel to the Type V line. It crosses the Type III line into the Type VI region (region between Types III and V) at a point where $\beta_1 \simeq 0.5$ and then moves in close to the log-normal line. Hence the Weibull distribution with $c > c_0$ will closely resemble the Pearson Type VI distribution when $\beta_1 \geq 0.5$, and for values $\beta_1 \simeq 1$ or greater it will also closely resemble the lognormal distribution. The skewness factor of the Weibull family has been studied at great length by Groeneveld (1986).

From (21.12) we obtain the moment generating function of $\log X$ to be

$$E[e^{t \log X}] = E[X^t] = \Gamma\left(\frac{t}{c} + 1\right). \qquad (21.15)$$

Hence the rth cumulant of $\log X$ is $\psi^{(r)}(1)/c^r$, where $\psi^{(r)}(\cdot)$ are the "polygamma functions" (derivatives of logarithm of gamma function); see Chapter 1, Section A2. In particular, we have

$$E[\log X] = -\frac{\gamma}{c}$$

$$\text{Var}(\log X) = \frac{\pi^2}{6c^2}. \qquad (21.16)$$

The variable $\log X$ has an *extreme value distribution* which is discussed in detail in Chapter 22.

The information generating function for the standard distribution in (21.10) is

$$T(u) = E\left[\left(p_X(X)\right)^u\right] = c^u \int_0^\infty x^{u(c-1)} e^{-ux^c} \, dx. \qquad (21.17)$$

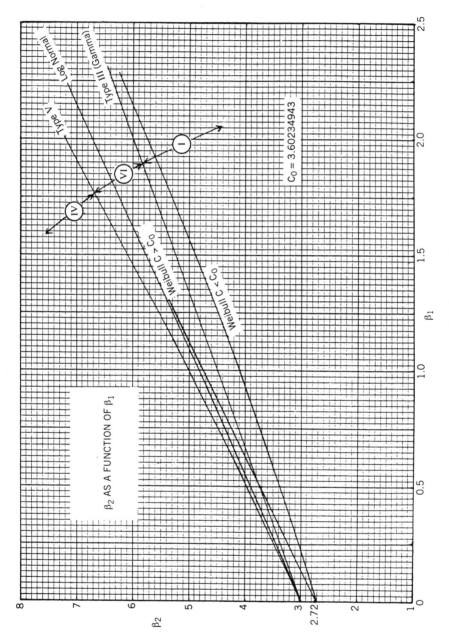

Figure 21.3 β_2 as a Function of β_1

Setting $ux^c = v$ (taking $u > 0$ and noting that $c > 0$), we obtain

$$T(u) = \frac{c^{u-1}\Gamma\{[1 + u(c - 1)]/c\}}{u^{\{1+u(c-1)\}/c}}, \tag{21.18}$$

from which the entropy of the distribution is

$$-T'(1) = \frac{(c - 1)\gamma}{c} - \log c + 1. \tag{21.19}$$

A detailed account of these and other properties of the Weibull distribution has been presented by Lehman (1963).

3 ORDER STATISTICS

Let $X_1, X_2, \ldots, X_n$ be independent and identically distributed Weibull random variables with density function in (21.3) and cumulative distribution function (21.4); further, let $X_1' \le X_2' \le \cdots \le X_n'$ denote the order statistics obtained from these n variables. The density function of the smallest order statistic X_1' is

$$p_{X_1'}(x) = n\{1 - F_X(x)\}^{n-1}p_X(x)$$

$$= \frac{nc}{\alpha}\left(\frac{x - \xi_0}{\alpha}\right)^{c-1}e^{-n\{(x-\xi_0)/\alpha\}^c}, \qquad x > \xi_0. \tag{21.20}$$

It is readily observed from (21.20) that X_1' is also distributed as a Weibull random variable, except that α is replaced by $\alpha n^{-1/c}$. In other words, the Weibull distribution is "closed under minima."

From the density function and joint density function of order statistics, exact explicit expressions for the single and the product moments of order statistics can be derived. For this purpose it is convenient to work with order statistics for a random sample from the standard Weibull distribution (21.10), since the corresponding results for the three-parameter distribution (21.3) can be obtained by using the linear transformation $\xi_0 + \alpha X_r'$.

The density function of X_r' $(1 \le r \le n)$ is

$$p_{X_r'}(x) = \frac{n!}{(r - 1)!(n - r)!}(1 - e^{-x^c})^{r-1}e^{-x^c(n-r+1)}cx^{c-1}, \qquad x > 0. \tag{21.21}$$

From (21.21), we obtain the kth raw moment of X'_r to be

$$
E\left[(X'_r)^k\right] = \frac{n!}{(r-1)!(n-r)!} \int_0^\infty x^k \{1 - e^{-x^c}\}^{r-1} e^{-x^c(n-r+1)} c x^{c-1}\, dx
$$

$$
= \frac{n!}{(r-1)!(n-r)!} \sum_{i=0}^{r-1} (-1)^i \binom{r-1}{i} \int_0^\infty e^{-x^c(n-r+i+1)} x^k c x^{c-1}\, dx
$$

$$
= \frac{n!}{(r-1)!(n-r)!} \Gamma\left(1 + \frac{k}{c}\right) \sum_{i=0}^{r-1} \frac{(-1)^i \binom{r-1}{i}}{(n-r+i+1)^{1+(k/c)}}.
$$

$$(21.22)$$

Using (21.22), which is due to Lieblein (1955), Weibull (1967b) tabulated means and variances of order statistics for $n = 5(5)20$ and $c^{-1} = 0.1(0.1)1.0$. Govindarajulu and Joshi (1968) similarly tabulated means and variances of all order statistics for n up to 10 and $c = 1, 2, 2.5, 3(1)10$. Harter (1970) presented a more extensive table of means of order statistics for samples of sizes up to 40 and $c = 0.5(0.5)4(1)8$. Recently Balakrishnan and Chan (1993a) have tabulated means and variances of all order statistics for n up to 20 and $c = \frac{1}{5}, \frac{1}{4}, \frac{1}{3}, \frac{1}{2}, 1.5(0.5)3, 4(2)10$.

An alternate simple recursive method of computing the single moments has been proposed by Balakrishnan and Joshi (1981). David and Shu (1978) showed that for an arbitrary distribution with distribution function $F_X(x)$,

$$
F_{X'_{r:n}}(x) = F_{X'_{r-1:n}}(x) - \binom{n}{r-1} \{F_X(x)\}^{r-1} \{1 - F_X(x)\}^{n-r+1}, \quad (21.23)
$$

with $F_{X'_{0:n}} \equiv 1$ for all x so that (21.23) is true for $r = 1$ as well. Further, for any arbitrary distribution we have the relation [David (1981, pp. 46–47); Arnold, Balakrishnan, and Nagaraja (1992, pp. 111–112)]

$$
(r-1)F_{X'_{r:n}}(x) + (n-r+1)F_{X'_{r-1:n}}(x) = nF_{X'_{r-1:n-1}}(x) \quad (21.24)
$$

which, together with (21.23), readily yields

$$
F_{X'_{r:n}}(x) = F_{X'_{r-1:n-1}}(x) - \binom{n-1}{r-1} \{F_X(x)\}^{r-1} \{1 - F_X(x)\}^{n-r+1}. \quad (21.25)
$$

Then, using [Parzen (1960, p. 212)]

$$
E\left[(X'_{r:n})^k\right] = k \int_0^\infty x^{k-1} \{1 - F_{X'_{r:n}}(x)\}\, dx
$$

and substituting for $F_{X'_{r:n}}(x)$ from (21.25), we get

$$E\left[(X'_{r:n})^k\right] = E\left[(X'_{r-1:n-1})^k\right] + k\binom{n-1}{r-1}J_k(n-r+1, r-1), \quad (21.26)$$

where

$$J_k(n-r+1, r-1) = \int_0^\infty x^{k-1}e^{-x^c(n-r+1)}\{1-e^{-x^c}\}^{r-1}\,dx. \quad (21.27)$$

These quantities can be computed recursively using

$$J_k(p,0) = \frac{\Gamma(k/c)}{cp^{k/c}},$$

$$J_k(p,m) = J_k(p,m-1) - J_k(p+1, m-1), \quad m \ge 1, \quad (21.28)$$

and then the single moments may be computed recursively from (21.26).
The joint density function of X'_r and X'_s $(1 \le r < s \le n)$ is

$$p_{X'_r, X'_s}(x, y) = \frac{n!}{(r-1)!(s-r-1)!(n-s)!}\{1-e^{-x^c}\}^{r-1}$$

$$\times\{e^{-x^c} - e^{-y^c}\}^{s-r-1}e^{-x^c}cx^{c-1}e^{-y^c(n-s+1)}cy^{c-1},$$

$$0 < x < y < \infty. \quad (21.29)$$

From (21.29) we obtain the product moment of X'_r and X'_s as

$$E[X'_r X'_s] = \int_0^\infty \int_0^y xy\, p_{X'_r, X'_s}(x, y)\, dx\, dy$$

$$= \frac{n!c^2}{(r-1)!(s-r-1)!(n-s)!}\sum_{i=0}^{r-1}\sum_{j=0}^{s-r-1}(-1)^{s-r-1-j+i}$$

$$\times\binom{r-1}{i}\binom{s-r-1}{j}$$

$$\times\int_0^\infty\int_0^y e^{-(i+j+1)x^c}e^{-(n-r-j)y^c}x^c y^c\, dx\, dy$$

$$= \frac{n!}{(r-1)!(s-r-1)!(n-s)!}\sum_{i=0}^{r-1}\sum_{j=0}^{s-r-1}(-1)^{s-r-1-j+i}$$

$$\times\binom{r-1}{i}\binom{s-r-1}{j}$$

$$\times\phi_c(i+j+1, n-r-j), \quad (21.30)$$

where $\phi_c(a, b)$ is Lieblein's ϕ-function defined by

$$\phi_c(a, b) = c^2 \int_0^\infty \int_0^y e^{-ax^c - by^c} x^c y^c \, dx \, dy. \tag{21.31}$$

Through a differential equation approach, Lieblein (1955) has shown that for $a \geq b$,

$$\phi_c(a, b) = \frac{\{\Gamma[1 + (1/c)]\}^2}{(ab)^{1 + (1/c)}} I_{a/(a+b)}\left(1 + \frac{1}{c}, 1 + \frac{1}{c}\right), \tag{21.32}$$

where $I_p(a, b)$ is incomplete beta function ratio defined by (see Chapter 1, Section A5)

$$I_p(a, b) = \frac{1}{B(a, b)} \int_0^p t^{a-1}(1 - t)^{b-1} \, dt, \qquad 0 < p < 1. \tag{21.33}$$

When $a < b$, $\phi_c(a, b)$ may be computed from the identity

$$\phi_c(a, b) + \phi_c(b, a) = \frac{\{\Gamma[1 + (1/c)]\}^2}{(ab)^{1 + (1/c)}}. \tag{21.34}$$

After computing the ϕ-function from (21.32) and (21.34), the product moments (and hence the covariances) of Weibull order statistics may be computed from the formula in (21.24). Weibull (1967b) presented covariances of all order statistics for $n = 5(5)20$ and $c^{-1} = 0.1(0.1)1.0$. Govindarajulu and Joshi (1968) tabulated covariances of order statistics for sample sizes up to 10 and $c = 1, 2, 2.5, 3(1)10$. Balakrishnan and Chan (1993a) prepared tables of covariances of all order statistics for n up to 20 and $c = \frac{1}{5}, \frac{1}{4}, \frac{1}{3}, \frac{1}{2}, 1.5(0.5)3, 4(2)10$; see also Adatia (1994) for some extended tables of covariances.

The tables of means, variances, and covariances of order statistics are necessary for the determination of best linear unbiased estimators of the parameters ξ_0 and α (when c is assumed to be known), as explained in the following section. We note here that Patel and Read (1975) and Patel (1975) discussed bounds on moments of linear functions of order statistics. Malik and Trudel (1982) derived the density function of quotient of two order statistics. Recurrence relations satisfied by the product moments of order statistics were derived by Khan et al. (1983). By noting from (21.12) that negative moments (of order $r > -c$) do exist for the standard Weibull random variable, Khan, Khan, and Parvez (1984) discussed the negative moments of all Weibull order statistics (of order greater than $-c$). Further relevant discussions on order statistics have been provided by Miller (1976), Harter (1988), and Shor and Levanon (1991).

4   METHODS OF INFERENCE

For the two-parameter Weibull distribution (with $\xi_0 = 0$) with probability density function

$$p_X(x) = \frac{c}{\alpha^c}x^{c-1}e^{-(x/\alpha)^c}, \qquad x > 0, \alpha > 0, c > 0, \qquad (21.35)$$

numerous papers have appeared over the years dealing with exact or approximate methods of inference for the shape parameter c and the scale parameter α. In this case the transformed variable $Y = \log X$ is easily seen to have the probability density function

$$p_Y(y) = ce^{c(y\log\alpha)}e^{-e^{c(y-\log\alpha)}}, \qquad -\infty < y < \infty, \alpha > 0, c > 0, \quad (21.36)$$

which is a Type 1 extreme-value distribution (a member of the location-scale family of distributions) with the location parameter $\eta = \log\alpha$ and the scale parameter $\delta = 1/c$. Due to this fact, given a sample from the two-parameter Weibull distribution (21.35), after using the logarithmic transformation on the data and estimating the parameters η and δ using some method or criterion, one can obtain estimators of the parameters α and c through the relationships $\alpha = e^{\eta}$ and $c = 1/\delta$. Of course some properties of the estimators of the parameters η and δ will be passed on to the transformed estimators of the parameters α and c (e.g., the maximum likelihood property), while some other properties may not be inherited by the transformed estimators of α and c (e.g., the unbiasedness property). Almost all the papers that deal with inference for the two-parameter Weibull distribution are based on this transformation idea. Since these results, in principle, were derived primarily in developing inference for the location and scale parameters η and δ of the Type 1 extreme-value distribution, they will be discussed in detail in Chapter 22. Other inference-related topics like tolerance limits and intervals and prediction limits and intervals are handled in Sections 5 and 6, respectively. Once again, most of the developments in these topics are also based on the transformation idea. As a result our primary focus in this section will be on methods of inference developed for the three-parameter Weibull distribution, or those results that do not hinge upon the transformation of the Weibull data to the extreme-value form.

4.1   Moment Estimation

By equating the first three moments of X to the first three sample moments and solving, it is possible to determine the moment estimators of ξ_0, α, and c. The value of the first moment ratio ($\sqrt{\beta_1}$) depends only on c, and once $\sqrt{\beta_1}$ is estimated from the sample coefficient of skewness ($\sqrt{b_1}$), c can be determined numerically. Using this estimated value of c, α is determined from the standard deviation, and finally ξ_0 is determined from the sample mean. If ξ_0 is known, then c can be estimated from the ratio (standard deviation)/(mean $- \xi_0$). A nomogram provided by Kotel'nikov (1964) for this purpose is presented in Section 8, Figure 21.6. Dubey (1966f, 1967b) provided tables of asymptotic relative efficiency of moment estimators.

Using the formula for the moments of X, it is possible to construct simple estimators of c^{-1}, applicable when ξ_0 is known. Menon (1963) proposed the estimator

$$
\widetilde{(c^{-1})} = \frac{\sqrt{6}}{\pi} \times \{\text{Sample standard deviation of } \log X_1, \log X_2, \ldots, \log X_n\}
$$

$$
= \frac{\sqrt{6}}{\pi} \times \sqrt{\frac{1}{n-1} \sum_{i=1}^{n} \left(\log X_i - \overline{\log X}\right)^2}, \tag{21.37}
$$

which is an asymptotically normal and unbiased estimator of c^{-1}. Its variance is $\{1.1 + O(n^{-1})\}(c^{-2}/n)$, and its asymptotic efficiency, relative to the Cramér-Rao lower bound, is 55%. Kagan (1965) constructed a generalization of Menon's estimator with a higher asymptotic efficiency.

Weibull (1967b) proposed that estimation of the parameters can be based on the first and second *moments about the smallest sample value*

$$
R_i = \frac{1}{n} \sum_{j=1}^{n} \left(X_j' - X_1'\right)^i, \qquad i = 1, 2. \tag{21.38}
$$

As $n \to \infty$, the expected value of R_2/R_1^2 tends to a function $f(c)$ of the shape parameter c. By solving the equation

$$
f(c^*) = \frac{R_2}{R_1^2}, \tag{21.39}
$$

an asymptotically unbiased estimator of c can be derived. The table presented below, taken from Weibull (1967b), gives some values of $f(c)$:

c	$f(c)$
0.0	1.0000
0.1	1.0145
0.2	1.0524
0.3	1.1093
0.4	1.1831
0.5	1.2732
0.6	1.3801
0.7	1.5045
0.8	1.6480
0.9	1.8124
1.0	2.0000

For $c \geq 0.5$, this estimator is markedly more accurate than the estimator based on central moments. The asymptotic efficiency (compared with MLEs) decreases from 97.6% when $c = 0.5$ to 60.8% when $c = 1.0$, while that of a central moment estimator (obtained by equating $\sqrt{\beta_1}$ to the sample value $\sqrt{b_1}$) decreases from 17.1% to 7.6% over the same range of values of c. Of course the moment estimator uses the third central moment, while R_2/R_1^2 in (21.39) uses only the first two moments, so one might expect the former to be relatively inaccurate.

The statistic

$$\frac{1}{\gamma}\left\{\log \alpha - \frac{1}{n}\sum_{j=1}^{n} \log X_j\right\} \tag{21.40}$$

is an unbiased estimator of c^{-1}, with variance

$$\left(c^2 n\right)^{-1}\frac{\pi^2}{6\gamma^2} \doteq 4.93\frac{c^{-2}}{n}, \tag{21.41}$$

where γ is Euler's constant. Menon (1963) pointed out that this estimator, apart from its unbiasedness, does not have very many desirable properties; for example, its efficiency is as low as 12%. On the other hand, the estimator of α derived from (21.40), given by

$$\tilde{\alpha} = \exp\left\{\frac{1}{n}\sum_{i=1}^{n} \log X_i + \gamma\left(\widetilde{c^{-1}}\right)\right\}, \tag{21.42}$$

is asymptotically unbiased, with variance

$$\mathrm{Var}(\tilde{\alpha}) = 1.2\frac{c^{-2}}{n} + c^{-2}O\left(n^{-3/2}\right). \tag{21.43}$$

This estimator has an asymptotic efficiency of 95%.

Moment estimation of the shape parameter c was also discussed by Mihram (1977). The properties of moment estimators for the three-parameter case have been discussed by many authors, including Newby (1980, 1984) and Cran (1988). A nomogram for estimating all three parameters has been prepared by Sen and Prabhashanker (1980). Tables that would facilitate the method of moments estimation of all three parameters have been presented by Blischke and Scheuer (1986). Bowman and Shenton (1983) have given series expressions for moments of these moment estimators.

4.2 Best Linear Unbiased Estimation

Let $X'_{r+1}, X'_{r+2}, \ldots, X'_{n-s}$ be a doubly Type II censored sample from the three-parameter Weibull distribution in (21.3), where the smallest r and the largest s lifetimes (out of n components based on a life test) were not observed. Let $Z'_{r+1}, Z'_{r+2}, \ldots, Z'_{n-s}$ be the corresponding doubly Type II censored sample from the standard Weibull distribution (21.10), where $Z'_i = (X'_i - \xi_0)/\alpha$ for $r + 1 \leq i \leq n - s$. Further, let us denote $E[Z'_i]$ by μ_i, $\mathrm{Var}(Z'_i)$ by $\sigma_{i,i}$, $\mathrm{Cov}(Z'_i, Z'_j)$ by $\sigma_{i,j}$, and write

$$\boldsymbol{\mu}^T = (\mu_{r+1}, \mu_{r+2}, \ldots, \mu_{n-s}),$$

$$\mathbf{1}^T = (1, 1, \ldots, 1)_{1 \times (n-r-s)},$$

$$\boldsymbol{\Sigma} = \left((\sigma_{i,j}; r + 1 \leq i, j \leq n - s) \right).$$

Explicit expressions for these quantities have already been presented in Section 3, wherein the available tables for these quantities have also been listed. Realize, however, that these depend on the value of the shape parameter c. With the assumption that the shape parameter c is known the BLUEs of the location (or threshold) and scale parameters ξ_0 and α can be derived as shown below.

With $X^T = (X'_{r+1}, X'_{r+2}, \ldots, X'_{n-s})$, upon minimizing the generalized variance

$$(X - \xi_0 \mathbf{1} - \alpha\boldsymbol{\mu})^T \boldsymbol{\Sigma}^{-1} (X - \xi_0 \mathbf{1} - \alpha\boldsymbol{\mu}) \qquad (21.44)$$

with respect to the unknown parameters ξ_0 and α, we obtain the BLUEs of ξ_0 and α as

$$\xi_0^* = \left\{ \frac{\boldsymbol{\mu}^T\boldsymbol{\Sigma}^{-1}\boldsymbol{\mu}\mathbf{1}^T\boldsymbol{\Sigma}^{-1} - \boldsymbol{\mu}^T\boldsymbol{\Sigma}^{-1}\mathbf{1}\boldsymbol{\mu}^T\boldsymbol{\Sigma}^{-1}}{(\mathbf{1}^T\boldsymbol{\Sigma}^{-1}\mathbf{1})(\boldsymbol{\mu}^T\boldsymbol{\Sigma}^{-1}\boldsymbol{\mu}) - (\boldsymbol{\mu}^T\boldsymbol{\Sigma}^{-1}\mathbf{1})^2} \right\} X$$

$$= \sum_{i=r+1}^{n-s} a_i X'_i, \qquad (21.45)$$

$$\alpha^* = \left\{ \frac{\mathbf{1}^T\boldsymbol{\Sigma}^{-1}\mathbf{1}\boldsymbol{\mu}^T\boldsymbol{\Sigma}^{-1} - \mathbf{1}^T\boldsymbol{\Sigma}^{-1}\boldsymbol{\mu}\mathbf{1}^T\boldsymbol{\Sigma}^{-1}}{(\mathbf{1}^T\boldsymbol{\Sigma}^{-1}\mathbf{1})(\boldsymbol{\mu}^T\boldsymbol{\Sigma}^{-1}\boldsymbol{\mu}) - (\boldsymbol{\mu}^T\boldsymbol{\Sigma}^{-1}\mathbf{1})^2} \right\} X$$

$$= \sum_{i=r+1}^{n-s} b_i X'_i. \qquad (21.46)$$

The variances and covariance of these estimators are given by

$$\text{Var}(\xi_0^*) = \alpha^2 \left\{ \frac{\boldsymbol{\mu}^T \boldsymbol{\Sigma}^{-1} \boldsymbol{\mu}}{(\mathbf{1}^T \boldsymbol{\Sigma}^{-1} \mathbf{1})(\boldsymbol{\mu}^T \boldsymbol{\Sigma}^{-1} \boldsymbol{\mu}) - (\boldsymbol{\mu}^T \boldsymbol{\Sigma}^{-1} \mathbf{1})^2} \right\} = \alpha^2 V_1, \quad (21.47)$$

$$\text{Var}(\alpha^*) = \alpha^2 \left\{ \frac{\mathbf{1}^T \boldsymbol{\Sigma}^{-1} \mathbf{1}}{(\mathbf{1}^T \boldsymbol{\Sigma}^{-1} \mathbf{1})(\boldsymbol{\mu}^T \boldsymbol{\Sigma}^{-1} \boldsymbol{\mu}) - (\boldsymbol{\mu}^T \boldsymbol{\Sigma}^{-1} \mathbf{1})^2} \right\} = \alpha^2 V_2, \quad (21.48)$$

$$\text{Cov}(\xi_0^*, \alpha^*) = -\alpha^2 \left\{ \frac{\boldsymbol{\mu}^T \boldsymbol{\Sigma}^{-1} \mathbf{1}}{(\mathbf{1}^T \boldsymbol{\Sigma}^{-1} \mathbf{1})(\boldsymbol{\mu}^T \boldsymbol{\Sigma}^{-1} \boldsymbol{\mu}) - (\boldsymbol{\mu}^T \boldsymbol{\Sigma}^{-1} \mathbf{1})^2} \right\} = \alpha^2 V_3. \quad (21.49)$$

The coefficients a_i and b_i and the values of V_1, V_2, and V_3 have been tabulated by various authors for different choices of the sample size n and values of the shape parameter c. For example, Weibull (1967a) presented tables for $n = 5(5)20$ and $c^{-1} = 0.1(0.1)1.0$. Govindarajulu and Joshi (1968) presented tables for sample size n up to 10 and $c = 1, 2, 2.5, 3(1)10$. Balakrishnan and Chan (1993b) prepared tables for n up to 20 and $c = \frac{1}{5}, \frac{1}{4}, \frac{1}{3}, \frac{1}{2}, 1.5(0.5)3, 4(2)10$.

Musson (1965) provided early discussion on the determination of the BLUEs of ξ_0 and α. Stump (1968) proposed nearly best unbiased estimators of ξ_0 and α as functions of order statistics. Further discussion and developments on the best linear unbiased estimators of the parameters ξ_0 and α obtained through least-squares theory have been provided by Friedman (1981), Friedman and Gertsbakh (1981), Engeman and Keefe (1982), and Yildirim (1990).

For the case when ξ_0 can be assumed to be zero, Vogt (1968) has constructed a median-unbiased estimator of c. Since $Y_j' = (X_j'/\alpha)^c$ ($j = 1, 2, \ldots, n$) are order statistics corresponding to independent standard exponential variables, it can be shown that

$$\Pr\left[\frac{Y_n'}{Y_1'} \leq w \right] = n \sum_{j=0}^{n-1} (-1)^j \binom{n-1}{j} \{ j(w-1) + n \}^{-1}.$$

There is a unique value of w (> 0), say $w_{1/2}$, such that the probability is equal to $\frac{1}{2}$. Then, since the events $Y_n'/Y_1' \leq w_{1/2}$ and

$$c \leq (\log X_n' - \log X_1')^{-1}(\log w_{1/2})$$

are identical, it follows that c is the median of the distribution of

$$(\log X_n' - \log X_1')^{-1}(\log w_{1/2}).$$

Some values of $\log_{10} w_{1/2}$, based on Vogt (1968), are shown in Table 21.2.

Table 21.2 Values of $g_n = \log_{10} w_{1/2}$ such that $g_n(\log_{10} X'_n - \log_{10} X'_1)^{-1}$ is a Median-Unbiased Estimator of c

n	2	3	4	5	6	7	8	9	10
g_n	0.477	0.802	1.009	1.159	1.278	1.375	1.458	1.530	1.594

For large samples, when ξ_0 is known to be zero, Dubey (1967c) suggested using the estimated 16.7% and 97.4% quantiles, y_1, y_2, respectively, and estimating c by the formula

$$c^* = 2.989 \left[\log\left(\frac{\hat{y}_2}{\hat{y}_1} \right) \right]^{-1}.$$

This estimator is asymptotically unbiased and normal, with approximate variance $0.916c^2/n$.

The construction of best linear unbiased estimators for c^{-1} and $\log \alpha$ using the first k order statistics has been studied by White (1965). For the special case when $k = 2$, the following estimators have been suggested by Leone, Rutenberg, and Topp (1960):

1. If ξ_0 is known to be zero, c is estimated by solving the equation

$$\frac{X'_2}{X'_1} = n\left(\frac{n}{n-1} \right)^{1/c} - n + 1.$$

2. If the value of ξ_0 is not known but that of c is known, then ξ_0 is estimated as

$$X'_1 - n^{-1}(X'_2 - X'_1)\left[\left(\frac{n}{n-1} \right)^{1/c} - 1 \right]^{-1}.$$

3. If the values of both α and c are known, then ξ_0 is estimated as

$$X'_1 - \alpha(n+1)^{-1/c}.$$

This is an unbiased estimator of ξ_0.

4. If the value of c is known, α is estimated as

$$\frac{1}{n}(X'_2 - X'_1)\frac{(n-1)^{1/c}}{1 - n^{-1/c}(n-1)^{1/c}\Gamma(c^{-1} + 1)}.$$

5. Finally, if the values of both ξ_0 and c are known, α may be estimated as

$$(X'_{\mu+1} - \xi_0)[\log\{(n+1)/(n-\mu)\}]^{-1/c}$$

using the single order statistic $X'_{\mu+1}$, where μ is the largest integer not exceeding $n(1 - e^{-1})$.

Note that a linear function of order statistics $\{X'_j\}$ is also a linear function of c^{-1}-th powers $\{Z'^{1/c}_j\}$ or order statistics corresponding to n independent exponentially distributed variables. If all the sample were available, then it would be best to use a linear function of the Z'_j's (i.e., of X'^c_j's). While this may be not so when incomplete sets of order statistics are used, it appears quite likely that if c is known, functions of form $\Sigma\lambda_j X'^c_j$ will be more useful than those of form $\Sigma\lambda_j X'_j$.

4.3 Asymptotic Best Linear Unbiased Estimation

By making use of the asymptotic approximations (see Chapter 12), we have

$$\mu_{n_i} \doteq F^{-1}(\lambda_i) + O\left(\frac{1}{n}\right) \tag{21.50}$$

and

$$\sigma_{n_i, n_j} \doteq \frac{\lambda_i(1 - \lambda_j)}{np\left(F^{-1}(\lambda_i)\right)p\left(F^{-1}(\lambda_j)\right)} + O\left(\frac{1}{n^2}\right), \tag{21.51}$$

where $p(\cdot)$ and $F^{-1}(\cdot)$ are the density function and the inverse cumulative distribution function of the standard Weibull distribution (with $\xi_0 = 0$ and $\alpha = 1$), and $n_i = [n\lambda_i] + 1$. Then the asymptotic best linear unbiased estimators (ABLUEs) of ξ_0 and α may be derived, based on k optimally selected order statistics (or quantiles), by determining the optimal values of $\lambda_1, \lambda_2, \ldots, \lambda_k$, subject to $0 < \lambda_1 < \lambda_2 < \cdots < \lambda_k < 1$, that minimize some norm of the variance-covariance matrix of the estimates.

Using the approximations in (21.50) and (21.51), we can write the asymptotic approximations for the variances and covariances of the BLUEs ξ_0^* and α^* in (21.47) through (21.49) as

$$\text{Var}(\xi_0^*) = \frac{\alpha^2}{n} \times \frac{V_{22}}{V_{11}V_{22} - V_{12}^2}, \tag{21.52}$$

$$\text{Var}(\alpha^*) = \frac{\alpha^2}{n} \times \frac{V_{11}}{V_{11}V_{22} - V_{12}^2}, \tag{21.53}$$

$$\text{Cov}(\xi_0^*, \alpha^*) = -\frac{\alpha^2}{n} \times \frac{V_{12}}{V_{11}V_{22} - V_{12}^2}. \tag{21.54}$$

In the above formulas

$$V_{11} = \sum_{i=1}^{k+1} \frac{\{p(F^{-1}(\lambda_i)) - p(F^{-1}(\lambda_{i-1}))\}^2}{\lambda_i - \lambda_{i-1}}, \qquad (21.55)$$

$$V_{22} = \sum_{i=1}^{k+1} \frac{\{F^{-1}(\lambda_i)p(F^{-1}(\lambda_i)) - F^{-1}(\lambda_{i-1})p(F^{-1}(\lambda_{i-1}))\}^2}{\lambda_i - \lambda_{i-1}}, \qquad (21.56)$$

$$V_{12} = \sum_{i=1}^{k+1} \frac{\{p(F^{-1}(\lambda_i)) - p(F^{-1}(\lambda_{i-1}))\}\{F^{-1}(\lambda_i)p(F^{-1}(\lambda_i)) - F^{-1}(\lambda_{i-1})p(F^{-1}(\lambda_{i-1}))\}}{\lambda_i - \lambda_{i-1}},$$

$$(21.57)$$

with $\lambda_0 = 0$, $\lambda_{k+1} = 1$, $p(F^{-1}(\lambda_0)) = F^{-1}(\lambda_0)p(F^{-1}(\lambda_0)) = p(F^{-1}(\lambda_{k+1})) = F^{-1}(\lambda_{k+1})p(F^{-1}(\lambda_{k+1})) = 0$. Then the ABLUEs ξ_0^{**} and α^{**} are derived by determining the optimal spacing $(\lambda_1, \lambda_2, \ldots, \lambda_k)$, subject to $0 < \lambda_1 < \lambda_2 < \cdots < \lambda_k$, which minimizes the determinant or the trace of the variance-covariance matrix of ξ_0^* and α^*. For example, if the determinant norm is used for this purpose, then we will need to determine $\lambda_1, \lambda_2, \ldots, \lambda_k$ that maximize the quantity $(V_{11}V_{22} - V_{12}^2)$.

Similarly, if the threshold parameter ξ_0 is unknown but the scale parameter α is known, then the ABLUE ξ_0^{**} is derived by determining $\lambda_1, \lambda_2, \ldots, \lambda_k$ to maximize V_{11} since in this case the asymptotic approximation for the variance of the BLUE ξ_0^* is given by

$$\mathrm{Var}(\xi_0^*) = \frac{\alpha^2}{nV_{11}}. \qquad (21.58)$$

On the other hand, if the threshold parameter ξ_0 is known but the scale parameter α is unknown, then the ABLUE α^{**} is derived by determining $\lambda_1, \lambda_2, \ldots, \lambda_k$ that maximize V_{22}, since in this case the asymptotic approximation for the variance of the BLUE α^* is given by

$$\mathrm{Var}(\alpha^*) = \frac{\alpha^2}{nV_{22}}. \qquad (21.59)$$

Estimation of the scale parameter α by the use of one optimally selected order statistic has been discussed by Quayle (1963) and Moore and Harter (1965). Clark (1964) has discussed the estimation of α by the use of k optimally selected order statistics. Moore and Harter (1966) have discussed the interval estimation of the parameter α based on one optimally selected order statistic. Murthy and Swartz (1975) discussed the estimation of parameters based on two order statistics. Chan, Cheng, and Mead (1974) discussed the simultaneous estimation of the parameters ξ_0 and α based on k optimally selected order statistics [also see Hassanein (1971)]. Hassanein, Saleh, and Brown (1984) provided estimators based on k optimally selected order statistics from complete as well as Type II censored samples. Zanakis and Mann (1982) proposed a simple percentile estimator for the shape parameter c when all three parameters are unknown.

For distribution (21.2), the cumulative distribution function is

$$F_X(x) = 1 - \exp\left[-\left(\frac{x - \xi_0}{\alpha}\right)^c\right].$$

This equation can be written

$$x = \xi_0 + \alpha\left[-\log\{1 - F_X(x)\}\right]^{1/c}.$$

Suppose that $\hat{X}_{p_i}, \hat{X}_{p_j}, \hat{X}_{p_k}$ are estimators of the values of X corresponding to $F_X(x) = p_i, p_j, p_k$, respectively. (In a large sample $\hat{X}_{p_i}$ would be the $[(n + 1)p_i]$-th order statistic.) Then the three equations

$$\hat{X}_{p_s} = \xi_0^* + \alpha^*\left[-\log(1 - p_s)\right]^{1/c^*}, \qquad s = i, j, k,$$

could be solved, yielding the estimators ξ^*, α^*, and c^*.

The equation for c^* is

$$\frac{\hat{X}_{p_i} - \hat{X}_{p_j}}{\hat{X}_{p_j} - \hat{X}_{p_k}} = \frac{\left[-\log(1 - p_i)\right]^{1/c^*} - \left[-\log(1 - p_j)\right]^{1/c^*}}{\left[-\log(1 - p_j)\right]^{1/c^*} - \left[-\log(1 - p_k)\right]^{1/c^*}}.$$

If it can be arranged that

$$-\log(1 - p_j) = \sqrt{\{-\log(1 - p_i)\}\{-\log(1 - p_k)\}},$$

then the above equation for c^* simplifies to

$$c^* = \frac{\frac{1}{2}\left\{\log\left[\{-\log(1 - p_k)\}/\{-\log(1 - p_j)\}\right]\right\}}{\log\left[\left(\hat{X}_{p_i} - \hat{X}_{p_j}\right)/\left(\hat{X}_{p_j} - \hat{X}_{p_k}\right)\right]}.$$

Dubey (1967c) has discussed the optimal choice of p_i, p_j, and p_k.

If the value of ξ_0 is known (e.g., zero), then a variety of methods of estimating c and α is available. The function $Y_j = 1 - \exp[-(X_j/\alpha)^c]$ is the probability integral transform of X_j if X_j has distribution (21.2) with $\xi_0 = 0$. If $Y'_1 \leq Y'_2 \leq \cdots \leq Y'_n$ are the corresponding order statistics, then

$$Y'_j = 1 - \exp\left[-\left(\frac{X'_j}{\alpha}\right)^c\right], \qquad j = 1, 2, \ldots, n,$$

and the Y_j''s are distributed as order statistics in a random sample from the standard uniform distribution over the interval 0 to 1 [see (26.1) of Chapter 26 with $a = h = \frac{1}{2}$], and so $E[Y_j'] = j/(n + 1)$.

Estimators α^*, c^* of α, c, obtained by minimizing

$$\sum_{j=1}^{n} \left[\log\{-\log(1 - Y_j')\} - \log\{-\log(1 - E[Y_j'])\} \right]^2$$

$$= \sum_{j=1}^{n} \left[c \log\left(\frac{X_j'}{\alpha}\right) - \log\left\{ -\log\left(\frac{n - j + 1}{n + 1}\right) \right\} \right]^2$$

with respect to c and α, are

$$c^* = \frac{\sum_{j=1}^{n}\left(K_j - \bar{K}\right)\left(\log X_j' - \overline{\log X}\right)}{\sum_{j=1}^{n}\left(\log X_j' - \overline{\log X}\right)^2},$$

where

$$K_j = \log\left[-\log\left(\frac{n - j + 1}{n + 1}\right) \right]$$

$$\bar{K} = n^{-1} \sum_{j=1}^{n} K_j$$

$$\overline{\log X} = n^{-1} \sum_{j=1}^{n} \log X_j'$$

$$\alpha^* = \left(\prod_{j=1}^{n} X_j' \right)^{1/n} \left[\prod_{j=1}^{n} \left\{ -\log\left(\frac{n - j + 1}{n + 1}\right) \right\} \right]^{-1/(nc^*)}.$$

These estimators were proposed by Gumbel (1958).

Estimators of similar form, with K_j replaced by $\log[\sum_{i=1}^{j}(n - i + 1)^{-1}]$, were proposed by Bain and Antle (1967), and with K_j replaced by

$$\log\left[-\log\frac{n - j + (3/2)}{n - j} \right]$$

by Miller and Freund (1965). [A further estimator was obtained by Bain and Antle (1967), replacing K_j by *minus* expected value of the $(n - j + 1)$-th order statistic of random samples of size n from the standard Type 1 extreme value distribution (22.26) of Chapter 22; see also Kimball (1960) and Mann (1968). Bain and Antle (1967) show that for all estimators of this form the distribution of c^*/c does not depend on any parameter (and so confidence intervals of c can be constructed from an observed value of c^*). They further found, from a Monte Carlo investigation, that all the estimators (with various values of K_j) just described are comparable with (and rather better than) Menon's estimator (21.40).

The c^*'s are biased estimators of c, but the bias could be removed by an appropriate corrective multiplier (equal to $\{E[c^*/c]\}^{-1}$). Bain and Antle give values for this multiplier.

Sometimes it is desired to estimate the "reliability," that is,

$$
R(x) = \Pr[X > x] = \exp\left[-\left(\frac{x - \xi_0}{\alpha}\right)^c\right]
$$

or, if ξ_0 be known to be zero, $\exp[-(x/\alpha)^c]$. Of course, if $\hat{a}, \hat{c}$ are maximum likelihood estimators of α, c, respectively, then $\exp[-(x/\hat{a})^{\hat{c}}]$ is a maximum likelihood estimator of $R(x)$.

4.4 Minimum Quantile Distance Estimation

In the case of the three-parameter Weibull distribution, the method proposed by Carmody, Eubank, and LaRiccia (1984) provides a minimum quantile distance estimator for the parameter $\boldsymbol{\theta} = (\xi_0, \alpha, c)$. Parr and Schucany (1980) presented a useful discussion on minimum distance estimation using the empirical cumulative distribution function, while LaRiccia (1982) gave a lucid account of the quantile domain approach.

Let $Q(\lambda) = -\log(1 - \lambda)$, $0 < \lambda < 1$, denote the quantile function of the standard exponential distribution. Then the quantile function corresponding to the three-parameter Weibull distribution in (21.3) is

$$
Q(\lambda; \boldsymbol{\theta}) = \xi_0 + \alpha\{Q(\lambda)\}^{1/c}. \tag{21.60}
$$

Define the sample quantile function as

$$
\tilde{Q}(\lambda) = X_i', \qquad \frac{i-1}{n} < \lambda \le \frac{i}{n}, \qquad i = 1, 2, \dots, n. \tag{21.61}
$$

For a given set of $k < n$ percentile points $\boldsymbol{\lambda} = \{\lambda_1, \lambda_2, \dots, \lambda_k\}$ with $0 < \lambda_1 < \lambda_2 < \cdots < \lambda_k < 1$, let

$$
\tilde{\boldsymbol{Q}}_{\lambda} = \left(\tilde{Q}(\lambda_1), \tilde{Q}(\lambda_2), \dots, \tilde{Q}(\lambda_k)\right)^T
$$

and

$$Q_\lambda(\theta) = (Q(\lambda_1;\theta), Q(\lambda_2;\theta), \ldots, Q(\lambda_k;\theta))^T,$$

where $Q(\lambda;\theta)$ and $\tilde{Q}(\lambda)$ are as defined in (21.60) and (21.61), respectively. Then the minimum quantile distance estimator of θ is the vector that minimizes the quadratic form

$$(\tilde{Q}_\lambda - Q_\lambda(\theta))^T W(\theta)(\tilde{Q}_\lambda - Q_\lambda(\theta)) \tag{21.62}$$

as a function of θ, where $W(\theta)$ is a user-defined $k \times k$ matrix of weights that may be chosen to depend on θ. Thus (21.62) provides an entire family of estimators $\hat{\theta}_\lambda(W)$, indexed by both λ and W. A good choice of W, suggested by Carmody, Eubank, and LaRiccia (1984), is

$$W^*(c) = H_\lambda(c) R_\lambda^{-1} H_\lambda(c), \tag{21.63}$$

where $R_\lambda^{-1} = \min(\lambda_i, \lambda_j) - \lambda_i \lambda_j$ and $H_\lambda(c)$ is the $k \times k$ diagonal matrix having its ith diagonal element to be $c(1 - \lambda_i)\{Q(\lambda_i)\}^{(c-1)/c}$. Since R_λ^{-1} has a tridiagonal inverse whose typical row has nonzero entries

$$-\frac{1}{\lambda_i - \lambda_{i-1}}, \quad \frac{\lambda_{i+1} - \lambda_{i-1}}{(\lambda_{i+1} - \lambda_{i-1})(\lambda_i - \lambda_{i-1})}, \quad -\frac{1}{\lambda_{i+1} - \lambda_i},$$

the elements of $W^*(c)$ in (21.63) can easily be evaluated.

Although the estimator $\hat{\theta}_\lambda(W)$ will not have a closed form, the estimator can be computed without much difficulty, using a standard minimization routine. The interesting thing about this method of estimation is that it requires no modifications if the sample is Type II right censored, left censored, or doubly censored. Carmody, Eubank, and LaRiccia (1984) have established some asymptotic properties of the estimator including the uniqueness, consistency, and asymptotic normality. Further they have made some suggestions on the selection of λ, since the variance-covariance matrix of the estimator is a function of the specific quantiles selected, even though the estimator $\hat{\theta}_\lambda(W^*)$ is optimal for any given λ.

4.5 Modified Moment Estimation

For the three-parameter Weibull distribution, Cohen, Whitten, and Ding (1984) suggested the use of the traditional method of moments equations $E[X] = \bar{X}$ and $\mathrm{Var}(X) = S^2$ along with the equation $E[X_1'] = X_1'$ for the simultaneous estimation of the parameters ξ_0, α, and c, and this is the method of modified moment estimation. Realizing from (21.13) and (21.22) that

$$E[X] = \xi_0 + \alpha\Gamma_1(c)$$

$$\text{Var}(X) = \alpha^2\{\Gamma_2(c) - \Gamma_1^2(c)\} \tag{21.64}$$

$$E[X_1'] = \xi_0 + \frac{\alpha}{n^{1/c}}\Gamma_1(c),$$

the estimating equations readily become

$$\xi_0 + \alpha\Gamma_1(c) = \bar{X}$$

$$\alpha^2\{\Gamma_2(c) - \Gamma_1^2(c)\} = S^2 \tag{21.65}$$

$$\xi_0 + \frac{\alpha}{n^{1/c}}\Gamma_1(c) = X_1',$$

where $\Gamma_k(c) = \Gamma[1 + (k/c)]$, $\bar{X} = (1/n)\sum_{i=1}^n X_i$ is the observed sample mean, $S^2 = [1/(n-1)]\sum_{i=1}^n (X_i - \bar{X})^2$ is the observed sample variance, and X_1' is the observed smallest order statistic. After a few simple algebraic manipulations, the three equations of (21.65) reduce to

$$\frac{S^2}{(\bar{X} - X_1')^2} = \frac{\hat{\Gamma}_2 - \hat{\Gamma}_1^2}{\{(1 - n^{-1/\hat{c}})\hat{\Gamma}_1\}^2} \tag{21.66a}$$

$$\hat{\xi}_0 = \frac{n^{1/\hat{c}}X_1' - \bar{X}}{n^{1/\hat{c}} - 1} \tag{21.66b}$$

$$\hat{\alpha} = \frac{n^{1/\hat{c}}(\bar{X} - X_1')}{(n^{1/\hat{c}} - 1)\hat{\Gamma}_1}, \tag{21.66c}$$

where $\hat{\Gamma}_j = \Gamma_j(\hat{c}) = \Gamma[1 + (j/\hat{c})]$. Equivalent expressions for $\hat{\xi}_0$ and $\hat{\alpha}$ that are convenient for use are

$$\hat{\alpha} = SC(\hat{c}) \quad \text{and} \quad \hat{\xi}_0 = \bar{X} - SD(\hat{c}), \tag{21.67}$$

where $C(c) = 1/\sqrt{\Gamma_2 - \Gamma_1^2}$ and $D(c) = \Gamma_1/\sqrt{\Gamma_2 - \Gamma_1^2}$.

With $\bar{X}$, S^2, and X_1' available from the sample data, (21.66a) need to be solved for $\hat{c}$, and subsequently $\hat{\xi}_0$ and $\hat{\alpha}$ can be determined from (21.66b) and (21.66c), respectively. Although this method of estimation is applicable for all values of the shape parameter c, Cohen, Whitten, and Ding (1984) pointed out that some computational difficulties may be encountered when $c < 0.5$ (a case not of great practical importance). Graphs of $\{\Gamma_2(c) - \Gamma_1^2(c)\}/$

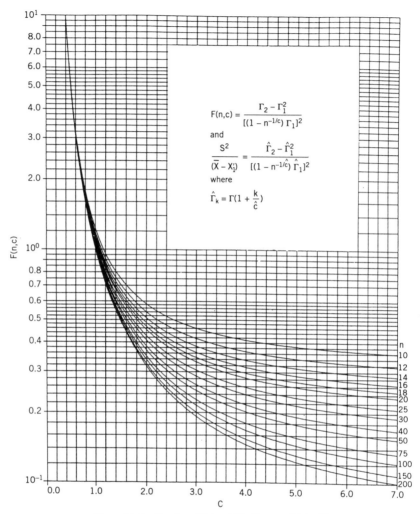

Figure 21.4   Graphs of $(\Gamma_2 - \Gamma_1^2)/[(1 - n^{-1/c})\Gamma_1]^2$

$\{(1 - n^{-1/c})\Gamma_1(c)\}^2$ are presented in Figure 21.4. Similarly graphs of $C(c)$ and $D(c)$ are presented in Figure 21.5.

Even though asymptotic variances and covariances for the maximum likelihood estimates are not strictly applicable for the modified moment estimates presented here, a simulation study carried out by Cohen and Whitten (1988) disclosed close agreement between simulated variances and corresponding asymptotic variances. However, when all three parameters ξ_0, α, and c have to be estimated, the asymptotic variances are valid only if $c > 2$. In fact, in order to avoid possible computational difficulties in practical applications, c needs to be greater than 2.2 approximately (see the next

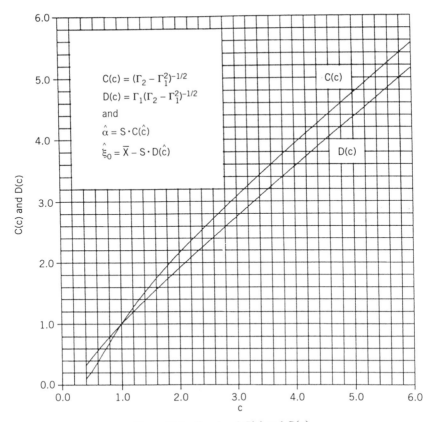

Figure 21.5 Graphs of $C(c)$ and $D(c)$

Section). For more details on this method of estimation, readers are referred to the books by Cohen and Whitten (1988) and Balakrishnan and Cohen (1991).

Instead of the simple modified moment equation $E[X_1'] = X_1'$, Cohen and Whitten (1982) considered some other possible equations:

$$E[F_X(X_1')] = F_X(X_1') \tag{21.68}$$

giving $-\log[n/(n + 1)] = [(X_1' - \xi_0)/\alpha]^c$, and

$$\text{median}(X) = \text{Sample median, } x_{\text{Med}} \tag{21.69}$$

giving $\xi_0 + \alpha(\log 2)^{1/c} = x_{\text{Med}}$. Cohen and Whitten (1982) have compared the performance of these estimators with the MLEs (discussed in Section 4.6) and some modified maximum likelihood estimators (discussed in Section 4.7).

4.6   Maximum Likelihood Estimation

The most usual situation is ξ_0 known (e.g., zero, without loss of generality) but c unknown, so both c and α must be estimated. Given sample values from a random sample of size n from the two-parameter Weibull distribution with probability density function

$$p_X(x) = \frac{c}{\alpha}\left(\frac{x}{\alpha}\right)^{c-1} e^{-(x/\alpha)^c}, \qquad x > 0, \tag{21.70}$$

the maximum likelihood estimators $\hat{c}$ and $\hat{\alpha}$ of c and α, respectively, satisfy the equations

$$\hat{\alpha} = \left\{\frac{1}{n}\sum_{i=1}^{n} X_i^{\hat{c}}\right\}^{1/\hat{c}} \tag{21.71}$$

and

$$\hat{c} = \left[\left\{\sum_{i=1}^{n} X_i^{\hat{c}} \log X_i\right\}\left\{\sum_{i=1}^{n} X_i^{\hat{c}}\right\}^{-1} - \frac{1}{n}\sum_{i=1}^{n} \log X_i\right]^{-1}. \tag{21.72}$$

If ξ_0 is not equal to zero, then each X_i is replaced by $X_i - \xi_0$ in the above equations. The value of $\hat{c}$ needs to be obtained from (21.72) and then used in (21.71) to obtain $\hat{\alpha}$. It should be noted that if c were in fact known to be equal to $\hat{c}$, then $\hat{\alpha}$ in (21.71) would be the maximum likelihood estimator of α.

If the parameter ξ_0 is also unknown, then the maximum likelihood estimates $\hat{c}$, $\hat{\alpha}$, and $\hat{\xi}_0$ satisfy the equations

$$\hat{\alpha} = \left\{\frac{1}{n}\sum_{i=1}^{n}\left(X_i - \hat{\xi}_0\right)^{\hat{c}}\right\}^{1/\hat{c}}, \tag{21.73}$$

$$\hat{c} = \left[\left\{\sum_{i=1}^{n}\left(X_i - \hat{\xi}_0\right)^{\hat{c}}\log\left(X_i - \hat{\xi}_0\right)\right\}\left\{\sum_{i=1}^{n}\left(X_i - \hat{\xi}_0\right)^{\hat{c}}\right\}^{-1}\right.$$
$$\left. - \frac{1}{n}\sum_{i=1}^{n}\log\left(X_i - \hat{\xi}_0\right)\right]^{-1}, \tag{21.74}$$

and

$$(\hat{c} - 1)\sum_{i=1}^{n}\left(X_i - \hat{\xi}_0\right)^{-1} = \hat{c}\hat{\alpha}^{-\hat{c}}\sum_{i=1}^{n}\left(X_i - \hat{\xi}_0\right)^{\hat{c}-1}. \tag{21.75}$$

If the value $\hat{\xi}_0$ satisfying (21.73)–(21.75) is larger than X_1', then it is the MLE of ξ_0. Otherwise, the MLE is $\hat{\xi}_0 = X_1'$ in which case (21.73) and (21.74) must be solved for $\hat{\alpha}$ and $\hat{c}$. It should be noted that the MLEs are "regular" (in the sense of having the usual asymptotic distribution) only for $c > 2$. If c is known to be in the interval $(0, 1)$, then X_1' is a "superefficient" estimator for $\hat{\xi}_0$. If the MLEs are regular, then Dubey (1965) has given for large n (when ξ_0 is known)

$$n \, \text{Var}(\hat{\alpha}) \doteq \left[1 + \frac{\{\psi(2)\}^2}{\psi'(1)}\right]\left(\frac{\alpha}{c}\right)^2 = 1.109\left(\frac{\alpha}{c}\right)^2,$$

$$n \, \text{Var}(\hat{c}) \doteq c^2[\psi'(1)]^{-1} = \frac{6c^2}{\pi^2} = 0.608c^2,$$

$$\text{Corr}(\hat{\alpha}, \hat{c}) \doteq \frac{\psi(2)}{\left[\psi'(1) + \{\psi(2)\}^2\right]^{1/2}} = 0.313,$$

where $\psi(\cdot)$ and $\psi'(\cdot)$ are digamma and trigamma functions (see Chapter 1, Section A2).

Haan and Beer (1967) suggested that (21.74) be solved for $\hat{c}$, for each of a series of trial values of $\hat{\xi}_0$. The value of $\hat{\alpha}$ is then easily obtained from (21.73) and the value of the likelihood can be calculated. An alternative method would be to compare the values of $\hat{\alpha}$ obtained from (21.73) and (21.75). Monte Carlo studies conducted by Aroian (1965) and Miller (1966) established that $\hat{c}$ is a biased estimator of c, the relative bias depending on sample size but not on c. For sample size 170, for example, the bias is about 0.8%.

Based on a Type II censored sample $X_1' \le X_2' \le \cdots X_k'$ with the largest $n - k$ observations censored, Cohen (1965) has given the maximum likelihood equations for estimators of ξ_0, α, and c. For example, if ξ_0 is known to be 0, then the equations are of form similar to (21.71) and (21.72) but with

$$n^{-1}, \quad \sum_{i=1}^{n} X_i^{\hat{c}}, \quad \sum_{i=1}^{n} X_i^{\hat{c}} \log X_i, \quad \sum_{i=1}^{n} \log X_i$$

replaced by

$$k^{-1}, \quad \sum_{i=1}^{k} X_i'^{\hat{c}} + (n-k) X_k'^{\hat{c}},$$

$$\sum_{i=1}^{k} X_i'^{\hat{c}} \log X_i' + (n-k) X_k'^{\hat{c}} \log X_k',$$

$$\sum_{i=1}^{k} \log X_i',$$

respectively. For samples of size 10, censored by omission of the five largest values, McCool (1966) presented estimated percentile points of the distribution of $\hat{c}/c$, based on sampling experiments. The estimates are based on five sets of 1000 samples each, for $c = 0.9, 1.1, 1.3, 1.5, 1.7$, respectively. It is noteworthy that the estimated value of $E[\hat{c}/c]$ is 1.37, indicating that $\hat{c}$ should be divided by 1.37 to obtain an approximately unbiased estimator of c. McCool also showed that the distribution of $\hat{c} \log(\hat{X}_p/X_p)$, where $\hat{X}_p$ is the maximum likelihood estimator of

$$X_p = \alpha[-\log(1 - p)]^{1/c}$$

does not depend on any of the parameters.

Since (21.73)–(21.75) do not yield explicit solutions for the estimates, Cohen (1965) suggested eliminating α^c from the first two equations to give

$$\left[\frac{\sum_{i=1}^{n}(X_i - \hat{\xi}_0)^{\hat{c}} \log(X_i - \hat{\xi}_0)}{\sum_{i=1}^{n}(X_i - \hat{\xi}_0)^{\hat{c}}} - \frac{1}{\hat{c}}\right]$$

$$-\frac{1}{n}\sum_{i=1}^{n} \log(X_i - \hat{\xi}_0) = 0. \tag{21.76}$$

When ξ_0 is known, (21.76) can be easily solved for $\hat{c}$. When ξ_0 is unknown, Cohen suggested the following approach: Select a first approximation $\hat{\xi}_0 < X_1'$, and solve (21.76) and (21.73) for $\hat{c}$ and $\hat{\alpha}$; then verify whether the likelihood equation for ξ_0 in (21.75) is satisfied for these values of the estimates, and iterate until a pair of estimates for ξ_0 is attained that is sufficiently close.

The asymptotic variance-covariance matrix for the MLEs $\hat{\xi}_0$, $\hat{\alpha}$ and $\hat{c}$ can be shown to be

$$\begin{bmatrix} \dfrac{\alpha^2}{n}\phi_{11} & \dfrac{\alpha^2}{n}\phi_{12} & \dfrac{\alpha}{n}\phi_{13} \\[2ex] & \dfrac{\alpha^2}{n}\phi_{22} & \dfrac{\alpha}{n}\phi_{23} \\[2ex] & & \dfrac{c^2}{n}\phi_{33} \end{bmatrix}, \tag{21.77}$$

where

$$\phi_{11} = \frac{\psi'(1)}{c^2 M}, \quad \phi_{22} = \frac{KC - J^2}{c^2 M},$$

$$\phi_{33} = \frac{C - \Gamma^2[2 - (1/c)]}{M},$$

$$\phi_{12} = -\frac{J\psi(2) + K\Gamma[2 - (1/c)]}{c^2 M},$$

$$\phi_{13} = \frac{J + \psi(2)\Gamma[2 - (1/c)]}{M},$$

$$\phi_{23} = \frac{C\psi(2) - J\Gamma[2 - (1/c)]}{M}. \tag{21.78}$$

Here

$$M = KC - 2J\psi(2)\Gamma\left(2 - \frac{1}{c}\right) - C\psi^2(2) - K\Gamma^2\left(2 - \frac{1}{c}\right) - J^2,$$

with

$$C = \left\{\Gamma\left(1 - \frac{2}{c}\right) + c\Gamma\left(2 - \frac{2}{c}\right)\right\}\frac{(c-1)}{c^2},$$

$$J = \Gamma\left(1 - \frac{1}{c}\right) - A\Gamma\left(2 - \frac{1}{c}\right),$$

$$K = \psi'(1) + \{\psi(2)\}^2,$$

$$A = 1 + \psi\left(2 - \frac{1}{c}\right).$$

The variance-covariance factors in (21.78) have been tabulated by Cohen and Whitten (1988, p. 48) and Balakrishnan and Cohen (1991, p. 125) for various values of the shape parameter $c > 2$.

Harter (1970, pp. 104–106) has presented similar expressions for the case where the available sample is doubly Type II censored. Harter has also given some tables of the variance-covariance factors for selected values of c and different levels of censoring at both ends. Lawless (1982) and Bain and Engelhardt (1991a) have also discussed iterative methods of determining the maximum likelihood estimates of the parameters. Cohen (1975), Lemon (1975), and Wingo (1973) have discussed the maximum likelihood estimation of parameters where the available sample is progressively censored.

Escobar and Meeker (1986a) presented the Fisher information matrix based on complete and censored data. Pertinent discussion on the maximum likelihood estimation for this case may also be seen in the paper by Smith (1985) discussing the problem in general for the class of nonregular cases. Zanakis and Kyparisis (1986) reviewed various maximum likelihood estimation methods available in literature for the three-parameter case. Zanakis (1979) also carried out a simulation study comparing various simplified estimators proposed for the three-parameter case.

By adapting Fisher's conditional approach to inference, Lawless (1978, 1980, 1982) has developed procedures for exact inference concerning the parameters and some functions of them such as quantiles. Although these exact conditional procedures have the advantage of not requiring extensive tables like the unconditional procedures, numerical integration is usually necessary for their application. For this reason DiCiccio (1987) proposed a method of approximate conditional inference based on normal approximations to the distributions of signed square roots of likelihood ratio statistics. By comparing the results with those of Lawless (1982), DiCiccio has shown that his approximate conditional method provides fairly accurate approximations to the exact method even in case of small samples. DiCiccio, Field, and Fraser (1990) have described two tail probability approximations for this problem in order to avoid high-dimensional integrals.

4.7 Modified Maximum Likelihood Estimation

Cohen and Whitten (1982) proposed modified maximum likelihood estimators (MMLEs) of ξ_0, α, and c as alternatives for use when the MLEs are likely to be unsatisfactory (when $c < 2$ or perhaps 2.2); these estimators also possess some advantages with respect to ease of computation, bias, and variance for all values of c. The various MMLEs proposed by Cohen and Whitten (1982) are based on replacing the likelihood equation for ξ_0 in (21.75) with alternate functional relationships of the following forms:

$$E[F_X(X_1')] = \frac{1}{n+1} \quad \text{reducing to}$$

$$-\log\left(\frac{n}{n+1}\right) = \frac{(X_1' - \xi_0)^c}{\alpha^c}, \tag{21.79}$$

$$E[X_1'] = X_1' \quad \text{reducing to}$$

$$\xi_0 + \left(\frac{\alpha}{n^{1/c}}\right)\Gamma\left(1 + \frac{1}{c}\right) = X_1', \tag{21.80}$$

$$E[X] = \bar{X} \quad \text{reducing to } \xi_0 + \alpha\Gamma\left(1 + \frac{1}{c}\right) = \bar{X}, \tag{21.81}$$

$$\text{Var}(X) = S^2 \quad \text{reducing to}$$

$$\alpha^2\left[\Gamma\left(1 + \frac{2}{c}\right) - \Gamma^2\left(1 + \frac{1}{c}\right)\right] = S^2, \tag{21.82}$$

$$E[X_{\text{Med}}] = X_{\text{Med}} \quad \text{reducing to}$$

$$\xi_0 + \alpha(\log 2)^{1/c} = X_{\text{Med}}. \tag{21.83}$$

Here, X_{Med} denotes the sample median. Then (21.73), (21.74), and one of (21.79)–(21.83) are solved jointly to obtain the MMLEs. Through a simulational study, Cohen and Whitten (1982) have also shown that the entries in the asymptotic variance-covariance matrix of the MLEs in (21.77) and (21.78) provide a reasonable approximation to variances and covariances of the MMLEs (when $c > 2$). Further discussion on the MMLEs may be found in Cohen and Whitten (1988), Balakrishnan and Cohen (1991), and Cohen (1991).

Wyckoff, Bain, and Engelhardt (1980) proposed the following procedure for estimating the parameters ξ_0, α, and c. X_1' is used as an initial estimate of ξ_0. An initial estimate of c is obtained by taking $\xi_0 = X_1'$ and using the estimator of c proposed by Dubey (1967c) given by [see Section 4.2]

$$\hat{c}_1 = \frac{2.989}{\log(X_k - X_1') - \log(X_h - X_1')}, \tag{21.84}$$

where X_k and X_h are the 94th and 17th sample percentiles, respectively. ξ_0 is reestimated from (21.80), with α replaced by $(\bar{X} - \xi_0)/\Gamma[1 + (1/c)]$ [from (21.81)]; $\hat{c}_1$ is substituted for c, and the resulting equation is solved for ξ_0. The solution is the new estimate for ξ_0, denoted by $\hat{\xi}_0$. The parameters α and c are then estimated by assuming $\xi_0 = \hat{\xi}_0$ and using the estimators given by Engelhardt and Bain (1977). Proceeding somewhat similarly, Kappenman (1985b) proposed the following estimators of ξ_0, α, and c:

$$\hat{\xi}_0 = \frac{X_1' - X_m\{\Gamma[1 + (1/\hat{c}_1)]/n^{1/\hat{c}_1}\}}{[1 - \{\Gamma[1 + (1/\hat{c}_1)]/n^{1/\hat{c}_1}\}]}, \tag{21.85}$$

$$\hat{c} = \frac{nk_n}{[s/(n-s)]\sum_{i=s+1}^{n}\log(X_i' - \hat{\xi}_0) - \sum_{i=1}^{s}\log(X_i' - \hat{\xi}_0)}, \tag{21.86}$$

and

$$\hat{\alpha} = \exp\left\{\frac{0.5772}{\hat{c}} + \frac{1}{n}\sum_{i=1}^{n}\log(X_i' - \hat{\xi}_0)\right\}, \tag{21.87}$$

where $\hat{c}_1$ is given by (21.84), $s = [0.84n]$ with $[\cdot]$ denoting the integer part, X_m is the 63rd sample percentile, and k_n is a constant that depends upon the sample size [Engelhardt and Bain (1977)]. Through a simulation study Kappenman (1985b) has displayed that the mean squared errors for the estimators of ξ_0 and c in (21.85) and (21.86) are always considerably smaller than those for the corresponding MMLEs; the same is true for the estimators of α, except for the case where $c = 0.5$.

4.8 Bayesian Estimation and Shrinkage Estimation

Considerable amount of work dealing with Bayes, empirical Bayes, and shrinkage estimation methods for the parameters of the Weibull distribution has been done since 1970. Volumes by Tsokos and Shimi (1977) and Martz and Waller (1982) will provide interested readers with many details and developments in this direction.

Soland (1966, 1967) initiated the Bayesian estimation of the Weibull parameters. Papadopoulos and Tsokos (1975) derived Bayesian confidence bounds for the Weibull model. By considering a two-parameter form under stochastic variation of the shape and scale parameters c and α, Tsokos and

Rao (1976) described the Bayesian analysis. While Kamat (1977) discussed the Bayesian estimation of the system reliability using Monte Carlo simulation, Martz and Lian (1977) considered the Bayes and empirical Bayes point as well as interval estimation of the reliability; see Bennett (1977) for further results on empirical Bayes estimates. Lingappaiah (1977) presented the Bayesian approach to the prediction problem based on complete and censored samples; also see Tziafetas (1987) for further discussion on this issue. By considering the two-parameter Weibull distribution, Erto (1982) derived practical Bayesian estimators for the parameters α and c. Bayesian estimation of the parameters as well as the reliability function for the three-parameter case has been discussed by Sinha (1986), Sinha and Guttman (1988), and Sinha and Sloan (1988). Similar discussions have been made by Sinha (1982, 1987) and Sinha and Sloan (1989) for mixtures of Weibull distributions. Singpurwalla (1988) and Singpurwalla and Song (1988) have presented a discussion on how to incorporate expert opinion in the reliability analysis based on Weibull distributions. Smith and Naylor (1987) compared the performance of the Bayesian estimators of the parameters ξ_0, α, and c with the corresponding maximum likelihood estimators. Canavos (1983) similarly compared the performance of the Bayesian and empirical Bayesian estimators of the parameters. Abdel-Wahid and Winterbottom (1987) proposed approximate Bayesian estimates for the reliability function and the hazard rate based on complete and censored data, while Dey and Kuo (1991) gave a new empirical Bayesian estimator based on censored data.

An experimenter involved in life-testing experiments becomes quite familiar with failure data and hence may often develop knowledge about some parameters of the distribution. In the case of Weibull distribution, for example, knowledge on the shape parameter c can be utilized to develop improved inference for the other parameters. If an initializing value θ_0 of a parameter θ is available, say, the shrunken estimator $k\hat{\theta} + (1 - k)\theta_0$ ($0 < k < 1$) for θ as proposed by Thompson (1968) is more efficient than the estimator $\hat{\theta}$ only if θ is close to θ_0 and is less efficient otherwise. This shrinkage estimation of the Weibull parameters has been discussed by a number of authors, including Singh and Bhatkulikar (1978), Pandey (1983, 1988), Pandey, Malik, and Srivastava (1989), Pandey and Singh (1984), Pandey and Upadhyay (1985, 1986), and Pandey and Singh (1993). For example, Singh and Bhatkulikar (1978) suggested performing a significance test of the validity of the prior value of c (which they took as 1). Pandey (1983) also suggested a similar preliminary test shrunken estimator for c. In a recent paper Pandey and Singh (1993) derived a test based on Type II censored data for the hypothesis $H_0 : c = c_0$ versus $H_1 : c \neq c_0$ (by minimizing the sum of the probabilities of the two types of error) and used it to propose another preliminary test shrunken estimator for c. The authors have shown that this estimator has higher efficiency and also Pitman's closeness than other shrunken estimators even in case of small sample sizes.

5   TOLERANCE LIMITS AND INTERVALS

For any arbitrary continuous distribution function $F_X(x)$, the proportion of the population that falls in the interval (l, u) is $F_X(u) - F_X(l)$. Suppose that L and U are random variables such that $L \le U$ with probability 1. Then the proportion of the population that falls in the random interval (L, U) will also be random and is equal to $F_X(U) - F_X(L)$. Suppose that L and U are chosen in such a way that

$$\Pr[F_X(U) - F_X(L) \ge \beta] = \gamma. \qquad (21.88)$$

That is, L and U are chosen so that at least $100\beta\%$ of the population falls in the random interval (L, U) with probability γ. This interval is called a *$100\gamma\%$ tolerance interval with proportion β*, and L and U are called the *tolerance limits*. The difference between a tolerance interval and a confidence interval is quite apparent. However, as in the case of confidence intervals, one can define one-sided tolerance limits and intervals. From (21.88) we define L to be a lower γ tolerance limit for proportion β (setting $U = \infty$) if

$$\Pr[1 - F_X(L) \ge \beta] = \gamma$$

or, equivalently,

$$\Pr[F_X(L) \le 1 - \beta] = \gamma. \qquad (21.89)$$

For a general location-scale family of distributions (Weibull distributions belong to this class when c is known), Dumonceaux (1969) demonstrated that tolerance limits can be constructed based on the pivotal quantity

$$T = \frac{\hat{\xi}_0 - \xi_0}{\hat{\alpha}} - \frac{\alpha}{\hat{\alpha}} F^{-1}(1 - \beta), \qquad (21.90)$$

where $F^{-1}(1 - \beta)$ is the inverse of the distribution function $F_X(x)$ (in the standardized form); that is, $F^{-1}(1 - \beta) = [-\log \beta]^{1/c}$, and $\hat{\xi}_0$ and $\hat{\alpha}$ are the chosen estimates of ξ_0 and α, respectively, determined from the given sample. In this case the pivotal quantity T in (21.90) becomes

$$T = \frac{\hat{\xi}_0 - \xi_0}{\hat{\alpha}} - (-\log \beta)^{1/c} \frac{\alpha}{\hat{\alpha}}. \qquad (21.91)$$

Now, suppose that t_γ is the lower γ percentage point of the distribution of the pivotal quantity T in (21.91); that is,

$$\Pr[T \le t_\gamma] = \gamma. \qquad (21.92)$$

Then it is easily verified from (21.89) that

$$L = \hat{\xi}_0 - t_\gamma \hat{\alpha} \tag{21.93}$$

forms a lower γ tolerance limit for proportion β.

Similarly from (21.88) we may define the tolerance limit U to be an *upper γ tolerance limit for proportion β* (setting $L = \xi_0$) if

$$\Pr[F_X(U) \geq \beta] = \gamma. \tag{21.94}$$

It can be constructed based on the pivotal quantity

$$T^* = -\frac{\hat{\xi}_0 - \xi_0}{\hat{\alpha}} + \frac{\alpha}{\hat{\alpha}} F^{-1}(\beta)$$

$$= -\frac{\hat{\xi}_0 - \xi_0}{\hat{\alpha}} + \frac{\alpha}{\hat{\alpha}} \{-\log(1 - \beta)\}^{1/c}. \tag{21.95}$$

Suppose that t_γ^* denotes the lower γ percentage point of the distribution of the pivotal quantity T^* in (21.95); that is,

$$\Pr[T^* \leq t_\gamma^*] = \gamma.$$

Then it is easily verified from (21.94) that

$$U = \hat{\xi}_0 + t_\gamma^* \hat{\alpha} \tag{21.96}$$

forms an upper γ tolerance limit for proportion β.

Tables of lower tolerance limits can also be used to determine lower confidence limits for the reliability function as originally observed by Lloyd and Lipow (1962, p. 204) in the case of normal distributions. Let $R_X(t) = \Pr[X \geq t] = 1 - F_X(t)$ denote the reliability at a specified time t. Then a lower confidence limit for the reliability function $R_X(t)$ can be obtained by setting $t = L$ and then determining for a specified γ the value of β that would make the statement in (21.92) to be true. The value of β thus determined would then be the lower confidence limit for the reliability function $R_X(t)$. To see this, let $\hat{R}_1$ denote the observed value of the chosen estimate $\hat{R}_X(t)$ for the given sample. Then, by setting

$$L = \hat{\xi}_0 - t_\gamma \hat{\alpha} = t,$$

we obtain

$$t_\gamma = -\frac{t - \hat{\xi}_0}{\hat{\alpha}} = -F^{-1}\left(1 - \hat{R}_1\right) = -\left(-\log \hat{R}_1\right)^{1/c}. \qquad (21.97)$$

Now, we need to choose β so as to satisfy

$$\Pr\left[\frac{\hat{\xi}_0 - \xi_0}{\hat{\alpha}} - \frac{\alpha}{\hat{\alpha}}(-\log \beta)^{1/c} \leq -\left(-\log \hat{R}_1\right)^{1/c}\right] = \gamma \qquad (21.98)$$

[see (21.91) and (21.92)]. If the reliability function $R_X(t) = \beta$, then $F^{-1}(1 - \beta) = (-\log \beta)^{1/c} = (t - \xi_0)/\alpha$ so that (21.98) becomes

$$\Pr\left[\frac{\hat{\xi}_0 - \xi_0}{\hat{\alpha}} - \frac{t - \xi_0}{\alpha}\frac{\alpha}{\hat{\alpha}} \leq -\left(-\log \hat{R}_1\right)^{1/c}\right] = \gamma \qquad (21.99)$$

iff
$$\Pr\left[\frac{t - \hat{\xi}_0}{\hat{\alpha}} \geq \left(-\log \hat{R}_1\right)^{1/c}\right] = \gamma,$$

or iff
$$\Pr\left[F\left(\frac{t - \hat{\xi}_0}{\hat{\alpha}}\right) \geq 1 - \hat{R}_1\right] = \gamma,$$

or iff
$$\Pr\left[1 - F\left(\frac{t - \hat{\xi}_0}{\hat{\alpha}}\right) = 1 - \hat{F}_X(t) = \hat{R}_X(t) \leq \hat{R}_1\right] = \gamma.$$

Thus we have $R_X(t) = \beta$ to be the value of the reliability function which makes (21.99) to be true, and hence β is the lower confidence limit for $R_X(t)$. Such a value exists and is independent of the unknown parameters ξ_0 and α (for the case when the shape parameter c is known), since Dumonceaux (1969) has shown that the distribution of $\hat{R}_X(t)$ depends only on the unknown parameter $R_X(t)$.

Thoman, Bain, and Antle (1970) discussed the determination of tolerance limits and the estimation of the reliability $R_X(t)$ and also confidence intervals for the reliability function using the maximum likelihood method. This discussion is for the two-parameter Weibull distribution (with $\xi_0 = 0$ or assumed to be known). Tables are also provided for the lower confidence limits of the reliability and hence the lower tolerance limits. Mann and Fertig (1973) presented similar tables based on the best linear invariant estimates of α and c (when $\xi_0 = 0$ or assumed to be known). The results of Thoman, Bain, and Antle (1970) were extended to the case of Type II censored samples by Billman, Antle, and Bain (1972), who also presented some necessary tables for the lower confidence limits of the reliability $R_X(t)$ when the censoring on the right is either 25% or 50%. For the two-parameter Weibull distribution, Lawless (1975) discussed the same issues using the conditional approach which does not require the construction of any tables and is applicable for complete as well as Type II censored samples. Mann and Fertig (1977) discussed the construction of tolerance limits and lower confidence limits on the reliability using the asymptotic linear estimators based on k optimally selected order statistics for the parameters α and c (when $\xi_0 = 0$). Mann (1977) also examined the adequacy of an F-approximation for a statistic involved in the derivation of lower confidence bounds on the percentiles of the distribution discussed earlier by Mann, Schafer, and Singpurwalla (1974), for Type II censored data. Mann (1978a, b) has made further investigation on calculating simple approximate small-sample lower tolerance bounds or confidence bounds for percentiles of the distribution based on a two-parameter Weibull accelerated life-testing model. Bain and Engelhardt (1981) provided some simple approximations for the distributions of pivotal quantiles (based on the MLEs) for the two-parameter Weibull model that are useful for construction of approximate confidence limits for the parameters α and c, and also in determining approximate tolerance limits and confidence limits on reliability. These involve the chi-square, t, and noncentral t distribution (see Chapters 18, 28, 31) approximations for the pivotal quantities. Based on the least-squares method and median plotting positions, Erto and Guida (1985) determined lower confidence limits for reliability (also useful for finding lower confidence limits for population quantiles), and presented the necessary tables for various sample sizes and different choices of Type II right censoring.

All of these developments are based on the two-parameter Weibull distribution (taking $\xi_0 = 0$) and hence, not surprisingly (as explained in the beginning of Section 4), use the logarithmic transformation of the Weibull data into the extreme-value form and then proceed with the location parameter $\eta = \log(\alpha)$ and the scale parameter $\delta = 1/c$. As mentioned earlier in Section 4, the Weibull distribution with unknown scale and shape parameters is rewritten in the familiar and convenient "location-scale form" of the extreme-value distribution that made all the above-mentioned derivations possible [with η and δ appearing in (21.90)–(21.99) in place of ξ_0 and α].

Johnson and Haskell (1984) considered the three-parameter Weibull distribution in (21.3) and discussed a large-sample approach for deriving lower tolerance bounds. This is based on a large-sample approximation for the distribution of the maximum likelihood estimates established by Johnson and Haskell (1983). By means of a simulation study, they observed that $n = 70$ is even too small for obtaining a good approximate normal distribution for the proposed approximate tolerance limits. The approximation is observed to be particularly weak when β is large (e.g., 0.95), with great improvement taking place when β is around 0.75.

6 PREDICTION LIMITS AND INTERVALS

For the two-parameter Weibull distribution (with $\xi_0 = 0$), Mann and Saunders (1969) used three specially selected order statistics to predict the minimum of a single future sample. Suppose that X'_r, X'_p, and X'_q are the three selected order statistics from the available sample of size n. Further suppose that Y'_1 is the smallest in a future sample of size N. Mann (1970) has presented tables for the factor v in the probability statement

$$\Pr\left[Y'_1 > X'_r + v(X'_p - X'_q)\right] = \gamma \qquad (21.100)$$

for $\gamma = 0.95$ and selected values of n, N, r, p, and q. If $Y'_{1(i)}$ denotes the minimum in the ith future sample, Hewett and Moeschberger (1976) showed that

$$\Pr\left[\bigcap_{i=1}^{k} \{Y'_{1(i)} > X'_r + v(X'_p - X'_q)\}\right] \geq \gamma^k, \qquad (21.101)$$

where the sizes $N_1, N_2, \ldots, N_k$ of the future samples all equal N.

In a repairable system suppose that n breakdowns have occurred and that the times of occurrence follow a Weibull process (see Section 12 for details). Engelhardt and Bain (1978) discussed the prediction of the time of the next breakdown, say, T_{n+1}, and in general the time T_{n+k} of the kth future breakdown. In the case of the prediction of T_{n+1}, they derived an exact closed-form prediction limit; for the general case of T_{n+k}, they discussed the exact limit and also an approximate prediction limit. The γ level lower confidence limit for T_{n+k} is a statistic $T_L = T_L(n, k, \gamma)$ such that

$$\Pr\left[T_L(n, k, \gamma) < T_{n+k}\right] = \gamma. \qquad (21.102)$$

The form of $T_L(n, k, \gamma)$ is

$$T_L(n, k, \gamma) = t_n \cdot \exp\left[\frac{y_{1-\gamma}}{(n-1)\hat{c}}\right], \qquad (21.103)$$

where

$$\hat{c} = \frac{n}{\sum_{i=1}^{n-1} \log(t_n/t_i)}$$

and $y_{1-\gamma}$ satisfies the equation in y:

$$1 - \gamma = \sum_{j=1}^{k} m_j \left\{ 1 - \left[1 + (n + j - 1)\frac{y}{n(n-1)} \right]^{-(n-1)} \right\},$$

with

$$m_j = \frac{(-1)^{j-1}(n + k - 1)!}{(n-1)!(k-j)!(j-1)!(n+j-1)!}.$$

For the case $k = 1$, (21.103) can be written as

$$T_L(n, 1, \gamma) = t_n \exp \left[\frac{\gamma^{-1/(n-1)} - 1}{\hat{c}} \right]. \qquad (21.104)$$

Assuming that a Type II right-censored sample is available from the two-parameter Weibull distribution (with $\xi_0 = 0$), Engelhardt and Bain (1979) used the logarithmic transformation to change to the Type 1 extreme-value distribution form (as explained in the first paragraph of Section 4) and then discussed the construction of prediction limits for the minimum and, in general, the jth smallest in a future sample of size N. These intervals are based on some closed-form estimates of the extreme-value parameters η ($= \log \alpha$) and δ ($= 1/c$) and F-approximations to the resulting pivotal quantities. Engelhardt and Bain (1982) subsequently presented a simplified approximation for the lower prediction limit for the smallest in a future sample of size N; this simplified limit is an explicit form in terms of the estimates of the extreme-value parameters η and δ and the variances and covariance of these estimates. Sherif and Tan (1978) have discussed this prediction problem when the available sample is Type II progressively censored. Fertig, Meyer, and Mann (1980), along the lines of Engelhardt and Bain (1979), discussed the prediction of a single future observation through pivotal quantities based on the best linear invariant estimators of the extreme-value parameters η and δ and presented the necessary tables of percentage points for different sample sizes and levels of censoring.

Wright and Singh (1981) considered the problem of predicting the sth-order statistic X_s', having observed the first r order statistics in a sample of size n from a two-parameter Weibull distribution (with $\xi_0 = 0$) when (1) n is fixed and (2) n is randomly distributed as a binomial variable. The strong assumption that the shape parameter c is known is made in this analysis. With the assumption that c is known, the prediction problem considered here becomes equivalent to that in the case of the scaled exponential distribution (see Chapter 19).

Adatia and Chan (1982) used the maximin estimator and the adaptive estimator of α, both of which are robust as explained briefly in Section 4, to construct prediction intervals for an order statistic X'_s, having observed the first r order statistics $X'_1, \ldots, X'_r$ from a sample of size n from the two-parameter Weibull distribution (with $\xi_0 = 0$). These authors assume that the shape parameter c is known to lie in the interval $[c_L, c_R]$ and then use the maximin or the adaptive estimator to propose "robust" predictors of X'_s. These predictors are taken to be linear functions of the observed order statistics $X'_1, X'_2, \ldots, X'_r$ with weights determined from the guaranteed efficiency required. Necessary tables have also been provided by Adatia and Chan (1982). Under the same setup Balasooriya and Chan (1983) carried out a robustness study of four types of predictors of X'_s, given $X'_1, \ldots, X'_r$, assuming that it is only known that the shape parameter c lies in an interval. For this comparative study they considered the following four predictors:

1. Best linear unbiased predictor

$$\hat{X}'_s(c) = \omega(c)^T V(c)^{-1} X + \hat{\alpha}(c) \left[E[Z'_s | c] - \omega(c)^T V(c)^{-1} E\left[\frac{X}{\alpha}\bigg| c\right] \right],$$

$$(21.105)$$

where $\hat{\alpha}(c)$ is the BLUE of α (see Section 4), Z'_i is the ith standardized order statistic $(= X'_i / \alpha)$, X denotes the column vector consisting of the given r order statistics, $V(c)$ is the variance-covariance matrix of $(Z'_1, \ldots, Z'_r)$, and $\omega(c)$ is the $r \times 1$ covariance matrix of $(Z'_1, \ldots, Z'_r)$ and Z'_s $(1 \le r < s \le n)$.

2. Best linear invariant predictor

$$\tilde{X}'_s(c) = \hat{X}'_s(c) + \hat{\alpha}(c) \left[E[Z'_s | c] - \omega(c)^T V(c)^{-1} E\left[\frac{X}{\alpha}\bigg| c\right] \right]$$

$$\times \frac{\text{Var}(\hat{\alpha}(c)/\alpha)}{1 + \text{Var}(\hat{\alpha}(c)/\alpha)}.\qquad (21.106)$$

3. Final linear unbiased predictor

$$X'^*_s(c) = X'_r + \hat{\alpha}(c)\{E[Z'_s | c] - E[Z'_r | c]\}. \qquad (21.107)$$

4. Cross-validatory predictive function

$$X'^{\text{cross}}_s(c; a) = X'_r + a\{E[Z'_s | c] - E[Z'_r | c]\}, \qquad (21.108)$$

where a depends on the sample and is given by

$$a = \frac{\sum_{i=s-r+1}^{r}\{X_i' - X_{i-(s-r)}'\}\{E[Z_i'|c] - E[Z_{i-(s-r)}'|c]\}}{\sum_{i=s-r+1}^{r}\{E[Z_i'|c] - E[Z_{i-(s-r)}'|c]\}^2}. \quad (21.109)$$

Balasooriya and Chan (1983) showed that the cross-validatory predictive function is the best in the sense of guaranteed efficiency; in addition this method does not require the knowledge of the covariances of Weibull order statistics or the inverse of an $r \times r$ matrix.

Pandey and Upadhyay (1986) considered the prediction of the smallest or the ith order statistic from a future sample of size N having observed a Type II right-censored sample $X_1', \ldots, X_r'$ from a sample of size n from the two-parameter Weibull distribution (with $\xi_0 = 0$). They assume that the scale parameter α is known and then develop prediction limits by using the preliminary test estimator of the unknown shape parameter c proposed by Pandey (1983) (also see Section 4).

As mentioned in Section 4.8, Tziafetas (1987) considered the Bayesian prediction of a single future observation, having observed a Type II right-censored sample from the two-parameter Weibull distribution (with $\xi_0 = 0$). Bayesian predictive intervals are derived after assigning Weibull and uniform (c_1, c_2) prior distributions for the scale parameter α and the shape parameter c, respectively. Nigm (1990) derived similar Bayesian prediction bounds for order statistics from the same sample and also for order statistics from a future sample, having observed a Type I censored sample up to time t. Based on a Type II censored sample $X_1', \ldots, X_r'$ from a sample of size n, Nigm (1989) had earlier discussed the derivation of Bayesian prediction bounds for the order statistic X_s' ($r < s \leq n$) from the same sample, and also for the order statistic Y_i' in a future sample of size N. For this development Nigm (1989) used an informative prior family for the parameters α^* and c of the bivariate form

$$\Pi(\alpha^*, c) = c^{2a}(\alpha^*)^{a+h/\phi(c)} e^{-cg} e^{-b\alpha^*\psi(c)}, \quad a > -1, g, h, b > 0, c, \alpha^* > 0,$$

where $\phi(\cdot)$ and $\psi(\cdot)$ are increasing functions of c. Here the parameter α^* is given by $\alpha^{-1/c}$ and the reparametrized Weibull density is $c\alpha^* x^{c-1} e^{-\alpha^* x^c}$. The same Bayesian prediction problem had earlier been discussed by Evans and Nigm (1980a, b) but using the noninformative prior for the parameters α^* and c given by

$$\Pi(\alpha^*, c) \propto \frac{1}{\alpha^* c^2}, \quad c > 0, \alpha^* > 0.$$

Dellaportas and Wright (1991) have described a numerical approach to this Bayesian prediction problem and the method of evaluating the posterior expectations. These authors, in addition to constructing prediction bounds for future lifetimes, also discuss the construction of the posterior distribution of the median lifetime.

This section deals only with the two-parameter Weibull distribution (with $\xi_0 = 0$ or known). The prediction problems for the three-parameter Weibull model in (21.3) seem to be unresolved and is certainly worth looking into.

7 RECORD VALUES

Let $X_1, X_2, \ldots$ be an i.i.d. sequence of standard Weibull random variables with density function in (21.10), and let $X_{U(1)}, X_{U(2)}, \ldots$ be the upper record values arising from this sequence (see Chapter 13, Section 5, for definitions). Then the density function of the nth upper record value $X_{U(n)}$ is

$$
\begin{aligned}
p_{X_{U(n)}}(x) &= \frac{1}{\Gamma(n)} \{-\log[1 - F_X(x)]\}^{n-1} p_X(x) \\
&= \frac{1}{\Gamma(n)} (x^c)^{n-1} e^{-x^c} cx^{c-1}, \qquad x > 0, \quad n = 1, 2, \ldots . \quad (21.110)
\end{aligned}
$$

Similarly the joint density function of $X_{U(m)}$ and $X_{U(n)}$ is

$$
\begin{aligned}
&p_{X_{U(m)}, X_{U(n)}}(x, y) \\
&= \frac{1}{\Gamma(m)\Gamma(n-m)} \{-\log[1 - F_X(x)]\}^{m-1} \frac{p_X(x)}{1 - F_X(x)} \\
&\qquad \times \{-\log[1 - F_X(y)] + \log[1 - F_X(x)]\}^{n-m-1} p_X(y) \\
&= \frac{1}{\Gamma(m)\Gamma(n-m)} (x^c)^{m-1} cx^{c-1} (y^c - x^c)^{n-m-1} e^{-y^c} cy^{c-1},
\end{aligned}
$$

$$
0 < x < y, \quad m = 1, 2, \ldots, n - 1. \quad (21.111)
$$

From (21.110), the kth raw moment of $X_{U(n)}$ is obtained to be

$$
E[X_{U(n)}^k] = \Gamma\left(n + \frac{k}{c}\right) \bigg/ \Gamma(n), \qquad k \geq 1, \quad n = 1, 2, \ldots . \quad (21.112)
$$

In particular, we have

$$E[X_{U(n)}] = \frac{\Gamma[n + (1/c)]}{\Gamma(n)},$$

$$\mathrm{Var}(X_{U(n)}) = \frac{\Gamma[n + (2/c)]}{\Gamma(n)} - \left\{\frac{\Gamma[n + (1/c)]}{\Gamma(n)}\right\}^2. \qquad (21.113)$$

Observe also from (21.110) that $X_{U(n)}^c$ is distributed as standard gamma with shape parameter n. Dallas (1982) has discussed some further distributional results on upper record values from the Weibull population.

Next, from the joint density function of $X_{U(m)}$ and $X_{U(n)}$ in (21.111), we obtain the product moment to be

$$E[X_{U(m)}X_{U(n)}] = \int_0^\infty \int_0^y xy f_{m,n}(x, y)\, dx\, dy$$

$$= \frac{\Gamma[m + (1/c)]}{\Gamma(m)} \times \frac{\Gamma[n + (2/c)]}{\Gamma[n + (1/c)]}, \qquad 1 \le m \le n - 1.$$

$$(21.114)$$

From (21.113) and (21.114) we immediately obtain

$$\mathrm{Cov}(X_{U(m)}, X_{U(n)}) = \frac{\Gamma[m + (1/c)]}{\Gamma(m)}\left\{\frac{\Gamma[n + (2/c)]}{\Gamma[n + (1/c)]} - \frac{\Gamma[n + (1/c)]}{\Gamma(n)}\right\}.$$

$$(21.115)$$

It is of interest to note here that the variance-covariance matrix $((\sigma_{i,j}))$ is of the form $\sigma_{i,j} = a_i b_j$, $i \le j$; this will make the inversion of the matrix possible explicitly (see Section 4.4) which will enable us to derive the best linear unbiased estimators of the parameters ξ_0 and α (when c is known) in an explicit form.

In addition to deriving the above results, Balakrishnan and Chan (1993c) also established the following simple recurrence relations satisfied by the single and the product moments of Weibull upper record values:

$$E\left[X_{U(n+1)}^k\right] = \left(1 + \frac{k}{nc}\right)E\left[X_{U(n)}^k\right], \qquad n \geq 1, \quad k = 1, 2, \ldots, \quad (21.116)$$

$$E\left[X_{U(m)}^k X_{U(m+1)}^l\right] = \frac{mc}{mc + k}E\left[X_{U(m+1)}^{k+l}\right], \qquad m \geq 1, \quad k, l = 1, 2, \ldots,$$
$$\hspace{11cm} (21.117)$$

$$E\left[X_{U(m)}^k X_{U(n)}^l\right] = \frac{mc}{mc + k}E\left[X_{U(m+1)}^k X_{U(n)}^l\right],$$
$$1 \leq m \leq n - 2, \quad k, l = 1, 2, \ldots, \quad (21.118)$$

$$E\left[X_{U(m)}^k X_{U(m+2)}^l\right] = \left(1 + \frac{l}{c}\right)E\left[X_{U(m)}^k X_{U(m+1)}^l\right]$$
$$- m\left\{E\left[X_{U(m+1)}^k X_{U(m+2)}^l\right] - E\left[X_{U(m+1)}^{k+l}\right]\right\},$$
$$m \geq 1, \quad k, l = 1, 2, \ldots, \quad (21.119)$$

$$E\left[X_{U(m)}^k X_{U(n+1)}^l\right] = \left(1 + \frac{l}{c(n - m)}\right)E\left[X_{U(m)}^k X_{U(n)}^l\right]$$
$$- \frac{m}{n - m}\left\{E\left[X_{U(m+1)}^k X_{U(n+1)}^l\right] - E\left[X_{U(m+1)}^k X_{U(n)}^l\right]\right\},$$
$$1 \leq m \leq n - 2, \quad k, l = 1, 2, \ldots . \quad (21.120)$$

Suppose that the original sequence of variables arise from the three-parameter Weibull population with density function in (21.3). Then the likelihood function based on the first n upper records is

$$L = \frac{p_X(x_1)}{1 - F_X(x_1)} \frac{p_X(x_2)}{1 - F_X(x_2)} \cdots \frac{p_X(x_{n-1})}{1 - F_X(x_{n-1})} p_X(x_n)$$

$$= \left(\frac{c}{\alpha}\right)^n \left\{\prod_{i=1}^{n}\left(\frac{x_i - \xi_0}{\alpha}\right)^{c-1}\right\} e^{-\{(x_n - \xi_0)/\alpha\}^c},$$

$$\xi_0 < x_1 < x_2 < \cdots < x_n. \quad (21.121)$$

Suppose the threshold parameter ξ_0 is known. If the shape parameter c is also known, then the MLE of α is easily observed to be

$$\hat{\alpha} = \frac{X_{U(n)} - \xi_0}{n^{1/c}}, \tag{21.122}$$

with the variance [from (21.113)]

$$\text{Var}(\hat{\alpha}) = \frac{\alpha^2}{n^{2/c}} \times \left[\frac{\Gamma[n + (2/c)]}{\Gamma(n)} - \left\{ \frac{\Gamma[n + (1/c)]}{\Gamma(n)} \right\}^2 \right]. \tag{21.123}$$

Suppose that the shape parameter c is unknown. Then the MLE's of α and c may be determined numerically from the likelihood function in (21.121), or from the likelihood function based on the first n record values from the extreme value distribution (for details, see Chapter 22). If the threshold parameter ξ_0 is also unknown, then the MLE's of ξ_0, α and c have to be determined numerically by solving simultaneously the three likelihood equations obtained from (21.121).

For the case when the shape parameter c is known, Balakrishnan and Chan (1993c) derived the BLUE's explicitly by making use of the explicit expressions of the means, variances, and covariances of record values presented earlier in (21.113) and (21.115). Because of the form of the variance-covariance matrix elements $\sigma_{i,j} = a_i b_j$, $i \leq j$ (as mentioned earlier), where

$$a_i = \frac{\Gamma[i + (1/c)]}{\Gamma(i)},$$

$$b_j = \frac{\Gamma[j + (2/c)]}{\Gamma[j + (1/c)]} - \frac{\Gamma[j + (1/c)]}{\Gamma(j)}, \tag{21.124}$$

the inverse of $((\sigma_{i,j}))$ can be explicitly written as [Graybill (1983, p. 198)]

$$\sigma^{1,1} = \frac{(c + 1)^2}{\Gamma[1 + (2/c)]},$$

$$\sigma^{i,i} = \frac{\Gamma(i)}{\Gamma[i + (2/c)]} \{c^2(2i^2 - 2i + 1) + c(4i - 2) + 1\},$$

$$i = 2, 3, \ldots, n - 1,$$

$$\sigma^{n,n} = \frac{\Gamma(n)}{\Gamma[n + (2/c)]} \frac{b_{n-1}}{b_n} (nc - c + 1)(nc - c + 2),$$

$$\sigma^{i,i+1} = -\frac{\Gamma(i)}{\Gamma[i + (2/c)]} ic(ic + 1), \qquad i = 1, 2, \ldots, n - 1,$$

$$\sigma^{i,j} = 0, \qquad |j - i| > 2. \tag{21.125}$$

For the case when ξ_0 and c are known, the BLUE of α becomes

$$\alpha^* = \frac{\Gamma(n)}{\Gamma[n + (1/c)]}(X_{U(n)} - \xi_0). \qquad (21.126)$$

Its variance is

$$\text{Var}(\alpha^*) = \alpha^2 \left\{ \frac{\Gamma(n)\Gamma[n + (2/c)]}{\Gamma^2[n + (1/c)]} - 1 \right\}. \qquad (21.127)$$

For large n, these are easily seen to be equivalent to the MLE in (21.122) and its variance in (21.123). Proceeding similarly for the case where both ξ_0 and α are unknown but the shape parameter c is known, Balakrishnan and Chan (1993c) derived the BLUE's of ξ_0 and α in explicit form, and also presented the variances and covariance of these estimators.

8   TABLES AND GRAPHS

Several tables and graphs relating to different methods of inference for the Weibull parameters have already been listed in Section 4. These will not be repeated here. Although the Weibull cumulative distribution function is available in an exact explicit form [(21.4)] that is easily computable, it is often convenient to have tables, or graphs, from which values can be obtained quickly without direct calculations.

For the standard Weibull case [$\xi_0 = 0$ and $\alpha = 1$ in (21.3)], Plait (1962) presented tables giving values of the probability density function ($cx^{c-1}e^{-x^c}$) to eight decimal places for $c = 0.1(0.1)3(1)10$ and also of the cumulative distribution function ($1 - e^{-x^c}$) to seven decimal places for $c = 0.1(0.1)4.0$. Dourgnon and Reyrolle (1966) also tabulated the cumulative distribution function (less extensively). Harter and Dubey (1967) presented values of the mean, variance, and first six standardized cumulant ratios $\kappa_r/\kappa_2^{r/2}$, $r = 3, \ldots, 8$, to eight decimal places for $c = 1.1(0.1)10.0$.

For the case where the threshold parameter ξ_0 is zero, by noting that $E[X] = \mu_1' = \alpha\Gamma[1 + (1/c)]$, the cumulative distribution function of X in (21.4) can be rewritten as

$$F_X(x) = 1 - \exp\left\{ -\left[\frac{x\Gamma[1 + (1/c)]}{\mu_1'} \right]^c \right\}, \qquad x \geq 0. \quad (21.128)$$

As it was indicated in Section 4, Kotel'nikov (1964) used (21.128) in constructing a nomogram for finding $F_X(x)$, given the mean μ_1' and the standard deviation σ. This nomogram is presented in Figure 21.6. One part of the

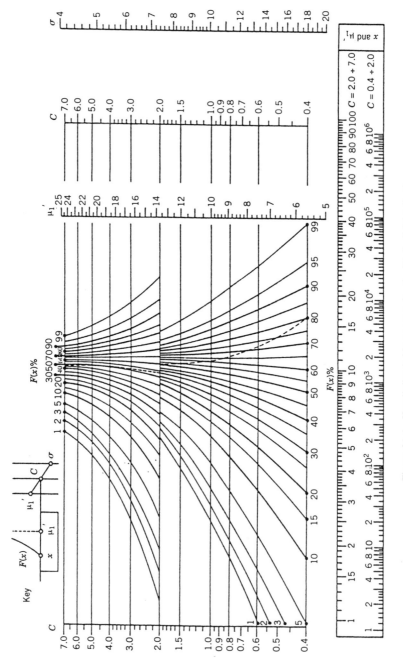

Figure 21.6 Kotel'nikov's Nomogram for the Weibull Distribution

676

nomogram determines the shape parameter c from given values of μ'_1 and σ. This is done by finding the intersection of the straight line joining the appropriate points on the μ'_1 and σ scales with the c scale. Then it is necessary to place the scale at the foot of the figure along the appropriate line μ'_1 and move it till the dotted line on the nomogram passes through the appropriate σ value on the scale. (Note that the edge of the scale to be used depends on c.) Then for any given value of x, the value of the cumulative distribution function $F_X(x)$ is the "$F_X(x)$ curve" passing through the value x on the appropriate edge of the scale. It is convenient in practice for the movable scale to be transparent.

Weibull probability paper can be constructed in several different ways [e.g., see Kao (1959); Nelson (1967); Plait (1962)]. Starting with the form of the cumulative distribution function in (21.4), we observe that

$$\log \log\{1 - F_X(x)\} = -c \log\left(\frac{x - \xi_0}{\alpha}\right), \qquad x > \xi_0,$$

or

$$\log \log\left\{\frac{1}{1 - F_X(x)}\right\} = c \log(x - \xi_0) - c \log \alpha, \qquad x > \xi_0,$$

Setting $\log \log\{1/[1 - F_X(x)]\}$ as w and $\log(x - \xi_0)$ as v, we obtain the linear relationship

$$w = cv - c \log \alpha, \qquad c > 0.$$

Therefore, if $1 - F_X(x)$ or $1/\{1 - F_X(x)\}$ is plotted against $x - \xi_0$ on a *log log versus log* paper, a straight line is observed. The slope of the line is $-c$ or c, and the intercept with the y-axis is $c \log \alpha$ or $-c \log \alpha$. This Weibull probability paper, as presented by Nelson (1967), is given in Figure 21.7.

Nelson and Thompson (1971) have provided a detailed discussion on the many different probability papers available for the Weibull analysis. Nelson (1972) also discussed the theory and application of hazard plotting based on censored data. The volume by D'Agostino and Stephens (1986) provides an authoritative account of graphical analysis of Weibull data and also of many formal goodness-of-fit methods such as EDF tests. Fowlkes (1987) has also presented theoretical quantile-quantile plots that provide some insight for users working with real-life data regarding the usage of the Weibull model for the data at hand.

Harter and Dubey (1967) used Monte Carlo simulations to construct extensive tables of the distribution of sample variances, $s^2 = \sum_{i=1}^{n}(x_i - \bar{x})^2/(n - 1)$, calculated from random samples from Weibull popu-

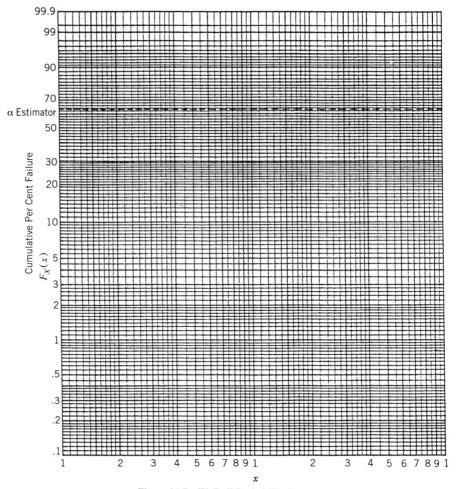

Figure 21.7   Weibull Probability Paper

lations, and also of the "Weibull-t" defined by

$$\frac{\sqrt{n}\,(\bar{x} - \mu_1')}{s}$$

constructed from i.i.d. samples from the Weibull distribution. They also constructed tables of the distribution of the sample mean $\bar{x}$ using a Cornish-Fisher expansion (see Chapter 12).

Details of tables relating to order statistics from Weibull populations have been presented in Section 3. As mentioned earlier, tables and graphs have been prepared by several authors in order to simplify the different methods of inference proposed in the literature; for the benefit of readers and users, a description of such tables and graphs may be seen in Section 4. Tables prepared for the determination of Weibull tolerance limits have been described in Section 5. Finally, tables relating to the prediction problems concerning the Weibull distribution are detailed in Section 6.

9 CHARACTERIZATIONS

It was already shown in Section 3 that when $X_1, X_2, \ldots, X_n$ are independent and identically distributed Weibull random variables, then the smallest order statistic X_1' is also distributed as Weibull. Dubey (1966a) established this to be a characterizing property of a Weibull distribution by showing that if X_1' is distributed as Weibull, then the common distribution of the X_i has to be a Weibull distribution. The proof is straightforward if one looks at the survival function of X_1' (see Chapter 12, Section 6).

By assuming that X and Y are independent nonnegative random variables, Arnold and Isaacson (1976) have shown that, for some pair (a, b) such that $0 < a, b < 1$,

$$\min(X, Y) \stackrel{\mathrm{d}}{=} aX \stackrel{\mathrm{d}}{=} bY$$

if and only if X and Y have Weibull distributions. More generally, Arnold and Isaascon have proved the following characterization result:

Let $X_1, X_2, \ldots, X_m$ be independent non-degenerate non-negative random variables. Suppose that

$$\min(X_1, X_2, \ldots, X_m) \stackrel{\mathrm{d}}{=} a_1 X_1 \stackrel{\mathrm{d}}{=} a_2 X_2 \stackrel{\mathrm{d}}{=} \cdots \stackrel{\mathrm{d}}{=} a_m X_m,$$

where $0 < a_i < 1$, $i = 1, 2, \ldots, m$. If, for that α for which $\sum_{i=1}^{m} a_i^{\alpha} = 1$, $\lim_{x \to 0+} F(x)/x^{\alpha} = c$, then the X_i's have Weibull distributions.

By generalizing the Cauchy functional equation $\Phi(x + y) = \Phi(x) + \Phi(y)$ to the form

$$\Phi\left((x^c + y^c)^{1/c}\right) = \Phi(x) + \Phi(y), \qquad c \neq 0,$$

assuming Φ to be left- or right-continuous, Wang (1976) established a characterization of the Weibull distribution as follows:

Let $c \neq 0$ and X be a non-degenerate non-negative random variable. Then, X has a two-parameter Weibull distribution if and only if X satisfies

$$\Pr\left(X > (x^c + y^c)^{1/c} \mid X > y\right) = \Pr(X > x) \qquad \text{for all } x, y > 0 \quad (21.129)$$

with $c > 0$. If $c < 0$, there is no random variable possessing property (21.129).

Note that this characterization is in the spirit of the memoryless property of the exponential distribution [see Chapter 19 for details and also Janardan (1978)].

Let $X_1, X_2, \ldots, X_n$ by n positive real random variables. For $a_j \geq 0$ and $c > 0$, let

$$
Z_j(a_j) = \begin{cases} X_j - a_j, & x_j > a_j, \\ 0, & x_j \leq a_j, \end{cases}
$$

and suppose that for some r between 1 and $n - 1$, $(Z_1, \ldots, Z_r)$ and $(Z_{r+1}, \ldots, Z_n)$ are independent. Then Cook (1978) proved a characterization result that for all $a_j \geq 0$, $E[\Pi_{j=1}^n Z_j(a_j)]$ depends on the parameters only through a function of $\Sigma_{j=1}^n a_j$ if and only if all X_j are mutually independent two-parameter Weibull random variables.

Let $X_{1:n}, X_{2:n}, \ldots, X_{n:n}$ be the order statistics obtained from n i.i.d. random variables with absolutely continuous strictly increasing distribution function $F(x)$. Janardan and Taneja (1979a) characterized the two-parameter Weibull distribution based on the following statements:

1. For any fixed r and two distinct numbers S_1 and S_2 and with $1 < r < S_1 < S_2 \leq n$, the distributions of

$$
V_i = \left(X_{S_i:n}^c - X_{r:n}^c \right)^{1/c} \quad \text{and} \quad W_i = X_{S_i - r:n-r}
$$

are identical for $i = 1$ and 2.
2. The statistics $D_1 = X_{1:n}$ and $D_2 = (X_{2:n}^c - X_{1:n}^c)^{1/c}$ are independent.
3. The statistics $U_j = (X_{j+1:n}^c - X_{j:n}^c)^{1/c}$, $1 \leq j \leq n - 1$, and $X_{i:n}$ ($1 \leq i \leq j$) are independent [also see Janardan and Schaeffer (1978)].
4. $X, X_1, \ldots, X_k$ are random variables satisfying $\Pr[X - \Sigma_{i=1}^k X_i > 0]$ is positive, and

$$
\Pr\left[X - \sum_{i=1}^k X_i > \left(a^c + \sum_{i=1}^k b_i^c \right)^{1/c} \middle| X - \sum_{i=1}^k X_i > \left(\sum_{i=1}^k b_i^c \right)^{1/c} \right]
$$
$$
= \Pr[X > a], \quad c > 0.
$$

Notice the similarity of condition **4** to (21.129).

Janardan and Taneja (1979b) have provided some further characterization results. One of these is based on the conditional expectation as follows: Let X be a nonnegative random variable with distribution function $F(x)$. Suppose that $E[X] < \infty$ and that y is a positive number. Then $F(x)$ is a two-parameter Weibull distribution if and only if

$$
E[X^c | X > y] = y^c + E[X] \qquad \text{for all } y, c > 0.
$$

Another condition based on which they characterize the Weibull distribution is

$$
E\left[\left(X - (s^c + t^c)^{1/c} \right)^+ \right] = \alpha E\left[(X - s)^+ \right] E\left[(X - t)^+ \right]
$$

for all s, t belonging to a dense subset of $(0, \infty)$, where $(X - u)^+$ denotes max$\{X - u, 0\}$. Roy (1984) established some characterization results for the generalized gamma distribution based on the Fisher-information-minimization and entropy-maximization principles which gives as a special case results for the Weibull distribution. Further, denoting the cumulative hazard function $-\log[1 - F(x)]$ by $H(x)$ and the hazard function by $h(x) = H'(x) = p(x)/[1 - F(x)]$, Roy and Mukherjee (1986) presented some characterizations for the two-parameter Weibull distribution based on the following statements:

1. $H(xy)H(1) = H(x)H(y)$ for all $x, y > 0$.
2. $H(xy)H(1) = H(x)H(y)$, with $H(1) > 0$, for at least one value of $y > 0$ (and not equal to 1), for all $x > 0$, and $xh(x)/H(x)$ is nondecreasing for $x > 0$.
3. $h(x)$ is strictly increasing with $h(0) = 0$ and $h(X)$ is distributed as two-parameter Weibull (shape parameter $c' > 1$) where $(1/c) + (1/c') = 1$.

Khan and Beg (1987) proved that the simple condition that $\text{Var}(X^c_{r+1:n}|X_{r:n} = x)$ does not depend on x characterizes the two-parameter Weibull distribution. This is, in principle, a direct consequence of a characterization of the exponential distribution based on the condition that $\text{Var}(X_{r+1:n}|X_{r:n} = x)$ is a constant, a result due to Beg and Kirmani (1978) (see also Chapter 19). Along similar lines Khan and Ali (1987) have characterized the Weibull distribution through the condition that

$$E[X^c_{r+1:n}|X_{r:n} = x] = x^c + \text{constant}$$

[see also El-Din, Mahmoud, and Youssef (1991)].

Ouyang (1987), Beg and Ali (1989), Moothathu (1990), and Chaudhuri and Chandra (1990) have provided some other characterizations of the Weibull distribution. For example, the characterization result presented by the first author is based on the conditional expectation of the function $g(X)$, given that $X \leq x$. Kamps (1991) has provided a characterization result for the Weibull distribution based on inequalities for moments of record values, with equality being satisfied if and only if the parent distribution is two-parameter Weibull. The proof follows easily from a simple application of Hölder's inequality.

Scholz (1990) established an interesting characterization of the three-parameter Weibull density function in (21.3): Let $C = \{(u, v, w): 0 < u < v < w < 1, \log(1 - u)\log(1 - w) = (\log(1 - v))^2\}$, and $F^{-1}(\cdot)$ denote the inverse of the distribution function in (21.4). Then it is easily verified from (21.4) that for some fixed t, namely $t = \xi_0$, the quantile relation

$$F^{-1}(u)F^{-1}(w) - (F^{-1}(v))^2 = t(F^{-1}(u) + F^{-1}(w) - 2F^{-1}(v))$$

holds for every $(u, v, w) \in C$. Scholz proves that this in fact is a characterizing relation for the three-parameter Weibull distribution in (21.4). He also has made some suggestions with regard to the usage of this characterization result in developing a goodness-of-fit procedure for the three-parameter Weibull distribution.

Shimizu and Davies (1981) have established the following two characterization results:

1. Let N be an integer valued random variable (independent of the X's) such that $N \geq 2$ with probability 1 and the distribution of $\log N$ has a finite mean and is not concentrated on a lattice $\rho, 2\rho, \ldots$ for any $\rho > 0$. Then the random variable $Y = N^{1/c} \cdot X_{1:N}$ is distributed as X if and only if X's are distributed as Weibull.

2. Let $m \geq 2$ be a fixed positive number, and let $Z_1, \ldots, Z_m$ be a set of positive random variables (independent of the X's) satisfying

$$\Pr\left[\sum_{i=1}^{m} Z_i^c = 1\right] = 1,$$

$$\Pr\left[\frac{\log Z_i}{\log Z_j} \text{ is irrational for some } i \text{ and } j\right] > 0.$$

Then the random variable $Y = \min\{X_1/Z_1, \ldots, X_m/Z_m\}$ is distributed as X if and only if the X's have Weibull distributions.

Roy (1984) presented a characterization of the generalized gamma (or generalized Weibull; see Section 12) for two independent nonnegative random variables X and Y based on the conditional distribution of X, given $Z = X/Y$, being also a generalized gamma distribution.

10   SIMULATION ALGORITHMS

Since the cumulative distribution function F of a three-parameter Weibull variable is of a closed explicit form [see (21.4)], it is easy to generate pseudorandom observations through the probability integral transformation approach. Specifically, letting

$$U = F(X; \xi_0, \alpha, c) = 1 - e^{-[(X - \xi_0)/\alpha]^c} \qquad (21.130)$$

and inverting this transformation, we obtain

$$X = \xi_0 + \alpha\{-\log(1 - U)\}^{1/c}. \qquad (21.131)$$

Thus, after generating a pseudo-random observation U from the uniform $(0, 1)$ population, the required Weibull observation X can be generated from (21.131) for the specified values of the parameters, ξ_0, α, and c. This simple simulational process is easily programmable in languages such as Fortran. However, within the context of GPSS, (21.131) cannot be directly implemented because GPSS does not contain any built-in functions, and in particular contains no logarithmic function and no exponentiation operator. Tadikamalla and Schriber (1977) have proposed a simulational algorithm in

GPSS that involves the following process:

1. A piecewise linear approximation to the inverse of the cdf (21.131) is formed.
2. This piecewise linear approximation is expressed in GPSS by defining a C-type ("continuous") function consisting of ordered pairs of points corresponding to the piecewise linear approximation.
3. When a random observation is needed, the function is invoked; in response, the GPSS processor evaluates the function's argument (which is a value returned by a 0–1 random number generator), then does a table lookup in the function defining set of ordered pairs, performs a linear interpolation across the interval that brackets the argument, and returns the interpolant value as the value of the function.

To improve the piecewise linear approximation involved, they suggested writing (21.131) equivalently as

$$X = \xi_0 + \alpha \exp\left[\left(\frac{1}{c}\right)\log\{-\log(1 - U)\}\right] \qquad (21.132)$$

and using a GPSS C-type function to determine a value for $\log\{-\log(1 - U)\}$ and another GPSS C-type function to determine a value for $\exp(argument)$.

Of course the process described above can be used for any distribution by introducing an appropriate piecewise linear approximation (even when the cdf is not available in an explicit form). In languages wherein the logarithmic function and the exponentiation operator are available, (21.132) can be implemented directly to simulate the required Weibull observation.

Another natural way to simulate a Weibull observation is to make use of any efficient exponential simulational algorithm. By realizing that $[(X - \xi_0)/\alpha]^c$ is distributed as standard exponential, we can simulate the required Weibull observation X by

$$X = \xi_0 + \alpha Z^{1/c}, \qquad (21.133)$$

where Z denotes the standard exponential pseudo-random observation already simulated. Note that the transformation in (21.133) is equivalent to the transformation in (21.132), since $-\log(1 - U)$ when $U \sim \text{uniform}(0, 1)$, is distributed as standard exponential Z (see Chapter 19).

Ishioka (1990) has discussed the generation of Weibull random numbers using the ratio of uniform pseudorandom observations. Ramberg and Tadikamalla (1974) have shown how gamma variates can be generated by using Weibull sampling mechanisms.

11 APPLICATIONS

As mentioned earlier in Sections 1 and 2, the fact that the hazard rate of the Weibull distribution is decreasing, constant, and increasing according as $c < 1$, $c = 1$, and $c > 1$ has made this distribution highly useful as a lifetime model. Naturally numerous articles appeared dealing with this particular type of application and the analysis of data resulting from such life tests.

Despite a warning issued by Gorski (1968) on the "Weibull euphoria," many more papers have appeared since 1970 exploring several new and interesting applications of the Weibull distribution. Van der Auwera, de Meyer, and Malet (1980), Barros and Estevan (1983), Dixon and Swift (1984), Conradsen, Nielsen, and Prahm (1984), and Tuller and Brett (1984) have all discussed the Weibull distribution in the analysis of wind speed. Pavia and O'Brien (1986) used the Weibull distribution to model wind speed over the ocean, while Carlin and Haslett (1982) applied the distribution to model wind power from a dispersed array of wind turbine generators. The Weibull distribution also found applications in analyzing rainfall and flood data. Zhang (1982), Boes (1989), and Nathan and McMahon (1990) considered the Weibull distribution in the analysis of flood data, while Wilks (1989) and Selker and Haith (1990) applied the distribution to model the rainfall intensity data.

The Weibull model was utilized in many analyses relating to health sciences. For example, Berry (1975) discussed the design of carcinogenesis experiments using the Weibull distribution. Dyer (1975) applied the distribution to analyze the relationship of systolic blood pressure, serum cholesterol, and smoking to 14-year mortality in the Chicago Peoples Gas Company; coronary and cardiovascular-renal mortality were also compared in two competing risk models in this study. Whittemore and Altschuler (1976) used the model in the analysis of lung cancer incidence in cigarette smokers by considering Doll and Hill's data for British physicians. Aitkin, Laird, and Francis (1983) applied the Weibull model in analyzing the Stanford heart transplant data. Christensen and Chen (1985) proposed a general noninteractive multiple toxicity model including the Weibull. While carrying out a Bayesian analysis of survival curves for cancer patients following treatment, Chen et al. (1985) utilized the Weibull distribution. Portier and Dinse (1987) made use of the Weibull distribution in their semiparametric analysis of tumor incidence rates in survival/sacrifice experiments. Rao, Talwalker, and Kundu (1991) applied the model to study the relative relapse rate of placebo versus 6-Mercaptopurine group of acute leukemia patients. Some further uses of the Weibull distribution in analysis relating to health issues may be found in Goldman (1984), Achcar, Brookmeyer, and Hunter (1985), Struthers and Farewell (1989), and Koehler and McGovern (1990). Ellingwood and Rosowsky (1991), while studying the duration of load effects in Load and Resistance Factor Design (LRFD) for wood construction, used the Weibull distribution function for short-term modulus of rupture, obtained by loading the beam to failure over a period of approximately 5 to 10 minutes.

In addition to the above-mentioned applications, the Weibull distribution also found important uses in a variety of other problems. For example, Fong, Rehm, and Graminski (1977) applied the distribution as a microscopic degradation model of paper. The uses of the distribution in analyzing hydrometeorological, metrological, and pharmacokinetic data have been illustrated by Wong (1977), Vodă (1978), and Schwenke (1987), respectively. A Weibull shelf life model for pharmaceutical problems was proposed by Ogden (1978). Rink et al. (1979) used the three-parameter Weibull distribution to quantify sweetgum germination data in genetic research. The application of the Weibull distribution to the analysis of the reaction time data has been introduced by Ida (1980). A role for the Weibull distribution in offshore oil/gas lease bidding problems has been demonstrated by Dyer (1981). While Berry (1981) used the distribution as a human performance descriptor, Newby and Winterton (1983) applied it to model the duration of industrial stoppages. Applications of the Weibull distribution in lumber industry have been discussed by Johnson and Haskell (1983). Rawlings and Cure (1985) introduced the Weibull function as a dose-response model to describe ozone effects on crop yields. Kanaroglou, Liaw, and Papageorgiou (1986) used the distribution in the analysis of migratory systems. The application of the Weibull model in the analysis of thunderstorm data has been displayed by Schuette, Salka, and Israelsson (1987). Homan, Pardo, and Trudeau (1987) discussed an application of Weibull modeling and discriminant analysis of alcoholism relapse among veterans. In order to evaluate the effect of temperature on cumulative germination of alfalfa, Bahler, Hill, and Byers (1989) considered both Weibull and logistic functions and compared the performance of the two. Lalla (1990) used the distribution to model the spell duration data. Gates (1985) and Braennaes (1986) have illustrated some further applications of the Weibull distribution. Quite interestingly Papastavridis (1987, 1988) and Chryssaphinou and Papastavridis (1990) established Weibull distribution as the limiting distribution for the reliability of consecutive k-out-of-n, consecutive k-within-m-out-of-n, and consecutive k-out-of-n: F systems, respectively.

It is of interest to mention here the works of Revfeim (1983, 1984) and Leese (1973). By assuming that the number of occurrences of an event in a fixed period of time has a Poisson distribution, with probability of r events in a fixed time T given by

$$p_r = \frac{(\rho T)^r e^{-\rho T}}{r!}$$

where ρ is the mean rate of occurrence of the event per unit of time in which T is measured, and then using a two-parameter Weibull distribution, Revfeim

(1984) obtained the maximum value distribution (in a stochastic form) as

$$G(x) \simeq \exp\left\{-\rho T \exp\left(\frac{-x^c}{\alpha}\right)\right\}.$$

He used this stochastic model to analyze extreme annual rainfall data.

Some other applications of the Weibull distribution have been described earlier in Section 1.

12 RELATED DISTRIBUTIONS

As pointed our earlier, the Weibull distribution in (21.4), when $c = 1$, becomes the two-parameter exponential distribution

$$F_X(x) = 1 - e^{-(x-\xi_0)/\alpha}, \qquad x > \xi_0,$$

which has been discussed in Chapter 19. For the case where $c = 2$, the Weibull distribution in (21.4) becomes the two-parameter Rayleigh distribution

$$F_X(x) = 1 - e^{-[(x-\xi_0)/\alpha]^2}, \qquad x > \xi_0,$$

which has already been discussed in Chapter 18.

If X has a Weibull distribution in (21.4), then the probability density function of $Y = -c \log[(X - \xi_0)/\alpha]$ is given by

$$p_Y(y) = e^{-e^{-y}}e^{-y}, \qquad -\infty < y < \infty,$$

which is one of the extreme value distributions that will be discussed in detail in Chapter 22. As mentioned on several occasions in this chapter, this transformation forms a basis for some methods of estimation of the parameters c and α (when ξ_0 is known).

When $\xi_0 = 0$, the probability density function of the random variable X can be conveniently written in the form

$$p_X(x) = c\theta x^{c-1}e^{-\theta x^c}, \qquad x > 0, c > 0, \tag{21.134}$$

where $\theta = \alpha^{-c}$. If c is fixed but α varies so that the parameter θ has a gamma probability density function

$$p_\theta(t) = \frac{\delta^p}{\Gamma(p)}t^{p-1}e^{-\delta t}, \qquad \theta > 0; p, \delta > 0,$$

then the probability density function of the random variable X can be written as

$$p_X(x) = \frac{\delta^p}{\Gamma(p)} cx^{c-1} \int_0^\infty t^p e^{-t(x^c+\delta)} \, dt$$

$$= \frac{p\delta^p cx^{c-1}}{(x^c + \delta)^{p+1}}, \qquad x > 0. \tag{21.135}$$

Note that the transformation X^c (of this *compound Weibull* random variable) has a Pareto distribution (see Chapter 20). This was observed by Dubey (1968). It is also of interest to note that the density in (21.135) is Burr's Type III distribution with a scale parameter δ (see Chapter 12, Section 4). Proceeding on similar lines, Harris and Singpurwalla (1968) derived compound Weibull distributions by assigning for the parameter $\theta = \alpha^{-c}$ (1) uniform, (2) two-point, and (3) two-parameter gamma distributions. Harris and Singpurwalla (1969) discussed the estimation of parameters in such Weibull distributions with random scale parameters.

Rodriguez (1977) made an interesting connection between Burr's Type XII distribution (see Chapter 12) and the Weibull family. Specifically, he observed that the Weibull curve is the lower bound for Burr's Type XII distribution in the Pearson $(\sqrt{\beta_1}, \beta_2)$ plane, and he explained it as follows: For a Burr Type XII random variable X, we have

$$\Pr\left[X \le \left(\frac{1}{k}\right)^{1/c} x\right] = 1 - \left(1 + \frac{x^c}{k}\right)^{-k}$$

$$= 1 - \exp\left\{-k \log\left(1 + \frac{x^c}{k}\right)\right\}$$

$$= 1 - \exp\left[-k\left\{\frac{x^c}{k} - \frac{1}{2}\left(\frac{x^c}{k}\right)^2 + \cdots\right\}\right]$$

$$\to 1 - e^{-x^c} \qquad \text{as } k \to \infty.$$

He also noted that Burr's Type XII distribution can be obtained as a smooth mixture of Weibull distributions.

On reflecting the Weibull distribution with density (21.3) about the vertical axis $x = \xi_0$, the density function and the distribution function of the reflected distribution become [Cohen (1973)]

$$p_{X_R}(x) = \frac{c}{\alpha}\left(\frac{\xi_0 - x}{\alpha}\right)^{c-1} e^{-[(\xi_0-x)/\alpha]^c}, \qquad x < \xi_0, \alpha > 0, c > 0,$$

$$F_{X_R}(x) = e^{-[(\xi_0-x)/\alpha]^c}, \qquad x < \xi_0, \alpha > 0, c > 0,$$

respectively. When any skewed distribution is reflected about a vertical axis, the sign of the skewness is reversed. Accordingly, the "reflected Weibull distribution" will be positively skewed for $c > c_0$ and negatively skewed for $c < c_0$ (see Section 2). With the Weibull distribution in (21.3) being the third asymptotic distribution of smallest values as discussed by Gumbel (1958), the reflected Weibull distribution given above is actually the third asymptotic distribution of largest values or the Fisher-Tippett Type III distribution of largest values.

By reflecting the Weibull density function in (21.3) onto the left of ξ_0 (a mirror image of the function on the right), Balakrishnan and Kocherlakota (1985) defined the *double Weibull distribution* with probability density function

$$p_X(x) = \frac{c}{2\alpha}\left|\frac{x - \xi_0}{\alpha}\right|^{c-1} e^{-|(x-\xi_0)/\alpha|^c}, \qquad -\infty < x < \infty. \quad (21.136)$$

The density function in (21.136) is symmetric about ξ_0. [For the case $c = 1$ it becomes the Laplace or double exponential density, which will be studied in Chapter 24.] For this distribution Balakrishnan and Kocherlakota (1985) discussed the properties of order statistics and derived, in particular, explicit expressions for means, variances, and covariances of order statistics. Using these quantities, they derived the BLUEs of the parameters ξ_0 and α, assuming the shape parameter c to be known, based on complete samples and presented the necessary tables for sample sizes up to 10. These results were extended by Dattatreya Rao and Narasimham (1989) to the case of Type II censored sample. Vasudeva Rao, Dattatreya Rao, and Narasimham (1991) have also derived optimal linear estimators for α based on absolute values of order statistics (when ξ_0 is known).

Zacks (1984) introduced a three-parameter *Weibull-exponential* distribution having a distribution function

$$F_X(x; \lambda, c, \tau) = 1 - e^{-\lambda x - [\lambda(x-\tau)^+]^c}, \qquad x \geq 0,$$

where $Y^+ = \max(0, Y)$, $\lambda > 0$ (scale parameter), $c \geq 1$ (shape parameter), and $\tau \geq 0$ (change-point parameter). This family possesses a nondecreasing hazard rate function given by

$$h_X(x; \lambda, c, \tau) = \begin{cases} \lambda & \text{if } 0 \leq x < \tau, \\ \lambda + \lambda^c c(x - \tau)^{c-1} & \text{if } x \geq \tau, \end{cases}$$

which is simply a superposition (for $x \geq \tau$) of a Weibull hazard rate on the constant hazard rate of an exponential distribution. It is clear that as the change-point parameter $\tau \to \infty$, the distribution approaches the exponential distribution with parameter λ. As Zacks (1984) aptly pointed out, this family is suitable for modeling systems that, after a certain length of time, enter a wear-out phase in which components of the system have an increasing hazard rate. Adaptive Bayesian inference has also been developed by Zacks for the parameter τ.

Mention should be made here of the piecewise Weibull model discussed by Nelson (1982).

Stacy and Mihram's (1965) generalized gamma distribution with the probability density function

$$p_X(x) = \frac{c}{\Gamma(\beta)\alpha^c\beta}(x - \xi_0)^{c\beta-1}e^{-[(x-\xi_0)/\alpha]^c}, \qquad x \geq \xi_0 \geq 0,$$

and cumulative distribution function

$$F_X(x) = \frac{1}{\Gamma(\beta)}\Gamma\left(\beta; \left(\frac{x - \xi_0}{\alpha}\right)^c\right),$$

where $\Gamma(\beta; y) = \int_0^y e^{-t}t^{\beta-1}\,dt$ is the incomplete gamma function, is a generalization of the three-parameter Weibull distribution in (21.3); the three-parameter Weibull pdf in (21.3) is a special case of this distribution when the second shape parameter $\beta = 1$. A discussion of this distribution is presented in Chapter 17 (as it is a generalization of the three-parameter gamma distribution as well). Cohen (1969) discussed this distribution as the four-parameter generalized Weibull distribution, while Arora (1974) studied in detail various characteristics of this generalized Weibull family.

Suppose that X_i $(i = 1, 2, \ldots, k)$ are independently distributed as three-parameter Weibull (ξ_0, α_i, c_i) with probability density function (21.3); then a random variable X that is distributed as X_i with probability π_i (such that $\pi_1 + \pi_2 + \cdots + \pi_k = 1$) is said to have a *finite-mixture Weibull distribution*. The density function of X is given by

$$p_X(x) = \sum_{i=1}^{k} \pi_i \frac{c_i}{\alpha_i}\left(\frac{x - \xi_0}{\alpha_i}\right)^{c_i-1}e^{-[(x-\xi_0)/\alpha_i]^{c_i}},$$

$$x > \xi_0; \alpha_i, c_i > 0, \quad (21.137)$$

and the cumulative distribution function is given by

$$F_X(x) = \sum_{i=1}^{k} \pi_i\{1 - e^{-[(x-\xi_0)/\alpha_i]^{c_i}}\}, \qquad x > \xi_0; \alpha_i, c_i > 0, \sum_{i=1}^{k} \pi_i = 1.$$

$$(21.138)$$

The case when $k = 2$ is commonly referred to as a *two-component mixture Weibull distribution* and has received considerable attention over the years from numerous researchers. For example, Kao (1959) suggested the use of this model in life-testing experiment involving electron tubes and proposed a graphical method of estimation of the parameters involved in this model. Rider (1961) discussed the method of moments estimation of the parameters of this mixed-Weibull model. Kaylan and Harris (1981) developed efficient numerical algorithms to derive the maximum likelihood estimates of the parameters involved in the finite-mixture Weibull model. Cheng and Fu (1982) have discussed further on estimation of mixed Weibull parameters. Bayesian inference has been developed for a variety of problems involving the two-component mixture Weibull distribution by many researchers including Ashour and Rashwan (1981), Ashour, Shoukry, and Mohamed (1983), Ashour (1987a), Sinha (1987), and Sinha and Sloan (1989). Of these, the first and the last articles dealt with the prediction problem, while Sinha (1987) dealt with the Bayesian estimation of the parameters and the associated reliability function of the model; see also Sinha (1982). Dykacz and Hennessey (1989) have explained how a two-component Weibull model can arise due to some natural features of a process. The usage of mixtures of Weibull distributions in order to estimate mixing proportions has been elaborated by Woodward and Gunst (1987). The special case of mixed Weibull-exponential model (a two-component mixture Weibull distribution in which one of the components has an exponential distribution, i.e., $c_1 = 1$) has also been studied in detail by Ashour (1985, 1987b). This author has discussed estimation of parameters from censored samples secured from this model and also the Bayesian estimation of the parameters involved. Mixtures of exponential distributions also could be considered as a special case of mixture Weibull distribution in (21.137) [Jewell (1982)], but more details on this specific case may be seen in Chapter 19. An interesting discussion of general mixture models and their role in survival analysis has been made by Farewell (1986).

In a similar effort Al-Hussaini and Abd-El-Hakim (1989) proposed the *Inverse Gaussian-Weibull mixture distribution* with density function

$$p_X(x; \underline{\theta}) = \pi p_1(x) + (1 - \pi)p_2(x), \qquad 0 \le \pi \le 1, \quad (21.139)$$

where $p_1(x)$ is the $IG(\mu, \lambda)$ density (see Chapter 15) given by

$$p_1(x) = \left(\frac{\lambda}{2\pi x^3}\right)^{1/2} \exp\left(-\frac{\lambda(x - \mu)^2}{\mu^2 x}\right), \qquad x > 0; \mu, \lambda > 0,$$

and $p_2(x)$ is the two-parameter Weibull density given by

$$p_2(x) = \frac{c}{\alpha}\left(\frac{x}{\alpha}\right)^{c-1} e^{-(x/\alpha)^c}, \qquad x > 0; \alpha, c > 0.$$

The parameter θ in (21.139) is the vector of all five parameters involved in the mixture model. The density in (21.139) can be either unimodal or bimodal. Since the hazard rate of the IG distribution is either upside-down bathtub or increasing and the hazard rate of the Weibull distribution is decreasing, constant, or increasing, the mixture model yields six different combinations of hazard rates; see Al-Hussaini and Abd-El-Hakim (1989). Thus the mixture model in (21.139) provides a very flexible model for reliability studies. Estimation of parameters for this model has been considered by Al-Hussaini and Abd-El-Hakim (1990).

The Weibull distribution in (21.3) can be restricted to lie in the interval (L, U); that is, the distribution is truncated below L and above U. In this case the distribution is referred to as the *doubly truncated Weibull distribution*, and its density function is given by

$$p_X(x) = \frac{c}{\alpha\{e^{-[(L-\xi_0)/\alpha]^c} - e^{-[(U-\xi_0)/\alpha]^c}\}} \left(\frac{x - \xi_0}{\alpha}\right)^{c-1} e^{-[(x-\xi_0)/\alpha]^c},$$

$$\xi_0 \leq L < x < U; c, \alpha > 0. \quad (21.140)$$

The special case of $L = \xi_0$ is referred as the *right-truncated Weibull distribution*, and its pdf is

$$p_X(x) = \frac{c}{\alpha\{1 - e^{-[(U-\xi_0)/\alpha]^c}\}} \left(\frac{x - \xi_0}{\alpha}\right)^{c-1} e^{-[(x-\xi_0)/\alpha]^c},$$

$$\xi_0 < x < U; c, \alpha > 0. \quad (21.141)$$

Similarly the special case of $U \to \infty$ is referred as the *left-truncated Weibull distribution*, and its pdf is

$$p_X(x) = \frac{c}{\alpha} \left(\frac{x - \xi_0}{\alpha}\right)^{c-1} e^{-[(x-\xi_0)/\alpha]^c + [(L-\xi_0)/\alpha]^c},$$

$$\xi_0 < L < x < \infty; c, \alpha > 0. \quad (21.142)$$

These truncated Weibull distributions have also been suggested for use in life-test analysis. Sugiura and Gomi (1985) presented the Pearson diagram of $(\sqrt{\beta_1}, \beta_2)$ for truncated Weibull distributions, while McEwen and Parresol (1991) presented expressions for the moments and summary statistics. Wingo (1988) described methods for fitting the right-truncated Weibull distribution in (21.141) to life-test and survival data; see also Martinez and Quintana (1991). Wingo (1989) discussed the role and properties of the left-truncated Weibull distribution in (21.142). A mention should be made of Mittal and Dahiya (1989) who have discussed the problem of estimation of parameters of a truncated Weibull distribution.

Nakagawa and Osaki (1975) defined a *discrete Weibull distribution* as a distribution $\{P_k\}_{k=0}^{\infty}$ such that

$$\sum_{j=k}^{\infty} P_j(q,c) = q^{k^c}, \qquad k = 0,1,2,\ldots, \; c > 0, 0 < q < 1. \quad (21.143)$$

In this case the probability mass function and the hazard rate are given by

$$P_k(q,c) = q^{k^c} - q^{(k+1)^c} \qquad (21.144)$$

and

$$h_k(q,c) = \frac{P_k(q,c)}{\sum_{j=k}^{\infty} P_j(q,c)} = 1 - q^{(k+1)^c - k^c}, \; k = 0,1,\ldots. \quad (21.145)$$

It is clear from (21.145) that this distribution has decreasing hazard rate if $0 < c < 1$, constant hazard rate if $c = 1$ (in fact it becomes the geometric distribution in this case), and increasing hazard rate if $c > 1$. Thus this distribution mimics the property of the continuous Weibull distribution and hence becomes a natural analogue of it. It is also easy to verify that if X has the discrete Weibull distribution in (21.143), then X^c has a geometric distribution (a property similar to the continuous Weibull distribution). Nakagawa and Osaki (1975) appropriately pointed out that this discrete Weibull distribution will be useful in modeling failures of some devices that often depend more on the total number of cycles than on the total time that they have been used (switching devices, rail tracks, tires, etc.). Stein and Dattero (1984) introduced another discrete Weibull distribution as the distribution whose hazard rate is given by

$$h_k(q,c) = \begin{cases} qk^{c-1}, & k = 1,2,\ldots,m, \\ 0, & k = 0 \text{ or } k > m, \end{cases} \qquad (21.146)$$

where m is given by

$$m = \begin{cases} \left[q^{-1/(c-1)} \right] & \text{if } c > 1, \\ +\infty & \text{if } c \le 1. \end{cases}$$

This discrete Weibull distribution has the following appealing properties: (1) the hazard rate behaves similarly to that of the continuous Weibull distribution, (2) it provides the exact lifetime distribution of a specific system, and (3) the lifetime converges in limit to that given by the continuous Weibull distribution. Ali Khan, Khalique, and Abouammoh (1989) have recently proposed a simple method of estimation of the parameters of these two discrete Weibull distributions and compared its efficiency with the method of moments estimates.

When X has a Weibull distribution with density function as in (21.3), the distribution of $1/X$ is referred as the *inverse Weibull distribution*. Its density function or distribution function can be written down easily from (21.3) and (21.4). Erto (1989) has discussed the properties of this distribution and its potential use as a lifetime model. The maximum likelihood estimation and the least-squares estimation of the parameters of the inverse Weibull distribution are discussed by Calabria and Pulcini (1990).

Starting with the Weibull density with unit median given by

$$p_W(x) = c(\log 2)x^{c-1}e^{-\log 2 \cdot x^c}, \tag{21.147}$$

Ekstrom (1973) used the usual Rice procedure [e.g., see Whalen (1971, pp. 103–105)] to derive the density function of the *Weibull-Rician distribution* as

$$p_{WR}(y) = \int_0^{2\pi} \left(cy\frac{\log 2}{2\pi} \right) \left\{ \frac{\exp\left[(-\log 2)(y^2 - 2Ay\cos\phi + A^2)^{c/2} \right]}{\left[y^2 - 2Ay\cos\phi + A^2 \right]^{(2-c)/2}} \right\} d\phi. \tag{21.148}$$

Here y is the envelope, A is the peak signal-to-median clutter ratio, and c is the parameter of the unit median Weibull density in (21.147). The integration on the RHS of (21.148) is not difficult to do numerically. Nilsson and Glisson (1980) derive this density function and the corresponding cumulative distribution function using conditional probability arguments, and in this method the cumulative distribution function is also expressed as a single integral.

A *Weibull process* is a useful model for phenomena that are changing over time. As noted by Ascher (1979), it is important to distinguish between a Weibull process, which models a repairable system, and the Weibull distribution (discussed so far), which models a nonrepairable system. Crow (1974) proposed a stochastic model in which the number of system failures is assumed to occur according to a nonhomogeneous Poisson process, $\{N(t): t \geq 0\}$, with mean value function of the form $m(t) = E[N(t)] = \lambda t^c$ and intensity function of the form $v(t) = dm(t)/dt = \lambda ct^{c-1}$. This is consistent with the empirical finding of Duane (1964) that the number of system failures and operating time are approximately linearly related (in the logarithmic scale), since $\log m(t) = \log \lambda + c \log t$ is a linear function of $\log t$. Another commonly used parametrization is $m(t) = (t/\theta)^c$, with $c < 1$ corresponding to improvement of the system and $c > 1$ corresponding to its deterioration. The Weibull process inherits its name due to the fact that in the sequence of successive failure times $T_1, T_2, \ldots$, the time to first failure T_1 has the Weibull distribution with hazard rate function $v(t) = c(t/\theta)^{c-1}/\theta$. Furthermore the conditional failure time T_n, given $T_1 = t_1, \ldots, T_{n-1} = t_{n-1}$, also has a Weibull distribution truncated below at t_{n-1}. More details on various results regarding Weibull processes can be obtained from a review article by

Engelhardt (1988). Moeller (1976) has presented a similar discussion on the Rasch-Weibull process.

Based on the Griffith theory of brittle fracture and the statistical theory of extreme values, Black et al. (1989) proposed the *brittle fracture distribution*, which is a three-parameter distribution that includes the Weibull and the exponential distributions as special cases. The cumulative distribution function of this distribution is

$$F_X(x; \alpha, \beta, r) = 1 - e^{-\alpha x^{2r} e^{-\beta/x^2}}, \qquad x > 0, \alpha > 0, \beta \geq 0, r > 0.$$

$$(21.149)$$

Here, X represents the breaking stress or strength. The corresponding probability density function is

$$p_X(x; \alpha, \beta, r) = 2\alpha x^{2r-1} \left(\frac{\beta}{x^2} + r \right) e^{-(\beta/x^2) - \alpha x^{2r} e^{-\beta/x^2}}, \qquad x > 0. \quad (21.150)$$

Black, Durham, and Padgett (1990) observed that the value of β affects the scale and that α and r both increase the variance as they increase.

For the case when $\beta = 0$ and $\delta = 2r$, (21.149) reduces to the Weibull distribution with shape parameter δ and scale parameter α. If $\beta = 0$ and $r = 1/2$, then (21.149) becomes an exponential distribution with mean $1/\alpha$. Since a closed-form expression does not exist for the moment generating function or the characteristic function of the brittle fracture distribution in (21.149), Gulati, Durham, and Padgett (1993) derived bounds for the kth moment ($k \geq 1$). While the lower bound turns out to be sharp, the upper bound does not. Black, Durham, and Padgett (1990) discussed estimation of the three parameters by a simple least-squares procedure and also by the maximum likelihood method. These authors have noted that the least-squares estimates could be used as good initial values for an iterative procedure for obtaining the maximum likelihood estimates; they also examined the behavior of these estimators by means of simulations.

Kopaev (1990, 1993) has considered a *modified Weibull distribution* with cumulative distribution function

$$F(x) = 1 - e^{-[(x/c)\exp\{\pm(x/c)\}^p/A]^B}, \qquad (21.151)$$

where $+(x/c)$ is taken for $p > 0$ and $-(x/c)$ for $p < 0$. Kopaev has discussed many properties of this distribution and also some methods of estimation of the parameters in (21.151).

BIBLIOGRAPHY

Abdel-Wahid, A. R., and Winterbottom, A. (1987). Approximate Bayesian estimates for the Weibull reliability function and hazard rate from censored data, *Journal of Statistical Planning and Inference*, **16**, 277–283.

Achcar, J. A., Brookmeyer, R., and Hunter, W. G. (1985). An application of Bayesian analysis to medical follow-up data, *Statistics in Medicine*, **4**, 509–520.

Adatia, A. (1994). Extended tables of covariances of order statistics from the Weibull distribution, *Preprint*, University of Regina, Regina, Saskatchewan.

Adatia, A., and Chan, L. K. (1982). Robust procedures for estimating the scale parameter and predicting future order statistics of the Weibull distribution, *IEEE Transactions on Reliability*, **31**, 491–498.

Adatia, A., and Chan, L. K. (1985). Robust estimators of the 3-parameter Weibull distribution, *IEEE Transactions on Reliability*, **34**, 347–351.

Aitkin, M., Laird, N., and Francis, B. (1983). A reanalysis of the Stanford heart transplant data (with discussion), *Journal of the American Statistical Association*, **78**, 264–292.

Al-Baidhani, F. A., and Sinclair, C. D. (1987). Comparison of methods of estimation of parameters of the Weibull distribution, *Communications in Statistics—Simulation and Computation*, **16**, 373–384.

Al-Hussaini, E. K., and Abd-El-Hakim, N. S. (1989). Failure rate of the inverse Gaussian-Weibull mixture model, *Annals of the Institute of Statistical Mathematics*, **41**, 617–622.

Al-Hussaini, E. K., and Abd-El-Hakim, N. S. (1990). Estimation of parameters of the inverse Gaussian-Weibull mixture model, *Communications in Statistics—Theory and Methods*, **19**, 1607–1622.

Ali Khan, M. S., Khalique, A., and Abouammoh, A. M. (1989). On estimating parameters in a discrete Weibull distribution, *IEEE Transactions on Reliability*, **38**, 348–350.

Antle, C. E., and Bain, L. J. (1988). Weibull distribution, In *Encyclopedia of Statistical Sciences*, **9**, S. Kotz, N. L. Johnson, and C. B. Read (editors), 549–556, New York: Wiley.

Archer, N. P. (1980). A computational technique for maximum likelihood estimation with Weibull models, *IEEE Transactions on Reliability*, **29**, 57–62.

Archer, N. P. (1982). Maximum likelihood estimation with Weibull models when the data are grouped, *Communications in Statistics—Theory and Methods*, **11**, 199–207.

Arnold, B. C., Balakrishnan, N., and Nagaraja, H. N. (1992). *A First Course in Order Statistics*, New York: Wiley.

Arnold, B. C., and Isaacson, D. (1976). On solutions to $\min(X, Y) \overset{d}{=} aX$ and $\min(X, Y) \overset{d}{=} aX \overset{d}{=} bY$, *Z. Wahrscheinlichkeitstheorie verw. Gebiete*, **35**, 115–119.

Aroian, L. A. (1965). Some properties of the conditional Weibull distribution, *Transactions of the Nineteenth Technical Conference of the American Society for Quality Control*, 361–368.

Arora, M. S. (1974). The 4-parameter generalized Weilbull family and its derivatives, *Abstract 711-62-6, Notices of the Institute of Mathematical Statistics*, A-240.

Ascher, H. (1979). Comments on "Models for Reliability of Repaired Equipment," *IEEE Transactions on Reliability*, **28**, 119.

Ashour, S. K. (1985). Estimation of the parameters of mixed Weibull-exponential models from censored samples, *Tamkang Journal of Mathematics*, **16**, No. **4**, 103–111.

Ashour, S. K. (1987a). Multi-censored sampling in mixed Weibull distribution, *Journal of the Indian Association for Productivity, Quality and Reliability*, **12**, No. 2, 51–56.

Ashour, S. K. (1987b). Bayesian estimation of mixed Weibull exponential in life testing, *Applied Stochastic Models and Data Analysis*, **3**, 51–57.

Ashour, S. K., and Jones, P. W. (1977). Shortest confidence intervals for the shape parameter of a Weibull distribution based on two adjacent failure times, *Technometrics*, **19**, 201–204.

Ashour, S. K., and Rashwan, D. R. (1981). Bayesian predictions for compound Weibull model, *Communications in Statistics—Theory and Methods*, **10**, 1613–1624.

Ashour, S. K., and Shalaby, O. A. (1982). Estimating sample size with Weibull failure, *Mathematische Operationsforschung und Statistik, Series Statistics*, **14**, 263–268.

Ashour, S. K., Shoukry, E., and Mohamed, T. (1983). Bayesian estimation for compound Weibull model, *Mathematische Operationsforschung und Statistik, Series Statistics*, **14**, 381–386.

Bahler, C., Hill, R. R., Jr., and Byers, R. A. (1989). Comparison of logistic and Weibull functions: The effect of temperature on cumulative germination of alfalfa, *Crop Science*, **29**, 142–146.

Bain, L. J. (1972). Inferences based on censored sampling from the Weibull or extreme-value distribution, *Technometrics*, **14**, 693–702.

Bain, L. J., and Antle, C. E. (1967). Estimation of parameters in the Weibull distribution, *Technometrics*, **9**, 621–627.

Bain, L. J., and Engelhardt, M. (1980a). Inferences on the parameters and current system reliability for a time truncated Weibull process, *Technometrics*, **22**, 421–426.

Bain, L. J., and Engelhardt, M. (1980b). Probability of correct selection of Weibull versus gamma based on likelihood ratio, *Communications in Statistics—Theory and Methods*, **9**, 375–381.

Bain, L. J., and Engelhardt, M. (1981). Simple approximate distributional results for confidence and tolerance limits for the Weibull distribution based on maximum likelihood estimators, *Technometrics*, **23**, 15–20.

Bain, L. J., and Engelhardt, M. (1986). Approximate distributional results based on the maximum likelihood estimators for the Weibull distribution, *Journal of Quality Technology*, **18**, 174–181.

Bain, L. J., and Engelhardt, M. (1991a). *Statistical Analysis of Reliability and Life-testing Models* (Second edition), New York: Marcel Dekker.

Bain, L. J., and Engelhardt, M. (1991b). Reliability test plans for one-shot devices based on repeated samples, *Journal of Quality Technology*, **23**, 304–311.

Balakrishnan, N., and Chan, P. S. (1993a). Extended tables of means, variances and covariances of Weibull order statistics, *Report*, McMaster University, Hamilton, Ontario.

Balakrishnan, N., and Chan, P. S. (1993b). Extended tables of the best linear unbiased estimators of the location and scale parameters of the Weibull distribution based on complete and Type-II censored data, *Report*, McMaster University, Hamilton, Ontario.

Balakrishnan, N., and Chan, P. S. (1993c). Record values from Rayleigh and Weibull distributions and associated inference, Presented at the *International Conference on Extremes and Applications*, Gaithersburg, MD.

Balakrishnan, N., and Cohen, A. C. (1991). *Order Statistics and Inference: Estimation Methods*, San Diego: Academic Press.

Balakrishnan, N., and Joshi, P. C. (1981). A note on order statistics from Weibull distribution, *Scandinavian Actuarial Journal*, 121–122.

Balakrishnan, N., and Kocherlakota, S. (1985). On the double Weibull distribution: Order statistics and estimation, *Sankhyā, Series B*, **47**, 161–178.

Balasooriya, U., and Chan, L. K. (1983). The prediction of future order statistics in the two-parameter Weibull distributions—A robust study, *Sankhyā, Series B*, **45**, 320–329.

Barlow, R. E., and Hsiung, J. H. (1983). Expected information from a life test experiment, *The Statistician*, **32**, 35–45.

Barlow, R. E., and Proschan, F. (1988). Life distribution models and incomplete data, In *Handbook of Statistics*, **7**, P. R. Krishnaiah and C. R. Rao (editors), 225–250, Amsterdam: North-Holland.

Barros, V. R., and Estevan, E. A. (1983). On the evaluation of wind power from short wind records, *Journal of Climate and Applied Meteorology*, **22**, 1116–1123.

Beg, M. I., and Ali, M. M. (1989). Characterization of the exponential and Weibull distributions, *Pakistan Journal of Statistics, Series B*, **5**, 287–291.

Beg, M. I., and Kirmani, S. N. U. A. (1978). Characterization of the exponential distribution by a weak homoscedasticity, *Communications in Statistics—Theory and Methods*, **7**, 307–310.

Bennett, G. K. (1977). Basic concepts of empirical Bayes methods with some results for the Weibull distribution, In *Theory and Applications of Reliability: With Emphasis on Bayesian and Nonparametric Methods*, **2**, C. P. Tsokos and I. N. Shimi (editors), 181–202, New York: Academic Press.

Berrettoni, J. N. (1964). Practical applications of the Weibull distribution, *Industrial Quality Control*, **21**, 71–79.

Berry, G. (1975). Design of carcinogenesis experiments using the Weibull distribution, *Biometrika*, **62**, 321–328.

Berry, G. L. (1981). The Weibull distribution as a human performance descriptor, *IEEE Transactions on Systems, Man, Cybernetics*, **11**, 501–504.

Bhattacharya, S. K. (1962). On a probit analogue used in a life-test based on the Weibull distribution, *Australian Journal of Statistics*, **4**, 101–105.

Bhattacharyya, G. K., and Soejoeti, Z. (1989). A tampered failure rate model for step-stress accelerated life test, *Communications in Statistics—Theory and Methods*, **18**, 1627–1643.

Billman, B. R., Antle, C. E., and Bain, L. J. (1972). Statistical inference from censored Weibull samples, *Technometrics*, **14**, 831–840.

Black, C. M., Durham, S. D., Lynch, J. D., and Padgett, W. J. (1989). A new probability distribution for the strength of brittle fibers, *Fiber-Tex 1989*, The Third Conference on Advanced Engineering Fibers and Textile Structures for Composites, *NASA Conference Publication 3082*, 363–374.

Black, C. M., Durham, S. D., and Padgett, W. J. (1990). Parameter estimation for a new distribution for the strength of brittle fibers: A simulation study, *Communications in Statistics—Simulation and Computation*, **19**, 809–825.

Blischke, W. R., Johns, M. V., Truelove, A. J., and Mundle, P. B. (1965). Estimation of the location parameters of the Pearson Type III and Weibull distributions in the non-regular case and other results in non-regular estimation, Aerospace Research Laboratories ARL 6-0233, Wright-Patterson Air Force Base, OH.

Blischke, W. R., and Scheuer, E. M. (1986). Tabular aids for fitting Weibull moment estimates, *Naval Research Logistics Quarterly*, **33**, 145–153.

Blom, G. (1958). *Statistical Estimates and Transformed Beta Variables*, New York: Wiley.

Boes, D. C. (1989). Regional flood quantile estimation for a Weibull model, *Water Resources Research*, **25**, 979–990.

Boland, P. J., Proschan, F., and Tong, Y. L. (1989). Crossing properties of mixture distribution, *Probability in the Engineering and Informational Sciences*, **3**, 355–366.

Bowman, K. O., and Shenton, L. R. (1983). Moment series for moment estimators of the parameters of a Weibull density, In *Proceedings of the Fourteenth Symposium on the Interface*, K. W. Keiner, R. S. Sacher, and J. W. Wilkinson (editors), 174–186, Berlin: Springer-Verlag.

Braennaes, K. (1986). Prediction in a duration model, *Journal of Forecasting*, **5**, 97–103.

Brown, G., and Wingo, D. R. (1975). Comments on "MLE of Weibull parameters by quasilinearization," *IEEE Transactions on Reliability*, **24**, 158–159.

Brownlee, J. (1923). *Tracts for Computers 9*, Cambridge: Cambridge University Press.

Bugaighis, M. M. (1988). Efficiencies of MLE and BLUE for parameters of an accelerated life-test model, *IEEE Transactions on Reliability*, **37**, 230–233.

Bunday, B., and Al-Mutwali, I. (1981). Direct optimization for calculating maximum likelihood estimates of parameters of the Weibull distribution, *IEEE Transactions on Reliability*, **30**, 367–369.

Bury, K. V. (1975). *Statistical Models in Applied Science*, New York: Wiley.

Cacciari, M., and Montanari, G. C. (1985). An approximative method for calculating parameters and tolerance limits for percentiles of a Weibull distribution valid for progressive censored tests, *Metron*, **43**, No. 2, 67–84. (In Italian)

Cacciari, M., and Montanari, G. C. (1987). A method to estimate the Weibull parameters for progressively censored tests, *IEEE Transactions on Reliability*, **36**, 87–93.

Calabria, R., and Pulcini, G. (1990). On the maximum likelihood and least squares estimation in the inverse Weibull distribution, *Statistica Applicata*, **2**, 53–66.

Canavos, G. C. (1983). A performance comparison of empirical Bayes and Bayes estimators of the Weibull and gamma scale parameters, *Naval Research Logistics Quarterly*, **30**, 465–470.

Carlin, J., and Haslett, J. (1982). The probability distribution of wind power from a dispersed array of wind turbine generators, *Journal of Climate and Applied Meteorology*, **21**, 303–313.

Carmody, T. J., Eubank, R. L., and LaRiccia, V. N. (1984). A family of minimum quantile distance estimators for the three-parameter Weibull distribution, *Statistische Hefte*, **25**, 69–82.

Chan, L. K., Cheng, S. W. H., and Mead, E. R. (1974). Simultaneous estimation of location and scale parameters of the Weibull distribution, *IEEE Transactions on Reliability*, **23**, 335–341.

Chandra, N. K., and Chaudhuri, A. (1990a). On testimating the Weibull shape parameter, *Communications in Statistics—Simulation and Computation*, **19**, 637–648.

Chandra, N. K., and Chaudhuri, A. (1990b). On the efficiency of a testimator for the Weibull shape parameter, *Communications in Statistics—Theory and Methods*, **19**, 1247–1259.

Chao, A., and Hwang, S. -J. (1986). Comparison of confidence intervals for the parameters of the Weibull and extreme value distributions, *IEEE Transactions on Reliability*, **35**, 111–113.

Charernkavanich, D., and Cohen, A. C. (1984). Estimation in the singly truncated Weibull distribution with an unknown truncation point, *Communications in Statistics—Theory and Methods*, **13**, 843–857.

Chaudhuri, A., and Chandra, N. K. (1989). A test for Weibull populations, *Statistics & Probability Letters*, **7**, 377–380.

Chaudhuri, A., and Chandra, N. K. (1990). A characterization of the Weibull distribution, *Journal of the Indian Association for Productivity, Quality and Reliability*, **15**, 69–72.

Chen, W. C., Hill, B. M., Greenhouse, J. B., and Fayos, J. V. (1985). Bayesian analysis of survival curves for cancer patients following treatment, In *Bayesian Statistics*, **2**, J. M. Bernardo, M. H. DeGroot, D. V. Lindley, and A. F. M. Smith (editors), 299–328, Amsterdam: North-Holland.

Chen, W. W. S. (1987). Testing gamma and Weibull distribution: A comparative study, *Estadistica*, **39**, 1–26.

Cheng, K. F., and Chen, C. H. (1988). Estimation of the Weibull parameters with grouped data, *Communications in Statistics—Theory and Methods*, **17**, 325–341.

Cheng, K. F., and Shun, J. Y. (1989). Testing equality of two Weibull distributions with jointly type II censored samples, *Bulletin of the Institute of Mathematics, Academia Sinica*, **17**, 143–155.

Cheng, R. C. H., and Amin, N. A. K. (1983). Estimating parameters in continuous univariate distributions with a shifted origin, *Journal of the Royal Statistical Society, Series B*, **45**, 394–403.

Cheng, R. C. H., and Iles, T. C. (1983). Confidence bands for cumulative distribution functions of continuous random variables, *Technometrics*, **25**, 77–86.

Cheng, R. C. H., and Iles, T. C. (1987). Corrected maximum likelihood in non-regular problems, *Journal of the Royal Statistical Society, Series B*, **49**, 95–101.

Cheng, R. C. H., and Iles, T. C. (1988). One-sided confidence bands for cumulative distribution functions, *Technometrics*, **30**, 155–159.

Cheng, S. W., and Fu, J. C. (1982). Estimation of mixed Weibull parameters in life testing, *IEEE Transactions on Reliability*, **31**, 377–381.

Christensen, E. R., and Chen, C. -Y. (1985). A general noninteractive multiple toxicity model including probit, logit, and Weibull transformations, *Biometrics*, **41**, 711–725.

Christofferson, R. D., and Gillette, D. A. (1987). A simple estimator of the shape factor of the two-parameter Weibull distribution, *Journal of Climate and Applied Meteorology*, **26**, 323–325.

Chryssaphinou, O., and Papastavridis, S. G. (1990). Limit distribution for a consecutive-*k*-out-of-*n*: *F* system, *Advances in Applied Probability*, **22**, 491–493.

Clark, E. C., and Williams, G. T. (1958). Distribution of the members of an ordered sample, *Annals of Mathematical Statistics*, **29**, 862–870.

Clark, L. J. (1964). Estimation of the scale parameter of the Weibull probability density function by the use of one order and of M order statistics, Unpublished thesis, Air Force Institute of Technology, Wright-Patterson Air Force Base, Dayton, OH.

Cobb, E. B. (1989). Estimation of the Weibull shape parameter in small-sample bioassay, *Journal of Statistical Computation and Simulation*, **31**, 93–101.

Cohen, A. C. (1965). Maximum likelihood estimation in the Weibull distribution based on complete and on censored samples, *Technometrics*, **7**, 579–588.

Cohen, A. C. (1969). A four-parameter generalized Weibull distribution and its inference, NASA Contractor Report No. 61293, Contract NAS 8-11175, Marshall Space Flight Center, Huntsville, AL.

Cohen, A. C. (1973). The reflected Weibull distribution, *Technometrics*, **15**, 867–873.

Cohen, A. C. (1975). Multi-censored sampling in the three parameter Weibull distribution, *Technometrics*, **17**, 347–352.

Cohen, A. C. (1991). *Truncated and Censored Samples: Theory and Applications*, New York: Marcel Dekker.

Cohen, A. C., and Whitten, B. J. (1982). Modified maximum likelihood and modified moment estimators for the three-parameter Weibull distribution, *Communications in Statistics—Theory and Methods*, **11**, 2631–2656.

Cohen, A. C., and Whitten, B. J. (1988). *Parameter Estimation in Reliability and Life Span Models*, New York: Marcel Dekker.

Cohen, A. C., Whitten, B. J., and Ding, Y. (1984). Modified moment estimation for the three-parameter Weibull distribution, *Journal of Quality Technology*, **16**, 159–167.

Conradsen, K., Nielsen, L. B., and Prahm, L. P. (1984). Review of Weibull statistics for estimation of wind speed distributions, *Journal of Climate and Applied Meteorology*, **23**, 1173–1183.

Cook, L. (1978). Characterization of exponential and Weibull distributions through truncated statistics, *Scandinavian Actuarial Journal*, 165–168.

Couture, D. J., and Martz, H. F., Jr. (1972). Empirical Bayes estimation in the Weibull distribution, *IEEE Transactions on Reliability*, **R-21**, 75–83.

Cran, G. W. (1988). Moment estimators for the 3-parameter Weibull distribution, *IEEE Transactions on Reliability*, **37**, 360–363.

Crow, L. H. (1974). Reliability analysis for complex, repairable systems, In *Reliability and Biometry: Statistical Analysis of Life Lengths*, F. Proschan and R. J. Serfling (editors), 379–410, Philadelphia: SIAM.

Crow, L. H. (1982). Confidence interval procedures for the Weibull process with applications to reliability growth, *Technometrics*, **24**, 67–72.

Crowder, M. (1985). A distributional model for repeated failure time measurements, *Journal of the Royal Statistical Society, Series B*, **47**, 447–452.

Crowder, M. (1990). On some nonregular tests for a modified Weibull model, *Biometrika*, **77**, 499–506.

D'Agostino, R. B., and Stephens, M. A. (eds.) (1986). *Goodness-of-fit Techniques*, New York: Marcel Dekker.

Dallas, A. C. (1982). Some results on record values from the exponential and Weibull law, *Acta Mathematica of Academy of Sciences of Hungary*, **40**, 307–311.

Dattatreya Rao, A. V., and Narasimham, V. L. (1989). Linear estimation in double Weibull distribution, *Sankhyā, Series B*, **51**, 24–64.

David, H. A. (1981). *Order Statistics* (Second edition), New York: Wiley.

David, H. A., and Moeschberger, M. L. (1978). *The Theory of Competing Risks*, New York: Macmillan.

David, H. A., and Shu, V. S. (1978). Robustness of location estimators in the presence of an outlier, In *Contributions to Survey Sampling and Applied Statistics: Papers in Honor of H. O. Hartley*, H. A. David (editor), 235–250, New York: Academic Press.

De Braganca Pereira, B. (1984). On the choice of a Weibull model, *Estadistica*, **36**, 157–163.

Dellaportas, P., and Wright, D. E. (1991). Numerical prediction for the two-parameter Weibull distribution, *The Statistician*, **40**, 365–372.

Dey, D. K., and Kuo, L. (1991). A new empirical Bayes estimator with type II censored data, *Computational Statistics & Data Analysis*, **12**, 271–279.

DiCiccio, T. J. (1987). Approximate inference for the generalized gamma distribution, *Technometrics*, **29**, 33–40.

DiCiccio, T. J., Field, C. A., and Fraser, D. A. S. (1990). Approximations of marginal tail probabilities and inference for scale parameters, *Biometrika*, **77**, 77–95.

Dixon, J. C., and Swift, R. H. (1984). The directional variation of wind speed and Weibull probability parameters, *Atmospheric Environment*, **18**, 2041–2047.

Dourgnon, F., and Reyrolle, J. (1966). Tables de la fonction de répartition de la loi de Weibull, *Revue de Statistique Appliquée*, **14**, No. 4, 83–116.

Duane, J. T. (1964). Learning curve approach to reliability monitoring, *IEEE Transactions on Aerospace*, **2**, 563–566.

Dubey, S. D. (1965). Asymptotic properties of several estimators of Weibull parameters, *Technometrics*, **7**, 423–434.

Dubey, S. D. (1966a). Characterization theorems for several distributions and their applications, *Journal of Industrial Mathematics*, **16**, 1–22.

Dubey, S. D. (1966b). Comparative performance of several estimators of the Weibull parameters, *Proceedings of the Twentieth Technical Conference of the American Society for Quality Control*, 723–735.

Dubey, S. D. (1966c). Some test functions for the parameters of the Weibull distributions, *Naval Research Logistics Quarterly*, **13**, 113–128.

Dubey, S. D. (1966d). On some statistical inferences for Weibull laws, *Naval Research Logistics Quarterly*, **13**, 227–251.

Dubey, S. D. (1966e). Hyper-efficient estimator of the location parameter of the Weibull laws, *Naval Research Logistics Quarterly*, **13**, 253–264.

Dubey, S. D. (1966f). Asymptotic efficiencies of the moment estimators for the parameters of the Weibull laws, *Naval Research Logistics Quarterly*, **13**, 265–288.

Dubey, S. D. (1966g). Transformations for estimation of parameters, *Journal of the Indian Statistical Association*, **4**, 109–124.

Dubey, S. D. (1967a). Normal and Weibull distributions, *Naval Research Logistics Quarterly*, **14**, 69–79.

Dubey, S. D. (1967b). Revised tables for asymptotic efficiencies of the moment estimators for the parameters of the Weibull laws, *Naval Research Logistics Quarterly*, **14**, 261–267.

Dubey, S. D. (1967c). Some percentile estimators for Weibull parameters, *Technometrics*, **9**, 119–129.

Dubey, S. D. (1967d). Monte Carlo study of the moment and maximum likelihood estimators of Weibull parameters, *Trabajos de Estadistica*, **18**, 131–141.

Dubey, S. D. (1968). A compound Weibull distribution, *Naval Research Logistics Quarterly*, **15**, 179–188.

Dumonceaux, R. H. (1969). Statistical inferences for location and scale parameter distributions, *Ph.D. thesis*, Rolla, MO: University of Missouri.

Dyer, A. R. (1975). An analysis of the relationship of systolic blood pressure, serum cholesterol, and smoking to 14-year mortality in the Chicago Peoples Gas Company Study. Part I: Total mortality in exponential—Weibull model, Part II: Coronary and cardiovascular-renal mortality in two competing risk models, *Journal of Chronic Diseases*, **28**, 565–578.

Dyer, D. (1981). Offshore oil/gas lease bidding and the Weibull distribution, In *Statistical Distributions in Scientific Work*, **6**, C. Taillie, G. P. Patil, and B. A. Baldessari (editors), 33–45, Dordrecht: Reidel.

Dykacz, J. M., and Hennessey, J. C. (1989). An example of a two-part Weibull model prompted by some natural features of a process, *ASA Proceedings of Social Statistics Section*, 79–83.

Ekstrom, J. (1973). The detection of Steady Targets in Weibull Cluster, Presented at *IEEE Conference on Radar—Present and Future*, London, Publication No. **105**.

El-Din, M. M. M., Mahmoud, M. A. W., and Youssef, S. E. A. (1991). Moments of order statistics from parabolic and skewed distributions and a characterization of Weibull distribution, *Communications in Statistics—Simulation and Computation*, **20**, 639–645.

Ellingwood, B., and Rosowsky, D. (1991). Duration of load effects in LRFD for wood construction, *Journal of Structural Engineering*, **117**, 584–596.

Ellis, W. C., and Rao Tummala, V. M. (1983). Bayesian approach to maximum likelihood estimation of parameters of Weibull distribution, *ASA Proceedings of Business & Economic Statistics Section*, 574–575.

Ellis, W. C., and Rao Tummala, V. M. (1986). Minimum expected loss estimators of the shape & scale parameters of the Weibull distribution, *IEEE Transactions on Reliability*, **35**, 212–213.

Emoto, S. E., and Matthews, P. C. (1990). A Weibull model for dependent censoring, *Annals of Statistics*, **18**, 1556–1577.

Engelhardt, M. (1975). On simple estimation of the parameters of the Weibull or extreme-value distribution, *Technometrics*, **17**, 369–374.

Engelhardt, M. (1988). Weibull processes, In *Encyclopedia of Statistical Sciences*, **9**, S. Kotz, N. L. Johnson, and C. B. Read (editors), 557–561, New York: Wiley.

Engelhardt, M., and Bain, L. J. (1973). Some complete and censored sampling results for the Weibull or extreme-value distribution, *Technometrics*, **15**, 541–549.

Engelhardt, M., and Bain, L. J. (1974). Some results on point estimation for the two-parameter Weibull or extreme-value distribution, *Technometrics*, **16**, 49–58.

Engelhardt, M., and Bain, L. J. (1977). Simplified statistical procedures for the Weibull or extreme-value distribution, *Technometrics*, **19**, 323–332.

Engelhardt, M., and Bain, L. J. (1978). Prediction intervals for the Weibull process, *Technometrics*, **20**, 167–170.

Engelhardt, M., and Bain, L. J. (1979). Prediction limits and two-sample problems with complete or censored Weibull data, *Technometrics*, **21**, 233–238.

Engelhardt, M., and Bain, L. J. (1982). On prediction limits for samples from a Weibull or extreme-value distribution, *Technometrics*, **24**, 147–150.

Engeman, R. M., and Keefe, T. J. (1982). On generalized least squares estimation of the Weibull distribution, *Communications in Statistics—Theory and Methods*, **11**, 2181–2193.

Engeman, R. M., and Keefe, T. J. (1985). Two-step estimators of the scale parameter of the Weibull distribution, *Computers and Biomedical Research*, **18**, 391–396.

Erto, P. (1982). New practical Bayes estimators for the 2-parameter Weibull distribution, *IEEE Transactions on Reliability*, **31**, 194–197.

Erto, P. (1989). Genesis, properties and identification of the inverse Weibull lifetime model, *Statistica Applicata*, **1**, 117–128. (In Italian)

Erto, P., and Guida, M. (1985). Tables for exact lower confidence limits for reliability and quantiles, based on least-squares of Weibull parameters, *IEEE Transactions on Reliability*, **34**, 219–223.

Escobar, L. A., and Meeker, W. Q., Jr. (1986a). Elements of the Fisher information matrix for the smallest extreme value distribution and censored data, *Applied Statistics*, **35**, 80–86.

Escobar, L. A., and Meeker, W. Q., Jr. (1986b). Planning accelerated life tests with Type II censored data, *Journal of Statistical Computation and Simulation*, **23**, 273–297.

Evans, I. G., and Nigm, A. M. (1980a). Bayesian prediction for two-parameter Weibull lifetime models, *Communications in Statistics—Theory and Methods*, **9**, 649–658.

Evans, I. G., and Nigm, A. M. (1980b). Bayesian 1-sample prediction for the 2-parameter Weibull distribution, *IEEE Transactions on Reliability*, **29**, 410–413.

Farewell, V. T. (1986). Mixture models in survival analysis: Are they worth the risk? *Canadian Journal of Statistics*, **14**, 257–262.

Farewell, V. T., and Prentice, R. L. (1977). A study of distributional shape in life testing, *Technometrics*, **19**, 69–76.

Fearn, D. H., and Nebenzahl, E. (1991). On the maximum likelihood ratio method of deciding between the Weibull and gamma distributions, *Communications in Statistics—Theory and Methods*, **20**, 579–593.

Fertig, K. W., and Mann, N. R. (1980). Life-test sampling plans for two-parameter Weibull populations, *Technometrics*, **22**, 165–177.

Fertig, K. W., Meyer, M. E., and Mann, N. R. (1980). On constructing prediction intervals for samples from a Weibull or extreme value distribution, *Technometrics*, **22**, 567–573.

Fisher, R. A., and Tippett, L. H. C. (1928). Limiting forms of the frequency distribution of the largest or smallest member of a sample, *Proceedings of the Cambridge Philosophical Society*, **24**, 180–190.

Flygare, M. E., Austin, J. A., and Buckwalter, R. M. (1985). Maximum likelihood estimation for the 2-parameter Weibull distribution based on interval-data, *IEEE Transactions on Reliability*, **34**, 57–59.

Fong, J. T., Rehm, R. G., and Graminski, E. L. (1977). Weibull statistics and a microscopic degradation model of paper, *Journal of the Technical Association of the Pulp and Paper Industry*, **60**, 156–159.

Fowlkes, E. B. (1987). *Folio of Distributions: A Collection of Theoretical Quantile-Quantile Plots*, New York: Marcel Dekker.

Franck, J. R. (1988). A simple explanation of the Weibull distribution and its applications, *Reliability Review*, **8**, No. 3, 6–9.

Fréchet, M. (1927). Sur la loi de probabilité de l'écart maximum, *Annales de la Société Polonaise de Mathematique, Cracovie*, **6**, 93–116.

Freimer, M., Mudholkar, G. S., and Lin, C. T. (1989). Extremes, extreme spacings and outliers in the Tukey and Weibull families, *Communications in Statistics—Theory and Methods*, **18**, 4261–4274.

Freudenthal, A. M., and Gumbel, E. J. (1954). Minimum life in fatigue, *Journal of the American Statistical Association*, **49**, 575–597.

Friedman, L. (1981). Parameter estimation in a minimum-type model by the least-squares method—A Monte Carlo study, *Communications in Statistics—Theory and Methods*, **10**, 463–487.

Friedman, L., and Gertsbakh, I. B. (1980). Maximum likelihood estimation in a minimum-type model with exponential and Weibull failure models, *Journal of the American Statistical Association*, **75**, 460–465.

Friedman, L., and Gertsbakh, I. B. (1981). Parameter estimations in a minimum-type scheme, *Communications in Statistics—Theory and Methods*, **10**, 439–462.

Fukuta, J. (1963). Estimation of parameters in the Weibull distribution and its efficiency, Research Report No. 13, Faculty of Engineering, Gifu University, Japan.

Fuller, W. A. (1991). Simple estimators for the mean of skewed populations, *Statistica Sinica*, **1**, 137–158.

Galambos, J. (1981). Failure time distributions: Estimates and asymptotic results, In *Statistical Distributions in Scientific Work*, **5**, C. Taillie, G. P. Patil, and B. A. Baldessari (editors), 309–317, Dordrecht: Reidel.

Galetto, F. (1988). Comment on "New practical Bayes estimators for the 2-parameter Weibull distribution," *IEEE Transactions on Reliability*, **37**, 562–565.

Gallagher, M. A., and Moore, A. H. (1990). Robust minimum-distance estimation using the 3-parameter Weibull distribution, *IEEE Transactions on Reliability*, **39**, 575–580.

Gates, D. J. (1985). On the optimal composition of electricity grids with unreliable units: Solvable models, *Advances in Applied Probability*, **17**, 367–385.

Gibbons, D. I., and Vance, L. C. (1981). A simulation study of estimators for the 2-parameter Weibull distribution, *IEEE Transactions on Reliability*, **30**, 61–66.

Gibbons, D. I., and Vance, L. C. (1983). Estimators for the 2-parameter Weibull distribution with progressively censored samples, *IEEE Transactions on Reliability*, **32**, 95–99.

Gittus, J. H. (1967). On a class of distribution functions, *Applied Statistics*, **16**, 45–50.

Glaser, R. E. (1984). Estimation for a Weibull accelerated life testing model, *Naval Research Logistics Quarterly*, **31**, 559–570.

Gnedenko, B. V. (1943). Sur la distribution limite du terme maximum d'une série aléatoire, *Annals of Mathematics*, **44**, 423–453.

Gnedenko, B. V., Belyaev, Yu. K., and Solov'ev, A. D. (1965). *Mathematical Methods in Reliability Theory*, Moscow. (In Russian. English Translation, New York: Academic Press, 1968.)

Goldman, A. I. (1984). Survivorship analysis when cure is a possibility: A Monte Carlo study, *Statistics in Medicine*, **3**, 153–163.

Gorski, A. C. (1968). Beware of the Weibull euphoria, *IEEE Transactions on Reliability*, **17**, 202–203.

Govindarajulu, Z., and Joshi, M. (1968). Best linear unbiased estimation of location and scale parameters of Weibull distribution using ordered observations, *Statistical Applications Research, JUSE*, **15**, 1–14.

Graybill, F. A. (1983). *Matrics with Applications in Statistics* (Second edition), Belmont, CA: Wadsworth.

Groeneveld, R. A. (1986). Skewness for the Weibull family, *Statistica Neerlandica*, **40**, 135–140.

Gross, A. J., and Clark, V. A. (1975). *Survival Distributions: Reliability Applications in the Biomedical Sciences*, New York: Wiley.

Gross, A. J., and Lurie, D. (1977). Monte Carlo comparisons of parameter estimators of the 2-parameter Weibull distribution, *IEEE Transactions on Reliability*, **26**, 356–358.

Gulati, S., Durham, S. D., and Padgett, W. J. (1993). C401. On bounds for moments of the brittle fracture distribution, *Journal of Statistical Computation and Simulation*, **47**, 105–108.

Gumbel, E. J. (1958). *Statistics of Extremes*, New York: Columbia University Press.

Haan, C. T., and Beer, C. E. (1967). Determination of maximum likelihood estimators for the three parameter Weibull distribution, *Iowa State Journal of Science*, **42**, 37–42.

Hahn, G. J., Godfrey, J. T., and Renzi, N. A. (1960). Weibull density computer programs, General Electric Company Report No. 60GL235.

Hahn, G. J., and Meeker, W. Q., Jr. (1983). Product life data analysis and some of its hazards, *Chemical Technology*, **13**, 282–284.

Hallinan, A. J., Jr. (1993). A review of the Weibull distribution, *Journal of Quality Technology*, **25**, 85–93.

Harris, C. M., and Singpurwalla, N. D. (1968). Life distributions derived from stochastic hazard functions, *IEEE Transactions on Reliability*, **17**, 70–79.

Harris, C. M., and Singpurwalla, N. D. (1969). On estimation in Weibull distributions with random scale parameters, *Naval Research Logistics Quarterly*, **16**, 405–410.

Harter, H. L. (1970). *Order Statistics and Their Use in Testing and Estimation*, **2**, Washington, DC: Government Printing Office.

Harter, H. L. (1986). Realistic models for system reliability, In *Reliability and Quality Control*, A. P. Basu (editor), 201–207, Amsterdam: North-Holland.

Harter, H. L. (1988). Weibull, log-Weibull and gamma order statistics, In *Handbook of Statistics*, **7**, P. R. Krishnaiah and C. R. Rao (editors), 433–466, Amsterdam: North-Holland.

Harter, H. L., and Dubey, S. D. (1967). Theory and tables for tests of hypotheses concerning the mean and the variance of a Weibull population, Aerospace Research Laboratories ARL 67-0059, Wright-Patterson Air Force Base, OH.

Harter, H. L., and Moore, A. H. (1965a). Point and interval estimators, based on *m* order statistics, for the scale parameter of a Weibull population with known shape parameter, *Technometrics*, **7**, 405–422.

Harter, H. L., and Moore, A. H. (1965b) Maximum-likelihood estimation of the parameters of gamma and Weibull populations from complete and from censored samples, *Technometrics*, **7**, 639–643. (Correction, *Ibid.*, **9**, 195.)

Harter, H. L., and Moore, A. H. (1967). Asymptotic variances and covariances of maximum likelihood estimators, from censored samples, of the parameters of Weibull and gamma populations, *Annals of Mathematical Statistics*, **38**, 557–570.

Harter, H. L., and Moore, A. H. (1976). An evaluation of exponential and Weibull test plans, *IEEE Transactions on Reliability*, **25**, 100–104.

Harter, H. L., Moore, A. H., and Wiegand, R. P. (1985). Sequential tests of hypotheses for system reliability modeled by a 2-parameter Weibull distribution, *IEEE Transactions on Reliability*, **34**, 352–355.

Hassanein, K. M. (1971). Percentile estimators for the parameters of the Weibull distribution, *Biometrika*, **58**, 673–676.

Hassanein, K. M., and Legler, W. K. (1975). On minimum variance stratification for estimating the mean of a Weibull population, *Scandinavian Actuarial Journal*, 207–214.

Hasssanein, K. M., Saleh, A. K. Md. E., and Brown, E. F. (1984). Quantile estimates in complete and censored samples from extreme-value and Weibull distributions, *IEEE Transactions on Reliability*, **33**, 370–373.

Haugh, L. D., and Delano, J. (1984). The effect of successive censoring and truncation on inference for the Weibull distribution: A Monte Carlo study, *ASA Proceedings of Statistical Computing Section*, 233–238.

Hentzschel, J. (1989). Parametric estimation of censored samples: The Weibull distribution, *Mathematische Operationsforschung und Statistik, Series Statistics*, **20**, 383–395. (In German)

Hewett, J. E., and Moeschberger, M. L. (1976). Some approximate simultaneous prediction intervals for reliability analysis, *Technometrics*, **18**, 227–230.

Hirose, H. (1991). Percentile point estimation in the three-parameter Weibull distribution by the extended maximum likelihood estimate, *Computational Statistics & Data Analysis*, **11**, 309–331.

Hobbs, J. R., Moore, A. H., and Miller, R. M. (1985). Minimum-distance estimation of the parameters of the 3-parameter Weibull distribution, *IEEE Transactions on Reliability*, **34**, 495–496.

Homan, S. M. (1989). A comparison of plotting rules under L_1 and L_2 estimation of the Weibull scale and shape parameters in situations of small samples with possible censoring and outliers, *Communications in Statistics—Simulation and Computation*, **18**, 121–143.

Homan, S. M., Pardo, E. S., and Trudeau, M. (1987). An application of Weibull modeling and discriminant analysis of alcoholism relapse among veterans, *ASA Proceedings of Social Statistics Section*, 222–224.

Hsieh, P. I., Lu, M. -W., and Frohman, T. F. (1987). Simulation study on the Weibull distribution, *Reliability Review*, 7, No. 4, 53–56.

Husler, J., and Schuepbach, M. (1986). On simple block estimators for the parameters of the extreme-value distribution, *Communications in Statistics—Simulation and Computation*, **15**, 61–76.

Ida, M. (1980). The application of the Weibull distribution to the analysis of the reaction time data, *Japanese Psychological Research*, **22**, 207–212.

Ishioka, T. (1990). Generating the Weibull random numbers using the ratio of uniform deviates, *Journal of the Japanese Society for Quality Control*, **20**, 127–129. (In Japanese)

Ishioka, T., and Nonaka, Y. (1991). Maximum likelihood estimation of Weibull parameters for two independent competing risks, *IEEE Transactions on Reliability*, **40**, 71–74.

Jaech, J. L. (1964). Estimation of Weibull distribution shape parameter when no more than two failures occur per lot, *Technometrics*, **6**, 415–422.

Jaech, J. L. (1968). Estimation of Weibull parameters from grouped failure data, Presented at the *American Statistical Association Annual Meeting at Pittsburgh*, August 1968.

Janardan, K. G. (1978). A new functional equation analogous to Cauchy-Pexider functional equation and its application, *Biometrical Journal*, **20**, 323–328.

Janardan, K. G., and Taneja, V. S. (1979a). Characterization of the Weibull distribution by properties of order statistics, *Biometrical Journal*, **21**, 3–9.

Janardan, K. G., and Taneja, V. S. (1979b). Some theorems concerning characterization of the Weibull distribution, *Biometrical Journal*, **21**, 139–144.

Janardan, K. G., and Schaeffer, D. J. (1978). Another characterization of the Weibull distribution, *Canadian Journal of Statistics*, **6**, 77–78.

Jewell, N. P. (1982). Mixtures of exponential distributions, *Annals of Statistics*, **10**, 479–484.

Johns, M. V., and Lieberman, G. J. (1966). An exact asymptotically efficient confidence bound for reliability in the case of the Weibull distribution, *Technometrics*, **8**, 135–175.

Johnson, L. G. (1968). The probabilistic basis of cumulative damage, *Transactions of the Twenty-second Technical Conference of the American Society of Quality Control*, 133–140.

Johnson, N. L. (1966). Cumulative sum control charts and the Weibull distribution, *Technometrics*, **8**, 481–491.

Johnson, R. A., and Haskell, J. H. (1983). Sampling properties of estimators of a Weibull distribution of use in the lumber industry, *Canadian Journal of Statistics*, **11**, 155–169.

Johnson, R. A., and Haskell, J. H. (1984). An approximate lower tolerance bound for the three parameter Weibull applied to lumber property characterization, *Statistics & Probability Letters*, **2**, 67–76.

Kagan, A. M. (1965). Zamechaniya o razdyelyayushchikh razbyeniakh, *Trudy Matematichskogo Instituta imeni Steklova*, **79**, 26–31 (In Russian. English translation by the American Mathematical Society, 1968.)

Kalbfleisch, J. D., and Lawless, J. F. (1988). Estimation of reliability in field-performance studies (with discussion), *Technometrics*, **30**, 365–836.

Kamat, S. J. (1977). Bayesian estimation of system reliability for Weibull distribution using Monte Carlo simulation, In *Theory and Applications of Reliability: With Emphasis on Bayesian and Nonparametric Methods*, **2**, C. P. Tsokos and I. N. Shimi (editors), 123–132, New York: Academic Press.

Kamps, U. (1988). Distance measures in a one-parameter class of density functions, *Communications in Statistics—Theory and Methods*, **17**, 2013–2019.

Kamps, U. (1991). Inequalities for moments of order statistics and characterizations of distributions, *Journal of Statistical Planning and Inference*, **27**, 397–404.

Kanaroglou, P., Liaw, K. -L., and Papageorgiou, Y. Y. (1986). An analysis of migratory systems. II: Operational framework, *Environment and Planning*, Series A, **18**, 1039–1060.

Kao, J. H. K. (1958). Computer methods for estimating Weibull parameters in reliability studies, *Transactions of IRE—Reliability and Quality Control*, **13**, 15–22.

Kao, J. H. K. (1959). A graphical estimation of mixed Weibull parameters in life-testing electron tubes, *Technometrics*, **1**, 389–407.

Kappenman, R. F. (1981). Estimation of Weibull location, *Journal of Statistical Computation and Simulation*, **13**, 245–254.

Kappenman, R. F. (1982). On a method for selecting a distributional model, *Communications in Statistics—Theory and Methods*, **11**, 663–672.

Kappenman, R. F. (1983). Parameter estimation via sample reuse, *Journal of Statistical Computation and Simulation*, **16**, 213–222.

Kappenman, R. F. (1985a). A testing approach to estimation, *Communications in Statistics—Theory and Methods*, **14**, 2365–2377.

Kappenman, R. F. (1985b). Estimation for the three-parameter Weibull, lognormal, and gamma distributions, *Computational Statistics & Data Analysis*, **3**, 11–23.

Kappenman, R. F. (1988). A simple method for choosing between the lognormal and Weibull models, *Statistics & Probability Letters*, **7**, 123–126.

Kapur, K. C., and Lamberson, L. R. (1977). *Reliability in Engineering Design*, New York: Wiley.

Kaylan, A. R., and Harris, C. M. (1981). Efficient algorithms to derive maximum-likelihood estimates for finite exponential and Weibull mixtures, *Computers and Operations Research*, **8**, 97–104.

Keating, J. P. (1984). A note on estimation of percentiles and reliability in the extreme-value distribution, *Statistics & Probability Letters*, **2**, 143–146.

Khan, A. H., and Ali, M. M. (1987). Characterization of probability distributions through higher order gap, *Communications in Statistics—Theory and Methods*, **16**, 1281–1287.

Khan, A. H., and Beg, M. I. (1987). Characterization of the Weibull distribution by conditional variance, *Sankhyā, Series A*, **49**, 268–271.

Khan, A. H., Khan, R. U., and Parvez, S. (1984). Inverse moments of order statistics from Weibull distribution, *Scandinavian Actuarial Journal*, 91–94.

Khan, A. H., Parvez, S., and Yaqub, M. (1983). Recurrence relations between product moments of order statistics, *Journal of Statistical Planning and Inference*, **8**, 175–183.

Khirosi, S., and Mieko, N. (1963). On the graphical estimation of the parameter of the Weibull distribution from small samples, *Bulletin of the Electrotechnical Laboratory*, **27**, 655–663.

Khirosi, S., Sideru, T., and Minoru, K. (1966). On the accuracy of estimation of the parameters of the Weibull distribution from small samples, *Bulletin of the Electrotechnical Laboratory*, **30**, 753–765.

Kimball, B. F. (1960). On the choice of plotting positions on probability paper, *Journal of the American Statistical Association*, **55**, 546–560.

Kingston, J. V., and Patel, J. K. (1981). Interval estimation of the largest reliability of k Weibull populations, *Communications in Statistics—Theory and Methods*, **10**, 2279–2298.

Klein, J. P., and Basu, A. P. (1981). Weibull accelerated life tests when there are competing causes of failure, *Communications in Statistics—Theory and Methods*, **10**, 2073–2100.

Klein, J. P., and Basu, A. P. (1982). Accelerated life tests under competing Weibull causes of failure, *Communications in Statistics—Theory and Methods*, **11**, 2271–2286.

Kleyle, R. (1978). Approximate lower confidence limits for the Weibull reliability function, *IEEE Transactions on Reliability*, **27**, 153–160.

Koehler, K. J., and McGovern, P. G. (1990). An application of the LFP survival model to smoking cessation data, *Statistics in Medicine*, **9**, 409–421.

Kopaev, B. V. (1990). A modified Weibull distribution and its properties, *Nadezhnost' i Kontrol Kachestva*, **4**, 48–51.

Kopaev, B. V. (1993). Applications of the weighing method for determination of estimators of parameters of a distribution law, *Zavodskaya Laboratoria*, **10**, 53–57.

Kotel'nikov, V. P. (1964). A nomogram connecting the parameters of Weibull's distribution with probabilities, *Teoriya Veroyatnostei i ee Primeneniya*, **9**, 743–746. (in Russian. English translation, **9**, 670–673.)

Kübler, H. (1979). On the fitting of the three parameter distributions lognormal, gamma and Weibull, *Statistische Hefte*, **20**, 68–125.

Lalla, M. (1990). Parameter estimation of the cumulative distribution function of spell duration, *Statistica*, **50**, 269–284. (In Italian)

Lamberson, L. R., and DeSouza, D. I., Jr. (1987). Bayesian Weibull estimation, *ASQC Technical Conference Transactions*, 497–506.

LaRiccia, V. N. (1982). Asymptotic properties of weighted L^2 quantile distance estimators, *Annals of Statistics*, **10**, 621–624.

Lawless, J. F. (1972). Confidence intervals for the parametres of the Weibull distribution, *Utilitas Mathematica*, **2**, 71–87.

Lawless, J. F. (1973a). On the estimation of safe life when the underlying life distribution is Weibull, *Technometrics*, **15**, 857–865.

Lawless, J. F. (1973b). Conditional versus unconditional confidence intervals for the parameters of the Weibull distribution, *Journal of the American Statistical Association*, **68**, 665–669.

Lawless, J. F. (1975). Construction of tolerance bounds for the extreme-value and Weibull distributions, *Technometrics*, **17**, 255–262.

Lawless, J. F. (1978). Confidence interval estimation for the Weibull and extreme-value distributions (with discussion), *Technometrics*, **20**, 355–368.

Lawless, J. F. (1980). Inference in the generalized gamma and log gamma distributions, *Technometrics*, **22**, 409–419.

Lawless, J. F. (1982). *Statistical Models and Methods for Lifetime Data*, New York: Wiley.

Lawrence, K. D., and Shier, D. R. (1981). A comparison of least squares and least absolute deviation regression models for estimating Weibull parameters, *Communications in Statistics—Simulation and Computation*, **10**, 315–326.

Lee, L., and Lee, S. K. (1978). Some results on inference for the Weibull process, *Technometrics*, **20**, 41–46.

Lee, S. M., Bell, C. B., and Mason, A. L. (1988). Discrimination, signal detection, and estimation for Weibull-type Poisson processes, *IEEE Transactions on Information Theory*, **34**, 576–580.

Leese, M. N. (1973). Use of censored data in the estimation of Gumbel distribution parameters for annual maximum flood series, *Water Resources Research*, **9**, 1534–1542.

Lehman, E. H. (1963). Shapes, moments and estimators of the Weibull distribution, *IEEE Transactions on Reliability*, **12**, 32–38.

Lemon, G. H. (1975). Maximum likelihood estimation for the three parameter Weibull distribution based on censored samples, *Technometrics*, **17**, 247–254.

Lemon, G. H., and Wattier, J. B. (1976). Confidence and "A" and "B" allowable factors for the Weibull distribution, *IEEE Transactions on Reliability*, **25**, 16–19.

Leone, F. C., Rutenberg, Y. H., and Topp, C. W. (1960). *Order Statistics and Estimators for the Weibull Distribution*, Case Statistical Laboratory Publication No. 1026, Cleveland, OH: Case Institute of Technology.

Lieblein, J. (1955). On moments of order statistics from the Weibull distribution, *Annals of Mathematical Statistics*, **26**, 330–333.

Lingappaiah, G. S. (1977). Bayesian approach to prediction in complete and censored samples from the Weibull population, *Metron*, **35**, 167–180.

Lingappaiah, G. S. (1983). A study of shift in parameter of a Weibull life test model, *Industrial Mathematics*, **33**, 139–148.

Lingappaiah, G. S. (1990). Inference in life tests based on a Weibull model when an outlier is present, *Journal of the Indian Association for Productivity, Quality and Reliability*, **15**, 1–10.

Lloyd, D. K., and Lipow, M. (1962). *Reliability: Management, Methods, and Mathematics*, Englewood Cliffs, NJ: Prentice-Hall.

Looney, S. W. (1983). The use of the Weibull distribution in bioassay, *ASA Proceedings of Statistical Computing Section*, 272–277.

Makino, T. (1984). Mean hazard rate and its application to the normal approximation of the Weibull distribution, *Naval Research Logistics Quarterly*, **31**, 1–8.

Malik, H. J., and Trudel, R. (1982). Probability density function of quotient of order statistics from the Pareto, power and Weibull distributions, *Communications in Statistics—Theory and Methods*, **11**, 801–814.

Malik, M. A. K. (1975). A note on the physical meaning of the Weibull distribution, *IEEE Transactions on Reliability*, **24**, 95.

Mandelbaum, J., and Harris, C. M. (1982). Parameter estimation under progressive censoring conditions for a finite mixture of Weibull distributions, In *Optimization in Statistics*, S. H. Zanakis and J. S. Rustagi (editors), 239–260, Amsterdam: North-Holland.

Mann, N. R. (1966). Exact three-order-statistics confidence bounds on reliability parameters under Weibull assumptions, Aerospace Research Laboratories Report ARL 67-0023, Wright-Patterson Air Force Base, OH.

Mann, N. R. (1967). Tables for obtaining the best linear invariant estimates of parameters of the Weibull distribution, *Technometrics*, **9**, 629–645.

Mann, N. R. (1968). Point and interval estimation procedures for the two-parameter Weibull and extreme-value distributions, *Technometrics*, **10**, 231–256.

Mann, N. R. (1970). Warranty periods based on three ordered sample observations from a Weibull population, *IEEE Transactions on Reliability*, **19**, 167–171.

Mann, N. R. (1971). Best linear invariant estimation for Weibull parameters under progressive censoring, *Technometrics*, **13**, 521–533.

Mann, N. R. (1977). An F approximation for two-parameter Weibull and lognormal tolerance bounds on possibly censored data, *Naval Research Logistics Quarterly*, **24**, 187–196.

Mann, N. R. (1978a). Calculation of small-sample Weibull tolerance bounds for accelerated testing, *Communications in Statistics—Theory and Methods*, **7**, 97–112.

Mann, N. R. (1978b). Weibull tolerance intervals associated with moderate to small survival proportions for use in a new formulation of Lanchester combat theory, *Naval Research Logistics Quarterly*, **25**, 121–128.

Mann, N. R., and Fertig, K. W. (1973). Tables for obtaining Weibull confidence bounds and tolerance bounds based on best linear invariant estimates of parameters of the extreme-value distribution, *Technometrics*, **15**, 87–101.

Mann, N. R., and Fertig, K. W. (1975). Simplified and efficient point and interval estimators for Weibull parameters, *Technometrics*, **17**, 361–368.

Mann, N. R., and Fertig, K. W. (1977). Efficient unbiased quantile estimators for moderate-size complete samples from extreme-value and Weibull distributions; confidence bounds and tolerance and prediction intervals, *Technometrics*, **19**, 87–94.

Mann, N. R., and Saunders, S. C. (1969). On evaluation of warranty assurance when life has a Weibull distribution, *Biometrika*, **59**, 615–625.

Mann, N. R., Schafer, R. E., and Singpurwalla, N. D. (1974). *Methods for Statistical Analysis of Reliability and Life Data*, New York: Wiley.

Marquina, N. (1979). A simple approach to estimating Weibull parameters, *ASA Proceedings of Statistical Computing Section*, 365–367.

Martinez, S., and Quintana, F. (1991). On a test for generalized upper truncated Weibull distributions, *Statistics & Probability Letters*, **12**, 273–279.

Martz, H. F., Jr., and Lian, M. G. (1977). Bayes and empirical Bayes point and interval estimation of reliability for the Weibull model, In *Theory and Applications of Reliability: With Emphasis on Bayesian and Nonparametric Methods*, **2**, C. P. Tsokos, and I. N. Shimi (editors), 203–234, New York: Academic Press.

Martz, H. F., Jr., and Waller, R. A. (1982). *Bayesian Reliability Analysis*, New York: Wiley.

McCool, J. I. (1966). Inference on Weibull percentiles and shape parameter from maximum likelihood estimates, SFK Industries Research Laboratory Report No. AL68PO23, King of Prussia, PA.

McCool, J. I. (1974). Inferential techniques for Weibull populations, Aerospace Research Laboratories ARL 74-0180, Wright-Patterson Air Force Base, OH.

McCool, J. I. (1976). Estimation of Weibull parameter with competing-mode censoring, *IEEE Transactions on Reliability*, **25**, 25–31.

McCool, J. I. (1979). Estimation of Weibull shape parameter for samples of size 2, *IEEE Transactions on Reliability*, **28**, 158–160.

McEwen, R. P., and Parresol, B. R. (1991). Moment expressions and summary statistics for the complete and truncated Weibull distribution, *Communications in Statistics—Theory and Methods*, **20**, 1361–1372.

McWilliams, T. P. (1989). Economic control chart designs and the in-control time distributions: A sensitivity study, *Journal of Quality Technology*, **21**, 103–110.

Meeker, W. Q., Jr. (1984). A comparison of accelerated life test plans for Weibull and lognormal distributions and type I censoring, *Technometrics*, **26**, 157–171.

Meeker, W. Q., Jr. (1986). Planning life tests in which units are inspected for failure, *IEEE Transactions on Reliability*, **35**, 571–578.

Meeker, W. Q., Jr. (1987). Limited failure population life tests: Application to integrated circuit reliability, *Technometrics*, **29**, 51–65.

Meeker, W. Q., Jr., and Nelson, W. (1976a). Weibull percentile estimates and confidence limits from singly censored data by maximum likelihood, *IEEE Transactions on Reliability*, **25**, 20–24.

Meeker, W. Q., Jr., and Nelson, W. (1976b). Optimum accelerated life-tests for the Weibull and extreme value distributions, *IEEE Transactions on Reliability*, **25**, 321–332.

Meeker, W. Q., Jr., and Nelson, W. (1977). Weibull variances and confidence limits by maximum likelihood for singly censored data, *Technometrics*, **19**, 473–476.

Mendenhall, W. (1958). A bibliography on life testing and related topics, *Biometrika*, **45**, 521–543.

Mendenhall, W., and Lehman, E. H. (1960). An approximation to the negative moments of the positive binomial useful in life testing, *Technometrics*, **2**, 227–242.

Menon, M. V. (1963). Estimation of the shape and scale parameters of the Weibull distribution, *Technometrics*, **5**, 175–182.

Mihram, G. A. (1975). A generalized extreme-value density, *South African Statistical Journal*, **9**, 153–162.

Mihram, G. A. (1977). Weibull shape parameter: Estimation by moments, *ASQC Technical Conference Transactions*, 315–322.

Miller, D. R. (1976). Order statistics, Poisson processes and repairable systems, *Journal of Applied Probability*, **13**, 519–529.

Miller, D. W. (1966). Degree of normality of maximum likelihood estimates of the shape parameter of the Weibull failure distribution, Unpublished thesis, St. Louis, MO: Washington University.

Miller, I., and Freund, J. E. (1965). *Probability and Statistics for Engineers*, Englewood Cliffs, NJ: Prentice-Hall.

Mittal, M. M., and Dahiya, R. C. (1989). Estimating the parameters of a truncated Weibull distribution, *Communications in Statistics—Theory and Methods*, **18**, 2027–2042.

Moeller, S. K. (1976). The Rasch-Weibull process, *Scandinavian Journal of Statistics*, **3**, 107–115.

Moore, A. H., and Harter, H. L. (1965). One-order-statistic estimation of the scale parameters of Weibull populations, *IEEE Transactions on Reliability*, **14**, 100–106.

Moore, A. H., and Harter, H. L. (1966). Point and interval estimation, from one-order statistic, of the location parameter of an extreme-value distribution with known scale parameter, and of the scale parameter of a Weibull distribution with known shape parameter, *IEEE Transactions on Reliability*, **15**, 120–126.

Moore, A. H., Harter, H. L., and Antoon, D. F. (1981). Confidence intervals and tests of hypotheses for the reliability of a 2-parameter Weibull distribution, *IEEE Transactions on Reliability*, **30**, 468–470.

Moothathu, T. S. K. (1990). A characterization property of Weibull, exponential and Pareto distributions, *Journal of the Indian Statistical Association*, **28**, 69–74.

Mukherjee, S. P., and Roy, D. (1987). Failure rate transform of the Weibull variate and its properties, *Communications in Statistics—Theory and Methods*, **16**, 281–291.

Mukherjee, S. P., and Sasmal, B. C. (1984). Estimation of Weibull parameters using fractional moments, *Calcutta Statistical Association Bulletin*, **33**, 179–186.

Murthy, V. K., and Swartz, G. B. (1975). Estimation of Weibull parameters from two-order statistics, *Journal of the Royal Statistical Society*, Series B, **37**, 96–102.

Musson, T. A. (1965). Linear estimation of the location and scale parameters of the Weibull and gamma distributions by the use of order statistics, Unpublished Thesis, Air Force Institute of Technology, Wright-Patterson Air Force Base, Dayton, OH.

Nakagawa, T. (1986). Periodic and sequential preventive maintenance policies, *Journal of Applied Probability*, **23**, 536–542.

Nakagawa, T., and Osaki, S. (1975). The discrete Weibull distribution, *IEEE Transactions on Reliability*, **24**, 300–301.

Nathan, R. J., and McMahon, T. A. (1990). Practical aspects of low-flow frequency analysis, *Water Resources Research*, **26**, 2135–2141.

Nelson, L. S. (1967). Weibull probability paper, *Industrial Quality Control*, **23**, 452–453.

Nelson, P. R. (1979). Control charts for Weibull processes with standards given, *IEEE Transactions on Reliability*, **28**, 283–288.

Nelson, W. (1972). Theory and application of hazard plotting for censored failure data, *Technometrics*, **14**, 945–966.

Nelson, W. (1982). *Applied Life Data Analysis*, New York: Wiley.

Nelson, W. (1985). Weibull analysis of reliability data with few or no failures, *Journal of Quality Technology*, **17**, 140–146.

Nelson, W., and Meeker, W. Q., Jr., (1978). Theory for optimum accelerated censored life tests for Weibull and extreme value distributions, *Technometrics*, **20**, 171–178.

Nelson, W., and Thompson, V. C. (1971). Weibull probability papers, *Journal of Quality Technology*, **3**, 45–50.

Newby, M. J. (1980). The properties of moment estimators for the Weibull distribution based on the sample coefficient of variation, *Technometrics*, **22**, 187–194. (Correction, *Ibid.*, **24**, 90.)

Newby, M. J. (1984). Properties of moment estimators for the 3-parameter Weibull distribution, *IEEE Transactions on Reliability*, **33**, 192–195.

Newby, M. J., and Winterton, J. (1983). The duration of industrial stoppages, *Journal of the Royal Statistical Society, Series A*, **146**, 62–70.

Nigm, A. M. (1989). An informative Bayesian prediction for the Weibull lifetime distribution, *Communications in Statistics—Theory and Methods*, **18**, 897–911.

Nigm, A. M. (1990). Prediction bounds using type I censoring for the Weibull distribution, *Mathematische Operationsforschung und Statistik, Series Statistics*, **21**, 227–237.

Nilsson, A. A., and Glisson, T. H. (1980). On the derivation and numerical evaluation of the Weibull-Rician distribution, *IEEE Transactions on Aerospace and Electronic Systems*, **16**, 864–867.

Ofosu, J. B. (1990). Design of experiments for selection from gamma and Weibull populations, *Pakistan Journal of Statistics, Series A*, **6**, 71–84.

Ogden, J. E. (1978). A Weibull shelf-life model for pharmaceuticals, *ASQC Technical Conference Transactions*, 574–580.

Ostrouchov, G., and Meeker, W. Q., Jr. (1988). Accuracy of approximate confidence bounds computed from interval censored Weibull and lognormal data, *Journal of Statistical Computation and Simulation*, **29**, 43–76.

Ouyang, L. Y. (1987). On characterizations of probability distributions based on conditional expected values, *Tamkang Journal of Mathematics*, **18**, 113–122.

Öztürk, A., and Korukoğlu, S. (1988). A new test for the extreme value distribution, *Communications in Statistics—Simulation and Computation*, **17**, 1375–1393.

Padgett, W. J., and Spurrier, J. D. (1990). Shewhart-type charts for percentiles of strength distributions, *Journal of Quality Technology*, **22**, 283–288.

Palisson, F. (1989). Determination of the parameters of the Weibull model using an actuarial method, *Revue de Statistique Appliquée*, **37**, No. 4, 5–39. (In French)

Panchang, V. G., and Gupta, R. C. (1989). On the determination of three-parameter Weibull MLE's, *Communications in Statistics—Simulation and Computation*, **18**, 1037–1057.

Pandey, B. N., Malik, H. J., and Srivastava, R. (1989). Shrinkage testimators for the shape parameter of Weibull distribution under Type II censoring, *Communications in Statistics—Theory and Methods*, **18**, 1175–1199.

Pandey, B. N., and Singh, K. N. (1984). Estimating the shape parameter of the Weibull distribution by shrinkage towards an interval, *South African Statistical Journal*, **18**, 1–11.

Pandey, M. (1983). Shrunken estimators of Weibull shape parameter in censored samples, *IEEE Transactions on Reliability*, **32**, 200–203.

Pandey, M. (1987). Bayes estimator of survival probability from randomly censored observations with Weibull distribution of time to death, *Biometrical Journal*, **29**, 491–496.

Pandey, M. (1988). A Bayesian approach to shrinkage estimation of the scale parameter of a Weibull distribution, *South African Statistical Journal*, **22**, 1–13.

Pandey, M., Ferdous, J., and Uddin, Md. B. (1991). Selection of probability distribution for life testing data, *Communications in Statistics—Theory and Methods*, **20**, 1373–1388.

Pandey, M., and Singh, U. S. (1993). Shrunken estimators of Weibull shape parameter from Type-II censored samples, *IEEE Transactions on Reliability*, **42**, 81–86.

Pandey, M., and Upadhyay, S. K. (1983). Selection of Weibull shape parameter, based on adaptive estimation, *IEEE Transactions on Reliability*, **32**, 205–208.

Pandey, M., and Upadhyay, S. K. (1985). Bayes shrinkage estimators of Weibull parameters, *IEEE Transactions on Reliability*, **34**, 491–494.

Pandey, M., and Upadhyay, S. K. (1986). Approximate prediction limit for Weibull failure based on preliminary test estimator, *Communications in Statistics—Theory and Methods*, **15**, 241–250.

Papadopoulos, A. S., and Tsokos, C. P. (1975). Bayesian confidence bounds for the Weibull failure model, *IEEE Transactions on Reliability*, **24**, 21–26.

Papastavridis, S. G. (1987). A limit theorem for the reliability of a consecutive-k-out-of-n system, *Advances in Applied Probability*, **19**, 746–478.

Papastavridis, S. G. (1988). A Weibull limit for the reliability of a consecutive k-within-m-out-of-n system, *Advances in Applied Probability*, **20**, 690–692.

Park, W. J. (1983). Percentiles of pooled estimates of Weibull parameters, *IEEE Transactions on Reliability*, **32**, 91–94.

Parr, W. C., and Schucany, W. R. (1980). Minimum distance and robust estimation, *Journal of the American Statistical Association*, **75**, 616–624.

Parzen, E. (1960). *Modern Probability Theory and Its Applications*, New York: Wiley.

Patel, J. K. (1975). Bounds on moments of linear functions of order statistics from Weibull and other restricted families, *Journal of the American Statistical Association*, **70**, 670–674.

Patel, J. K., and Read, C. B. (1975). Bounds on conditional moments of Weibull and other monotone failure rate families, *Journal of the American Statistical Association*, **70**, 238–244.

Patel, S. R., and Patel, N. P. (1989). On shrinkage type Bayes estimators of the scale parameter of Weibull distribution, *Metron*, **47**, 351–360.

Pavia, E. J., and O'Brien, J. J. (1986). Weibull statistics of wind speed over the ocean, *Journal of Climate and Applied Meteorology*, **25**, 1324–1332.

Pinder, J. E., III, Wiener, J. G., and Smith, M. H. (1978). The Weibull distribution: A new method of summarizing survivorship data, *Ecology*, **59**, 175–179.

Plait, A. (1962). The Weibull distribution—with tables, *Industrial Quality Control*, **19**, 17–26.

Portier, C. J., and Dinse, G. E. (1987). Semiparametric analysis of tumor incidence rates in survival/sacrifice experiments, *Biometrics*, **43**, 107–114.

Quayle, R. J. (1963). Estimation of the scale parameter of the Weibull probability density function by use of one-order statistic, Unpublished thesis, Air Force Institute of Technology, Wright-Patterson Air Force Base, Dayton, OH.

Quesenberry, C. P., and Kent, J. (1982). Selecting among probability distributions used in reliability, *Technometrics*, **24**, 59–65. (Correction, *Ibid.*, **24**, 171.)

Raja Rao, B., and Talwalker, S. (1989). Bounds on life expectancy for the Rayleigh and Weibull distributions, *Mathematical Biosciences*, **96**, 95–115.

Ramberg, J. S., and Tadikamalla, P. R. (1974). An algorithm for generating gamma variates based on the Weibull distribution, *American Institute of Industrial Engineers Transactions*, **6**, 257–260.

Rao Tummala, V. M. (1980). Minimum expected loss estimators of reliability and shape parameter of Weibull distribution, *Industrial Mathematics*, **30**, 61–67.

Rao, B. R., Talwalker, S., and Kundu, D. (1991). Confidence intervals for the relative relapse rate of placebo vs. 6-Mercaptopurine group of acute leukemia patients, *Biometrical Journal*, **33**, 579–598.

Ratnaparkhi, M. V., and Park, W. J. (1986). Lognormal distribution: Model for fatigue life and residual strength of composite materials, *IEEE Transactions on Reliability*, **35**, 312–315.

Rawlings, J. O., and Cure, W. W. (1985). The Weibull functions as a dose-response model to describe ozone effects on crop yields, *Crop Science*, **25**, 807–814.

Revfeim, K. J. A. (1983). On the analysis of extreme rainfalls, *Journal of Hydrology*, **22**, 107–117.

Revfeim, K. J. A. (1984). Generating mechanisms of, and parameter estimators for, the extreme value distribution, *Australian Journal of Statistics*, **26**, 151–159.

Rider, P. R. (1961). Estimating the parameters of mixed Poisson, binomial, and Weibull distributions by the method of moments, *Bulletin de l'Institut International de Statistique*, **39**, 225–232.

Rigdon, S. E. (1990). The effect of assuming a homogeneous Poisson process when the true process is a power law process, *Journal of Quality Technology*, **22**, 111–117.

Rigdon, S. E., and Basu, A. P. (1988). Estimating the intensity function of a Weibull process at the current time: Failure truncated case, *Journal of Statistical Computation and Simulation*, **30**, 17–38.

Rigdon, S. E., and Basu, A. P. (1989). The power law process: A model for the reliability of repairable systems, *Journal of Quality Technology*, **21**, 251–260.

Rigdon, S. E., and Basu, A. P. (1990). Estimating the intensity function of a power law process at the current time: Time truncated case, *Communications in Statistics—Simulation and Computation*, **19**, 1079–1104.

Ringer, L. J., and Sprinkle, E. E., III. (1972). Estimation of the parameters of the Weibull distribution from multicensored samples, *IEEE Transactions on Reliability*, **R-21**, 46–51.

Rink, G., Dell, T. R., Switzer, G., and Bonner, F. T. (1979). Use of the [three-parameter] Weibull function to quantify sweetgum germinating data, *Silvae Genetica*, **28**, 9–12.

Robinson, J. A. (1983). Bootstrap confidence intervals in location-scale models with progressive censoring, *Technometrics*, **25**, 179–187.

Rodriguez, R. N. (1977). A guide to the Burr type XII distribution, *Biometrika*, **64**, 129–134.

Rosen P., and Rammler, B. (1933). The laws governing the fineness of powdered coal, *Journal of the Institute of Fuels*, **6**, 29–36.

Rousu, D. N. (1973). Weibull skewness and kurtosis as a function of the shape parameter, *Technometrics*, **15**, 927–930.

Roy, D. (1984). A characterization of the generalized gamma distribution, *Calcutta Statistical Association Bulletin*, **33**, 137–141. (Correction, *Ibid.*, **34**, 125.)

Roy, D., and Mukherjee, S. P. (1986). A note on characterizations of the Weibull distribution, *Sankhyā, Series A*, **48**, 250–253.

Salvia, A. A. (1979). Consonance sets for 2-parameter Weibull and exponential distribution, *IEEE Transactions on Reliability*, **28**, 300–302.

Saylor, F. (1977). A Monte Carlo comparison of the method-of-moments to the maximum-likelihood estimates of Weibull parameters for CAS data, In *Theory and Applications of Reliability: With Emphasis on Bayesian and Nonparametric Methods*, **2**, C. P. Tsokos and I. N. Shimi (editors), 517–530, New York: Academic Press.

Schafer, D. B. (1975). A note on a simple test function for the Weibull distribution location parameter, *Scandinavian Actuarial Journal*, 1–5.

Schafer, R. E., and Angus, J. E. (1979). Estimation of Weibull quantiles with minimum error in the distribution function, *Technometrics*, **21**, 367–370.

Schneider, H. (1989). Failure-censored variables-sampling plans for lognormal and Weibull distributions, *Technometrics*, **31**, 199–206.

Schneider, H., and Weissfeld, L. A. (1989). Interval estimation based on censored samples from the Weibull distribution, *Journal of Quality Technology*, **21**, 179–186.

Scholz, F. -W. (1990). Characterization of the Weibull distribution, *Computational Statistics & Data Analysis*, **10**, 289–292.

Schuepbach, M., and Huesler, J. (1983). Simple estimators for the parameters of the extreme-value distribution based on censored data, *Technometrics*, **25**, 189–192.

Schuette, T., Salka, O., and Israelsson, S. (1987). The use of the Weibull distribution for thunderstorm parameters, *Journal of Climate and Applied Meteorology*, **26**, 457–463.

Schulz, W. (1983). The planning of sample sizes for maximum likelihood estimation of the parameters of the Weibull distribution, *Biometrical Journal*, **25**, 783–789.

Schwenke, J. R. (1987). Using the Weibull distribution in modeling pharmacokinetic data, *ASA Proceedings of the Biopharmaceutical Section*, 104–109.

Selker, J. S., and Haith, D. A. (1990). Development and testing of single-parameter precipitation distributions, *Water Resources Research*, **26**, 2733–2740.

Sen, I., and Prabhashanker, V. (1980). A nomogram for estimating the three parameters of the Weibull distribution, *Journal of Quality Technology*, **12**, 138–143.

Sherif, A., and Tan, P. (1978). On structural predictive distribution with type II progressively censored Weibull data, *Statistische Hefte*, **19**, 247–255.

Shier, D. R., and Lawrence, K. D. (1984). A comparison of robust regression techniques for the estimation of Weibull parameters, *Communications in Statistics—Simulation and Computation*, **13**, 743–750.

Shimizu, R., and Davies, L. (1981). General characterization theorems for the Weibull and the stable distributions, *Sankhyā, Series A*, **43**, 282–310.

Shor, M., and Levanon, N. (1991). Performances of order statistics CFAR, *IEEE Transactions on Aerospace and Electronic Systems*, **27**, 214–224.

Singh, J., and Bhatkulikar, S. G. (1978). Shrunken estimation in Weibull distribution, *Sankhyā, Series B*, **39**, 382–393.

Singh, V. P., Cruise, J. F., and Ma, M. (1990). A comparative evaluation of the estimators of the Weibull distribution by Monte Carlo simulation, *Journal of Statistical Computation and Simulation*, **36**, 229–241.

Singpurwalla, N. D. (1988). An interactive PC-based procedure for reliability assessment incorporating expert opinion and survival data, *Journal of the American Statistical Association*, **83**, 43–51.

Singpurwalla, N. D., and Song, M. S. (1988). Reliability analysis using Weibull life-time data and expert opinion, *IEEE Transactions on Reliability*, **37**, 340–347.

Sinha, S. K. (1982). Reliability estimation with Type I censoring of a mixture of Weibull distributions, Publications of Institute of Statistics, University of Paris IV, **27**, No. 2, 75–92.

Sinha, S. K. (1986). Bayes estimation of the reliability function and hazard rate of a Weibull failure time distribution, *Trabajos de Estadistica*, **37**, 47–56.

Sinha, S. K. (1987). Bayesian estimation of the parameters and reliability function of a mixture of Weibull life distributions, *Journal of Statistical Planning and Inference*, **16**, 377–387.

Sinha, S. K., and Guttman, I. (1988). Bayesian analysis of life-testing problems involving the Weibull distribution, *Communications in Statistics—Theory and Methods*, **17**, 343–356.

Sinha, S. K., and Sloan, J. A. (1988). Bayes estimation of the parameters and reliability function of the 3-parameter Weibull distribution, *IEEE Transactions on Reliability*, **37**, 364–369.

Sinha, S. K., and Sloan, J. A. (1989). Prediction inverval for a mixture of Weibull failure-time distributions: A Bayesian approach, *South African Statistical Journal*, **23**, 119–130.

Sirvanci, M. (1984). An estimator for the scale parameter of the two parameter Weibull distribution for type I singly right-censored data, *Communications in Statistics—Theory and Methods*, **13**, 1759–1768.

Sirvanci, M., and Yang, G. (1984). Estimation of the Weibull parameters under Type I censoring, *Journal of the American Statistical Association*, **79**, 183–187.

Smith, R. L. (1985). Maximum likelihood estimation in a class of nonregular cases, *Biometrika*, **72**, 67–90.

Smith, R. L., and Naylor, J. C. (1987). A comparison of maximum likelihood and Bayesian estimators for the three-parameter Weibull distribution, *Applied Statistics*, **36**, 358–369.

Smith, R. M. (1977). Some results on interval estimation for the two parameter Weibull or extreme-value distribution, *Communications in Statistics—Theory and Methods*, **6**, 1311–1322.

Soland, R. M. (1966). Use of the Weibull distribution in Bayesian decision theory, Research Analysis Corporation Technical Paper RAC-TP-225, McLean, VA.

Soland, R. M. (1967). Bayesian analysis of the Weibull process with unknown scale parameter, *IEEE Transactions on Reliability*, **17**, 84–90.

Splitstone, D. (1967). Estimation of the Weibull shape and scale parameters, Masters thesis, Ames: Iowa State University, Iowa.

Srinivasan, R., and Wharton, R. M. (1975). Confidence bands for the Weibull distribution, *Technometrics*, **17**, 375–380.

Srivastava, J. N. (1987). More efficient and less time-consuming censoring designs for life testing, *Journal of Statistical Planning and Inference*, **16**, 389–413.

Srivastava, J. N. (1989). Advances in the statistical theory of comparison of lifetimes of machines under the generalized Weibull distribution, *Communications in Statistics—Theory and Methods*, **18**, 1031–1045.

Stacy, E. W., and Mihram, G. A. (1965). Parameter estimation for a generalized gamma distribution, *Technometrics*, **7**, 349–358.

Stauffer, H. B. (1979). A derivation for the Weibull distribution, *Journal of Theoretical Biology*, **81**, 55–63.

Stein, W. E., and Dattero, R. (1984). A new discrete Weibull distribution, *IEEE Transactions on Reliability*, **33**, 196–197.

Stone, G. C., and Rosen, H. (1984). Some graphical techniques for estimating Weibull confidence intervals, *IEEE Transactions on Reliability*, **33**, 362–369.

Struthers, C. A., and Farewell, V. T. (1989). A mixture model for time to AIDS data with left truncation and an uncertain origin, *Biometrika*, **76**, 814–817.

Stump, F. B. (1968). Nearly best unbiased estimation of the location and scale parameters of the Weibull distribution by the use of order statistics, Master's thesis, Air Force Institute of Technology, Wright-Patterson Air Force Base, Dayton, OH.

Sugiura, N., and Gomi, A. (1985). Pearson diagrams for truncated normal and truncated Weibull distributions, *Biometrika*, **72**, 219–222.

Suzuki, K. (1985). Estimation of lifetime parameters from incomplete field data, *Technometrics*, **27**, 263–271.

Suzuki, Y. (1988). On selection of statistical models, In *Statistical Theory and Data Analysis*, **2**, K. Matusita (editor), 309–325, Amsterdam: North-Holland.

Tadikamalla, P. R., and Schriber, T. J. (1977). Sampling from the Weibull and gamma distributions in GPSS, *Simuletter*, **9**, No. 1, 39–45.

Tang, D. (1989). Confidence interval estimation for the coefficient of variation of the Weibull distribution, *Chinese Journal of Applied Probability and Statistics*, **5**, 276–282. (In Chinese)

Tate, R. F. (1959). Unbiased estimation: Functions of location and scale parameters, *Annals of Mathematical Statistics*, **30**, 341–366.

Taylor, J. A., and Jakeman, A. J. (1985). Identification of a distributional model, *Communications in Statistics—Simulation and Computation*, **14**, 497–508.

Thoman, D. R., Bain, L. J., and Antle, C. E. (1970). Reliability and tolerance limits in the Weibull distribution, *Technometrics*, **12**, 363–371.

Thompson, J. R. (1968). Some shrinkage techniques for estimating the mean, *Journal of the American Statistical Association*, **63**, 113–123.

Tiago de Oliveira, J. (1981). Statistical choice of univariate extreme models, In *Statistical Distributions in Scientific Work*, **6**, C. Taillie, G. P. Patil, and B. A. Baldessari (editors), 367–387, Dordrecht: Reidel.

Tiago de Oliveira, J., and Littauer, S. B. (1976). Mean square invariant forecasters for the Weibull distribution, *Naval Research Logistics Quarterly*, **23**, 487–512.

Tseng, S. -T., and Wu, H. -J. (1990). Selecting, under type-II censoring, Weibull populations that are more reliable, *IEEE Transactions on Reliability*, **39**, 193–198.

Tsokos, C. P., and Rao, A. N. V. (1976). Bayesian analysis of the Weibull failure model under stochastic variation of the shape and scale parameters, *Metron*, **34**, 201–218.

Tsokos, C. P., and Shimi, I. N. (eds.) (1977). *Theory and Applications of Reliability: With Emphasis on Bayesian and Nonparametric Methods*, **1** and **2**, New York: Academic Press.

Tuller, S. E., and Brett, A. C. (1984). The characteristics of wind velocity that favor the fitting of a Weibull distribution in wind speed analysis, *Journal of Climate and Applied Meteorology*, **23**, 124–134.

Tziafetas, G. N. (1987). On the construction of Bayesian prediction limits for the Weibull distribution, *Mathematische Operationsforschung und Statistik*, *Series Statistics*, **18**, 623–628.

Vance, L. C. (1979). Estimating parameters of a Weibull distribution, *ASQC Technical Conference Transactions*, 113–118.

Van der Auwera, L., de Meyer, F., and Malet, L. M. (1980). The use of the Weibull three-parameter model for estimating mean wind power densities, *Journal of Applied Meteorology*, **19**, 819–825.

Vasudeva Rao, A., Dattatreya Rao, A. V., and Narasimham, V. L. (1991). Optimum linear unbiased estimation of the scale parameter by absolute values of order statistics in the double exponential and the double Weibull distributions, *Communications in Statistics—Simulation and Computation*, **20**, 1139–1158.

Vodă, V. G. (1978). Concerning an application of the Weibull distribution to reliability aspects in metrology, *Metrologia Aplicata*, **25**, 123–124.

Vogt, H. (1968). Zur Parameter- und Prozentpunktschätzung von Lebensdauerverteilungen bei kleinem Stichprobenumfang, *Metrika*, **14**, 117–131.

Wallenius, K. T., and Korkotsides, A. S. (1990). Exploratory model analysis using CDF knotting with applications to distinguishability, limiting forms, and moment ratios to the three parameter Weibull family, *Journal of Statistical Computation and Simulation*, **35**, 121–133.

Wang, Y. H. (1976). A functional equation and its application to the characterization of the Weibull and stable distributions, *Journal of Applied Probability*, **13**, 385–391.

Weibull, W. (1939a). A statistical theory of the strength of material, Report No. 151, Ingeniörs Vetenskaps Akademiens Handligar, Stockholm.

Weibull, W. (1939b). The phenomenon of rupture in solids, Report No. 153, Ingeniörs Vetenskaps Akademiens Hadligar, Stockholm.

Weibull, W. (1951). A statistical distribution of wide applicability, *Journal of Applied Mechanics*, **18**, 293–297.

Weibull, W. (1967a). Estimation of distribution parameters by a combination of the best linear order statistics and maximum likelihood, Air Force Materials Laboratory Technical Report AFML-TR-67-105, Wright-Patterson Air Force Base, Dayton, OH.

Weibull, W. (1967b). Moments about smallest sample value, Air Force Materials Laboratory Technical Report AFML-TR-67-375, Wright-Patterson Air Force Base, Dayton, OH.

Whalen, A. D. (1971). *Detection of Signals in Noise*, New York: Academic Press.

White, J. S. (1964). Least-squares unbiased censored linear estimation for the log Weibull (extreme value) distribution, *Journal of Industrial Mathematics*, **14**, 21–60.

White, J. S. (1965). Linear estimation for the log Weibull distribution, General Motors Research Publication GMR-481.

White, J. S. (1966). A technique for estimating Weibull percentage points, General Motors Research Publication GMR-572.

White J. S. (1967). Estimating reliability from the first two failures, *General Motors Research Publication GMR-669*.

Whittemore, A., and Altschuler, B. (1976). Lung cancer incidence in cigarette smokers: Further analysis of Doll and Hill's data for British physicians, *Biometrics*, **32**, 805–816.

Whitten, B. J., and Cohen, A. C. (1981). Percentiles and other characteristics of the four-parameter generalized gamma distribution, *Communications in Statistics—Simulation and Computation*, **10**, 175–219. (Correction, *Ibid.*, **10**, 543.)

Wilks, D. S. (1989). Rainfall intensity, the Weibull distribution, and estimation of daily surface runoff, *Journal of Applied Meteorology*, **28**, 52–58.

Wingo, D. R. (1973). Solution of the three-parameter Weibull equations by constrained modified quasilinearization (progressively censored samples), *IEEE Transactions on Reliability*, **22**, 96–102.

Wingo, D. R. (1988). Methods for fitting the right-truncated Weibull distribution to life-test and survival data, *Biometrical Journal*, **30**, 545–551.

Wingo, D. R. (1989). The left-truncated Weibull distribution, *Statistische Hefte*, **30**, 39–48.

Wilson, R. B. (1965). Two notes on estimating shape parameters, RAND Corporation Memorandum RM-4459-PR.

Wong, P. G., and Wong, S. P. (1982). A curtailed test for the shape parameter of the Weibull distribution, *Metrika*, **29**, 203–209.

Wong, R. K. W. (1977). Weibull distribution, iterative likelihood techniques and hydrometeorological data, *Journal of Applied Meteorology*, **16**, 1360–1364.

Woo, J., and Lee, K. H. (1983). Estimation of the scale parameter in the Weibull distribution based on the quasi-range, *Journal of the Korean Statistical Society*, **12**, 69–80.

Woodward, W. A., and Gunst, R. F. (1987). Using mixtures of Weibull distributions to estimate mixing proportions, *Computational Statistics & Data Analysis*, **5**, 163–176.

Wright, W. P., and Singh, N. (1981). A prediction interval in life-testing: Weibull distribution, *IEEE Transactions on Reliability*, **30**, 466–467.

Wyckoff, J., Bain, L. J., and Engelhardt, M. (1980). Some complete and censored sampling results for the three-parameter Weibull distribution, *Journal of Statistical Computation and Simulation*, **11**, 139–151. (Correction, *Ibid.*, **13**, 149.)

Yildirim, F. (1990). Least squares estimation of the Weibull parameters, *Pakistan Journal of Statistics, Series A*, **6**, 93–104.

Zacks, S. (1984). Estimating the shift to wear-out systems having exponential-Weibull life distributions, *Operations Research*, **32**, 741–749.

Zanakis, S. H. (1979). A simulation study of some simple estimators for the three-parameter Weibull distribution, *Journal of Statistical Computation and Simulation*, **9**, 101–116.

Zanakis, S. H., and Kyparisis, J. (1986). A review of maximum likelihood estimation methods for the three-parameter Weibull distribution, *Journal of Statistical Computation and Simulation*, **25**, 53–73.

Zanakis, S. H., and Mann, N. R. (1982). A good simple percentile estimator of the Weibull shape parameter for use when all three parameters are unknown, *Naval Research Logistics Quarterly*, **29**, 419–428.

Zhang, Y. (1982). Plotting positions of annual flood extremes considering extraordinary values, *Water Resources Research*, **18**, 859–864.

Abbreviations

ABLUE	asymptotically best linear unbiased estimator
ARE	asymptotic relative efficiency
BAEE	best affine equivariant estimator
BAN	best asymptotically normal
BLIE	best linear invariant estimator
BLUE	best linear unbiased estimator
cdf	cumulative distribution function
GGC	generalized gamma convolution
GIG	generalized inverse Gaussian distribution
GPD	generalized Pareto distribution
IG	inverse Gaussian distribution
LBIG	length biased inverse Gaussian distribution
MLE	maximum likelihood estimator
MME	(method of) moments estimator
MMME	modified (method of) moments estimator
MPE	maximum probability estimator
MPS	maximum product-of-spacings
MSE	mean square error
MVUE	minimum variance unbiased estimator
$N(\mu, \sigma^2)$	normal distribution with mean μ and variance σ^2
P(X)(.)	Pareto distribution of X-th kind (X = I, II, or III)
pdf	probability density function
$\Pr[E]$	probability of event E
PWM	probability-weighted moment
UMVUE	uniform(ly) minimum variance unbiased estimator

Author Index

The purpose of this Index is to provide readers with quick and easy access to the contributions (pertinent to this volume) of any individual author, and not to highlight any particular author's contribution.

725

Subject Index

For readers' convenience, lists of all continuous and discrete distributions that are referred to in this volume have also been included in this Index.

WILEY SERIES IN PROBABILITY
AND MATHEMATICAL STATISTICS

ESTABLISHED BY WALTER A. SHEWHART AND SAMUEL S. WILKS

Editors

Vic Barnett, Ralph A. Bradley, Nicholas I. Fisher, J. Stuart Hunter, J. B. Kadane, David G. Kendall, Adrian F. M. Smith, Stephen M. Stigler, Jozef L. Teugels, Geoffrey S. Watson

Probability and Mathematical Statistics

*Now available in a lower priced paperback edition in the Wiley Classics Library.

*Now available in a lower priced paperback edition in the Wiley Classics Library.

*Now available in a lower priced paperback edition in the Wiley Classics Library.